Beavers

Ecology, Behaviour, Conservation, and Management

Frank Rosell

*Department of Natural Sciences and Environmental Health,
University of South-Eastern Norway, Norway*

Róisín Campbell-Palmer

*Independent Beaver Ecologist,
UK and Republic of Ireland*

OXFORD
UNIVERSITY PRESS

OXFORD

UNIVERSITY PRESS

Great Clarendon Street, Oxford, OX2 6DP,
United Kingdom

Oxford University Press is a department of the University of Oxford.
It furthers the University's objective of excellence in research, scholarship,
and education by publishing worldwide. Oxford is a registered trade mark of
Oxford University Press in the UK and in certain other countries

First Edition published in 2022

Impression: 1

Published in the United States of America by Oxford University Press
198 Madison Avenue, New York, NY 10016, United States of America

British Library Cataloguing in Publication Data
Data available

Library of Congress Control Number: 2021924426

ISBN 978–0–19–883504–2 (hbk.)
ISBN 978–0–19–883505–9 (pbk.)

DOI: 10.1093/oso/9780198835042.001.0001

Printed and bound by
CPI Group (UK) Ltd, Croydon, CR0 4YY

Links to third party websites are provided by Oxford in good faith and
for information only. Oxford disclaims any responsibility for the materials
contained in any third party website referenced in this work.

Preface

Both species of beavers (the Eurasian beaver [*Castor fiber*] and the North American beaver [*C. canadensis*]) have followed similar histories, from near extinction, largely through hunting, to their recovery being viewed as somewhat of a conservation success story. Beavers have been widely reintroduced across Europe and North America for species restoration, biodiversity, and ecological benefits. Many countries have recently reintroduced beavers in Europe, demonstrating that the beaver's role as a keystone engineer is well understood, with proven abilities to increase the complexity and biodiversity of freshwater ecosystems.

During the last 20 years, there has been a huge increase in the number of scientific papers published about both beaver species, revealing precisely how unique they are. There is no other animal quite like the beaver, which, like humans, has the ability to adapt its surroundings to suit its needs. This makes it a fascinating and exciting animal to learn about and watch. What animals other than humans can act as engineers, forest workers, carpenters, masons, creators of habitats, and nature managers? Beavers build their own houses, with indoor pools, often having winter and summer residences. They have their own 'freezer box' where they store food during cold winter months. They have even been observed using tools, brandishing sticks in territorial displays. Their dams act as bridges for a variety of animals to use, and they create wetlands. They can restore entire forests by selectively clearing trees and allowing natural regeneration, thereby encouraging plant diversity, which in turn supports a whole host of other species.

This book will, for the first time, bring together themes and latest research from behaviour, ecology, conservation, and management for both species. New methods such as GPS and tri-axial accelerometers attached to beavers have moved the field forward in the last few years. In this book, we have focused on scientific publications but have also included results from some important theses, accessible reports, and writings from those working and living closely with these animals.

Many people have assisted in and provided input to this book, so in no particular order we would like to thank Shea Allison Sundstøl for help with literature and translating some Norwegian texts, Fyodor Fyodorov for help with some Russian papers, Alexander J.A. Briggs, Harry Hirst for help with tables and Endnote, and especially Rachel Hinds for help with 'everything'. Without Rachel's help this book would have been much more work and stressful. Specially we thank the following people for reviewing the chapters: Göran Hartman (Chapters 1 and 2), Derek Gow (Chapters 2, 9, and 11), Duncan Halley (Chapter 2 and many questions!), Alexander Saveljev (Chapter 3), Hanna Kavli Lodberg-Holm (Chapters 4–8), Glynnis Hood (Chapters 4 and 5), Ken Tape (Chapters 4 and 5), Peter Busher (Chapters 6 and 7), Howard Parker (Chapter 7.5), Tom Gable (Chapter 8), Alan Puttock (Chapters 4 and 9), Kent Woodruff

(Chapter 10), and Gerhard Schwab (Chapters 10 and 11). Special thanks also go to the editors at Oxford University Press.

Finally, Frank wishes to thank his wife Frid Elisabeth Berge and stepdaughter Yrja Skjærum for their patience and support during the writing of this book. Róisín would also like to thank her family, in particular her mum for proofreading, beyond the call of duty, and her sister for her skill and patience with the illustrations throughout the book.

Frank Rosell and Róisín Campbell-Palmer
September 2021
Bø in Telemark
and Pitlochry/Belfast

Contents

CHAPTER 1

Introducing the beaver

1.1 A buck-toothed wonder

Early written descriptions of beavers (*Castor* spp.) are not kind. The openings of Lewis Henry (Morgan)'s iconic writings in 1868, introducing the beaver to a European and colonial American audience, clearly place the beaver at the lower end of mammalian evolution. With descriptions that include 'a coarse vegetable feeder' whose 'clumsy proportions render him slow', 'inferior to the carnivorous and even the herbivorous animals' when comparing to both other land and water mammals, seem somewhat harsh. However, he could not help but be impressed by the beaver's architectural skills, dedicating most of his book to these. Again perhaps somewhat unfairly, the American naturalist and historian Earl Hilfiker (1991) described the beaver's appearance as 'definitely not impressive. It is the things he does rather than his appearance that make him one of the most widely recognized forms of North American wildlife'. More recently, Frances (Backhouse)'s book *Once They Were Hats* (2015) opens with discussing the beaver's image problem: 'a chubby rodent with goofy buckteeth and a tail that looks like it was run over by a tractor tire'. She goes on to elegantly argue for the beaver's rightful place in history, their incredible influences on our ecosystems, and why we should respect this fascinating animal with its uniquely adapted biology and natural history. Saying that, if the reports of some of the objects found in beaver constructions are to be believed, from lengths of pipe, steals from nearby firewood stores, fence posts, beer bottles, drinks cans, tyres, and even a prosthetic leg (Goldfarb, 2018), beavers are also practical recyclers with potentially a great sense of humour!

Beavers are unique, oversized, semi-aquatic rodents with distinctive features such as their flat scaly tail, webbed hind feet, prominent teeth, and luxurious fur (Figure 1.1), and they exhibit specialized behaviours such as damming and tree felling that can transform landscapes. Because of such adaptations, few species, bar humans, can so readily modify their surrounding environments if left to their own devices.

Beavers have a long history of being utilized and eradicated by humans—for food, various body parts, and of course their highly valued fur (see Chapter 2) but also indirectly with their activities providing habitats for numerous species, thereby providing foraging opportunities and ecological benefits such as water storage in times of drought (see Chapter 9). They are a highly adaptable species and can modify many types of natural, cultivated (Schwab et al., 1994), and urban habitats (Pachinger and Hulik, 1999) to suit their needs. Although beavers can also establish in brackish water (Pasternack et al., 2000), especially at higher population density, they are typically a freshwater species, occupying a wide range of freshwater systems including ponds, streams, marshes, rivers, lakes, and even agricultural drainage systems. They thrive in areas stretching from sea level (0 m a.s.l.) to mountain areas (upto 3,500 m a.s.l., including the Rocky Mountains, Colorado), though preferring low-gradient watercourses (Novak, 1987; Osmundson and Buskirk, 1993).

Box 1.1 describes the classification of beavers (*Castor fiber* and *C. canadensis*). Both species play a crucial role in wetland ecology and species biodiversity and can provide important ecosystem services such as habitat creation and water management. This is challenging in modern, often heavily modified landscapes. The history of the beaver represents

Beavers: Ecology, Behaviour, Conservation, and Management. Frank Rosell and Róisín Campbell-Palmer, Oxford University Press.
© Frank Rosell and Róisín Campbell-Palmer 2022. DOI: 10.1093/oso/9780198835042.003.0001

Figure 1.1 Beavers are unique semi-aquatic rodents, highly social in family groups, living in actively defended territories against other beaver families. (Photo supplied courtesy of Michael Runtz.)

important lessons in conservation, as both species were on the verge of extinction solely through human activities. These biodiversity and ecological services benefit that beaver activities can generate are only now being recognized in more modern times (Rosell et al., 2005; Stringer and Gaywood, 2016). Their restoration has offered exciting opportunities for habitat and biodiversity restoration if we are prepared to tolerate, accept and even embrace their activities.

1.2 All in the name

The formal scientific name, Latin *'castor'* and Greek *'kastor'*, is thought to originate from the Sanskrit *'kasturi'* meaning musk, though it has also been suggested that the Greeks called it *Castor* from *gastro*, the stomach, given their rounded appearance (Martin, 1892). When formally naming the beaver (1758), recognizing only one species at the time, Carl Linnaeus basically named it *'beaver beaver'*, as the genus name *'fiber'* means beaver in Latin (Poliquin, 2015). Kuhl formally described and named the North American beaver in 1820, over two centuries after some of the first fur trading posts were established in Canada (Martin, 1892). The genus name *'canadensis'* represents the North American geographic range (Long, 2000).

The Old Norse for beaver was *'bjorr'*, leading to *'bjur'*, *'bur'*, and *'björ'* in Old Scandinavian. In today's more modern languages, the Norwegian *'bever'*, Swedish *'bäver'* (first appearing in Swedish texts from the sixteenth century), Danish *'bæver'*, Dutch *'bever'*, and German *'biber'* are especially similar. The Old English term *'beofor'* has at various points also been spelt *'befor'*, *'byfor'*, *'befer'*, and *'bever'*—all presumed to have their origins in the Old Teutonic term *'bebru'*, a general reference to a brown animal. Yet another similar word is *'bhebhrú'* meaning brown water animal in Old Aryan (Long, 2000). During the Middle Ages in Britain, *'bever'* or *'bevor'* were words used to describe a drink or snack between meals, leading to the word beverage. The verb to *'bever'* was to tremble or shake. At other times, *'beaver'* or *'bever'* was a type of face guard on military helmets and even undergrowth associated with hedges was called *'beaver'* or *'beever'* (Long, 2000).

1.3 A robust rodent

Beavers are often described as a 'robust' rodent. They are streamlined and thought of as more agile in the water, though are sizeable, chunky animals especially when viewed on land. Both species are remarkably similar in morphology, with body shapes often described as teardrop with short, stumpy limbs. Beaver morphology reflects their adaptations to a semi-aquatic lifestyle (see Chapter 3). Their large webbed hind feet are specialized for swimming, providing forward thrust to quietly propel them through the water, along with a unique tail with a developed caudal muscle attachment to the vertebrae. Their tail is dorsal-ventrally flattened, pretty much hairless, with skin patterning often described as scaly in appearance. Beaver tails (see more in Chapter 3) are fairly large affairs, though slight variability is evident across adults and according to body condition.

Beavers are dominantly brown in coloration, with dark-grey tails, for example accounting for 91.8% of animal observed in Karelia, Russia (Danilov et al., 2011a), though pelage colour can range from almost blonde to reddish-brown to black, with even white individuals known (Baker and Hill, 2003). Very light-coloured beavers have been recorded, but albinism is rare (Novak, 1987). Native Americans attached significant value to the skins of white beavers, which were often made into medicine bags (Martin, 1892). Partial albinism or 'spotted' beavers have been recorded in

North America and Russia, with white appearing as irregular patches especially on the stomach and hind feet (Lovallo and Suzuki, 1993). Isolated beaver populations in Central Asia have also been observed with white spotting on their ventral sides (Busher, 2016). Completely black beavers are more common than white or spotted variations, and these often fetched the highest prices in early fur exploitation in Canada (Martin, 1892). Up until the 1970s it was reported that all beavers residing in the Luga watershed, Leningrad region, were composed of only black individuals (Danilov and Kan'shiev, 1983). Out of 350 North American beavers trapped in Karelia and the Leningrad region, only two were nearly black, with the rest varying in coloration from light- to dark-brown (Danilov and Kan'shiev, 1983; Kanshiev, 1998). Export lists of beaver pelts from sixteenth-century Stockholm note colour variations, with black pelts being the most expensive, suggesting reintroductions from Norway may have lost some of this colour variation during genetic bottlenecking (G. Hartman, pers. comm.). Brown pelage is therefore presumed to be the dominant genetic trait for both species, given the scarcity of other colours and brown offspring being born to black parents (Danilov and Kan'shiev, 1983).

Typical body dimensions vary due to a range of factors, such as time of year and habitat quality, and age class should also be considered (Tables 1.1 and 1.2; Figure 1.2). Newborn kits tend to weigh between 380 and 620 g (mean 525 g in captive Eurasian beavers; Żurowski, 1977) and typically reach between 7 and 9 kg by the end of their first year (Ognev, 1947). Adults ($\geq$ 2–3 years) on average weigh around 18 kg but can reach 26+ kg, though general body dimensions are less variable with age (Grinnell et al., 1937; Leege and Williams, 1967; Aleksiuk and Cowan, 1969; Parker et al., 2012). Mass is used to distinguish subadults (between $\geq$ 17 and $\leq$ 19.5 kg) and adults $\geq$ 3 years ($\geq$ 19.5 kg) (Rosell et al., 2010). Rarer examples of Eurasian beavers weighing 29–35 kg have been trapped in Russia, including a 36-kg female (Yazan, 1964; Solov'yov, 1973; Danilov et al., 2011a). North American adults weighing between 24 and 28 kg are common; more rarely maximum body weights of 37–39 kg have been recorded in North

Table 1.1 Reported average adult body dimensions of the two species (note 2-year olds are included, thus lowering the mean mass).

Parameter	Eurasian	References	North American	References
Body weight	17.8 kg	Danilov et al. (2011a)	17.2 kg	Jenkins and Busher (1979); Baker and Hill (2003); Danilov et al. (2011a) (North American in Karelia)
Body length	80.5 cm	Danilov et al. (2011a)	76.8 cm	Jenkins and Busher (1979); Baker and Hill (2003); Danilov et al. (2011a)
Tail length	26.3–30 cm	Danilov et al. (2011a)	25.8–32.5 cm	Grinnell et al. (1937); Davis (1940); Osborn (1953); Jenkins and Busher (1979); Baker and Hill (2003) Danilov et al. (2011a);
Tail width	13cm	Danilov et al. (2011a)	9–20 cm	Grinnell et al. (1937); Davis (1940); Osborn (1953); Jenkins and Busher (1979); Baker and Hill (2003); Danilov et al. (2011a)

Table 1.2 Age class body dimension breakdown of the Norwegian beaver (Rosell and Pedersen, 1999).

Age class	Body length (cm)	Tail length (cm)	Tail width (middle point) (cm)
One-year olds	70–80	17–23	5–8
Two-year olds	90–100	23–27	8–10
Adult	100–110	27–31	10–12

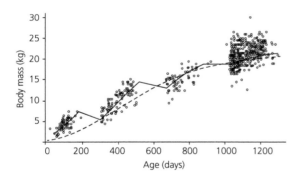

Figure 1.2 Body weight with beaver age. (Reprinted from Campbell, R. 2010. Demography and life history of the Eurasian beaver *Castor fiber*. PhD thesis, University of Oxford.)

America (Grinnell et al., 1937; Schorger, 1953) and even a 44-kg individual (Seaton-Thomson, 1909). Adult body lengths can vary, and though there may be inconsistences with measuring methods, 100–120 cm has been recorded in North American beavers (Grinnell et al., 1937; Osborn, 1953; Jenkins and Busher, 1979).

1.4 The two beavers

The collective term 'beavers' recognizes that there are two species of beavers alive today. Although people have been quite challenged to physically distinguish between them, genetic evidence clearly determines the Eurasian beaver (*C. fiber*) from the North American or Canadian beaver (*C. canadensis*). Both modern species are incredibly similar in appearance and behaviour which can make them hard to distinguish in the field (Rosell et al., 2005; Danilov et al., 2011a). Whilst some differences in skull morphology were first described by Cuvier (1825), historically most zoologists considered that all beavers were either one species or two subspecies (Morgan, 1868). It was not until differences in the number of chromosomes determined fairly late on the existence of the two distinct species: the Eurasian beaver has 48 pairs of chromosomes, whereas the North American has 40 pairs (Lavrov and Orlov, 1973), which clearly distinguishes them as separate species. Captive experiments investigating whether these would interbreed took place, but although copulations were recorded, no hybrid offspring resulted (Lavrov and Orlov, 1973). More recent genetic analysis in developing a rapid DNA assay between the two beaver species determined that SNP positions 1971 and 2473 in the 16-s mitochondrial gene are fixed for nucleotides C/A in Eurasian beavers and G/T in North American beavers, respectively (McEwing et al., 2014). These fixed differences are ideal for species identification purposes and could be used to develop a quick field test.

To the experienced observer, subtle differences in pelage colouring (such as more buff-coloured cheeks and rarity of black individuals in North American beavers) and differences in tail shape (generally slightly rounded, more oval shaped in North American whereas straighter edged, parallel sides of the Eurasian have been noted) have been reported, though setting consistent defining species standards can be complicated (Danilov, 1995). Internal differences in skull morphology are evident; at least seven differences have been reported, including nasal bone structure, depression of basioccipital, and shape of nostril and foramen magnum (Miller, 1912; Ognev, 1947; Figure 1.3a and b). These of course cannot be used in field assessments, and many of these physical differences are only relevant on post mortem examination. One curious and seemingly reliable difference appears to be the colour and viscosity of their anal gland secretions (AGS) (Figure 1.4). Examination of these is a quick and reliable method to determine the sex of a beaver, as external sexually dimorphic features are often lacking across both species (Rosell and Sun, 1999).

Many authors have investigated and postulated over dissimilarities between the species (Table 1.3)

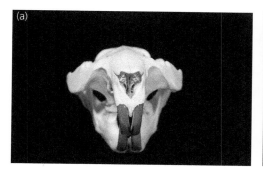 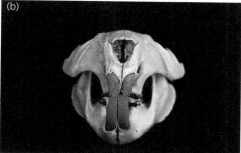

Figure 1.3a and b Differences in skull morphology exist between the two beaver species, the Eurasian (a, on the left) North American beaver (b, on the right), note the more triangular rostrum in the Eurasian. (Photos supplied courtesy of Michael Runtz.)

Figure 1.4 Anal gland secretion (AGS) colour and viscosity vary but can be reliably used to identify species and sexes. (Photo supplied courtesy of Frank Rosell.)

Table 1.3 Main reported differences between the two extant beaver species (Miller, 1912; Ognev, 1947; Lavrov, 1980; Lavrov, 1983; Rosell and Sun, 1999; Rosell et al., 2005; Danilov et al., 2011a; Müller-Schwarze, 2011).

Feature	Eurasian	North American
Genetic		
Chromosome number	48	40
Cranium		
Skull volume	Smaller	Larger
Nasal opening	Triangular	Quadrangular, slightly shorter below than above
Nasal bones	Extend beyond nasal processes of premaxillae	Do not extend beyond nasal processes of premaxillae
Depression between auditory bullae in the lower basioccipital region	Broad and rounded	Ovate
Pterygoid process	Large, 4–6 mm wide	Thin, < 2 mm wide
Least depth of rostrum behind the incisors	Greater than distance from gnathion to end of infraorbital foramen	Nearly equal to distance from gnathion to end of infraorbital foramen
Occipital foramen	Vertically elongated	Horizontally elongated
Foramen magnum	Rounded	Triangular
Cranium width in front postorbital processes	Nearly equals greatest breadth of nasal bones	Greater than greatest breadth of nasal bones
Mandible		
Mandible angular process	Elongately rounded, moderately massive	Short, with rounded edge and very massive
Coronoid process	Strongly bent backwards	Sharpened, bent backwards
Depression between coronoid and angular processes	Prominent	Shallow
Internal		
Uterus masculinus	Present and more consistently identifiable	Not always present and highly variable in form and shape
Anal glands	Larger volume	Smaller volume
Anal gland secretions	Female: greyish-white, paste-like Male: yellow to light-brownish, more fluid	Female: whitish to light-yellow, runny Male: brown and viscous, darker than Eurasian
Tail vertebrae	Narrower, less developed processes	Broader, more developed processes. Deeply bifurcated laminae
Crus bones	Shorter	Longer
External		
Fur	Longer hollow medulla	Shorter hollow medulla
Tail shape	Parallel at midpoint, width 47% of length	Broader across midpoint, width 56% of length
Vertical posture	Assumed less often due to crus bone morphology	Assumed more often due to longer crus bones
Life history		
Sexual maturity	Later typical	Earlier possible
Average fetus number	~2.5	~4.0
Average litter size	1.9–3.1	3.2–4.7
Average family group	3.8 ± 1.0	5.2 ± 1.4
Behavioural		
Dam building	Similar	Similar—greater falsely reported
Lodges	Bank lodges numerous	Free-standing lodges numerous
Scent mounds	Smaller	Some 'giant' mounds recorded

and whether these permit one to have a competitive edge over the other, but general conclusions appear unified (for example, Danilov, 1995; Dewas et al., 2012; Parker et al., 2012; Frosch et al., 2014). Over the years, various studies have reported ecological and life history trait differences. Such comparison studies are particularly relevant where both species are meeting along several fronts, such as in parts of Finland and Russia. For example, studies of adjacent populations of both species living in north-western Russia and Ruusila stated that North American beavers built more dams and stick-type lodges (Danilov and Kan'shiev, 1983), other North American biologists claiming at one point that the greater building activities of North American beavers assisted in giving them a competitive edge (Hilfiker, 1991; Müller-Schwarze, 2011). More recent investigations have reviewed a wide range of ecological features across numerous populations of both species and not found any significant differences, including diet, habitat use, and construction types (Danilov et al., 2011b; Parker et al., 2012). The latest conclusions are that both species build the equivalent frequency and degree of dams and lodges under the same habitat conditions in south Karelia (Danilov and Fyodorov, 2015). Therefore, the niche overlap of both beavers is considered as virtually complete unless further information from sympatric populations comes to light (Parker et al., 2012).

Typically, North American beavers were often presumed to be bigger, raising concerns that this may give them a competitive advantage in body size and aggression in territorial disputes, therefore enabling them to outcompete the native Eurasian. However, body length and masses are highly comparable (Danilov et al., 2011a). It has long been believed that the North American beaver becomes sexually mature faster and has higher fecundity rates (Müller-Schwarze, 2011), with the combined effect of higher fetus numbers (*C. c.* ~4.0, *C. f.* ~2.5) and mean litter sizes (*C. c.* 3.2–4.7, *C. f.* 1.9–3.1). Logically this leads to differences in mean family group sizes (*C. c.* 5.2 ± 1.4, *C. f.* 3.8 ± 1.0) (see Chapter 6). This was thought to give North American beavers a competitive edge and

to explain their more rapid population expansion compared to Eurasian beavers in Finland (Nummi, 2001). However, this is contested by other researchers; moreover, as the two species meet in various parts of Finland and Russia, there are no clear winners and the picture is more complex (see Chapter 2). Actual differences in age of sexual maturity may not be accurately distinguished but rather confounded with data on first age at reproduction, in turn more influenced by population density as opposed to fundamental species differences in reproductive biology. It is therefore not clear if the two beaver species would eventually coexist or be excluded by the other (Petrosyan et al., 2019). Interestingly, several studies report immunophysiological differences between the species, also demonstrated where they both occupy the same habitat. North American beavers appear to have a higher susceptibility to tularaemia and to be a significant reservoir (Mörner, 1992), whereas the Eurasian beaver is reported only sporadically a host (Girling et al., 2019). During tularaemia outbreaks among wild rodents in Voronezh province, Russia, between 1943 and 1945, over half of the North American beavers held at the research centre died, whilst local Eurasian beavers held at the same facility were unaffected (Avrorov and Borisov, 1947).

1.5 Fossil beavers

All rodents share a common ancestor around 57–76 million years ago (myr), with the early beaver-like animals diverging from the scaly-tailed squirrel *Anomalurus* and first appearing around 54 myr (Horn et al., 2011). Therefore, the animal we know today has an incredibly long evolutionary history, though questions still exist around its evolution and closest relatives, as rodent phylogenetic relationships are still difficult to decide (Korth, 1994; Horn et al., 2011). Today's beavers are evolutionary distinct—the only remaining members of the once much larger and diverse family of Castoridae, which dates back nearly 40 myr. The diverse fossil taxa were thought to include up to 30 genera, with more than 100 species at one point (McKenna and

Bell, 1997; Korth and Samuels, 2015; Mörs et al., 2016; Li et al., 2017). In China alone at least ten species of extinct beaver spanning eight genera have been determined from sediments spanning the Early Oligocene to the Pleistocene (Yang et al., 2019), originating from a mouse-related clade, which contains several families including Anomaluridae (scaly-tailed squirrels), Geomyidae (burrowing rodents like gophers), Dipodidae (such as jumping mice), Heteromyidae (such as burrowing rodents, e.g., kangaroo rats), Muridae (true mice), and Pedetidae (springhares) (Huchon et al., 2002; Adkins et al., 2003; Huchon et al., 2007; Blanga-Kanfi et al., 2009). Some debate remains as to whether beavers are more closely related to the scaly-tailed squirrels (Horn et al., 2011) or the Geomyidae (Blanga-Kanfi et al., 2009). The extinct Eutypomyidae are proposed to be the closest related group to Castoridae (Wahlert, 1977). Castoridae varied greatly, from small burrowers around 1 kg in size such as *Palaeocastor* spp. found in the Late Oligocene and Early Miocene to the bear-sized giant beavers of the Pleistocene (Korth, 2001; Rybczynski, 2007).

The origin of the Castoridae, genus *Agnotocastor* (Stirton, 1935), is found in North America at the end of the Eocene (37 myr) representing species that could produce, store, and dispense castoreum (Korth, 2001; Rybczynski et al., 2010). These were also found in Asia and France (Hugueney and Escuillie, 1996) in the Oligocene, suggesting they originated in North America and then radiated out into Asia and wider Eurasia, but this remains debated (Horn et al., 2014). Fossil remains of Castoridae have been found in the Middle East from around the Lower Oligocene and Upper Miocene (Turnbull, 1975). Either way, this occurred via the Beringia isthmus. This was an Arctic land bridge, existing throughout most of the Cenozoic, and although subject to climatic change it enabled 'faunal interchange' and mammalian dispersal between Eurasia and North America (Beard and Dawson, 1999; Gladenkov et al., 2002). This is demonstrated by fossil beaver finds in the Yushe basin, China, which are characterized by long lineages of *Dipoides*, *Trogontherium*, and *Sinocastor* (considered a subgenus of *Castor*), the majority dating back to the Pliocene, with some *Dipoides*

species occurring in the Late Miocene (Xu et al., 2017). These early castorids then gave rise to the burrowing Palaeocastorinae (Martin, 1987; Hugueney and Escuillie, 1996; Korth, 2001; Korth and Rybczynski, 2003). The first animals closely related to beavers and often described as the direct ancestors of contemporary beavers of Eurasia are the genus *Steneofiber* (Geoffroy, 1803), appearing in the Late Oligocene (Hugeney, 1975; Lavrov, 1983; Savage and Russell, 1983). Though they were around the size of marmots (genus *Marmota*), they share morphological similarities, especially in molar structures (Lavrov, 1983). Fossil remains of a family found in France, presumed as ten individuals, were discovered in close proximity and displayed various teeth development, from worn adult teeth to erupting premolars, indicating family structure and breeding patterns parallel to extant beaver species (Hugueney and Escuillie, 1996). The genus *Castor* in Europe has been dated back to between 10 and 12 million years (Lavrov, 1983).

It is thought that up to 30 genera once existed up until the Miocene in Eurasia and up to the Late Pleistocene in the rest of the northern hemisphere (Korth, 2001; Rybczynski, 2007; Rook and Angelone, 2013), so that beavers in the past were much more diverse than today. The palaeocastorine beavers tend to refer to those fossorial beavers restricted to North America and represent an Early Oligocene–Miocene radiation (Stefen, 2014). Around 30 myr palaeocastorines diverged into approximately 15 species (Martin, 1987); this radiation was thought to be associated with a climatic shift, the appearance of more open, grass-dominated habitats, and burrowing adaptations (Strömberg, 2002; Samuels and Valkenburgh, 2009). Table 1.4 shows the approximate beaver timeline based on fossil finds.

It was not until the Early Miocene (24 myr) that the familiar traits of modern beavers—wood cutting and swimming—evolved (Rybczynski, 2007). Prior to this, ancestral beavers tended to be small burrowers (a bit larger that a prairie dog) and are thought to have originated from *Palaeocastor* in the Late Oligocene and Early Miocene (~25 myr) which were adapted to fossorial habits in more upland and arid habitats not associated with

Table 1.4 Approximate fossil beaver timeline (years ago) (Pilleri et al., 1983; Müller-Schwarze, 2011).

Period	Pleistocene	Tertiary				
		Pliocene	Miocene	Oligocene	Eocene	Palaeocene
Years before present	1.9 million–8,000	5.3–1.9 million	23.9–5.3 million	33.7–23.9million	55–33.7 million	66–55 million
Key events	Both extant species coexisted with giant forms	*Castor* arrives in North America, speciation of extant species	Wood cutting and swimming evolve	*Palaeocastorine* species divergence	Common rodent ancestor exists. Split from *Anomalurus*	
North America	*Castoroides ohioensis* *Castor canadensis*	*Dipoides* *Amblycastor* *Castor* *Eucastor*	*Eucastor*	*Palaeocastor* *Agnotocastor*	Earliest beaver-like fossils recorded in California, Germany, and China	
Eurasia	*Castor fiber* *Trogontherium*	*Trogontherium* *Steneofiber* *Castor* spp.	*Castor* spp. *Palaeomys* *Steneofiber eseri* (not adapted to aquatic lifestyle)	*Steneofiber fossor* (underground lifestyle)		

wetlands. Canonical analyses of a wide range of extinct beavers have demonstrated that skull morphology was highly adapted for digging behaviours, though these fossorial features demised around 20 myr (Samuels and Valkenburgh, 2009). Some of these burrowing beavers (three species are associated: *Palaeocastor fossor*, *P. magnus*, and *P. barbouri*) dug unusual, deep, helical burrows, ending in an inclined living chamber (Martin and Bennett, 1977). When first discovered, these structures were described as silicified sponges or extinct plants and named '*Daimonelix*' (Barbour, 1892; Barbour, 1895). Later, Cope (1893) and Fuchs (1983) proposed these were rodent burrows, with plant material being roots of various plants gnawed during burrow construction, which was later validated (Peterson, 1906; Schultz, 1942). Several thousand of these unusual structures ('devil's corkscrews'; Figure 1.5), which can descend to a depth of nearly 3 m and were often found in clusters or 'towns', have now been identified in North America (Martin, 1994).Various theories have been proposed as to why these ancient beavers dug such energetically expensive and complex burrows requiring so much effort to construct (Meyer, 1999). This unusual shape was proposed as an efficient space utilization design (Martin and Bennett, 1977), though this was later ruled out as neighbouring burrows are clustered together apparently randomly (Meyer, 1999). Neither did they seem to offer increased predator protection, as the remains of both predated beavers and various predators have been found in these burrows (Martin and Bennett, 1977; Martin, 1994). In the end they concluded that excavated deep, helical burrow systems in arid grasslands maintained a more consistent subsurface temperature and humidity and may even have trapped some water. All burrowing beaver species (up to eight coexisting species known) disappear from the fossil record around the same time, ~20 myr, with those utilizing a more semi-aquatic lifestyle becoming more successful (Hugueney and Escuillie, 1996).

The *Eucastor* gave rise to *Dipoides*, another former member of the family Castoridae, sharing the trait of tree exploitation and possessing the wood-cutting abilities of the genus *Castor*, and therefore our modern beavers (Rybczynski, 2007). Though

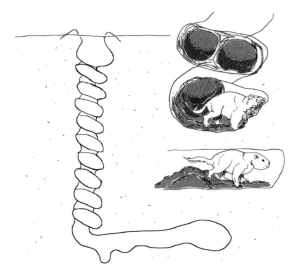

Figure 1.5 'Devil's corkscrew' burrows of ancient fossorial beaver ancestors. (Illustration provided courtesy of Rachael Campbell-Palmer, redrawn after Martin & Bennett, 1977.)

Castor and *Dipoides* share a common wood-cutting ancestor, thought to have been a burrower, which can be traced back at least 24 myr, they are not close relatives but share a semi-aquatic clade (Rybczynski, 2007). Their respective gnawing marks on fossilized tree remains may be difficult to distinguish without careful examination. *Dipoides* are considered less advanced, producing smaller and more overlapping cuts due to their smaller and more round incisors, compared to the more evolved chisel-like ones seen in beavers today (Tedford and Harington, 2003). The wood-cutting abilities of *Castor* are more efficient as their straight edge incisors produce larger cuts, taking fewer bites to fell equivalent sizes of woody material than *Dipoides* with their strongly curved incisors (Rybczynski, 2008). Fossil remains of intertwined cut sticks have also been found in association with them, implying 'nest-type' structures that could be evidence of early lodge and/or dam building activities (Tedford and Harington, 2003). Evidence of dam building behaviours and impacts on geomorphology in extinct Castoridae appears lacking, though wood cutting pre-existed modern beavers (Plint et al., 2020).

Modern beavers (genus *Castor*) first appeared during the Late Miocene and Pliocene (11–2.5 myr) as one common species around the Palaearctic region (Xu, 1994; Rekovets et al., 2009; Rybczynski et al., 2010) from their close relative species *Steneofiber* (Xu, 1994; Rybczynski, 2007; Flynn and Jacobs, 2008). *Castor praefiber* (Depéret, 1897) for example is considered an intermediate species in *Castor fiber* evolution thought to have first appeared in the early Pliocene (Rekovets et al., 2009). The genus *Castor* is believed to have emerged in Eurasia and then penetrated into North America via the Bering land bridge (Lavrov, 1983; Lindsay et al., 1984; Xu, 1994; Hugueney and Escuillie, 1996; Flynn and Jacobs, 2008) during the Pliocene around 4.9–6.6 myr (Lindsay et al., 1984; Xu, 1994). The earliest *C. fiber* is known from is the Early Pleistocene (Barisone et al., 2006; Rekovets et al., 2009), and it is thought to have overlapped with *Castor plicidens* (Cuenca-Bescós et al., 2015).

Trogontherium was a congener of modern beavers; these were the giant beavers of Eurasia, though widely distributed throughout the Palaearctic, comprising three species, *T. minutum*, *T. minus*, and the largest *T. cuvieri*, in the Upper Pliocene (Mayhew, 1978; Fostowicz-Frelik, 2008). Their fossil remains have been discovered regularly with those of the genus *Castor* in both Europe and Asia from the Early Pleistocene (~2.4–0.13 myr), and the recent finding of a specimen in China has extended its extinction date to the Late Pleistocene (Yang et al., 2019). It appears that modern beavers lived alongside or were possibly locally extirpated by the slightly larger *Trogontherium cuvieri*, as the prevalence of the two forms at archaeological sites demonstrates an inverse relationship (Mayhew, 1978). Additionally, this places the survival of this giant beaver as overlapping with Pleistocene people and therefore a candidate with the other extinctions of large Ice Age mammals caused by human activities (Yang et al., 2019). Figure 1.6a and b shows comparison sizes.

More recent genetic studies have determined that divergence between our two modern species occurred around 7.5 myr (Horn et al., 2011). The Bering land bridge would have permitted animal movements between Eurasia and North America. After the land bridge disappeared this most likely triggered the speciation into *C. fiber* and *C. canadensis* as they became completely isolated from each other (Horn et al., 2011). The Eurasian beaver is therefore thought to be around twice as old, dating back around 2 myr. Genetic differences and lack of hybridization are evident, but their biology,

(a)

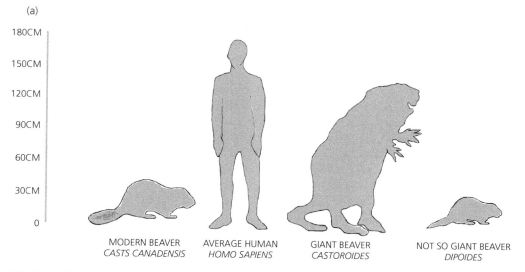

Figure 1.6a Modern beaver size in relation to *Homo sapiens*, their giant ancestor *Castoroides*, and *Dipoides*. (Illustration provided courtesy of Rachael Campbell-Palmer, redrawn after Scott Woods, Western University, Canada.)

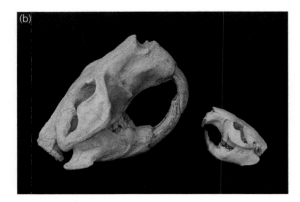

Figure 1.6b Giant beaver replica skull in comparison to beavers today. (Photo supplied courtesy of Michael Runtz.)

morphology, and physiology remain remarkably similar, along with their shared ancient, coevolved parasites, the beaver beetle *Platypsyllus castoris* and stomach nematode *Travassosius rufus* (Lavrov, 1983; see Chapter 8).

Today's beavers are the second largest species of rodent in the world (the largest being the capybara, *Hydrochoerus hydrochaeris*, from South America), although the now-extinct giant beavers (genus *Castoroides*) really capture the imagination. It is believed that giant beavers and modern beavers shared most of the same range, with fossils found from Alaska to Florida (Kurtén and Anderson, 1980). These giant beavers existed in North America during the Pleistocene and were one of the last megafauna (typically defined as animals with body weights > 44 kg; Martin, 1984) to go extinct near the end of this epoch—thought to be due to a combination of climatic and anthropogenic impacts (Boulanger and Lyman, 2014; Cooper et al., 2015). One such giant beaver *Castoroides leiseyorum*, was thought to reach adult body sizes of 2.5 m in length and weigh between 150 and 200 kg (Kurtén and Anderson, 1980) and was known to exist throughout the southeastern USA (Parmalee and Graham, 2002). Another, *Castoroides ohioensis*, thought to have been one of the last of the giant beavers, disappeared around 10,000 years ago (Boulanger and Lyman, 2014). There are conflicting theories on giant beaver ecology, especially regarding their tree cutting and dam and lodge building abilities, as little evidence of these structures exist (Rybczynski,

2008), though their large body size and short limbs are thought to have made them poorly adapted for terrestrial life (Plint et al., 2019).

However, very recent palaeodietary studies have used stable carbon and nitrogen isotope analysis of bone collagen to confirm these beavers were likely to be cold-tolerant and highly dependent on submerged and floating macrophytes, a significant factor allowing *C. canadensis* to coexist, as they would have exhibited complementary dietary niches (Plint et al., 2019). This dietary analysis did not support tree material consumption, which is consistent with their dental morphology (Rinaldi et al., 2009), both indicating giant beavers had to rely on significant existing wetlands (Plint et al., 2019). Fossil finds display rounded incisors with blunt tips, and this has prompted researchers to believe these teeth were used to cut off and grind coarse swamp vegetation rather than trees (Kurtén and Anderson, 1980).

Evidence for the last known giant beaver populations is concentrated in the Great Lake Basins, northern USA, and Ontario, Canada, where they are thought to have hung on before their final extinction (Boulanger and Lyman, 2014). Although there is evidence of giant beavers and humans overlapping in these areas for ~1,000 years, no evidence of them hunting these animals currently exists (Boulanger and Lyman, 2014). It is therefore concluded that changes to warmer and drier climatic conditions resulted in suitable wetland habitat loss through reduction in glacial melt water and sediment infilling, with associated changes in woody vegetation and giant beavers being reduced to small, isolated populations that eventually died out completely (Plint et al., 2019); on the other hand, *Castor* spp. possessed incisors, allowing tree felling, taking advantage of woody tree species, and would have had the ability to create new habitats, giving them a competitive edge (Plint et al., 2019).

1.6 Modern beavers

Extant beavers were first named and described as *Castor fiber* by Linnaeus (1758), while Kuhl first named and described *Castor canadensis* (1820). Fossil evidence of Eurasian beavers has been found in Italy, Spain, Turkey, Syria, Iraq, Israel, and Iran, but the species is thought to have gone extinct in southern Eurasia by the Late Holocene (Linstow, 1908; Legge and Rowley-Conwy, 1986; Barisone et al., 2006). For example, fossil evidence of Eurasian beaver from Spain documents their presence from the Early Pleistocene ~1.4 myr, but then there is a great absence of remains during the Middle Pleistocene when human occupation intensified and suitable habitat most likely became scarcer (Cuenca-Bescós et al., 2015). Both beaver species remained widespread in suitable freshwater habitats throughout the northern hemisphere successfully until human populations grew and began to exploit them, beginning in Eurasia (see Chapter 2). The mass exploitation of beavers, to the point of near complete extinction, has impacted on both their modern distribution and subspecies classification. For both species the usage of subspecies names is complicated by inconsistent application in the literature, with some names not following the rules of the International Code of Zoological Nomenclature and based on contested scientific data such as differences only in relict population survival location (Gabrys and Wazna, 2003). Table 1.5 shows the historical classification of subspecies for both extant beaver species.

The greatest impact on Eurasian beavers, with the most rapid period of decline, occurred in the nineteenth century, by the end of which this species was on the verge of becoming extinct and reduced to a handful of populations in fragmented refugia left after the fur trade, thought to number 1,200 individuals overall (Veron, 1992; Nolet and Rosell, 1998). As a side note, in some early taxonomical descriptions the Eurasian beaver was classified as two species—the eastern beaver, *C. fiber*, and the western beaver, *C. albicus*—by Matschie (1907) based on craniological differences (Lavrov, 1981; Lavrov, 1983). Most zoologists at that time, however, recognized only two contemporary species, the North American and a single Eurasian species (Figure 1.7a Eurasian and Figure 1.7b North American), so this third species (*C. albicus*) was rejected and placed in subspecies classification, which themselves underwent several debates (Gabryś and Ważna, 2003). Ancient beaver DNA

analysis does not provide any evidence to support defined substructure categories, instead forming part of a continuous clade (Horn et al., 2014; Marr et al., 2018), though divergence in mtDNA haplotypes is evident (eastern and western phylogroups), caused by population retreat into glacial refugia during the last Ice Age (~25,000 years ago) (Durka et al., 2005). So enough defined, it was recommended they be managed as separate evolutionary significant units (ESU) (Durka et al., 2005). ESUs are largely defined as reciprocally monophyletic mtDNA units exhibiting significant divergence of allele frequencies at nuclear markers and in regard to conservation management can suggest sourcing for reintroductions (Moritz, 1994; Frosch et al., 2014). Ancient beaver populations were pretty much continuous across the whole of Eurasia and although the two main lineages were apparent, so too was a higher degree of haplotype diversity, and later differences determined in the remaining relict populations were not as stark (Horn et al., 2014). Comparison of DNA from fossil beavers with modern beavers (using samples ranging from several hundred to 11,000 years old) demonstrates Eurasian beavers have suffered a significant genetic bottleneck, losing at least a quarter of their unique haplotypes (Horn et al., 2014). This loss of genetic diversity occurred during the Holocene, when human populations expanded (Horn et al., 2014), with the most recent impact and distribution strongly linked with human activities (Halley et al., 2020).

These nine relict populations, characterized by low genetic variability and a low proportion of polymorphic loci (Ellegren et al., 1993; Babik et al., 2005; Ducroz et al., 2005), were previously considered to be distinct subspecies based on morphological skull measurements, disjunct distribution (Lavrov, 1981; Heidecke, 1986; Frahnert, 2000), and mitochondrial differences, including proportion of assignment to eastern or western clades (Durka et al., 2005). The Belarus refuge has more recently been determined to be the most genetically diverse compared to all the other relict populations, given its larger population size and more complex distribution remaining at isolated locations across several water basins while it passed through this genetic bottlenecking (Munclinger et al., in prep).

Table 1.5 Historical subspecies classification for both extant beaver species. Note many of these are no longer formally recognized as a subspecies or referred to as a fur trade refugia (Gabryś and Ważna, 2003; Pelz-Serrano, 2011).

Species	Subspecies	Region	References
C. fiber	C. f. albicus	Germany (Elbe), Poland	Matschis (1907)
	C. f. galliae	France (Rhone)	Geoffroy (1803)
	C. f. fiber	Norway	Linneaus (1758)
	C. f. belarusicus	Belarus, northern Ukraine	Heidecke (1986)
	C. f. orientoeuropaeus	Russia (Voronez), Belarus	Lavrov (1981)
	C. f. pohlei	Western Siberia, Urals	Serebrennikov (1929)
	C. f. tuvinicus	South-central Siberia	Lavrov (1969)
	C. f. birulai	Southwest Mongolia, China	Serebrennikov (1929)
	C. f. vistulanus	Vistula, Poland	Matschis (1907)
C. canadensis	C. c. acadicus	New Brunswick, New England, Nova Scotia, Quebec	Bailey and Doutt (1942)
	C. c. baileyi	Humboldt River, Nevada	Nelson (1927)
	C. c. belugae	Yukon, Alaska	Taylor (1916)
	C. c. caecator	Newfoundland	Kuhl (1820); Banngs (1913)
	C. c. canadensis	Canada, British Columbia	
	C. c. carolinensis	North Carolina, Louisiana, Mississippi	Rhoads (1898)
	C. c. concisor	Colorado, Mexico	Warren and Hall (1939)
	C. c. duchesnei	Duchesne River, Utah	Durrant and Crane
	C. c. frondator	Rio San Pedro, Mexico border	(1948) Mearns (1898)
	C. c. idoneus	Oregon, Washington	Jewett and Hall (1940)
	C. c. labradorensis	Labrador rivers	Bailey and Doutt (1942)
	C. c. leucodontus	Vancouver Island, British Columbia	Gray (1869)
	C. c. mexicanus	New Mexico, Texas	Bailey (1913)
	C. c. michiganensis	Michigan, Wisconsin, Minnesota Saskatchewan, Manitoba,	Bailey (1913)
	C. c. missouriensis	Missouri, Dakota Washington, British Columbia,	Bailey (1919)
	C. c. pacificus	Idaho	Benson (1933)
	C. c. pallidus	Raft River, Utah	Durrant and Crane
	C. c. phaeus	Pleasant Bay, Alaska	(1948)
	C. c. repentinus	Grand Canyon, Arizona	Heller (1909)
	C. c. rostralis	Red Butte Canyon, Utah	Goldman (1932)
	C. c. sagittatus	British Columbia, Yukon, Idaho Shasta Mountains, California	Durrant and Crane (1948)
	C. c. shastensis	California	Benson (1933)
	C. c. subauratus	Big Wood River, Idaho	
	C. c. taylori	Cummings Creek, Texas	Taylor (1916)
	C. c. texensis		Taylor (1912) Davis (1939) Bailey (1905)

Since the 1900s, beaver numbers have recovered throughout much of their former European range as a result of a combination of legal protection (species and habitat), reduction in hunting pressure and increased regulation, land-use shifts including farmland abandonment, proactive reintroductions/ translocations, and natural recolonizations (Deinet et al., 2013; see Section 2.8.1). Genetic analysis of mitochondrial DNA and major histocompatibility complex (MHC) DRB gene sequences demonstrates low diversity within these refugia populations, though distinctions between them do exist (Babik

Figure 1.7a Adult Eurasian beaver (*Castor fiber*). (Photo supplied courtesy of Frank Rosell.)

Figure 1.7b Adult North American beaver (*Castor canadensis*). (Photo supplied courtesy of Jan Herr.)

et al., 2005; Ducroz et al., 2005; Durka et al., 2005). Despite large geographical distances, there was no significant genetic differentiation between Mongolian and eastern Russian beaver populations (Ducroz et al., 2005). No major variations in haplotypes have been found within central European relict populations (Durka et al., 2005). However, some skull morphometric analyses suggest some evidence for relict population differentiation (Frahnert, 2000) and behavioural discrimination studies indicate lesser recognition and reactions to anal gland secretions between, *C. f. fiber* (relict Norwegian) and *C. f. albicus* (relict German), different Eurasian subspecies (Rosell and Steifetten, 2004). Whilst distinctive eastern (Russia, Belarus, Siberia, and Mongolia) and western (relict France, Germany, and Norway) clades based on genetic differences based on mtDNA have been found (Durka et al., 2005; Horn et al., 2014), as for many European mammals following the last Ice Age, hybridization zones are also apparent. Recent genetic studies dismiss these differences as being significant enough to warrant subspecies classification, as the remnants of a once much more diverse, mixed, and expansive population, now relegated to the artefact of human hunting and recent anthropogenic genetic bottlenecking rather than previously separated subspecies (Horn et al., 2014). This has been supported by nuclear and mitochondrial analyses (Frosch et al., 2014; Horn et al., 2014; Senn et al., 2014). Recent genetic analysis has concluded that much of Europe and Russia is now populated by admixed beavers, resulting in increased genetic diversity leading to viable and successfully expanding populations,

indicating that outbreeding depression is not a significant impact (Munclinger et al., in prep). Some authors still argue for the recognition of certain subspecies/populations, for example Siberian beavers, *C. f. pohlei*, defined as possessing a specific haplotype marker (Saveljev and Lavrov, 2016), and *C. f. tuvinicus* and *C. f. birulini* populations in Mongolia and China given their long period of isolation (Munclinger et al., in prep).

Genetic screening of modern-day Eurasian beaver populations demonstrates that the degree of mixing between eastern and western lineages is already so advanced (Frosch et al., 2014; Senn et al., 2014) across several areas in Eurasia that it seems pointless to try and maintain this former glaciation and geographically induced clade separation, especially as anthropogenic translocations continue and naturally population expansion leads to secondary contact and mixing across their native range (Frosch et al., 2014). Recent genetic analysis of the current population determined that mitochondrial DNA (mtDNA) and nuclear microsatellites reflected the composition of the founder animals and that admixture zones occurred (Minnig et al., 2016). Recent recovery through both natural spread and reintroductions demonstrates that beavers from these relict populations are meeting and admixing (Frosch et al., 2014). Poland for example is a modern-day mixing zone, where as a result of natural range expansion and multiple translocations, admixed populations now exist, with both lineages represented (Biedrzycka et al., 2014).

One of the most detailed genetic analyses of the Eurasian beaver genome was undertaken by Senn et al. (2014), through the identification of 306 single nucleotide polymorphisms (SNP) genotyping derived from restriction site associated DNA (RAD) sequencing data (Senn et al., 2013) comparing samples from across Europe. Several key outcomes were established: the Norwegian *C. f. fiber* and French *C. f. galliae* are distinct populations with no evidence of mixing with any other populations; there is no evidence of *C. f. albicus* remaining as a separate identity with a high degree of introgression with other populations; *C. f. belorussicus* and *C. f. orientoeuropaeus* from Belarus and Russia share a common genetic cluster; and beavers from Bavaria and Switzerland clearly display multiple genetic origins and represent admixed populations (Senn et al., 2014). After they were extirpated at the start of the nineteenth century, reintroductions of 141 individuals to Switzerland occurred between 1956 and 1977. These beavers were sourced from the Rhone valley in France (*C. f. galliae*) released into the Rhone and Rhine catchment areas in the west, beavers from the Telemark region in Norway (*C. f. fiber*), and beavers from the Voronezh province in Russia (*C. f. orientoeuropaeus*) were released into the eastern Rhine catchment (Stocker, 1985), though haplotypes of French, Norwegian, and German (*C. f. albicus*), due to secondary contact, were later determined genetically (Minnig et al., 2016).

It has been clearly demonstrated that genetic diversity is highest in populations involving the reintroduction of mixed sourced animals, compared to those composed solely of remaining fur trade refugia (Frosch et al., 2014; Senn et al., 2014). In addition, these studies provided evidence that the genetic difference between the previous eastern and western clades was not as distinct as the haplotype data suggested and subsequent mixing is fairly extensive as beaver populations meet along varying fronts. However, in some populations relatedness between individuals is still high, e.g. parts of Britain (Campbell-Palmer et al., 2020) and Switzerland where some closely related individuals were found up to 50 km away (Minnig et al., 2016) (Figure 1.8).

A very similar story occurred with the North American beaver; at one time 24+ subspecies were recognized (Jenkins and Busher, 1979). Previously widespread across Canada and North America,

from the Arctic to the Rio Grande, until they were commercially removed by the fur trade, *C. c. acadicus*, *C. c. canadensis*, *C. c. carolinensis*, and *C. c. missouriensis* were considered to be the most widespread North American subspecies (Hall, 1981). Repeated and widespread reintroduction and restoration projects have led to mixing of many of the more isolated and potentially discrete populations (Baker and Hill, 2003). The pattern of hunting, then subsequent reintroductions and mixing has essentially made many of these categorizations redundant. Some flexible mating strategies of beavers, including extra-pair mating (Crawford et al., 2008) and equal dispersal between the sexes (Sun et al., 2000), suggest they are socially monogamous but opportunistically promiscuous, which should promote mixing and genetic diversity. However, genetic structure comparison analysis has been undertaken between different Illinois populations, one acting as a small population with single family units, whilst the other had larger family numbers and size, with multiple breeding adults on a less linear system (Sun, 2003; Crawford et al., 2008). These mating and ecological differences had clear influences on genetic population structure—for example, groups of animals in non-linear systems proving more difficult to scent mark and defend, and multiple families presenting more opportunities for interactions (Crawford et al., 2009). They also found evidence of female philopatry; however, parental genetic analysis was suggestive of dispersal between populations and promiscuous mating systems (Crawford et al., 2008). Therefore, some gene flow was maintained, even though families functioned as fairly distinct breeding units (Crawford et al., 2009; see also Chapter 6). The authors conclude that despite reintroductions occurring previously to Illinois (Pietsch, 1956), beaver populations are acting as fairly isolated units with some limited dispersal between them, a fact to consider in longer-term beaver population management. Overall, North American beavers are now once again present in all the states and provinces they previously occupied, with numbers still growing (see Section 2.8.2).

To try and unravel the genome of the North American beaver, a beaver volunteer called Ward, descended from wild stock originating in Quebec, was undertaken by geneticists in Ontario (Lok et al.,

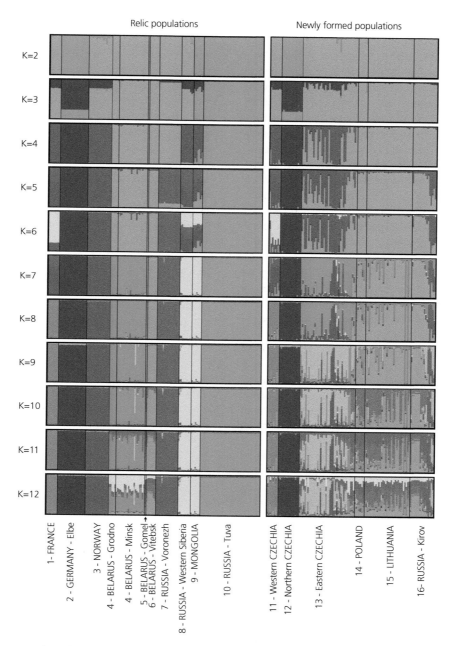

Relic populations   Newly formed populations

Figure 1.8 Structure analysis of microsatellite data for K (number of assumed groups) from 2 to 12. Newly formed populations display a greater degree of admixing in contrast to relict populations. Newly formed populations are viable and expanding and typically display higher genetic diversity and hybrid vigor, outbreeding depression does not appear to be a significant issue. (Figure supplied courtesy of Pavel Munclinger.)

2017). Despite being almost made extinct, with remaining populations small and dispersed, high levels of genetic diversity appear to have been retained with no strong indication of recent genetic bottleneck, known to have occurred ~100 years ago (Pelz-Serrano et al., 2009; Lok et al., 2017). Microsatellite studies have found mean heterozygosity ranging from 66.3 to 74% in tested populations, though the degree of gene flow possible amongst all the various geographic locations is questioned (Crawford et al., 2009; Lok et al., 2017). Beavers sampled (n=117) from Alabama, Arizona, Maine, Minnesota, South Carolina, Texas, and Wisconsin and tested across nine microsatellites found them to be highly polymorphic, again finding mean heterozygosity around 74%, with Alabama displaying the lowest and Texas populations the highest (Pelz-Serrano, 2011). Therefore, the North American beaver has not appeared to have lost the same extent of genetic diversity as the Eurasian beaver.

Analysis of mitochondrial DNA does indicate that past climatic and geological events, predominantly ice sheet coverage, determined phylogenetic relationships, with evidence of multiple Pleistocene refugia, including in Texas and the Rocky and Appalachian Mountains (Pelz-Serrano, 2011). Rapid population recovery, potentially due to the greater expanses of remaining and less human-populated landscapes along with larger remaining population size, as compared to Europe, could explain the lesser loss of heterozygosity. There have been investigations on whether this mass persecution, followed by a series of translocation and reintroduction efforts to compensate for this loss serve to enable the evaluation of these human impacts on current genetic diversity of populations today. Genetic analysis (based on nine microsatellites) of seven North American geographic areas subject to historic translocations (Alabama, Arizona, Maine, Minnesota, South Carolina, Texas, and Wisconsin) concluded that high genetic diversity at the mitochondrial DNA has remained, though this doesn't mean that other parts of the genome have not been negatively impacted, and high gene flow rates among some of these geographic areas are lacking (Pelz-Serrano et al., 2009). For example, Russian populations of Eurasian beaver demonstrated a high degree of polymorphism at the microsatellite but not MHC loci, whilst both were extremely low in Scandinavian beavers (Ellegren et al., 1993). The Alabama population for example seem composed of alleles from both the Texas and South Carolina populations (probably due to historical translocations); therefore current populations are a result of ancestral populations, admixed with modern translocations, though natural dispersal is also contributing to low levels of genetic flow as these populations do not exist in complete genetic isolation (Pelz-Serrano et al., 2009). Early, interregional translocations have most certainly aided the preservation of genetic diversity and recovery of this species, though that is not to say some genetic distinctions and specialized local adaptations have not been lost.

1.7 Mistaken identity

Even for such a distinctive species it is surprising how often and how variable misidentification of beavers can be. A number of semi-aquatic mammals utilize and simultaneously overlap occupation of freshwater riparian habitats (Hood, 2020). Identifying swimming animals in particular seems to cause some confusion. This is not entirely surprising as regular and clear viewing of beavers can still be an elusive experience to many. Not only may beavers be misidentified as other species, but also other species, especially those commonly utilizing swimming and burrowing near to freshwater environments, can often be labelled as beavers. Particularly where beavers are relatively new recolonizers of areas, any swimming mammal (and sometimes even a duck!) can be enthusiastically assigned as a *Castor* species. Depending on which continent you are in (and recognizing that some of these species are now common introduced invasive animals), otters (*Lutra lutra*), coypu or nutria (*Myocastor coypus*), muskrat (*Ondatra zibethicus*), and even water voles (*Arvicola amphibius*) may be commonly mistaken (Figure 1.9a–d). In North America, beavers, coypu, and muskrat are important furbearing semi-aquatic mammals, which can be confused for each other by a casual observer as their ranges commonly overlap.

Figure 1.9a Both the Eurasian (*Lutra lutra*) and North American (*Lutra canadensis*) otter can be confused for beavers, especially when swimming, but otters move quite differently, with much more slender bodies and completely different tails. (Photo supplied courtesy of Kirsty Taylor-Wilson.)

Figure 1.9b The coypu or nutria (*Myocastor coypus*) looks the most similar to beavers apart from their tails, but they are completely unrelated. (Photo supplied courtesy of Leopold Kanzler.)

Figure 1.9c Swimming muskrats (*Ondatra zibethicus*) can look especially like swimming beaver kits, but out of water their tail is clearly different. (Photo supplied courtesy of Leopold Kanzler.)

Figure 1.9d Though much smaller, swimming water voles (*Arvicola amphibious*) have also been mistaken for beaver kits. (Photo supplied courtesy of Kirsty Taylor-Wilson.)

A naming confusion may come with the mountain beaver (*Aplodontia rufa*) also known as a sewellel, found in western North America and which of course is not a beaver at all. They are small (~20–46 cm long and weighing 0.8–1.2 kg), burrow-dwelling rodents, with a short furry tail which is largely non-visible, living in forest with dense understories, and confusingly not aquatic or found in the mountains (Kays and Wilson, 2010). They eat a range of vegetation, including bark and small branches, which is thought to be how they got their name (Long, 2000).

It is perhaps the beaver's tail, a common and distinctive feature of both species, that tends to set it apart from other species—not only in shape and size but also in the fact it is generally hairless bar

the odd, short and scattered bristles. Their tails' fish-scale-like appearance was once falsely accredited to their pescatarian diet, and this is still a surprisingly commonly held belief. Perhaps the most similar looking species is the coypu, which both in and out of the water is uncannily comparable. This South American species has been introduced to both North America and parts of Europe as a productive, furbearing mammal. This confusion with beavers occurs to such an extent that many media stories concerning beavers are often illustrated with photos of coypu. The key difference is undoubtedly the tail, being round, long, thin, and sparsely covered with bristle hair; whereas the beaver tail is uniquely beaver! On land these two animals move completely differently and hopefully if you are seeing them out of the water then the tail should be immediately obvious.

Beavers are often described as cumbersome on land, walking on all fours with the tail dragging behind in a purposeful direction but not at great speed, unless they feel in immediate danger. Coypu are more agile and can almost be described as being able to scuttle on land, rather than plod along. Coypu are also much smaller, with average weights of ~5.4 kg, though they can reach 9.1 kg, and body lengths of ~61 cm (LeBlanc, 1994), being more similar to a 1-year-old beaver. Facially they have a more truncated snout than beavers and possess noticeable white whiskers around the muzzle. They share webbed hind feet, though in coypu the first four toes are connected whilst the outer toe is free, as opposed to being fully webbed in beavers; therefore tracks, especially the hind paws, can be differentiated. They also possess prominent orange incisor teeth which can range into a deeper, more red-orange than beavers in coloration (LeBlanc, 1994). Field signs (especially burrows and feeding signs) from both these species can also be hard to distinguish and can overlap. Their scats are also distinct from beavers as they produce dark green to almost black cylindrical faeces (~5 cm long) marked by deep parallel grooves (LeBlanc, 1994). Life history traits also vary, with coypu being shorter lived, reaching sexual maturity at a much younger stage (~4 months), and having multiple litters throughout the year (LeBlanc, 1994). Lactating coypus have more visible nipples than beavers, which are set high on the female's side.

Muskrats and beavers can be confused. Though native to North America, muskrats have been introduced to parts of Europe and share the same aquatic habitats. It is proposed that some degree of commensal relationship may exist as they sometimes co-inhabit beaver burrows and lodges, and the muskrat nest or 'push-up' can look like a mini-beaver lodge (MacArthur and Aleksiuk, 1979). Misidentification can happen especially with a swimming animal, though muskrats are significantly smaller (overall length 46–61 cm with average weights of 1.1 kg) than beavers and coypu (Miller, 1994). They could be mistaken for a beaver kit, though they can swim much faster. Facially the muskrat's snout tends to be more pointed, with no visible white whiskers. Again, the tail is distinctly different from beavers, being scaly and laterally flattened but long and thin in shape and when swimming makes a clear undulating motion. Coypu and beaver tails on the other hand are still while surface swimming. Muskrat and water vole feeding signs on green vegetation (as opposed to woody vegetation) could be mistaken for beaver feeding (Strachan et al., 2011); however, alternative field signs in close proximity should aid identification. When out of the water and eating, for example, their digits are visually more elongated with longer, often lighter nails than beavers. Their tracks are significantly smaller than both beaver and coypu, and they possess skin folds between the toes of their hind feet (partially webbed) as opposed to clear webbing between toes (Miller, 1994).

On occasions, otters have been confused for beavers and vice versa. there are few physical similarities, though they do share a similar semi-aquatic lifestyle and are often seen swimming in the same freshwater habitats (Woodroffe, 2007). Either the excitement of knowing beavers may be recolonizing an area, mixed with at times poorer viewing conditions, such as seeing a swimming animal in the dusk, can fool the observer. Many of their behavioural and habitat use traits differ, due to their carnivore and vegetarian diets respectively. Otters are faster, agile, energetic swimmers capable of changing direction and speed abruptly, with elongated bodies and prominent whiskers (e.g. Erlinge, 1968), whereas beavers tend to be more set on their course—yacht to tanker comparisons come to mind! Otters have been reported to use abandoned beaver burrows and sometimes even active lodges (Vorel, 2001). Otters tend to hold their heads higher out of the water while swimming, whereas when beavers swim on the surface only the upper part of their head (eyes and above) are visible.

References

Adkins, R. M., Walton, A. H., and Honeycutt, R. L. 2003. Higher-level systematics of rodents and divergence time estimates based on two congruent nuclear genes. *Molecular Phylogenetics and Evolution*, 26, 409–420.

Aleksiuk, M. and Cowan, I. M. 1969. Aspects of seasonal energy expenditure in the beaver (*Castor canadensis* Kuhl) at the northern limit of its distribution. *Canadian Journal of Zoology*, 47, 471–481.

Avrorov, A. and Borisov, B. P. 1947. Cited in Lavrov, L.S. 1983. Evolutionary development of the genus *Castor* and taxonomy of the contemporary beavers of Eurasia. *Acta Zoologica Fennica*, 174, 87–90.

Babik, W., Durka, W., and Radwan, J. 2005. Sequence diversity of the MHC DRB gene in the Eurasian beaver (*Castor fiber*). *Molecular Ecology*, 14, 4249–4257.

Backhouse, F. 2018. *Once They Were Hats: In Search of the Mighty Beaver*. Toronto: ECW Press.

Bailey, V. 1905. Biological survey of Texas. *Northern American Fauna*, 25, 1–222.

Bailey, V. 1913. Two new subspecies of North American beavers. *Proceedings of the Biological Society of Washington*, 26, 191–193.

Bailey, V. 1919. A new subspecies of beaver from North Dakota. *Journal of Mammalogy*, 1, 31–32.

Bailey, V. and Doutt, J. K. 1942. Two new beavers from Labrador and New Brunswick. *Journal of Mammalogy*, 23, 86–88.

Baker, B. and Hill, E. 2003. Beaver (*Castor canadensis*). In: Feldhamer, G. A., Thompson, B. C., and Chapman, J. A. (eds.) *Wild Mammals of North America: Biology, Management and Conservation*. Baltimore: Johns Hopkins University Press.

Banngs, O. 1913. The land mammals of Newfoundland. *Bulletin Harvard Museum Comparative Zoology*, 54, 507–516.

Barbour, E. H. 1895. Progress made in the study of Daemonelix. *Proceedings of Nebraska Academic Scientific Publications*, 5, 1894–1895.

Barbour, I. H. 1892. Notice of new gigantic fossils. *Science*, 19, 99–100.

Barisone, G., Argenti, P., and Kotsakis, T. 2006. Plio-Pleistocene evolution of the genus *Castor* (Rodentia, Mammalia) in Europe: *C. fiber* Plicidens of Pietrafitta (Perugia, Central Italy). *Geobios*, 39, 757–770.

Beard, K. C. and Dawson, M. R. 1999. Intercontinental dispersal of Holarctic land mammals near the Paleocene/Eocene boundary; paleogeographic, paleoclimatic and biostratigraphic implications. *Bulletin de la Société Géologique de France*, 170, 697–706.

Benson, S. B. 1933. A new race of beaver from British Columbia. *Journal of Mammalogy*, 14, 320–325.

Biedrzycka, A., Konior, M., Babik, W., Świsłocka, M., and Ratkiewicz, M. 2014. Admixture of two phylogeographic lineages of the Eurasian beaver in Poland. *Mammalian Biology*, 79, 287–296.

Blanga-Kanfi, S., Miranda, H., Penn, O., Pupko, T., Debry, R. W., and Huchon, D. 2009. Rodent phylogeny revised: analysis of six nuclear genes from all major rodent clades. *Biomed Central Evolutionary Biology*, 9, 1–12.

Boulanger, M. T. and Lyman, R. L. 2014. Northeastern North American Pleistocene megafauna chronologically overlapped minimally with Paleoindians. *Quaternary Science Reviews*, 85, 35–46.

Busher, P. 2016. Family Castoridae (beavers). In: Wilon, D. E., Lacher, T. E., and Mittermeier, J. R. A. (eds.) *Handbook of the Mammals of the World; Logomorphs and Rodents I*. Barcelona: Lynx Edicions.

Campbell-Palmer, R., Senn, H., Girling, S., Pizzi, R., Elliott, M., Gaywood, M. and Rosell, F. 2020. Beaver genetic surveillance in Britain. *Global Ecology and Conservation*, 24, e01275.

Cooper, A., Turney, C., Hughen, K. A., Brook, B. W., Mcdonald, H. G., and Bradshaw, C. J. A. 2015. Abrupt warming events drove late Pleistocene Holarctic megafaunal turnover. *Science*, 349, 602–606.

Cope, E. D. 1893. A supposed new order of gigantic fossils from Nebraska. *American Naturalist*, 27, 559–560.

Crawford, J. C., Liu, Z., Nelson, T. A., Nielsen, C. K., and Bloomquist, C. K. 2008. Microsatellite analysis of mating and kinship in beavers (*Castor canadensis*). *Journal of Mammalogy*, 89, 575–581.

Crawford, J. C., Liu, Z., Nelson, T. A., Nielsen, C. K., and Bloomquist, C. K. 2009. Genetic population structure within and between beaver (*Castor canadensis*) populations in Illinois. *Journal of Mammalogy*, 90, 373–379.

Cuenca-Bescós, G., Ardevol, J. R., Morcillo-Amo, Á., Galindo-Pellicena, M. Á., Santos, E., and Costa, R. M. 2015. Beavers (Castoridae, Rodentia, Mammalia) from the Quaternary sites of the Sierra de Atapuerca, in Burgos, Spain. *Quaternary International*, 433, 263–277.

Cuvier, G. 1825. *Recherches sur les Ossemens Fossiles, où l'on Rétablit les Caractères de Plusieurs Animaux dont les Révolutions du Globe Ont Détruit les Espèces: 4*. Paris: Dufour et d'Ocagne.

Danilov, P. 1995. Canadian and Eurasian beavers in Russian North-West (distribution, number, comparative ecology). *Proceedings of the 3rd Nordic Beaver Symposium*, Helsinki, Finland.

Danilov, P. I. and Fyodorov, F. V. 2015. Comparative characterization of the building activity of Canadian and European beavers in northern European Russia. *Russian Journal of Ecology*, 46, 272–278.

Danilov, P. I. and Kan'shiev, V. Y. 1983. State of populations and ecological characteristics of European (*Castor fiber* L.) and Canadian (*Castor canadensis* Kuhl). Beavers in the northwestern USSR. *Acta Zoologica Fennica*, 174, 95–97.

Danilov, P., Kanshiev, V., and Fyodorov, F. 2011a. Differences of the morphology of the North American and Eurasian beavers in Karelia. In: Sjöberg, G. and Ball,

J. P. (eds.) *Restoring the European Beaver: 50 Years of Experience*. Sofia-Moscow: Pensoft.

Danilov, P., Kanshiev, V. Y., and Fedorov, F. 2011b. History of beavers in eastern Fennoscandia from the Neolithic to the 21st century. In: Sjøberg, G. and Ball, J. P. (eds.) *Restoring the European Beaver: 50 Years of Experience*. Sofia-Moscow: Pensoft.

Davis, W. B. 1939. *The Recent Mammals of Idaho*. Caldwell: Caxon Printers.

Davis, W. B. 1940. Critical notes on the Texas beaver. *Journal of Mammalogy*, 21, 84–86.

Deinet, S., Ieronymidou, C., Mcrae, L., Burfield, I. J., Foppen, R. P., Collen, B., and Böhm, M. 2013. *Wildlife Comeback in Europe: The Recovery of Selected Mammal and Bird Species. Final Report to Rewilding Europe by ZSL*. London: ZSL: Birdlife International and the European Bird Census Council.

Depéret, C. 1890. *Les Animaux Pliocènes du Roussillon*. Amsterdam: Swets and Zeitlinger.

Dewas, M., Herr, J., Schley, L., Angst, C., Manet, B., Landry, P., and Catusse, M. 2012. Recovery and status of native and introduced beavers *Castor fiber* and *Castor canadensis* in France and neighbouring countries. *Mammal Review*, 42, 144–165.

Ducroz, J.-F., Stubbe, M., Saveljev, A. P., Heidecke, D., Samjaa, R., Ulevičius, A., Stubbe, A., and Durka, W. 2005. Genetic variation and population structure of the Eurasian beaver *Castor fiber* in eastern Europe and Asia. *Journal of Mammalogy*, 86, 1059–1067.

Durka, W., Babik, W., Ducroz, J. F., Heidecke, D., Rosell, F., Samjaa, R., Saveljev, A., Stubbe, A., Ulevicius, A., and Stubbe, M. 2005. Mitochondrial phylogeography of the Eurasian beaver *Castor fiber* L. *Molecular Ecology*, 14, 3843–3856.

Durrant, S. D. and Crane, H. S. 1948. Three new beavers from Utah. *Publications of the Museum of Natural History University of Kansas*, 1, 407–417.

Ellegren, H., Hartman, G., Johansson, M., and Andersson, L. 1993. Major histocompatibility complex monomorphism and low levels of DNA fingerprinting variability in a reintroduced and rapidly expanding population of beavers. *Proceedings of the National Academy of Sciences*, 90, 8150–8153.

Erlinge, S. 1968. Food studies on captive otters *Lutra lutra* L. *Oikos*, 19, 259–270.

Flynn, L. J. and Jacobs, L. L. 2008. Castoridae. In: Janis, C. M., Gunnell, G. F., and Uheh, M. D. (eds.) *Evolution of Tertiary Mammals of North America*. Cambridge: Cambridge University Press.

Fostowicz-Frelik, Ł. 2008. First record of *Trogontherium cuvieri* (Mammalia, Rodentia) from the Middle Pleistocene of Poland and review of the species. *Geodiversitas*, 30, 765–778.

Frahnert, S. 2000. Ontogenetic changes of cranial proportions in beaver, *Castor fiber* L., 1758 (Rodentia, Castoridae): taxonomical implications and functional aspects. *Bonner Zoologische Beiträge*, 49, 131–153 (in Russian).

Frosch, C., Kraus, R. H. S., Angst, C., Allgöwer, R., Michaux, J., Teubner, J., and Nowak, C. 2014. The genetic legacy of multiple beaver reintroductions in central Europe. *PLoS ONE*, 9, e97619.

Fuchs, T. 1983. Ueber die natur von daimonelix barbour. *Naturhist Hofmus, Wien*, 7, 91–94.

Gabryś, G. and Ważna, A. 2003. Subspecies of the European beaver *Castor fiber* Linnaeus, 1758. *Acta Theriologica*, 48, 433–439.

Geoffroy, E. 1803. *Catalogue des Mammiféres du Muséum National d'Histoire Naturelle*. Paris: Muséum National d'Histoire Naturelle.

Girling, S. J., Naylor, A., Fraser, M., and Campbell-Palmer, R. 2019. Reintroducing beavers *Castor fiber* to Britain: a disease risk analysis. *Mammal Review*, 49, 300–323.

Gladenkov, A. Y., Oleinik, A. E., Marincovich Jr, L., and Barinov, K. B. 2002. A refined age for the earliest opening of Bering Strait. *Palaeogeography, Palaeoclimatology, Palaeoecology*, 183, 321–328.

Goldfarb, B. 2018. Why two countries want to kill 100,000 beavers. https://www.washingtonpost.com/news/animalia/wp/2018/08/09/why-two-countries-want-to-kill-100000-beavers/ [Accessed 10/09/2018].

Goldman, E. A. 1932. A new beaver from Arizona. *Journal of Mammalogy*, 13, 266–267.

Gray, J. E. 1869. On the white-toothed American beaver. *Journal of Natural History*, 4, 293.

Grinnell, J., Dixon, J. S., and Linsdale, J. M. 1937. *Fur-Bearing Mammals of California, their Natural History, Systematic Status and Relations to Man*. Berkeley: University of California Press.

Hall, E. R. 1981. *The Mammals of North America*. New York: John Wiley.

Halley, D., Saveljev, A., and Rosell, F. 2020. Population and distribution of Eurasian beavers (*Castor fiber*). *Mammal Review*, 51, 1–24.

Heidecke, D. 1986. Aspekte des artenschutzes am beispiel biber Eurasiens. *Hercynia Nf Leipzig*, 22, 146–161.

Heller, E. 1909. The mammals. In: Grinnell, J., Stephens, F., Dixon, J., and Heller, E. (eds.) *Birds and Mammals of the 1907 Alexander Expedition to Southeastern Alaska*. Berkeley: University of California Press.

Hilfiker, E. L. 1991. *Beavers, Water, Wildlife and History, Interlaken*. New York: Windswept Press.

Hood, G. A. 2020. *Semi-Aquatic Mammals: Ecology and Biology*. Baltimore: Johns Hopkins University Press.

Horn, S., Durka, W., Wolf, R., Ermala, A., Stubbe, A., Stubbe, M., and Hofreiter, M. 2011. Mitochondrial genomes reveal slow rates of molecular evolution and the timing of speciation in beavers (*Castor*), one of the largest rodent species. *PLoS ONE*, 6, e14622.

Horn, S., Prost, S., Stiller, M., Makowiecki, D., Kuznetsova, T., Benecke, N., Pucher, E., Hufthammer, A. K., Schouwenburg, C., Shapiro, B., and Hofreiter, M. 2014. Ancient mitochondrial DNA and the genetic history of Eurasian beaver (*Castor fiber*) in Europe. *Molecular Ecology*, 23, 1717–1729.

Huchon, D., Madsen, O., Sibbald, M. J., Ament, K., Stanhope, M. J., Catzeflis, F., De Jong, W. W., and Douzery, E. J. 2002. Rodent phylogeny and a timescale for the evolution of glires: evidence from an extensive taxon sampling using three nuclear genes. *Molecular Biology and Evolution*, 19, 1053–1065.

Huchon, D., Chevret, P., Jordan, U., Kilpatrick, C. W., Ranwez, V., Jenkins, P. D., Brosius, J., and Schmitz, J. 2007. Multiple molecular evidences for a living mammalian fossil. *Proceedings of the National Academy of Sciences*, 104, 7495–7499.

Hugeney, M. 1975. Les Castoridae (Mammalia, Rodentia) dans l'Oligocene d'Europe. *International Centre Natural Research Science*, 218, 791–804.

Hugueney, M. and Escuillie, F. 1996. Fossil evidence for the origin of behavioral strategies in early Miocene Castoridae, and their role in the evolution of the family. *Paleobiology*, 22, 507–513.

Jenkins, S. H. and Busher, P. E. 1979. *Castor canadensis* Kuhl, 1820. *Mammalian Species*, 120, 1–8.

Jewett, S. G. and Hall, E. R. 1940. A new race of beaver from Oregon. *Journal of Mammalogy*, 21, 87–89.

Kanshiev, V. Y. 1998. The peculiarities of European (*Castor fiber* L.) and Canadian (*Castor canadensis* Kuhl) beavers' morphology in the north-west regions of Russia. *Zoological Journal*, 77, 222–230 (in Russian).

Kays, R. W. and Wilson, D. E. 2010. *Mammals of North America*. Princeton: Princeton University Press.

Korth, W. W. 1994. *The Tertiary Record of Rodents in North America*. New York: Plenum Press.

Korth, W. W. 2001. Comments on the systematics and classification of the beavers (Rodentia, Castoridae). *Journal of Mammalian Evolution*, 8, 279–296.

Korth, W. W. and Rybczynski, N. 2003. A new, unusual Castorid (Rodentia) from the earliest Miocene of Nebraska. *Journal of Vertebrate Paleontology*, 23, 667–675.

Korth, W. W. and Samuels, J. X. 2015. New rodent material from the John Day Formation (Arikareean, Middle Oligocene to Early Miocene) of Oregon. *Annals of Carnegie Museum*, 83, 19–84.

Kuhl, H. 1820. *Beiträge zur Zoologie und Vergleichenden Anatomie*. Frankfurt am Main: Verlag Der Hermannschen Buchandlung.

Kurtén, B. and Anderson, E. 1980. *Pleistocene Mammals of North America*. New York: Columbia University Press.

Lavrov, L. S. 1969. A new subspecies of the European beaver (*Castor fiber* L.) from the Enissei River upper flow. *Zoologicheski Zhurnal*, 48, 456–457 (in Russian).

Lavrov, L. S. 1980. *Beaver*. Moscow: Nauka.

Lavrov, L. S. 1981. *Beavers of the Palearctic*. Voronezh: Voronezh State University.

Lavrov, L. S. 1983. Evolutionary development of the genus *Castor* and taxonomy of the contemporary beavers of Eurasia. *Acta Zoologica Fennica*, 174, 87–90.

Lavrov, L. S. and Orlov, V. N. 1973. Karyotypes and taxonomy of modern beavers (*Castor*, Castoridae, Mammalia). *Zoologische Zhurnal*, 52, 734–742.

Leblanc, D. J. 1994. *Nutria. The Handbook: Prevention and Control of Wildlife Damage*, 16. http://digitalcommons.unl.edu/icwdmhandbook/16.

Leclerc, G. L. 1807. *Buffon's Natural History*. London: W. Strahan and T. Cadell.

Leege, T. A. and Williams, R. M. 1967. Beaver productivity in Idaho. *Journal of Wildlife Management*, 31, 326–332.

Legge, A. J. and Rowley-Conwy, P. A. 1986. The beaver (*Castor fiber* L.) in the Tigris-Euphrates Basin. *Journal of Archaeological Science*, 13, 469–476.

Li, L., Li, Q., Lu, X., and Ni, X. 2017. Morphology of an Early Oligocene beaver *Propalaeocastor irtyshensis* and the status of the genus *Propalaeocastor*. *PeerJ*, 5, e3311.

Lindsay, E. H., Opdyke, N. D., and Johnson, N. M. 1984. Blancan-Hemphillian land mammal ages and late Cenozoic mammal dispersal events. *Annual Review of Earth and Planetary Sciences*, 12, 445–488.

Linneaus, C. 1758. *Systema Naturae*. Holmiae: Salvii.

Linstow, O. V. 1908. Die Verbreitung des Bibers im Quartär. *Abhandlungen und Berichte aus dem Museum Fuer Natur-und Heimatkunde Magdeburg*, 1, 246–346.

Lok, S., Paton, T. A., Wang, Z., Kaur, G., Walker, S., Yuen, R. K., Sung, W. W., Whitney, J., Buchanan, J. A., and Trost, B. 2017. De novo genome and transcriptome assembly of the Canadian beaver (*Castor canadensis*). *G3: Genes, Genomes, Genetics*, 7, 755–773.

Long, K. 2000. *Beavers: A Wildlife Handbook*. Colorado: Johnson Books.

Lovallo, M. and Suzuki, M. 1993. Partial albinism in two related beavers, *Castor canadensis*, in Central Wisconsin. *Canadian Field-Naturalist*, 107, 229.

MacArthur, R. A. and Aleksiuk, M. 1979. Seasonal micro-environments of the muskrat (*Ondatra zibethicus*) in a northern marsh. *Journal of Mammalogy*, 60, 146–154.

Marr, M. M., Brace, S., Schreve, D. C., and Barnes, I. 2018. Identifying source populations for the reintroduction of the Eurasian beaver, *Castor fiber* L. 1758, into Britain: evidence from ancient DNA. *Scientific Reports*, 8, 2708.

Martin, L. D. 1987. Beavers from the Harrison Formation (Early Miocene) with a revision of Euhapsis. *Dakoterra*, 3, 73–91.

Martin, L. D. 1994. *The Devil's Corkscrew*. https://www.naturalhistorymag.com/htmlsite/editors_pick/1994_04_pick.html.

Martin, L. D. and Bennett, D. K. 1977. The burrows of the Miocene beaver *Palaeocastor*, western Nebraska, USA. *Palaeogeography, Palaeoclimatology, Palaeoecology*, 22, 173–193.

Martin, P. S. 1984. Prehistoric overkill: the global model. In: Martin, P. S. and Kliein, R. G. (eds.) *Quaternary Extinctions*. Tucson: University of Arizona Press.

Martin, T. H. 1892. *Castologia. History and Traditions of the Canadian Beaver*. Montreal: W.M. Drysdale and Company.

Matschie, P. 1907. Die Wissenschaftliche Bezeichnung der Sogenannten Altai-Hirsche. *Sitzungsbericht der Gesellschaft Naturforschender Freunde zu Berlin*, 8, 221–228.

Mayhew, D. F. 1978. Reinterpretation of the extinct beaver *Trogontherium* (Mammalia, Rodentia). *Philosophical Transactions of the Royal Society of London B: Biological Sciences*, 281, 407–438.

McEwing, R., Frosch, C., Rosell, F., and Campbell-Palmer, R. 2014. A DNA assay for rapid discrimination between beaver species as a tool for alien species management. *European Journal of Wildlife Research*, 60, 547–550.

McKenna, M. C. and Bell, S. K. 1997. *Classification of Mammals: Above the Species Level*. New York: Columbia University Press.

Mearns, E. A. 1898. Preliminary diagnoses of new mammals of the genera *Sciurus*, *Castor*, *Neotoma*, and *Sigmodon*, from the Mexican border of the United States. *Proceedings of the United States National Museum*, 20, 501–505.

Meyer, R. C. 1999. Helical burrows as a palaeoclimate response: Daimonelix by Palaeocastor. *Palaeogeography, Palaeoclimatology, Palaeoecology*, 147, 291–298.

Miller, G. S. 1912. *Catalogue of the Mammals of Western Europe (Exclusive of Russia) in the Collection of the British Museum*. London: British Museum.

Miller, J. E. 1994. *Muskrats. The Handbook: Prevention and Control of Wildlife Damage*, 15. http://digitalcommons.unl.edu/icwdmhandbook/15.

Minnig, S., Angst, C., and Jacob, G. 2016. Genetic monitoring of Eurasian beaver (*Castor fiber*) in Switzerland and implications for the management of the species. *Russian Journal of Theriology*, 15, 20–27.

Morgan, L. H. 1868. *The American Beaver and his Works*. Philadelphia: J.B. Lippincott and Company.

Moritz, C. 1994. Applications of mitochondrial DNA analysis in conservation: a critical review. *Molecular Ecology*, 3, 401–411.

Mörner, T. 1992. *Liv Och Död Bland Vilda Djur*. Stockholm, Sellin & Partner.

Mörs, T., Tomida, Y., and Kalthoff, D. C. 2016. A new large beaver (Mammalia, Castoridae) from the early Miocene of Japan. *Journal of Vertebrate Paleontology*, 36, e1080720.

Müller-Schwarze, D. 2011. *The Beaver: Its Life and Impact*, 2nd edn. New York: Cornell University Press.

Munclinger, P., Syruckova, A., Nahlovsky, J., Durka, W., Saveljev, A. P., Rosell, F., Samjaa, R., Stubbe, A., Stubbe, M., Ulevicius, A., Samjaa, R., Yanuta, G. & Vorel, A. Recovery in the melting pot: complex origins and restored genetic diversity in newly-established Eurasian beaver (Rodentia: Castoridae) populations. Submitted to Biological Journal of the Linnean Society.

Nelson, E. 1927. Description of a new subspecies of beaver. *Proceedings of the Biological Society Washington*, 40, 125–126.

Nolet, B. A. and Rosell, F. 1998. Comeback of the beaver *Castor fiber*: an overview of old and new conservation problems. *Biological Conservation*, 83, 165–173.

Novak, M. 1987. Beaver. In: Novak, M., Baker, J. A., Obbard, M. E., and Mallock, B. (eds.) *Wild Furbearer Management and Conservation in North America*. Toronto: Queens Printer for Ontario.

Nummi, P. 2001. *Canadian beaver (Castor canadensis). Alien Species in Finland*. Helsinki: Ministry of the Environment.

Ognev, S. I. 1947. *Animals of USSR and Adjacent Countries. Rodents*. Moscow: Leningrad.

Osborn, D. J. 1953. Age classes, reproduction, and sex ratios of Wyoming beaver. *Journal of Mammalogy*, 34, 27–44.

Osmundson, C. L. and Buskirk, S. W. 1993. Size of food caches as a predictor of beaver colony size. *Wildlife Society Bulletin (1973–2006)*, 21, 64–69.

Pachinger, K. and Hulik, T. 1999. Beavers in an urban landscape. In: Busher, P. E. and Dzięciołowski, R. M. (eds.) *Beaver Protection, Management, and Utilization in Europe and North America*. Boston: Springer.

Parker, H., Nummi, P., Hartman, G., and Rosell, F. 2012. Invasive North American beaver *Castor canadensis* in Eurasia: a review of potential consequences and a strategy for eradication. *Wildlife Biology*, 18, 354–365.

Parmalee, P. W. and Graham, R. W. 2002. Additional records of the giant beaver, *Castoroides*, from the mid-south: Alabama, Tennessee, and South Carolina. In: Emry, R. J. (ed.) *Cenozoic Mammals of Land and Sea: Tributes to the Career of Clayton E. Ray*. Washington, DC: Smithsonian Institution Press.

Pasternack, G. B., Hilgartner, W. B., and Brush, G. S. 2000. Biogeomorphology of an upper Chesapeake Bay river-mouth tidal freshwater marsh. *Wetlands*, 20, 520–537.

Pelz-Serrano, K. 2011. Molecular phylogeography of the American beaver (*Castor canadensis*): implications for management and conservation. PhD thesis, University of Arizona.

Pelz-Serrano, K., Munguia-Vega, A., Piaggio, A. J., Neubaum, M., Munclinger, P., Pártl, A., Van Riper Iii, C., and Culver, M. 2009. Development of nine new microsatellite loci for the American beaver, *Castor canadensis*

(Rodentia: Castoridae), and cross-species amplification in the European beaver, *Castor fiber*. *Molecular Ecology Resources*, 9, 551–554.

Peterson, O. A. 1906. Description of new rodents. In: Holland, W. J. (ed.) *Memoirs of the Carnegie Museum*. Pittsburgh: Authority of the Board of Trustees of the Carnegie Institute of Biodiversity.

Petrosyan, V. G., Golubkov, V. V., Zavyalov, N. A., Khlyap, L. A., Dergunova, N. N., and Osipov, F. A. 2019. Modelling of competitive interactions between native Eurasian (*Castor fiber*) and alien North American (*Castor canadensis*) beavers based on long-term monitoring data (1934–2015). *Ecological Modelling*, 409, 1–15.

Pietsch, L. R. 1956. The beaver in Illinois. *Transaction of the Illinois State Academy of Science*, 49, 193–201.

Pilleri, G., Gihr, M., and Kraus, C. 1983. Vocalization and behaviour in the young beaver (*Castor canadensis*). *Investigations on Beavers*, 1, 67–80.

Plint, T., Longstaffe, F. J., and Zazula, G. 2019. Giant beaver palaeoecology inferred from stable isotopes. *Scientific Reports*, 9, 1–12.

Poliquin, R. 2015. *Beaver*. London: Reaktion Books.

Rekovets, L., Kopij, G., and Nowakowski, D. 2009. Taxonomic diversity and spatio-temporal distribution of late Cenozoic beavers (Castoridae, Rodentia) of Ukraine. *Acta Zoologica Cracoviensia—Series A: Vertebrata*, 52, 95–105.

Rhoads, S. N. 1898. Contributions to a revision of the North American beavers, otters and fishers. *Transactions of the American Philosophical Society*, 19, 417–439.

Rinaldi, C., Martin, L., Cole Iii, T., and Timm, R. 2009. Occlusal wear morphology of giant beaver (*Castoroides*) lower incisors: functional and phylogenetic implications. *Abstracts Society of Vertebrate Palaeontology*, 29, 171.

Rook, L. and Angelone, C. 2013. Just a few: rodents and lagomorphs in the Plio-Pleistocene fossil record of the Upper Valdarno basin. *Italian Journal of Geosciences*, 132, 98–103.

Rosell, F. and Pedersen, K. V. 1999. *Bever [The Beaver]*. Oslo: Landbruksforlaget.

Rosell, F. and Steifetten, Ø. 2004. Subspecies discrimination in the Scandinavian beaver (*Castor fiber*): combining behavioral and chemical evidence. *Canadian Journal of Zoology*, 82, 902–909.

Rosell, F. and Sun, L. X. 1999. Use of anal gland secretion to distinguish the two beaver species *Castor canadensis* and *Castor fiber*. *Wildlife Biology*, 5, 119–123.

Rosell, F., Bozser, O., Collen, P., and Parker, H. 2005. Ecological impact of beavers *Castor fiber* and *Castor canadensis* and their ability to modify ecosystems. *Mammal Review*, 35, 248–276.

Rosell, F., Zedrosser, A., and Parker, H. 2010. Correlates of body measurements and age in Eurasian beaver from Norway. *European Journal of Wildlife Research*, 56, 43–48.

Rybczynski, N. 2007. Castorid phylogenetics: implications for the evolution of swimming and tree-exploitation in beavers. *Journal of Mammalian Evolution*, 14, 1–35.

Rybczynski, N. 2008. Woodcutting behavior in beavers (Castoridae, Rodentia): estimating ecological performance in a modern and a fossil taxon. *Paleobiology*, 34, 389–402.

Rybczynski, N., Ross, E. M., Samuels, J. X., and Korth, W. W. 2010. Re-evaluation of sinocastor (Rodentia: Castoridae) with implications on the origin of modern beavers. *PLoS ONE*, 5, e13990.

Samuels, J. X. and Valkenburgh, B. V. 2009. Craniodental adaptations for digging in extinct burrowing beavers. *Journal of Vertebrate Paleontology*, 29, 254–268.

Savage, R. J. G. and Russell, D. E. 1983. *Mammalian Paleofaunas of the World*. Reading, MA: Addison-Wesley Publishing Company.

Saveljev, A. P. and Lavrov, V. L. 2016. About possible ways of genes penetration from West Siberian beavers *Castor fiber pohlei* into Austria. *Russian Journal of Theriology*, 15, 78–79.

Schorger, A. 1953. Large Wisconsin beaver. *Journal of Mammalogy*, 34, 260–261.

Schultz, C. B. 1942. *A Review of the Daimonelix Problem*. Lincoln: University of Nebraska Studies Science and Technology.

Schwab, G., Dietzen, W., and Lossow, G. V. 1994. Biber in Bayern-Entwicklung Eines Gesamtkonzepts Zum Schutz des Bibers. Schriftenreihe Bayer. *Landesamt Für Umweltschutz*, 28, 9–44.

Seaton-Thomson, E. 1909. *Life-Histories of Northern Animals: An Account of the Mammals of Manitoba*. New York: Charles Scribner's Sons.

Senn, H., Ogden, R., Cezard, T., Gharbi, K., Iqbal, Z., Johnson, E., Kamps-Hughes, N., Rosell, F., and Mcewing, R. 2013. Reference-free SNP discovery for the Eurasian beaver from restriction site-associated DNA paired-end data. *Molecular Ecology*, 22, 3141–3150.

Senn, H., Ogden, R., Frosch, C., Syrůčková, A., Campbell-Palmer, R., Munclinger, P., Durka, W., Kraus, R. H. S., Saveljev, A. P., Nowak, C., Stubbe, A., Stubbe, M., Michaux, J., Lavrov, V., Samiya, R., Ulevicius, A., and Rosell, F. 2014. Nuclear and mitochondrial genetic structure in the Eurasian beaver (*Castor fiber*)—implications for future reintroductions. *Evolutionary Applications*, 7, 645–662.

Serebrennikov, M. 1929. Review of the beavers of the Palaearctic region (*Castor*, Rodentia). *Proceedings of the Academy of Sciences USSR Leningrad*, 1929, 271–276.

Solov'yov, V. A. 1973. Beavers in Komi, USSR. *Syktyvkar*, 126 (in Russian).

Stefen, C. 2014. Cranial morphology of the Oligocene beaver *Capacikala gradatus* from the John Day Basin and comments on the genus. *Palaeontologia Electronica*, 17, 1–29.

Stirton, R. A. 1935. A review of Tertiary beavers. *University of California Publications in Geological Science*, 23, 391–458.

Stocker, G. 1985. The beaver (*Castor fiber* L.) in Switzerland. Biological and ecological problems of re-establishment. *Swiss Federal Institute of Forestry Research Reports*, 242, 1–49.

Strachan, R., Moorhouse, T., and Gelling, M. 2011. *Water Vole Conservation Handbook*. Oxford: Wildlife Conservation Research Unit, University Of Oxford.

Stringer, A. P. and Gaywood, M. J. 2016. The impacts of beavers *Castor* spp. on biodiversity and the ecological basis for their reintroduction to Scotland, UK. *Mammal Review*, 46, 270–283.

Strömberg, C. A. 2002. The origin and spread of grass-dominated ecosystems in the late Tertiary of North America: preliminary results concerning the evolution of hypsodonty. *Palaeogeography, Palaeoclimatology, Palaeoecology*, 177, 59–75.

Sun, L. 2003. Monogamy correlates, socioecological factors, and mating systems in beavers. In: Reichard, U. H. and Boesch, C. (eds.) *Monogamy: Mating Strategies and Partnerships in Birds, Humans and Other Mammals.* Cambridge: Cambridge University Press.

Sun, L., Müller-Schwarze, D., and Schulte, B. A. 2000. Dispersal pattern and effective population size of the beaver. *Canadian Journal of Zoology*, 78, 393–398.

Taylor, W. P. 1912. The beaver of west central California. *University of California Publications in Zoology*, 10, 167–169.

Taylor, W. P. 1916. The status of the beavers of western North America with a consideration of the factors in their speciation. *University of California Publications in Zoology*, 12, 413–495.

Tedford, R. H. and Harington, C. R. 2003. An Arctic mammal fauna from the early Pliocene of North America. *Nature*, 425, 388–390.

Veron, G. 1992. Biogeographical history of the European beaver *Castor fiber* (Rodentia, Mammalia). *Mammalia*, 56, 87–108.

Vorel, A. 2001. Bobr Evropsky (*Castor fiber*) Na Labi A Katerinském Potoce. Katedra Ekologie Lf Czu, Praha. Diplomova Prace, Nepubl.

Wahlert, J. H. 1977. Cranial foramina and relationships of *Eutypomys* (Rodentia, Eutypomyidae). *American Museum Novitates*, 2626, 1–8.

Warren, E. R. and Hall, E. R. 1939. A new subspecies of beaver from Colorado. *Journal of Mammalogy*, 20, 358–362.

Woodroffe, G. 2007. *The Otter*. London: The Mammal Society.

Xu, X. 1994. Evolution of Chinese Castoridae in rodent and lagomorph families of Asian origins and diversification. In: Tomida, Y., Li, C. K., and Setoguchi, T. (eds.) *Natural Science Museum Monograph*. Tokyo: Natural Science Museum.

Xu, X., Li, Q., and Flynn, L. J. 2017. *The Beavers (Castoridae) of Yushe Basin. Late Cenozoic Yushe Basin, Shanxi Province, China: Geology and Fossil Mammals*. Dordrecht: Springer.

Yang, Y., Li, Q., Fostowicz-Frelik, Ł., and Ni, X. 2019. Last record of *Trogontherium cuvieri* (Mammalia, Rodentia) from the late Pleistocene of China. *Quaternary International*, 513, 30–36.

Yazan, Y. P. 1964. About some morphological and ecological changes of beavers as a result of their reacclimatization in Pechora-Ylychsky reserve. *Proceedings of Pechora, Ylychsky State Nature Reserve*, 11, 75–82 (in Russian).

Żurowski, W. 1977. *Rozmnazanie sie bobrow europejskichw warunkach fermowych. Rozprawy habilitacyjne, Zeszyt 7.* Polska Akademia Nauk, Instytut Genetyki HodowliZwierzat, pp. 31–42 (in Polish).

CHAPTER 2

Utilization and distribution of beavers

2.1 An ancient relationship

Humans have an ancient relationship with beavers (*Castor* spp.) which has ranged broadly over time from the supply of food and fur, through the inspiration of religious and cultural beliefs, to a contemporary recognition of the species' critical role as a maintainer of ecological balance. This often-forgotten association is seldom recognized at a time when their modern populations are now just beginning to recover. Without exaggeration, beavers can be afforded the distinction of being one of the key wild animal species that have influenced history (Coles, 2006). Many of the Native American indigenous people have a long history of utilising beavers as a source of food and warmth, with legends relating to beavers, and even keeping more amiable individuals as pets (Schorger, 1965; Dolin, 2010). Beavers were a commonly hunted mammal dominant in the diets of Neolithic and Early Bronze Age humans across the East Baltic region (Daugnora and Girininkas, 1995; Antanaitis-Jacobs et al., 2009). They are also depicted in prehistoric art. Their behaviour and castoreum (their unique scent from the castor sacs; chapter 3.2.9) were discussed by the Roman naturalist Pliny the Elder (*c.* 77 BCE); Strabo (*c.* 63 BCE–24 CE), a Greek geographer and philosopher, states castoreum (which was actively traded in Roman times) from Pontus, where it was called the '*Pontic dog*' (now northern Turkey, near the Black Sea) was superior to that from Spain; and beavers are also included in the medieval Iranian text of *Bestiary* (1297–1298).

Strange theories regarding their behaviours and biology were commonly held. Early North American explorers told exaggerated tales of their building capabilities. They described them using their tails as spatulas to plaster their homes and spoke of their organized colonies where strict laws and social structures were maintained. While some individuals ruled, others had set jobs such as guards, ditch diggers, or carpenters (Denys, 1672). Across their range, similar beliefs about beavers as nature's 'workaholics' are held by people, while sayings such as 'to be as busy as a beaver', 'eager beaver', and 'beavering away' refer directly to their obvious work ethic. In contradiction, it was also thought that their industrial traits could result in confusion, conflict, and selfishness in their determination to finish their tasks (Lake-Thom, 1997)!

Wars have been waged, laws enshrined, religions spread, and new countries established on the glossy backs of beavers. The complex interweaving of the fur trade and colonial expansion is best exemplified in history by the formation of Canada and colonization of North America (Gerstell, 1985; Dolin, 2010). The *Beaver Wars* of the eighteenth century were predominantly fought between the early British and French colonialists in an effort to secure the most lucrative trapping areas and trading routes.

In 1975, Canada received Royal assent and, acknowledging its beaver escutcheon, its symbol of sovereignty incorporates the beaver's heritage and economic and cultural importance (Lok et al., 2017). The beaver is depicted on countless emblems, seals, and government and military badges; for example, the public seal of New Netherland used from 1623 (Martin, 1892). Though the Canadian five cent famously depicts the beaver on one of its faces

Beavers: Ecology, Behaviour, Conservation, and Management. Frank Rosell and Róisín Campbell-Palmer, Oxford University Press.
© Frank Rosell and Róisín Campbell-Palmer 2022. DOI: 10.1093/oso/9780198835042.003.0002

(a)

Figure 2.1a The Canadian five cent coin depicts the beaver in recognition of its economic and cultural importance. (Illustration provided courtesy of Rachael Campbell-Palmer.)

(b)

Figure 2.1b The illegal $5 'beaver coin'. (Illustration provided courtesy of Rachael Campbell-Palmer, drawn from photo of coin.)

(c)

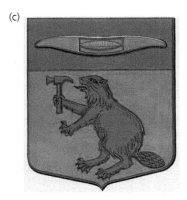

Figure 2.1c The crest of Bjursås a municipality of the Swedish city of Härnösand, city seal depicting the hard work ethic of the beaver.

(Figure 2.1a), it was not the first coin to bear the beaver image. In 1849, the gold $5 'beaver coin', showing a beaver standing on a log on one side (Figure 2.1b), was produced by a private mint following growing frustration at the US government's lack of establishment of a countrywide currency system; it was later deemed illegal (Crutchfield, 2018). The first Canadian postage stamp issued in 1851, the '3p beaver', depicts a beaver rather than Queen Victoria, which was a significant break from the customary practice of the British Empire at that time (Lok et al., 2017). As such it was the first animal to ever appear on an official stamp (Poliquin, 2015).

In other countries and states, the beaver appears on crests and coats of arms, such as the London School of Economics, and the state flag of Oregon,

on the reverse of which a golden beaver is depicted, the state animal. In *Fairbairn's Book of Crests of the Families of Great Britain and Ireland*, an extensive listing categorizing all surnames known at the time, the beaver appears on the crest of 25 families (Fairbairns, 1892, Vol. 1, plate 134; Box 2.1). More recently, *Amik* (originating from the Anishinaabe word for beaver) was the beaver mascot chosen for the Montreal Summer Olympics of 1976, while the logo for the National Parks of Canada (Parks Canada) is a beaver—symbolic of hard work, self-reliance, peacefulness, and an ability to tackle challenges!

2.1.1 Early human interactions

The largest rodent ever to have existed in North America and now extinct, the giant beaver (*Castorides ohioensis*) is believed to have appeared at the dawn of the Pleistocene and has been embedded in Native American legends (Martin, 1867; Powell, 1948; see Section 2.2). However, historical evidence of both modern species is widespread. Extensive beaver burrows have been identified from the Late Palaeolithic site of Grabow in the floodplain of North Germany (Tolksdorf et al., 2017). Prehistoric digs in Somerset, England, have determined that humans were attracted to beaver sites, building plank walkways in shared habitats, with beaver bone remains found in nearby caves from 12,700 to 8500 BCE (Coles, 2006).

┌─────────────────────────────────────┐
Box 2.1 Beaver crest family names
└─────────────────────────────────────┘

Family surnames including beaver in their crest (Fairbairns, 1892).

Alexander, Baynham, Beaver, Beevor, Bell, Besook, Bevers, Beynham, Brookes, Brooks, Coram, Corham, Danskine, Dimsdale, Eaton, Fenwick, Howel, Howell, Maclagan, McLagan, Molineux, Sadleyr, Symcock, Symcott, Trowell.

Common surnames deriving from beaver in some form include:

Beverley or *Beverly* 'beaver stream' in Old English and derived from a settlement in England.

Bieber 'beaver' German and Jewish origin, could also be a nickname for a hard worker.

Bjurström (Swedish spelling) or *Bjurstrøm* (Norwegian and Danish spelling) are Nordic surnames derived from *bjur* 'beaver' and *ström* 'river'.

One '*gost*' (a title awarded only by the tsar to represent a very wealthy merchant), Vasili Bobr, is known for building a brick church in the 1480s near to Red Square in Moscow today, presumed to have made his money from beaver pelts, given his surname (Monahan, 2016).

Three main human-related activities found in beaver remains have been described (Lebreton et al., 2017). Evidence of various cut marks on the bones demonstrates (1) the use of a sharp-edged tool, producing small perpendicular marks as evidence of skinning; (2) disarticulation via deep marks, made again by a sharp tool to separate limbs from the axial skeleton; and (3) defleshing, evident through short incisions in a regular pattern left by axe-shaped tools to separate meat from the bone. Butchery at numerous prehistory human sites such as Dalmeri, Italy (Fiore et al., 2001), Kettig, Germany (Baales, 2002), and multiple sites in Finland (Forstén and Lahti, 1976) clearly indicates that beavers were consumed quite normally alongside many other animals. A burnt beaver tooth from this same period has been found at Medzhibozh in the Ukraine (Stepanchuk and Moigne, 2016). In the Middle Palaeolithic, human cut marks have been identified on beaver remains in Grotta San Bernadino, Italy (Fiore et al., 2004) and at Taubach and Lehringen in Germany (Gaudzinski, 2004). Multiple beaver remains, including burnt and cut bones have been

dated from the Early Holocene at several sites in France (Dibble et al., 2009; Rendu, 2010; Silmak et al., 2010). Just over 5% of total mammal remains in the Rhone Valley and Danube basin from this time have been identified as beaver (Kind, 2009). In northern European Mesolithic sites, beaver bone fragments are much more numerous, for example they comprise 30–60% of mammal remains at some Estonian (Veski et al., 2005) and Russian (Chaix, 2003) sites. At one Russian location, a beaver skull with the remains of a harpoon head has been uncovered (Zhillin, 2004). In North America near Lake Huron, butchered beaver bones with clear knife cuts have been aged as being 3,700 years old and the remains of a beaver pelt which was once wrapped around a copper axe has been aged to 2,500 years, found in a burial mound (Backhouse, 2018).

In the Neolithic period, northern European tribes were also creating symbolic representations of beavers. Petroglyphs (rock carvings) of beavers have been found at Lake Onega and near the White Sea, in northwest Russia (Danilov, 1976). These representations are believed to have been drawn from a hunter's perspective, looking down on a swimming beaver from above (Danilov et al., 2011; Figure 2.2a and b). Further beaver remains at the Etruscan sanctuary of Podere Ortaglia, in Italy, are believed to have been part of a ritualistic offering to Artemis, the goddess of the hunt (Sorrentino and Landini, 2005). Whilst beavers are featured as part of cultural belief systems, occurring widely in Native American totem carvings as important animal spirits (Marcuzzi, 1986; Figure 2.3).

2.1.2 Beaver place names

Apart from Hawaii, every other American state and Canadian province possesses place names relating to beavers (Backhouse, 2018). Nearly 3,500 geographical locations with 'beaver' in their name have been recorded, including 61 swamps, 331 lakes, and 1,373 creeks (Long, 2000). Wherever beavers have been numerous they were an important feature of value in the landscape and were as such recognized by people.

A similar pattern of place name records exists in Europe. The Old French word for beaver being '*bièvre*' is recorded in towns such as Monthou-sur-Bièvre

(a)   (b)

Figure 2.2a and b Representations of rock carving depicting a swimming beaver from a hunter's perspective looking down. (Illustrations provided courtesy of Rachael Campbell-Palmer, redrawn after Savvateev, 1970.)

and Bièvres, whilst Beverley in England was named through the combination of '*beofor*' and '*leac*' which are the Old English words for beaver and stream respectively (Figure 2.4). In Germany, the municipality of Biberbach means beaver creek and there are also two Biber rivers elsewhere. The Polish word for beaver '*bóbr*' is recorded in the River Bóbr and the towns of Bober and Bóbrka (Poliquin, 2015). Similarly, the villages Bobr, Bobroviníky, and Bobrová can all be found in the Czech Republic. The Nordic '*bjur*' is found in numerous place names including Bjurälven (beaver river), site of the first beaver reintroduction in Sweden 1922. Throughout the whole of Sweden there are numerous derived place names including Bjurängen, Bjursås, Bjurbäck, and Bjuön.

In Karelian, Finnish, Sámi, and Vespian the word for beaver is '*majova*', '*majava*', '*mádjit*', and '*maji*', all of which sound very similar. Various place names, including the rivers Maija and Maya, the lakes Maima, Maimjärvi, and Maijezero, and numerous Karelia villages such as Maiguba and Mayaniemi (Danilov et al., 2011), take their root from these. Other names have proved more controversial. For example, the confirmed Old Welsh name for beaver is '*llostlydan*' meaning broadtail, which appears in the 'Laws of Hywel Dda' known as the Medieval Welsh Law Codes, dating back to 940 CE (Charles-Edwards, 1989), though no place names associated with this are known today. From around the seventeenth century the Welsh term for beaver more commonly becomes '*afanc*', also spelt to a water monster (Aybes and Yalden, 1995), variably described as anything from a crocodile-like creature to a giant beaver and even a hybrid between the two (A. Leow-Dyke, pers. comm.)!

2.2 Myths, folklore, and religious beliefs

Many folklore and numerous myths are associated with the beaver, including counting as a fish meal on religious holidays but also eating fish themselves, working in huge teams to build dams, and self-sacrificing body parts (often mistakenly referred to as their testicles, whilst meaning their castor sacs) to appease hunters (Wilsson, 1971; Poliquin, 2015; Figure 2.5a). Medieval writers, such as Dante Alighieri, believed beavers not only ate fish but also could lure them by waving their tails in the water and releasing an attractive fatty substance (Holbrook, 1902; Figure 2.5b), with Buffon, a French natural historian, even stating a beaver's scaly tail was caused by fish eating. Other myths such as an ability to predict the weather may be based on the biological realities of extra fur growth and larger food caches in harsh winter conditions. Many such as sleeping with their tails dipped in the water to detect its changing levels or their tail slapping leading to thunder (Newman, 1985) relate more to fairy tales than having a scientific basis, but wherever beavers exist, there are stories.

Figure 2.3 The beaver is often represented in the Native American belief system as a creator of life. (Photo supplied courtesy of Alicia Leow-Dyke.)

'*afangc*' or '*avanke*', with place names including Llyn-yr-Afanc, meaning beaver pool, and Bedd yr Afanc, meaning Grave of the Afanc, still evident (Coles, 2019). In early Welsh the term '*afanc*' refers

Figure 2.4 The English town of Beverley, named after the Old English for beaver and stream. (Photo supplied courtesy of Derek Gow.)

Many Native American tribes considered beavers to be superior animals, with a central role in their belief regarding the creation of the world. 'Beaver-man' was believed by indigenous Alaskans to have redesigned the animals we know today from their previous forms (Long, 2000). The Sioux have four creative gods, one of which is the beaver. The Amikonas 'People of the Beaver', an Algonquin tribe of Lake Huron, have long-standing myths about the giant 'great original father' beaver, which they claim to be descended from (Martin, 1892). *Castoroides ohioensis* was a giant beaver species which occurred in the Pleistocene when humans and the now extinct megafauna over-lapped in time?. The Cherokee thought that giant beavers helped the Great Spirit to create the Earth (Rue, 1964). The northeastern Algonquin tales talk of *Quahbeet*, the giant beaver, whose tail slapping produced thunder (Martin, 1892). The super myth-ical figure *Gluskap* was common in many tales of epic chases and hunts of giant beavers (Beck, 1966). Many of these credit the formation of the Great Lakes to the giant beaver when it dammed the Saint Lawrence River. *Gluskap* also created all ani-mals, including people whom he made last and then presented to the other creatures. Those who

showed no respect to his newest of beings he reduced in size by a stroke of his hand. While this worked for some, the crafty beavers were swift to escape and another epic chase began. Numerous bays and islands were formed from the paddle strokes of *Gluskap* chasing the beavers in his canoe and from the rocks he threw at them during these battles. Slight variations in this story and topographical fea-tures change, but its telling is fairly consistent amongst the indigenous people of this region (Beck, 1972). The Dene tribe of the Great Slave and Artillery Lakes tell stories of *Xachogh*, a hunter who killed giant beavers for food, while the Pocumtuck tribe believes the Pocumtuck hill range of the Connecticut River valley to be the petrified remains of a giant beaver that tried to eat people when it ran out of fish. The people called on the spirit *Hobomock* to chase the giant beaver back into the water and kill it (Poliquin, 2015). The commonality of giant beaver tales among various northeastern Algonquin tribes overlapped with the distribution of Pleistocene remains of this extinct animal represent a fossil mem-ory of shared existence (Beck, 1972).

The Cherokee credit the beaver for creating the antlers that are now borne by deer, which they won after a contest with the rabbit as to who was the

Figure 2.5a Old females or slave beavers could be recognized as they have bald backs as they were used for transporting trees. Johannes Bjurberg illustration, 1687, University of Uppsala.

Figure 2.5b Notice beavers are fishing with their tails in the water in this illustration of beaver hunting from the '*History of the Nordic People*', Olaus Magnus around 1550.

faster animal (Lake-Thom, 1997). With variations, many tribes believe that in the beginning the Earth was covered in water and it was the beaver that dived to bring up mud from which the Great Spirit Manitou formed the land. Manitou removed their power of speech as a punishment. One Cheyenne clan believes the father of all humans to be a great white beaver who retains the power to gnaw through the wooden post that holds up their world if he is angered. As a result, touching the skin or eating the meat of ordinary beavers would make you sick (Long, 2000).

In the USA Midwest, it was believed that human spirits were trapped inside beavers and that these would return to their human forms when they were killed. The Crow thought they were reincarnated humans (Rue, 1964) and as a result not all Native American tribes will hunt or eat beavers. So strong was this consideration for the Blackfeet that they refused to trade in their furs with the European colonists. Mistassini Cree ideology holds the beaver in higher regard than other animals (Cox, 1988). Utilizing and respecting their entire carcass required that even the often-purposeless forelegs were decorated and hung in trees to prevent their con-

sumption by scavengers (Speck, 1923; Tanner, 1979). Other tribes believed beaver bones should be burnt as a mark of respect. The Tlingits thought that feeding beaver meat to their dogs would result in the dead beaver complaining to the spirits of the living beavers with regard to their fate and that as a result the spirit animals would punish their hunters with reduced yields in the future.

The discovery of beaver artefacts at religious sites and graves from ancient European sites also indicates that the species retained an almost holy status for early people (Wilsson, 1968). Some of the first written accounts of beavers and their activities were provided by Giraldus Cambrensis 'Gerald of Wales'. In 1188, this historian and Cambro-Norman archdeacon described beaver castles and other more curious aspects of their behaviours. He stated that when moving large logs, they would work in a team, with some lying on their backs while others loaded logs onto their stomachs. They would then drag their load back to the water. A widespread European medieval literary cycle or fable *Reynard the Fox*, first appearing in the later twelfth century, includes a series of animal characters with defined personalities, including Bockert the beaver, considered wise and well educated, who acts as the secretary and translator at the court the trickster fox is called to (G. Hartman, pers. comm.). One of the first natural history accounts of North American beavers was written by Denys (1672). He described their collective workings and told how they came together in highly regimented units of over 400 individuals where labour divisions (lumberjacks, diggers, masons, etc.) were strictly enforced and overseen by commanders (Denys, 1672).

One of the oldest and most commonly held myths relating to beavers is believed to originate from one of Aesop's legendary fables, first written down in 1484 (Caxton, 1484). Aesop was believed to be an African native enslaved in ancient Greece around 620–564 BCE, whose fables have been widely translated and told for centuries throughout the world. The beaver fable describes how a beaver chased by a huntsman and his hounds will bite off and offer up its castor sacs, which were often mistaken for testicles, as they know the hunter will spare their life if they sacrifice these organs valued for their castoreum. Several varying morals were taken from this tale over time. The original has been translated as 'if only people would take the same approach and agree to be deprived of their possessions in order to live lives free from danger: no one, after all, would set a trap for someone already stripped to the skin' (Gibbs, 2002); in another, that 'wise [men] know to sacrifice possessions over their life' (Calabritto, 2002). The Roman satirist Juvenalis (AD 60–130) likens the behaviour of the character Catullus on saving his ship and crew from near-certain death as that of a beaver: 'For when the hold was half of water, and the waves rocked the hull from side to side, so that the white-haired skipper, with all his skill, could bring no succour to the labouring mast, he resolved to compound with the winds like the beaver, who gives up one part of his body that he may keep the rest: so conscious is he of the drug which he carries in his groin. "Overboard with everything!" shouted Catullus.' Medieval and Christian moral overtones link the tale with casting away of 'vices and shameless acts' and therefore sin, and castration was practiced by some priests at the time (Curley, 1979). This belief of beaver self-castration came from classical Greek naturalists, passing into Roman folklore, Christian teachings

(popularized in the medieval period by Isidore of Seville and the Latin *Bestiary*) (Raye, 2014), and numerous ancient natural history texts throughout Eurasia, until published anatomical studies by early French naturalists such as Guillaume Rondelet (around 1566) and Moyse Charas (around 1668) dismissed its possibility, clearly distinguishing between testes and castor sacs (Poliquin, 2015).

A similar story of the self-sacrificing nature of beavers is told by the Midwestern Omaha tribe: when a beaver family has no food to offer a guest, their youngest child will on its own behalf offer itself for dinner. As long as its bones were respected and not broken, the beaver child could then regenerate and return good as new. If, however, the guest was careless or dishonest while eating and cracked even a small bone, when the young beaver reappeared it would have a broken toe, and as a result all beavers carry two split nails on a single hind toe as a reminder of this possibility (Backhouse, 2018).

In medieval times the Catholic Church classed beavers as *'aquatilia'*, along with other water-dwelling animals like seals (Phocidae), otters (*Lutra lutra*), and turtles (Chelonia). These creatures were all believed to be more fish-like in nature and as such could be consumed during religious holidays and periods of penitence when meat was forbidden. In parts of Italy, beaver remains related to the diets of seventh- to ninth-century monks have been found in a number of monastic complexes such as San Vincenzo al Volturno, Molise and the Piazza della Signoria in Rome (Salari et al., 2019). The eleventh-century German monk Ekkehard IV has been credited with writing *'sit benedicta fibri caro piscis voce salubri'*, roughly translated as blessed be the fish-like flesh of the beaver.

2.3 Beaver territories as resources

It is worth noting that beaver activities in themselves have multifunctional uses for humans and historically have provided an important resource. In modern times their habitat modifications have often been used to justify reintroduction programmes, encourage wetland restoration, and implement a sustainable process of ecosystem services (see Chapters 9 and 11).

Beaver-generated environments generate a broad range of resources readily utilized by humans (Coles, 2006). Early hunter gatherers were drawn to wetlands as these were full of prey (game, water fowl, fish, and small mammals) and complex environments. At a time when there were many large predators in the wider environment, islands surrounded by complex channels were preferentially sought. The vast amount of dead wood generated by the beavers provided a ready fuel resource when dried, which ranged from larger logs cut from felled trunks to the near wood powder generated by gnawed chips. Charred material with beaver tooth marks has been identified and found at a number of archaeological sites (Coles, 2006). Further beaver-cut material of appropriate dimensions was variously used for fence posts, shelters, door stops, paper weights, and even walking sticks (Coles, 2006). The rod-straight regrowth of coppiced tree stumps from species such as willow (*Salix* spp.) must surely have prompted people to initially utilize and then replicate the production of this resource which was essential for wattle fencing, basket weaving, and trap construction (Backhouse, 2018).

Beaver dams can be significant structures and humans have been recorded using them as walkways through wetlands. Perhaps one of the most significant functions of beaver dam systems in our modern world is water retention. This natural function has significant ecological benefits (see Chapter 9), which are now increasingly linked to important economic and conservation gains. The renowned beaver ecologist Glynnis Hood has documented through her numerous studies the hydrological capabilities of beaver-generated environments (see Chapter 9). The impoundment of water behind their main dam structures coupled with a wider retention capacity in extensive networks of excavated canals hold huge volumes of water under drought conditions. This function directly benefited cattle ranchers whose stock could still drink in beaver ponds and graze on the lush availability of associated vegetation in otherwise arid Canadian landscapes (Hood and Bayley, 2008; Hood, 2011).

Beaver meadows, formed from the combination of silts and sediments layered behind their dams over millennia, have also been widely used by humans as environments for cultivation through much of the beaver's former range. As the beaver was hunted out of Eurasia and North America, not only did this leave 'a wealth of place names which no longer made much sense' but ever-expanding colonists used their abandoned dams as sites for water mills and the land left behind broken-down dams was used as rich fertile fields for crops or farming livestock (Cronon, 1983). Henry Wansey, an English traveller to the eastern USA, recognized the high hay production gained from farming these beaver meadows when he noted in 1794 that, 'it is a fortunate circumstance to have purchased lands where these industrious animals have made a settlement' (Wansey, 1798; Backhouse, 2018). In Sweden, a traditional superstition states if you rub a beaver's incisor along your axe then the clearing you make with it in the wood would stay open, most likely an interpretation of people noticing that beaver meadows remain open for long periods of time (G. Hartman, pers. comm.).

2.4 Cultural value of beaver teeth and bones

Entire beaver mandibles or their single long sharp incisor teeth were commonly used as woodworking tools during the Mesolithic. While the mandibles were employed as scrappers or chisels (Ewersen, 2006/2007; Zhillin, 2004), the incisors were typically joined to a simple wooden shaft with pitch and have been found in lowland European sites (Schacht and Bogen, 2001). Beaver incisors were used for carving various artefacts by Northwest Native Americans (Stewart, 1973). Numerous tribes including the Ingalik of Alaska continued such practices well into the twentieth century (Osgood, 1940), while, in particular, Pacific Northwest tribes, such as the Clayoquot of British Columbia, made dice from their molars, decorating them with carvings and colours for gambling games (Dorsey, 1901).

Their large long incisors were widely sought for ornamental pendants, which have been found in Mesolithic sites on the eastern Baltic and in western Russia (Schmölcke et al., 2017). The high value of these pendants for their symbolic and spiritual significance was demonstrated at a large burial site

(~6,400–6,000 BCE) on Lake Onega in northwestern Russia, where many women had been buried with them (O'Shea and Zvelebil, 1984). Beaver ornaments have also been discovered in the graves of children and women in Anglo-Saxon (sixth to seventh centuries) graves in central and eastern England (Coles, 2006). Several theories to explain their symbolism have been considered. Variations in the wear patterns associated with the bodies of growing children and pregnant women may indicate a belief rooted in medical folklore linked to dental health (Meaney, 1981). Alternatively, given the resemblance of the Latin for chastity *castitas* with *castor*, these pendants may represent the virginal status of those they were buried with (Blair, 2005). The Cherokee believed that when a child lost a baby tooth, they had to run around their home four times saying 'beaver put a new tooth into my jaw' before throwing it onto the roof (Long, 2000). Beaver tooth pendants were also worn by Vikings (Schmölcke et al., 2017), while the Romans in Gaul had beaver incisors as amulets, a custom continued in some parts of modern Italy to ward off the evil eye (Pilleri, 1986).

Astragalus (ankle bone) pendants were found in Mesolithic graves in Latvia (Eriksson et al., 2003) and their wider use is of interest as their symbolic value from the Estonian Vikings through to the early French Christians (sixth to seventh centuries) is likely to be different. The commonality of their use in graves either suggests a protective meaning or may be reflective of social status (Schmölcke et al., 2017). Beaver bones have also been found as suspected funeral offerings of German tribes (Pilleri, 1986).

The full symbolic value of the widespread discoveries of beaver teeth and astragalus has resulted in considerable speculation regarding their use, which ranges from common items such as trap markers (Jonuks, 2005) to economic tokens (Luik, 2010), oracles and amulets, and gambling tokens (Nikulina and Schmölcke, 2007; Nikulina and Schmölcke, 2008), and even proves the existence of 'beaver cults' (Fehner, 1963). Beaver teeth were also powdered and made into soup-like substances to be consumed for the prevention of a range of diseases, though castoreum was believed to have the most healing powers.

2.5 Castoreum and its creative uses

Though not used today for medicinal purposes, castoreum has long been employed in traditional medicine. In 500 BCE Hippocrates, the Greek 'Father of Medicine', was credited with describing diseases as natural rather than caused by the gods (Garrisson, 1966) and he fully recognized the healing properties of castoreum (Martin, 1892). Celsus (*c.* 30 CE), a Roman encyclopaedist, details the medicinal value of castoreum in his writing *De Medicina*, whilst Solomon III, King of Israel, is said to have obtained castoreum from Spanish beavers to cure his headaches (Westcott, 1989). Numerous early medical texts, originating with Baccius's *Castrologica* (1685), pronounce now somewhat outlandish claims for the medicinal properties of castoreum. Castoreum is believed to have analgesic and analeptic properties (Leung and Foster, 1996) and has been utilized in the treatment of a range of conditions including all sorts of fevers, toothache, preventing insomnia, stomach complaints, hysteria, and even insanity. Historical testimonies towards its power have included the curing of ear-ache and even deafness, tumours of the liver, gout, sciatica, lethargic people, maladies of the spleen, and a girl with complete memory loss, destroying fleas, stopping hiccoughs, both inducing and preventing sleep (presumably different doses!), as an antidote to the stings of scorpions and spiders, and even as doing 'much good to mad people' by clearing the brain (Martin, 1892). In fairness, castoreum may possess some analgesic properties, being mainly derived from beavers' ingestion of the salicin in the bark of the willow, meadowsweet (*Filipendula ulmaria*), and poplar (*Populus* spp.) trees which then concentrate in their castor sacs (Rainsford, 2016). Where present, beavers will readily select and consume meadowsweet plants with great enthusiasm. Acetylsalicylic acid produced from salicin derived from meadowsweet (at that time its botanical name was *Spiraea ulmaria*) led to the development of aspirin by Felix Hoffmann in 1897 (Jeffreys, 2008). However, no modern medicinal uses for castoreum are recognized (Figure 2.6).

It was also used in perfumes, soaps, and creams and as an oil base and fixative agent in cosmetics to increase the length of time scent lasts, as was ambergris from sperm whales and civet cat musk

Figure 2.6 Castoreum and castor sacs were believed to have significant medicinal and economic value, whilst being key to chemical communication in beavers. (Photo supplied courtesy of Dmitry Sventitsky.)

(Opdyke, 1973; Aftel, 2004), and it was believed to be an aphrodisiac by some (Mertin, 2003). The Algonquins, amongst others, at times when tobacco was rare or out of season, would substitute plants to which they added castoreum as flavour, with a reportedly soothing effect (Martin, 1892). Later, twentieth-century cigarette companies such as Philip Morris dusted dried castoreum over their tobacco, giving a distinguishing scent to brands such as Camel and Winston historically (Berenstein, 2018). Carolus Linnaeus reported that Laplanders mixed snuff with castoreum (Pilleri, 1984), and it was burnt in lamps by the Romans who believed its fumes to be an abortifacient (Müller-Schwarze, 1992). In the Middle Ages castoreum was applied to active hives by bee-keepers to deter predators and improve honey production (Pilleri, 1984; Marcuzzi, 1986).

Until more recently, extracts from castoreum were used to enhance food flavouring, often in association with vanilla-based foods and beverages (Burdock, 1995; Burdock, 2007). Castoreum was still sold in Norwegian pharmacies until around the 1970's. While this 'beaver butt juice' has been largely replaced by synthetic versions, its extract (CAS No. 8023-83-4, FEMA No. 2261) remains approved for human consumption in both Europe and North America (by the Council of Europe and the Food & Drug Administration Flavour & Extract Manufacturers Association; Burdock, 2007) and may still be present in some 'natural' vanilla flavourings (Berenstein, 2018). '*Bäversnaps*' is still locally available in parts of Sweden, with real castoreum giving it a unique taste and apparently not a liqueur for the fainthearted (Figure 2.7).

2.6 Beavers as food

In both Europe and North America there is a long tradition of beavers being eaten by humans. The Vikings of South Estonia commonly consumed them (Paaver, 1956). Beaver was a fairly familiar food resource to many Native American tribes, roasting them whole or cutting into strips to dry in the sun and use like modern jerky, or making into long-lasting pemmican cakes by pounding beaver meat with fat and berries (Dolin, 2010). In the seventeenth century English colonists in Plymouth and Massachusetts (Fischer et al., 1997) followed suit, as did the Sahtu Dene and Metis of the Northwest Territories (Appavoo et al., 1991). In modern-day North America, Norway, Sweden, Finland, Poland, Lithuania, and Latvia they are still a widespread game species, with hunters and trappers typically allowed to keep meat for their own consumption (Jankowska et al., 2005; Razmaité et al., 2011;

Figure 2.7 '*Bäversnaps*' liquor flavoured with castoreum, available in Russia. (Photo supplied courtesy of Frank Rosell.)

Strazdina et al., 2015), though illegal hunting no doubt occurs elsewhere. In medieval times their aquatic lifestyle, well-scaled tail, and duck-webbed feet were enough to warrant their classification as fish-like. As mentioned earlier, this interpretation allowed their consumption during religiously dedicated, Christian fasting, 'meat-free' Lenten food days (Gessner and Forer, 1563).

Beaver meat is high in protein (35 g per 100 g), essential amino acids and minerals, and polyunsaturated fatty acids (Jankowska et al., 2005; Razmaité et al., 2011; Razmaité et al., 2019). The nutritional value of cooked beaver meat has been compared with that of bear (*Ursus americanus*), boar (*Sus scrofa*), and cattle (*Bos taurus*) (US Department of Agriculture, 2006). Beaver meat averages 212 kcal per 100 g and is higher in protein and vitamin B_{12}

than the others. An adult beaver properly butchered can provide over 5.5 kg of meat, which is equivalent to a similarly sized roe deer (*Capreolus capreolus*) (Jankowska et al., 2005). The high fat content (40–60%; (Aleksiuk, 1970) of beaver tails especially during the early winter months would have been an important resource during leaner times. A professionally butchered beaver carcass is around 48.6% (n=17, 18.846 ± 3.596 kg) of the whole dead animal weight and typically composed of ~62.8% meat, 22.4% bones, and 14.5% fat (Jankowska et al., 2005).

2.7 The fur trade: 'brown or soft gold'

Humans have utilized furs from prehistoric times to protect themselves from the elements but also for decoration. The ancient Egyptians used vegetables to dye furs that they traded from the Arabians and Phoenicians, whilst the Romans imported pelts from Germany (Dolin, 2010). A vast array of animal pelts were traded across the whole of Eurasia in medieval times through commercial networks, with increasingly skilled products produced (e.g. Fisher, 1943; Veale, 2003). Ancient trade in beaver fur (Figure 2.8) is documented across Eurasia, with scattered documents in multiple languages and country records depicting a similar tale. In Britain, for example, records from a Neolithic site in Norfolk (Healy, 1996), an Iron Age site in Cambridgeshire (Evans and Serjeantson, 1988), an eighth-/ninth-century location in York (O'Connor, 1991), and another from the tenth/eleventh century in Suffolk (West, 1985) all demonstrate the prolonged use of beaver pelts. As these grew rarer, their value rose to 120 pence per beaver, as its dimensions were much greater than similar animals, for example 12 pence for an otter and 24 pence for a pine marten (*Martes martes*) (Coles, 2006). In 940 a law stated that only Royal garments could be made from their skins (Spriggs, 1998; Yalden, 1999). Later in the Middle Ages, beaver pelts were being imported to Britain from many sources including northern Eurasia and even Portugal, afterwards replaced by excessive imports from North America (Veale, 2003). Similar laws came later in mainland Europe; for example, in 1857 beaver hunting in Bavaria and Saxony was only permitted on behalf of the Royal Court as they were so scarce (Marcuzzi, 1986).

Figure 2.8 Beaver fur was a much sought-after commodity, which nearly saw both species become extinct. (Photo supplied courtesy of Robert Needham.)

Therefore, long before any settlement of North America by the European colonists and exploitation of beaver escalating in the infamous Beaver Wars, significant trade routes existed throughout Eurasia, with Siberian river routes linking Russia and China (Halley and Schwab, 2020). The Russian fur trade with China prevailed from the tenth to the eighteenth centuries, with huge markets selling every fur imaginable, which ultimately and eventually saw the demise of the Eurasian beaver in this unquenchable quest for fur (Halley and Schwab, 2020). Viatka, an old Novgorod human settlement, was founded in Perm (eastern Russia) by Russian trappers hunting beavers and squirrels (*Sciurus vulgaris*) for Russian merchants (Fisher, 1943). From the mid-fourteenth century Moscow dominated the fur market, being the predominant supplier to western Europe and the Ottoman Empire (Richards, 2014). So exhaustive was this early fur trading, by the Middle Ages European furriers, who for centuries had followed the fashion trends, were dealing in a

dwindling resource. By the 1600s the European fur supply was shrinking, with all furbearing mammals in serious decline (Veale, 2003). Species of exceptional value such as martens were increasingly rare by the thirteenth century across Europe. In eastern Prussia around 1400, beaver hunting was already being limited, with exclusive rights only accorded to specific hunters. By this time, a beaver was worth twice as much as an otter (*Enhydra lutris*) and by 1591 a beaver and bear (*Ursus arctos*) pelt were receiving similar bounties (Marcuzzi, 1986). Records of fur prices in Moscow in 1595 indicate that a beaver pelt had a value of 0.9 rubles, with one sable pelt being worth 0.716 though this is much smaller than a beaver pelt (Fisher, 1943). In the Moscow fur markets beaver was a popular fur, though indigenous people were paid only a fraction of what the furs were sold for, often costing more than a bear skin (Hellie, 1999). Around 22 furbearing species were commonly sold here, with beaver furs generally involved in only 5.6% of transactions (between 1600 and 1725) (Hellie, 1999).

This saw the Russian exploitation increase and the conquest of Siberia mainly due to sable fur. Centuries later an almost identical history followed in North America on the back of beaver pelts which were often nicknamed 'brown gold' or 'soft gold', given their high value, portability, and the fact they could be exchanged for almost everything (Monahan, 2016). In the quest for further hunting grounds, early Russian hunters traded with indigenous peoples, often offering the latter blankets, metal utensils, knives, flour, tea, and especially alcohol in return for supplying them with furs, though this trade could hardly be described as equitable (Richards, 2014). The mid-fifteenth century onwards saw a major and aggressive commercially based expansion of Russia into Siberia, beginning as sporadic raids and accumulating in domination of trade routes and ports, a policy of 'gathering of the Russian lands' eastwards (Gibson, 1969). A band of an estimated 800 Cossacks captured Sibir, the capital of the Siberian khante, in 1581, collecting tributes which they sent back to Moscow, including 2,000 beaver pelts which they had collected in less than 2 years (Fisher, 1943). In half a century, this rate of expansion saw the founding of Verkhoturye (1598) on the eastern edge of the Urals, followed by the navigation of the Bering Strait in 1648 (Gibson, 1969). Native populations were small and highly scattered across this vast range, so easily dominated by Russians with their advanced weapons.

As Russians established and colonized Siberia, a tax or *iasak* or *yasak* (fur tribute) was demanded to be paid in sable furs or their fixed ratio equivalent in other furs per adult male, the Siberian sable (*Martes zibellina*) being the most prized fur (Richards, 2014). Payment of this fur tax and allegiance to the Tsar were also encouraged by well-armed Cossacks, by the practice of taking hostages (Gibson, 1969). The native populations and hunters also had to compete with private Russian fur traders, often from long families of experienced trappers known as *promyshlenniks* who depleted their native hunting grounds seeking personal wealth. Like North America, Siberia was often viewed as providing a limitless source of high-quality fur, offering adventure, freedom, and wealth to those willing to seek this lifestyle, with secondary incentives of gold and silver, walrus and even mammoth ivory (Gibson, 1969). As with the fate to later befall Native Americans, the native Siberian populations not only suffered with the demands placed to meet this tax but were also exposed to diseases (e.g. small pox, measles) and saw their cultural and social systems collapse, with many forming a dependency on traded goods such as flour and alcohol and losing possession of their land (Richards, 2014). By around 1627, the sables and beavers were reportedly 'hunted out' from areas around Mangazeia, Turukhansk, lower Yenesei, and Tunguska (Fisher, 1943). In the 1690s the Chinese market opened to the west and so a seemingly insatiable demand opened to the Russians dominating the trade in furs. As this demand increased and beavers became more highly valued, some Sami communities recognized this and insisted any hunter killing beavers should pass their revenue to the community instead of retaining it themselves, thereby decreasing resource competition and social conflicts around concepts of greed and envy (G. Hartman, pers. comm.).

The beaver had a significant place in the early economy of Native Americans who lived in forests and practically used all of its resources. Skins were historically believed to have medicinal properties and could be applied to heal rheumatism and gout (Plantain, 1976), or were tanned to produce a leather for clothing and bags (Dolin, 2010), while the tail fat treated epilepsy, nerve conditions, convulsions, and asthma (Long, 2000). Beaver body fat was made into ointments to help prevent frostbite (Martin, 1892). The Apsaroke boiled down tail membranes and mixed these with hide scrapings to produce a glue which was strong enough to attach strings to bows (Long, 2000). However, the most valuable body part remained the pelt and to be buried in beaver skins or to give them as presents was a sign of honour and friendship in many tribes (Wilsson, 1968).

Perhaps the first significant European traders in North American beaver pelts were the French Normans who were dealing, and in part integrating, with native people long before the first trading posts and trappers arrived. They would purchase 'castor gras d'hiver' (fat beaver) pelts that had been cured by native peoples, worn over the winter and sold in the spring. As the demand for beaver pelts expanded, further explorations were made inland and French settlements begin to flourish. King Henry IV of France actively supported and encouraged the French fur trade in North America (Dolin, 2010). Whilst early attempts to colonize failed, as they often included people deported for religious or political beliefs and criminal activities with fewer incentives to endure, others considered more 'respectable' and 'civilized' (those with trades and professions) began to settle. The first trading posts were built on Île de Sable in 1598 and in 1600 at Tadoussac (Poliquin, 2015).

This informal hold on the New World fur trade became more formalized for France in 1603 with the establishment of a royal charter and a trading post near modern Québec by the French explorer Samuel de Champion. This move resulted in further inland expansion which claimed large areas of 'virgin beaver territory' around the Great Lakes, Lake Superior, and James Bay (Dugmore, 1914). The Dutch had long traded for furs with Russia, though as the stories of boundless fur resources in the 'New World' returned and hearing that furs could readily be traded for 'trinkets' such as glassware and duffel cloth with native populations they too sought to expand west in the search for 'tariff-free' furs (Dolin, 2010). As pressure grew from establishing of the first English (Jamestown in 1607) and Dutch (Manhattan Island in 1624) colonies, the French claim to a large part of the North American continent on the basis of their fur explorations weakened. As the English and Dutch colonies grew and developed they too began to trade in furs, resulting in a rise in tensions and conflicts with the French (Dolin, 2010). One significant trading item was shell beads, known as *wampum, peag,* or *sewan,* either white (cut from inner whelk shells) or the more valuable black (dark purple) cut from quahogs, a type of clam, and used in many ceremonial roles such as peace offerings or jewellery (Woodward, 1980; Dolin, 2010). Learning the significance of these beads as a form of currency between Native American tribes, the Dutch readily traded tools to assist in their production, such as drills and polishing tools (Weeden, 1884; Dolin, 2010).

While the colonial powers consolidated their influence, vast numbers of furs were sent back to Europe and China. In time, a combination of overhunting coupled with habitat destruction to make way for settlements and farming decimated any neighbouring beaver populations and forced a process of travel ever deeper into new territory to acquire new hunting grounds. Typically, most of the trapping was undertaken by the native peoples, with the colonists trading items such as blankets, mirrors, axes, guns, and alcohol for the pelts of various animals. At the Hudson's Bay Company's trading post in Fort Albany in 1733, trade was measured in 'Made Beaver' (MB), one adult beaver skin in prime condition (Hudson's Bay Company, 2016; Table 2.1); other pelts were also measured in MB and in 1703 this was equivalent to one black bear (*Ursus americanus*), pelt, two foxes (*Vulpes fulvus*), four raccoons (*Procyon lotor*), eight pairs of moose hooves (*Alces alces*), or 5 lb of goose feathers (Hudson's Bay Company, 2016). The pursuit of ever greater numbers of pelts combined with new trapping technologies and overexploitation, leading to profound shifts in tribal cultures turning away from more traditional hunting methods and lifestyles which were more appropriately in balance with nature. This process caused numerous intertribal conflicts and ultimately more colonial wars (Long, 2000).

Table 2.1 Standard of trade from the Hudson's Bay Company at Fort Albany, 1733. A trade item was set as one Made Beaver (MB), and other pelts could be traded, for example two otter pelts could equal 1 MB.

Trade item	Amount	Trade item	Amount	Trade item	Amount
Beads (coloured)	¾ lb	Breeches	1 pair	Ice chisels	2
Kettle, brass	1	Combs	2	Knives	8
Lead, black	1 lb	Egg boxes	4	Looking glasses	2
Gun powder	1½ lb	Feathers, red	2	Tobacco boxes	2
Shot	5 lb	Fish hooks	20	Needles	12
Sugar	2 lb	Fire steels	4	Powder horns	2
Tobacco, Brazil	2 lb	Files	1	Plain rings	6
Thread	1 lb	Flints	20	Stone rings	3
Ditto Leaf or Roll	1½ lb	Guns	1 for 10–12 pelts	Mocotagans (curved knives)	2
Vermilion	1½ oz	Pistols	1 for 4 pelts	Sword blades	2
Brandy	1 gallon	Gun worms	1	Spoons	4
Broad cloth	2 yard	Gloves, yarn	1	Shirts	2
Blankets	1	Goggles	2	Shoes	1 pair
Flannel	2 yard	Handkerchiefs	1	Stockings	1 for 1¼ pelts
Gartering	2 yard	Hats, laced	1	Sashes, worsted	2
Awl blades	12	Hatchets	2	Thimbles	6
Net lines	2	Scrapers	2		
Buttons	12 dozen	Hawk bells	8		

Source: Reproduced from Hudson's Bay Company (2016).

In 1670, the English made a significant strategic move when a Royal Charter was signed by Charles II affording rights over the area known as Rupert's Land. This acquisition led directly to the formation of the Hudson's Bay Company. The Company's dominion over this vast area, along with abundant resources, meant a significant loss to the French and marked their decline in dominance. A series of trading posts created by the Company went on to trade with tribes such as the Swampy Cree for another three centuries. At various points, its ownerships and land rights were challenged by the French between 1686 and 1714 and by the Swedish, and the English regained control in the end. The Dutch were also a significant fur-trading nation and established trading posts such as Fort Orange, near Albany (Dolin, 2010).

The fur wars have become famous in the history of the western exploration which led to the settlement of Canada and North America. Often bloody conflicts were fought at various times between the native peoples, France, England, Holland, Sweden, Spain, and Portugal. Dolin (2010) and Hilfiker (1991) graphically detail the history of these long, complicated, and political fightings and attribute the widespread use of the English language, laws, and customs throughout North America to the quest for the beaver's high-value fur. After expulsion of the Dutch throughout the 1700–1800s, the English and French waged several wars for the control of territories and therefore trade in beaver furs. These included King William's War (1689–1697), Queen Anne's War (1702–1713), King George's War (1744–1748), and the French–Indian War (1754–1763). While this last conflict culminated in the French being defeated, other major players then entered the fur trade. The North West Fur Company was formed as a syndicate of fur-trading groups in 1783, the Spanish established the Missouri Company in 1794, and the Russians founded the Russian–American Fur Company in 1799. Eventually the British were also ousted and the American colonists declared themselves independent, setting up their own fur-trading companies (Dolin, 2010). The American Fur Company formed in 1808, the Pacific Fur Company in 1810, and the Rocky Mountain Fur Company in 1822 (Westcott, 1989). The ferocious competition, including the purposeful creation of 'fur deserts' in some regions to sabotage exploitation by other companies through the aim to eliminate all fur bearers, combined with ruthless overexploitation generated by these organizations directly resulted in the exhaustion of ever greater tracts of beaver habitat (Ott, 2003).

Prior to the early nineteenth century, beavers were largely killed using traditional methods such as deadfall traps and snares, hunting with dogs, and staking out lodges, burrows, or ice hole exits before spearing or bludgeoning the emerging or dug-out animals (Long, 2000; Dolin, 2010). Guns were available for use in hunting, though not typically used for gathering pelts, while various iron trap designs existed from as early as 1633 (Novak, 1987). This changed when a new tooth-jawed, steel leghold trap was invented and made commercially available in approximately 1823 by Sewell Newhouse, New York (Gerstell, 1985). This was a highly effective and portable killing device, as documented by individual trappers. James O. Pattre, for example, recorded trapping 450 beavers between December 1824 and March 1825, with up to 60 beavers being taken in a single day through the use of 40 traps (Grescham, 1944). Castoreum derived from dead beavers was used to assist this process. Native trappers had been recorded by Chaplin Charles Woolley as using it as a lure for their traps around the New York area in the 1670s (Poliquin, 2015). Other mixes using it in combination with various tree barks, nutmeg, clove, cinnamon, and whiskey were also employed (Godman, 1826). This lethal combination of castoreum as a bait and the simple steel trap removed tens of millions of beavers from the North American landscape.

The peak trade in beaver pelts occurred during the first four decades of the nineteenth century. This was the era of the 'mountain men'(1825–1840), generally white male trappers, recruited and equipped by the fur companies to undertake risky and long treks into the western mountains, living an independent existence trapping and processing beavers (Dolin, 2010). They met periodically at annual *rendezvous* to sell and trade goods (Long, 2000) and were largely responsible for the extirpation of many remoter beaver populations. Overexploitation prompted a crash in the North American beaver population. While in 1871, the Hudson's Bay Company traded

174,000 beaver pelts, this dropped to 102,745 16 years later and in 1891 was further reduced to 57,260 (Seton, 1953). The supply in beaver fur was diminishing and a long list of other species, including otter (*Lutra canadensis*), marten (*Martes americana*), black bear, timber wolf (*Canis lupus*), raccoon, wolverine (*Gulo gulo*), bobcat (*Lynx rufus*), moose, caribou (*Rangifer tarandus*), and wild turkey (*Meleagris gallopavo*), had also been drastically reduced. At a similar time, the newly opened silk trade with China, around the 1830s, saw fashions shift. Coypu (*Myocastor coypus*) fur was introduced in an attempt to try and substitute some of the previous beaver demand (Martin, 1892). The trade in 'brown gold' dwindled and decimated populations were left as scattered fragments. Buffalo (*Bison bison*) became the next big exploitable American commodity, leading on to what is now known as the 'Age of Extermination' in the later nineteenth century, which saw now seemingly obscene numbers of animals killed for food, fur, and sport (Dolin, 2010). This also marked a turning point in sparking the conservation movement in recognition of this destruction and need for sustainable management for future utilization.

2.7.1 Beaver hats

While beaver pelts have multiple uses, as cradle blankets, clothing, and protective bags for musical instruments (Coles, 2006), tobacco, or medicine (Long, 2000), by far their most famous use was in felt production for hat making. Beaver felt is incredibly strong and resilient to all weather types. It keeps its pressed shape longer than other furs, tends not to tear easily, and is smooth. The pliability of beaver felt saw it modelled into many differing styles, following the changing fashions of the times.

The French protestant Huguenots of the fifteenth and sixteenth century exhibited a great vogue for felt hats, which was swiftly followed by the European upper classes, a must-have 'fashion necessary' which lasted nearly 200 years (Crean, 1962; Dolin, 2010). Beaver felt rose rapidly in popularity and the headwear constructed from this commodity known simply as 'beavers' was accompanied by a whole etiquette of wear (Westcott, 1989). They were expensive and owning them was a show of wealth. Records exist of these valuable items being gifted in wills. There was also a trade in counterfeits, and cheaper versions distinguished from pure beaver felt hats, such as 'castor' headwear (a mix of beaver and other fur such as rabbit *Oryctolagus cuniculus*), were around three times cheaper (Poliquin, 2015). Trade monopolies were an important business and in 1613 the hat and felt makers' guild of London lobbied Parliament in an attempt to ban any foreign imports. In 1638, the British Parliament passed a law ensuring the purity of these hats so that they were only permitted to be made from 'beaver stuff' or 'beaver wool' (Martin, 1892). England alone exported 69,500 beaver hats in 1700 which escalated to 500,000 beaver hats by 1760, but in total an astounding 21 million beaver and felt hats combined were exported from England in the course of ~70 years (Carlos and Lewis, 2010). Beaver hat theft was severely punished and in 1675 the Old Bailey, London, recorded 222 people being convicted of beaver-related thefts. Twelve were whipped, seventy-three transported to the New World, and forty-five condemned to death for crimes of this sort by the end of the nineteenth century (Poliquin, 2015).

2.8 Historic and current distribution range

From great historic lows both beaver species have followed a very similar path of modern recovery. This has been achieved through a combination of legal protection, the creation of hunting regulations, and proactive restoration and reintroduction programmes. All have combined to contribute to a modern acceleration in their natural spread.

In Europe around the twelfth century beavers had been substantially reduced, and by the sixteenth they were all but nearly completely extinct (Table 2.2), from being once widespread throughout freshwater habitats from the Arctic to the Mediterranean and from the Atlantic to the Pacific coasts, excluding areas north of the Arctic tundra, islands of Ireland, Iceland, Sicily, Corsica, Crete, the Balearics, Malta, Sardinia, and the very southern fringes of Greece and Italy, and they are not known across southern and South-East Asia (Halley and Schwab, 2020). Eurasian beavers where known in antiquity (typically 8th century BC to 6th century AD) but then disappeared early from the Mediterranean rim and southern areas of their range such as Greece, Portugal, Spain, Turkey, Azerbaijan, and Iraq (Boessneck, 1974).

Early protective decrees and charters in Europe often assigned ownership of beavers to the state or church as they became a scarce commodity; for example a Prussian royal edict mentions beavers in 1714, whilst another declared fines for those hunting Elbe beavers in 1725 (Martin, 1892). Some of the first national beaver protection laws had already been enacted in Norway in 1845. Hunting was banned in Sweden in 1873 at a time when they were already thought to be extinct. However, a much earlier medieval (mid-fourteenth-century) law from the county of Östergötland is roughly translated as 'he who kills a beaver and destroys his house, gives the beaver back to the landowner, along with three marks. If he takes the beaver from common land, he owns the beaver and will not stand trial'; this gives insight on how valued the beaver was at this time (G. Hartman, pers. comm.).

By the twentieth century more countries where beavers still remained recognized their plight and began to take an interest in their restoration. Although many early reintroductions were largely motivated to restore the fur trade, others that followed in more modern times did so as a realization of the species' pivotal role in wetland ecology and began to take root in their own right. Today the Eurasian beaver is thought to number ~1.5 million individuals (Halley et al., 2020). It is thought to have increased in range by 550% since the 1950/1960s, with one of the greatest species abundance increases and highest annual growth rates (~11% annually) of all recovering species in Europe (Deinet et al., 2013). Some of the most rapid expansions have occurred between 2000 and 2011, in western and south-central Europe, southern Russia, and west and central Siberia (Halley et al., 2020). This astounding return is due to conservation programmes and it is still enshrined as a European Protected Species throughout much of Europe, even though it is now considered to be of 'least concern' on the International Union for Conservation of Nature (IUCN)'s Red List, on both a global and European scale, with future range expansion and population increases expected (Batbold et al., 2008; IUCN, 2011).

It has been estimated that prior to the European settlement of North America there could have been anywhere between 60 and 400 million beavers in existence (Seton, 1929; Naiman et al., 1986). After centuries of relentless overhunting, largely for felt hats (Bryce, 1904), they were critically threatened across their whole range by the 1880s and had disappeared from the eastern USA (Baker and Hill, 2003). An eventual concern over these impacts on beaver populations began a movement towards more controlled harvesting and conservation, with trapping regulations implemented in 1899. One of the first people to really appreciate the scale of their destruction and the need to implement conservation

Table 2.2 Relict fur trade refugia populations in Eurasia.

Area of refugia	Estimated minimum population size	References
Belarus/Ukraine (Dnieper River system)	< 300	Zharkov and Sokolov (1967)
France (Lower Rhone delta)	30	Richard (1985)
Germany (middle Elbe)	200	Heidecke and Hörig (1986)
Mongolia/China (Urungu)	< 100–150	Lavrov and Lu (1961)
Norway (Telemark)	60–100	Collet (1897)
Russia (Konda-Sosva)	300	Lavrov and Lavrov (1986)
Russia (Upper Yenesei)	30–40	Lavrov and Lavrov (1986)
Russia (Voronezh)	70	Lavrov and Lavrov (1986)

Source: © Halley, D.J., Saveljev, A.P. and Rosell, F. Mammal Review published by Mammal Society and John Wiley & Sons Ltd.

efforts (all be it to protect fur resources) was George Simpson, the governor of the Hudson's Bay Company (Poliquin, 2015). Although he could not enforce a hunting ban he reduced the availability of steel traps, offered higher prices for alternative fur, closed trading posts in trapped-out areas, and opened others where more stable populations still remained.

2.8.1 Eurasian beaver distribution

The modern distribution of the Eurasian beaver is a curious combination of natural range expansion from the small relict populations that received protection and multiple reintroduction and translocation projects (Figure 2.9). Their current range, including the most recent reintroductions to Britain (there is no evidence that beavers ever colonized Ireland), extends from large populations in Norway and Sweden, with smaller populations in Finland, and a recently reintroduced population in Denmark, before forming a generally contiguous band across central Europe from Germany to Poland, the Baltic States, Belarus, Ukraine, Siberia, and into Russia. Their western and south-central European current range has extended greatly with exponential growth evident (Halley et al., 2020). A further scattering of small, distinct populations to the east and south of this European central range is mainly associated with reintroduction projects. Other populations exist on the periphery of their main Russian range and in central Asia. The Russian Federation range is vast and expanding, with recent estimates of around 630,000 animals (Borisov, 2011; Russian Fur Union, 2016). The vast majority of the Federation and the Asian landmass remains unoccupied, although three key nature reserves have been specifically created to restore beaver populations at Berezinsky, Voronezh, and Kondo-Sosvinsky (Zharkov and Sokolov, 1967).

The very eastern range of the Eurasian beaver is found in Mongolia and the western Chinese province of Xinjiang. Also known as the Sino-Mongolian beaver, it is represented by a small threatened relict population and often given subspecies status *C. f. biruli* given its isolation length and geographic restriction as mainly occurring within the Ulungur River watershed (Chu and Jiang, 2009). In 2007,

population estimates suggested 508–645 beavers existed, many occurring in the Bulgan Beaver Nature Reserve; however, they are likely to be decreasing due to logging and land reclamation impacts throughout the catchment (Chu and Jiang, 2009), human settlements, and intensive agriculture along the riparian zone and are recognized as requiring conservation efforts. Small numbers of these beavers have been successfully translocated to other sites such as the Bulgan and Chovd Rivers by the Mongolian Wildlife Management Authority (Halley and Rosell, 2002; Shar, 2005). More recently these authorities imported Bavarian beavers to supplement breeding efforts, as reproductive rates in *C. f. biruli* are reported to be low and they have been difficult to breed in captivity (Lu, 1981).

Reintroductions and translocations have been key in the highly successful restoration of this species (see Table 2.3 for full details). The very first ever known beaver reintroduction occurred in what is now modern Czech Republic, then a province of Bohemia, part of the Austrian Empire, in 1800–1810 at Třeboňsko, and a population survived there until 1876 (Vorel and Korbelová, 2016). Modern reintroductions from 1970 to 1990, along with natural immigrations from Bavaria have resulted in today's population, with Austrian beavers in turn populating adjacent regions in Hungary, the Czech Republic, Switzerland, Italy, and Slovenia (Halley and Rosell, 2002).

Throughout the rest of Europe natural spread and translocations have been significant and most of these have been well documented (Halley and Rosell, 2002; Halley and Rosell, 2003; Halley et al., 2020; Table 2.3). In Norway and Sweden, given pelt trade numbers, beavers were still abundant in the seventeenth century, but overhunting in the eighteenth century saw the last original Swedish populations recorded in the 1870s (Curry-Lindahl, 1967). Norway legally protected its remaining beavers in 1845 and so a small fur-trade refugia remained. In certain countries such as the Czech Republic, beavers were effectively made extinct twice; by the seventeenth century they were rare, becoming extinct in the middle of the eighteenth century through hunting, then natural recolonization saw them reoccur, only to be hunted again to extinction by the end of the nineteenth century—effectively

Table 2.3 Protection status, reintroduction effort history and current population estimates of Eurasian beavers in Eurasia.

Country	Extirpation	Protection	Reintroduction and/or translocations	Most recent population estimate	References/comments
Austria	1869	–	1970–1990	7,600	Kollar and Seiter (1990); Universität für Bodenkultur Wien (2017)
Belarus	Remnant	1922	1948	51,100	National Statistical Committee of the Republic of Belarus (2018)
Belgium	1848	–	1998–1999	2,200–2,400	Dewas et al. (2012); www.beverwerkgroep.be 2019
Bosnia and Herzegovina	?	?	2006	140	Trbojević and Trbojević (2016)
Bulgaria	1750–1850	?	–	–	Boev and Spassov (2019)
China	Remnant	1991	–	600	Chu and Jiang (2009)
Croatia	1857?	–	1996–1998	10,000	Grubešić et al. (2001); Kralj (2014); Cavrić (2016); Tomljanović et al. (2018)
Czech Republic	Mid-eighteenth century; 1876[a]	–	1800–1810, 1991–1992, 1996 > 6,000	10,000	Jitka Uhlikova, pers. comm.; Vorel et al. (2012); Vorel and Korbelová (2016)
Denmark	c. 500 BCE[b]	–	1999, 2009–2011	216–252	Asbirk (1998); Elmeros (2017); Taidal (2018)
England	Eighteenth century?	–	2016[c]	c. 150[c]	D. Gow, pers. comm., 2019; Coles (2006)
Estonia	1841	–	1957	18,000	Timm, Estonian Environmental Information Centre, pers. comm.; Veeroja and Mannil (2018); Estonian Hunters Association (2019)
Finland	1868	1868	1935–1937, 1995	3,300–4,500[d]	Lahti and Helminen (1969); Natural Resources Institute Finland, Natural Resources Institute Finland (LUKE) (2019)
France	Remnant	1909	1959–1995	>14,000	Dewas et al. (2012); Office Nationale de la Chasse et le Faune Sauvage (2018)
Germany	Remnant	1910	1936–1940, 1966–1989, 1999–2000	35,000	Schwab et al. (1994); G. Schwab, pers. comm., 2018
Hungary	1865	–	1980–2006	14,600–18,300	Bajomi (2011); Bajomi et al. (2016); Čanády et al. (2016)
Italy	1541	–	Proposed	1	Nolet (1997); Martina (2018)
Kazakhstan	1915	–	1963–1986	5,500	Karagoyshin (2000); Saveljev (2005)
Latvia	1870s	–	1927–1952, 1975–1984	100,000–150,000	Belova et al. (2016)
Liechtenstein	?	?	–	50	Fasel (2018)
Lithuania	1938	–	1947–1959	121,000	Belova et al. (2016); Ulevičius, pers. comm., 2019
Luxembourg	?	?	–	c. 75	Herr et al. (2018)
Moldova	?	?	–	?	Status uncertain
Mongolia	Remnant	?	1959–2002, 2012, 2018	800	Samiya (2013); Adiya et al. (2015); Saveljev et al. (2015)
Netherlands	1826	–	1988–2000	2,300–3,800	Kurstjens and Niewold (2011); Dijkstra et al. (2018); Dijkstra (2019)
Norway	Remnant	1845	1925–1932, 1952–1965	>80,000	Rosell and Pedersen (1999); Parker and Rosell (2003); F. Rosell, pers. comm., 2020

Poland	1844	1923	1943–1949, 1975–2000	124,622	Janiszewski and Misiukiewicz (2012); Rozkrut (2018)
Portugal	c. 1450	–	–	0	Antunes (1989)
Romania	1824?	–	1998–1999	2,145–2,250	Troidl and Ionescu (1997); Ionescu et al. (2006); Pașca et al. (2018)
Russian Federation	Remnant	1922	1927–1933, 1934–1941, 1946–2005	700,000	Ognev (1963); Federal State Statistics Service, Russian Research Institute of Game Management and Fur Farming 2019; and see text
Northwest FD				152,800[e]	Russian Fur Union (2016)
Central FD				153,750	Russian Fur Union (2016)
Volga FD				168,070	Russian Fur Union (2016)
South FD				7,660	Russian Fur Union (2016)
Ural FD				58,100	Russian Fur Union (2016)
Siberian FD				80,780	Russian Fur Union (2016)
Far-East FD				800[e]	Russian Fur Union (2016)
Scotland	Sixteenth century	–	2009	c. 319–547	Coles (2006); Campbell-Palmer et al. (2018)
Serbia	1903?	–	2003–2004	240	Ćirović, pers. comm., 2012; Smeraldo et al. (2017)
Slovakia	1858	–	1995	7,700–9,600	Čanády et al. (2016)
Slovenia	1750?	?	–	300–400	M. Grubešić, pers. comm.; Deberšek (2012); Juršič et al. (2017)
Spain	Seventeenth century	1980s	2003	450–650[f]	Cena et al. (2004); Sáenz de Buruaga (2017)
Sweden	1871	1873	1922–1939	130,000	Freye (1978); Hartman (1994); Hartman (1995); Belova et al. (2016)
Switzerland	1820	–	1956–1977	2,800	C. Angst, pers. comm., 2012; BAFU (2016)
Ukraine	Remnant	1922		46,000	Safonov and Pavlov (1973); Matsiboruk (2013a); Matsiboruk (2013b)
Wales	Sixteenth century	–	Feasibility study completed	c. 15	Coles (2006); A. Jones, pers. comm., 2009; D. Gow, pers. comm., 2019; R. Campbell-Palmer and A. Leow-Dyke, pers. comm., 2020

Minimum population estimate

1,487,000[e]

Source: Reprinted from Halley et al. (2020).

FD, Federal district.[a]

[a] 1876 extirpation of animals descended from 1800–1810 reintroduction in southern Bohemia (Vorel and Korbelová, 2016).

[b] Based on subfossil remains. Philological evidence from place names suggests a remnant may have survived as late as the eleventh century.

[c] First licenced release, to River Otter, to supplement a population apparently deriving from escapes. Populations on other rivers are apparently the result of escapes from fenced enclosures. Population figure is for wild-living beavers only.

[d] Also c. 10,300–19,100 *C. canadensis*. Source: https://www.luke.fi/en/natural-resources/game-and-hunting/beavers/.

[e] Rounded to nearest thousand. Population estimate includes some *C. canadensis*; Russian Northwest FD < 20,000, Far East FD < 200.

[f] Navarra region only.

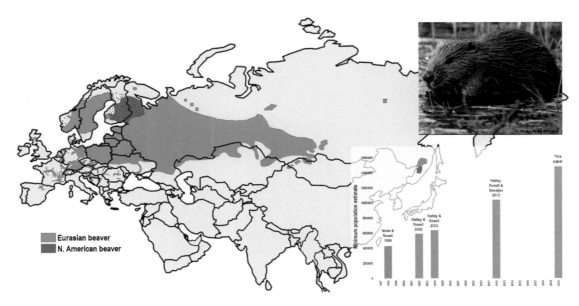

Figure 2.9 Beaver distribution (both species) in Europe and Asia in 2020. The graph shows minimum total Eurasian beaver (*Castor fiber*) population estimates 1998-2020 with sources indicated (© Halley, D.J., Saveljev, A.P. and Rosell, F. Mammal Review published by Mammal Society and John Wiley & Sons Ltd.)

going through two genetic bottlenecks (Ernst et al., 2017). The first wave of beaver reintroductions in Europe began in the 1920s–1950s, largely for the purpose of fur resource restoration (Halley and Schwab, 2020), from humble beginnings involving two Eurasian individuals trapped in pits in Aust-Agder, Norway, and moved to Sweden where beavers had previously died out in the 1870s (Nolet and Rosell, 1998; Vevstad, 1989). Early beaver translocations (involving ~40 individuals) in 1925–32 and 1952–65 around Norway mostly failed due to a low number of individuals (two to six) being released at each event (Myrberget, 1967). This marked the first of a series of beaver translocations from southern Norway, Telemark region, where beavers had survived as one of the few isolated fur trade refuge populations. This relict population also grew due to protective laws.

Today Eurasian beavers are found throughout the country, right up to the Arctic Circle, and by 2011 the population was estimated to have recovered to around 70,000 individuals (Rosell and Parker, 2011). A further 80 Norwegian animals were released across 19 locations in Sweden from 1922 to

1939 (Hartman, 1995). Latest Swedish populations are thought to number 130,000 individuals (Belova et al., 2016). Norwegian beavers were also reintroduced to Finland (1935–37), Switzerland, Latvia, and Scotland (Owesen, 1979; Rosell and Parker, 1995; Gaywood, 2015). Beavers had been completely lost from Switzerland by the early nineteenth century; today's successful population exists entirely due to reintroduction projects which, beginning in 1956, saw 141 individuals released at 30 sites along the Rhone and Rhine river systems (Stocker, 1985). These reintroductions saw animals sourced from the French part of the Rhone system, Scandinavia, and Russia (Stocker, 1985). From Switzerland, they naturally colonized the German state of Baden-Württemberg, growing to a population of at least 350 by 2004, including immigration from the River Danube (Banzhaf, 2004).

Hunting bans on beavers were instigated in Germany in 1910, enabling some recovery, which was then supplemented with translocations starting in 1936 and again in 1973, the 1980s, and 1990s (Heidecke, 1984; Nolet, 1997). One of the most successful reintroduction projects occurred in the

Danube river basin beginning with beaver releases to the Inn and Isar tributaries in Bavaria in 1966, Austria in 1976, the Czech Republic in 1991, Croatia in 1996, Romania in 1998, and Serbia in 2004 (Grubešić et al., 2006; Ionescu et al., 2006; Parz-Gollner and Vogl, 2006). Beavers were generally extinct in Bavaria in 1867 (Weinzierl, 1973). The translocations to Bavaria involved Russian animals and later included Polish, French, and Norwegian individuals but not those from neighbouring east Germany, where the fur trade relict population on the River Elbe had survived, purely due to political reasons (Weinzierl, 1973; Schwab and Schmidbauer, 2003). Bavarian beavers were so successful in breeding and range expansion that years later they have provided stock for numerous reintroductions and population augmentation projects across Europe, including Croatia in 1996–1998 (Grubešić et al., 2012; Cavrić, 2016), Romania in 1998–2003 (Paşca et al., 2018), Serbia in 2004–2005 (Smeraldo et al., 2017), and Bosnia in 2005–2006 (Trbojević and Trbojević, 2016). All of these have seen successful restoration.

France retained a very small relict population in the Rhone delta; through strict protection and within-country translocations (273 individuals) this has now greatly expanded in range (an estimated 60% of French streams are now occupied) and number to an estimated 14,000–16,000 beavers (Dubrulle and Catusse, 2012). Austria documented its last beaver shot dead in 1863, on the Danube floodplain near Fishamend (Sieber and Bauer, 2001). Around 40 Eurasian and an estimated 12 North American beavers were released in Austria from Polish, Russian, and Swedish stock from 1976 to 1985 and, along with natural immigration from Bavaria, this saw rapid and successful establishment, with an estimated 3,300 individuals by 2011 (Sieber, 1999; Parz-Gollner and Hölzler, 2012; see Section 2.8.3). No North American beavers have been determined in this population (Kautenburger and Sander, 2008). Other countries, such as the Czech Republic, saw successful, natural recolonization of beavers from neighbouring countries, namely Germany (Kostkan, 1999). A patchier picture was evident in Hungary, where beavers first reappeared again in the Szigetköz region around 1985, most likely naturalized recolonization from Austria (Schwab and

Lutschinger, 2001; Haarberg, 2007), though in other parts of the country small numbers of German-sourced beavers were released in the early 1990s (Bajomi, 2011). A further twist in this tale was that some of these animals released in Tisza Lake were retrospectively proved to be North American beavers and thereby removed. This, along with known escapes from captive collection, put a question mark over Hungarian beaver restoration projects, and although a limited number of North American beavers have been shot dead, it is believed that this species has not established (Haarberg, 2007).

Almost 17,000 beavers, have been translocated within the former Soviet Union, mainly between 1927 and 2004, to sites throughout Russia, Ukraine, Belarus, Latvia, Estonia, Lithuania, and Kazakhstan (Saveljev, 2003; Halley et al., 2012). The largest number of relict populations was retained, though isolated, in Belarus and different pockets in Russia; therefore natural recolonization also occurred and played an important role. Like other countries, Poland lost its beavers in the early nineteenth century, with restoration efforts beginning in the 1940s sourcing beavers from the Baltic region of Russia and bred individuals in farms specifically established to restore this species (Żurowski and Kasperczyk, 1988).

Britain represents the very western extent of the Eurasian beaver range. Evidence of beavers in southern Britain include archaeological remains from Saxon and Norman times (Coles, 2006), Old English place names, and written first-hand evidence from Gerald of Wales in 1188 of the last beavers know in one river in Wales. One of the last Scottish records tells of them around Loch Ness, in *The History and Chronicles of Scotland* (Boece, 1527), and the more recent discovery and radiocarbon-dating of a beaver-gnawed stick from Northumberland demonstrates they remained in the Tyne, northern England, until the fourteenth century (Manning et al., 2014), indicating they survived longer in the north. Although it was believed that the species had been lost from southern Britain and then hunted to extinction in Scotland by the sixteenth century (Kitchener and Conroy, 1997), intriguing evidence suggests that small, isolated populations most likely lingered until much later (Coles, 2006). In Wales a fourteenth-century Welsh poem (Halley et al., 2009) and in England a sixteenth-century English bounty specifically listing beavers records their presence, while as late as 1789 payments were still being made for their heads in Yorkshire (Coles, 2006). This theory has been rebutted by others (Raye, 2014), who suggest that accounts of bounty payments indicate no evidence of widespread culling (Lovegrove, 2007), that medieval art from this time depicted beavers castrating themselves which was a common folklore (i.e. not drawn from real life), medieval cooking texts were often copied, and a developed international trade in furs, including beavers, existed (Veale, 2003). Regardless, in a haphazard manner, beavers have now returned to Britain, via small-scale official release projects but mainly through either accidental or assisted unofficial releases; and though populations remain small and scattered, they could easily be approaching 2000 individuals.

The Ebro/Aragon watershed in Spain represents the current most southern extent of the Eurasian beaver, after Bavarian origin animals were unofficially released here in 2003 (Cena et al., 2004; Halley et al., 2012), though recent sightings have been reported in the Italian media on a headstream of the Danube; if proved correct these would be the first in 450 years (Halley et al., 2020). Mixed trapping effort has occurred in the Ebro system at cost and great political heat as the Spanish authorities obtained permission from the EU Habitats Directive to remove these animals which they considered an illegal release, even though beavers where already present in the Basque region (Echegaray et al., 2018, Halley et al., 2020). Needless to say, population data are sketchy but beaver populations have remained and expansion is evident; a similar situation may be witnessed in Italy going forward. Belgium's current population has also descended from unofficial releases, with 101 individuals released at several sites in 1998–2000, with additional natural spread from Germany and the Netherlands (Verbeylen, 2003).

Archaeologists have unearthed Eurasian beaver remains to prove their former occurrence in some surprising countries including Syria and Iraq (Danford and Alston, 1880; Harrison, 1972; Hatt, 1959). In 1835 Eurasian beavers were discovered on the Khabur and Euphrates Rivers (Ainsworth, 1838), where they are presumed to have remained in these countries up until the late nineteenth century. There have been invalidated claims of beaver presence in Turkey namely based on two observations, one dating back to the 1970s on the Yumurtalik marshes; however, these remain unconfirmed (Halley et al., 2012). Greece officially remains beaver free, though early reintroduction feasibility assessments may be in discussion (Halley et al., 2020).

2.8.2 North American beaver native distribution

Before the fur trade wars, the North American beaver's range was vast, expanding from Alaska (except along the Arctic Slope from Point Hope east to the Canadian border), mainland Canada (from below the northern tundra and Northwest Territories), to northern Mexico including the Colorado River and Rio Grande, and some coastal streams along the Gulf of Mexico (Hakala, 1952; Leopold, 1959; Baker and Hill, 2003). Before the Europeans arrived in North America and the commercial beaver harvest began in the 1600s, an estimated 60–400 million beavers were thought to have potentially existed (Seton, 1929). By the early 1900s the results of this astronomical hunting and habitat loss became evident, with the beaver population significantly declined or extirpated across Canada and the USA, with <100,000 thought to exist (Boyle and Owens, 2007). This realization, along with the passing of the Federal Aid in Wildlife Restoration Act in 1937, saw conservation efforts begin to restore beaver populations, including trapping regulations, wetland habitat protection, and proactive translocations (White et al., 2015). Early reintroductions and translocations started in the 1900s, though they took little account of the genetics, health, and any local adaptations (Müller-Schwarze, 2011).

Today the North American beaver has recovered to occupy much of this former native range, though in parts some populations are still small and restricted, largely due to habitat loss such as wetland drainage for agriculture (Hall, 1981; Larson and Gunson, 1983; Figure 2.10a). An estimated 195,000–260,000 km^2 of wetland has been converted to agricultural land-use in the USA since 1834 (Naiman et al., 1988). In the very arid southwestern USA, beavers tend only to be present on the larger rivers and streams; they are also generally absent from much of South Carolina and the Florida peninsula (Novak, 1987; Baker and Hill, 2003; Boyle and Owens, 2007; Figure 2.10). After the significant losses of the Beaver Wars era, curious records of scattered populations occurred in Mexico, including along the Cañón de Guadalupe at 5,000 ft (1,500 m) above sea level (Baird, 1859), Lower Río Colorado in Valle de Mexicali (Mellink and Luevano, 1998), Río Sonora, Río Bavispe, and Río San Pedro (Mearns, 1898; Burt, 1938; Gallo-Reynoso et al., 2002). Similar to the Eurasian beaver, North American beavers are absent from the tundra regions and are not present in these parts of Alaska and Canada.

In Canada, significant beaver declines since the late nineteenth century were recognized and in 1938 all beaver trapping was banned in Ontario, only resuming again in 1947 through registered trap lines with annual quotas once numbers had recovered. Following this, annual trapping numbers have increased from 120,000 in the 1950s, to 150,000 in the 1960s, and to 209,000 in 1979–80 (Robinson and Bolen, 1984). The latest figures of beaver pelts produced place Ontario highest, with 49,962 pelts recorded in 2009, whilst total annual figures for the whole of Canada were 147,685 in 2007, 152,782 in 2008, and 139,220 in 2009 (Statistics Canada, 2009). In the USA, 154,762 beavers were harvested for fur

in 2017, falling from 357,532 recorded in 2006 (AFWA, 2019), with many more controlled lethally (White et al., 2015). Populations across North America continue to increase and expand, particularly boosted after a wave of reintroductions in the 1950s, and there are now estimated between 6 and 12 million individuals (Naiman et al., 1988).

The extent and therefore recovery of beavers varied across the states. For example, Nebraska retained an acceptable proportion of its beavers, though the beaver became protected there in 1907. These populations recovered successfully through natural breeding and recolonization, so that the annual harvest in 2002–3 was 15,000, with beavers remaining common and widespread throughout this state (Boyle and Owens, 2007). Contrastingly, Wyoming almost lost all its beavers in 40 years of intensive fur trapping between 1820 and 1860 in a simultaneously depressing and impressive effort (Olson and Hubert, 1994). The hunting loss, in combination with habitat loss through intensive livestock grazing, lead to a state law protecting beavers in 1899 (Collins, 1993). Natural recovery was assisted with translocations from other states and was so successful that by 1958 beavers were reclassified as furbearers again and regulated harvesting was permitted (Olson and Hubert, 1994; Boyle and Owens, 2007). Kansas had an estimated historic

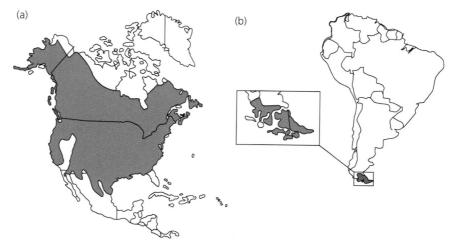

Figure 2.10 Distribution of North American beavers across North America (a) and South America (b). Map credits to Clker-Free-Vector-Images/Pixabay. Redrawn by Norith Eckbo after Anderson et al. (2009); Busher, P.E. (2016); Pietrek et al. (2017); Tape et al. (2018).

population of 50,000, which had almost disappeared by the 1800s due to commercial harvesting (Henderson, 1960). An expedition in northern Kansas in 1899–1900 apparently only found one beaver 'colony' remaining (Cockrum, 1952). In 1927, the beaver became protected by law in Kansas, with destruction of beaver lodges and dams prohibited and fur dealers requiring licences (Henderson, 1960). The population recovered to an estimated 1,500 individuals by 1937 (Wall, 1939) and, to supplement this, beavers were translocated by the state wildlife agency to several sites between 1930 and 1950, resulting in an estimated state population close to 23,000 beavers by 1959 (Henderson, 1960).

Translocation occurred similarly throughout North America, with beavers being moved from Canada and from state to state in the USA. By around the early 1800s the North American beaver had pretty much disappeared from Canada, from New Hampshire around 1865, Pennsylvania by 1890, and Kansas by 1907, amounting to <100,000 remaining in scattered and isolated populations across the whole of North America (Burroughs, 1961; Moloney, 1967; Hammond, 1993). By the mid-1800s some of the fur-trading companies had already extrapolated the impact of the beaver trade on remaining populations and began the implementation of limited trapping and even translocation of beavers to new areas to allow some form of population recovery (Long, 2000). In some states, small numbers of beavers remained, such as Kansas and Wisconsin, and beaver trapping was banned so that the populations recovered through natural recovery. Shifts in trapping efforts to other animals and changes in fashion trends enabled the natural recovery of remaining and scattered populations. Beaver populations are now considered as recovered across much of Canada, barring the Canadian and Arctic tundra as yet (Ingle-Sidorowicz, 1982).

Some of the earliest attempts to restore North American beavers saw small numbers of animals trapped and released to New York State firstly between 1901 and 1906, to California in 1924, and to Missouri in 1928 (Tappe, 1942; Hill, 1987). Yellowstone National Park was created in 1872 and this included protection for the few remaining beavers present, some of which were later translocated to several sites in the Adirondack Mountains, New

York state (Radford, 1907). These animals joined those starting one of the earliest North American beaver translocations. The original 20 founders were all wild trapped in Ontario and released between 1901 and 1906; proving successful, Yellowstone animals were later added. A series of reintroductions followed in areas where beavers had been extirpated including Massachusetts between 1928 and 1932 (Shaw, 1984; DeGraaf and Yamasaki, 2001), South Carolina between 1939 and 1941 (South Carolina Department of Natural Resources, 2010), and interstate translocations in Arizona between 1940 and 1950 (Carrillo et al., 2009). These early projects did not tend to maintain any genetic distinctions, sourcing animals widely and moving them between states, with beaver restoration to California involving locally native animals but also those brought in from Idaho and Oregon for example (Lynn, 1949). In California, 1,208 beavers were released over 274 sites during the 1930s and 1940s (Lynn, 1949), with 466 beavers released in North Dakota (Hibbard, 1958), and 76 beavers released over 40 sites in Washington state according to unpublished Okanogan Wenatchee National Forest records (Pollock et al., 2015). Table 2.4 shows the known translocations and introductions of beavers across North America between 1901 and 1948.

By the mid-1950s beavers started to make a more significant recovery and their distribution gained momentum. Many of these early reintroductions proved successful to the point that hunting bans tended to be lifted. By the 1970s it was estimated that around 15 million beavers were thought to exist and had a presence on all the major watersheds (Hill, 1987). Across much of its native range the North American beaver has recovered remarkably well, with populations naturally expanding, with some being controlled on human–wildlife conflict grounds (see Chapter 11). Other populations have remained small and are actively managed for conservation to increase their numbers, such as those in Arizona and northern Mexico (Pelz-Serrano et al., 2005; Carrillo et al., 2009).

2.8.3 North American beaver introductions

The North American beaver has been introduced historically to areas outside of its native range to

Table 2.4 Beaver translocations and reintroductions in North America as reported in the literature. C. c. = *Castor canadensis*.

Year	Source location	Source subspecies given	Release location	Resident subspecies given	References
1901–1906	Ontario, Canada	*C. c. acadicus*	Adirondacks, New York	*C. c. acadicus*	Radford (1907)
1907	Yellowstone, Wyoming	*C. c. missouriensis*	Adirondacks, New York	*C. c. acadicus*	Radford (1907)
1928–1932	Unknown	Unknown	Massachusetts	*C. c. acadicus*	Shaw (1984); DeGraaf and Yamasaki (2001)
1939–1941	Georgia	*C. c. carolinensis*	South Carolina	*C. c. carolinensis*	South Carolina Department of Natural Resources (2010)
1939	Wisconsin	*C. c. canadensis*	North Carolina	*C. c. carolinensis*	Woodward et al. (1985)
1957	Alabama	*C. c. carolinensis*	North Carolina	*C. c. carolinensis*	Woodward et al. (1985)
1940–1951	Alabama	*C. c. carolinensis*	Throughout Alabama	*C. c. carolinensis*	Beshears (1967)
1940–1950	Springville and Pima, Arizona	*C. c. frondator*	Chiricahua Mountains, Arizona	*C. c. frondator*	Carrillo et al. (2009)
1942–1944	Unknown	Unknown	Northern Black Hills, South Dakota	*C.c. missouriensis*	Harris and Aldous (1946)
1915–1937	Utah	*C. c. duchesnei, pallidus, rostralis*	Throughout Utah	*C. c. duchesnei, pallidus, rostralis*	Utah Division of Wildlife Resources (2010)
1940–1944	Snelling, California	*C. c. subauratus*	Mather Station, California	*C. c. subauratus*	Hensley (1946)
1985–1999	New Mexico	*C. c. frondator, mexicanus*	Zuni Indian Reservation, New Mexico	*C. c. frondator, mexicanus*	Albert and Trimble (2000)
1930–1950	Kansas	*C. c. missouriensis*	Throughout Kansas	*C. c. missouriensis*	Henderson (1960)
1937–1943	Unknown	Unknown	Western Iowa	*C. c. missouriensis*	Bowles (1970)
1933–1940	Unknown	Unknown	West Virginia	*C. c. canadensis*	Bailey (1954)
1917	Wisconsin	*C. c. canadensis*	Pennsylvania	*C. c. acadicus*	Jackson (1961)
1920–1936	Wisconsin	*C. c. canadensis*	Throughout Wisconsin	*C. c. canadensis*	Jackson (1961)
1935	Wisconsin	*C. c. canadensis*	Indiana	*C. c. carolinensis*	Indiana Department of Natural Resources (2011)
1939–1948	Unknown	Unknown	Ohio	*C. c. carolinensis*	Chapman (1949)
1920–1930	Wisconsin	*C. c. canadensis*	Kentucky	*C. c. carolinensis*	Barbour and Davis (1974)
1942–1943	Missouri	*C. c. missouriensis, carolinensis*	Tennessee	*C. c. carolinensis*	Goodpaster and Hoffmeister (1952)
1934	Unknown	*C. c. canadensis*	Arkansas	*C. c. carolinensis*	Sealander (1956)
1943–1945	Unknown	*C. c. carolinensis*	Arkansas	*C. c. carolinensis*	Sealander (1956)
1938–1948	Southeastern Louisiana	*C. c. texensis, carolinensis*	Western Louisiana	*C. c. carolinensis*	Elsey and Kinler (1996)
1930s	Mississippi	*C. c. carolinensis*	Throughout Mississippi	*C. c. carolinensis*	Mississippi Fish and Wildlife Foundation (2010)

continued

Table 2.4 Continued

Year	Source location	Source subspecies given	Release location	Resident subspecies given	References
1914	Unknown	Unknown	Connecticut	*C. c. acadicus*	Wilson (2001)
1926–1930	Unknown	Unknown	New Hampshire	*C. c. acadicus*	New Hampshire Fish And Game (2011)
1920–1930	New York and Maine	*C. c. acadicus*	Vermont	*C. c. acadicus*	Vermont Fish and Wildlife Department (2004)
1929–1936	Wisconsin	*C. c. canadensis*	Illinois	*C. c. carolinensis*	Pietsch (1956)
1999	Temecula, California	*C. c. repentinus*	Texas	*C. c. mexicanus, texensis*	Longcore et al. (2007)
1932–1948	Pennsylvania	*C. c. canadensis*	Virginia	*C. c. canadensis, carolinensis*	Longcore et al. (2007)
1932–1948	Michigan	*C. c. michiganensis*	Virginia	*C. c. canadensis, carolinensis*	Longcore et al. (2007)
1932–1948	New York	*C. c. acadicus*	Virginia	*C. c. canadensis, carolinensis*	Longcore et al. (2007)
1932–1948	New Hampshire	*C. c. acadicus*	Virginia	*C. c. canadensis, carolinensis*	Engle (1954)
1932–1948	Maine	*C. c. acadicus*	Virginia	*C. c. canadensis, carolinensis*	Engle (1954)
1932–1948	West Virginia	*C. c. canadensis*	Virginia	*C. c. canadensis, carolinensis*	Engle (1954)

Source: Karla Pelz Serrano, PhD, 2011, University of Arizona. Molecular phylogeography of the American beaver (Castor canadensis): implications for management and conservation. https://repository.arizona.edu/bitstream/ handle/10150/223320/azu_etd_12001_sip1_m.pdf?sequence=1#page=22.

bolster beaver populations and establish fur resources, though escapes from captive collections explain more recent records. A risk assessment on the invasion of North American beavers written for the European Union identifies three main pathways for their introduction: intentional releases, captive escapes from zoos and fur farms, and natural dispersal from existing populations (Hollander et al., 2017).

Some of the earliest documented introductions were into northeast Poland in 1926 and the River Goryn (formerly Poland and now Ukraine) in 1933 (Lahti and Helminen, 1974). North American beavers have since become extinct in both (Ukraine by 1957 and Poland by 1979 [Parker et al., 2012]). Some releases are varyingly believed to have been attempted in northern Norway, though none were successful (Parker et al., 2012). Originally seven individuals were sent from New York State and released into Finland between 1935 and 1937, alongside 17 Eurasian beavers from Norway, to supplement beaver restoration efforts. This occurred prior to the scientific confirmation that they were indeed two separate species (Parker et al., 2012). Beavers spread quickly throughout most of the southern part of Finland and crossed into Russia in the 1950s (Ivanov, 1975), augmented by further releases in Finland and neighbouring Karelia. They were thought to number ~8,000 individuals by 2003 (Danilov et al., 2011), presenting huge challenges to wildlife managers (Ermala et al., 1999). Today an estimated 13,000 North American beavers are found in Karelia, Russia and >24,000 in Finland (Brommer et al., 2017).

North American populations exist in northwest Russia largely on the Amur and in Kamchatka (Saveljev and Safonov, 1999; Halley and Rosell, 2003). As mentioned, these have appeared through colonization from Finland and known releases (Segal and Orlova, 1961; Nolet, 1997). Successful natural dispersal was facilitated by numerous waterbodies with a high degree of connectivity, assisted by deliberate and repeated translocations to beaver-free areas between 1960 and the 1980s. For example, over 7 years, 115 beavers were moved to unoccupied areas in central Karelia (Danilov et al., 2011).

In Argentina, as a 'one-off' release, North American beavers were introduced to Isla Grande de Tierra del Fuego in 1946, the largest island in this archipelago lying off the Chile/Argentine coast. This introduction is commonly documented as the release of 25 pairs of beavers undertaken to 'enrich native fauna' and establish a fur industry (Daciuk, 1978), though interestingly the hunting of beavers was not officially permitted in Argentina until 1981 (Lizarralde, 1993). Beavers never existed in South America; therefore this introduction differs significantly from other introductions in Europe, for example where the Eurasian beaver already or previously existed. As a consequence, beaver introduced here has resulted in significant changes to the native flora and fauna communities that have never evolved alongside beavers and their activities (Lizarralde et al., 2004; Anderson et al., 2006; Moorman et al., 2006; see Chapter 11). In addition to beavers, the fur industry introduced muskrat (*Ondatra zibethicus*) and American mink (*Neovison vison*). Beaver-modified habitats, particularly the transformation of fast-flowing streams into lentic ponds, encouraged muskrat occupation; additionally, muskrats play a significant role in mink diets (Crego et al., 2016). Therefore, beavers can facilitate the colonization and survival of both these invasive non-natives.

Since the initial release, beaver populations have expanded and dispersed to Chile across the Beagle Channel, first colonizing several islands from the 1960s (Navarino, Hoste, Picton, Nueva, Lennox, and Dawson Island), with the first reports of their presence on mainland Chile around the early 1990s (Lizarralde, 1993; Skewas et al., 2006; Wallem et al., 2007; Figure 2.10b).

A further small number of European countries have sporadically recorded minor numbers of North American beavers with escapes from captive collections being the most likely cause in recent times. There are thought to have been several releases over the years in Germany, including North Rhine-Westphalia in 1981 and 1989 and Rhineland-Palatinate in 1994 (Dewas et al., 2012). There are suspicions that some of the beavers brought from Polish beaver farms in Popel'no and released into Bavaria in 1966 and 1972 may have included North American beavers, as both species at one time were used at these breeding centres before the two species were recognized (Lahti and Helminen, 1974).

Bavarian beavers have been repeatedly used as source population for numerous reintroductions across Europe and some analysis suggests they could have potentially and unknowningly encouraged the spread of North American beavers in central and western Europe (Petrosyan et al., 2019).

Three North American beavers in 1977 were documented to have escaped from a zoo in France. The decision was made to remove this non-native with 24 animals removed by 1985 (Rouland, 1985). Similar releases and escapes from captivity in Austria (1978 and 1986) and Hungary (1990s) are also now believed to be extinct and not represented in current populations (Sieber, 2000; Halley and Rosell, 2002).

Luxembourg did not officially reintroduce beavers but waited for natural colonization along shared watercourses with surrounding countries. The first beaver activity was noted in early 2000 (Schley et al., 2001), expanding to other sites over the following years. However, the finding of a dead North American individual along the Luxembourg/German border initiated a wider survey of beaver species identification across western Europe and confirmed their wider presence in both these countries (Dewas et al., 2012).

References

Adiya, Y., Samiya, R., and Shar, S. 2015. History of translocation and building of breeding unit for beavers in Mongolia. Beavers—from genetic variation to landscape-level effects in ecosystems. *Proceedings of the 7th International Beaver Symposium Book of Abstracts.* Voronezh: Scientific Bulletin of the Uzhgorod University.

Aftel, M. 2004. *Essence and Alchemy: A Natural History of Perfume.* Salt Lake City: Gibbs Smith.

AFWA. 2019. US Furbearer Harvest Statistics Database 1970–2017. https://www.fishwildlife.org/afwa-inspires/furbearer-management.

Ainsworth, W. 1838. *Researches in Assyria, Babylonia, and Chaldaea: Forming Part of the Labours of the Euphrates Expedition.* London: John W. Parker.

Albert, S. and Trimble, T. 2000. Beavers are partners in riparian restoration on the Zuni Indian Reservation. *Ecological Restoration,* 18, 87–92.

Aleksiuk, M. 1970. The function of the tail as a fat storage depot in the beaver (*Castor canadensis*). *Journal of Mammalogy,* 51, 145–148.

Anderson, C. B., Griffith, C. R., Rosemond, A. D., Rozzi, R., and Dollenz, O. 2006. The effects of invasive North American beavers on riparian plant communities in Cape Horn, Chile: do exotic beavers engineer differently in sub-Antarctic ecosystems? *Biological Conservation,* 128, 467–474.

Anderson, C. B., Pastur, G. M., Lencinas, M. V., Wallem, P. K., Moorman, M. C., and Rosemond, A. D. 2009. Do introduced North American beavers *Castor canadensis* engineer differently in southern South America? An overview with implications for restoration. *Mammal Review,* 39, 33-52

Antanaitis-Jacobs, I., Richards, M., Daugnora, L., Jankauskas, R., and Orginc, N. 2009. Diet in early Lithuanian prehistory and the new stable isotope evidence. *Archaeologia Baltica,* 12, 12–30.

Antunes, M. T. 1989. *Castor fiber* na Gruta do Caldeirão. Existência, distribuição e extinção do *castor* em Portugal. *Ciências da Terra,* 10, 23–40.

Appavoo, D. M., Kubow, S., and Kuhnlein, H. V. 1991. Lipid composition of indigenous foods eaten by the Sahtu (Hareskin) Dene-Metis of the Northwest Territories. *Journal of Food Composition and Analysis,* 4, 107–119.

Asbirk, S. 1998. *Forvaltningsplan for Bæver.* Copenhagen: Skov- og Naturstyrelsen.

Aybes, C. and Yalden, D. W. 1995. Place-name evidence for the former distribution and status of wolves and beavers in Britain. *Mammal Review,* 25, 201–226.

Baales, M. 2002. Der spätpaläolithische Fundplatz Kettig. Untersuchungen zur siedlungsarchäologie der Federmesser-Gruppen am Mittelrhein. *Bulletin de la Société Préhistorique Française,* 101, 365–366.

Backhouse, F. 2018. *Once They Were Hats: In Search of the Mighty Beaver.* Toronto: ECW Press.

BAFU 2016. *Konzept Biber Schweiz. Vollzugshilfe des BAFU zum Biber management in der Schweiz.* St. Gallen: Federal Office for the Environment.

Bailey, R. W. 1954. Status of beaver in West Virginia. *Journal of Wildlife Management,* 18, 184–190.

Baird, S. F. 1859. Mammals of the boundary. *Report on the US and Mexican Boundary Survey, under the order of Lieut. Col. W.H. Emory,* 2, 1–62.

Bajomi, B. 2011. *Reintroduction of the Eurasian beaver (Castor fiber) in Hungary.* Budapest: Danube Parks Network of Protected Areas, Directorate of Duna-Dráva National Park.

Bajomi, B., Bera, M., Czabán, D., and Gruber, T. 2016. Eurasian beaver re-introduction in Hungary. In: Soorae, P. S. (Ed.) *Global Re-introduction Perspectives: 2016. Case-Studies from Around the Globe.* Gland: IUCN/SSC Re-introduction Specialist Group. Abu-Dhabi: UAE: Switzerland and Environment Agency.

Baker, B. and Hill, E. 2003. Beaver (*Castor canadensis*). In: Feldhamer, G. A., Thompson, B. C. and Chapman, J. A.

(Eds.) *Wild Mammals of North America: Biology, Management and Conservation*. Baltimore: Johns Hopkins University Press.

Banzhaf, D. 2004. Der Baumeister. *Badische Zeitung*, 28 August 2004.

Barbour, R. W. and Davis, W. H. 1974. *Mammals of Kentucky*. Lexington, University of Kentucky Press.

Batbold, J., Batsaikhan, N., Shar, S., Amori, G., Hutterer, R., Kryštufek, B., Yigit, N., Mitsain, G., and Muñoz, L. 2008. *Castor fiber*. IUCN Red List of Threatened Species. Gland: IUCN.

Beck, H. P. 1966. *Gluskap the Liar & Other Indian Tales*. Freeport: B. Wheelwright Company.

Beck, J. C. 1972. The giant beaver: a prehistoric memory? *Ethnohistory*, 19, 109–122.

Belova, O., Sjöberg, G., Ecke, F., Lode, E., and Thorell, D. 2016. *Förvaltningavbäveri Östersjöregionen—enöversiktavaktuellkunskap, metoderochutvecklingsbehov*. WAMBAF final document. https://www.skogsstyrelsen.se/en/wambaf/.

Berenstein, N. 2018. A history of flavouring food with beaver butt juice. https://www.vice.com/en_us/article/a3m885/a-history-of-flavoring-food-with-beaver-butt-juice?utm_source=vicefbus.

Beshears, W. W. 1967. Status of the beaver in Alabama. *Proceedings of the 1st Alabama Beaver Symposium*. Montgomery: Department of Conservation and Alabama Forest Product Association, pp. 2–6.

Bjurberg, J. 1687. *Castor breviter delineatus*. Thesis, University of Uppsala

Blair, J. 2005. *The Church in Anglo-Saxon Society*. Oxford: Oxford University Press.

Boece, H. 1527. *The History and Chronicles of Scotland*. Edinburgh: W. and C. Tait.

Boessneck, J. 1974. Ergänzungen zur einstige Verbreitung des Bibers, *Castor fiber* (Linne 1758). *Säugetierkundliche Mitteilungen*, 22, 83–88.

Boev, Z. and Spassov, N. 2019. Past distribution of *Castor fiber* in Bulgaria: fossil, subfossil and historical records (Rodentia: Castoridae). *Lynx, Series Nova*, 50, 37–49.

Borisov, B. P. 2011. Beaver. Status of hunting resources in Russian Federation in 2008–2010: information analytical materials. In: *Hunting Animals in Russia: Biology, Protection, Resources and Rational Use*. Moscow: Fizicheskaya Kul'tura.

Bowles, J. B. 1970. Historical record of some Iowan mammals. *Transactions of the Kansas Academy of Science*, 73, 419–430.

Boyle, S. and Owens, S. 2007. *North American beaver (Castor canadensis): a technical conservation assessment*. USDA Forest Service, Rocky Mountain Region, 50. http://www.fs.fed.us/r2/projects/scp/assessments/north-americanbeaver.pdf.

Brommer, J. E., Alakoski, R., Selonen, V., and Kauhala, K. 2017. Population dynamics of two beaver species in Finland inferred from citizen-science census data. *Ecosphere*, 8, e01947.

Bryce, G. 1904. *The Remarkable History of the Hudson's Bay Company*. New York: Burt Franklin Press.

Burdock, G. A. 1995. *Fenaroli's Handbook of Flavour Ingredients*. Boca Raton: CRC Press.

Burdock, G. A. 2007. Safety assessment of castoreum extract as a food ingredient. *International Journal of Toxicology*, 26, 51–55.

Burroughs, R. 1961. *The Natural History of the Lewis and Clark Expedition*. East Lansing: Michigan State University Press.

Burt, W. H. 1938. Faunal relationships and geographic distribution of mammals in Sonora, Mexico. *Miscellaneous Publication of the Museum of Zoology University of Michigan*, 39, 1–77.

Busher, P.E. 2016. Family Castoridae (Beavers). Pp. 150-168 in: Wilson D.E., Lacher, T.E., Jr. & Mittermeier, R.A. eds. Handbook of Mammals of the World, Vol. 6 Lagomorphs and Rodents I. Lynx Editions, Barcelona.

Calabritto, M. 2002. *Medical and Moral Dimensions of Feminine Madness: Representing Madwomen in the Renaissance. Forum Italicum*, pp. 26–52. London: SAGE.

Campbell-Palmer, R., Puttock, A., Graham, H., Wilson, K., Schwab, G., Gaywood, M., and Brazier, R. 2018. *Survey of the Tayside area beaver population 2017–2018*. Inverness: Scottish Natural Heritage.

Čanády, A., Krišovský, P., Bajomi, B., Huber, A., Czabán, D., and Olekšák, M. 2016. Is new spread of the European beaver in Pannonian basin an evidence of the species recovery? *European Journal of Ecology*, 2, 44–63.

Carlos, A. M. and Lewis, F. D. 2010. *Commerce by a Frozen Sea: Native Americans and the European Fur Trade*. Philadelphia: University of Pennsylvania Press.

Carrillo, C. D., Bergman, D. L., Taylor, J., Nolte, D., Viehoever, P., and Disney, M. 2009. An overview of historical beaver management in Arizona. *Proceedings of the Wildlife Damage Management Conference*, 13, 234–242.

Cavrić, K. 2016. The population of European beaver (*Castor fiber* L.) in Croatia. PhD thesis, Strossmayer University of Osijek.

Caxton, W. 1484. *The History and Fables of Aesop*. Westminster: William Caxton.

Cena, J. C., Alfaro, I., Cena, A., Itoitz, U., Berasategui, G., and Bidegain, I. 2004. *Castor* Europeo en Navarra y La Rioja. *Galemys*, 16, 91–98.

Chaix, L. 2003. A short note on the Mesolithic fauna from Zamostje 2 (in Russia). In: Kindgren, L. H., Knutsson, D., Loeffler, D., and Åkerland, A. (eds.) *Mesolithic on the Move. Proceedings of the 6th International Conference on the Mesolithic in Europe*, 2000, Stockholm. Oxford: Oxbow Books.

Chapman, F. B. 1949. The beaver in Ohio. *Journal of Mammalogy*, 30, 174–179.

Charles-Edwards, T. 1989. *The Welsh Laws*. Cardiff: University of Wales Press.

Chu, H. and Jiang, Z. 2009. Distribution and conservation of the Sino-Mongolian beaver *Castor fiber* birulai in China. *Oryx*, 43, 197–202.

Cockrum, E. L. 1952. Mammals of Kansas. *University of Kansas Museum of Natural History Publications*, 7, 1–303.

Coles, B. 2006. *Beavers in Britain's Past*. Oxford: Oxbow Books.

Coles, B. J. 2019. *Review—Avanke, Bever, Castor: The Story of Beavers in Wales*. Thorventon: Wetland Archaeology Research Project.

Collet, R. 1897. Bœveren I Norge, dens Utbredelsen og Levemaade 1896. *Bergens Museums Aarbog*, 1, 1–139.

Collins, T. 1993. *The Role of Beaver in Riparian Habitat Management. Habitat Extension Bulletin*. Cheyenne: Wyoming Game and Fish Department.

Cox, B. A. 1988. *Native People, Native Lands: Canadian Indians, Inuit and Metis*. Ottawa: McGill-Queen's University Press.

Crean, J. F. 1962. Hats and the fur trade. *Canadian Journal of Economics and Political Science*, 28, 373–386.

Crego, R. D., Jiménez, J. E., and Rozzi, R. 2016. A synergistic trio of invasive mammals? Facilitative interactions among beavers, muskrats, and mink at the southern end of the Americas. *Biological Invasions*, 18, 1923–1938.

Cronon, W. 1983. *Changes in the Land: Indians, Colonists, and the Ecology of New England*. New York: Hill and Wang.

Crutchfield, J. A. 2018. *It Happened in Oregon: Stories of Events and People that Shaped the Beaver State History*. Guilford, CN: Globe Pequot Press.

Curley, M. J. 1979. *Physiologus: A Medieval Book of Nature Lore*. Chicago: University of Chicago Press.

Curry-Lindahl, K. 1967. The beaver, *Castor fiber* Linnaeus, 1758 in Sweden—extermination and reappearance. *Acta Theriologica*, 12, 1–15.

Daciuk, J. 1978. Notas faunisticas y bioecológicas de Peninsula Valdés y Patagonia, IV. Estado actual de las especies de mamíferos introducidos en la Region Araucana (Rep. Argentina) y grado de coacción ejercido en algunos ecosistemas surcordilleranos. *Anales de Parques Nacionales*, 14, 104–130.

Danford, C. G. and Alston, E. R. 1880. On the mammals of Asia Minor, part II. *Proceedings of the Scientific Meeting of the Zoological Society London*, 1880, 50–64.

Danilov, P. I. 1976. *The History of Beaver Spread in Karelia. Ecology of Birds and Mammals in the North-West of USSR*. Petrozavodsk: ??.

Danilov, P., Kanshiev, V. Y., and Fedorov, F. 2011. History of beavers in eastern Fennoscandia from the Neolithic to the 21st century. In: Sjøberg, G. and Ball, J. P. (eds.) *Restoring the European Beaver: 50 Years of Experience*. Sofia-Moscow: Pensoft.

Daugnora, L. and Girininkas, A. 1995. Neolithic and Bronze Age mixed farming and stock breeding in the traditional Baltic culture-area. *Archeologia Baltica*, 1, 43–51.

Deberšek, B. 2012. Beaver territories (*Castor fiber*) in the region of Koroška, Slovenia. Graduate thesis, Environmental Protection College Velenje.

Degraaf, R. M. and Yamasaki, M. 2001. *New England Wildlife: Habitat, Natural History, and Distribution*. Lebanon, NH: University Press of New England.

Deinet, S., Ieronymidou, C., Mcrae, L., Burfield, I. J., Foppen, R. P., Collen, B., and Böhm, M. 2013. Wildlife comeback in Europe: the recovery of selected mammal and bird species. *Final Report to Rewilding Europe by ZSL*. London: ZSL: BirdLife International and the European Bird Census Council.

Denys, N. 1672. *The Description and Natural History of the Coasts of North America*. Translated by W.F. Ganong, 1908. Toronto: The Champlain Society.

Dewas, M., Herr, J., Schley, L., Angst, C., Manet, B., Landry, P., and Catusse, M. 2012. Recovery and status of native and introduced beavers *Castor fiber* and *Castor canadensis* in France and neighbouring countries. *Mammal Review*, 42, 144–165.

Dibble, H. L., Berna, F., Goldberg, P., McPherron, S. P., Mentzer, S., Niven, L., Richter, D., Sandgathe, D., Théry-Parisot, I., and Turq, A. 2009. A preliminary report on Pech de l'Azé IV, layer 8 (Middle Paleolithic, France). *PaleoAnthropology*, 2009, 182–219.

Dijkstra, V. 2019. Bever populatie blijftgroeien. https://www.zoogdiervereniging.nl/nieuws/2019/beverpopulatie-blijft-groeien.

Dijkstra, V., Polman, E., and van Oene, M. 2018. Verspreidingsonderzoek Bever en Otter. *Telganger*, October, 2–5.

Dolin, E. J. 2010. *Fur, Fortune and Empire*. New York: Norton and Company.

Dorsey, G. A. 1901. Games of the Makah Indians of Neah Bay. *American Antiquarian and Oriental Journal (1880–1914)*, 23, 69.

Dubrulle, P.-M. and Catusse, M. 2012. Où en est la colonisation du *Castor* en France? *Faune Sauvage*, 297, 24–35.

Dugmore, A. R. 1914. *The Romance of the Beaver*. Philadelphia: J.B. Lippincott.

Echegaray, J., Soto, J., Pérez de Obanos, C., and Artika, E. 2018. Any hope for the long-term conservation of beavers in Spain? New law insights. *Poster, 8th International Beaver Symposium*, Denmark.

Elmeros, M. 2017. *Bestandsudvikling og udbredelse af bæver i Jylland i foråret 2017. Note from DCE*. Aarhus: National Centre for Environment and Energy, Aarhus University [in Danish].

Elsey, R. M. and Kinler, N. 1996. Range extension of the American beaver, *Castor canadensis*, in Louisiana. *The Southwestern Naturalist*, 41, 91–93.

Engle, J. W. J. 1954. Present status of the beaver in Virginia. *Virginia Wildlife*, 15, ??.

Eriksson, G., Lõugas, L., and Zagorska, I. 2003. Stone Age hunter–fisher–gatherers at Zvejnieki, northern Latvia: radiocarbon, stable isotope and archaeozoology data. *Before Farming*, 1, 1–25.

Ermala, A., Lahti, S., and Vikberg, P. 1999. Bäverstammen ökar fortfarande—fangsten redan närmare 2500 bävrar. *Jägaren*, 4, 28–31.

Ernst, M., Putnová, L., Štohl, R., and Matoušková, J. 2017. Comparison of Czech and Latvian beaver population by microsatellite analysis and genetic differences between *Castor fiber* and *Castor canadensis*. *Transylvanian Review*, 25, 5339.

Estonian Hunters Association. 2019. Aasta loom 2019—kobras. http://www.ejs.ee/aasta-loom-2019-kobras/.

Evans, C. and Serjeantson, D. 1988. The backwater economy of a fen-edge community in the Iron Age: the Upper Delphs, Haddenham. *Antiquity*, 62, 360–370.

Ewersen, J. 2006/2007. Ein vergessenes steinzeitliches Gerät? Beobachtungen an Unterkiefern von Bibern. *Offa*, 63/64, 197–208.

Fairbairns, J. 1892. *Fairbairn's Book of Crests of the Families of Great Britain and Ireland*. Edinburgh: T.C. and E.C. Jack.

Fasel, M. 2018. Der Biber in Liechtenstein—Bestand und Verbreitung im Winter 2017/2018. Amt für Umwelt, Fürstentum Liechtenstein. https://www.llv.li/files/au/bestanderhebung-biber-liechtenstein-201718.pdf.

Fehner, M. 1963. Predmety yazytsheskogo kul'ta. Yaroslavskoye Povolzhye X–XI vv. Po materyalam Timerevskogo, Mihhailovskogo i Petrovskogo mogil'nikov (Moskva). ??.

Fiore, I., Tagiacozzo, A., and Cassoli, P. F. 2001. Ibex exploration at Dalmeri rockshelter (TN) and 'specialised hunting' in the sites of the Eastern Alps during the Tardiglacial and the Early Holocene. *Prehistoria Alpina*, 34, 173–183.

Fiore, I., Gala, M., and Tagliacozzo, A. 2004. Ecology and subsistence strategies in the Eastern Italian Alps during the Middle Palaeolithic. *International Journal of Osteoarchaeology*, 14, 273–286.

Fischer, J. B., McArthur, L. H., and Petersen, J. B. 1997. Continuity and change in the food habits of the seventeenth-century English colonists in Plymouth and Massachusetts Bay. *Ecology of Food and Nutrition*, 36, 65–93.

Fisher, R. H. 1943. *The Russian Fur Trade 1550–1700*. Berkeley: University of California Press.

Forstén, A. and Lahti, S. 1976. Postglacial occurrence of the beaver (*Castor fiber* L.) in Finland. *Boreas*, 5, 155–161.

Freye, H. A. 1978. *Castor fiber* Linnaeus, 1758—Europäischer Biber. In: Niethammer, J. and Krapp, F. (eds.) *Handbuch der Säugetiere Europas*. Wiesbaden: Akademische Verlagsgesellschaft.

Gallo-Reynoso, J.-P., Suárez-Gracida, G., Cabrera-Santiago, H., Coria-Galindo, E., Egido-Villarreal, J., and Ortiz, L. C. 2002. Status of beavers (*Castor canadensis frondator*) in Rio Bavispe, Sonora, Mexico. *The Southwestern Naturalist*, 47, 501–504.

Garrisson, F. H. 1966. *History of Medicince*. Philadelphia: W.B. Saunders Company.

Gaudzinski, S. 2004. A matter of high resolution? The Eemian Interglacial (OIS 5e) in north-central Europe and Middle Palaeolithic subsistence. *International Journal of Osteoarchaeology*, 14, 201–211.

Gaywood, M. J. 2015. *Beavers in Scotland: A Report to the Scottish Government*. Inverness: Scottish Natural Heritage.

Gerstell, C. 1985. *The Steel Trap in North America*. Harrisburg, PA: Stackpile Books.

Gessner, C. and Forer, C. 1563. *Thierbuch, das ist ein kurtze beschreybung aller vierfüssigen so wohl zahmer als wilder Thieren*. Zurich.

Gibbs, L. 2002. Aesop: 'The Beaver and his Testicles'. *Aesop's Fables*. Oxford: Oxford University Press.

Gibson, J. R. 1969. *Feeding the Russian Fur Trade. Provisionment of the Okhotsk Seaboard and the Kamchatka Peninsula 1639–1856*. Tennessee: University of Wisconsin Press, Kingsport Press.

Godman, J. D. 1826. *American Natural History*. Philadelphia: H.C. Carey and I. Lea.

Goodpaster, W. W. and Hoffmeister, D. F. 1952. Notes on the mammals of western Tennessee. *Journal of Mammalogy*, 33, 362–371.

Grescham, B. 1944. *The Beaver*. Winnipeg: The Hudson's Bay House.

Grubesić, M., Kusan, V., and Krapinec, K. 2001. Monitoring of beaver (*Castor fiber* L.) population distribution in Croatia. In: Czech, A. and Schwab, G. (eds.) The European Beaver in a New Millennium. *Proceedings of the 2nd European Beaver Symposium*, Bialowieza, Poland. Krakow: Carpathian Heritage Society.

Grubešić, M., Glavaš, M., Margaletić, J., Pernar, R., Ančić, M., and Krapinec, K. 2006. A decade of the beaver (*Castor fiber* L.) in Croatia. *Poster extract, Proceedings of the 4th European Beaver Symposium/3rd Euro-American Beaver Symposium*, 11–14 September, Freisig, Germany, pp. 11–14.

Grubešić, M., Margaletić, J., Bjedov, L., Tomljanović, K., and Vucelja, M. 2012. Beaver in Croatia—20 Years later. *Proceedings of the 6th International Beaver Symposium*, Ivanić-Grad, Croatia.

Haarberg, O. 2007. *Amit a hódról tudni érdemes. Az eurázsiai hód Magyarországon – visszatelepítés, védelem és állományszabályozás*. Budapest: WWF Magyarország.

Hakala, J. B. 1952. The life history and general ecology of the beaver (*Castor canadensis* Kuhl) in interior Alaska. MSc thesis, University of Alaska.

Hall, E. R. 1981. *The Mammals of North America*. New York, John Wiley.

Halley, D. and Rosell, F. 2002. The beaver's reconquest of Eurasia: status, population development and management of a conservation success. *Mammal Review*, 32, 153–178.

Halley, D. J. and Rosell, F. 2003. Population and distribution of European beavers (*Castor fiber*). *Lutra*, 46, 91–101.

Halley, D. and Schwab, G. 2020. Eurasian population and distribution: the past, present and future. *Presentation, Beaver CON 2020*, Baltimore, 3–5 March 2020.

Halley, D. J., Jones, A. L., Chesworth, S., Hall, C., Gow, D., Jones-Parry, R., and Walsh, J. 2009. The reintroduction of the Eurasian beaver *Castor fiber* to Wales: an ecological feasibility study. *NINA Report*, 457, 1–66.

Halley, D., Rosell, F., and Saveljev, A. 2012. Population and distribution of Eurasian beaver (*Castor fiber*). *Baltic Forestry*, 18, 168–175.

Halley, D., Saveljev, A., and Rosell, F. 2020. Population and distribution of Eurasian beavers (*Castor fiber*). *Mammal Review*, 51, 1–24.

Hammond, L. 1993. Marketing wildlife: the Hudson's Bay Company and the Pacific Northwest, 1821–49. *Forest and Conservation History*, 37, 14–25.

Harris, D. and Aldous, S. E. 1946. Beaver management in the northern Black Hills of South Dakota. *Journal of Wildlife Management*, 10, 348–353.

Harrison, D. L. 1972. *The Mammals of Arabia*. London, Ernest Benn.

Hartman, G. 1994. Long-term population development of a reintroduced beaver (*Castor fiber*) population in Sweden. *Conservation Biology*, 8, 713–717.

Hartman, G. 1995. Patterns of spread of a reintroduced beaver *Castor fiber* population in Sweden. *Wildlife Biology*, 1, 97–104.

Hatt, R. T. 1959. *The Mammals of Iraq*. Ann Arbor: Museum of Zoology, University of Michigan.

Healy, F. 1996. The Fenland Project, Number 11: the Wissey embayment: evidence for pre-Iron Age occupation accumulated prior to the Fenland Project. *East Anglian Archaeology*, 78, 1–194.

Heidecke, D. 1984. Untersuchungen zur Ökologie und Populationsentwicklung des Elbebibers (*Castor fiber albicus* Matschie, 1907). I: Biologische und populationsökologische Ergebnisse. Zoologische Jahrbücher. *Abteilung für Systematik, Ökologie und Geographie der Tiere*, 111, 1–41.

Heidecke, D. and Hörig, K. J. 1986. Bestands- und Schutzsituasjon des Elbebibers. *Halle and Magdeburg*, 23, 1–14.

Hellie, R. 1999. *The Economy and Material Culture of Russia, 1600–1725*. Chicago: University of Chicago Press.

Henderson, F. R. 1960. *Beaver in Kansas. State Biological Survey and Museum of Natural History Miscellaneous Publication Number 26*. Lawrence: University of Kansas.

Hensley, A. L. 1946. A progress report on beaver management. *California Fish and Game*, 32, 87–99.

Herr, J., Schley, C., Gonner, A., Arendt, G., Biver, A., and Bombardella, M. 2018. Aktions- und Management plan für den Umgang mit Bibern in Luxemburg. *Technischer Bericht der Naturverwaltung betreffend Wildtiermanagement und Jagd*, 6, 1–40.

Hibbard, E. A. 1958. Movements of beaver transplanted in North Dakota. *Journal of Wildlife Management*, 22, 209–211.

Hilfiker, E. L. 1991. *Beavers, Water, Wildlife and History*. Interlaken, NY: Windswept Press.

Hill, E. P. 1987. Beaver resortation. In: Kallman, H., Agee, C. P., Goforth, W. G. and Linduska, J. P. (eds.) *Restoring America's Wildlife, 1937–1987: The First 50 Years of the Federal Aid in Wildlife Restoration (Pittman-Robertson) Act*. Washington, DC: United States Department of the Interior; US Fish and Wildlife Service.

Holbrook, R. T. 1902. *Dante and the Animal Kingdom*. New York: Columbia University Press.

Hollander, H., van Duinen, G. A., Branquart, E., de Hoop, L., de Hullu, P. C., Matthews, J., van der Velde, G., and Leuven, R. S. E. W. 2017. Risk assessment of the alien North American beaver (*Castor canadensis*). https://repository.ubn.ru.nl/bitstream/handle/2066/169023/169023.pdf.

Hood, G. A. 2011. *The Beaver Manifesto*. Canada: Rocky Mountain Books.

Hood, G. A. and Bayley, S. E. 2008. Beaver (*Castor canadensis*) mitigate the effects of climate on the area of open water in boreal wetlands in western Canada. *Biological Conservation*, 141, 556–567.

Hudson's Bay Company. 2016. History Foundation. Standard of Trade. http://www.hbcheritage.ca/history/fur-trade/standard-of-trade.

Indiana Department of Natural Resources 2011. *DNR: Beaver. News Releases and Publications. Species Information*. Indianapolis: Indiana Department of Natural Resources.

Ingle-Sidorowicz, H. M. 1982. Beaver increase in Ontario. Result of changing environment. *Mammalia*, 46, 167–176.

Ionescu, G., Ionescu, O., Ramon, J., Pasca, C., Popa, M., Sarbu, G., Scurtu, M., and Visan, D. 2006. 8 years after beaver reintroduction in Romania. *Poster abstract, Proceedings of the 4th European Beaver Symposium/3rd Euro-American Beaver Congress*, 11–14 September, Freisig, Germany.

IUCN 2011. *The IUCN Red List of Threatened Species*. http://www.iucnredlist.org/.

Ivanov, P. 1975. The North American beaver on the Karelian Isthmus of Leningradoblast. *Proceedings of the Voronezh State Nature Reserve*, 21, 114–120 (in Russian).

Jackson, H. H. T. 1961. *Mammals of Wisconsin.* Madison: University of Wisconsin Press.

Janiszewski, P. and Misiukiewicz, W. 2012. Bóbr europejski *Castor fiber.* Warszawa: BTL Works.

Jankowska, B., Żmijewski, T., Kwiatkowska, A., and Korzeniowski, W. 2005. The composition and properties of beaver (*Castor fiber*) meat. *European Journal of Wildlife Research*, 51, 283–286.

Jeffreys, D. 2008. *Aspirin: The Remarkable Story of a Wonder Drug.* New York: Bloomsbury Publishing.

Jonuks, T. 2005. Archaeology of religion—possibilities and prospects. *Eesti Arheoloogia Ajakiri*, 9, 32–59.

Juršič, K., Zupancic, K., Set, J., and Mazinjanin, K. 2017. Ocena številčnosti populacije evrazijskega bobra *Castor fiber* Linnaeus, 1758 na reki Krki in njenih pritokih v letu 2017. *Natura Sloveniae*, 19, 29–46 (English summary).

Karagoyshin, Z. M. 2000. Ecology of beaver in Western Kazakhstan and its economic value. PhD thesis, Almaty, Institute of Zoology CS MES RK.

Kautenburger, R. and Sander, A. C. 2008. Population genetic structure in natural and reintroduced beaver (*Castor fiber*) populations in Central Europe. *Animal Biodiversity and Conservation*, 31, 25–35.

Kind, C.-J. 2009. The Mesolithic in southwest Germany. *Preistoria Alpina*, 44, 137–145.

Kitchener, A. and Conroy, J. 1997. The history of the Eurasian beaver *Castor fiber* in Scotland. *Mammal Review*, 27, 95–108.

Kollar, H. P. and Seiter, M. 1990. *Biber in den Donau-Auen östlich von Wien. Eine erfolgreiche Wiederansiedlung.* Vienna: Verein für Ökologie und Umweltforschung.

Kostkan, V. 1999. The European beaver (*Castor fiber*) population growth in the Czech Republic. In: Hollmann, H. (ed.) *Proceedings of the 3rd International Symposium on Semiaquatic Mammals and their Habitats*, 25–27 May, Universität Osnabrück, Germany.

Kralj, T. 2014. State and distribution of beaver (*Castor fiber* L.) in Croatia. BSc thesis, University of Zagreb.

Kurstjens, G. and Niewold, F. 2011. *De verwachte ontwikkelingen van de beverpopulatie in Nederland: naar een bevermanagement.* Beek-Ubbergen & Doesburg: Kurstjens Ecologisch Adviesbureau & Niewold Wildlife Infocentre.

Lahti, S. and Helminen, M. 1969. History of reintroductions and present population status of the beaver in Finland. *Suomen Riista*, 21, 67–75.

Lahti, S. and Helminen, M. 1974. The beaver *Castor fiber* (L.) and *Castor canadensis* (Kuhl) in Finland. *Acta Theriologica*, 19, 177–189.

Lake-Thom, B. 1997. *Spirits of the Earth: A Guide to Native American Nature, Symbols, Stories and Ceremonies.* New York: Plume.

Larson, J. and Gunson, J. 1983. Status of the beaver in North America. *Acta Zoologica Fennica*, 174, 91–93.

Lavrov, L. S. and Lavrov, V. L. 1986. Verteilung und Anzahl ursprunlicher und aborigener Biberpopulation und den USSR. *Zoologische Abhandlungen Dresden*, 41, 105–109.

Lavrov, L. and Lu, H.-T. 1961. Present conditions and ecological peculiarities of beavers (*Castor fiber* L.) in natural colonies in Asia. *Vestnik Leningradskogo Universiteta*, 9, 72–83.

Lebreton, L., Moigne, A.-M., Filoux, A., and Perrenoud, C. 2017. A specific small game exploitation for Lower Paleolithic: the beaver (*Castor fiber*) exploitation at the Caune de l'Arago (Pyrénées-Orientales, France). *Journal of Archaeological Science: Reports*, 11, 53–58.

Leopold, A. S. 1959. *Wildlife of Mexico, the Game Birds and Mammals.* Berkeley: University of California Press.

Leung, A. Y. and Foster, S. 1996. *Encyclopaedia of Common Natural Ingredients Used in Food, Drugs and Cosmetics.* New York: John Wiley & Sons.

Lizarralde, M. S. 1993. Current status of the introduced beaver (*Castor canadensis*) population in Tierra del Fuego, Argentina. *Ambio*, 22, 351–358.

Lizarralde, M., Escobar, J., and Deferrari, G. 2004. Invader species in Argentina: a review about the beaver (*Castor canadensis*) population situation on Tierra del Fuego ecosystem. *Interciencia*, 29, 352–356.

Lok, S., Paton, T. A., Wang, Z., Kaur, G., Walker, S., Yuen, R. K., Sung, W. W., Whitney, J., Buchanan, J. A., and Trost, B. 2017. De novo genome and transcriptome assembly of the Canadian beaver (*Castor canadensis*). *G3: Genes, Genomes, Genetics*, 7, 755–773.

Long, K. 2000. *Beavers: A Wildlife Handbook.* Colorado: Jognson Books.

Longcore, T., Rich, C., and Müller-Schwarze, D. 2007. Management by assertion: beavers and songbirds at Lake Skinner (riverside county, California). *Environmental Management*, 39, 460–471.

Lovegrove, R. 2007. *Silent Fields.* Oxford: Oxford University Press.

Lu, H. 1981. Distribution and ecological characteristics of beaver *Castor fiber* birulai in China. *Journal of Shandong University*, 4, 1–7 (in Chinese with English abstract).

Luik, H. 2010. Beaver in the economy and social communication of the inhabitants of south Estonia in the Viking age (800–1050 AD). In: Pluskowski, A., Kunst, G. K., Kucer, M., Bietak, M., and Hein, I. (eds.) *Bestial Mirrors: Using Animals to Construct Human Identities in Medieval Europe. Animals as Material Culture in the Middle Ages.* Vienna: Vienna Institute for Archaelolgical Science, University Wien.

Lynn, A. 1949. *Final Report Project California 34-D-2 Beaver Transplanting*. Sacramento: Bureau of Game Conservation, California Division of Fish and Game.

Magnus, O. 1555. *Historia de gentibus septentrionalibus (History of the Nordic People)*.

Manning, A. D., Coles, B. J., Lunn, A. G., Halley, D. J., Ashmole, P., and Fallon, S. J. 2014. New evidence of late survival of beaver in Britain. *The Holocene*, 24, 1849–1855.

Marcuzzi, G. 1986. The relations man–beaver. In: Pilleri, G. (ed.) *Investigations on Beavers*. Bern: Brain Anatomy Institute Bern, Vammalan Kirjapaino.

Martin, P. S. 1867. Pleistocene overkill. *Natural History*, 76, 32–38.

Martin, T. H. 1892. *Castologia, or History and Traditions of the Canadian Beaver*. Montreal: W.M. Drysdale and Company.

Martina, G. 2018. Il ritorno del castoro nelle foreste friulane, l'eccezionale avvistamento dopo 450 anni. https://messaggeroveneto.gelocal.it/udine/cronaca/2018/11/30/news/il-ritorno-del-castoro-nelle-foreste-friulane-l-eccezionale-avvistamento-dopo-450-anni-1.17515503?refresh_ce.

Matsiboruk, P. V. 2013a. Impacts of the population of European beaver (*Castor fiber* L.) on forest ecosystems of Ukrainian Polessye: environmental assessment. PhD thesis, University of Ukraine.

Matsiboruk, P. V. 2013b. The influence of the beaver population on the forest irrigation and drainage systems in Ukrainian Polessye. *Scientific Bulletin of the National Forestry University of Ukraine (Lviv)*, 23, 102–110 (in Ukrainian with Russian and English summary).

Meaney, A. L. 1981. Anglo-Saxon amulets and curing stones. *British Archaeology Reports British Series*, 96.

Mearns, E. A. 1898. Preliminary diagnoses of new mammals of the genera *Sciurus, Castor, Neotoma*, and *Sigmodon*, from the Mexican border of the United States. *Proceedings of the United States National Museum*, 20, 501–505.

Mellink, E. and Luevano, J. 1998. Status of beavers (*Castor canadensis*) in Valle de Mexicali, México. *Bulletin of the Southern California Academy of Sciences*, 97, 115–120.

Mertin, B. 2003. Castoreum—das Aspirin des Mittelalters. *Denisia*, 9, 47–51.

Mississippi Fish and Wildlife Foundation 2010. Managing beaver ponds for waterfowl. In: J. L. Cummins and A. Boyles (eds.) *Wildlife Mississippi*. Jackson: Mississippi Fish and Wildlife Foundation.

Moloney, F. X. 1967. *Fur Trade in New England*. Hamden: Archon Books.

Monahan, E. 2016. *The Merchants of Siberia*. Ithaca: Cornell University Press.

Moorman, M. C., Anderson, C. B., Gutiérrez, A. G., Charlin, R., and Rozzi, R. 2006. Watershed conservation and aquatic benthic macroinvertebrate diversity in the Alberto D'Agostini National Park, Tierra del Fuego, Chile. *Anales Instituto Patagonia*, 2006. Universidad de Magallanes, 41–58.

Müller-Schwarze, D. 1992. Castoreum of beaver (*Castor canadensis*): function, chemistry and biological activity of its components. In: Doty, R. L. and Müller-Schwarze, D. (eds.) *Chemical Signals in Vertebrates*. New York: Plenum.

Müller-Schwarze, D. 2011. *The Beaver: Its Life and Impact*, 2nd edn. New York: Cornell University Press.

Myrberget, S. 1967. The Norwegian population of beaver *Castor fiber*. *Papers of the Norwegian State Game Research Institute*, 2, 1–40.

Naiman, R. J., Melillo, J. M., and Hobbie, J. E. 1986. Ecosystem alteation of boreal forest streams by beaver (*Castor canadensis*). *Ecology*, 67, 1254–1269.

Naiman, R. J., Johnston, C. A., and Kelley, J. C. 1988. Alteration of North American streams by beaver. *BioScience*, 38, 753–762.

Natural Resources Institute Finland (Luke). 2019. Beavers. https://www.luke.fi/en/natural-resources/game-and-hunting/beavers/.

New Hampshire Fish and Game 2011. Beaver (*Castor canadensis*). Concord: NH Fish and Game Department.

Newman, P. C. 1985. *Company of Adventures*. Toronto: Penguin.

Nikulina, E. A. and Schmölcke, U. 2007. Spiel mit dem Knochen. *Abenteuer Arch*, 5, 12–15.

Nikulina, E. A. and Schmölcke, U. 2008. Les osselets, ancêtres du jeu de dés. *Pour la Science*, 365, 40–43.

Nolet, B. A. 1997. *Management of the beaver (Castor fiber): towards restoration of its former distribution and ecological function in Europe*. Strasboug: Report Council of Europe.

Nolet, B. A. and Rosell, F. 1998. Comeback of the beaver *Castor fiber*: an overview of old and new conservation problems. *Biological Conservation*, 83, 165–173.

Novak, M. 1987. Beaver. In: Novak, M., Baker, J. A., Obbard, M. E., and Mallock, B. (eds.) *Wild Furbearer Management and Conservation in North America*. Toronto: Queens Printer for Ontario.

O'Connor, T. P. 1991. *Bones from 46–54 Fishergate. The Archaeology of York*. London: Council for British Archaeology.

O'Shea, J. and Zvelebil, M. 1984. Reconstructing the social and economic organisation of prehistoric foragers in northern Russia. *Journal of Anthropology and Archeology*, 3, 1–40.

Office Nationale de la Chasse et le Faune Sauvage 2018. *Synthèse Nationale Annuelle de l'Activité du Reseau Castor*. Vincennes: ONCFS.

Ognev, S. I. 1963. Mammals of the Union of Soviet Socialist Republics. Volume V—Rodents. Tel Aviv: Israeli Programme of Scientific Translations.

Olson, R. A. and Hubert, W. A. 1994. *Beaver: Water Resources and Riparian Habitat Manager*. Laramie: University of Wyoming.

Opdyke, D. 1973. Monographs on fragrance raw materials. *Food and Cosmetics Toxicology*, 11, 855–876.

Osgood, C. 1940. *Ingalik Material Culture*. New Haven: Yale University Press.

Ott, J. 2003. 'Ruining' the rivers in the Snake Country: the Hudson's Bay Company's fur desert policy. *Oregon Historical Quarterly*, 104, 166–195.

Owesen, A. 1979. *I beverskog*. Oslo: Gylendal Norsk.

Paaver, K. 1956. Esialgne kokkuvóte Róuge linn use ja asula arheoloogilisel kaevamisel 1951 –1956 a kogutud osteoloogilise materjali määramisulemustest. Tallinn: Manuscript in the Institute of History, Tallinn University.

Parker, H. and Rosell, F. 2003. Beaver management in Norway: a model for continental Europe? *Lutra*, 46, 223–234.

Parker, H., Nummi, P., Hartman, G., and Rosell, F. 2012. Invasive North American beaver *Castor canadensis* in Eurasia: a review of potential consequences and a strategy for eradication. *Wildlife Biology*, 18, 354–365.

Parz-Gollner, R. and Hölzler, G. 2012. Numbers, distribution and recent beaver conflicts in Austria. *Poster abstract, Proceedings of the 4th European Beaver Symposium/3rd Euro-American Beaver Congress*, 11–14 September 2006, Freisig, Germany.

Parz-Gollner, R. and Vogl, W. 2006. Numbers, distribution and recent beaver conflicts in Austria. Poster abstract. *Proceedings of the 4th European Beaver Symposium/3rd Euro-American Beaver Congress*, 11–14 September, Freisig, Germany,.

Paşca, C., Popa, M., Ionescu, G., Vişan, D., Gridan, A., and Ionescu, O. 2018. Distribution and dynamics of beaver (Castor fiber) population in Romania. *Presentation, 8th International Beaver Symposium*, Denmark, 18–20 September 2018. https://8ibs.dk/media/191209/2-pasca_8ibs.pdf.

Pelz-Serrano, K., Guevara, E. P., and González, C. A. L. Habitat and conservation status of the beaver in the Sierra San Luis Sonora, Mexico. *Connecting Mountain Islands and Desert Seas: Biodiversity and Management of the Madrean Archipelago II and 5th Conference on Research and Resource Management in the Southwestern Deserts*, 11–15 May 2004, Tucson, Arizona. US Department of Agriculture, Forest Service, Rocky Mountain Research Station, 429.

Petrosyan, V. G., Golubkov, V. V., Zavyalov, N. A., Khlyap, L. A., Dergunova, N. N., and Osipov, F. A. 2019. Modelling of competitive interactions between native Eurasian (*Castor fiber*) and alien North American (*Castor canadensis*) beavers based on long-term monitoring data (1934–2015). *Ecological Modelling*, 409, 1–15.

Pietrek A. G., Boor, G. K. H., and Morris, W. F. 2017. How effective are buffer zones in managing invasive beavers in Patagonia? A simulation study. Biodiversity and Conservation, 26,2591-2605.

Pietsch, L. R. 1956. The beaver in Illinois. *Transaction of the Illinois State Academy of Science*, 49, 193–201.

Pilleri, G. 1984. *Investigations on Beavers*. Berne: Institute of Brain Anatomy, University of Berne.

Pilleri, G. 1986. *Investigations in Beavers*. Waldau-Berne: Brain Anatomy Institute.

Plantain, P. H. 1976. Rôle et utilization du castor (*Castor fiber* et *Castor canadensis*) dans la pharmacopée historique et actuelle. *Ethnosciences*, 143.

Poliquin, R. 2015. *Beaver*. London: Reaktion Books.

Pollock, M. M., Lewallen, G., Woodruff, K., Jordan, C. E., and Castro, J. M. 2015. *The Beaver Restoration Guidebook: Working with Beaver to Restore Streams, Wetlands, and Floodplains*. Portland, OR: United States Fish and Wildlife Service.

Powell, L. H. 1948. *The Giant Beaver Castoroides in Minnesota*. No. 2. St Paul: Science Museum, St Paul Institute.

Radford, H. V. 1907. *History of the Adirondack beaver. Annual reports of the Forest, Fish and Game Commission of the state of New York for 1904–1905–1906*. Albany, NY: Forest, Fish and Game Commission.

Rainsford, K. D. 2016. Occurrence, properties and synthetic developments of the salicylates. In: Rainsford, K. D. (ed.) *Aspirin and Related Drugs*. London: CRC Press.

Raye, L. 2014. The early extinction date of the beaver (*Castor fiber*) in Britain. *Historical Biology*, 27, 1029–1041.

Razmaité, V., Šveistienė, R., and Švirmickas, G. J. 2011. Compositional characteristics and nutritional quality of Eurasian beaver (*Castor fiber*) meat. *Czech Journal of Food Sciences*, 29, 480–486.

Razmaité, V., Pileckas, V., and Juškiene, V. 2019. Effect of muscle anatomical location on fatty acid composition of beaver (*Castor fiber*) females. *Czech Journal of Food Sciences*, 37, 106–111.

Rendu, W. 2010. Hunting behavior and Neanderthal adaptability in the Late Pleistocene site of Pech-de-l'Azé I. *Journal of Archaeological Science*, 37, 1798–1810.

Richard, P. B. 1985. Peculiarities on the ecology and management of the Rhodanian beaver (*Castor fiber* L.). *Zeitschrift für Angewandte Zoologie*, 72.

Richards, J. F. 2014. *The World Hunt: An Environmental History of the Commodification of Animals*. Berkeley: University of California Press.

Robinson, W. L. and Bolen, E. G. 1984. *Wildlife Ecology and Management*. New York: Macmillan.

Rosell, F. and Parker, H. 1995. *Beaver Management: Present Practice and Norway's Future Needs*. Bø i Telemark: Telemark University College.

Rosell, F. and Parker, H. 2011. A population history of the beaver *Castor fiber fiber* in Norway. In: Sjøberg, G. and Ball, J. P. (eds.) *Restoring the European Beaver: 50 Years of Experience*. Sofia-Moscow: Pensoft.

Rosell, F. and Pedersen, K. V. 1999. *Bever [The Beaver]* Oslo: Landbruksforlaget.

Rouland, P. 1985. Les castors canadiens de la Puisaye. *Bulletin Mensuel de l'Office National de la Chasse*, 91, 35–40.

Rozkrut, D. 2018. *Statistical Yearbook of Forestry*. Warsaw: Statistics Poland.

Rue, L. L. 1964. *The World of the Beaver*. New York: J.B. Lippincott.

Russian Fur Union 2016. *Overview of the Population of the Main Species of Fur Animals in the Russian Federation (2005–2015)*. Moscow: Russian Fur Union (in Russian).

Sáenz de Buruaga, P. 2017. El castor (*Castor fiber*) en el País Vasco. Presentation. *5o Congreso Nacional sobre Especies Exóticas Invasoras*, Girona.

Safonov, V. G. and Pavlov, M. P. 1973. The beaver (*Castor fiber* L.). In: Kiris, I. D. (ed.) *Acclimatization of Game Mammals and Birds in the USSR*. Kirov: Volgo-Vyatka Publications.

Salari, L., Masseti, M., and Silvestri, L. 2019. Late Pleistocene and Holocene distribution history of the Eurasian beaver in Italy. *Mammalia*, 84, 259–277.

Samiya, R. 2013. *Central Asian Beaver* Castor fiber *birulai Serebrennikov, 1929*. Ulaanbaatar: Mongolian Red Book.

Saveljev, A. P. 2003. Biological peculiarities of aboriginal and artificially created beaver populations in Eurasia and their significance for the resource management strategy. Dissertation, Russian Research Institution of Game Management and Fur Farming of RAAS.

Saveljev, A. P. 2005. Beavers in Asia: status, conservation, and management. *IX International Mammalogical Congress*, Sapporo, Japan.

Saveljev, A. P. and Safonov, V. G. 1999. The beaver in Russia and adjoining countries. Recent trends in resource changes and management problems. In: Busher, P. and Dzieciolowski, R. (eds.) *Beaver Protection, Management and Utilization in Europe and North America*. New York: Kluwer Academic/Plenum Publishers.

Saveljev, A., Shar, S., Scopin, A., Otgonbaatar, M., Soloviev, V., Putincev, N., and Lhamsuren, N. 2015. Introduced semiaquatic mammals in the Uvs Nuur Hollow (Current distribution and ecological vectors of naturalization). *Russian Journal of Biological Invasions*, 6, 37–50.

Schacht, S. and Bogen, C. 2001. Neue Ausgrabungen auf dem mesolithisch-neolithischen Fundplatz 17am Latzig-See bei Rothenklempenow, Lkr. Uecker-Randow. *Archäologische Berichte aus Mecklenburg-Vorpommern*, 8, 5–21.

Schley, L., Schmitz, L., and Schanck, C. 2001. First record of the beaver *Castor fiber* in Luxembourg since at least the 19th century. *Lutra*, 44, 41–42.

Schmölcke, U., Groß, D., and Nikulina, E. A. 2017. Bears and Beavers: 'The Browns' in daily life and spiritual world. In: Eriksen, B. V., Abegg-Wigg, A., Bleile, R., and Ickerodt, U. (eds.) *Interaction Without Borders. Exemplary Archaeological Research at the Beginning of the 21st Century*. Schleswig: Schleswig-Holsteinische Landesmuseen Schloss Gottorf.

Schorger, A. W. 1965. The beaver in early Wisconsin. *Transactions of the Wisconsin Academy of Science, Arts and Letters*, 54, 147–179.

Schwab, G. and Lutschinger, G. 2001. The return of the beaver (*Castor fiber*) to the Danube watershed. In: Czech, A. and Schwab, G. (eds.) The European Beaver in a New Millennium. *Proceedings of the 2nd European Beaver Symposium*, 27–30 September 2000, Bialowieza, Poland. Krakow: Carpathian Heritage Society.

Schwab, V. G. and Schmidbauer, M. 2003. Beaver (*Castor fiber* L., Castoridae) management in Bavaria. *Denisia*, 9, 99–106.

Schwab, G., Dietzen, W., and Lossow, G. V. 1994. Biber in Bayern-Entwicklung eines Gesamtkonzepts zum Schutz des Bibers. Schriftenreihe Bayer. *Landesamt für Umweltschutz*, 28, 9–44.

Sealander, J. A. 1956. A provisional check-list and key to the mammals of Arkansas (with annotations). *American Midland Naturalist*, 56, 257–296.

Segal, A. N. and Orlova, S. A. 1961. The beaver appearance in Karelia. *Zoological Journal*, 40, 1580–1583 (in Russian).

Seton, E. T. 1929. *Lives of Game Animals*. New York: Doubleday, Doran and Company.

Seton, E. T. 1953. *Lives of Game Animals: An Account of Those Land Animals in America, North of the Mexican Border, Which Are Considered 'Game,' Either Because They Have Held the Attention of Sportsmen, or Received the Protection of the Law*. Branford, CT: Doubleday.

Shar, S. 2005. Distribution of beavers in Chovd River. *Kyzyl*, 7, 316–317.

Shaw, S. P. 1984. *The Beaver in Massachusetts*. Boston: Massachusetts Department of Conservation Research Bulletin.

Sieber, J. 1999. City beavers in downtown Vienna. In: Hollmann, H. (ed.) *Proceedings of the 3rd International Symposium on Semiaquatic Mammals and their Habitats*, 25–27 May, Universität Osnabrück, Germany.

Sieber, J. 2000. Attempting beaver management in Austria. In: Czech, A. and Schwab, G. (eds.) *The European Beaver in a New Millennium. Proceedings of the 2nd European Beaver Symposium*, Bialowieza, Poland. Krakow: Carpathian Heritage Society.

Sieber, J. and Bauer, K. 2001. Eurpäischer und Kanadischer Biber. In: Spitzenberger, F. (ed.) *Grüne Reihe des BMLFUW Bd. 13*. Vienna.

Silmak, L., Lewis, J. E., Crégut-Bonnoure, E., Metz, L., Ollivier, V., André, P., Chrzavzez, J., Giraud, Y., Jeannet,

M., and Magnin, F. 2010. Le Grand Abri aux Puces, a Mousterian site from the Last Interglacial: paleogeography, paleoenvironment, and new excavation results. *Journal of Archaeological Science*, 37, 2749–2761.

Smeraldo, S., di Febbraro, M., Ćirović, D., Bosso, L., Trbojević, I., and Russo, D. 2017. Species distribution models as a tool to predict range expansion after reintroduction: a case study on Eurasian beavers (*Castor fiber*). *Journal for Nature Conservation*, 37, 12–20.

Sorrentino, C. and Landini, L. 2005. Una deposizione in contest sacrificale di *Sus scrofa* L. da Podere Ortaglia/Peccioli. In: Tagliacozzo, A., Fiore, I., Marconi, S., and Tecchiati, U. (eds.) *Atti del 5° Convegno Nazionsale di Archeozoologia*, Osiride, Rovereto.

South Carolina Department of Natural Resources 2010. *Life's Better Outdoors. Beaver in South Carolina*. Columbia: South Carolina Department of Natural Resources.

Speck, F. G. 1923. Mistassini hunting territories in the Labrador Peninsula. *American Anthropologist*, 25, 452–471.

Spriggs, J. A. 1998. The British beaver: fur, fact and fantasy. In: Cameron, E. (ed.) *Leather and Fur. Aspects of Early Medieval Trade and Technology*. London: Archetype Publications.

Statistics Canada. 2009. Table 32-10-0293-01 Number and value of pelts produced. https://doi.org/10.25318/3210029301-eng.

Stepanchuk, V. and Moigne, A.-M. 2016. MIS 11-locality of Medzhibozh, Ukraine: archaeological and paleozoological evidence. *Quaternary International*, 409, 241–254.

Stewart, H. 1973. *Indian Artifacts of the Northweat Coast*. Seattle: University of Washington Press.

Stocker, G. 1985. The beaver (*Castor fiber L.*) in Switzerland. Biological and ecological problems of re-establishment. *Swiss Federal Intitute of Forestry Research Reports*, 242, 1–49.

Strazdina, V., Sterna, V., Jemeljanovs, A., Jansons, I., and Ikauniece, D. 2015. Investigation of beaver meat obtained in Latvia. *Agronomy Research*, 13, 1096–1103.

Taidal, L. 2018. Dobbelt så mange bævere svømmer nu rundt i Nordsjælland. https://www.tv2lorry.dk/artikel/dobbelt-saa-mange-baevere-svoemmer-nu-rundt-i-nordsjaelland.

Tanner, A. 1979. *Bringing Home Animals: Religious Ideology and Mode of Production of the Mistassini Cree Hunters*. New York: St Martins Press.

Tape K.D., Jones, B.M., Arp, C.D., Nitze, I. and Grosse, G. 2018. Tundra be dammed: beaver colonization of the Arctic. *Global Change Biology*, 24, 4478-4488.

Tappe, D. T. 1942. The status of beavers in California. *Game Bulletin*, 3, 1–60.

Tolksdorf, J. F., Turner, F., Veil, S., Bittmann, F., and Breest, K. 2017. Beaver (*Castor fiber*) activity in an archaeological context: a Mid-Holocene beaver burrow feature and a Late-Holocene ecofact at the Late Palaeolithic Grabow site, northern Germany. *Journal of Wetland Archaeology*, 17, 36–50.

Tomljanović, K., Margaletić, J., Vucelja, M., and Grubešić, M. 2018. Beaver in Croatia—20 years after. *Presentation, 8th International Beaver Symposium*, 18–20 September 2018, Denmark. https://8ibs.dk/media/191209/2-pasca_8ibs.pdf.

Trbojević, I. and Trbojević, T. 2016. Distribution and population growth of Eurasian beaver (*Castor fiber* Linnaeus, 1758) in Bosnia and Herzegovina 10 years after reintroduction. *Glasnik Šumarskog fakulteta Univerziteta u Banjoj Luci*, 25, 51–60.

Troidl, C. and Ionescu, G. 1997. Beaver project Romania: a reintroduction with special focus on anthropic factors. *Proceedings of the European Beaver Symposium*, 1997, pp. 15–19.

Universität Für Bodenkultur Wien. 2017. Biber in Österreich. https://boku.ac.at/dib/iwj/forschung/projekte-aktuelle-informationen/der-biber-castor-fiber-in-oesterreich/biberverbreitung-und-bestand/biber-in-oesterreich/.

US Department of Agriculture 2006. Food Composition Database. https://ndb.nal.usda.gov/.

Utah Division of Wildlife Resources 2010. Utah Beaver Management Plan: 2010–2020. Salt Lake City: Utah Division of Wildlife Resources.

Veale, E. M. 2003. *The English Fur Trade*. London: London Record Society.

Veeroja, R. and Mannil, P. 2018. *Status of Game Populations in Estonia and Proposal for Hunting in 2018*. Tartu: Estonian Environmental Agency.

Vermont Fish and Wildlife Department 2004. Best management practices for resolving human–beaver conflicts in Vermont. Montpelier: Vermont Department of Environmental Conservation.

Veski, S., Heinsalu, A., Klassen, V., Kriiska, A., Lõugas, L., Poska, A., and Saluäär, U. 2005. Early Holocene coastal settlements and palaeoenvironment on the shore of the Baltic Sea at Pärnu, southwestern Estonia. *Quaternary International*, 130, 75–85.

Vevstad, A. 1989. Fangstmannsminne, Aslak Harstveit fortel til Andreas Vevstad. Åmli: Vest-Agder Norway (in Norwegian).

Vorel, A. and Korbelová, J. 2016. *Handbook for Coexisting with Beavers*. Prague: University of Life Sciences Prague.

Vorel, A., Šafář, J., and Šimůnková, K. 2012. Recentní rozšíření bobra evropského (*Castor fiber*) v České republice v letech 2002–2012 (Rodentia: Castoridae) [Recent expansion of *Castor fiber* in the Czech Republic during 2002–2012]. *Lynx*, 43, 149–179.

Wall, R. 1939. *Seventh biennial report of Forestry, Fish and Game Commission of the State of Kansas, for the period ending June 30, 1938*. Topeka: Kansas Forestry, Fish and

Game Commission (now Kansas Department of Wildlife and Parks).

Wansey, H. 1798. *An Excursion to the United States of North America in the Summer of 1794*. London, Salisbury.

Weeden, W. B. 1884. *Indian Money as a Factor in New England Civilization*. Baltimore: N. Murray, publication agent, Johns Hopkins University.

Weinzierl, H. 1973. *Projekt Biber*. Stuttgart: Frankh'sche Verlagsbuchhandlung.

West, S. 1985. West Stow the Anglo-Saxon village. *East Anglian Archaeology*, 24, 1–186.

Westcott, F. 1989. *The Beaver: Nature's Master Builder*. Canada: Apocrypha Corporation.

White, H. B., Decker, T., O'Brien, M. J., Organ, J. F., and Roberts, N. M. 2015. Trapping and furbearer management in North American wildlife conservation. *International Journal of Environmental Studies*, 72, 756–769.

Wilson, J. M. 2001. Beavers in Connecticut: Their Natural History and Management. Hartford: Connecticut Department of Environmental Protection, Bureau of Natural Resources, Wildlife Division.

Wilsson, L. 1968. *My Beaver Colony*. New York: Doubleday.

Wilsson, L. 1971. Observations and experiments on the ethology of the European beaver (*Castor fiber* L.). *Viltrevy*, 8, 160–203.

Woodward, D. K. 1980. *Wampum. A paper presented to the Numismatic and Antiquarian Society of Philadelphia*. Albany: Munsell.

Woodward, D. K., Hazel, R. B., and Gaffney, B. P. 1985. Economic and environmental impacts of beavers in North Carolina. *Proceedings of the 2nd Eastern Wildlife Damage Control Conference*, 22–25 September, North Carolina State University.

Yalden, D. 1999. *The History of British Mammals*. London: T. and A.D. Poyser.

Zharkov, I. V. and Sokolov, V. E. 1967. The European beaver (*Castor fiber* Linnaeus, 1758) in the Soviet Union. *Acta Theriologica*, 12, 27–46.

Zhillin, M. G. 2004. *The Environment and Economy of the Mesolithic Population of the Center and North-West of the Forest Zone of Eastern Europe*. Moscow: Academia (in Russian).

Żurowski, W. and Kasperczyk, B. 1988. Effects of reintroduction of European beaver in the lowlands of the Vistula basin. *Acta Theriologica*, 33, 325–338.

CHAPTER 3

Beaver morphology and physiology

3.1 Body form and keeping water out

On land, the beaver (*Castor* spp.) is a large (Figure 3.1), plodding animal that rarely waddles far from the water's edge. Beavers possess a range of unique features and specialized organs that enable them to adapt to their semi-aquatic lifestyle and, while potentially not the most agile of swimming animals, they spend large amounts of time in the water foraging and maintaining their territory. They are able swimmers, with typical surface swimming speeds of 0.8 m/s (0.6 m/s when submerged) (Allers and Culik, 1997), if determined they can move distances on land at speeds of 1.5 km/h (Nolet and Rosell, 1994). Their sense organs such as eyes, ears, and nose are linear and set high on their head so that they are exposed whilst swimming and can still be utilized (Figures 3.2a and b). Histological studies of selected beaver organs largely determined they are remarkably similar to other rodents, but they do possess a number of adaptive characteristics to aid their semi-aquatic lifestyle (Dolka et al., 2014), whilst the significant difference in form and function of their fore and hind limbs effectively illustrates the compromises of an amphibious lifestyle (Figure 3.3).

The beaver's body is recognizably rodent but stout, horizontally compressed, heavily muscled, and with small extremities. They are described as 'drop-shaped', being broader at the hips than the shoulders, tapering towards the nose and head, with the short, thick neck and shoulders appearing almost continuous with each other (Baker and Hill, 2003; Figure 3.2). Their fore legs are much shorter than their hind legs. This large rotund body size is a good compromise for life on the land dragging heavy branches but is also fairly streamline in

the water, with this shape being comparable to seals (Pinnipeds), with a hydrodynamic index of streamlining of 4.8 (Reynolds, 1993).

Along with the likes of water voles (*Arvicola amphibious*), capybara (*Hydrochoerus hydrochaeris*), and hippopotamus (Hippopotamidae), beavers are semi-aquatic. Whilst morphologically they are more typically terrestrial mammal in body form, they possess specialized adaptations for the aquatic environment. They have an additional eyelid, the nictitating membrane, which protects the eye from underwater debris, prevents eye fluids from being washed away, and enables them to keep their eyes open underwater (Wilsson, 1971; Lancia and Hodgdon, 1984). Beaver nostrils are situated higher than most rodents and automatically constrict when in contact with water (Warren, 1927). External ear flaps are much reduced in size and they have hair inside the ears traps air to further reduce water entry. Their epiglottis, located inside the nasal cavity, can be moved, preventing water from entering the larynx and trachea; reducing their ability to breathe through their mouths (Coles, 1970). They can close off the oral cavity using the raised tongue base (Figure 3.4) and closing their hairy, fur-lined lips behind their incisors at the diastema, thereby preventing unwanted material or water being swallowed while gnawing or foraging underwater (Morgan, 1868; Coles, 1970). This raised tongue also allows them to chew with their molars without swallowing water (Coles, 1970). Beavers have no external reproductive organs, just a cloaca which acts as a single opening for excretion, reproduction, and scent marking for both sexes. This may also reduce infections from the aquatic environment (Müller-Schwarze, 2011).

Beavers: Ecology, Behaviour, Conservation, and Management. Frank Rosell and Róisín Campbell-Palmer, Oxford University Press.
© Frank Rosell and Róisín Campbell-Palmer 2022. DOI: 10.1093/oso/9780198835042.003.0003

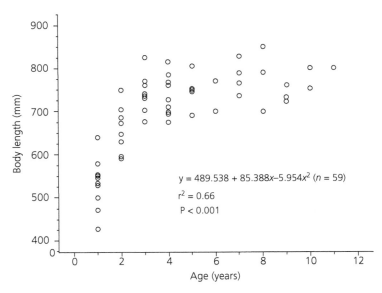

Figure 3.1 Regression plot of Eurasian beaver body length (minus tail length) for females culled in Telemark County, Norway (13th March and 15th May, 1997-1999). Reprinted from Parker et al., 2017. Journal of Zoology 302, p238.

Figure 3.2a Beaver sense organs for smell, sight and hearing are all placed high and in a linear manner on the head so can be used whilst surface swimming (Photo supplied courtesy of Jorn Van Den Bogaert.)

Figure 3.2b Beaver body shape is remarkably streamline making them efficient underwater swimmers with the main thrust coming from the hindlegs. Notice eyes are open underwater, protected by a nictitating membrane (Photo supplied courtesy of Frank Rosell.)

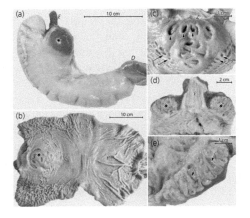

Figure 3.3 The fore (a) and hind digits (b) of beavers differ significantly, representing their different functions, mainly grasping and digging, and swimming. (Photo supplied courtesy of Jon Cracknell and Frank Rosell.)

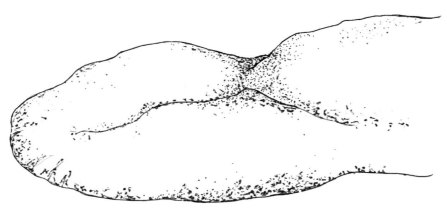

Figure 3.4 Raised tongue to enable chewing and foraging underwater without ingesting lots of water. (Illustration provided courtesy of Rachael Campbell-Palmer, adapted from Long, 2000.)

3.1.1 Fur

Beaver fur is incredibly thick and unfortunately this has been heavily exploited as a luxurious item for humans, almost making both species extinct (see Chapter 2). The tail and feet are the only areas of beavers that are not heavily furred, comprising about 30% of their surface area (Long, 2000). Beaver fur is specialized, consisting of two different hair types: long, coarse guard hairs (~5–6 cm/2–2.4 inches long, being longest along the beaver's back) and shorter (2–3 cm/0.8–1.2 inches long), dense, soft underfur (Müller-Schwarze, 2011). The guard hairs have nearly ten times the diameter of the underfur hairs (Baker and Hill, 2003). Interestingly, it is the outer guard hairs that account for much of the coloration (predominantly a range of browns from reddish to grey-brown and yellow-browns, and even black or white in smaller numbers), while the undercoat tends to remain a more uniform grey. North American beavers (*Castor canadensis*) generally have denser fur than Eurasian beavers (*C. fiber*), though they display fewer differences in hair length variation, for example hairs on their back and stomach are less different in length (Zamakhaeva, 1985).

The description of beaver fur by Hilfiker (1991) captures its multifunctional purpose nicely—a warm overcoat, raincoat, life jacket, and protective covering in fights. The underfur is very dense, so that when a beaver submerges in water, a layer of air is trapped next to the skin which helps repel water and improve insulation (Scholander et al., 1950). Therefore, beaver fur does not become saturated and it dries very quickly once out of the water. Beaver fur has good insulating properties, especially in water, with the fur on the belly (23,000 underfur hairs/cm^2) almost twice as dense as the fur on the back (12,000 hairs/cm^2) (Novak, 1987). They undergo a main moult in autumn each year (August–September). One of the most detailed and comprehensive studies on Eurasian beaver fur investigated seasonal, age, and geographical differences from 11 populations across eastern Europe and Siberia (Zamakhaeva, 1982). Beavers from Voronezh and Belarus have especially been used as source populations for numerous central and eastern translocation projects. Belarusian beavers have thicker fur than those of Voronezh (Table 3.1). Additionally, 25–30 years after first releases, this study demonstrated that descendants had even thicker fur in the more northern parts of Russia and Siberia. With changes in length and fineness of underfur and ratios between underfur coverage, these beavers have also adapted to colder conditions.

Beavers spend a lot of time grooming to maintain fur quality and have a specifically adapted grooming claw (split nail), which they use to comb their fur (Hamilton and Whitaker, 1979; Figure 3.5). The incisors are also used during grooming and frequent grooming is essential to ensure insulation and buoyancy (Fish et al., 2002). Beaver whiskers are used as sensory tools, particularly in dark

Table 3.1 Summary of hair densities differences thousands (tsd) (per cm²) between Eurasian beaver populations.

Density of hair (tsd/cm²)	Voronezh	Belarus	Populations arising from Voronezh source			Populations arising from Belarus source			Hybrids of both populations
			St Petersburg	Novosibirs	Krasnoyars	Smolensk	Arkhangelsk	Komi	Kirov
Back	16.3	22.5	16.4	18.0	18.2	21.7	16.9	24.2	15.9
Belly	20.4	27.1	20.8	26.4	25.2	29.9	29.6	29.4	24.6

Figure 3.5 Beavers have a specialized grooming claw, used to keep their fur free of dirt and parasites. (Photo supplied courtesy of Jon Cracknell.)

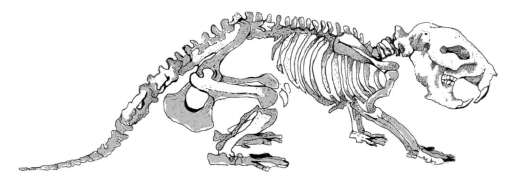

Figure 3.6 Beaver skeletal structure. (Illustration provided courtesy of Rachael Campbell-Palmer, adapted from Coles 2006.)

or murky water, to avoid objects and can give a sense of the strength of water currents (Müller-Schwarze, 2011).

3.1.2 Skeleton and skull

Relative to other mammals of a similar length, the skeleton of beavers is significantly larger (Baker and Hill, 2003). The entire skeleton is composed of 273 bones, 38 of which are associated with the head and 215 within the body trunk, tail (the beaver has 55 vertebrae), and extremities (Morgan, 1868; Figure 3.6). Based on studies of epiphyseal fusion, it has been concluded that beaver skeleton does not obtain full maturity until 12 years of age (Fandén, 2005). Beaver skull and jaw bones are also much larger, thicker, and more robust to support strong facial musculature and process trees, especially hardwoods (Morgan, 1868). The upper lip is emarginated whilst the lower lip is loose so that the incisors are often visible (Morgan, 1868). Beavers have 14 pairs of ribs, which are slender as are the five bones of the sternum (Morgan, 1868). In both species, the bones of the extremities tend to be strong and thick, though it is reported that North American beavers have longer tibias and fibula (shin bones), enabling a greater range of bipedal locomotion than the Eurasian (Hollander et al., 2017).

Beaver skulls are large and thick, with a low and elongated profile (almost 2.5 times its height), flattened on top but with a clear sagittal crest which splits rostrally; characteristic features include wide zygomatic bones with large arc and large mandible bones (Nowicki et al., 2019; Figure 3.7a). This skull morphology, high jaw joint, and large jaw-closing muscle set, as well as the incisor tip enable the beaver to produce a very effective and efficient bite (Cox and Baverstock, 2016; Stefen et al., 2016). The bony openings of the auditory canals, wide orbital bones, and wide, wedge-like nasal cavity with numerous nasal bones making up the key sense organs beavers rely on are clearly seen (Nowicki et al., 2019). The foramen magnum is quite large, the opening connecting the skull with the vertebral canal.

A whole range of skull morphometric characteristics (e.g. nasal, coronoid fossa, and mandibular length) have been studied widely in both species, particularly to investigate differences between the species (Chapter 1.4), ageing techniques (Chapter 10.4.4), and geographical variations (Lavrov, 1979; Lavrov, 1983). For example, higher craniometric variations in Eurasian beaver throughout Europe have been determined, especially in relation to mandibular size and shape, with beavers originating from relict German (Elbe) populations having large mandibles and tooth rows (Kitchener and Lynch, 2000). More recent comparisons between Ukrainian and Polish populations found eight significant cranial differences, with the Polish populations having longer skulls whilst the Ukrainian were wider with longer nasal bones (Teleky et al., 2018).

North American beavers tend to have shorter skull lengths than Eurasian, which are considered to possess more heavily built skulls (Pilleri, 1983). Typical length of skull measurements range from 12.1 to 14.6 cm in adult North American beavers, while zygomatic width varies from 8.7 to 10.8 cm (Grinnell et al., 1937; Osborn, 1953; Jenkins and Busher, 1979). No significant sex differences in skull dimensions have been determined in North American beavers (Dillman and Barnett, 1984). However, recent studies on Eurasian beavers in Poland suggest a slight degree of sexual dimorphism relating to the length of the viscerocranium, greatest nasal bone length, cranial base length, height of foramen magnum, and angular height of the neurocranium (Nowicki et al., 2019). Polish studies of Eurasian beavers determined average total skull lengths of 13.05 cm in females and 14.0 cm in males and zygomatic widths of 9.62 cm in females and 10.39 cm in males (Nowicki et al., 2019).

3.1.3 Teeth

Typical beaver dentation (Figures 3.7a and b) in the maxilla and mandible consists of 20 teeth, with the dental pattern the same for the upper and lower jaw, consisting of a pair of incisors (I), a pair of premolars (P), and three pairs of molars (M) [2 × (I 1/1, C 0/0, P 1/1, M 3/3) = 20] (Figure 3.7a). Supernumerary teeth in beavers are rare. This has only been recorded for premolar teeth in a very small number of individuals (2% of 142 North American beaver skulls examined [Pilleri, 1983). In every case these teeth were much smaller, lying below the masticatory plane, and were bilateral in occurrence. Therefore, their actual use for eating is doubtful. The premolar and molar teeth are highly similar, with a masticatory surface formed from transverse enamel folds, producing a crushing and grinding surface (Nowicki et al., 2019). The dentine between these layers of enamel is worn faster so as to create higher enamel ridges, with each molar being slightly curved, creating two concave surfaces (Morgan, 1868). The first molars are the largest and longest, while the last ones are the smallest (Morgan, 1868). Beavers have a large diastema (space) between the premolars and incisors.

Their incisors are extensively developed, strong, semi-circular, and broad. Such self-sharpening inci-

(a)

(b)

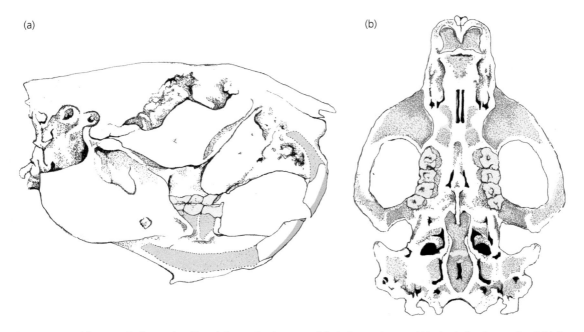

Figure 3.7a and b Beaver skull, ventral and lateral view, noting the extent of the incisor teeth roots within the skull and proportion visible in the buccal cavity (Illustration provided courtesy of Rachael Campbell-Palmer redrawn with permission from Pelagic Publishers and further adapted from Kitchener 2001.)

sors evolved for gnawing are a characteristic of the order Rodentia. As rodents, beavers possess four prominent, orange incisor teeth—only much larger! The incisors have open roots and therefore grow continually, whereas their molars tend to stop growing when the animal reaches around two years old. The incisors grow faster and are highly curved (Troszynski, 1977) with the majority remaining with the mandible itself. Growth rates of 0.39cm per month have been estimated, through measuring beaver skulls with unchecked growing incisors (Rosell and Kile, 1998). Incisor growth is kept in check for appropriate feeding behaviours through gnawing and grinding action. The lower incisors are larger and less curved and their root begins much further back in the mandible, whereas the upper incisors have a much shorter root. These lower incisors are particularly well adapted to cutting and the tip is optimized by the wedge angle (Stefen et al., 2016). Beavers have large superficial masseter and temporalis muscles. The incisor bite force is between 550 and 740 Newtons (kg × m/s²)

(80 kg equals around 785 Newtons). This bite force is much higher than would be expected from the beaver's body mass (202 Newtons) or incisor dimensions (334 Newtons) and is likely related to the very high percentage (> 96%) of bite force directed along the long axis of the lower incisor (Cox and Baverstock, 2016).

Variability in beaver teeth dimensions (length and width) across both beaver species ($n=77$) has been investigated (Stefen, 2009). Beaver teeth size seems to increase linearly with age up to around 4.2 years, after which little increase in tooth size occurs. There is large variation in teeth size over all ages of both species, though North American beaver teeth were found to be slightly smaller than Eurasain. Another study investigated the 'hardness' of beaver teeth, determining that upper incisors possess softer dentine than lower ones. This hard, orange enamel is mineralized with iron and calcium (Wilson and Ruff, 1999). Difference in erosion rates between soft dentine and harder enamel is well documented in rodents, leading to their sharp 'chisel-like' incisor teeth. Within the incisors, dentine hardness was least graduating towards the middle of the tooth and greatest at the periphery (Osborn, 1969). The average thickness of enamel tends to be greater in the mandible incisors at an average of 0.34 mm, whilst the maxilla enamel was ~0.29 mm in adult Slovakian beavers (Teleky et al., 2018).

3.1.4 Legs and feet

Whilst on land, beavers walk using all four feet, being generally slow and cumbersome. Their gait is often described as a slow waddle, unless pushed when for short distances they can reach a 'lumbering quadruped gallop' (Hilfiker, 1991). They often rise onto their hind legs when felling trees or gnawing branches or for short distances when carrying materials in their fore arms, using their tail to stabilize themselves (Carlson and Welker, 1976). Beaver feet vary greatly between the front and back (Figure 3.3a and b) and are unique amongst rodents in having five toes on each foot (Long, 2000). Each of these digits ends in strong and sturdy claws, which are shorter and more rounded on the hind feet compared to the fore paws. The upper surface of each digit is covered with short hairs, whilst the undersides are hairless, and covered with a tough black epidermis (Morgan, 1868).

The fore arms are mainly used for walking, digging, carrying, and manipulating earth and vegetation; they are not used for swimming, apart from in very young animals. The front paws are very dexterous and strongly clawed (only very rudimentarily webbed) and can grasp and manipulate objects. They can carry remarkably heavy objects such as stones by holding them close to their body and tucked under their chin with their fore arms (Figure 3.8). Beavers are strong and capable diggers, forming burrows, chambers, and canals as well as gathering earth and mud for various constructions (see chapter 4.2).

The hind legs are much bigger and more heavily muscled, with two separate lower-leg bones, not fused like rats and mice. The back feet are fully webbed, providing a large surface area for propulsion in the water. The toes on the hind feet are particularly long, ~20 cm in adults (Baker and Hill, 2003). Another unique feature is the presence of split or double nails, also known as the grooming claw, on the second outer toe of each hind foot. Both medial (inside) toes on each hind foot have double nails which grow perpendicularly to each other. A beaver can move these claws to some extent so that the split claw acts like a clamp during grooming (Long, 2000; see Figure 3.5).

Figure 3.8 Beavers can walk short distances on their back legs, especially when carrying materials. The skeletal structure of American beavers is thought to lend itself more readily to bipedal locomotion compared to the Eurasian. (Photo supplied courtesy of Michael Runtz.)

3.1.5 The tail

Beavers are renowned for their unique tails, which are not only curious to look at but also multifunctional on both land and in water, capable of an impressive range of movement (Mahoney and Rosenberg, 1981; Figure 3.9). The tail has several uses: for signalling a warning sign to other family members to alert them to danger (see Section 7.4.2); as a balancing aid when the beaver is raised on its back legs while gnawing trees or carrying mud (see Section 5.2.2); as a swimming tool and in assisting kits (see Section 6.8); in thermoregulation (see Section 6.2 on sleeping behaviours); and as a fat storage unit (see Section 10.3.2 on body condition).

The tail is made up of two very different parts: a hairy, conical muscular base and a distinctive distal section which is longer, flat, and broader, making up most of the tail. This latter section consists mainly of thick, fatty, collagenous tissue (subcutaneous), which acts as an important fat store and so

Figure 3.9 The beaver tail is perhaps one of its most defining features. (Photo supplied courtesy of Rachel Hinds.)

varies in dimension according to body condition (Aleksiuk, 1970). The outer epidermis is covered with diamond-shaped keratin scales (thick stratum lucidum), with sparse, short, thick hairs, under which is a dense, white adipose tissue layer acting as a fat depot and insulation layer (Dolka et al., 2014). Numerous capillary vessels are found in the dermis layer, along with nerve endings and sparse hair follicles with their associated sebaceous glands (Dolka et al., 2014). The rest of the tail consists of the coccygeal vertebrae which have numerous large tendons from more caudally situated muscles in the upper tail section (Carlson and Welker, 1976). The upper, furry part of the tail is structurally very different, being significantly more muscular. Three major muscle groups (a dorsal *sacrococcygeus dorsalis*; a ventral *sacrococcygeus ventralis*; and a short ventral *coccygeus*) lie along four to seven coccygeal vertebrae (Carlson and Welker, 1976). The muscular arrangement in the beaver tail is a complex system with extensive intermeshing muscle fibres between muscles groups which enables fine control of distal parts by a proximal muscle mass (Mahoney and Rosenberg, 1981). Beavers can rotate their tails in different directions and along various planes, i.e. horizontally, vertically, and in rolling motions, allowing 'a four-way rudder', thanks to their specialized muscle groups (Carlson and Welker, 1976). Not only does this result in powerful thrusting movements during swimming and diving, but also both beaver species will 'tail slap' as a warning behaviour, lifting their tail above the water surface and bringing it down sharply, producing a noise that travels remarkably well (Aleksiuk, 1970; Rosell et al., 2005; Müller-Schwarze, 2011; chapter 7.4.2).

3.1.6 The cloaca

One unique feature of the beaver is the 'cloaca' (Figure 3.10), which is not a true vertebrate cloaca but a common chamber for the various excretions and secretions (Müller-Schwarze, 2011). Thus, beavers do not have any external sex organs (see Section 3.2.8). In addition to the anus, urethra, and genital openings, they have a pair of castor sacs (see Section 3.2.9) and a pair of anal glands

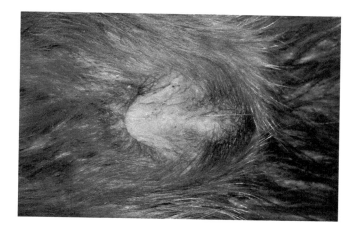

Figure 3.10 Both sexes have a cloaca rather than external sex organs. (Photo supplied courtesy of Jon Cracknell.)

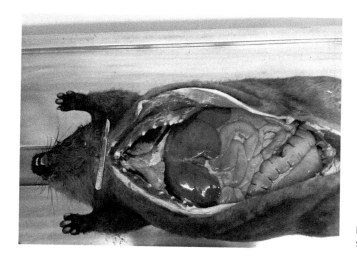

Figure 3.11 Partial exposure of internal organs. (Photo supplied courtesy of Romain Pizzi.)

(see Section 3.2.10) located in a shared outer opening on the ventral side at the base of the tail behind the pelvis. The anal glands are located posterior to the castor sacs. These scent organs produce two main sources of scent: castoreum from the castor sacs and secretions from the anal glands (AGS) (see below and Section 7.2.2 on territorial defence). The two anal glands are secretory glands, which open into the cloaca, while the two castor sacs are pockets, lined with a non-secretory tissue (Svendsen, 1978; Walro and Svendsen, 1982; Valeur, 1988), which open separately into the cloaca.

3.2 Internal organs

Figure 3.11 shows the autopsy of a beaver, cut open to display the internal organs.

3.2.1 Brain

Beaver brains are not thought to be particularly specialized, whether compared to other rodents or displaying adaptations to their lifestyle (Müller-Schwarze, 2011). Dorsally beaver brains are oval in aspect. Though they have bigger brains than other rodents, they are smaller and less complex than the larger capybara for example. In contrast, the outer surface of the cerebral neocortical is lissencephalic (smooth) and unfissured, whereas capybara have a more fissured cortex (Carlson and Welker, 1976).

Brain mass has been recorded in North American beavers as 41 g for a 11.7-kg individual, 30 g for a 15-kg, 45 g for a 17-kg, and 52 g for a 22.5-kg (Pilleri, 1983, 1984). Brain to body mass ratios are sometimes used as a reflection on intelligence, in terms of the ability to problem solve, with the encephalization quotient (EQ) being a measure of actual to expected brain size (Müller-Schwarze, 2011). The EQ of the North American beaver is given at 0.9 whereas the Eurasian was determined as 0.8, with muskrat (*Ondatra zibethicus*) 0.68 and coypu (*Myocastor coypus*) 0.78, making beavers higher than typical terrestrial rodents but lower than comparable arboreal species (Pilleri, 1984), though they do possess a higher cerebrum to hypothalamus length, with the ratio ranging from 0.20 to 0.24 in a study of 12 North American beavers (Pilleri, 1983). Some believe this is a crude measure of intelligence. The cerebrum is certainly responsible for the co-ordination of locomotion, which the beaver scores higher in compared to other rodents, though the cerebrum of arboreal rodents tends to be larger (Pilleri, 1983; Müller-Schwarze, 2011).

A physiological study of the cerebral motor cortex of beavers found stimulation tended to elicit patterns of movement rather than strictly control specific body parts (Langworthy and Richter, 1938). A series of brain section studies undertaken in beavers relate to their semi-aquatic and more nocturnal lifestyles (Szteyn and Galert, 1979), with their motor cortex comparable to other highly specialized rodents such as porcupines (infraorder Hystricidae) (Langworthy and Richter, 1933). In examination of the *nucleus corporis geniculati lateralis pars dorsalis* this was found to be structurally quite primitive in beavers, leading to the conclusion that the optic organ is not well developed. In contrast, the *nucleus corporis geniculati medialis* was more robustly formed, indicating that the sense of hearing in beavers is more developed than their sense of vision. In general, large areas of a beaver's brain seem evolved to

process auditory and somatic sensory information (Carlson and Welker, 1976).

3.2.2 Digestive organs

As typical members of the Rodentia, the digestive system of beavers consists of a relatively simple stomach with an enlarged caecum (Kitts et al., 1957; Vispo and Hume, 1995; Figure 3.12). They are classified as monogastric hind-gut (caecum) colon fermenters based on their gut morphology (Hume, 1989; Vispo and Hume, 1995), meaning the caecum is the main site for microbial plant carbohydrate digestion. The whole digestive tract constitutes between 9 and 13% of total body mass (Vispo and Hume, 1995; Bełżecki et al., 2018). They are completely herbivorous, possessing a digestive system that can handle large amounts of materials that are considered difficult to digest (see Section 3.4.1 and Chapter 5). The digestive capacity of a typical adult beaver has been measured at 3,458 g wet weight of food per day, whilst the digestive organ capacity is 1,484 g of wet weight of food at a time. Beavers typically empty their digestive tracts 2.33 times per day (Belovsky, 1984).

3.2.2.1 Digestion of food in the mouth and stomach

Digestion begins in the mouth where the food is physically chewed into small pieces and further broken down by saliva enzymes. Beaver salivary glands are well developed, with the parotid and submaxillary glands particularly large, covering the front and sides of the neck (Morgan, 1868). The stomach is a unilocular, C-shaped sac with a characteristic cardiac gland which is a highly developed, secretory gland, located near the oesophageal entrance on the lesser curvature of the stomach (Nasset, 1953; Vispo and Hume, 1995; Ziółkowska et al., 2014). The adult stomach has a capacity of 350–1,400 cm^3 (Janiszewski et al., 2017), roughly measuring 28–40 cm along the greater curvature, with the perimeter at the top end or corpus ranging from 22 to 30 cm and at the bottom end or pylorus ranging from 15 to 18 cm (Bełżecki et al., 2018). The cardiogastric gland (CGG) looks like a flattened football and is 6–7 cm long, 4 cm wide, and about 2 cm thick and weighs 60 g, with multiple openings

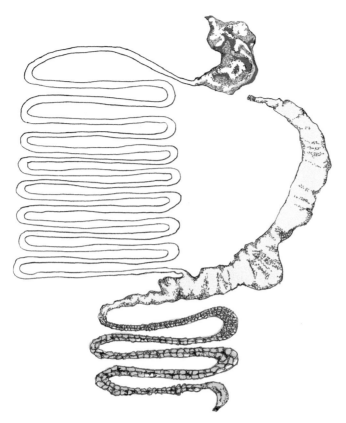

Figure 3.12 Beaver digestive system and stomach gland. Beavers have elongated C-shaped caecums to facilitate digestion of a high-cellulose diet. (Illustration provided courtesy of Rachael Campbell-Palmer, redrawn with permission from Pelagic Publishers.)

to release gastric fluids (Nasset, 1953). This gland is found in other mammals with high-fibre diets and it secretes enzymes that reduce cellulose to starch. The gland is connected to the stomach cavity by 36–40 orifices on its luminal surface. Twelve of the orifices are large and the remaining are small (Figure 3.13). Another unusual feature of the beaver's stomach is the presence of specific mucus that coats the luminal surface of the stomach. The mucus likely acts as a barrier against physical injuries to the stomach, including those caused by the beaver's diet, or might form a niche for microorganisms and provide optimal conditions for their development (Ziółkowska et al., 2014).

After food is processed in the stomach it moves to the small intestine. The small intestine is approximately six times the beaver's body size and can reach 534–629 cm in North American beavers and

400 cm in Eurasian beavers (Vispo and Hume, 1995; Bełżecki et al., 2018). In the small intestine nutrients are absorbed before the remaining digested matter is sent on to the caecum.

3.2.2.2 Digestion in the stomach–intestinal canal

Other features to enable beavers to more efficiently digest highly woody diets are their large intestines and very large caecum which is around twice the size of their stomach (Müller-Schwarze, 2011). Similar to the rumen in ruminants, the enlarged (nearly 40 cm long in adults), C-shaped caecum 'fermentation chamber' is located in the uppermost part of the large intestine (Nasset, 1953). Its main function is to enable beavers to digest coarse-fibered plant material. A large caecum is an adaptation in herbivores that only have one stomach. Near the entrance to the caecum, the food is divided into a

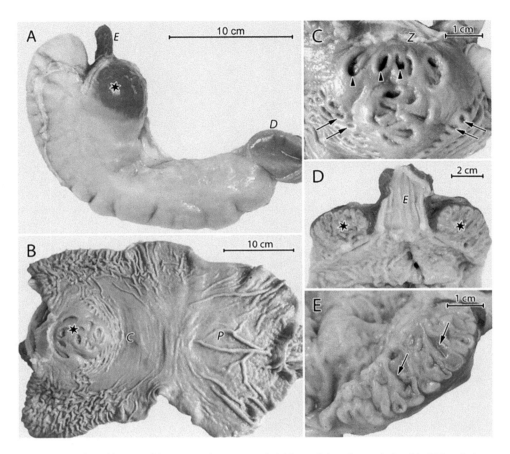

Figure 3.13 Gross morphological features of the European beaver stomach. (A) External view of stomach. Asterisk: CGG on the lesser curvature; E: oesophagus; D: duodenum. (B) Mucosa after cutting the stomach along the greater curvature. Note differences in mucosal colour between corpus (C) and pylorus (P). Asterisk: CGG. (C) CGG viewed from the stomach lumen. Note large crater-like (arrow heads) and small (arrows) orifices. Z: gastroesophageal junction. (D) Stomach cut along the lesser curvature. Note continuity between muscular coat of CGG and muscularis of oesophagus. Asterisk: CGG; E: oesophagus. (E) Section of CGG showing its internal organization. Arrows: branched tubes that open on luminal mucosal surface. (Reprinted from Ziółkowska, N., Lewczuk, B., Petryński, W., Palkowska, K., Prusik, M., Targońska, K., Giżejewski, Z. and Przybylska-Gornowicz, B. 2014. Light and electron microscopy of the European beaver (*Castor fiber*) stomach reveal unique morphological features with possible general biological significance. *PLoS ONE*, 9, e94590 under CC BY 4.0 License.)

coarse, fibrous mass that continues out of the large intestine as faeces and a more finely textured portion that is sent into the caecum for further digestion.

The caecum contains a substantial population of microorganisms (bacteria and protozoa; see Section 8.5.3.1) due to the anaerobic conditions and is the main site of cellulose digestion (Currier et al., 1960; see Section 3.4.1). The proportion of large-sized particles (> 4 mm) are higher here than in the small intestine and colon. The food components influence the activity of microorganisms, and the percentage of fine-sized particles (> 0.5 mm) is also higher in the summer than autumn when more woody shoots are digested (Bełżecki et al., 2018; Miltko et al., 2019). After processing in the caecum and prior to defecation, plant material is finally sent to the large intestine and its capacity corresponds to twice the volume of the stomach (Vecherskii et al., 2006). In comparative studies of intestinal length of Eurasian (*n*=15) and North American (*n*=13) beavers living in similar habitats on tributaries of the

Amur River, Russian Far East (Saveljev, 1984), intestinal length of North American kits was determined to be 12.5 times longer than body length and 11.4 times longer in adult animals, whilst in Eurasian individuals this ratio did not exceed 10.6 and 9.3, respectively (Saveljev, 1984). Such differences in evolutionarily close species living in the same habitat conditions appear to reveal species characteristics. It has been argued that this enhanced development of the intestine, especially the colon, in North American beavers is representative of the general direction of evolution of herbivorous mammals from protein-lipid to fiber digestion (Vorontsov, 1967), and therefore they can be considered more evolved (Saveljev, 1989).

Beavers produce two types of faeces. One faeces originates in the caecum and another passes directly through the small and large intestines. The intestinal faeces are short and thick often marble-shaped, with a hint of a point at one end. They are usually light brown and resemble sawdust; they are about 2 cm thick and 2–4 cm long. The consistency is similar to that of Lagomorph faeces because they are also primarily made up of coarse plant parts, which can be identified on the surface. Normally, beavers defecate intestinal faeces in the water and do not defecate within their lodge or burrows. As for all caecotrophs, the caecal faeces is reingested to enable the beavers to absorb the proteins (volatile fatty acids) and vitamins released by the caecum symbiont bacteria during the cellulose breakdown process. It has been estimated that up to 19% of a beaver's energy requirements could be met through the ingestion of these volatile fatty acids (Hoover and Clarke, 1972).

3.2.3 Liver

As for all mammals, livers are multifunctional organs that assist in blood storage, digestion, metabolism, and detoxification. Beavers' livers are long and flattened, with two principal lobes and two smaller lobes (Morgan, 1868). They can store large amounts blood which assists in longer dives by storing oxygen (Dolka et al., 2014). Histological studies of 21 beavers found numerous binucleated hepatocytes (Dolka et al., 2014). These cells store glycogen (energy reserve) and this could be an adaptation to increase seasonal energy demands or hypoxia when diving (Allers and Culik, 1997; Pescador et al., 2010; Dolka et al., 2014).

3.2.4 Spleen and thymus

Beaver spleens are very similar to other rodents, being small, long and linear, functioning as part of the immune system and to filter blood (Morgan, 1868). Like other rodents they possess more connective

Figure 3.14 Beaver intestinal faeces, often defecate in the water where it quickly breaks down so not often seen (Photo supplied courtesy of Frank Rosell.)

tissue in their fibrous capsule as opposed to smooth muscle found in other mammals, meaning they are more likely to keep a consistent size and colour (Valli et al., 2002; Dolka et al., 2014). Beaver spleens exhibit numerous lymphoid nodules and therefore typically this represents a defensive type of spleen rather than a storage type as they have a lack of blood-storing properties (Dolka et al., 2014). The authors speculate this may reflect an adaptation to an aquatic lifestyle and explain their tendency to avoid long periods on land, though they admit this requires further investigation. Beaver spleens are also capable of temporarily doing without red blood cells during the dive response and have the ability for aerobic metabolism (Cabanac et al., 1997).

The beaver thymus consists of *pars cervicalis* and *pars thoracalis*, which is unusual in most species apart from the water shrew (*Neomys fodiens*) (Bazan, 1955). The cervical part forms two lobes, with a distinctive structure, located near the thyroid, while the thoracic part covers the lungs and heart and tends to be unpaired, loose in structure, and irregularly shaped (Wyrzykowska and Wyrzykowski, 1979). In most other mammals, the smaller *pars thoracalis* disappears during development (Nickel and Seiferle, 1976).

3.2.5 Heart, blood vessels, and blood chemistry

Beaver heart anatomy is more similar to aquatic than land mammals, with diving thought to be the main evolutionary pressure responsible for this. Their cardiac muscle is also comparable to aquatic mammals (Dolka et al., 2014), though their hearts are relatively small, which is indicative of animals that do not generally exert themselves for long periods of time (McKean and Carlton, 1977). Heart weights represent 0.40% of their body weight, putting them in a similar range to terrestrial animals (Bisaillon, 1982). Beavers' hearts do have a more unusual right atrium which is divided into two areas—one on either side of the ascending aorta (Bisaillon, 1982). Typical heart rates for an active beaver are around 120 beats/minute, dropping to around 100 beats/minute when resting on land, 75 beats/minute when sleeping, and around 61 beats/minute when diving (Swain et al., 1988).

Beaver blood vessels are typically mammalian (Bisaillon, 1981; Bisaillon, 1982), many blood characteristics being similar to other rodents (Stevenson et al., 1959). Both beaver species appear to possess a well-developed carotid artery (Pilleri, 1983; Frąckowiak and Śmiełowski, 1998), which is lacking in some other rodents (Bugge, 1974). This differentiation in cephalic arteries offers a clear taxonomic distinction of beavers from Sciuromorpha rodents (Bugge, 1974; Frąckowiak and Śmiełowski, 1998). Significant haematological differences include lower red blood cell or erythrocyte numbers in both species (Kitts et al., 1958; Girling et al., 2015) and high cholesterol levels in North American beavers, almost double those of other rodents (Stevenson et al., 1959). Beaver erythrocyte counts are similar to guinea pigs (*Cavia porcellus*), below 5 million/mm^3, both being much lower than most other mammals including other rodents (Kitts et al., 1958). The average diameter (7.6 μm ± 1.2 S.D., range 6.0–9.7 μm) of these red blood cells was greater than those of other mammals (Kitts et al., 1958). Packed cell volumes have been found to differ slightly between the species, being lower in Eurasian (range of 0.33–0.43 L/L; Girling et al., 2015) compared to North American (0.4–0.46 L/L; Kitts et al., 1958) species. White blood cell counts are similar between both species, reported as 7–18.2 × 10^9/L in North American beavers (Kitts et al., 1958) and 6.9–13.4 × 10^9/L in Eurasian animals (Girling et al., 2015).

Cholesterol blood values for Eurasian beavers on the other hand were found to be largely within ranges published for other rodents, including coypu (Jenkins, 2008; Martino et al., 2012). As for humans, diet and genetic differences can influence cholesterol levels. Blood protein levels, mainly composed of haemoglobin, though this also includes albumin, fibrinogen, and globulin, are significantly lower in summer than autumn, most likely reflective of increased demands for growth and breeding (Stevenson et al., 1959). Beavers appear to have higher levels of volatile fatty acids, produced via bacterial action in the caecum during fat metabolism (Stevenson et al., 1959). All other variations in haematology and serum biochemistry parameters appear to be consistent across both species and similar to those typical for other rodents, though slight variations have been noted between beaver populations, most likely reflective of dietary and environmental differences (Girling et al., 2015). Tables 3.2 and 3.3 give mean values for haematological parameters and serum biochemistry in beavers, respectively.

Table 3.2 Typical mean values for haematological parameters in adult beavers (ISIS, 2002; Girling et al., 2015).

Parameter	Eurasian beaver (mean ± SD)	North American beaver (mean and reference interval)
Red cell count	3.69 ± 0.25 (×10^{12}/L)	3.74 (×10^{12}/L) 3.01–4.59
Haemoglobin	12.98 ± 0.98 (g/dL)	12.7 (g/dL) 10.3–15.0
Packed cell volume	0.37 ± 0.03 (L/L)	41.2% ± 2.5 S.E.
Mean cell volume	99.71 ± 5.74 (fl)	96.9 (fl) 79.1–117.1
Mean cell haemoglobin	35.25 ± 2.05 (pg)	33.9 (pg) 29.1–39.3
Mean cell haemoglobin concentration	35. 25 ± 0.60 (g/dL)	34.7 (g/dL) 26.4–40.8
Platelets	216.50 ± 49.86 (×10^9/L)	0.212 (×10^{12}/L) 0.011–0.478
White cell count	10.45 ± 2.09 (×10^9/L)	10.75 (×10^9/L) 4.96–20.28
Neutrophils segmented	7.50 ± 1.72 (×10^9/L)	5.77 (×10^9/L) 0.23–11.48
Neutrophils banded	0.08 ± 0.18 (×10^9/L)	0.102 (×10^9/L)
Lymphocytes	2.43 ± 1.04 (×10^9/L)	30% (% of total leucocyte count) and 29.8% 23.0–41.0 (Kitts et al., 1958)
Monocytes	0.33 ± 0.30 (×10^9/L)	3.3% and 3.5% 1.0–7.0 (Kitts et al., 1958)
Eosinophils	0.08 ± 0.09 (×10^9/L)	2.4% (1.0–5.0) (Kitts et al., 1958)
Basophils	0 (×10^9/L)	0.1% (0.0–1.0) and 0.6% (0.0–2.0) (Kitts et al., 1958)

Source: Reprinted from Girling, S. J., Campbell-Palmer, R., Pizzi, R., Fraser, M. A., Cracknell, J., Arnemo, J. and Rosell, F. 2015. Haematology and serum biochemistry parameters and variations in the Eurasian beaver (*Castor fiber*). *PLoS ONE*, 10, e0128775 under CC BY 4.0 License.

3.2.6 Lungs

The beaver respiratory system is similar to that of mice and rats (Muridae family) (Dolka et al., 2014). In examinations of rodent lung morphology, beavers have fairly unspecialized lungs (Wallau et al., 2000). The right lung has two lobes, while the left lung has four (Morgan, 1868). Their bronchioles are short and terminal, do not contain any cartilage or glands, and end in alveolar ducts (Dolka et al., 2014). They have an impressively efficient exchange ability, enabling them to exchange about 75% of the air in their lungs when they breathe (Irving and Orr, 1935; McKean and Carlton, 1977). The mean total lung capacity for beavers has been recorded as 60.5 ± 4.8 ml/kg (McKean and Carlton, 1977).

3.2.7 Kidneys

Beaver kidneys consist of the cortex, medulla, and medullary pyramid (Dolka et al., 2014). In mammals, the renal medulla varies according to habitat conditions (Beuchat, 1996), and although beavers differ from other rodents such as rats and mice, they are similar to guinea-pigs (Schmidt-Nielsen and Pfeiffer, 1970), with features linked to aquatic habitats (Al-Kahtani et al., 2004). Beaver kidneys have only short loops of Henle, with thin descending and thick ascending limbs and shallow renal papilla (Dolka et al., 2014). Therefore their ability to produce concentrated urine is low (Jamison, 1987); some of this dilute urine can become concentrated in the castor sacs to become castoreum (Müller-Schwarze, 2011).

3.2.8 Reproductive organs

In beavers, all sex organs are internal, which makes sense when spending so much time in cold water. Females have two pairs of nipples located in the thoracic region. These are only visibly obvious if the female is in later pregnancy and/or lactating (Figure 3.15). Otherwise there is no obvious external distinguishing feature between the sexes. The morphology and immunohistochemical properties of mammary glands in beavers are highly similar to that described for other mammals, particularly rats and mice (Franke-Radowiecka et al., 2016). There

Table 3.3 Typical mean values for serum biochemistry parameters in adult beavers (ISIS, 2002; Girling et al., 2015).

Parameter	Eurasian beaver (mean ± SD)	North American beaver (mean ± SD)
Alanine transferase	36.83 ± 7.28 (IU/L)	2 (U/L) 7– 45
Aspartate aminotransferase	74.80 ± 12.59 (IU/L)	69 (U/L) 34–139
Alkaline phosphatase	366.23 ± 135.72 (IU/L)	204 (U/L) 47–600
Bile acids	3.54 ± 1.13 (µmol/L)	–
Total bilirubin	4.41 ± 0.88 (µmol/L)	2.90 (µmol/L) 0.0–7.0
Total protein	64.89 ± 3.78 (g/L)	64 (g/L) 47– 79
Albumin	36.43 ± 1.70 (g/L)	4.37% ± 0.44 (Stevensen et al., 1959)
Alpha 1 globulin	2.70 ± 0.66 (g/L)	All globulin 3.19% ± 0.73
Alpha 2 globulin	10.32 ± 1.45 (g/L)	Albumin/globulin ratio 1.46 ± 0.36
Beta globulin	7.85 ± 1.16 (g/L)	(Stevensen et al., 1959)
Gamma globulin	7.38 ± 1.81 (g/L)	
Cholesterol	2.71 ± 0.65 (mmol/L)	2.03 (mmol/L) 0.96–3.29
Triglycerides	0.99 ± 0.49 (mmol/L)	0.40 (mmol/L) 0.13–1.22
Non-esterified fatty acids (NEFA)	1006.89 ± 164.67 (mmol/L)	
Sodium	133.57 ± 3.37 (mmol/L)	132 (mmol/L) 124–143
Potassium	4.44 ± 0.58 (mmol/L)	4.5 (mmol/L) 3.30–6.50
Sodium:potassium ratio	30.56 ± 3.74 (mmol/L)	30.40 19.10–41.90
Chloride	84.94 ± 2.65 (mmol/L)	89 (mmol/L) 82–98
Urea	5.99 ± 2.02 (mmol/L)	7.4 (mmol/L) 3.20–12.90
Creatinine	88.69 ± 21.62 (mmol/L)	109 (mmol/L) 53–177
Total calcium	2.54 ± 0.19 (mmol/L)	2.5 (mmol/L) 2.20–2.80
Phosphorus	2.88 ± 1.65 (mmol/L)	2.02 (mmol/L) 1.02–3.48
Magnesium	1.51 ± 0.15 (mmol/L)	0.96 (mmol/L) 0.32–1.23

Source: Reprinted from Girling, S. J., Campbell-Palmer, R., Pizzi, R., Fraser, M. A., Cracknell, J., Arnemo, J. and Rosell, F. 2015. Haematology and serum biochemistry parameters and variations in the Eurasian beaver (*Castor fiber*). *PLoS ONE*, 10, e0128775 under CC BY 4.0 License.

Figure 3.15 Late-pregnancy and lactating females have four visible nipples, facilitating visual sexing of animals only during such time. (Photo supplied courtesy of Einar Hugnes.)

are no distinct differences in the distribution and immunohistochemical characteristics of nerve fibres of the mammary gland between juvenile and adult females (Franke-Radowiecka et al., 2016).

Beaver ovary size and shape vary greatly amongst individuals, even depending on breeding condition and body weight, though ovaries are smaller in immature animals (Provost, 1962). The surface of the ovaries appears smooth, unless the animal is in oestrous when it can appear bumpy in texture (Provost, 1962). Beaver follicles are estimated at being around 5–6.5 mm in diameter at time of ovulation, which happens within a short time frame (Provost, 1962).

The male genital system is well formed, even from newborns (Doboszyńska and Żurowski, 1981). Male beavers possess a baculum or os penis; palpitation for this structure in sexing is more reliable than feeling for testes, but can be difficult and inconsistent especially in younger individuals (see Section 10.4.3 on sexing). They have relatively small testes (see Section 6.5 on mating strategies), around half the predicted size for similar mammals with similar body mass (Kenagy and Trombulak, 1986), which are retained within the abdominal cavity, so they do not possess a scrotum. In adults these are around 2.7–3.2 cm long and 1.5–1.8 cm wide (Doboszyńska and Żurowski, 1981). The testis and epididymis are covered by *lamina visceralis tunica vaginalis* and during the mating season the testis can sometimes be palpated, but otherwise all these structures project into the diverticulum of the abdominal cavity (Doboszyńska and Żurowski, 1981). Overall, their testes represent around 0.05% of their body weight (Osborn, 1953).

This relatively small testis size in rodents tends to be accompanied by highly derived sperm, usually lacking the hook-like sperm head typically found throughout the Rodent superfamily Muroidea (Bierla et al., 2007). Sperm morphology in beavers represents a paddle-shaped head, though this can be variable, with a small, wedge-shaped nucleus, i.e. broader at the base and more pointed at the head (Bierla et al., 2007; Figure 3.16). Another unusual characteristic is that their sperm heads tend to have a large vacuole at the anterior region of

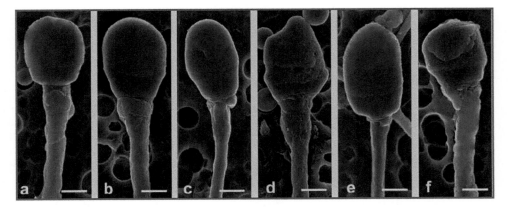

Figure 3.16 Scanning electron microscope (SEM) image of spermatozoa of the Eurasian beaver obtained from the cauda epididymidis and ampulla. Note variability in overall sperm shape, although in most it is paddle-shaped. Bar line: a–f = 1.1 μm. (Bierla, J. B., Gizejewski, Z., Leigh, C. M., Ekwall, H., S.derquist, L., Rodriguez-Martinez, H., Zalewski, K., and Breed, W. G., 2007. Sperm morphology of the Eurasian beaver, Castor fiber: an example of a species of rodent with highly derived and pleiomorphic sperm populations. Journal of Morphology, 268, 683–689. © 2007 Wiley-Liss, Inc.)

the nucleus, for reasons not yet determined (Bierla et al., 2007). The acrosome is very large which becomes highly folded as sperm mature in the testis, whilst the tail is short (~50 μm) compared to other rodents (Bierla et al., 2007). Living in family groups, this social monogamy accompanied with low levels of intermale sperm competition are hought to account for small testis size and sperm morphology in both beaver species (Bierla et al., 2007).

Males of both beaver species have a bicornuate vestigial uterus, historically referred to as the '*uterus masculinus*', first noted in Eurasian beavers (Brandt, 1829; Hinze, 1950; Figure 3.17). More accurately these are remnants of the paramesonephric (Müllerian) ducts, which in female embryos develop into the uterus and in males generally degenerate (Meier et al., 1998). Typically found running parallel to the *ducti deferentes*, their histology is comparable to that of the female uterus (Conaway, 1958). Unlike the Eurasian beaver, gross anatomy studies of North American males determined that these structures were highly variable in shape and size between individuals and even absent in some (Conaway, 1958; Meier et al., 1998). Most typically these structures comprised two horns, which then fuse into a single tube-like structure, cystic in nature, and although lying alongside, it does open into the urethra (Conaway, 1958). Of 24 North American male beavers studied in Ontario, 83% possessed distinct uterus-like structures (Meier et al., 1998). Variations included only one horn or no horns

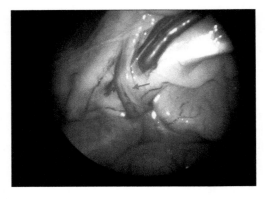

Figure 3.17 The *uterus masclius* (red arrow) seems to be present in the majority of male Eurasian beavers, but presence is more variable in North American males. The ductus deference is mark by green arrow. (Photo supplied courtesy of Romain Pizzi.)

being present, or an absent body, with all structures displaying variable lengths (Conaway, 1958; Meier et al., 1998). Paramesonephric duct remains are not found in all male North American beavers (Meier et al., 1998), as previously reported (Conaway, 1958). The reason for this difference between the species remains unknown. One theory proposed is that their chromosomal difference may influence the hormonal feedback loop by which degeneration of the Müllerian ducts typically occurs (Meier et al., 1998).

Box 3.1 describes the physiology of pregnancy and reproduction in the Eurasian beaver.

Box 3.1 Physiology of pregnancy and reproduction in the Eurasian beaver

The physiology of beaver reproduction and pregnancy has been investigated recently. Sex- and season-dependent changes in plasma concentrations of gonadotropins, gluco-corticoids, and sex steroids have been demonstrated (Chojnowska et al., 2015; Czerwińska et al., 2015; Chojnowska et al., 2019). Plasma follicle-stimulating hormone levels are higher at the end of lactation and during sexual inactivity compared to during pregnancy, while testosterone and progesterone peak during pregnancy (Chojnowska et al., 2015). Cortisol secretions are highest in males during the post-breeding (offspring rearing) season, with corticosterone and adrenocorticotropic hormones generally constant in both sexes regardless of season (Czerwińska et al., 2015). The expression of leptin mRNA in the hypothalamic–pituitary–gonadal axis (endocrine glands associated with development, reproduction, and ageing in animals) and in the myometrium (the middle layer of the uterine wall) are markedly higher in July than in April for females (Chojnowska et al., 2017). The expression of genes and protein of leptin receptor in the structure of the hypothalamic–pituitary–adrenal and hypothalamic–pituitary–gonadal axes as well as

in the uterus of the beaver has been demonstrated (Chojnowska et al., 2019).

Numerous studies by Lipka et al. (Lipka et al., 2017a; Lipka et al., 2017b; Lipka et al., 2018; Lipka et al., 2019) aimed to characterize crucial genes associated with pregnancy outcome and placental development have attempted to identify the mechanisms underlying effective reproduction in the beaver. They found that the number of foetuses present affects the expression profile in the beaver subplacental transcriptome (the set of all RNA molecules in one cell or a population of cells). They also determined that placental aspartic proteinase levels influence efficient implantation, placenta development, and pregnancy maintenance. Their results provided a broad-based characterization of the global expression pattern of the beaver placental transcriptome, which in turn should improve understanding of the crucial pathways relevant to proper placenta development and successful reproduction. The expression of orexin genes and proteins varies seasonally and between sexes and this suggests that the receptors are associated with the beaver's reproductive activity and adaptations to changing environmental conditions (Czerwińska et al., 2017).

3.2.9 Castor sacs

Beavers of both sexes produce a unique viscous fluid in two castor sacs, known as castoreum (see Sections 2.5 for human use and 7.2.2.1 for chemical composition and use in scent marking). The castor sacs are 'a pair of pouch-like structures' located between the kidneys and bladder (Müller-Schwarze, 2011: Figure 3.18), being more similar to bags of skin surrounded by a covering of cross-striped muscles, which allow conscious contractions leading to expulsion of the liquid substance from within the sac. They are not true glands in the anatomic sense since no secretory cells are found in these sacs. Their anatomy and placement in relation to the urethra allows urine to pass into the sacs and go through a chemical process (Figure 3.19). A coating of concentrated urine inside the folded inside skin will cause a gradual formation of castoreum, a substance that is yellowish-grey to slightly brownish. It becomes darker brown when air-dried. Castoreum should be considered as concentrated urine, which enhances the smell of ordinary urine

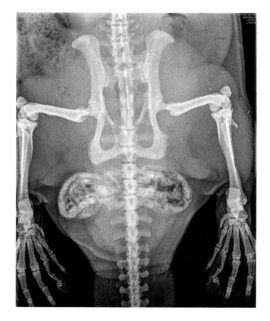

Figure 3.18 X-ray clearly showing castor sacs, with anal glands appearing associated lighter masses. Note also the presence of the os-penis bone as a floating small bone. (Photo supplied courtesy of Kathryn Perrin.)

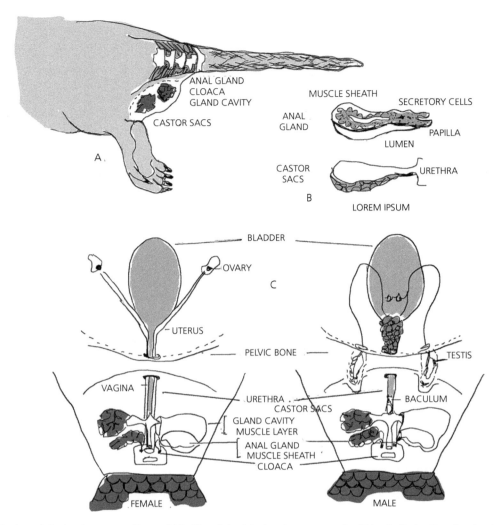

Figure 3.19 Anal gland and castor sac of beaver. (a) Position of glands in the subcutaneous cavity; (b) longitudinal section of anal and castor sacs; (c) anatomical relationships of anal and castor sacs to other structures in the beaver. (Redrawn by Rachael Campbell-Palmer from Gerald E. Svendsen, Castor and Anal Glands of the Beaver (*Castor canadensis*), *Journal of Mammalogy* 1978; 59 (3): 618–620, doi:10.2307/1380239. Reprinted by permission of Oxford University Press on behalf of American Society of Mammalogists.)

when used for territorial scent marking (Svendsen, 1978; Walro and Svendsen, 1982; Valeur, 1988; see Section 7.2.2.3).

In kits, the castor sacs are empty and not fully developed; they mainly contain only urine at this stage, with a very sparse coating of concentrated urine (Walro and Svendsen, 1982). During the second year, the castor sacs grow and become firmer as they are gradually filled up with concentrated urine that is layered onto the folded mucosa. In their third year, a significant increase in growth occurs in the sacs, which allows a beaver to scent mark more when dispersing and ultimately settling down in its own territory. The relative weight of the sacs increases from 0% of their body weight at birth to about 0.4% in their first autumn, winter, and second summer. During their third year, when the animal is 2 years old, they develop quickly and comprise about 1% of their body weight (Wilsson, 1971). This most likely reflects their increasing use in scent marking and territorial behaviour with age. When beavers are fully mature, the castor sacs will have reached their maximum volume capacity. Typically, in adult animals castor sacs weigh between 150 and 200 g in spring when their marking activity is at it is highest, measured at ~60 g in North American beavers at other times, around 0.3% of their body

weight, with each sac ~8–11 cm long and 5 cm wide (Wilsson, 1971; Müller-Schwarze, 2011; Rosell and Schulte, 2004).

Sexual dimorphism in castor sacs is variably reported, with any differences thought to arise due to intersexual differences in territorial scent marking (see Section 7.2.2.3). Eurasian males scent mark more and therefore have smaller castor sacs, due to higher flushing rates. Young females (< 4 years old) had average castor sac weights of 95 g compared to 79 g for males (Rosell and Schulte, 2004; Figure 3.20

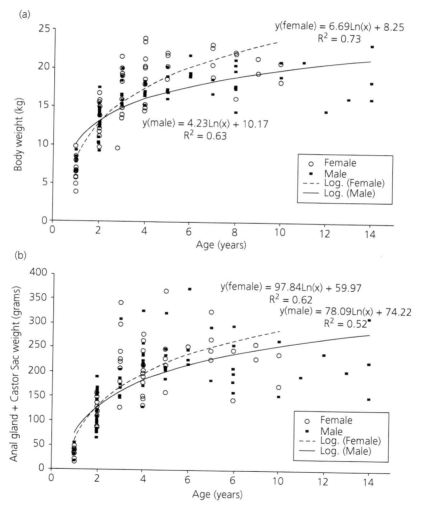

Figure 3.20 a) Body weight (kg) of male and female Eurasian beavers plotted against age in years as determined from tooth analysis, b) Scent structure weight (g) for anal glands plus castor sacs plotted against age. The best-fit log line is shown for each sex separately. a) Anal gland weight and b) castor sac weight ($\log_{10}$ g of weight) of male and female Eurasian beavers plotted against body weight ($\log_{10}$ kg of weight).

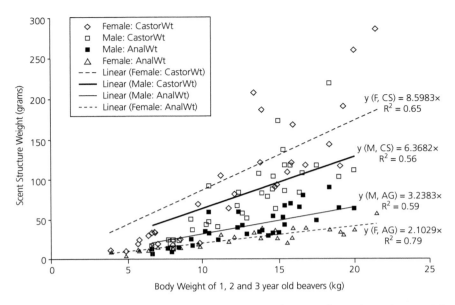

Figure 3.21 Anal gland and castor sac weights (g) plotted against body weight (kg) for Eurasian beavers in age class years 1, 2, and 3 as determined from tooth analysis. The best-fit line is shown for males (n = 33) and females (n = 29) for each scent-structure type. All fits were significant (P < 0.001). Intercepts were set to 0. (Rosell, F. and Schulte, B. A. 2004. Sexual dimorphism in the development of scent structures for the obligate monogamous Eurasian beaver (*Castor fiber*), Journal of Mammalogy, 85, 1138-1144 by permission of Oxford University Press).

and Figure 3.21). Studies of North American beavers have determined that females have slightly smaller castor sacs, by around 9 g on average (Bollinger et al., 1983), although previous studies found no sex differences (Svendsen, 1978).

3.2.10 Anal glands

The anal glands are true glands, found below the castor sacs, surrounded by a pale pink membrane, and have two openings on either side of the anal opening. They are thick-walled bladders with groups of glandular tissue attached to the epithelial walls of the bladders (Svendsen, 1978; Walro and Svendsen; 1982, Valeur, 1988). The glands have a restricted pathway that ends in one papilla on each and normally are turned inward towards the wall of the cloaca but are exposed during scent marking. The papillae (small, pale pink-red tips) are covered with a few brush-like hairs. When expelling the secretion from the anal glands, the papillae are forced outwards and the secretion is emitted when the hairs are in contact with a solid object (see Chapter 7.2.2.2). This has been observed in captive

(Wilsson, 1971) and in free-living beavers both inside (Müller-Schwarze and Herr, 2003) and outside their lodges (Rosell and Bergan, 1998). The anatomical structure of the anal glands is different between the two beaver species (Sokolov and Shchennikov, 1990)—the inner surface and wall thickness of the anal gland and location and shape of the inner cavity, as shown in Figure 3.22.

The anal glands were well developed in prenatal kits and relatively large and turgid with sebum at birth (Walro and Svendsen, 1982). Again, differing from the castor sacs, they are relatively well formed in a beaver's first autumn. This may indicate that they may have an important function soon after birth, most likely in connection with recognition of individuals. It also could be that they just develop faster embryologically since they are different tissue compared with the castor sacs (P. Busher, pers. comm.). The two anal glands of an adult animal normally weigh between 60 and 90 g in the spring when marking activity is at its highest level and at this point they are 6–7 cm long and 4 cm wide (Rosell and Schulte, 2004; Figure 3.20). Beavers also exhibit sexual dimorphism in the development of

Eurasian beaver

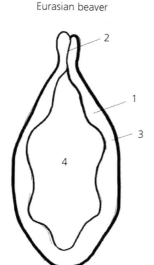

North American beaver

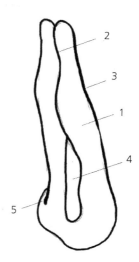

Figure 3.22 Structure of the anal glands of adult beavers showing (1) glandular tissue, (2) secretory duct, (3) muscular membrane, and (4) inner cavity. In Eurasian beavers, the inner cavity is broad and occupies most of the cell. The inner surface of the gland is covered by tubercles and the wall is only 1–3 mm thick. In the North American beaver, however, the inner cavity is much narrower and displaced in the distal area; also the inner surface has evenly distributed sebaceous glands and the wall is much thicker at 4–10 mm. (5) A fold on the outer wall of the gland is only present in North American beavers. (Redrawn by Hanna Wakeford from Sokolov, V. Y. and Shchennikov, G. N. 1990. Identification of the species *Castor fiber* and *C. canadensis* (Rodentia, Castoridae) by the structure of the anal glands. *Zoologiceskij Zurnal*, 69, 158–159.)

the size of the anal glands and this difference is thought to have arisen because of intersexual differences in territorial scent marking (Rosell and Schulte, 2004). Male anal glands are larger than the female's, mainly as they scent mark more and their glands therefore produce more AGS. Young females (< 4 years old) had average anal gland weights of 26 g compared to 40 g for males (Rosell and Schulte, 2004). North American females' anal glands averaged 4 g smaller than in equivalent-aged males. Also, the weight and volume varied significantly among spring (March to June), summer (July to October), and winter (November to February) (Bollinger et al., 1983).

3.3 Sense organs

3.3.1 Eyes and vision underwater

Beavers have disproportionally small eyes relative to their body size, with their vision being near-sighted and generally thought to be more related to motion detection. This is further supported by them

having a narrow optic nerve, especially in relation to size of their eye (Pilleri, 1983). They tend not to rely on sight but do possess a relatively large field of vision, with an angle of 65 degrees between their optical and body axis, representative of being prey animals (Ballard et al., 1989). The beaver eye is not especially adapted for underwater vision (Müller-Schwarze, 2011). Their retina lacks any light-reflecting crystals (*tapetum lucidum*), present in practically all other nocturnal mammals, and therefore at low light levels beaver eyesight is particularly poor (Hut et al., 2012; Rodriguez-Ramos Fernandez and Dubielzig, 2013).

Ballard et al. (1989) investigated the aerial and aquatic visual acuity in Canadian otters (*Lutra canadensis*) and beavers; even though they share the same habitats, otters are presumed to have much superior vision in both environments, largely to facilitate their piscivorous diet. They found that the beaver iris was similar to other rodents, being slender, heavily pigmented, and with a rudimentary dilator with most of the muscle cells located near the pupillary ruff. Similarly to small-eyed nocturnal mammals, the beaver possesses small ciliary muscles composed of few slender muscle fibers. In more truly aquatic birds and mammals, sphincter and dilator muscles are highly developed to enable them to adjust the curvature of their lens to refract light as they change from terrestrial and aquatic environments (Ballard et al., 1989). Beaver eyes are set high on their head, more so than for other rodents and semi-aquatic mammals, and this is an adaptation for living in water but not to assist aquatic vision. The authors conclude that beavers do have limited vision in water but rely on other senses, with their eyes being mainly evolved for aerial vision. This was confirmed in later visual acuity studies determining that due to the shorter posterior nodal distance when the eye was underwater, the retina resolution was lower, and additionally the image is disturbed by the retina itself (hypermetropia) (Mass and Supin, 2017; Figure 3.23).

3.3.2 Ears and hearing

Beaver external ear flaps are surprisingly small, but their hearing is significantly more developed compared to their eyesight, which makes sense when you're mostly active in the dark. Beavers possess large (relative to mammals of comparative size) auditory canals (especially in relation to their skull size) that end in enlarged auditory cavities, with well-developed acoustic nerves (Pilleri, 1983). It is thought that they can also hear vibrations and sounds underneath the water relatively well (Rosell and Pedersen, 1999). Valvular flaps in the ear prevent water entering the auditory canals during swimming and diving. Beavers are reactive to unexpected noises and hearing plays a role in predator detection. Given their use of varying vocalizations (see Chapter 7.4.1), hearing seems to have some relation to social behaviours.

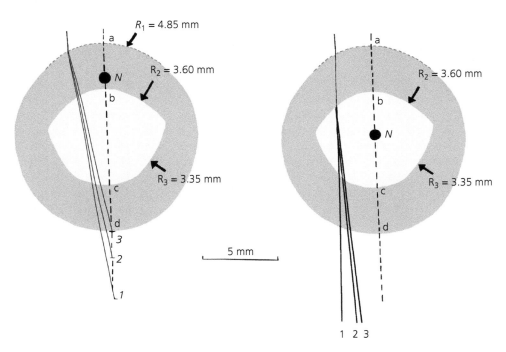

Figure 3.23 Schemes of light ray paths in the beaver eye in (left) aerial and (right) aquatic environments. Radiuses of curvature: R_1 external corneal surface; R_2 front lens surface; R_3 rear lens surface. Points a–d are axial points of corneal surface, front and rear lens surfaces, and eyecup bottom, respectively. 1, 2, 3 are light ray paths after light passing through corneal surface and front and back surfaces of lens, respectively. N = the nodal point which in air was in close proximity to the lens front surface, whilst in water was closer to the lens surface. (Reprinted by permission from Springer Nature Customer Service Centre GmbH: Springer, Physiology Estimates of Underwater and Aerial Visual Acuity in the European Beaver *Castor fiber* L. Based on Morphological Data, Mass, A.M. & Supin, A.Ya. © 2017.)

3.3.3 Nose and smell

Of all the senses, smell is the most critical and chemical communication is highly developed in beavers, playing a key role in social interactions (Sun and Müller-Schwarze, 1997; see Chapter 6) and territorial behaviours (see Section 7.2.2). This highly developed sense of smell enables them to sense predators (see Section 5.2.1) and to locate and select food items (see Section 5.2.2), with high-value items such as aspen being detected several hundreds of metres away (Doucet et al., 1994). Chemical signaling and scents are also factors often over-looked in beaver conservation, captive care, and mitigation techniques, though their use in trapping has an incredibly long history (Campbell-Palmer and Rosell, 2010; see Chapter 10).

The nose is the most immediately obvious feature, the majority being fur covered, though the bare nasal extremity between the nostrils is moist and hairless. The nostrils are lateral, and round when expanded, changing to a sub-triangular or crescentic form when contracted (Morgan, 1868). Beavers breathe through their noses, as opposed to their mouths. The nostrils enable air to enter the respiratory tract, though on contact with water these automatically constrict to prevent water entry (see Section 3.4.2). In mammals, the olfactory system primarily involves the olfactory bulb, a region of the fore brain which olfactory nerves target and where information processing occurs (Doty, 2001). This distinct area is readily recognizable and can be measured and compared (Pihlström, 2008). The olfactory bulb in beavers is not especially developed or different compared to other mammals. Their olfactory bulb to cerebrum length ratio is given as 0.35, which is similar to the Alpine marmot (*Marmota marmot*) and European red squirrel (*Sciurus vulgaris*, 0.34) (Pilleri, 1983). Another key component in the physiology of olfaction is the accessory olfactory system, or the vomeronasal system (sometimes also referred to as the Jacobson's organ); these are paired, tube- or pouch-like structures filled with fluid within the nasal cavity involving communication between the oral and nasal cavities (Pihlström, 2008). In mammals, this is surrounded by cartilage or bone, so that there is a separate airstream passing over the olfactory receptor cells with their own neural pathways (Su et al., 2009). Beavers posses a vomeronasal organ which is often linked with pheromone detection but is not always present in all mammal species (e.g. Estes 1972; Brennan and Keverne 2004; see section 7.2.2.3).

3.4 Significant physiological processes

3.4.1 Nutrient uptake

In terms of efficiency, beavers digest 30–33% of ingested cellulose and 44% of protein (Currier et al., 1960; Hoover and Clarke, 1972). The relatively long retention time of many plant species in the digestive tract translates into a need to eat large amounts of food and then less food intake once there is no room for additional food. For example, red maple is less digestible and has less energy content (Fryxell and Doucet, 1993), taking 30–50 hours (mean 37–42 hours, but it may take as long as 84 hours) to pass through a beaver's digestive tract compared to aspen which takes 10–20 hours (mean 14–16 hours) (Fryxell and Doucet, 1993; Fryxell et al., 1994). White willow bark has a retention time of 12 hours (Nolet et al., 1995) and alder tissue is retained for 37 hours (Fryxell et al., 1994). The faster the food passes through the beaver's system, the higher the rate of nutrient absorption. Therefore, beavers cannot meet their energy requirements if they eat red maple exclusively (Fryxell and Doucet, 1993) and a beaver may starve to death in a forest of alders even though it is eating as much as it can process each day (Fryxell, 1999). The digestive content tends to increase in the caecum and colon in winter months when poor-quality food dominates the beaver's diet. The digestive tract parameters of beavers vary based on the composition of available forage (Bełżecki et al., 2018).

The expanded caecum is the main site of cellulose digestion (Currier et al., 1960), where the enzyme cellulose severs the glycosidic bonds between glucose molecules that comprise cellulose (Vispo and Hume, 1995). The caecum contains microorganisms that are capable of decomposing cellulose (Gruninger et al., 2016), which is often the obstacle for herbivores when extracting nutrients from plant cells. These commensal microbes can assist in up to 32% of beaver cellulose digestion (Buech, 1984).

Cellulose is an important energy source and beavers convert woody plants into nutrition in an effective way (Wong et al., 2016; Wong et al., 2017; Armstrong et al., 2018). Microbial nitrogen fixation is also assumed to play a major role in nitrogen nutrition of beavers. Nitrogen-fixing bacteria are found in all parts of the large intestine. The dominant microorganisms are bacteria of the genera *Bacteroides*, *Alcaligenes*, and *Flavobacterium*. Nitrogen fixation by bacteria also serves as a source of biological nitrogen for the beavers (Vecherskii et al., 2006; Vecherskii et al., 2009). Bacteria and single-celled animals break down plant material into more easily digestible components, such as carbohydrates and fatty acids, which is accompanied by methane and carbon dioxide gas as by-products.

Because the fermentation chamber is behind the majority of the intestines' absorptive portion (the small intestine), some of the nutrients that the microorganisms produce, including vitamins and proteins, are lost in the faeces and not utilized by the animal. However, beavers commonly consume their own faeces from their caecum and, therefore, do not lose these nutrients (Wilsson, 1971; Hoover and Clarke, 1972; Vispo and Hume, 1995). Beavers ingest specially produced green faeces directly from their anus which, when reingested, maximizes nutrient uptake and allows digestion and absorption of bacterial proteins and vitamins. This type of faeces is very different in colour, consistency, and shape from other faecal matter. It is darker and softer and does not have the usual round form. In hares (*Lepus* spp.) this type of faeces is coated in a mucus capsule and swallowed whole. Those from beavers are uncoated and are chewed before being swallowed (Buech, 1984). Ingestion of these faeces occurs regularly as part of the beaver's daily activity. All beavers, except unweaned kits, display such coprophagy and it occurs on land and usually inside their lodges (Wilsson, 1971). It has been postulated that fixed atmospheric nitrogen is incorporated in the beaver body in the form of bacterial protein as a result of coprophagy (Vecherskii et al., 2009).

Beavers have very long small intestines which constitute the longest part of their digestive tract, resulting in a high capacity for absorption (Vispo and Hume, 1995). During the second passage of materials through the small intestine, recovered nutrients are absorbed. Fifty-five percent of smaller particles pass after 10–14 hours and 88% pass after 40 hours. By comparison, only 6% of eaten bark is digested after 11 hours (first defecation after a meal) but 88% is digested by the second defecation on the second day. Passage time for 100% of both original food and reingested material is about 60 hours (Buech, 1984).

3.4.2 Swimming and the diving reflex

As semi-aquatic mammals, beaver anatomy and morphology must accommodate quadruped locomotion on land along with swimming and diving. The tail tends to be functional as a rudder, though it can be used in fast, submerged swimming as an additional downward thrust especially if the beaver feels threatened. Beaver hind feet are fully webbed to the distal end of each of the five toes. They only engage the hind legs for swimming and the large webbed feet provide a large surface area for propulsion. Swimming pattern locomotion differs depending on if an animal is swimming below or at the water's surface. During submerged or subsurface swimming beavers engage both their hind webbed feet simultaneously, while at the surface they engage them alternately (Allers and Culik, 1997). When diving or when faster swimming is required beavers will engage a fully submerged swimming pattern. Most swimming locomotion is at the surface; this is typically slower but associated with increased manoeuvrability. Fish (1993) describes this surface paddling stroke as a modification of the terrestrial gait, with the power-generating phase of the stroke along with the posterior motion of the webbed foot generating a drag force that is translated into forward thrust. The foot is then repositioned during the recovery phase and the toes contracted to reduce the webbed area so as to reduce drag (Fish, 1993).

Though not commonly undertaken, beavers can dive for up to 15 minutes in forced dives without becoming asphyxiated (Irving and Orr, 1935). The diving reflex is an automatic, vasoconstriction response which is triggered when nerve endings around the nose and mouth area come into contact with water (Müller-Schwarze, 2011). This causes a redistribution of blood flow to the most vital organs such as the brain, heart, lungs, and adrenal glands and reductions to others like muscle tissue that are less likely to be damaged by an oxygen shortage and can tolerate higher carbon dioxide (CO_2) concentrations (Irving, 1937; McKean, 1982). As diving mammals, beavers exhibit a range of interacting adaptations to cope with aerobic metabolism during diving, including oxygen storage capacity, reducing demands for oxygen, reducing blood flow, and behavioural changes (Snyder, 1983). Oxygen storage in diving mammals typically includes variations in lung, blood, and tissue capacity, along with adaptations to tolerate and recover from periods of shortage and higher CO_2 concentrations (Snyder, 1983). Variations in lung oxygen storage occur in diving mammals, though in beavers these are mainly comparable with terrestrial mammals (McKean and Carlton, 1977), unlike sea otters (*Enhydra lutris*) which possess significantly greater total lung volume capacity (Lenfant et al., 1970; Snyder, 1983). Even though beavers do not have greater lung volumes than terrestrial mammals, they dive following inspiration and therefore their lungs do provide some of the oxygen storage needed (Clausen and Ersland, 1968).

Beaver blood oxygen storage capacity is not especially impressive—on average 8.3 ml/O_2/kg (Kitts et al., 1958; Clausen and Ersland, 1968)—being significantly less than for truly aquatic mammals (McKean and Carlton, 1977). Their mean blood volume constitutes 6.5 ± 0.8% of their body mass, with haemoglobin 12.4 ± 1.5 g/100 ml and myoglobin 1.2 ± 0.3 g/100 mg (McKean and Carlton, 1977). Muskrats, with incredibly similar diving behaviours and habitat selection, have an average blood oxygen storage capacity of 20.8 ml/O_2/kg, more than double that of beavers (Irving, 1939; Aleksiuk and Frohlinger, 1971). However, beavers can tolerate high concentrations of CO_2 in their tissues (Irving, 1937). In beavers, haemoglobin affinity for oxygen is generally high (McKean and Carlton, 1977). In their histological studies of beaver hearts, Dolka et al. (2014) determined a larger amount of binucleated cardiomyocytes, which, although to be further studied, could be indicative of higher metabolic activity, increased RNA production, and protein synthesis, which

could be an aquatic adaptation, though it has also been argued that as burrowing rodents, having high oxygen affinities compared to other terrestrial mammals may actually have been a precursor later utilized in their diving capabilities (Snyder, 1983).

Beaver muscles (60% body mass; McKean and Carlton, 1977) exhibit greater oxygen storage capacity than other rodents and almost twice that of terrestrial mammals, but this is still significantly lower than comparable marine divers (Lenfant et al., 1970; Snyder, 1983). During diving, arterial oxygen content fell from 16% to 3% volume after 4 minutes, with venous oxygen dropping from 8 to 2% (McKean, 1982), while cardiac output fell from 0.12 ± 0.01 to 0.03 ± 0.01 ml/min/kg (McKean, 1982). These adaptations optimize reduced oxygen use, conserve oxygen stores for the respiration of most-dependent organs, and enable beavers to remain underwater for longer (Irving and Orr, 1935; Feldhamer et al., 2007).

Bradycardia will also occur, reducing heart rate by around half, with typical resting heart rates on land of 116 ± 38 beats/minute and submerged rates of 50 beats/minute (Gilbert and Gofton, 1982), while a swimming beaver would display heart rate drops from 125 ± 41 to 67 beats/minute when it dived (Gilbert and Gofton, 1982). In forced submersion experiments (beavers restrained to a tipping board that could submerge animals underwater), heart rate changes occurred after a few beats to up until 10 seconds before bradycardia was initiated (McKean, 1982). After a dive, beaver heart rates typically remained around 75–90 beats/minutes for several minutes to 'repay the oxygen debt' incurred (Irving, 1937; Gilbert and Gofton, 1982).

Beaver livers demonstrate high catalase activity, whilst their hearts have high lactic dehydrogenase and various antioxidant enzyme activities, meaning these tissues have high aerobic capacity (Sergina et al., 2015). Catalase activity is higher in beaver heart, liver, and skeletal muscles compared to muskrats (Sergina et al., 2015), which have shorter diving time capacity of 9–12 minutes (Scholander, 1940). Overall, beaver heart, lungs, and spleen have a high ability for aerobic metabolism (Sergina et al., 2015).

3.4.3 Thermoregulation

Beavers do not hibernate and therefore must maintain constant body temperatures all year round, regardless of environmental and climatic conditions (Figure 3.24). To maintain homeostasis requires an ability to lose excess heat and also prevent heat loss as required. On land, adult beaver body temperature generally lies between 37.1 and 37.8°C, but this varies with activity levels and season (MacArthur, 1989; Reynolds, 1994). During the summer, they tend to vary by 1.4°C, while this can be by up to 2.5°C in winter (Smith et al., 1991). However, Dyck and MacArthur (1992) contest this finding; in their study, mean abdominal temperature remained around 37°C throughout the year.

Spending extended periods in water, swimming and diving, especially in northern hemisphere freshwater bodies which are not typically the warmest, makes thermoregulation vital to enable occupancy of these habitats. Many warm-blooded animals living in cold-water environments have evolved adaptations to retain body heat. Beavers have incredibly thick fur; even so this is estimated to account for only 24% of beavers' total insulation capacity, whilst body fat accounts for the rest (Scholander et al., 1950). The lowest critical air temperature, which causes a resting beaver to shiver to stay warm (myogenic heat production), is around 0–2°C for the North American beaver (MacArthur and Krause, 1989). In water, beavers display abdominal cooling at water temperatures between 2 and 20°C (MacArthur and Dyck, 1990). Kits lose heat faster than adults, as demonstrated by MacArthur and Dyck (1990), with adults immersed for 40 minutes dropping body temperature by a maximum of 2°C, whilst kits dropped up to 7°C after 20 minutes. Kits will compensate for this through behavioural thermoregulation, i.e. spending less time in the water (kits 4–7 months old: 5–10 minutes; adults: 14–23 minutes; MacArthur and Dyck, 1990). Behaviours by other family members towards kits, such as grooming and transporting them to and from lodges, has also been postulated to assist in their thermoregulation (MacArthur and Dyck, 1990).

To increase the water-repellent properties of their fur beavers have been observed rubbing AGS into their fur when grooming to give it a slightly oily coating in order to help fur remain water-resistant and trap air in the dense underfur. Despite these observations, it is still possible to read in the literature that the beaver does not transfer its AGS to its fur to make it waterproof. Some authors, however, believe they do not need these compounds since their fur is so dense already. Typically, when beavers sit back on their tails, the anal glands will protrude slightly out of the cloaca, where they then rub their fore feet against them, before transferring AGS to other parts of the body. The water-resistant properties of fur are lost if beavers are prevented from using their fore legs to groom or if their anal glands are surgically removed even though the grooming movement patterns are still in place (Wilsson, 1971; Walro and Svendsen, 1982).

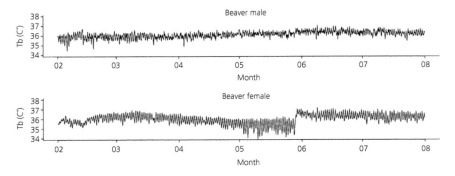

Figure 3.24 Mean body core temperature for an adult male (top graph) and pregnant female depicting a clear change during parturition (May 29th) ranging from February to July 2016. Reprinted from Mayer, M., Robstad, C., Pi Serra, E., Hohwieler, K. et al., Exposure of the Eurasian beaver (*Castor fiber*) towards hunters during the spring hunt. Reprinted with permission from USN, 2017.

Beavers possess two types of heat exchangers—the *venae comitantes* and the *rete mirabile*. In the superficial chest muscles, beavers exhibit a *venae comitantes*, which is of course insulated by outer fur and so was originally thought not to significantly function in thermoregulation (Cutright and McKean, 1979). However, thermal images of beavers at various ambient temperatures found that at higher ambient temperatures more heat loss occurs via the back rather than beaver tails as this is where fat coverage is thinnest (Zahner and Müller, 2006). Beaver tails are vital to assist in thermoregulation, serving in heat exchange through a counter-current arrangement of blood vessels, the *rete mirabile* (Grinnell et al., 1937; Carlson and Welker, 1976; Cutright and McKean, 1979; Thomsen et al., 2007). This is a complex network of blood vessels starting at the base of the tail that have a counter-current arrangement of arteries, with warm blood from the body core transferring heat to a closely laying arrangement of veins returning colder blood from the extremities back towards the heart (Cutright and McKean, 1979). The other significant, uninsulated structures regularly exposed to water are the large hind feet, which function in a similar manner as the tail as these counter-current heat exchangers are also present in the hind legs.

It is estimated that beavers can lose up to 20–25% of their generated body heat via the tail by diverting more blood here when hot (Steen and Steen, 1965), though this can be reduced to around 2% in the winter when beavers need to conserve heat (Marchand, 1996). Beavers do not pant to lose heat when environmental temperatures are increased. Most of the tail is hairless, with a thin layer of skin between scales, not only increasing flexibility but also facilitating heat loss. A lack of arrector pili muscles has been recorded (Dolka et al., 2014), which suggests the thick layer of subcutaneous adipose acts to prevent heat loss (Khamas et al., 2012). Beavers given access to dip their tails into water did not experience hyperthermia when temperatures were artificially raised to 25°C (Steen and Steen, 1965). Several studies have reported the presence of arteriovenous anastomoses in tail dermis, which have a function in temperature regulation (Steen and Steen, 1965; Müller-Schwarze, 2011; Dolka et al., 2014). Melanin-containing cells have

been found in the tails of both Eurasian (Dolka et al., 2014) and North American (Jenkins and Busher, 1979) beavers. Tail pigmentation is thought to be associated with UV absorption rates. Tail colour changes with age in beavers, being darker in young animals and getting lighter with age (Jenkins and Busher, 1979).

3.4.4 Seasonal changes and surviving winter

Beavers possess a range of specializations to survive winter and prolonged periods with limited growth seasons, especially at northern latitudes. These can be quite significant seasonal changes in behaviours and physiology, including food caching (Slough, 1978), lodge construction and group nesting (Buech et al., 1989; Stephenson, 1969), reduced activity (Lancia et al., 1982), winter foraging (Novakowski, 1967), seasonal thickening of fur (Scholander et al., 1950), reducing core body temperature (Smith et al., 1991), fat deposition, and varying thyroid activity (Aleksiuk and Cowan, 1969a; Aleksiuk, 1970). Beavers do not hibernate, a common response to prolonged periods of significantly reduced energy availability at high latitudes, though they can demonstrate metabolic depression during winter (Aleksiuk and Cowan, 1969b). These authors examined seasonal changes in splenic, perirenal, and subcutaneous fat deposits in immature beavers. A common pattern is of lean body fat during the summer months, followed by increasing levels of fat deposition during autumn, high and stable levels through winter, which then steadily fall as fat is readily mobilized coming into the spring. This definite seasonal fat content pattern is representative of hibernating mammals. Thyroid activity is age-related in beavers, being highest in kits and decreasing steadily with age, which is a common pattern for other mammals (Aleksiuk and Cowan, 1969b). Within this there is also a seasonal pattern of thyroid activity, rising steadily from late spring, peaking in late summer, before falling in autumn, and remaining low over winter (Aleksiuk and Cowan, 1969b). These fat deposit and thyroid activity patterns show adaptations to vegetation growing seasons, especially non-woody vegetation, with summer being the optimal time for aquatic, emergent and semi-emergent plant growth, and therefore an absence of body fat indicates energy requirements are meet through feeding. Seasonal changes in beaver weights and tail dimensions have been investigated and demonstrate a constant pattern of gains in summer (May–September) and losses during winter (November–April) (Smith and Jenkins, 1997; Figure 3.25). This isn't completely clear-cut. Although differences between families (e.g. local resources and family composition) and years (e.g. winter severity) were demonstrated, individuals from yearlings and older animals repeatedly show this seasonal change. Kits from that year, however, increase in weight and tail size during summer and winter periods, suggesting they are still utilizing resources (Smith et al., 1991), whereas weight loss of older family members during winter is evidence of energy deficits and potentially even periods of non-eating (Novakowski, 1967).

3.4.5 Living underground

Millions of years of evolution have given the beaver the ability to adapt to the extreme conditions of an underground existence and to live on minimal available oxygen levels, as displayed by an experiment by Owesen (1979). He placed a large wooden crate and tunnel leading to a body of water, to simulate a beaver lodge. Measuring several parameters, he was able to investigate the sensitivity of beavers to low oxygen levels with correspondingly high levels of CO_2. The oxygen content of standard air is around 21%; even when lodge oxygen levels fell to an incredibly low 4–5% beavers did not respond like typical mammals (e.g. displaying increased breathing or any signs of hyperventilating). The atmosphere generally has a concentration of 0.03% CO_2. In the cramped chambers where several beavers are breathing simultaneously one must assume that the concentration of CO_2 would be much higher. In North America, CO_2 levels ranging from

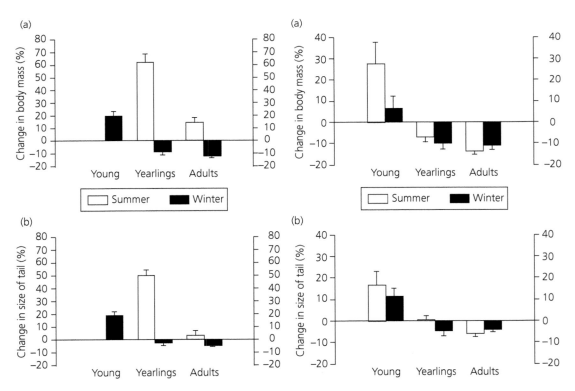

Figure 3.25 Seasonal weight changes. Reprinted from Smith, D. W. and Jenkins, S. H. Seasonal change in body mass and size of tail of northern beavers, *Journal of Mammalogy*, 1997, 78, 869–876 by permission of Oxford University Press.

1.5 to 1.8% have been measured in some active lodges, but the concentration varied from 0.03 to 1.8% (Dyck and MacArthur, 1993). To simulate conditions within a lodge where air quality becomes gradually worse, concentration of CO_2 was slowly increased over a 5-hour period (Owesen, 1979). The beavers did not leave the lodge until the concentration reached 16.5%! This concentration should be dangerous for any animal and the beaver should not have survived, but the beavers were unaffected.

References

Al-Kahtani, M. A., Zuleta, C., Caviedes-Vidal, E., and Garland, J. T. 2004. Kidney mass and relative medullary thickness of rodents in relation to habitat, body size, and phylogeny. *Physiological and Biochemical Zoology*, 77, 346–365.

Aleksiuk, M. 1970. The function of the tail as a fat storage depot in the beaver (*Castor canadensis*). *Journal of Mammalogy*, 51, 145–148.

Aleksiuk, M. and Cowan, I. M. 1969a. Aspects of seasonal energy expenditure in the beaver (*Castor canadensis* Kuhl) at the northern limit of its distribution. *Canadian Journal of Zoology*, 47, 471–481.

Aleksiuk, M. and Cowan, I. M. 1969b. The winter metabolic depression in Arctic beavers (*Castor canadensis* Kuhl) with comparisons to California beavers. *Canadian Journal of Zoology*, 47, 965–979.

Aleksiuk, M. and Frohlinger, A. 1971. Seasonal metabolic organization in the muskrat (*Ondatra zibethica*). I. Changes in growth, thyroid activity, brown adipose tissue, and organ weights in nature. *Canadian Journal of Zoology*, 49, 1143–1154.

Allers, D. and Culik, B. M. 1997. Energy requirements of beavers (*Castor canadensis*) swimming underwater. *Physiological Zoology*, 70, 456–463.

Armstrong, Z., Mewis, K., Liu, F., Morgan-Lang, C., Scofield, M., Durno, E., Chen, H. M., Mehr, K., Withers, S. G., and Hallam, S. J. 2018. Metagenomics reveals functional synergy and novel polysaccharide utilization loci in the *Castor canadensis* fecal microbiome. *International Society for Microbial Ecology Journal*, 12, 2757–2769.

Baker, B. and Hill, E. 2003. Beaver (*Castor canadensis*). In: Feldhamer, G. A., Thompson, B. C., and Chapman, J. A. (eds.) *Wild Mammals of North America: Biology, Management and Conservation*. Baltimore: Johns Hopkins University Press.

Ballard, K., Sivak, J., and Howland, H. 1989. Intraocular muscles of the Canadian river otter and Canadian beaver and their optical function. *Canadian Journal of Zoology*, 67, 469–474.

Bazan, I. 1955. Badania nad zmiennoscia aparatu plciowego I grasicy u rzesorka rzeczka (*Neomys fodiens fodiens* Schreb.). *Annals University Mariae Curie-Sklodowska*, 9, 213–259.

Belovsky, G. E. 1984. Summer diet optimization by beaver. *American Midland Naturalist*, 111, 209–222.

Bełżecki, G., Miltko, R., Kowalik, B., Demiaszkiewicz, A. W., Lachowicz, J., Giżejewski, Z., Obidziński, A., and McEwan, N. R. 2018. Seasonal variations of the digestive tract of the Eurasian beaver *Castor fiber*. *Mammal Research*, 63, 21–31.

Beuchat, C. A. 1996. Structure and concentrating ability of the mammalian kidney: correlations with habitat. *American Journal of Physiology—Regulatory, Integrative and Comparative Physiology*, 271, R157–R179.

Bierla, J. B., Gizejewski, Z., Leigh, C. M., Ekwall, H., Söderquist, L., Rodriguez-Martinez, H., Zalewski, K., and Breed, W. G. 2007. Sperm morphology of the Eurasian beaver, *Castor fiber*: an example of a species of rodent with highly derived and pleiomorphic sperm populations. *Journal of Morphology*, 268, 683–689.

Bisaillon, A. 1981. Gross anatomy of the cardiac blood vessels in the North American beaver (*Castor canadensis*). *Anatomischer Anzeiger*, 150, 248.

Bisaillon, A. 1982. Anatomy of the heart in the North American beaver (*Castor canadensis*). *Anatomischer Anzeiger*, 151, 381–391.

Bollinger, K. S., Hodgdon, H. E., and Kennelly, J. J. 1983. Factors affecting weight and volume of castor and anal glands of beaver (*Castor canadensis*). *Acta Zoologica Fennica*, 174, 115–116.

Brandt, J. F. 1829. *Medizinsche Zoologie*. Berlin: Hirschwald.

Brennan, P. A. and Keverne, E. B. 2004. Something in the air? New insights into mammalian pheromones. *Current Biology*, 14, R81–R89.

Buech, R. R. 1984. Ontogeny and diurnal cycle of fecal reingestion in the North American beaver (*Castor canadensis*). *Journal of Mammalogy*, 65, 347–350.

Buech, R. R., Rugg, D. J., and Miller, N. L. 1989. Temperature in beaver lodges and bank dens in a near-boreal environment. *Canadian Journal of Zoology*, 67, 1061–1066.

Bugge, J. 1974. The cephalic arterial system in insectivores, primates, rodents and lagomorphs. *Acta Anatomica*, 62, 1–160.

Cabanac, A., Folkow, L. P., and Blix, A. S. 1997. Volume capacity and contraction control of the seal spleen. *Journal of Applied Physiology*, 82, 1989–1994.

Campbell-Palmer, R. and Rosell, F. 2010. Conservation of the Eurasian beaver *Castor fiber*: an olfactory perspective. *Mammal Review*, 40, 293–312.

Carlson, M. and Welker, W. 1976. Some morphological, physiological and behavioral specializations in North

American beavers (*Castor canadensis*). *Brain, Behavior and Evolution*, 13, 302–326.

Chojnowska, K., Czerwinska, J., Kaminski, T., Kaminska, B., Panasiewicz, G., Kurzynska, A., and Bogacka, I. 2015. Sex-and seasonally related changes in plasma gonadotropins and sex steroids concentration in the European beaver (*Castor fiber*). *European Journal of Wildlife Research*, 61, 807–811.

Chojnowska, K., Czerwinska, J., Kaminski, T., Kaminska, B., Kurzynska, A., and Bogacka, I. 2017. Leptin plasma concentrations, leptin gene expression, and protein localization in the hypothalamic–pituitary–gonadal and hypothalamic–pituitary–adrenal axes of the European beaver (*Castor fiber*). *Theriogenology*, 87, 266–275.

Chojnowska, K., Czerwinska, J., Kaminski, T., Kaminska, B., Kurzynska, A., and Bogacka, I. 2019. Leptin/leptin receptor system in the regulation of reproductive functions and stress response in the European beaver. *Current Zoology*, 65, 197–203.

Clausen, G. and Ersland, A. 1968. The respiratory properties of the blood of two diving rodents, the beaver and the water vole. *Respiration Physiology*, 5, 350–359.

Coles, R. W. 1970. Pharyngeal and lingual adaptations in the beaver. *Journal of Mammalogy*, 51, 424–425.

Conaway, C. H. 1958. The uterus masculinus of *Castor canadensis*. *Journal of Mammalogy*, 39, 97–108.

Cox, P. G. and Baverstock, H. 2016. Masticatory muscle anatomy and feeding efficiency of the American Beaver, *Castor canadensis* (Rodentia, Castoridae). *Journal of Mammalian Evolution*, 23, 191–200.

Currier, A., Kitts, W. D., and Cowan, I. M. 1960. Cellulose digestion in the beaver (*Castor canadensis*). *Canadian Journal of Zoology*, 38, 1109–1116.

Cutright, W. J. and McKean, T. 1979. Countercurrent blood vessel arrangement in beaver (*Castor canadensis*). *Journal of Morphology*, 161, 169–175.

Czerwińska, J., Chojnowska, K., Kamiński, T., Bogacka, I., Panasiewicz, G., Smolińska, N., and Kamińska, B. 2015. Plasma glucocorticoids and ACTH levels during different periods of activity in the European beaver (*Castor fiber* L.). *Folia Biologica*, 63, 229–234.

Czerwińska, J., Chojnowska, K., Kaminski, T., Bogacka, I., Smolinska, N., and Kaminska, B. 2017. Orexin receptor expression in the hypothalamic–pituitary–adrenal and hypothalamic–pituitary–gonadal axes of free-living European beavers (*Castor fiber* L.) in different periods of the reproductive cycle. *General and Comparative Endocrinology*, 240, 103–113.

Dillman, B. V. and Barnett, R. J. 1984. Skull size in beavers from California to Alaska. *The Murrelet*, 65, 78–79.

Doboszyńska, T. and Żurowski, W. 1981. Anatomical studies of the male genital organs of the European beaver. *Acta Theriologica*, 26, 331–340.

Dolka, I., Giżejewska, A., Giżejewski, Z., Kluciński, W., and Kołodziejska, J. 2014. Histological evaluation of selected organs of the Eurasian beavers (*Castor fiber*) inhabiting Poland. *Journal of Veterinary Medicine*, 44, 378–390.

Doty, R. L. 2001. Olfaction. *Annual Review of Psychology*, 52, 423–452.

Doucet, C. M., Walton, R. A., and Fryxell, J. M. 1994. Perceptual cues used by beavers foraging on woody plants. *Animal Behaviour*, 47, 1482–1484.

Dyck, A. P. and Macarthur, R. A. 1992. Seasonal patterns of body temperature and activity in free-ranging beaver (*Castor canadensis*). *Canadian Journal of Zoology*, 70, 1668–1672.

Dyck, A. P. and Macarthur, R. A. 1993. Seasonal variation in the microclimate and gas composition of beaver lodges in a boreal environment. *Journal of Mammalogy*, 74, 180–188.

Estes, R. D. 1972. The role of the vomeronasal organ in mammalian reproduction. *Mammalia*, 36, 315–341.

Fandén, A. 2005. Ageing the beaver (*Castor fiber* L.): a skeletal development and life history calendar based on epiphyseal fusion. *Archaeofauna*, 14, 199–213.

Feldhamer, G. A., Drickamer, L. C., Vessey, S. H., Merritt, J. F., and Krajewski, C. 2007. *Mammalogy: Adaptation, Diversity, Ecology*. Baltimore: John Hopkins University Press.

Fish, F. E. 1993. Influence of hydrodynamic-design and propulsive mode on mammalian swimming energetics. *Australian Journal of Zoology*, 42, 79–101.

Fish, F. E., Smelstoys, J., Baudinette, R. V., and Reynolds, P. S. 2002. Fur doesn't fly, it floats: buoyancy of pelage in semi-aquatic mammals. *Aquatic Mammals*, 28, 103–112.

Frąckowiak, H. and Śmiełowski, J. 1998. Cephalic arteries in the European beaver *Castor fiber*. *Acta Theriologica*, 43, 219–224.

Franke-Radowiecka, A., Giżejewski, Z., Klimczuk, M., Dudek, A., Zalecki, M., Jurczak, A., and Kaleczyc, J. 2016. Morphological and neuroanatomical study of the mammary gland in the immature and mature European beaver (*Castor fiber*). *Tissue and Cell*, 48, 552–557.

Fryxell, J. M. 1999. Functional responses to resource complexity: an experimental analysis of foraging by beavers. In: Collen, P. (ed.) *Herbivores: Between Plants and Predators*. Oxford: Blackwell Scientific.

Fryxell, J. M. and Doucet, C. M. 1993. Diet choice and the funcional response of beavers. *Ecology*, 74, 1297–1306.

Fryxell, J., Vamosi, S., Walton, R., and Doucet, C. 1994. Retention time and the functional response of beavers. *Oikos*, 71, 207–214.

Gilbert, F. F. and Gofton, N. 1982. Heart rate values for beaver, mink and muskrat. *Comparative Biochemistry and Physiology Part A: Physiology*, 73, 249–251.

Girling, S. J., Campbell-Palmer, R., Pizzi, R., Fraser, M. A., Cracknell, J., Arnemo, J., and Rosell, F. 2015. Haematology and serum biochemistry parameters and variations in the eurasian beaver (*Castor fiber*). *PLoS ONE*, 10, e0128775.

Grinnell, J., Dixon, J. S., and Linsdale, J. M. 1937. *Fur-Bearing Mammals of California, their Natural History, Systematic Status and Relations to Man*. Berkeley: University of California Press.

Gruninger, R. J., McAllister, T. A., and Forster, R. J. 2016. Bacterial and archaeal diversity in the gastrointestinal tract of the North American beaver (*Castor canadensis*). *PLoS ONE*, 11, e0156457.

Hamilton, W. J. J. and Whitaker, J. O. 1979. Mammals in the eastern United States. *Journal of Mammalogy*, 22, 250–263.

Hilfiker, E. L. 1991. *Beavers, Water, Wildlife and History*. Interlaken, NY: Windswept Press.

Hinze, G. 1950. Der Biber. Körperbau und Lebensweise Verbreitung und Geschichte. Berlin: Akademie.

Hollander, H., van Duinen, G. A., Branquart, E., de Hoop, L., de Hullu, P. C., Matthews, J., van der Velde, G., and Leuven, R. S. E. W. 2017. Risk assessment of the alien North American beaver (*Castor canadensis*). https://repository.ubn.ru.nl/bitstream/handle/2066/169023/169023.pdf.

Hoover, W. H. and Clarke, S. D. 1972. Fiber digestion in the beaver. *Journal of Nutrition*, 102, 9–15.

Hume, I. D. 1989. Nutrition of marsupial herbivores. *Proceedings of the Nutrition Society*, 48, 69–79.

Hut, R. A., Kronfeld-Schor, N., van der Vinne, V., and de la Iglesia, H. 2012. In search of a temporal niche: environmental factors. In: Kalsbeek, A., Merrow, M., Roenneberg, T., and Foster, R. G. (eds.) *Progress in Brain Research*. Amsterdam: Elsevier.

Irving, L. 1937. The respiration of beaver. *Journal of Cell Comparative Physiology*, 9, 437–451.

Irving, L. 1939. Respiration in diving mammals. *Physiological Reviews*, 19, 112–134.

Irving, L. and Orr, M. 1935. The diving habits of the beaver. *Science*, 82, 569.

ISIS 2002. Reference values for captive North American beaver, *Castor canadensis*, both sexes and all ages combined (CD-ROM). St Paul, MN: ISIS.

Jamison, R. L. 1987. Short and long loop nephrons. *Kidney International*, 31, 597–605.

Janiszewski, P., Kolasa, S., and Strychalski, J. 2017. The preferences of the European beaver *Castor fiber* for trees and shrubs in riparian zones. *Applied Ecology and Environmental Research*, 15, 313–327.

Jenkins, J. R. 2008. Rodent diagnostic testing. *Journal of Exotic Pet Medicine*, 17, 16–25.

Jenkins, S. H. and Busher, P. E. 1979. *Castor canadensis* Kuhl, 1820. *Mammalian Species*, 120, 1–8.

Kenagy, G. J. and Trombulak, S. C. 1986. Size and function of mammalian testes in relation to body size. *Journal of Mammalogy*, 67, 1–22.

Khamas, W. A., Smodlaka, H., Leach-Robinson, J., and Palmer, L. 2012. Skin histology and its role in heat dissipation in three pinniped species. *Acta Veterinaria Scandinavica*, 54, 46.

Kitchener, A. 2001. *Beavers*. Whittet Books Ltd, UK.

Kitchener, A. and Lynch, J. M. 2000. A morphometric comparison of the skulls of fossil British and extant European beavers, *Castor fiber*. *Scottish Natural Heritage Review*, 127.

Kitts, W. D., Bose, R. J., Wood, A. J., and Cowan, I. M. 1957. Preliminary observations on the digestive enzyme system of the beaver (*Castor canadensis*). *Canadian Journal of Zoology*, 35, 449–452.

Kitts, W. D., Robertson, M. C., Stephenson, B., and Cowan, I. M. 1958. The normal blood chemistry of the beaver (*Castor canadensis*): A. Packed-cell volume, sedimentation rate, hemoglobin, erythrocyte diameter, and blood cell counts. *Canadian Journal of Zoology*, 36, 279–283.

Lancia, R. A. and Hodgdon, H. E. 1984. Beavers. In: Macdonald, D. W. (ed.) *The Encyclopedia of Mammals*. New York: Facts on File.

Lancia, R. A., Dodge, W. E., and Larson, J. S. 1982. Winter activity patterns of two radio-marked beaver colonies. *Journal of Mammalogy*, 63, 598–606.

Langworthy, O. R. and Richter, C. P. 1933. The cerebral motor cortex of the porcupine. *Journal fur Psychologie und Neurologie*, 45, 138–142.

Langworthy, O. R. and Richter, C. P. 1938. A physiological study of cerebral motor cortex and decerebrate rigidity in the beaver. *Journal of Mammalogy*, 19, 70–77.

Lavrov, L. S. 1979. Species of beavers (*Castor*) of the Palaeartics. *Zoologicheskii Zhurnal*, 58, 88–96.

Lavrov, L. S. 1983. Evolutionary development of the genus *Castor* and taxonomy of the contemporary beavers of Eurasia. *Acta Zoologica Fennica*, 174, 87–90.

Lenfant, C., Johansen, K., and Torrance, J. D. 1970. Gas transport and oxygen storage capacity in some pinnipeds and the sea otter. *Respiration Physiology*, 9, 277–286.

Lipka, A., Majewska, M., Panasiewicz, G., Bieniek-Kobuszewska, M., and Szafranska, B. 2017a. Gene structure of the pregnancy-associated glycoprotein-like (PAG-L) in the Eurasian beaver (*Castor fiber* L.). *Functional & Integrative Genomics*, 17, 599–605.

Lipka, A., Paukszto, L., Majewska, M., Jastrzebski, J., Myszczynski, K., Panasiewicz, G., and Szafranska, B. 2017b. Identification of differentially expressed placental transcripts during multiple gestations in the Eurasian beaver (*Castor fiber* L.). *Reproduction, Fertility and Development*, 29, 2073–2084.

Lipka, A., Panasiewicz, G., Majewska, M., Paukszto, L., Bieniek-Kobuszewska, M., and Szafranska, B. 2018. Identification of placental aspartic proteinase in the Eurasian beaver (*Castor fiber* L.). *International Journal of Molecular Sciences*, 19, 1229.

Lipka, A., Paukszto, L., Majewska, M., Jastrzebski, J. P., Panasiewicz, G., and Szafranska, B. 2019. De novo characterization of placental transcriptome in the Eurasian beaver (*Castor fiber* L.). *Functional & Integrative Genomics*, 19, 1–15.

Long, K. 2000. *Beavers. A Wildlife Handbook*. Colorado: Jognson Books.

MacArthur, R. A. 1989. Energy metabolism and thermoregulation of beaver (*Castor canadensis*). *Canadian Journal of Zoology*, 67, 651–657.

MacArthur, R. A. and Dyck, A. P. 1990. Aquatic thermoregulation of captive and free-ranging beavers (*Castor canadensis*). *Canadian Journal of Zoology*, 68, 2409–2416.

MacArthur, R. A. and Krause, R. E. 1989. Energy requirements of freely diving muskrats (*Ondatra zibethicus*). *Canadian Journal of Zoology*, 67, 2194–2200.

Mahoney, J. M. and Rosenberg, H. I. 1981. Anatomy of the tail in the beaver (*Castor canadensis*). *Canadian Journal of Zoology*, 59, 390–399.

Marchand, P. J. 1996. *Life in the Cold*. Lebanon, NH: University Press of New England.

Martino, P. E., Aráuz, S. M., Anselmino, F., Cisterna, C. C., Silvestrini, M. P., Corva, S., and Hozbor, F. A. 2012. Hematology and serum biochemistry of free-ranging nutria (*Myocastor coypus*). *Journal of Zoo and Wildlife Medicine*, 43, 240–247.

Mass, A. and Supin, A. Y. 2017. Estimates of underwater and aerial visual acuity in the European beaver *Castor fiber* L. based on morphological data. *Doklady Biological Sciences*, 473, 35–38.

McKean, T. 1982. Cardiovascular adjustments to laboratory diving in beavers and nutria. *American Journal of Physiology—Regulatory, Integrative and Comparative Physiology*, 242, R434–R440.

McKean, T. and Carlton, C. 1977. Oxygen storage in beavers. *Journal of Applied Physiology*, 42, 545–547.

Meier, C., Partlow, G. D., Fisher, K. R., and Rennie, B. 1998. Persistent paramesonephric ducts (masculine uterus) in the male North American beaver (*Castor canadensis*). *Canadian Journal of Zoology*, 76, 1188–1193.

Miltko, R., Kowalik, B., Kędzierska, A., Demiaszkiewicz, A., McEwan, N., Obidziński, A., and Bełżecki, G. 2019. Effect of seasonal diet composition changeson the characteristics of the gastrointestinal tract contents of the Eurasian beaver (*Castor fiber*). *Journal of Animal and Feed Sciences*, 28, 392–397.

Morgan, L. H. 1868. *The American Beaver and His Works*. Philadelphia: J.B. Lippincott.

Müller-Schwarze, D. 2011. *The Beaver: Its Life and Impact*. New York: Cornell University Press.

Müller-Schwarze, D. and Herr, J. 2003. Are they 'listening in the dark'? Behaviour of beavers in their lodge. *Lutra*, 46, 255–257.

Nasset, E. S. 1953. Gastric secretion in the beaver (*Castor canadensis*). *Journal of Mammalogy*, 34, 204–209.

Nickel, E. S. and Seiferle, E. 1976. *Lehrbuch der Anatomie der Haustiere*, 3. Singhofen: Paul Parey, pp. 292–302.

Nolet, B. A. and Rosell, F. 1994. Territoriality and time budgets in beavers during sequential settlement. *Canadian Journal of Zoology*, 72, 1227–1237.

Nolet, B. A., van der Veer, P. J., Evers, E. G. J., and Ottenheim, M. M. 1995. A linear programming model of diet choice of free-living beavers. *Netherlands Journal of Zoology*, 45, 315–337.

Novak, M. 1987. Beaver. In: Novak, M., Baker, J. A., Obbard, M. E., and Mallock, B. (eds.) *Wild Furbearer Management and Conservation in North America*. Toronto: Queens Printer for Ontario.

Novakowski, N. S. 1967. The winter bioenergetics of a beaver population in northern latitudes. *Canadian Journal of Zoology*, 45, 1107–1118.

Nowicki, W., Brudnicki, W., Kirkiłło-Stacewicz, K., Skoczylas, B., and Wach, J. 2019. Craniometric features of the population of European beaver (*Castor fiber* L., 1758). *Slovenian Veterinary Research*, 56.

Osborn, D. J. 1953. Age classes, reproduction, and sex ratios of Wyoming beaver. *Journal of Mammalogy*, 34, 27–44.

Osborn, J. W. 1969. Dentine hardness and incisor wear in the beaver (*Castor fiber*). *Cells Tissues Organs*, 72, 123–132.

Owesen, A. 1979. *I Beverskog*. Oslo: Gylendal Norsk Forlag.

Pescador, N., Villar, D., Cifuentes, D., Garcia-Rocha, M., Ortiz-Barahona, A., Vazquez, S., Ordoñez, A., Cuevas, Y., Saez-Morales, D., and Garcia-Bermejo, M. L. 2010. Hypoxia promotes glycogen accumulation through hypoxia inducible factor (HIF)-mediated induction of glycogen synthase 1. *PLoS ONE*, 5, e9644.

Pihlström, H. 2008. Comparative anatomy and physiology of chemical senses in aquatic mammals. In: Thewissen, J. G. M. and Nummela, S. (eds.) *Sensory Evolution on the Threshold: Adaptations in Secondarily Aquatic Vertebrates*. Berkeley: University of California Press.

Pilleri, G. 1983. Central nervous system, cranio-cerebral topography and cerebral hierarchy of the Canadian beaver (*Castor canadensis*). In: Pilleri, G. (ed.) *Investigations on Beavers*. Berne: Brain Anatomy Institute.

Pilleri, G. 1984. *Investigations on Beavers*. Berne: Institute of Brain Anatomy, University of Berne.

Provost, E. E. 1962. Morphological characteristics of the beaver ovary. *Journal of Wildlife Management*, 26, 272–278.

Reynolds, P. 1993. Size, shape, and surface area of beaver, *Castor canadensis*, a semiaquatic mammal. *Canadian Journal of Zoology*, 71, 876–882.

Reynolds, P. S. 1994. Time-series analyses of beaver body temperatures. In: Lange, N., Ryan, L., Billard, L., Brillinger, D., Conquest, L., and Greenhouse, J. (eds.) *Case Studies in Biometry*. Toronto: John Wiley & Sons.

Rodriguez-Ramos Fernandez, J. and Dubielzig, R. R. 2013. Ocular comparative anatomy of the family Rodentia. *Veterinary Ophthalmology*, 16, 94–99.

Rosell, F. and Bergan, F. 1998. Free-ranging Eurasian beavers, *Castor fiber*, deposit anal gland secretion when scent marking. *Canadian Field-Naturalist*, 112, 532–535.

Rosell, F. and Kile, N.B. 1998. Abnormal incisor growth in Eurasian beaver. *Acta Theriologica*, 43, 329-332.

Rosell, F. and Pedersen, K. V. 1999. *Bever [The Beaver]*. Oslo: Landbruksforlaget.

Rosell, F. and Schulte, B. A. 2004. Sexual dimorphism in the development of scent structures for the obligate monogamous Eurasian beaver (*Castor fiber*). *Journal of Mammalogy*, 85, 1138–1144.

Rosell, F., Bozser, O., Collen, P., and Parker, H. 2005. Ecological impact of beavers *Castor fiber* and *Castor canadensis* and their ability to modify ecosystems. *Mammal Review*, 35, 248–276.

Saveljev, A. P. 1984. Peculiarities of the digestive system of Canadian and European beavers (Castoridae, Rodentia). *Rodents. Proceedings of the 6th All-Union Conference*, Leningrad.

Saveljev, A. P. 1989. Comparative biological characteristic of the European and Canadian beavers in the USSR (adaptive changes during acclimatization). Dissertatsiya Kandidata Biologicheskikh Nauk [PhD Dissertation in Zoology], Moscow.

Schmidt-Nielsen, B. and Pfeiffer, E. W. 1970. Urea and urinary concentrating ability in the mountain beaver *Aplodontia rufa*. *American Journal of Physiology—Legacy Content*, 218, 1370–1375.

Scholander, P. F. 1940. Experimental Investigations on the Respiratory Function in Diving Mammals and Birds. Hvalradets Skrifter no. 22. Oslo: I kommisjon hos Jacob Dybwad.

Scholander, P. F., Walters, V., Hock, R., and Irving, L. 1950. Body insulation of some arctic and tropical mammals and birds. *Biological Bulletin*, 99, 225–236.

Sergina, S., Antonova, E., Ilyukha, V., Łapiński, S., Lis, M., Niedbała, P., Unzhakov, A., and Belkin, V. 2015. Biochemical adaptations to dive-derived hypoxia/reoxygenation in semiaquatic rodents. *Comparative Biochemistry and Physiology Part B: Biochemistry and Molecular Biology*, 190, 37–45.

Slough, B. G. 1978. Beaver food cache structure and utilization. *Journal of Wildlife Management*, 42, 644–646.

Smith, D. W. and Jenkins, S. H. 1997. Seasonal change in body mass and size of tail of northern beavers. *Journal of Mammalogy*, 78, 869–876.

Smith, D. W., Peterson, R. O., Drummer, T. D., and Sheputis, D. S. 1991. Over-winter activity and body temperature patterns in northern beavers. *Canadian Journal of Zoology*, 69, 2178–2182.

Snyder, G. K. 1983. Respiratory adaptations in diving mammals. *Respiration Physiology*, 54, 269–294.

Sokolov, V. Y. and Shchennikov, G. N. 1990. Identification of the species *Castor fiber* and *C. canadensis* (Rodentia, Castoridae) by the structure of the anal glands. *Zoologiceskij Zurnal*, 69, 158–159.

Steen, I. and Steen, J. B. 1965. Thermoregulatory importance of the beaver's tail. *Comparative Biochemistry and Physiology*, 15, 267–270.

Stefen, C. 2009. Intraspecific variability of beaver teeth (Castoridae: Rodentia). *Zoological Journal of the Linnean Society*, 155, 926–936.

Stefen, C., Habersetzer, J., and Witzel, U. 2016. Biomechanical aspects of incisor action of beavers (*Castor fiber* L.). *Journal of Mammalogy*, 97, 619–630.

Stephenson, A. B. 1969. Temperatures within a beaver lodge in winter. *Journal of Mammalogy*, 50, 134–136.

Stevenson, A. B., Kitts, W. D., Wood, A. J., and Cowan, I. M. 1959. The normal blood chemistry of the beaver (*Castor canadensis*). B. Blood glucose, total protein, albumin, globulin, fibrinogen, non-protein nitrogen, amino acid nitrogen, creatine, creatinine, cholesterol, and volatile fatty acids. *Canadian Journal of Zoology*, 37, 9–14.

Su, C.-Y., Menuz, K., and Carlson, J. R. 2009. Olfactory perception: receptors, cells, and circuits. *Cell*, 139, 45–59.

Sun, L. and Müller-Schwarze, D. 1997. Sibling recognition in the beaver: a field test for phenotype matching. *Animal Behaviour*, 54, 493–502.

Svendsen, G. E. 1978. Castor and anal glands of the beaver (*Castor canadensis*). *Journal of Mammalogy*, 59, 618–620.

Swain, U. G., Gilbert, F., and Robinette, J. D. 1988. Heart rates in the captive, free-ranging beaver. *Comparative Biochemistry and Physiology Part A: Physiology*, 91, 431–435.

Szteyn, S. and Galert, D. 1979. Nuclei of the geniculate bodies in the European beaver. *Acta Theriologica*, 10, 109–114.

Teleky, J., Melnik, O., Tóth, T., Rajský, D., Kremeň, J., Petrovová, E., and Flešárova, S. 2018. Craniometry of the Slovak northeastern beavers (*Castor fiber*) in comparison with the Ukrainian and Polish populations and contribution to the knowledge of the enamel thickness of beaver's incisors. *Biologia*, 73, 379–387.

Thomsen, L. R., Campbell, R. D., and Rosell, F. 2007. Tool-use in a display behaviour by Eurasian beavers (*Castor fiber*). *Animal Cognition*, 10, 477–482.

Troszynski, W. 1977. Changes in the form of incisors as an age criterion for beaver (*Castor fiber* L.). *Zoologica Poloniae*, 26, 167–175.

Valeur, P. 1988. Beverens territorial-adferd som populasjons-regulerende faktor. *Fauna*, 41, 20–34.

Valli, V. E., McGarth, J. P., and Chu, I. 2002. Hematopoietic system. In: Haschek, W., Rousseau, C. G., and Wallig, M. A. (eds.) *Handbook of Toxicology Pathology*. New York: Academic Press.

Vecherskii, M. V., Naumova, E. I., Kostina, N. V., and Umarov, M. M. 2006. Some specific features of nitrogen fixation in the digestive tract of the European beaver (*Castor fiber*). *Doklady Biological Sciences*, 411, 452–454.

Vecherskii, M. V., Naumova, E. I., Kostina, N. V., and Umarov, M. M. 2009. Assimilation of biological nitrogen by European beaver. *Biology Bulletin*, 36, 92–95.

Vispo, C. and Hume, I. D. 1995. The digestive tract and digestive function in the North American porcupine and beaver. *Canadian Journal of Zoology*, 73, 967–974.

Vorontsov, N. N. 1967. *The Evolution of the Digestive System of Rodents*. Novosibirsk: Nauka Publishing.

Wallau, B. R., Schmitz, A., and Perry, S. F. 2000. Lung morphology in rodents (Mammalia, Rodentia) and its implications for systematics. *Journal of Morphology*, 246, 228–248.

Walro, J. M. and Svendsen, G. E. 1982. Castor sacs and anal glands of the North American beaver (*Castor canadensis*): their histology, development, and relationship to scent communication. *Journal of Chemical Ecology*, 8, 809–819.

Warren, E. R. 1927. *The Beaver: Its Work and its Ways*. Baltimore: Williams & Wilkins Company.

Wilson, D. E. and Ruff, S. 1999. *The Smithsonian Book of North American Mammals*. Washington, DC: Smithsonian Institution Press in association with the American Society of Mammalogists.

Wilsson, L. 1971. Observations and experiments on the ethology of the European beaver (*Castor fiber* L.). *Viltrevy*, 8, 160–203.

Wong, M. T., Wang, W., Lacourt, M., Couturier, M., Edwards, E. A., and Master, E. R. 2016. Substrate-driven convergence of the microbial community in lignocellulose-amended enrichments of gut microflora from the Canadian beaver (*Castor canadensis*) and North American moose (*Alces americanus*). *Frontiers in Microbiology*, 7, 1–13.

Wong, M. T., Wang, W., Couturier, M., Razeq, F. M., Lombard, V., Lapebie, P., Edwards, E. A., Terrapon, N., Henrissat, B., and Master, E. R. 2017. Comparative metagenomics of cellulose- and poplar hydrolysate-degrading microcosms from gut microflora of the Canadian beaver (*Castor canadensis*) and North American moose (*Alces americanus*) after long-term enrichment. *Frontiers in Microbiology*, 8, 1–14.

Wyrzykowska, K. and Wyrzykowski, Z. 1979. Structure and topography of the thymus in the European beaver. *Acta Theriologica*, 24, 399–404.

Zahner, V. and Müller, R. 2006. Thermoregulation—a main function of the beaver tail. *Abstracts of the 4th European Beaver Symposium/3rd Euro-American Beaver Congress*, 11–14 September, Freising, Germany, p. 69.

Zamakhaeva, N. M. 1982. On the geographical variability of the fur of a Eurasian beaver. In: Bibikov, D. I. and Grakov, N. N. (eds.) *Hunting Theriology*. Moscow: Nauka Publishing.

Zamakhaeva, N. M. 1985. About the fur of the Canadian beaver. *Hunting and Hunting Husbandry*, 5, 12–13.

Ziółkowska, N., Lewczuk, B., Petryński, W., Palkowska, K., Prusik, M., Targońska, K., Giżejewski, Z., and Przybylska-Gornowicz, B. 2014. Light and electron microscopy of the European beaver (*Castor fiber*) stomach reveal unique morphological features with possible general biological significance. *PLoS ONE*, 9, e94590.

CHAPTER 4

Habitat use and constructions

4.1 Habitat selection: where can we find beavers?

Beavers (*Castor* spp.) have relatively simple requirements: water for locomotion, movement and protection; vegetation, particularly woody species, for forage, damming, and shelter construction; and suitable land access for feeding and shelter (Macfarlane et al., 2017). They can survive under an impressive diversity of freshwater landscape conditions, ranging from tundra (Tape et al., 2018), boreal forest (Naiman et al., 1988), and steppe (Pietrek et al., 2017), to desert (Andersen and Shafroth, 2010; Barela and Frey, 2016). This riparian zone, or water and land interface, allows them to forage in relative safety from predators and construct protective structures. Beavers prefer areas of deeper water, mainly for protection from predators, but will select smaller wetlands or build lodges close to the bank to shorten distances to forage and construction materials (Francis et al., 2017). Tree availability tends to limit beaver distribution, especially in the northern latitude. In regions where extreme winters do not occur, beavers are capable of surviving with limited tree coverage such as bogs that also lack open water (Rebertus, 1986). Beavers can exist in areas of high altitude, and live at elevations between 2,300 and 3,500 m (Osmundson and Buskirk, 1993) (Figure 4.1).

4.1.1 Important habitat factors

Various studies have investigated habitat selection in an attempt to identify and model key criteria, which then can be applied in potential conflict situations with human land-use (Swinnen et al., 2017; Alakoski et al., 2019; Hood, 2020). It is important to

be aware that such studies have used different sets of variables and methodologies (Hartman, 1996). The identification of key habitat indicators often differs between studies, depending on the object being modelled. For example, some studies focus on family territory mapping versus dam location, while others focus on the geomorphological characteristics of the region. The scale of the study area is also important, with most focusing on a local scale (Touihri et al., 2018). Four orders of hierarchy are commonly used in beaver habitat selection studies (Johnson, 1980):

1) Geographical distribution of the species (first-order).
2) Home range or territory (second-order).
3) Utilization of habitat patches within these home ranges/territories (third-order).
4) Preference for particular food items within these habitat patches (fourth-order).

Habitat selection by beavers has been extensively investigated through comparisons of beaver occupancy or density with habitat characteristics (e.g. geomorphology, soil, and vegetation) (Davis et al., 2016; Bergman et al., 2018; Holland et al., 2019; Wang et al., 2019; Hood, 2020). Second- and third-order habitat selection by North American beavers in northern Alabama demonstrated that herbaceous wetlands were more important habitat components than open water bodies (Wang et al., 2019). In a literature review of 12 major studies of beaver habitat in North America, the dominant habitat factors affecting beaver occurrence or abundance were identified as stream gradient, watershed size, and the presence of riparian hardwoods (broadleaved trees such as oak [*Quercus* spp.], ash, American mountain ash [*Sorbus Americana*], or

Beavers: Ecology, Behaviour, Conservation, and Management. Frank Rosell and Róisín Campbell-Palmer, Oxford University Press.
© Frank Rosell and Róisín Campbell-Palmer 2022. DOI: 10.1093/oso/9780198835042.003.0004

Figure 4.1 Beavers can exist in areas of high altitude, such as here in the Rocky Mountains, Colorado. at ca. 2,900 m above sea level. (Photo supplied courtesy of Frank Rosell.)

beech [*Fagus* spp.]) (Touihri et al., 2018), whereas other reviews assessing habitat quality of beaver release sites found that water depth, river width, bank composition, as well as height, slope gradient, and riparian vegetation cover were the most important criteria (Macdonald et al., 1995; Macdonald et al., 2000). Another review stated that a combination of river flow and lake-level regime, river channel characteristics, and food availability were important habitat characteristics for beavers (Gurnell, 1998). The habitat selection of beavers in an expanding population in south-central Sweden was evaluated, along with the variation in population density and habitat selection (Hartman, 1996). Finding the tortuosity of the riverbank and a dominant cover of grasses and forbs were the factors most influencing beaver occupation.

4.1.2 Preferred and potential habitats

When considering the habitat choice of a 'choosy generalist' like the beaver (Jenkins, 1980; Vorel et al., 2015), it is important to be aware of preferred habitats, along with potential habitats, because beavers can persist outside their ideal habitats (Hartman, 1996). For example, beavers tend to avoid landscapes

dominated by coniferous and mixed pine–hardwood habitats but will utilize these areas under high population densities. An area's suitability is also dependent on how long beavers have resided at the location and their population size. Studies of habitat selection in high-density populations do not necessarily provide accurate information on general beaver habitat preference, but rather provide insights into the full range of potential habitats that beavers may utilize, as well as which habitats are unusable for long-term occupation (Hartman, 1996).

The selection of preferred habitat variables has effectively been demonstrated to vary according to stage of colonization (John et al., 2010). During the initial phase of colonization, there are numerous areas of unoccupied optimal habitat in which habitat selection appeared chaotic. In the intermediate phase, beavers occupied the optimal sites, whilst during the long-term phase, animals expand into suboptimal habitats such as agricultural and urban areas. In southern Patagonia, Argentina, the beaver presence was predictable at the intermediate phase and the most significant variable to predict their presence was the river sinuosity index, i.e. the curve of the river (> 50% chance of occupation at sites with a sinuosity index higher than 1.51) (Davis et al., 2016). This variable alone explained 74% of data variability. Greater river sinuosity tends to increase with decreasing elevation (Cardenas, 2009), which also leads to slower water flow that favours beaver dam construction on suitably sized watercourses (Howard and Larson, 1985).

Beaver habitat selection was studied over a 24-year period following their reintroduction to mountain meadows in the Absaroka-Beartooth Wilderness north of Yellowstone National Park by the US Forest Service in 1986 (Scrafford et al., 2018). Willow (*Salix* spp.) cover and height were positively associated with longevity of beaver territories, but numerous influential variables such as secondary channels, sinuosity, stream depth, and sandbar width also existed. In Telemark, Norway, most of the habitat variables investigated referred to as being key for expanding beaver populations were also significant in populations near their carrying capacities (Pinto et al., 2009). Beavers selected narrower river widths, accessible banks (slope < 36 degrees), soft bank substrate, and predominantly deciduous tree cover.

As a generalist herbivore, beavers can adjust food choices based on availability (Northcott, 1971; Histøl, 1989). Resource availability is heterogeneous over space and time, such that in different locations or time periods varying factors may limit an animal's potential to occupy a site (Hunter and Price, 1992). In many localities, beavers therefore act like rotational crop farmers (within the territory at high densities but between territories at low densities). They can work an area intensely for two to five years and then let it rest and recover for two to ten years before potentially returning to work it again (Hyvönen and Nummi, 2008; Macfarlane et al., 2017; Labrecque-Foy et al., 2020). This is typical in relatively marginal sites in mature populations where beavers consume broadleaf woody resources more quickly than these species can regenerate and, following resource exhaustion and abandonment, can only reoccupy the site after a period of regeneration (Halley et al., 2013). If suitable habitat and vegetation are present, then typical foraging range lies within 10–30 m from the shoreline, though on rarer occasions maximum distances of > 200 m are possible (e.g. Belovsky 1984; Raffel et al. 2009).

In the upper Missouri River in southwest Montana, USA, habitat selection by North American beavers starting new families in novel areas was investigated (Ritter et al., 2019). Beavers settled down in areas with low-gradient stream segments, high canopy of woody riparian vegetation, relatively narrow stream channels, high channel complexity in the form of side channels and tributaries, and wetland types corresponding to low-lying areas directly adjacent to the stream.

4.1.3 Hydrology

Water is a basic requirement and factors such as water speed, water fluctuations, water depth, and water salinity are all important for habitat selection for beavers. They prefer slow-moving mature rivers over faster-flowing, steep-gradient streams (Hartman, 1996; John et al., 2010), but they may be found throughout major rivers with large rapids, such as in the Grand Canyon, Colorado (Breck et al., 2012). Typical stream gradients where beavers are found are ~1% (0–6%), but they may utilize steeper gradients (up to 15%) under high population density and/or in mountainous landscapes (Retzer et al., 1956; Beier and Barrett, 1987). Steep gradients increase water velocity, which may destroy dams and food caches (the main winter food resource and building material for beavers; see Section 4.3) (Taylor, 1970). Water bodies with shores that experience significant wave action and/or large fluctuations in water levels are also less favoured (Smith and Peterson, 1991; Hartman, 1996). Smith and Peterson (1991) suggested that total annual water fluctuation should not exceed 1.5 m and winter drawdown should not exceed 0.7 m. Beavers therefore tend to prefer a water depth of at least 0.6–1.0 m throughout the year so shelter entrances remain submerged underwater, thereby ensuring protection against terrestrial predators and safe access to any food cache. Beavers might dam a suitable watercourse if the depth is less than this to obtain sufficient and stable water levels around residential structures, if the watercourse allows (Hartman and Törnlöv, 2006; Barela and Frey, 2016; see Section 4.4). Beavers also construct dams to flood areas to facilitate access to suitable food and building materials so that they can forage within the safety of water.

Figure 4.2 Exposed lodge entrance with deep furrows visible outside due to a drop in water level. (Photo supplied courtesy of Frank Rosell.)

Despite these generalizations, numerous examples of beavers occupying extreme conditions exist. They can cope with extreme fluctuations in water levels (i.e. between 6 and 7 m annually) by building lodges on higher ground over the course of a few days or by resting on floating wood (Nitsche, 2001; Kurstjens and Bekhuis, 2003). In Norway, a beaver family persisted in a lake with 4.6 m annual fluctuation, which exposed the lodge entrance every spring (Kvamme, 1985; Figure 4.2). Some dispersing subadults (see Section 6.9) may try to spend the winter in watercourses with depths as low as 20 cm, which indicates that beavers do not always need open water to survive over winter. Evidence of current or previous history of beaver activity was found in 200 of 481 marshes (42%) that lacked open water in north-central Minnesota, USA (Rebertus, 1986). In harsh winters, beaver survival is reduced in places with low water levels because freezing restricts access to food caches. In such places, the beavers may gnaw their way through the roof of the lodge to access surrounding vegetation.

Beavers can swim extended distances to disperse through saltwater (Anderson et al., 2009; Halley et al., 2013). They cannot establish themselves in areas with high saltwater concentrations because they are not biologically adapted to effectively excrete salt. Beavers need to drink freshwater daily and to rid their pelage of salt, which impacts on condition and buoyancy of fur in saltwater environments. Other significant challenges associated with living in tidal zones include wave action, resulting in the destruction of the lodge or the food cache. Waves also present a dispersal challenge through increased energy requirements. Numerous beaver cadavers have been recovered in coastal zones in Scotland and parts of England, indicating beavers do attempt to disperse through such routes around freshwater estuaries but survival is limited by tidal action and ability to cope with prolonged saltwater exposure. Beavers may, however, occupy tidal zones with low saltwater concentrations (23% seawater and salinity < 8 ppt) (Hood, 2012), i.e. in brackish water (a salinity < 35 ppt) (Nolet and Baveco, 1996; Rosell and Nolet, 1997; Pasternack et al., 2000).

4.1.4 Vegetation, dam, and intrinsic modelling approaches

Woody vegetation is another basic requirement often mentioned in the literature, but no habitat suitability index (HSI) has so far been developed with good predictive qualities over the entire range of the beaver (subarctic to subtropics). The most broadly used beaver HSI considers woody vegetation a key limiting factor (Allen et al., 1982). Several studies have found that woody vegetation variables are important to beaver occupancy or density—see for example Slough and Sadleir (1977) and Scrafford et al. (2018)—but others did not find woody vegetation to be important (Hartman, 1996; Cox and Nelson, 2009; Pietrek et al., 2017). In climates that experience prolonged winters below freezing, access to woody browse in order to establish a food cache is an essential limiting factor in survival and population growth (see Section 4.3).

Additionally, available information about beaver habitat suitability is largely based on short-term studies in stream habitats that focused on woody vegetation and not aquatic plant suitability, even though aquatic plants may comprise over half of beavers' annual diets (see Section 5.6). One exception is a study that investigated the persistence and density of beaver occupancy in 23 lakes over a 50-year period at Isle Royale, Michigan, USA (Bergman et al., 2018). The results from this study showed that total macrophyte (aquatic plants) cover was associated with persistence and density of beaver occupancy in lakes and, less importantly, lake perimeter, sinuosity, and surface area. Percent cover of floating-leaved aquatic plants was a leading predictor of beaver family density in lakes, independently explaining 72% of the variation in the persistence of beaver occupancy. Bays and inlets created by sinuous shorelines provide beavers with needed protection from wind and ice (Shelton, 1966; Allen et al., 1982), but they also improve habitat conditions for aquatic plants by reducing exposure to turbulence (Spence, 1967; Lacoul and Freedman, 2006). These local HSI models consistently show that both woody and aquatic vegetation are important for beavers. Historically some landscapes' vegetation data are not mapped at fine enough spatial resolution to allow for landscape-scale HSI models to be applied over large spatial extents (Dittbrenner et al., 2018). However, this is changing with the increasign availability of high-resolution remote sensing datasets freely or cheaply available over large spatial extents. Such data

(i.e. those used by Graham et al., 2020) can now capture thin strips of riparian vegetation that form core beaver habitat in intensively managed European landscapes but were often overlooked in historic landcover maps.

An alternative to HSI models is intrinsic models (they generally use both intrinsic and extrinsic predictors) which use geomorphic variables that are less prone to change through time. The variables used in HSI models are often good predictors of current or historical beaver presence but fail to identify areas that may become suitable if transformed by beavers into high-quality habitat through restoration actions or management changes. Therefore, they are less useful in areas below carrying capacity or areas altered by anthropogenic impacts (Dittbrenner et al., 2018). In Washington state, USA, a beaver intrinsic potential model was successfully developed (Dittbrenner et al., 2018). The model could predict where high-quality habitat currently exists and where colonization will occur as population levels increase or if management changes are made. However, the model missed adjacent palustrine wetlands and the authors speculated that additional factors may dictate use of the model in other regions, as demonstrated in Table 4.1. A habitat model for Scotland has been developed based on locating woodland areas (> 0.5 ha) within 50 m of freshwater streams with a gradient < 15%, though authors acknowledge this cut-off could be replaced with a gradual classification to improve the model (Stringer et al., 2018). Reaches containing woodland and wider than 6 m are classed as unlikely to be dammed, again with an acknowledgement that the spatial distribution of dams is complex. This model is generally proving to be effective, though it does occasionally fail to identify dispersing beaver activity and field signs in discontinuous habitat containing patchy woodland < 0.5 ha.

The Beaver Restoration Assessment Tool has been developed in conjunction with the development of a Beaver Vegetation Index (BVI) in North America landscapes to determine the capacity for a river system to support beaver dams (Macfarlane et al., 2017). This constitutes a rules-based fuzzy inference system which allows for any uncertainty associated with generalist foraging behaviours of beavers.

Table 4.1 Summary of general and regional beaver habitat suitability models identifying environmental variables for predicting potential beaver occupation in the USA. Note that some studies focused on specific variable categories (e.g. vegetation) for the purpose of their study objectives.

Habitat quality variables	Allen (1982)	Retzer et al. (1956)	Howard and Larson (1985)	Barnes and Mallik (1997)	Suzuki and McComb (1998)	McComb et al. (1990)	Pollock et al. (2004)	Cox and Nelson (2009)	Anderson and Bonner (2014)	Macfarlane et al. (2017)	All studies
Focus area: region/state	USA	Rocky Mountains	MA	ON	OR	OR	WA	IL	WV	UT	$n=10$
Intrinsic											27
Valley width		X			X		X	X			4
Stream length	X										1
Stream gradient	X		X		X	X	X	X	X	X	8
Stream depth and width			X		X	X	X			X	5
Stream bank steepness		X									1
Stream substrate						X					1
Stream power/flood risk		X		X			X			X	4
Basin size, perennial flow			X				X			X	3
Extrinsic											17
Vegetation composition	X	X	X			X			X		5
Vegetation density		X			X			X			3
Canopy cover	X			X	X	X					4
Canopy height	X										1
Stem diameter	X										1
Habitat and vegetation area	X									X	2
Shoreline development ratio	X										1

Source. After Dittbrenner et al. (2018), reprinted under the Creative Commons CC0 public domain dedication. Additional recent European examples include Stringer et al. (2018) and Graham et al. (2020).

Fuzzy inference offers a way to deal with uncertainties generated by geographic information systems (GIS) datasets covering large areas, which are of course less precise than actual field surveys, which in turn may impact on traditional statistical approaches requiring high-precision data, though this does offer a pragmatic approach, enabling general conclusions to be drawn from lower-precision datasets, which in itself can be highly practical for assessing large tracts of land and future likely trends (Adriaenssens et al., 2004). As an evolution of this approach (Graham et al., 2020), nationally available landcover and hydrological datasets have been used in Britain to develop a Beaver Forage Index (BFI) model, identifying habitat for beaver foraging, and a Beaver Dam Capacity (BDC) model, classifying suitability of river reaches for dam construction (Graham et al., 2020). This has enabled dam potential location and number to be estimated over a catchment scale. These models have been successfully evaluated against field survey data (collected from wild beaver populations in Scotland, England, and Wales) using a categorical binomial Bayesian framework to calculate probability of foraging and dam construction.

4.1.5 Urbanized landscapes

Beavers successfully adapt to human-dominated landscapes (Swinnen et al., 2017) and increased human activity and land-use (Nix et al., 2018; Figure 4.3). In an urban landscape with dense human populations, the most highly selected habitats were the forest drainage canals, followed by rivers, lakes, and brooks. The least selected habitats were the field drainage canals due to lack of woody vegetation (Ulevičius et al., 2011), but beavers are commonly found living in field ditches with very limited tree coverage in Bavaria and Scotland. Beavers can also habituate to heavy road traffic and railways. Interestingly, beavers in a saturated area in Telemark, Norway, tended to select areas close to roads (Steyaert et al., 2015), most likely being attracted to nutritious roadside vegetation (Maringer and Slotta-Bachmayr, 2006). Beavers can also occupy recreational areas such as golf courses and swimming lakes. Beavers are found in numerous cities across Eurasia and North America, including Bratislava, Munich, Stockholm, Boston, and Toronto. However, some studies have found that locations subjected to a high degree of human disturbance often do not support long-term beaver settlements.

Figure 4.3 Beaver in an urban environment. (Photo supplied courtesy of Leopold Kanzler.)

In Bratislava, two-thirds of the animals stayed at their sites for only one or two years (Pachinger and Hulik, 1999).

A beaver habitat classification system for use in urban areas was developed and the authors suggested that their system, with slight modifications for local conditions, can be used in other urban areas (Pachinger and Hulik, 1999). It can also be used to predict which currently unoccupied areas may be suitable for beavers in the future and how increased human development will impact potential beaver habitat. In the Netherlands, beavers occupy a 50-km² estuary with many canals. A few years after they were reintroduced, as many as 3,000 boat visitors have daily visited the area during the summer season, resulting in beavers moving from the most visited areas to other less visited areas where boat traffic was not allowed (Nolet, 1994).

4.1.6 Local to ecoregional scale

In Poland, habitat selection by Eurasian beaver (*Castor fiber*) across a spatial gradient from local (within the family territory) to broad ecoregional scale was evaluated (Zwolicki et al., 2019). Beavers preferred habitats with high availability of woody plants, such as shrubs, and avoided anthropogenically modified habitats, particularly arable lands. Zwolicki et al. (2019) observed a decreasing woody plant cover with increasing distance from the centre (lodge, den, or winter food cache) of the territory, which suggests that beaver habitat preferences depend on the assessment of both the abundance and spatial distribution of preferred habitat elements. Geographical scale explained 46.7% of the variation in habitat composition, while local beaver density explained only 10.3% of this variability. Their results indicated that Eurasian beaver habitat selection was affected by different scale-related phenomena related to central-place foraging behaviour (see Section 5.2.1). The distribution of woody plants within a beaver family territory was clumped and non-random; the cover of woody plants declined with distance from the family centre, while anthropogenic landcover increased. Therefore, settlement decisions were based on the assessment of both the total amount and the spatial distribution of resources at the potential future territory scale.

Finally, habitat selection occurred independently across the largest spatial scale studied (e.g. between watersheds), probably due to the limited natal dispersal range of the animals.

4.1.7 Individual variation

Habitat selection is a context-dependent mechanism and is dependent on not only external factors but also internal states which can affect the behaviour and decisions of an individual. In Telemark, the variation in habitat selection was stronger (approximately six times) between territories than among individual beavers or years (Steyaert et al., 2015). The family size and the presence of kits, but not sex, explained individual variation in habitat selection. Adults with kits and/or larger families tended to exhibit low risk-taking behaviour. Adults avoided areas with a high risk of encountering humans (roads, buildings, and agricultural lands) and stayed close to their main lodge to carry out parental care required to successfully raise their offspring. There are individual and family differences, since some beavers live in the heart of cities, right next to roads, and/or often come out where dog walkers are frequent (for example in Devon and Tayside in the UK), whereas others are highly secretive and relocate or disappear when out and about if they sense a human.

4.1.8 Differences in habitat use between the two beaver species

There have been many speculations about the difference in habitat use between the two beaver species due to their general non-overlapping distribution. Only more recently have they been found in the same area. In Finland, habitat use by the two beaver species was compared in their core range and in a smaller area where they now live sympatrically (Alakoski et al., 2019). The authors hypothesized that the two species would differ in their habitat use in relation to agriculture and urban areas, because the native Eurasian beaver tends to live in more agriculture-dominated landscapes, whereas the North American beaver typically lives in more forest-dominated areas with a sparse human population. Their main results showed that the native

Eurasian beaver's core areas and territories were located closer to agriculture and the volume of birch (*Betula* spp.) was slightly greater near their lodge than those of the invasive North American beaver. The average distance to the nearest agricultural area was approximately 300 m for the Eurasian beaver and more than 1,200 m for the North American beaver. They conjectured that there may be a species-specific difference in their behaviour. Thus, Alakoski et al. (2019) speculated that management favoring birch forest patches within agriculture areas might benefit Eurasian beavers over the invasive North American beaver, but see Nelner and Hood (2011). Findings in Alakoski et al. (2019) indicate that habitat use of both species is quite flexible and it remains questionable how effective this approach would be. Other methods may be more effective in controlling the invasive species (see Parker et al. (2012; see Chapter 11).

4.2 Beaver-made constructions

Beavers have a unique ability to modify their environment by actively constructing burrows, lodges, lairs (a nest-like resting and sleeping place, dams, and food caches. They also build scent mounds (see Section 7.2.2.3) and construct forage trails and canals.

No other mammal, except humans, has such a broad range of construction behaviour and field signs (Table 4.2). The construction activity of the two species is practically identical, with no species-specific features, and is evidently determined by geomorphological and hydrographic settings in the habitat (Danilov et al., 2011; Danilov and Fyodorov, 2015).

Table 4.2 Fields signs: features and possible confusions.

Field signs	Identification	Possible confusion
Teeth marks	Distinct teeth marks, individual incisor scrapes identifiable. Ridges persist over years and can be felt for by running a finger over.	
Felled and gnawed trees	Felling cut approx. 45°. Distinct teeth marks. Wood chipping around base of tree.	At a distance trees fallen and snapped in wind can be mistaken for felled trees. Bark stripping by deer, Lagomorphs, and sheep may be mistaken for gnawing by beavers.
Peeled woody material	Bark peeling, either fully or partially. Teeth marks very often visible. Cut ends distinct at 45°.	
Angled cuts on vascular stems	Clean cut at 45°. Rarely frayed edges.	Lagomorphs. Water vole (*Arvicola amphibius*).
Grazed lawns/crops	Areas of lawns/crops grazed with clear foraging trails leading to and from water.	Lagomorphs. Vole spp.
Feeding stations	Piles of peeled sticks and/or mixed vegetation (e.g. bracken, reeds, aquatic plants) all distinctively cut. Varying quantity (2–100+) and sizes of cut sticks/plant material. Age assessed by coloration of sticks (very fresh is light in colour gradually getting darker with age).	Natural stick piles/debris.
Forage trails	Flattened vegetation pathways. Always leading from the water. Any other beaver feeding signs along the trail.	Other mammals such as otters, badgers and deer.
Lodges and burrows	Large piled structure made from sticks, mud, and vegetation. Varies greatly in size and shape. Fresh construction activity prevalent in autumn and spring. Fresh signs include wet/damp mud or vegetation.	Burrows can be very hard to find, as the entrance will be underwater. Can be found in low water conditions.
Dams	Structure built at right angles to a watercourse. Interwoven from sticks, mud, stones, and vegetation.	Debris dams can be distinguished by lack of uniformity of beaver-cut material.

continued

Table 4.2 Continued

Field signs	Identification	Possible confusion
Canals	Narrow channel excavated, often originating from forage trails. Radiate from the primary water body. Excavated debris deposited on canal sides.	Artificial drainage ditches in forestry and agricultural land.
Scent mounds	Small piles of mud or detritus, near water's edge and territorial boundaries, marked with distinctive scent.	
Faeces	Not often seen as beavers defecate in water and faeces quickly break down. Can persist longer in slow-moving water around lodges, dams, and feeding stations.	
Tracks	Hind paws are large and long. Possible to see print of web of skin connecting toes on hind feet. Distorted tracks possible due to tail being dragged through the tracks.	Distorted tracks may be confused with other riparian mammals.

Source: After Campbell-Palmer et al. (2015), reprinted by permission of Pelagic Publishing.

Figure 4.4 Beavers sleeping on their back inside the lodge. (Photo supplied courtesy of Odd Arne Olderbakk/NRK.)

4.2.1 Shelters: burrows, lodges, and lairs

Burrows and lodges provide favourable environmental conditions, protect from water and predators (Żurowski, 1992), and act as focal points of a beaver's territory. Both burrows and lodges normally contain an entrance tunnel or tunnels starting underwater, a feeding chamber at water level, and a sleeping chamber above water level (Figure 4.4), but they may have more than one of each (Wilsson, 1971).

4.2.1.1 Burrows

Beavers are excellent burrowers, an often-overlooked behaviour as the true extent is not often seen unless the burrow collapses or water levels drop (see Chapter 11). They can live their whole life in these structures. They prefer natural crevices in the bank but will also dig burrows in environments with friable soil and where banks are sufficiently high (Wilsson, 1971; Żurowski, 1992). In the Rhone valley in France, 95% of beaver families live predominantly in burrows throughout the year (Richards, 1983), whereas in northwest Russia (Pskov and Novgorod regions) nearly 50% of all beavers live in lodges, while further north in the Lapland Reserve only 10% live in lodges (Danilov, 1995). Dispersing beavers will often dig burrows and spend the winter there. Dispersers on their own do not need a large lodge or burrow. In established populations, these singletons often settle down later and in marginal habitats. These burrows are usually occupied only until the spring when they move on to better habitats and/or find a mate.

Burrows are commonly invisible above ground (unless the top partially collapses, which can lead to management conflicts; see Chapter 11), although entrances can sometimes be observed where water visibility is clear or where water levels are unusually low. In narrow watercourses, such burrows are most likely to be found by viewing them from the opposite bank or from a canoe in clearer, wider, and deeper watercourses (Campbell-Palmer et al., 2015). Figure 4.5a is an illustration of a burrow, with a bank lodge (Figure 4.5b) and stick lodge provided for comparison (Figure 4.5c).

Beavers dig burrows with their fore feet and can display burrowing behaviours from 2 months of age onwards (Novak, 1987). When substrate is sufficiently loosened through digging with their front feet, they push it backwards using their hind legs by using a series of forward and backward movements to push material out of the tunnel (Wilsson, 1971). The preferred place for digging burrows is under roots of large trees (10–15 m high), which allows them to use the strong root system as a framework to prevent the burrow from collapsing (Aeschbacher and Pilleri, 1983; Fustec et al., 2003). The entrances and sleeping chambers of these burrows are free of branches. The passages inside are usually 0.8–11 m long, 30–40 cm wide, and 25–50 cm high. In some instances, the passageways within these burrows form enormous underground labyrinths; at the extreme end they can be between 100 and 200 m in length (Wilsson, 1971; Danilov and Fyodorov, 2015). Beavers invest much effort in making passages and tunnels even and flat (Wilsson, 1971).

Sometimes burrows act as access routes from the water's edge to food sources further up on the bank. At times, upper sections of these tunnels may partially collapse, exposing the burrows and chambers beneath. If the beavers still reside in the burrow when this occurs, they will often respond by reinforcing these areas with sticks and mud (Campbell-Palmer et al., 2015). Burrows may also collapse when an area is flooded after heavy rainfall or following pressure (footfall, machinery, etc. from above). Dead beavers

(a)

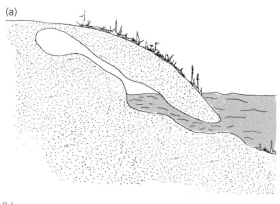

(b)

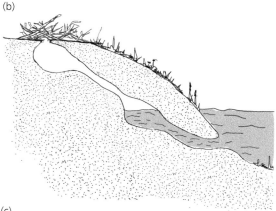

(c)

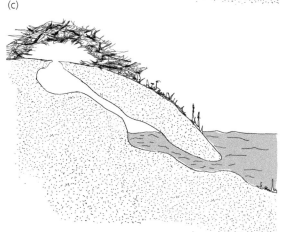

Figure 4.5a–c. Illustrations of a burrow (a), a bank lodge (b), and stick lodge (4c). (Illustration supplied courtesy of Rachael Campbell-Palmer.)

have been recorded within a partially collapsed burrow network. A dead beaver was found in a collapsed burrow most likely due to heavy rainfall and a sandy soil substrate, with suffocation presumed as the main cause of death (Thomsen et al., 2004).

Beavers also make another type of burrow—a short unchambered burrow—sometimes referred to as 'day rests'. These are hollows dug out in the bank but with a substrate roof. They usually extend only 1 m or so into the bank. They are located near the water's edge and may be multiple within a territory, providing a safe retreat an animal can just glide into if spooked. These types of burrows are more commonly made by dispersing subadults, though established members of a family will also construct them, especially during spring.

4.2.1.2 Lodges

Beavers tend to select lodge locations based on water depth and bank characteristics (Hartman, 1996). The absence of sandbank and canopy cover appear to be good predictors for lodge settling in the Loire River, France (Fustec et al., 2003). In a North American beaver study on a prairie river in eastern South Dakota, USA, slope of the riverbank was the most important physical factor, with horizontal vegetation cover being the most important terrestrial vegetative factor influencing lodge-site selection (Dieter and McCabe, 1989).

There are three main types of lodge structures common across both species. The first type is an island lodge (freestanding domed lodge or atoll), which is usually a freestanding structure surrounded by water, often as a pool or pond (Figure 4.6). In places where the banks are low, such as in marshy areas, beavers cannot dig a burrow into a bank and may, therefore, construct upwards. The water level is usually more stable around island lodge locations, as otherwise beavers would become flooded out. The entire burrow and chamber system is contained within the sticks and mud construction above ground level. These are gnawed out from the material gathered and placed by the animals. This type of lodge can also be established when beavers dam a small stream to create a pool, which then expands to surround the lodge. Beavers add to the lodge as the pond deepens. This type of lodge appears to be more common in North America than in Europe, although the reasons are unknown.

The second type of lodge is a bank lodge and is located on the water's edge, built with some sticks and formed around a burrow dug into the bank (Figure 4.7). The sleeping chamber is dug out in the soil. Bank dens are found along steeper edges with loose soil or sand, usually along streams and rivers.

Figure 4.6 Freestanding lodge, which is more commonly found in North America. (Photo supplied courtesy of Tom Gable.)

Figure 4.7 Bank lodge with chambers and tunnels dug out from the soil. (Photo suppled courtesy of Frank Rosell.)

Figure 4.8 Stick lodge extending from the shoreline. (Photo supplied courtesy of Frank Rosell.)

The tunnels and sleeping chambers are dug out from loose materials as described above.

The third type of lodge is a stick lodge (Figure 4.8). It is created from material extending from the shore-line. The chambers are contained above ground within the sticks and mud, i.e. the chambers are gnawed out in the pile of sticks.

Beavers often build the lodge they use during winter in the northeast to northwest facing part of its chosen location whenever possible. In these areas the sunlight will melt the ice on and around the lodge quicker than on a lodge built in the southern facing parts. Earlier snow melt gives the animals greater freedom of movement in the late winter and better access to food on land. The direction of the prevailing winds is also important. In larger lakes that have choppy waters and free-floating ice in the spring, beavers will choose to build their lodge in sheltered locations such as within the shelter of bays and inlets or behind small islands (Rosell and Pedersen, 1999).

Beavers occasionally build adjacent lodges, most likely to have more room. There can be a tunnel system between the lodges, but they can also have separate entrances. One report from Canada in the 1700s mentions a very unusual beaver lodge. It was located on a small island and had roughly a dozen rooms under one roof (Pilleri, 1960). There were only underwater connections between the rooms. Indigenous hunters killed 12 older animals and 25 young and immature animals from the lodge, but there were others that got away, indicating that the lodge had more than 37 occupants. It would seem that there were different families, each with its own apartment and private entrance (Rosell and Pedersen, 1999).

In active territories where beaver families build several lodges, beavers can often choose between a summer and a winter lodge. They may have summer lodges in places where the water is too shallow for winter occupancy, while the winter lodges are placed in areas with deeper water. Beavers can also move to a different summer lodge if there are more aquatic plants and herbs near the summer lodge than near the winter lodge. Winter lodges will often be located near deciduous forests, allowing the beaver to maximize its utilization of woody food resources during winter when other types of vegetation are inaccessible or absent. Beavers that build their winter lodges in areas with strong currents, for instance where a tributary feeds into or runs out of the river, will also have greater freedom of movement, as the water often freezes later in these areas. The ice will also be thinner here, allowing the beaver to more easily break and come to land. At other locations, the family group may use all lodges during summer but regroup into a single lodge in early autumn and winter. The family group appears to be

less cohesive in the summer months, despite using the same territory.

Lodge construction

A beaver lodge is often a very solid construction. It can occasionally take a long time before an individual or group of beavers begin building a lodge. Many beavers will build a lodge during their first autumn at a new location, whereas others wait until they have lived in the same place for as long as 8 years. Lone individuals or pairs without kits tend to utilize burrows rather than build lodges; often it is the impending birth of offspring that triggers lodge construction (Żurowski, 1992), as they now need a more solid and safe shelter. Building behaviour can also be initiated by pregnant females just ahead of giving birth. A substantial lodge for the winter can be built in as little as two nights (Müller-Schwarze, 2011). Though the time to build a lodge can be highly variable and any family can add to the size of a lodge over a number of breeding seasons. Beavers seem to be born with the ability to build structures, so even a juvenile can build a large lodge on its own in its first autumn (Wilsson, 1971). Beaver lodges can be constantly occupied for several decades. If the beaver site is abandoned, newcomers will often reuse lodges and/or burrows constructed by other beavers, though forming additional constructions is not uncommon.

Wilsson (1971) describes a gradation of constructions from burrow to lodge, depending on bank substrate and water levels. To create a lodge, beavers first tend to dig a burrow or excavate a cleft in the bank next to the water. In rocky environments or artificially reinforced riverbanks, beavers may adapt any available shallow shelves or exploit any cracks or crevices. Construction of lodges tends to begin with beavers assembling a collection of branches and mud in a large mound. This can occur both on land and in shallow water. When beavers build or repair a lodge, they generally use chewed-off branches or materials that they do not consider food (see Section 5.1). A wide range of objects have been recorded in beaver lodges including fishing rods, rubber boots, glass bottles, plastic bags, pipes, and fence posts. A beaver lodge in a coastal marsh area in Louisiana, USA, was constructed partly of commercial lumber, possibly made available by recent hurricanes (Elsey et al., 2015). Large branches are gripped at one end with the front teeth and dragged into the lodge. The branches are placed through shoving them into position using energetic body movements with the head. When collecting building materials, smaller branches are lifted into the mouth with one fore leg until the mouth is full (Wilsson, 1971). Twigs, soil, clay, mud, small rocks, bark, and plant material are piled up by the beaver using its fore paws and pressed between their chins (Figures 3.8). Snow can also be used to insulate the lodges late in the autumn. When a beaver transports loose insulating materials, it may first place some twigs between its jaws to create a larger carrying surface. The finer materials are then pressed together with its hands, nose, or chin. Once the constructed pile is large enough, the beaver will dig its way in through the bank under the pile and chews out chambers and tunnels within. Another strategy is to first excavate an underwater entrance and tunnel that ends at the water's edge, then assemble materials over the tunnel and chew out structures within (Wilsson, 1971).

The construction of bank lodges begins with a beaver digging an underwater tunnel up to the bank, preferably near a site that has been used as a feeding area, so that there is already debarked branches and logs at the water's edge and on the ground. A 'roof' over the entrances is built of twigs and insulated with soil, mud, or clay dug up from the riverbed. This 'roof' can be as thick as 15–20 m in diameter. The burrow is excavated in the same order as the bank lodge, but beavers dig out the chambers in the soil. From here, there is often a connection to a system of tunnels in the surrounding ground, when the area is suitable. Not all bank lodges have a 'roof' built over the entrance and occasionally there are only branches lain over the sleeping chamber further up the riverbank. If the 'roof' gets worn away by ice for example, the water-filled portions of the passages and feeding chambers can be eroded away. Animals in bank lodges that lose their 'roof' in the winter have limited chances of survival due to the difficulties of finding another suitable shelter under these harsh winter conditions. Because the 'roof' is so well insulated, the water in the submerged parts of the passages and feeding chambers does not freeze. And because the 'roof's' rear section

is not insulated with finer materials, there is good air circulation even in the parts of the tunnels that lie deep under the surface (Wilsson, 1971).

Beaver lodges in southern Europe can be very different from those in the north. Beavers in the Netherlands, for example, do not always insulate their lodges with finer materials and one can even see through the branches right into the sleeping chamber. This form of lodge construction is most likely because of the milder climate and lack of predators in the Netherlands. The animals here will also use lodges with openings located above the water level that lead directly to the sleeping chamber and these lodges might lack swimming pools and feeding chambers. In other parts of Europe we can find lodges built entirely from mud and roots, because there are no trees nearby (Rosell and Pedersen, 1999).

Lodge structure: exterior

The beaver lodge is often the largest and most obvious construction. They can vary greatly in shape and size depending on the surrounding habitat and number of occupants. Some dwellings can be very obvious structures at the water's edge, while others mainly consist of few external features and go largely unnoticed. The outward appearance of a lodge again varies depending upon the shape of the water's edge and materials utilized. No two beaver lodges are alike, even when they share the same construction method or are made by the same family within the same territory (Wilsson, 1971).

Beaver lodges can be impressively large, especially if it has been inhabited for several years. They can be > 4 m high and have a perimeter of 30 m. Most beaver lodges are smaller than this size and only 1–2 m high, 2–4 m wide, and 8–10 m long. In areas where the water level can fall significantly, the entrance to the lodge is lengthened to ensure that the entrance is always under the surface of the water. Some lodges can be as much as 12 m long (Zavyalov, 2011). Some lodges can easily be missed because they are very small or located in secluded locations, such as under bushes or between boulders. A beaver lodge can also be confused with watercourse debris or dumped material. This misidentification can occasionally have negative consequences for beavers. Lodges have been removed as accidentally mistaken for 'brush piles' or debris dams and even put on fire. Checks for characteristic beaver-cut woody material and teeth ridges can establish the difference (Rosell and Parker, 1995).

At the top of the lodge the walls are often much thinner and function as a ventilation channel (Wilsson, 1971). Even when the walls of the lodges are frozen solid in the winter, the ventilation channels remain open, probably because of the warmth from deeper within the lodge (Salvensen, 1934). When temperatures fall below −20°C and there is little wind, it is possible to see clouds of steam rising from active lodges and it can look as if the beaver has lit a fire inside the lodge to stay warm.

In spring, especially when heavy rains wash off parts of the outer mud surface or floods partially destroy a lodge, beavers may display increased lodge repair behaviour. Lodge maintenance behaviours are most prevalent during the autumn as beavers prepare for winter. If a beaver family has one or more established lodges, they tend to focus efforts on repairing one of them. In a fewer occasions they might also start repairing more than one to have a choice of places to spend the winter. Some beaver families maintain three lodges simultaneously and there can be as many as five lodges within a family's area (Zavyalov, 2011). Signs of recent activity include wet mud, sticks/branches, mosses, or aquatic plants deposited on the outer surface. Branches with green leaves that have not yet wilted are also good indicators of fresh activity (Rosell and Parker, 1995; Campbell-Palmer et al., 2015; see Chapter 11).

Lodge structure: from the entrance to the interior

Most lodges have two entrances (range 1–5), which begin below water and are seldom visible unless water levels drop significantly and/or water is clear enough to allow visibility of the submerged bank or obvious runs or furrows in the sediment outside the entrance. If the water level drops enough that it is absent from the lodge entrance in spring, beavers may move to another lodge (Smith and Peterson 1991). The entrances must be large enough to allow the animals to drag branches into the lodges. They generally have a diameter of about 40–50 cm and deep furrows in the lake- or riverbed are often visible just outside the entrance to the lodge.

Following the entrance tunnel into a lodge, the first area encountered is a feeding chamber, which has its own in-built swimming pool. The beaver will often sit in this feeding area to eat the forage it has carried in (Figure 4.9). When the chamber is full of debarked branches, the beaver will carry them out in large bunches and dump them in the water a short distance away from the lodge's entrance(s). Beavers will also use the feeding chamber as a place to let the excess water drain from their fur before entering the upper sleeping chambers. These chambers are normally located at least 10 cm above water level but can be much higher in areas with fluctuating water levels. The internal floor of these upper chambers tends to be lined with a bed of dry material such as bark, finely split wood shreds (chips), or grass, which the beavers use to create their own

Figure 4.9 A beaver eating a branch in the feeding area of the lodge. (Photo supplied courtesy of Odd Arne Olderbakk/ NRK.)

bedding material. Bedding material draws out the remaining moisture from the beaver's fur and will also absorb any moisture in the lodge (Wilsson, 1971).

The size of the sleeping chamber varies according to the number of animals inhabiting the lodge but is usually about 1.3 m long, 1.0 m wide, and 0.6 m high. The walls surrounding the sleeping chamber can be a metre thick. As the size of the family increases, beavers increase the size of the sleeping chamber so that there is room for all (Wilsson, 1971). This is also where any kits are born. The walls of the sleeping chamber are fairly smooth, where beavers have dug away the soil and gnawed off any branches protruding into the walls of the chamber. Beavers may occasionally use moss/vegetation to insulate the chamber. Such mosses also contain trace amounts of acids with disinfecting properties. When there are numerous animals in a lodge there can be several sleeping chambers, with one or two (or more) shared tunnel systems. However, in some cases they use the feeding chamber as a sleeping chamber. In areas where the water level can increase by significant amounts, they can also have a 'security sleeping chamber' further up the bank (Wilsson, 1971).

The lodge insulates beavers extremely well from outside temperatures. Despite large daily fluctuations in air temperature during January and February in northern latitudes, the lodge temperature remains stable, with one study in Algonquin Provincial Park, Ontario, Canada, measuring only 0.8°C separating the average minimum and maximum temperatures of 0.8°C and 1.6°C, respectively. Outside temperatures were considerably lower, with an average minimum and maximum of –21.0 and –6.8°C, respectively (Stephenson, 1969). In winter, thin-walled lodges are 7–8°C cooler than thicker-walled ones. In summer, a thinner wall provides the advantages of better ventilation (Buech et al., 1989). In North America, instances with significantly higher CO_2 and lower O_2 in active versus abandoned lodges in the winter have been reported (Dyck and MacArthur, 1993). The beaver's respiratory demands (see Chapter 3), combined with its limited activity (due to frozen water; see Section 6.2), have likely contributed to the observed differences in gas concentration between active and abandoned lodges in the winter. This wall difference might be

the reason why beavers change between summer and winter lodges. Soil and water temperature can also influence the chamber air temperature (Buech et al., 1989). Interestingly, bank lodges are cooler in summer than open-water (island) lodges (Buech et al., 1989). Hence, onshore shelters may offer better protection against heat stress during hot weather. Under similar conditions, both species build their lodges (and dams) with equal frequencies. Therefore, the building activity of beavers is a response to habitat conditions rather than a species-specific behaviour governing the manifestation of the building instinct (Danilov and Fyodorov, 2015).

4.2.1.3 Lairs (day nests) and other shelters

Beavers may form lairs under vegetation or on ledges. These lairs can be found close to the water's edge or several metres inland and are usually hollows in the ground with or without shredded woody material. They are usually made during spring and summer by beavers at northern latitudes but may also be used during the winter by beavers at southern latitudes. Beavers in the Biesbosch in the Netherlands commonly use lairs and often five to ten such lairs are found within a family's territory. The whole family can use these lairs, but they are most often used by only one animal (young or adult). In France, for example, lairs are often temporary and observed during floods since the water level is above the sleeping chamber (Richard, 1985).

Sometimes beavers use existing topographical features as shelters, including caves (McAlpine, 1977; Gore and Baker, 1989; Østbye et al., 1993), cavities between boulders (Grater, 1936), abandoned mine tunnels (Alcorn, 1934), and road culverts. They can also construct a food cache and dams inside caves (McAlpine, 1977) and evidence of breeding in such locations has been reported. Most beavers using these more temporary features are most likely to be subadults, perhaps dispersing from their natal families and requiring a shelter that needed minimal effort to maintain (Gore and Baker, 1989).

4.3 The food cache

Beavers spend a lot of time preparing for the winter to be able to survive in harsh climates. Not only do they build and repair lodges and dams but also

Figure 4.10 Large food cache extending into a river outside the lodge entrance. (Photo supplied courtesy of Frank Rosell.)

in northern environments they fill their 'larders' with food for the winter. They spend the autumn collecting woody material, which they store as a submerged cache outside their principal lodge or burrow in which they intend to overwinter (Wilsson, 1971; Hartman and Axelsson, 2004; Figure 4.10). A family will usually have just one food cache, but on rare occasions they can have two. Initiation of food cache-building usually begins from the end of September to the end of November (Busher, 2003; Hartman and Axelsson, 2004). For example, in Sweden and Lithuania beavers typically start food-caching between 28 September and 11 October, whereas in Norway this tends to be around 12–18 October (Busher et al., 2020), though in more extreme/northerly climates they have been reported caching as earlier as mid-August in northwestern Siberia (Vasin 2001) and late-August in northern Canada (Aleksiuk 1970). The majority of food caches are finished by the end of October or in early November. Timelines vary, though there appears to be a strong incentive and stimulus to begin food caching according to local climatic conditions (Busher et al., 2020). Usually, either the adult male or female initiates food cache construction (Hodgdon and Lancia, 1983). Adult pairs are responsible for the majority of food cache assembly, but kits can

contribute to the food cache during late autumn (Hodgdon and Lancia, 1983; Patenaude, 1984).

What causes the animals to begin building a food cache? Food caching behaviour is innate (Roberts, 1937). A beaver pair trapped in France in February and later released in Switzerland was able to build both a lodge and a food cache during their first autumn in their new surroundings. How did they know that they needed to build a food cache that first winter in a new climate? One explanation could be that the animals had come to a new environment and they built the food cache to be on the safe side, since they had no experience of winter in their new surroundings. External factors, such as temperature, can also initiate food cache construction (Semyonoff, 1957; Wilsson, 1971; Hodgdon, 1978). Furthermore, the quality of available food and nutritional status can be important, since food-caching behaviour is activated after the change from a summer to winter diet and after a period of extensive, energetically costly construction work. In Nordic countries, there are indications that the animals that build food caches start after the first frost.

The food cache is protected from rotting by the frigid water. Thick snow or frozen surface water means beavers can swim under the ice from their lodge to collect the stored branches and eat them through the winter. The branches are then eaten inside the lodge, near the swimming pool, which is eventually filled with peeled sticks. They can also gnaw holes through the frozen water surfaces and eat the branches on top of the ice. The discarded sticks are regularly removed from the lodge and can be found in large piles outside the lodge in the spring. They can then be used again, for example for repairing lodges or dams (Slough, 1978).

4.3.1 Food cache construction

Normally beavers start a food cache by building a 'raft' of 'inedible' branches. These 'inedible' branches stick out of the lodge and over the surface of the water. Many people, therefore, mistake the food cache for an unfinished section of the lodge. Under the 'raft' there is a collection of more appetising species such as aspen (*Populus tremuloides*) and willow. The 'raft' will slowly become inundated with water, which makes the materials trapped underneath sink to the bottom (Roberts, 1937; Slough, 1978). The branches also contain sap which ensures that they sink. The best food is securely stored under the ice, while the 'raft' freezes in the ice and becomes unavailable, so typically their less-preferred food species are found in the raft section. The branches are interwoven with smaller twigs and eventually the cache becomes very dense.

Food cache construction can also begin with beavers setting branches in the waterbed bottom in front of the lodge. The thickest end is driven into the sediments and branches are attached to those that are already stuck. Finally, beavers place the largest branches and logs at the top, which weighs down the entire food cache. There are also lodges with just a few small branches stuck in the sediments, invisible from over the water's surface. The food cache can, on rare occasions, be attached to the dam and in places where there is a strong current between stones. Sometimes the large-crowned tree and associated branches from a tree felled by the beaver can function as an extra food cache. The food cache can include aquatic plants such as water lilies (e.g. *Nuphar* spp. and *Nymphaea* spp.) and stalks of grains. Incomplete utilization of the food cache can be left in place to submerge for the following winter (Slough, 1978). The material

in caches that remains unused can, on occasion, regenerate by sending out new side shoots, particularly where it contains willow if the water is shallow enough.

Should the food cache disappear, for example because of spring floods, autumn ice, or it being removed as part of conflict mitigation (however, there may be a questionable welfare issue if food caches are purposively removed during late fall and winter), beavers will start work on a new one if they can. There have been occasions where a family has had to rebuild its food cache as many as four times in a single autumn (Rosell and Pedersen, 1999).

4.3.2 Plant species in the food cache

Beavers cache different plant species, both preferred and unpreferred species, at different locations. A cafeteria-style feeding experiment was conducted in Quabbin Reservation, Massachusetts, USA, during the time when the beavers were caching food (Busher, 1996). Witch hazel (*Hamamelis virginiana*) was consistently cached more often than other species, especially in early autumn. During late autumn they cache branches of all species, including red maple (*Acer rubrum*), witch hazel, red oak (*Quercus rubra*), black cherry (*Prunus seritona*), yellow birch (*B. lutea*), and Eastern white pine (*Pinus strobus*). Pine branches that are not preferred as food items can be stored to stabilize the construction of the food cache. In northeastern Poland, the food cache included four or five species (with one exception) (Dzięciołowski and Misiukiewicz, 2002). These were willow, birch, alder (*Alnus glutinosa*), trembling aspen, and mountain ash (*Sorbus aucuparia*), occurring in all four habitats studied (i.e. river, lake, farmland, and oligotrophic small reservoir). Willow alone constituted 62.5% of the total number of branches. Willow together with birch, hazel (*Corylus avellana*), alder, and trembling aspen amounted to 97% of all branches in caches. Most branches were 0.5–1 m long and < 3 cm thick at their base. In Algonquin Provincial Park, Ontario, Canada, beavers preferred trembling aspen, beaked hazel (*Corylus cornuta*), maple (*A. saccharum* and *A. rubrum*), and serviceberry (*Amelanchier arborea*) as food cache material (Doucet et al., 1994a). Shrubs of winterberry (*Ilex verticilla*) and blueberry (*Vaccinium* spp.) can be included in the food cache

(Busher, 1991). Many food caches are not completely browsed. The species most commonly left unbrowsed are alder and conifer (Slough, 1978).

On the Tongue River in Montana, a marked differences in cache composition between two study areas were observed and suggested that this possibly resulted in differences in fertility of the breeding females (Swenson and Knapp 1980). The low rate of ovum loss in one of the areas may be due to high food and habitat quality. In this area, 91% of the food items were willow, while in the other area only 51% were willow and this area also had 4% occurrence of corn stalks, which suggests that quality food was in shorter supply here.

4.3.3 Factors affecting the food cache size

The food caches vary in size, from a few branches to 100 m^3 of condensed constructed material (Busher, 2003; Hartman and Axelsson, 2004; Gorshkov and Gorshkov, 2011), and might be larger in colder climates (e.g. the mean was 4.6 m^3 in Voronezh in Russia) (Djoshkin and Safonov, 1972). Branch volume estimates of between 1 and 2 m^3 per day or 7 and 14 m^3 per week being added to food caches have been recorded (Busher et al., 2020). Food caches can be smaller and less well constructed on small tributaries than on large rivers (Wilsson, 1971). The occurrence, time of initiation, and finite size of food cache were related to watercourse characteristics in a population of Eurasian beavers in Sweden (Hartman and Axelsson, 2004). Water depth in front of the lodge was a strong predictor of occurrence of food caches but a weak predictor of finite cache size. Recent analysis across multiple studies determined that food caches where not found in water depths of less than 1 m or more than 2.3 m (Busher et al., 2020). Beavers on wide waters seem to begin constructing food caches earlier and have larger caches. This is probably due to the slower currents, which allow the surrounding water to freeze sooner. Wider watercourses will also generally be ice covered for longer than narrower watercourses.

Factors that can affect whether the beaver builds a food cache or not and the relative size of the cache are as follows:

- Latitude and altitude (lodge/burrow ice coverage longevity). Beavers in colder climates (northern latitudes) build food caches more

often than beavers in mild climates (Hartman and Axelsson, 2004). Beavers that have the opportunity to collect fresh food during the winter when there are ice-free conditions generally have smaller food caches or will not create food caches at all. In locations that lack extreme seasonality, for example in France and in the southern USA, beavers generally do not build food caches (Blanchet, 1959; Brzyski and Schulte, 2009).
- Depth of water outside the shelter (Lavsund, 1989) suggested that the stream provided little room for a sufficient food cache because of the low water depth in combination with a thick ice layer. The water depth was on average 1.1 m (range 0.5–2.3 m) by settlement with a food cache and 0.7 m (range 0.3–1.5 m) by those lacking a food cache (Hartman and Axelsson, 2004).
- Number of animals in the family. Some studies suggest a link between family size and cache size (Djoshkin and Safonov, 1972; Easter-Pilcher, 1990; Vasin, 2001), while another study found no association (Osmundson and Buskirk, 1993). Recent analysis indicates there is no strong correlation when comparing food cache sizes and presence or absence of kits. For example, mean cache volume in November in families with kits was 16.31 m^3 and without kits was 17.28 m^3 and in December 30.1 m^3 and 34.9 m^3, respectively (Busher et al., 2020).
- Experience through many years at the same location. Variability in food cache initiation may also reflect individual behavioural plasticity, with little variation observed between years for the same family, whereas between-family variation is much less consistent.
- The below-ice availability of aquatic plant roots or other vegetation that does not need to be collected. The roots of water lilies can often be extremely important food for beavers during winter (Shelton, 1966; Ray et al., 2001). Beavers with a lot of aquatic vegetation, e.g. cattail (*Typha* spp.), water lilies, or pondweed (*Lemna* spp.), often do not build food caches (Gorshkov and Gorshkov, 2011).

4.3.4 Invisible food caches

Not all food caches are easily visible from outside the lodge. Beavers can pull branches into the burrow or lodge or in other places where branches cannot be seen from the water's edge (Semyonoff, 1957; Slough, 1978). Beavers can also occasionally stick branches into the lodge from below the water's surface. The majority of families that do not have visible food caches are found in streams and larger ponds. In one study conducted in Telemark, Norway, 25% of the families surveyed had not built a visible food cache. In Belarus, where average temperatures are below freezing from November to March, the number was 40% (Zavyalov, 2011). Families on larger bodies of water have plenty of space for building food caches; it is possible that the food has been driven into the bed of the lake and is therefore simply not visible from above the surface. Families in marshes can store branches under the edge of the marshes.

4.4 The dam

One thing that beavers and humans have in common is the ability to build complex dams to impound water. While humans need the help of multiple technical experts to complete a dam, a solitary beaver can manage by itself (Figure 4.11). Beaver dams are the main structure they are typically famous for and by which they can modify whole landscapes (Lapointe St-Pierre et al., 2017; Swinnen et al., 2018) and dramatically increase wetland ecosystems (Hood and Bayley, 2008; see Chapter 9). However, importantly

beavers do not always build dams. Dam building is a high-cost activity and would only justify the investment in building and maintaining where resources exist to sustain a beaver territory. Dam formation by beavers involves two processes: (1) intentional construction through the selection and placement of wood and sediment (or other material), which facilitates (2) the passive capture and accretion of suspended wood and sediment (Blersch and Kangas, 2014). Beavers can live in the same place for several years before deciding to build a dam.

Andersen and Shafroth (2010) state that dam construction is an innate behaviour, while Wilsson's (1971) study of semi-captive beavers found they typically begin to exhibit building behaviour as juveniles (at ~1 year of age), i.e. during their second autumn.

4.4.1 Dam selection: where do beavers build dams?

Numerous factors help predict dam-building behaviour, such as watercourse width, water depth, watercourse gradient, watershed size, and valley floor width (Retzer et al., 1956; Barnes and Mallik, 1997; Suzuki and McComb, 1998; Hartman and Törnlöv, 2006; Jakes et al., 2007; Petro et al., 2018; Swinnen et al., 2018; Figure 4.12). Dam-building behaviour is so closely associated with beavers that habitat selection (see Section 4.1) and

Figure 4.11 Dam with an incorporated lodge. (Photo supplied courtesy of Tom Gable.)

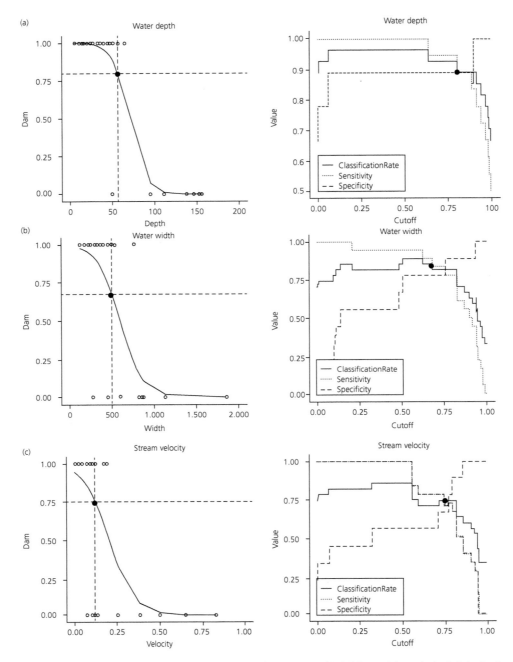

Figure 4.12 Binomial distribution graphs of the five environmental conditions, measured in Belgium and the Netherlands (July–October 2013), that are determinant of a waterway being suitable for dam construction by beavers, together with specificity-sensitivity graphs of each binomial model: (a) water depth (in centimetres, cm) at 10 m downstream of the dam, burrow, or lodge; (b) water width (cm) at 10 m downstream of the dam or at the burrow or lodge; (c) stream velocity (cm/second); (d) average bank height (cm) from 50 m downstream to 50 m upstream of the dam, burrow, or lodge; and (e) distance from dam, burrow, or lodge to nearest woody vegetation (m). Response variables are presence of a dam (1) or absence of a dam (0). Threshold values are indicated by dashed lines. (Swinnen, K. R. R., Rutten, A., Nyssen, J., and Leirs, H., 2018. Environmental factors influencing beaver dam locations. *Journal of Wildlife Management*, 83, 356–364. © 2018 The Wildlife Society (Swinnen et al., 2018).)

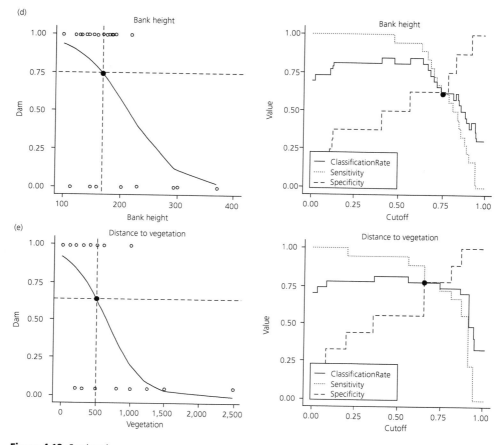

Figure 4.12 Continued

dam-site selection sometimes have been regarded as synonyms (McComb et al., 1990; Barnes and Mallik, 1997; Suzuki and McComb, 1998).

Macfarlane et al. (2017) modelled the capacity of riverscapes to support beaver dams. The model predicted the potential for beaver dams throughout 40,561 km of stream in Utah, USA, and portions of surrounding states, predicting an overall network capacity of 356,294 dams at an average capacity of 8.8 dams/km. They validated the performance of the predictive ability of their model by comparing the predictions with 2,852 known dams across 1,947 km of streams. The model showed excellent

agreement with observed dam densities where beaver dams were present. Beaver dam distribution on perennial streams and rivers was related to the distribution of preferred riparian vegetation, local flow regime, and stream gradient. More recently, Graham et al. (2020) adapted the dam capacity model of Macfarlane et al. (2017) to work in European landscapes classifying suitability of river reaches for dam construction and estimating the location and number of dams at the catchment scale. Model outputs produced by Graham et al. (2020) showed a strong relationship to validation datasets of dam locations in Scotland and England.

The following variables were considered to be appropriate for dam-site selection in a study in northwestern Quebec, Canada: stream gradient, dam upstream catchment area, stream order, channel geomorphic unit (stream width, depth, or cross-sectional area), floodplain size, riverbank slopes, and substrate type (Tremblay et al., 2017). The presence, abundance, size, and distribution of deciduous stems frequently used by beavers, either as food or as building material, might also be crucial variables (Fryxell and Doucet, 1991; Barnes and Mallik, 1996). Several modelling studies concluded that food availability was not an important factor in the selection of dam-building sites (Howard and Larson, 1985; Beier and Barrett, 1987; McComb et al., 1990; Barnes and Mallik, 1997). However, in Michigan's Jordan River Watershed, USA, the majority of dam sites were located in close proximity to aspen stands, i.e. 53% of dams were within 50 m of stands (Martin et al., 2015). A Canadian study determined that the density of woody vegetation with a diameter from 1.5 to 4.4 cm along the waterway was also important for the choice of dam location (Barnes and Mallik, 1996).

Other predictors related to beaver dam abundance include the presence of anthropogenic constructions (Jensen et al., 2001; Jakes et al., 2007), although no association between the number of beaver dams and the number of intersections between roads and streams was found in a study area in Quebec, Canada (Lapointe St-Pierre et al., 2017). In northern New Mexico, USA, biotic factors that might limit the occurrence of dam-building beavers were evaluated (Small et al., 2016). It was found that dam-building beavers were exceptionally rare on public lands managed for cattle grazing.

The presence of clay and silt deposits is often associated with the construction of beaver dams in a forest landscape in Quebec in Canada (Lapointe St-Pierre et al., 2017). Beavers use clay and silt substrate as building material; they usually do not tend to build dams on rocky substrate (Slough and Sadleir, 1977). The mean stream gradient was the top variable influencing dam abundance in Quebec, and these results showed that dam abundance at the landscape-scale increased with hardwood cover whereas it decreased with non-forested landcover.

There was also evidence of substantial variability of the effect of variables such as hardwood cover and non-forested cover among the six ecoregions studied (Lapointe St-Pierre et al., 2017).

Beavers in Europe rarely build dams when they occupy lakes, gravel pits, or wide rivers (>6 m) with friable banks and an abundance of easily accessible food (Hartman and Törnlöv, 2006). For example, in an area in Telemark, Norway, no dams exist in any of the main river canals that are 20–150 m wide and typically between 4 and 10 m deep. However, in some extreme situations dams are built by North American beavers in rivers up to 46 m wide (Müller and Watling, 2016). Beaver dams are most frequently present on streams 3–4 m wide in Oregon, USA and no dams on streams with widths > 10 or < 2 m (Suzuki and McComb, 1998). Beavers living on narrower watercourses in Sweden (< 6 m wide and < 0.7–1 m deep), however, often build dams and can create extensive systems of multiple impoundments (Hartman and Törnlöv, 2006). Likewise, in the central Oregon Coast Range, *ca.* > 90% of dams were on watercourses < 6 m wide (Suzuki and McComb, 1998). Stream flow must be sufficient to allow for ponded deep water year-round but also low enough that dams are not frequently breached by floods (Macfarlane et al., 2017). The slope of the watershed is also important. If it is too steep (> 18%), beavers cannot maintain dams (Retzer et al., 1956). Streams on landslides in southern Colorado and New Mexico, USA, have a greater portion of their gradients below which geologic and ecological literature suggest is a reasonable upper threshold (12%) for beaver dam maintenance (Krueger and Johnson, 2016). The majority (87%) of beaver dams identified in another study in Michigan's Jordan River watershed, USA, fell within the typical slope range for beaver habitation (1–4%) and few dams (4%) were located in areas of relatively extreme slopes (> 8%) (Martin et al., 2015).

Dams are also constructed at the outlets of existing bodies of water. Culverts with and without dams in northwestern Quebec were compared and showed that catchment surface, cumulative length of all local streams within a 2-km-radius, and road embankment height had a negative effect on the probability of dam construction on culverts, while flow and culvert diameter ratio had a positive effect (Tremblay et al., 2017).

The incidence of damming varies according to the characteristics of a watershed. Many studies identify that there is a larger proportion of dams on first- and second-order streams than on higher-stream orders (Snodgrass, 1997; McCullough et al., 2004; Martin et al., 2015). Under certain environmental conditions dam building can be common and families often construct more than one dam along a stretch of watercourse (Hartman and Törnlöv, 2006). By building dams, beavers can change a small watercourse into a series of smaller ponds. One waterway in Poland reported 24 dams in a 1.3-km reach (Żurowski, 1989) and 16 dams/km were found in Quebec, Canada (Naiman et al., 1988). It is also common to find two or three dams close together, as they better control the flow of water in a drainage system. One of the dams can function as the main dam while the others are support dams. 'Support dams' reduce the pressure of the water reaching the main dam, thereby reducing the loss of water that runs through the dam. Dams are most often built downstream from the lodge; the main dam (where the lodge is located) is always located downstream. On one river system in Norway, ~10% of beaver territories had actively maintained dams (Parker and Rønning, 2007). In a Polish study, 19.5% of beaver territories had active damming (Żurowski and Kasperczyk, 1986), while 19% of all beaver families in Bavaria had dams (Zahner, 2001). Russian studies report figures ranging from 19 to 53% (Danilov and Kan'shiev, 1983). In North America an average of 3.2 beaver dams per territory were found in Minnesota, USA (Longley and Moyle, 1963; Johnston and Naiman, 1990). In other areas there may be as many as 40 dams per territory (Müller-Schwarze, 2011).

4.4.2 Why do beavers build dams?

Beavers do not typically build dams unless they wish to retain, stabilize, and manage water levels. If required water depths are not met by naturally occurring riparian conditions, beavers are likely to build dams to increase depth to ensure a constant water level and reduce the flow of water. The most important parameter to correctly (with 97% accuracy) classify territories as dam territory or control territory is water depth. It is very likely that beavers will build dams if the water depth is < 68 cm in summer (Swinnen et al., 2018).

Dam building tends to occur in late summer and autumn, but beavers repair dams throughout the year whenever necessary (Müller-Schwarze, 2011; Raškauskaité and Šimkevičius, 2017). The legendary Swedish beaver researcher Lars Wilsson undertook a number of behavioural experiments on captive beavers, concluding that it was the sound of running water that stimulated dam-building activities (Wilsson, 1971). When he played a recording of running water through a speaker located behind a wall of the beavers' pool, beavers built a dam along that wall. This belief has often become folklore, with newer research demonstrating that it is the amount of water at the site that is the decisive factor in triggering dam-building activities (Żurowski, 1992). Other factors triggering such activity include pregnancy and the presence of kits (Buech, 1995). Larger families rebuild dams faster (Raškauskaité and Šimkevičius, 2017). Once the water level behind the dam has reached a certain point, beavers stop working on the dam even if water starts to run over the top. It appears that a depth of ~70 cm behind the dam is the 'target', though higher dams have been observed, especially on streams with high banks. Falling below a critical point will trigger beavers to repair leaks in the dam.

The increased water surface and depth has a number of advantages for the beaver as follows:

- It enables the beaver to build a secure place to live, with entrance(s) below the surface year-round, thereby providing protection against predators (Hodgdon and Hunt, 1966; Müller-Schwarze, 2011). Beaver dams increase the water level by an average of 46–47 cm in central Belgium and southern Netherlands and Sweden, respectively (Hartman and Törnlöv, 2006; Swinnen et al., 2018).
- It provides a safer environment through which beavers move. Beavers strongly prefer to move through water as opposed to travelling overland, which may involve more energy expenditure

and leaves them vulnerable to predators (Müller-Schwarze, 2011; Gable et al., 2018).

- It increases the availability of broad-leafed trees near the water's edge, which shortens the distance they must travel on land to find food (Breck et al., 2003).
- It becomes easier to transport building materials, which can be used to build or repair the lodge or food cache (Rosell and Pedersen, 1999; Hartman and Axelsson, 2004).
- It prevents the pond from freezing solid, which can be a catastrophe for the beaver (Rosell and Pedersen, 1999; see Chapter 8.3.6).
- It increases chances of building a large enough food cache near the entrances(s) to their living space, which reduces the danger of their food getting stuck in the ice during the winter (Hartman and Axelsson, 2004; Müller-Schwarze, 2011; see also Section 4.3).
- It gives beavers the possibility of opening the dams during winter in order to gain air space under the ice and a larger water surface in which to swim. Beavers will, therefore, be able to swim longer distances under the ice to openings and this allows them to swim farther in the winter. It can also enable them to move between two ponds (Rosell and Pedersen, 1999).

The beaver's dam building can occasionally lead to the destruction of its own food resources. This happens when the broad-leaved trees near the water become submerged and die (Campbell et al., 2012), thus causing the food source to disappear, and the beaver must move to a new area.

4.4.3 Dam construction

Beavers skilfully integrate different materials into the tightly interwoven structure, often utilizing natural features such as a narrowing of a channel, fallen trees, or large rocks to assist the building (Doucet et al., 1994b; Barnes and Mallik, 1996; Baker and Hill, 2003) (Figure 4.13). Dam construction technique can vary according to the hydrology of a site. In standing, shallow water, beavers raise the water level gradually by depositing mud, leaves/vegetation, and small twigs. In larger streams, a solid base is required, and dams are built by placing sticks downstream, against the flow of water and pointing upstream at an angle of about 30 degrees. Beavers weigh down these poles with heavy stones, pushing up sedimentary material and stuffing vegetation between any gaps (Macdonald et al., 1995; Müller-Schwarze, 2011).

Figure 4.13 Structure of the downstream side of a dam. (Photo supplied courtesy of Frank Rosell.)

In Algonquin Provincial Park, Ontario, beavers preferentially used coniferous species and speckled alder in dam construction (Doucet et al., 1994a). Dams can also be constructed from various other material including herbaceous materials, mud, stones, and aquatic vegetation (Doucet et al., 1994b; Barnes and Mallik, 1996; Baker and Hill, 2003). Beavers often choose building materials they would not choose as food (Barnes and Mallik, 1996) and/ or de-bark edible woody vegetation first. Following dam selection, beavers will often start on the side of the river where the current is weakest, where the bank is higher, or where riverbanks are close together. A beaver builds first one side, then the other, before blocking the current in the middle, or it can build the dam evenly across the waterway. The strength of the current and width of the waterway determine the method that is used (Rosell and Pedersen, 1999). Logs and branches are placed in the direction of the current at a narrow point along the creek in such a way that these materials get stuck (Barnes and Mallik, 1996). The beaver can also use previously felled trees, larger stones, or an outcrop in the riverbed as a starting point.

The sizes of the pieces of woody material that compose the dam generally relate to browse selection and composition of the riparian forest. Beavers prefer stick sizes between 1 and 5 cm in diameter and up to 3 m in length (Johnston and Naiman, 1990; Barnes and Mallik, 1996; Haarberg and Rosell, 2006). In central Maryland, USA, beaver selection of stick diameter was a subset of the overall range of stick diameter available, which might be expected to correlate with certain physiological constraints of the beaver (e.g. front paw grasping radius or maximum jaw range of motion) or optimization behavioural programs (e.g. avoiding smallest-diameter wood because of the low return on overall mass of wood per unit effort expended) (Blersch and Kangas, 2014).

Beavers use other types of materials if they do not have access to their preferred building material or need to supplement them. Peat and moss, soil, clay, mud, and stones are used to create an increasingly strong, even, solid, and watertight construction. Where the riverbed is rocky, beavers will shove smaller stones together on the upstream side, while larger stones are carried onto the dam where they roll down and increase the support of the

downstream side. A beaver can lift stones weighing up to 8 kg (Rosell and Pedersen, 1999), and two beavers have been observed moving a 15.9-kg stone (Richards, 1983). When wood is not in sufficient supply, beavers also build dams from rocks with diameter of up to 30 cm, combined with some wood. The largest stones used in central Yukon, USA, weighed approximately 4–5 kg (Jung and Staniforth, 2010). Small dams made from cereal or crop stalks have been found in Bavaria, Germany, and Scotland. Beaver dams have also been built exclusively from conifers, coal, stones, or grasses, but these are rare (Rosell and Pedersen, 1999). Dams are usually very strong and solid and can be used as 'bridges' by larger mammals, including moose (*Alces alces*), livestock and humans (see Chapter 2).

When moving building materials, the beaver swims with branches, twigs, and pieces of logs in its mouth. The thickest end is driven against the bottom of the waterway and the branches are attached to those already in the dam wall. As the dam grows, rocks are often positioned to help hold everything in place. On the downstream side of the dam, the beaver will attach large, long logs and branches. These logs and branches are placed parallel to the current and down to the original water level of the stream. To ensure that they are securely in place, a beaver will hold the branches with its fore legs while using its mouth to jam the branches into the riverbed. In this manner, the dam is secured against increasing water pressure. Beavers carefully place new logs in the dam; if obviously displeased, a beaver removes the log again and places it somewhere else until it is convinced that the log is used in the most effective manner (Wilsson, 1971).

The dam is usually reinforced from the upstream side, where logs and branches are carried by the current, but it can also be reinforced on the downstream side. Reinforcement is performed using excavated materials from the streambed, thus resulting in a deepening of the pond just above the dam. This approach helps to reduce the speed of the water and protects the dam. Much of the reinforcement work is performed underwater, but the beaver can also sit upright on the dam with its collection of building materials and shove them into place between the logs and branches using its front legs (Rosell and Pedersen, 1999).

When building dams, two animals can work together to move heavy objects. Although beavers generally work alone, they do cooperate. In her book *Beaversprite*, Dorothy Richards (1983) reports that she once observed two beavers working together to move a large, heavy piece of timber. They placed their teeth in strategic locations and, with their feet firmly planted on the ground, shoved as hard as they could. Hope Ryden, author of the book *The Lily Pond*, also recounts several occasions where two animals worked together (Ryden, 1990). She twice observed a pair of beavers cooperating to move a large log into the water and over the dam. On another occasion, a beaver struggling to remove a branch stuck in a pile of floating debris received help from another beaver after it began to grunt. The helper held the debris steady using its fore legs and teeth, while the first beaver was able to extract the branch it was after.

As a beaver dam grows, the difference between the two sides becomes apparent. The side facing against the current is built high and becomes very solid, so that water passage is reduced. The other side is more angled from sediment capture, which allows the dam to withstand the pressure of the water formed behind the dam. Longer dams are often curved to offer the greatest amount of support against the weight of the running water (Wilsson, 1971).

When complete, beaver dams are fronted on their upstream side by ramps of captured silt or specifically positioned debris. This design ensures that they are highly resilient to flood conditions. Non-beaver-related 'debris' dams often occur in streams where detritus is trapped against an obstacle in the watercourse. These dams can be distinguished from beaver dams by a lack of uniformity in their structure and an absence of beaver-cut sticks.

4.4.4 Dam dimensions

Dams can be as low as 20 cm or as high as 5 m and from 0.3 m to several hundred metres in length (Putnam, 1955; Müller-Schwarze, 2011). The longest recorded dam built by the Eurasian beaver was found on the Ostrovskaja River in Russia. It was 265 m long and between 0.1 and 1.6 m high (Zavyalov, 2011). In Telemark in Norway, a dam that was 3.5 m high was recorded (Collen and Gibson, 2001). In North America, beavers have built dams across valleys that are several hundred metres wide. One beaver dam in Wood Buffalo National Park in Alberta, Canada, was 850 m long and it is visible from space (Brown and Fouty, 2011). Thus, beaver dam dimensions are highly variable and dependent on the various factors mentioned in this chapter.

4.4.5 Dam maintenance and failure

Beavers can spend a significant amount of time maintaining dams, even on a daily basis. They repair or abandon a damaged dam, and functional dams might be abandoned and slowly break down in time when resident beavers die or emigrate from a territory (Andersen and Shafroth, 2010). Beavers rebuild dams sooner and faster during autumn than during spring (Raškauskaitė and Šimkevičius, 2017). Maintenance of active dams requires constant attention from resident beavers. In some cases, they build an overflow at each end, which can be enlarged during heavy rains and protects the dam from being destroyed. Once the rains subside, the beaver will probably need to rebuild to avoid water levels from falling below the burrow or lodge entrance and leaving the beavers exposed. Many dams that are damaged by spring floods are often, but not always, built up again. During summer, beavers tend to spend less time repairing dams and start these activities again during the autumn. Dam building in most cases can therefore be assumed to be related to preparation for winter.

In Arizona, USA, the dam failure–flood intensity relationship was investigated (experimental floods with peak discharge ranging from 37 to 65 m^3s^{-1}) and the major damage (breach ≥ 3m wide) occurred at ≥ 20% of monitored dams (Andersen and Shafroth, 2010). A similar or higher proportion of dams were moderately damaged. No relationship between damage to the dam and its size were detected or between flood discharge and the level of damage. Dam constituent material appeared to control the probability of major damage at low (attenuated) flood magnitude. Other studies report that beaver dams are washed out by flows of 12–325 m^3s^{-1} (Rutherford, 1953; Butler, 1989; Hillman, 1998;

Westbrook et al., 2006; Dalbeck and Weinberg, 2009). The age of the dam affects its resistance to failure in a complex manner. Wood components add weight as they lose buoyancy (become waterlogged), but bottom roughness and internal integrity weaken as the wood loses mass and rigidity through decomposition (Andersen and Shafroth, 2010).

Failure of beaver dams is a more common phenomenon (at least in North America) than often assumed in the literature (Butler and Malanson, 2005) and these failures might be influenced by the activities of other species. In two studies in Alberta, Canada, a beaver dam had been weakened or undermined by muskrats (*Ondatra zibethicus*) tunnelling under or into the dam (Hillman, 1998) and river otters (*Lontra canadensis*) had constructed tunnels through beaver dams (Reid et al., 1988). Beaver dams may also fail as the result of a variety of other processes such as high-intensity precipitation, rapid snowmelt, human destruction of portions of dams, and collapse of upstream dams (Butler and Malanson, 2005). The length of time that dams persist in the environment varies and can be relatively short, particularly if nearby food resources become depleted and the dams, therefore, are not worth maintaining (Rosell et al., 2005; Halley et al., 2009). Most beaver dams persist less than a decade (Gurnell, 1998), but 20% of the beaver families in Algonquin Provincial Park, Ontario, persisted beyond the 11-year study period (Fryxell, 2001). Some dams are maintained by several generations of beavers and have lasted ~150 years (Johnston, 2015).

4.5 Trails and canals

Beavers use specific trails when moving among their foraging spaces and transporting materials, which tend to become well-worn, eroded paths (Wilsson, 1971; Novak, 1987; Müller-Schwarze, 2011). These forage trails (or beaver runs, bankslides, or 'tunnel slides') can be a few to several hundred metres long (Meentemeyer et al., 1998; Figure 4.14). Following such trails from the shoreline will often lead to a foraging area and more beaver field signs. Beavers may purposefully modify these trails to make them easier to navigate by chewing through fallen tree trunks or digging through uneven substrate.

Grudzinski et al. (2019) reviewed why, where, and how beavers develop canals. They concluded that beavers mainly develop canals to increase accessibility to riparian resources, facilitate transport of harvested resources, and decrease predation

Figure 4.14 Long foraging trail leading to an aspen stand. (Photo supplied courtesy of Frank Rosell.)

risk. They classified beaver canals into three categories based on their landscape position: (1) extension canals, i.e. from a beaver pond or occupied wetland into a riparian area; (2) connector canals, i.e. connecting two hydrologically isolated aquatic environments; and (3) benthic canals, i.e. an excavated route at the bottom of a beaver pond or wetland (Figure 4.15).

Beaver canals are basically forage trails built through active digging or repeated trail use which then fill with water (Wilsson, 1971; Müller-Schwarze, 2011) and are usually made during autumn (Townsend, 1953). All canals require regular maintenance. The canals are often deepened through 'dredging' in which beavers shove any dredged material along the edge of the canal (Figure 4.16). The canals, even with very shallow water, provide safe access routes back to deeper water should they feel threatened while on the land. These canals also make the transportation of foraged material, especially branches, easier. Beavers dredge out channels in deeper water bodies, such as ponds, during drought periods. These channels provide waterways in an otherwise useless mudflat (Müller-Schwarze, 2011; Grudzinski et al., 2019).

Canal width is usually between 0.4 and 2.9 m and depth can range from 0.2 to 1.2 m (Zavyalov, 2011; Hood and Larson, 2015; Grudzinski et al., 2019). In some situations, the canals can be very long, for instance between different waterways. A network of canals can be as long as 1,290 m but is normally around 23 m long (Jasiulionis and Ulevičius, 2011; Hood and Larson, 2015). These canals create access corridors, which are also exploited as habitats by many other species, such as aquatic invertebrates and amphibians. In times of drought, beavers will create small dams or muddy impoundments in canals to maintain a basic water level, and these features afford a vital resource for a range of other wildlife (Hood and Larson, 2015; see Chapter 9).

Figure 4.15 Beaver pond (1) with a stick lodge (2). Dams (3) are built to regulate water levels to maintain sufficient depth. A feeding station (4) is visible as a pile of discarded sticks. The extension canal (5) provides access to a riparian area. A beaver burrow (6) can sometimes act as a tunnel to access foraging areas. The wetland (7) contains beaver-felled trees, characterized by the pencil-like shape. Benthic canals (8) are excavated in the bed of the waterbody. A beaver slide is also visible (9). By creating canals beavers can spend less time on land when accessing foraging areas, limiting risk of predation (10). The final canal type shown is a connector canal (11) which joins two previously isolated areas. (Illustration provided courtesy of Oskar Lacy Corral.)

Figure 4.16 Beaver-created canals. (Photo supplied courtesy of Frank Rosell.)

References

Adriaenssens, V., de Baets, B., Goethals, P. L., and de Pauw, N. 2004. Fuzzy rule-based models for decision support in ecosystem management. *Science of the Total Environment*, 319, 1–12.

Aeschbacher, A. and Pilleri, G. 1983. *Observation on the Building Behaviour of the Canadian Beaver (Castor canadensis) in Captivity. Investigations on Beaver*. Berne: Brain Anatomy Institute.

Alakoski, R., Kauhala, K., and Selonen, V. 2019. Differences in habitat use between the native Eurasian beaver and the invasive North American beaver in Finland. *Biological Invasions*, 21, 1601–1613.

Alcorn, G. D. 1934. Unusual home of beaver. *Murrelet*, 24, 82.

Allen, A. W. 1982. *Habitat Suitability Index Models: Beaver*. Washington, DC: US Fish and Wildlife Service.

Andersen, D. C. and Shafroth, P. B. 2010. Beaver dams, hydrological thresholds, and controlled floods as a management tool in a desert riverine ecosystem, Bill Williams River, Arizona. *Ecohydrology*, 3, 325–338.

Anderson, C. B., Pastur, G. M., Lencinas, M. V., Wallem, P. K., Moorman, M. C., and Rosemond, A. D. 2009. Do introduced North American beavers *Castor canadensis* engineer differently in southern South America? An

overview with implications for restoration. *Mammal Review*, 39, 33–52.

Anderson, J. T. and Bonner, J. L. 2014. Modeling habitat suitability for beaver (*Castor canadensis*) using geographic information systems. *Proceedings of the 4th International Conference on Future Environment and Energy*, Melbourne, Australia.

Baker, B. W. and Hill, E. 2003. *Beaver* (Castor canadensis*)*, 2nd edn. Baltimore: Johns Hopkins University Press.

Barela, I. A. and Frey, J. K. 2016. Habitat and forage selection by the American beaver (*Castor canadensis*) on a regulated river in the Chihuahuan Desert. *The Southwestern Naturalist*, 61, 286–293.

Barnes, D. M. and Mallik, A. U. 1996. Use of woody plants in construction of beaver dams in northern Ontario. *Canadian Journal of Zoology*, 74, 1781–1786.

Barnes, D. M. and Mallik, A. U. 1997. Habitat factors influencing beaver dam establishment in a northern Ontario watershed. *Journal of Wildlife Management*, 61, 1371–1377.

Beier, P. and Barrett, R. H. 1987. Beaver habitat use and impact in Truckee River Basin, California. *Journal of Wildlife Management*, 51, 794–799.

Belovsky, G. E. 1984. Summer diet optimization by beaver. *American Midland Naturalist*, 111, 209–222.

Bergman, B. G., Bump, J. K., and Romanski, M. C. 2018. Revisiting the role of aquatic plants in beaver habitat selection. *American Midland Naturalist*, 179, 222–246.

Blanchet, M. J. H. 1959. Was machen die 'Taler-Biber' an der Versoix. ??, 53, 119–123.

Blersch, D. M. and Kangas, P. C. 2014. Signatures of self-assembly in size distributions of wood members in dam structures of *Castor canadensis*. *Global Ecology and Conservation*, 2, 204–213.

Breck, S. W., Wilson, K. R., and Andersen, D. C. 2003. Beaver herbivory of willow under two flow regimes: a comparative study on the Green and Yampa rivers. *Western North American Naturalist*, 63, 463–471.

Breck, S. W., Goldstein, M. I., and Pyare, S. 2012. Site-occupancy monitoring of an ecosystem indicator: linking characteristics of riparian vegetation to beaver occurrence. *Western North American Naturalist*, 72, 432–441.

Brown, S. T. and Fouty, S. 2011. Beaver wetlands. *Lakeline*, 34–38.

Brzyski, J. R. and Schulte, B. A. 2009. Beaver (*Castor canadensis*) impacts on herbaceous and woody vegetation in southeastern Georgia. *American Midland Naturalist*, 162, 74–86.

Buech, R. R. 1995. Sex difference in behavior of beavers living in near-boreal lake habitat. *Canadian Journal of Zoology*, 73, 2133–2143.

Buech, R. R., Rugg, D. J., and Miller, N. L. 1989. Temperature in beaver lodges and bank dens in a near-boreal environment. *Canadian Journal of Zoology*, 67, 1061–1066.

Busher, P. E. 1991. *Food caching behaviour by the North American beaver,* Castor canadensis, *in western Massachusetts. Global Trends in Wildlife Management.* Krakow: Swiat Press.

Busher, P. E. 1996. Food caching behavior of beaver (*Castor canadensis*): selection and use of woody species. *American Midland Naturalist* 135, 343–348.

Busher, P. E. 2003. Food caching behaviour of the American beaver in Massachusetts. *Lutra*, 46, 139–146.

Busher, P. E., Mayer, M., Ulevičius, A., Samus, A., Hartman, G. and Rosell, F. 2020. Food caching behavior of the Eurasian beaver in Northern Europe. *Wildlife Biology*, 2020(3), 1-11.

Butler, D. R. 1989. The failure of beaver dams and resulting outburst flooding: a geomorphic hazard of the southeastern Piedmont. *The Geographical Bulletin*, 31, 29–38.

Butler, D. R. and Malanson, G. P. 2005. The geomorphic influences of beaver dams and failures of beaver dams. *Geomorphology*, 71, 48–60.

Campbell, R. D., Nouvellet, P., Newman, C., Macdonald, D. W., and Rosell, F. 2012. The influence of mean climate trends and climate variance on beaver survival and recruitment dynamics. *Global Change Biology*, 18, 2730–2742.

Campbell-Palmer, R., Gow, D., Needham, R., Jones, S., and Rosell, F. 2015. *The Eurasian Beaver.* Exeter: Pelagic Publishing.

Cardenas, M. B. 2009. Stream–aquifer interactions and hyporheic exchange in gaining and losing sinuous streams. *Water Resources Research*, 45, 1–13.

Collen, P. and Gibson, R. J. 2001. The general ecology of beavers (*Castor* spp.), as related to their influence on stream ecosystems and riparian habitats, and the subsequent effects on fish—a review. *Reviews in Fish Biology and Fisheries*, 10, 439–461.

Cox, D. R. and Nelson, T. A. 2009. Beaver habitat models for use in Illinois streams. *Illinois State Academy of Science*, 102, 55–64.

Dalbeck, L. and Weinberg, K. 2009. Artificial ponds: a substitute for natural beaver ponds in a central European highland (Eifel, Germany)? *Hydrobiologia*, 630, 49–62.

Danilov, P. 1995. Canadian and Eurasian beavers in Russian north-west (distribution, number, comparative ecology). *Proceedings of the 3rd Nordic Beaver Symposium*, Helsinki, Finland.

Danilov, P. I. and Fyodorov, F. V. 2015. Comparative characterization of the building activity of Canadian and European beavers in northern European Russia. *Russian Journal of Ecology*, 46, 272–278.

Danilov, P. I. and Kan'shiev, V. Y. 1983. State of populations and ecological characteristics of European (*Castor fiber* L.) and Canadian (*Castor canadensis* Kuhl.) beavers in the northwestern USSR. *Acta Zoologica Fennica*, 174, 95–97.

Danilov, P., Kanshiev, V., and Fyodorov, F. 2011. Characteristics of North American and Eurasian beaver ecology in Karelia. In: Sjøberg, G. and Ball, J. P. (ed.) *Restoring the European Beaver: 50 Years of Experience.* Sofia-Moscow: Pensoft.

Davis, E. F., Valenzuela, A. E. J., Murcia, S., and Anderson, C. B. 2016. Habitat use by invasive North American beaver during intermediate and long-term colonization periods in southern Patagonia. *Mastozoología Neotropical*, 23, 51–61.

Dieter, C. D. and McCabe, T. R. 1989. Factors influencing beaver lodge-site selection on a prairie river. *American Midland Naturalist*, 122, 408–411.

Dittbrenner, B. J., Pollock, M. M., Schilling, J. W., Olden, J. D., Lawler, J. J., and Torgersen, C. E. 2018. Modeling intrinsic potential for beaver (*Castor canadensis*) habitat to inform restoration and climate change adaptation. *PLoS ONE*, 13, e0192538.

Djoshkin, W. W. and Safonov, V. G. 1972. *Die Biber der alten und neuen Welt.* Wittenberg: A. Ziemsen.

Doucet, C. M., Adams, I. T., and Fryxell, J. M. 1994a. Beaver dam and cache composition: are woody species used differently? *Ecoscience*, 1, 268–270.

Doucet, C. M., Walton, R. A., and Fryxell, J. M. 1994b. Perceptual cues used by beavers foraging on woody plants. *Animal Behaviour*, 47, 1482–1484.

Dyck, A. P. and MacArthur, R. A. 1993. Daily energy requirements of beaver (*Castor canadensis*) in a simulated winter microhabitat. *Canadian Journal of Zoology*, 71, 2131–2135.

Dzięciołowski, R. and Misiukiewicz, W. 2002. Winter food caches of beavers *Castor fiber* in NE Poland. *Acta Theriologica*, 47, 471–478.

Easter-Pilcher, A. 1990. Cache size as an index to beaver colony size in northwestern Montana. *Wildlife Society Bulletin*, 18, 110–113.

Elsey, R. M., Platt, S. G. and Shirley, M. 2015. An unusual beaver (*Castor canadensis*) lodge in a Louisiana coastal marsh. *Southeastern Naturalist*, 14, N28–N30.

Francis, R. A., Taylor, J. D., Dibble, E., Strickland, B., Petro, V. M., Easterwood, C., and Wang, G. 2017. Restricted cross-scale habitat selection by American beavers. *Current Zoology*, 63, 703–710.

Fryxell, J. M. 2001. Habitat suitability and source–sink dynamics of beavers. *Journal of Animal Ecology*, 70, 310–316.

Fryxell, J. M. and Doucet, C. M. 1991. Provisioning time and central-place foraging in beavers. *Canadian Journal of Zoology*, 69, 1308–1313.

Fustec, J., Cormier, J.-P., and Lodé, T. 2003. Beaver lodge location on the upstream Loire River. *Comptes Rendus Biologies*, 326, 192–199.

Gable, T. D., Windels, S. K., Bruggink, J. G., and Barber-Meyer, S. M. 2018. Weekly summer diet of gray wolves (*Canis lupus*) in northeastern Minnesota. *American Midland Naturalist*, 179, 15–27.

Gore, J. A. and Baker, W. W. 1989. Beavers residing in caves in northern Florida. *Journal of Mammalogy*, 70, 677–678.

Gorshkov, D. and Gorshkov, Y. 2011. *Feeding Strategy of Beavers in Tatarstan Republic*. Sofia-Moscow: Pensoft.

Graham, H. A., Puttock, A., Macfarlane, W. W., Wheaton, J. M., Gilbert, J. T., Campbell-Palmer, R., Elliott, M., Gaywood, M. J., Anderson, K., and Brazier, R. E. 2020. Modelling Eurasian beaver foraging habitat and dam suitability, for predicting the location and number of dams throughout catchments in Great Britain. *European Journal of Wildlife Research*, 66, 1–18.

Grater, R. K. 1936. An unusual beaver habitat. *Journal of Mammalogy*, 17, 66.

Grudzinski, B. P., Cummins, H., and Vang, T. K. 2019. Beaver canals and their environmental effects. *Progress in Physical Geography: Earth and Environment*, 44, 189–211.

Gurnell, A. M. 1998. The hydrogeomorphological effects of beaver dam-building activity. *Progress in Physical Geography*, 22, 167–189.

Haarberg, O. and Rosell, F. 2006. Selective foraging on woody plant species by the Eurasian beaver (*Castor fiber*) in Telemark, Norway. *Journal of Zoology*, 270, 201–208.

Halley, D. J., Jones, A. L., Chesworth, S., Hall, C., Gow, D., Jones-Parry, R., and Walsh, J. 2009. The reintroduction of the Eurasian beaver *Castor fiber* to Wales: an ecological feasibility study. *NINA Report*, 457, 1–66.

Halley, D., Teurlings, I., Welsh, H., and Taylor, C. 2013. Distribution and patterns of spread of recolonising Eurasian beavers (*Castor fiber* Linnaeus 1758) in fragmented habitat, Agdenes peninsula, Norway. *Fauna Norvegica*, 32, 1–12.

Hartman, G. 1996. Habitat selection by European beaver (*Castor fiber*) colonizing a boreal landscape. *Journal of Zoology*, 240, 317–325.

Hartman, G. and Axelsson, A. 2004. Effect of watercourse characteristics on food-caching behaviour by European beaver, *Castor fiber*. *Animal Behaviour*, 67, 643–646.

Hartman, G. and Törnlöv, S. 2006. Influence of watercourse depth and width on dam-building behaviour by Eurasian beaver (*Castor fiber*). *Journal of Zoology*, 268, 127–131.

Hillman, G. R. 1998. Flood wave attenuation by a wetland following a beaver dam failure on a second order boreal stream. *Wetlands*, 18, 21–34.

Histøl, T. 1989. Sommerdiett hos bever *Castor fiber* L. i et utvalg av skogsvann i Vennesla kommune, Vest Agder. *Fauna*, 42, 96–103.

Hodgdon, H. 1978. *Social Dynamics and Behavior Within an Unexploited Beaver Population*. Amherst: University of Massachusetts.

Hodgdon, H. E. and Lancia, R. A. 1983. Behaviour of the North American beaver, *Castor canadensis*. *Acta Zoologica Fennica*, 174, 99–103.

Hodgdon, K. W. and Hunt, J. H. 1966. Beaver management in Maine. *Game Division Bulletin*, 3.

Holland, A. M., Schauber, E. M., Nielsen, C. K., and Hellgran, E. C. 2019. Occupancy dynamics of semi-aquatic herbivores in riparian systems in Illinois, USA. *Ecosphere*, 10, 1–17.

Hood, G. A. 2020. Not all ponds are created equal: long-term beaver (*Castor canadensis*) lodge occupance in a heterogeneous landscape. *Canadian Journal Zoology*, 98, 210–218.

Hood, G. A. and Bayley, S. E. 2008. The effects of high ungulate densities on foraging choices by beaver (*Castor canadensis*) in the mixed-wood boreal forest. *Canadian Journal of Zoology*, 86, 484–496.

Hood, G. A. and Larson, D. G. 2015. Ecological engineering and aquatic connectivity: a new perspective from beaver-modified wetlands. *Freshwater Biology*, 60, 198–208.

Hood, W. G. 2012. Beaver in tidal marshes: dam effects on low-tide channel pools and fish use of estuarine habitat. *Wetlands*, 32, 401–410.

Howard, R. J. and Larson, J. S. 1985. A stream habitat classification system for beaver. *Journal of Wildlife Management*, 49, 19–25.

Hunter, M. D. and Price, P. W. 1992. Playing chutes and ladders: heterogeneity and relative roles of bottom-up and top-down forces in natural communities. *Ecology*, 73, 724–732.

Hyvönen, T. and Nummi, P. 2008. Habitat dynamics of beaver *Castor canadensis* at two spatial scales. *Wildlife Biology*, 14, 302–308.

Jakes, A. F., Snodgrass, J. W., and Burger, J. 2007. *Castor canadensis* (beaver) impoundment associated with geomorphology of southeastern streams. *Southeastern Naturalist*, 6, 271–282.

Jasiulionis, M. and Ulevičius, A. 2011. Beaver impact on canals of land reclamation in two different landscapes. *Acta Zoologica Lituanica*, 21, 207–214.

Jenkins, S. H. 1980. A size–distance relation in food selection by beavers. *Ecology*, 61, 740–746.

Jensen, P. G., Curtis, P. D., Lehnert, M. E., and Hamelin, D. L. 2001. Habitat and structural factors influencing beaver interference with highway culverts. *Wildlife Society Bulletin*, 29, 654–664.

John, F., Baker, S., and Kostkan, V. 2010. Habitat selection of an expanding beaver (*Castor fiber*) population in central and upper Morava River basin. *European Journal of Wildlife Research*, 56, 663–671.

Johnson, D. H. 1980. The comparison of usage and availability measurements for evaluating resource preference. *Ecology*, 61, 65–71.

Johnston, C. A. 2015. Fate of 150 year old beaver ponds in the Laurentian Great Lakes region. *Wetlands*, 35, 1013–1019.

Johnston, C. A. and Naiman, R. J. 1990. Browse selection by beaver: effects on riparian forest composition. *Canadian Journal of Forest Research*, 20, 1036–1043.

Jung, T. S. and Staniforth, J. A. 2010. Unusual beaver, *Castor canadensis*, dams in central Yukon. *Canadian Field-Naturalist*, 124, 274–275.

Krueger, K. A. and Johnson, B. G. 2016. The effect of subalpine landslides on headwater stream gradient and beaver habitat. *Physical Geography*, 37, 344–360.

Kurstjens, G. and Bekhuis, J. 2003. Adaptation of beavers (*Castor fiber*) to extreme water level fluctuations and ecological implications. *Lutra*, 46, 147–151.

Kvamme, H. 1985. *Bever (Castor fiber L.) i østmarka ved Oslo: En populasjonsøkologisk undersøkelse etter en etsetting*. Oslo: Universitetet i Oslo.

Labrecque-Foy, J.-P., Morin, H., and Girona, M. M. 2020. Dynamics of territorial occupation by North American beavers in Canadian boreal forests: a novel dendroecological approach. *Forests*, 11, 1–12.

Lacoul, P. and Freedman, B. 2006. Environmental influences on aquatic plants in freshwater ecosystems. *Environmental Reviews*, 14, 89–136.

Lapointe St-Pierre, M., Labbé, J., Darveau, M., Imbeau, L., and Mazerolle, M. J. 2017. Factors affecting abundance of beaver dams in forested landscapes. *Wetlands*, 37, 941–949.

Lavsund, S. 1989. Winter activity of beaver (*Castor fiber*) in Sweden. *Abstracts: 5th International Theriological Congress*, 22–29 August, Rome.

Longley, W. and Moyle, J. B. 1963. *The Beaver in Minnesota*. St. Paul: Minnesota Department of Conservation, Division of Game and Fish, Section of Research and Planning.

Macdonald, D. W., Tattersall, F. H., Brown, E. D., and Balharry, D. 1995. Reintroducing the European beaver to Britain: nostalgic meddling or restoring biodiversity? *Mammal Review*, 25, 161–200.

Macdonald, D. W., Tattersall, F. H., Rushton, S., South, A. B., Rao, S., Maitland, P. and Strachan, R. 2000. Reintroducing the beaver (*Castor fiber*) to Scotland: a protocol for identifying and assessing suitable release sites. *Animal Conservation*, 3, 125–133.

Macfarlane, W. W., Wheaton, J. M., Bouwes, N., Jensen, M. L., Gilbert, J. T., Hough-Snee, N., and Shivik, J. A. 2017. Modeling the capacity of riverscapes to support beaver dams. *Geomorphology*, 277, 72–99.

Maringer, A. and Slotta-Bachmayr, L. 2006. A GIS-based habitat-suitability model as a tool for the management of beavers *Castor fiber*. *Acta Theriologica*, 51, 373–382.

Martin, S. L., Jasinski, B. L., Kendall, A. D., Dahl, T. A., and Hyndman, D. W. 2015. Quantifying beaver dam dynamics and sediment retention using aerial imagery, habitat characteristics, and economic drivers. *Landscape Ecology*, 30, 1129–1144.

McAlpine, D. F. 1977. Notes on cave utilization by beaver. *The NSS Bulletin*, 39, 90–91.

McComb, W. C., Sedell, J. R., and Buchholz, T. D. 1990. Dam-site selection by beavers in an eastern Oregon basin. *Great Basin Naturalist*, 50, 273–281.

McCullough, M. C., Harper, J. L., Eisenhauer, D. E., and Dosskey, M. G. 2004. Channel aggradation by beaver dams on a small agricultural stream in Eastern Nebraska. Self-Sustaining Solutions for Streams, Wetlands, and Watersheds, 12–15 September. *American Society of Agricultural and Biological Engineers*, 63.

Meentemeyer, R. K., Vogler, J. B., Hill, C., and Butler, D. 1998. The geomorphic influences of burrowing beavers on streambanks, Bolin Creek, North Carolina. *Zeitschrift fur Geomorphologie*, 42, 453–468.

Müller, G. and Watling, J. 2016. The engineering in beaver dams. *River Flow 2016: Eighth International Conference on Fluvial Hydraulics*, Saint Louis, USA.

Müller-Schwarze, D. 2011. *The Beaver: Its Life and Impact*. Ithaca: Cornell University Press.

Naiman, R. J., Johnston, C. A., and Kelley, J. C. 1988. Alteration of North American streams by beaver. *BioScience*, 38, 753–762.

Nelner, T. B. and Hood, G. A. 2011. Effect of agriculture and presence of American beaver *Castor canadensis* on winter biodiversity of mammals. *Wildlife Biology*, 17, 326–336.

Nitsche, K.-A. 2000. Behaviour of beavers (*Castor fiber albicus* Matschie, 1907) during the flood periods. In: Czech, A. and Schwab, G. (eds.) The European beaver in a new millennium. *Proceedings of the 2nd European Beaver Symposium*, 27–30 September, Bialowieza. Krakow: Carpathian Heritage Society, pp. 85–90.

Nix, J. H., Howell, R. G., Hall, L. K., and McMillan, B. R. 2018. The influence of periodic increases of human activity on crepuscular and nocturnal mammals: testing the weekend effect. *Behavioural Processes*, 146, 16–21.

Nolet, B. A. 1994. Return of the beaver to the Netherlands: viability and prospects of a re-introduced population. PhD thesis, University of Groningen.

Nolet, B. A. and Baveco, J. M. 1996. Development and viability of a translocated beaver *Castor fiber* population in the Netherlands. *Biological Conservation*, 75, 125–137.

Northcott, T. H. 1971. Feeding habits of beaver in Newfoundland. *Oikos*, 22, 407–410.

Novak, M. 1987. Beaver. In: Novak, M., Baker, J. A., Obbard, M. E., and Mallock, B. (eds.) *Wild Furbearer Management and Conservation in North America*. Toronto: Queens Printer for Ontario, pp. 283–304.

Osmundson, C. L. and Buskirk, S. W. 1993. Size of food caches as a predictor of beaver colony size. *Wildlife Society Bulletin (1973–2006)*, 21, 64–69.

Østbye, E. K., Tøsti, Ø. E., and Lauritzen, S. E. 1993. Limestone cave as residence for beaver *Castor fiber*. *Fauna*, 46, 16–21.

Pachinger, K. and Hulik, T. 1999. Beavers in an urban landscape. In: Busher, P. E. and Dzięciołowski, R. M. (eds.) *Beaver Protection, Management, and Utilization in Europe and North America*. Boston: Springer.

Parker, H. and Rønning, Ø. C. 2007. Low potential for restraint of anadromous salmonid reproduction by beaver *Castor fiber* in the Numedalslågen River catchment, Norway. *River Research and Applications*, 23, 752–762.

Parker, H., Nummi, P., Hartman, G., and Rosell, F. 2012. Invasive North American beaver *Castor canadensis* in Eurasia: a review of potential consequences and a strategy for eradication. *Wildlife Biology*, 18, 354–365.

Pasternack, G. B., Hilgartner, W. B., and Brush, G. S. 2000. Biogeomorphology of an upper Chesapeake Bay river-mouth tidal freshwater marsh. *Wetlands*, 20, 520–537.

Patenaude, F. 1984. The ontogeny of behavior of free-living beavers (*Castor canadensis*). *Zeitschrift für Tierpsychologie*, 66, 33–44.

Petro, V. M., Taylor, J. D., Sanchez, D. M., and Burnett, K. M. 2018. Methods to predict beaver dam occurrence in coastal Oregon. *Northwest Science*, 92, 278–289.

Pietrek, A. G., Escobar, J. M., Fasola, L., Roesler, I., and Schiavini, A. 2017. Why invasive Patagonian beavers thrive in unlikely habitats: a demographic perspective. *Journal of Mammalogy*, 98, 283–292.

Pilleri, G. 1960. Biber. *Umschau*, 14, 420–422.

Pinto, B., Santos, M. J., and Rosell, F. 2009. Habitat selection of the Eurasian beaver (*Castor fiber*) near its carrying capacity: an example from Norway. *Canadian Journal of Zoology*, 87, 317–325.

Pollock, M. M., Pess, G. R., Beechie, T. J., and Montgomery, D. R. 2004. The importance of beaver ponds to coho salmon production in the Stillaguamish river basin, Washington, USA. *North American Journal of Fisheries Management*, 24, 749–760.

Putnam, E. F. 1955. *Beaver Management and Ecology in Wyoming*. Cheyenne: Wyoming Game and Fish Commission.

Raffel, T. R., Smith, N., Cortright, C., and Gatz, A. J. 2009. Central place foraging by beavers (*Castor canadensis*) in a complex lake habitat. *American Midland Naturalist*, 162, 62–73.

Raškauskaitė, M. and Šimkevičius, K. 2017. Eurasian beaver (*Castor fiber* L) population in ASU science and teaching hunting area and beaver dams rebuild intensity. In: Raupelienė, A. (ed.) *Proceedings of the 8th International Scientific Conference on Rural Development*, Aleksandras Stulginskis University, pp. 769–774.

Ray, A. M., Rebertus, A. J., and Ray, H. L. 2001. Macrophyte succession in Minnesota beaver ponds. *Canadian Journal of Botany*, 79, 487–499.

Rebertus, A. J. 1986. Bogs as beaver habitat in north-central Minnesota. *American Midland Naturalist*, 116, 240–245.

Reid, D. G., Herrero, S. M., and Code, T. E. 1988. River otters as agents of water loss from beaver ponds. *Journal of Mammalogy*, 69, 100–107.

Retzer, J. L., Swope, H. M., Remington, J. D., and Rutherford, H. W. 1956. *Suitability of Physical Factors for Beaver Management in the Rocky Mountains of Colorado*. Denver: State of Colorado Department of Game and Fish.

Richard, P. B. 1985. Peculiarities on the ecology and management of the Rhodanian beaver (*Castor fiber* L.). *Zeitschrift für Angewandte Zoologie*, 72, 143–152.

Richards, D. 1983. *Beaversprite: My Years Building an Animal Sanctuary*. Interlaken, NY: Heart of the Lakes Publishing.

Ritter, T. D., Gower, C. N., and McNew, L. B. 2019. Habitat conditions at beaver settlement sites: implications for beaver restoration projects. *Restoration Ecology*, 28, 1–10.

Roberts, T. S. 1937. How two captive young beavers constructed a food pile. *Proceedings Minnesota Academy of Science Journal*, 5, 24–27.

Rosell, F. and Nolet, B. A. 1997. Factors affecting scent-marking behavior in Eurasian beaver (*Castor fiber*). *Journal of Chemical Ecology*, 23, 673–689.

Rosell, F. and Parker, H. 1995. *Beaver Management: Present Practice and Norway's Future Needs*. Bø i Telemark: Telemark University College.

Rosell, F. and Pedersen, K. V. 1999. *Bever [The Beaver]*. Oslo: Landbruksforlaget.

Rosell, F., Bozér, O., Collen, P., and Parker, H. 2005. Ecological impact of beavers *Castor fiber* and *Castor canadensis* and their ability to modify ecosystems. *Mammal Review*, 35, 248–276.

Rutherford, W. H. 1953. Effects of a summer flash flood upon a beaver population. *Journal of Mammalogy*, 34, 261–262.

Ryden, H. 1990. *Lily Pond: Four Years with a Family of Beavers*. New York: HarperCollins.

Salvensen, S. 1934. Air circulation in the beaver hut. *Journal of Mammalogy*, 15, 322–323.

Scrafford, M. A., Tyers, D. B., Patten, D. T., and Sowell, B. F. 2018. Beaver habitat selection for 24 yr since reintroduction north of Yellowstone National Park. *Rangeland Ecology & Management*, 71, 266–273.

Semyonoff, B. T. 1957. Beaver biology in winter in Archangel Province. *Russian Game Reports*, 1, 71–92.

Shelton, P. C. 1966. Ecological studies of beavers, wolves, and moose in Isle Royale National Park, Michigan. PhD thesis, Purdue University.

Slough, B. G. 1978. Beaver food cache structure and utilization. *Journal of Wildlife Management*, 42, 644–646.

Slough, B. G. and Sadleir, R. M. F. S. 1977. A land capability classification system for beaver (*Castor canadensis* Kuhl). *Canadian Journal of Zoology*, 55, 1324–1335.

Small, B. A., Frey, J. K., and Gard, C. C. 2016. Livestock grazing limits beaver restoration in northern New Mexico. *Restoration Ecology*, 24, 646–655.

Smith, D. W. and Peterson, R. O. 1991. Behavior of beaver in lakes with varying water levels in northern Minnesota. *Environmental Management*, 15, 395–401.

Snodgrass, J. W. 1997. Temporal and spatial dynamics of beaver-created patches as influenced by management practices in a south-eastern North American landscape. *Journal of Applied Ecology*, 34, 1043–1056.

Spence, D. H. N. 1967. Factors controlling the distribution of freshwater macrophytes with particular reference to the lochs of Scotland. *Journal of Ecology*, 55, 147–170.

Stephenson, A. B. 1969. Temperatures within a beaver lodge in winter. *Journal of Mammalogy*, 50, 134–136.

Steyaert, S., Zedrosser, A., and Rosell, F. 2015. Socioecological features other than sex affect habitat selection in the socially obligate monogamous Eurasian beaver. *Oecologia*, 179, 1023–1032.

Stringer, A. P., Blake, D., Genney, D. R., and Gaywood, M. J. 2018. A geospatial analysis of ecosystem engineer activity and its use during species reintroduction. *European Journal of Wildlife Research*, 64, 1–9.

Suzuki, N. and McComb, W. C. 1998. Habitat classification models for beaver (*Castor canadensis*) in the streams of the central Oregon Coast Range. *Northwest Science*, 72, 102–110.

Swenson, J. E. and Knapp, S. J. 1980. Composition of beaver caches on the Tongue River in Montana. *Prairie Naturalist*, 12, 33–36.

Swinnen, K. R. R., Strubbe, D., Matthysen, E., and Leirs, H. 2017. Reintroduced Eurasian beavers (*Castor fiber*): colonization and range expansion across human-dominated landscapes. *Biodiversity and Conservation*, 26, 1863–1876.

Swinnen, K. R. R., Rutten, A., Nyssen, J., and Leirs, H. 2018. Environmental factors influencing beaver dam locations. *Journal of Wildlife Management*, 83, 356–364.

Tape, K. D., Jones, B. M., Arp, C. D., Nitze, I., and Grosse, G. 2018. Tundra be dammed: beaver colonization of the Arctic. *Global Change Biology*, 24, 4478–4488.

Taylor, D. E. M. 1970. Growth, decline, and equilibrium in a beaver population at Sagehen Creek, California. PhD thesis, University of California.

Thomsen, L. R., Sharpe, F., and Rosell, F. 2004. Collapsing burrow causes death of a Eurasian beaver, *Castor fiber*. *Canadian Field-Naturalist*, 118, 434–435.

Touihri, M., Labbé, J., Imbeau, L., and Darveau, M. 2018. North American beaver (*Castor canadensis* Kuhl) key habitat characteristics: review of the relative effects of geomorphology, food availability and anthropogenic infrastructure. *Ecoscience*, 25, 9–23.

Tremblay, G., Valeria, O., and Imbeau, L. 2017. Characterisation of beaver habitat parameters that promote the use of culverts as dam construction sites: can we limit the damage to forest roads? *Forests*, 8, 494–507.

Townsend, J. E. 1953. Beaver ecology in western Montana with special reference to movements. *Journal of Mammalogy*, 34, 459–479.

Ulevičius, A., Kisielytė, N., and Jasiulionis, M. 2011. Habitat use and selectivity by beavers (*Castor fiber*) in anthropogenic landscape. *Ekologija*, 57, 47–54.

Vasin, A. 2001. Beavers of the north of West Siberia. In: Busher, P. E. and Gorshkov, Y. (eds.) *Proceedings of the 1st Euro-American Beaver Congress*, Transactions of Volga-Kama National Naure Zapovednik, pp. 51–60.

Vorel, A., Válková, L., Hamšíková, L., Maloň, J., and Korbelová, J. 2015. Beaver foraging behaviour: seasonal foraging specialization by a choosy generalist herbivore. *Behavioral Ecology and Sociobiology*, 69, 1221–1235.

Wang, G., McClintic, L. F., and Taylor, J. D. 2019. Habitat selection by American beaver at multiple spatial scales. *Animal Biotelemetry*, 7, 1–8.

Westbrook, C. J., Cooper, D. J., and Baker, B. W. 2006. Beaver dams and overbank floods influence groundwater–surface water interactions of a Rocky Mountain riparian area. *Water Resources Research*, 42, 1–12.

Wilsson, L. 1971. Observations and experiments on the ethology of the European beaver (*Castor fiber* L.). *Viltrevy*, 8, 160–203.

Zahner, V. 2001. Dam building by beaver (*Castor fiber*) and its impact on forest stands in South Germany. *Trudy Pervogo Evro-Amerikanskogo Kongressa Po Bobru*, 4, 119–126.

Zavyalov, N. A. 2011. Settlement history, population dynamics and the ecology of beavers (*Castor fiber* L.) in the Darwin Reserve. In: Sjøberg, G. and Ball, J. P. (eds.) *Restoring the European Beaver: 50 Years of Experience.* Sofia-Moscow: Pensoft.

Żurowski, W. 1989. Dam building activity of beavers on the mountainous streams. *Abstracts: 5th International Theriological Congress*, 22–29 August, Rome, pp. 316–317.

Żurowski, W. 1992. Building activity of beavers. *Acta Theriologica*, 37, 403–411.

Żurowski, W. and Kasperczyk, B. 1986. Characteristics of a European beaver population in the Suwałki Lakeland. *Acta Theriologica*, 31, 311–325.

Zwolicki, A., Pudełko, R., Moskal, K., Świderska, J., Saath, S., and Weydmann, A. 2019. The importance of spatial scale in habitat selection by European beaver. *Ecography*, 41, 187–200.

CHAPTER 5

The seasonal vegetarian

5.1 The opportunistic generalist

Without doubt beavers (*Castor* spp.) are the ultimate vegetarian; although certain species are favoured, the range of plants and parts they can utilize is vastly impressive, with little being off the menu. Beavers are opportunistic generalist herbivores that feed on bark (including bast fibre [also called phloem fibre or skin fibre] and cambium), twigs, shoots, and leaves of numerous woody plants, grass, herbs, ferns, aquatic plants, mushrooms, algae, fruits, and crops. Over 450 plant species have to date been recorded in their diets (Wilsson, 1971, Jenkins, 1980, Gorshkov and Gorshkov, 2011, Nitsche, 2017, Dezhkin et al., 1986, Hilfiker, 1991); see Table 5.1 for plants commonly eaten/consumed by beavers). This can even include plants that are poisonous or generally unpalatable for humans and other animals, such as bracken (*Pteridium aquilinum*) and Japanese knotweed (*Fallopia japonica*). Common plant defence mechanisms include toxins, thorns, and burning hairs, but these do not appear to be a hindrance to beavers.

Despite being herbivores, it is sometimes assumed that beavers predate and consume fish. This belief may be due to the fact that they share similar habitats with both fish and their predators, e.g. otters (Lutrinae) and thanks to stories such as *The Chronicles of Narnia*, where a pair of beavers famously cooks a meal of fish. This belief has caused controversies in some fishing sectors and led to beaver removal in some places. This misunderstanding is unfortunately persistent in some places and has been reinforced by the media, leading to the beaver's fish-eating habits becoming an enduring myth. There are rare reports of beavers infrequently scavenging on offered meat, fish, and even hard-boiled eggs in captivity (Rosell and Pedersen, 1999) and one case of a beaver biting into an intact salmon that was hanging from an angler's line (Gleason et al., 2005). A few records exist of wild individuals feeding on discarded Chinook salmon (*Oncorhynchus tshawytscha*) carcasses in south-central Alaska. This behaviour may be a fairly common strategy employed by beavers in Alaskan streams and rivers to take advantage of a superabundant high-energy food source (i.e. protein or fat) when these are seasonally available (Gleason et al., 2005). Although there are occasional records of fish consumption by beavers, this foraging behaviour remains extremely rare and by no means involves predation.

5.1.1 Seasonal variation

Beavers are opportunistic vegetarians and they will utilize what is available during the different seasons (see Section 10.2.2 for captive dietary issues). An investigation of 400 faecal samples from Danish beavers revealed seasonal variation in the utilization of woody and non-woody plants. Non-woody plants were preferred from June to September, while woody plants were preferred from November to May (Elmeros et al., 2003). Based on macro- and micro-histological analysis of 97 faeces, beavers in the Czech Republic generally consumed mostly deciduous trees and forbs (a herbaceous flowering plant) during the growing season (Krojerová-Prokešová et al., 2010; Figure 5.1). Consumption of grasses, aquatic plants, and field crops was negligible throughout the year, whilst in spring, beavers consumed mainly deciduous trees, and during summer and autumn the proportion of forbs significantly increased (Krojerová-Prokešová et al., 2010). In Mississippi, USA, over a one-year

Beavers: Ecology, Behaviour, Conservation, and Management. Frank Rosell and Róisín Campbell-Palmer, Oxford University Press.
© Frank Rosell and Róisín Campbell-Palmer 2022. DOI: 10.1093/oso/9780198835042.003.0005

Table 5.1 Wild plant types commonly eaten/consumed by beavers.

Common name	Plant type	Distribution	Latin name
Alder	Tree	North America and Europe	*Alnus glutinosa*
Alder buckthorn	Tree	North America and Europe	*Frangula alnus*
American white water lily	Aquatic plant	North America	*Nymphaea odorata*
Angelica	Flowering plant	North America and Europe	*Angelica sylvestris*
Apple tree	Tree	North America and Europe	*Malus* spp.
Ash	Tree	North America and Europe	*Fraxinus* spp.
Aspen	Tree	North America and Europe	*Populus* spp.
Aster	Flowering plant	North America and Europe	Asteraceae
Barley	Grass	North America and Europe	*Hordeum vulgare*
Basswood	Tree and shrub	North America and Europe	*Tila* spp.
Bedstraw	Herbaceous plant	North America and Europe	*Galium* spp.
Beech	Tree	North America and Europe	*Fagus* spp.
Bellflower	Flowering plant	North America and Europe	*Campanula* spp.
Bilberry	Shrub	North America and Europe	*Vaccinium myrtillus*
Birch	Tree	North America and Europe	*Betula* spp.
Bird cherry	Tree	Europe	*Prunus padus*
Bittersweet nightshade	Flowering plant	North America and Europe	*Solanum dulcamara*
Black cherry	Tree	North America and Europe	*Prunus serotina*
Black locust	Flowering plant	North America and Europe	*Robinia pseudoacacia*
Blackberry	Flowering plant	North America and Europe	*Rubus fruticosus*
Blackcurrant	Shrub	North America and Europe	*Ribes nigrum*
Blackthorn	Tree	North America and Europe	*Prunus spinosa*
Blueberry	Flowering plant	North America and Europe	*Cyanococcus*
Bog bean (buckbean)	Flowering plant	North America and Europe	*Menyanthes trifoliata*
Bog myrtle	Shrub	North America and Europe	*Myrica gale*
Bog-arum	Flowering plant	Europe	*Calla palustris*
Bottle/beaked sedge	Sedge	North America and Europe	*Carex rostrata*
Bracken	Fern	North America and Europe	*Pteridium aquilinum*
Branched bur-reed	Flowering plant	North America and Europe	*Sparganium erectum*
Broad buckler-fern	Fern	North America and Europe	*Dryopteris dilatata*
Broad-leaved pondweed	Aquatic plant	North America and Europe	*Potamogeton natans*
Broccoli	Flowering plant	North America and Europe	*Brassica oleracea*
Buckthorn	Trees and shrubs	North America and Europe	*Rhamnus* spp.
Butterbur	Flowering plants	North America and Europe	*Petasites* spp.
Cabbages	Flowering plant	North America and Europe	*Brassica oleracea*
Canary grass	Grass	North America and Europe	*Digraphis arundiancea*
Carrots	Flowering plant	North America and Europe	*Daucus carota*
Cattails	Flowering plant	North America and Europe	*Typha latifolia*
Cherry, plum, peach, almond	Tree	North America and Europe	*Prunus* spp.

continued

Table 5.1 Continued

Common name	Plant type	Distribution	Latin name
Christmas fern	Fern	North America	*Polystichum acrostichoides*
Cinquefoils	Flowering plants	North America and Europe	*Potentilla* spp.
Clovers	Flowering plant	North America and Europe	*Trifolium* spp.
Cocklebur	Flowering plant	Europe	*Xanthium albinum*
Common barberry	Flowering plant	North America and Europe	*Berberis vulgaris*
Common buckthorn	Tree	North America and Europe	*Rhamus cathartica*
Common club-rush	Flowering plant	North America and Europe	*Schoenoplectus lactustris*
Common cottongrass	Shrub	North America and Europe	*Eriophorum angustifolium*
Common dogwood	Tree	North America and Europe	*Cornus sanguinea*
Common elderberry	Tree	North America and Europe	*Sambucus nigra*
Common globeflower	Flowering plant	North America and Europe	*Trollius europaeus*
Common hedge hyssop	Shrub	North America and Europe	*Gratiola officinalis*
Common ivy	Flowering plant	North America and Europe	*Hedera helix*
Common mistletoe	Flowering plant	Europe	*Viscum album*
Common quillwort	Aquatic plant	Europe	*Isoetes lacustris*
Common reed	Reed	North America and Europe	*Phragmites australis*
Common snowberry	Flowering plant	North America	*Symphoricarpos albus*
Common spindle	Flowering plant	Europe	*Euonymus europaeus*
Common water-plantain	Flowering plant	North America and Europe	*Alisma plantago-aquatica*
Common yew	Flowering plant	North America and Europe	*Taxus baccata*
Conifers	Tree	North America and Europe	*Pinopsida*
Cow-parsnip	Flowering plant	North America and Europe	*Heracleum sibiricum*
Crowfoot	Flowering plants	North America and Europe	Ranunculaceae
Dandelion	Flowering plant	North America and Europe	*Taraxacum* spp.
Devil's-bit scabious	Flowering plant	Europe	*Succisa pratensis*
Downy burdock	Flowering plant	Europe	*Arctium tomentosum*
Dropwort	Flowering plants	North America and Europe	*Filipendula vulgaris*
Drummonds willow	Tree	North America	*Salix drummondiana*
Elm	Tree	North America and Europe	*Ulmus* spp.
Fenugreek	Flowering plant	Europe	*Trignonella foenum-graecum*
Fine-leaved water dropwort	Flowering plant	Europe	*Oenanthe aquatica*
Firs	Tree	North America and Europe	*Abies* spp.
Fremont cottonwood	Tree	North America	*Populus fremontii*
Garden loosestrife	Flowering plant	North America and Europe	*Lysimachia vulgaris*
Goldenrod	Flowering plant	North America and Europe	*Solidago virgaurea*
Grasses	Grass	North America and Europe	Poaceae
Greater water parsnip	Flowering plant	Europe	*Sium latifolium*
Ground elder	Tree	North America and Europe	*Aegopodium podagraria*
Ground ivy	Flowering plant	North America and Europe	*Glechoma hederacea*

Common name	Plant type	Distribution	Latin name
Guelder rose	Flowering plant	North America and Europe	*Viburnum opulus*
Hairy St John's wort	Europe	Europe	*Hypericum hirsutum*
Hawthorn	Trees and shrubs	North America and Europe	*Crataegus* spp.
Hazel	Tree	North America and Europe	*Corylus* spp.
Hemlock	Tree	North America and Europe	*Tsuga* spp.
Hemp-agrimony	Flowering plant	North America and Europe	*Eupatorium cannabinum*
Himalayan balsam	Flowering plant	North America and Europe	*Impatiens glandulifera*
Hornbeam	Tree	North America and Europe	*Carpinus* spp.
Horse chestnut	Tree	North America and Europe	*Aesculus hippocastanum*
Japanese honeysuckle	Flowering plant	North America and Europe	*Lonicera japonica*
Japanese knotweed	Herbaceous plant	North America and Europe	*Fallopia japonica*
Jeffrey pine	Tree	North America	*Pinus jeffreyi*
Juniper	Tree	Europe and North America	*Juniperus* spp.
Large-leafed dogwood	Tree	North America and Europe	*Cornus macrophylla*
Lawson cypress	Tree	North America and Europe	*Chamaecyparis lawsonia*
Legumes	Flowering plant	North America and Europe	Fabaceae
Lily of the valley	Flowering plant	North America and Europe	*Convallaria majalis*
Lizard's tail	Flowering plant	North America	*Saururus cernuus*
Long-beaked willow	Tree	North America	*Salix bebbiana*
Mackenzie's willow	Tree	North America	*Salix prolixa*
Maize	Grass	North America and Europe	*Zea mays*
Male fern	Fern	North America and Europe	*Dryopteris filix-mas*
Manyseed goosefoot	Flowering plant	North America and Europe	*Chenopodium polyspermum*
Maple	Trees	North America and Europe	*Acer* spp.
Marsh cinquefoil	Flowering plant	North America and Europe	*Potentilla palustris*
Marsh euphorbia	Flowering plant	Europe	*Euphorbia palustris*
Marsh horsetail	Herbaceous plant	North America and Europe	*Equisetum palustre*
Marsh marigold	Flowering plant	North America and Europe	*Caltha palustris*
Marsh purslane	Flowering plant	North America and Europe	*Ludwigia* spp.
Meadow buttercup	Flowering plant	North America and Europe	*Ranunculus acris*
Meadowsweet	Herbaceous plant	North America and Europe	*Filipendula ulmaria*
Milk parsley	Flowering plant	Europe	*Peucedanum palustre*
Mint/deadnettle	Flowering plant	North America and Europe	Lamiaceae
Mugwort	Flowering plant	North America and Europe	*Artemisia vulgaris*
Mulleins	Flowering plant	North America and Europe	*Verbascum* spp.
Narrow-leaf cottonwood	Tree	North America	*Populus angustifolia*
Nettle	Flowering plant	North America and Europe	*Urtica* spp.
Northern water hemlock	Flowering plant	North America and Europe	*Cicuta virosa*
Northern/eastern white cedar	Tree	North America	*Thuja occidentalis*

continued

Table 5.1 Continued

Common name	Plant type	Distribution	Latin name
Oak	Trees	North America and Europe	*Quercus* spp.
Oats	Grass	North America and Europe	*Avena sativa*
Pea	Flowering plant	North America and Europe	*Pisum sativum*
Pilewort	Flowering plant	Europe	*Ficaria verna*
Pine	Tree	North America and Europe	*Pinus* spp.
Pondweed	Aquatic plant	North America and Europe	*Potamogeton* spp.
Poplar	Trees	North America and Europe	*Populus* spp.
Potato	Flowering plant	North America and Europe	*Solanum tuberosum*
Purple marshlock	Flowering plant	North America and Europe	*Comarum palustre*
Quaking aspen	Tree	North America and Europe	*Populus tremuloides* and *P. tremula*
Rape	Flowering plant	North America and Europe	*Brassica napus*
Raspberry	Flowering plant	North America and Europe	*Rubus idaeus*
Red dogwood	Tree	North America and Europe	*Cornus sericea*
Red maple	Tree	North America and Europe	*Acer rubrum*
Reed grass	Grass	North America and Europe	*Calamagrostis lanceolatus*
Reed	Reed	North America and Europe	*Phragmites* spp.
Rosebay willowherb	Flowering plant	North America and Europe	*Epilobium angustifolium*
Rowan (mountain ash)	Tree	North America and Europe	*Sorbus aucuparia*
Rush	Flowering plants	North America and Europe	*Juncus* spp.
Rye	Grass	North America and Europe	*Secale cereale*
Sandbar willow	Tree	North America	*Salix exigua*
Saw sedge/swamp sawgrass	Sedge	North America and Europe	*Cladium mariscus*
Sawara cypress	Tree	North America and Europe	*Chamaecyparis pisifera*
Sedge	Sedge	North America and Europe	*Carex* spp.
Shining willow	Tree	North America	*Salix lucida*
Siberian squill	Flowering plant	Europe	*Scilla siberica*
Siberian pea tree	Tree	North America and Europe	*Caragana arborescens*
Silverberry	Flowering plant	North America and Europe	*Elaeagnus commatata*
Sitka willow	Tree	North America	*Salix sitchensis*
Small-leaved lime	Tree	North America and Europe	*Tila cordata*
Smartweed	Flowering plant	North America and Europe	*Polygonum* spp.
Snowy mespilus	Flowering plant	North America and Europe	*Amelanchier ovalis*
Sorrel	Herbaceous plant	North America and Europe	*Rumex acetosa*
Soybean	Flowering plant	North America and Europe	*Cephalanthus occidentalis*
Speckled alder/grey alder/hoary alder	Tree	North America and Europe	*Alnus incana*
Spruces	Tree	North America and Europe	*Picea* spp.
St John's wort	Flowering plant	North America and Europe	*Hypericum perforatum*
Stinging nettle	Flowering plant	North America and Europe	*Urtica dioica*
Sugar beet	Flowering plant	North America and Europe	*Beta vulgaris*

Common name	Plant type	Distribution	Latin name
Swede	Flowering plant	North America and Europe	*Brassica napobrassica*
Sweet flag	Flowering plant	North America and Europe	*Acorus calamus*
Sycamore	Trees	North America and Europe	*Acer pseudoplatanus*
Thistle	Flowering plant	North America and Europe	*Cirsium* spp.
Timothy-grass	Grass	North America and Europe	*Phleum pratense*
Turnip	Flowering plant	North America and Europe	*Brassica rapa*
Warty spindle	Shrub	Europe	*Euonymus verruosa*
Water avens	Flowering plant	North America and Europe	*Geum rivale*
Water horsetail	Herbaceous plant	North America and Europe	*Equisetum fluviatile*
Water lobelia	Aquatic plant	North America and Europe	*Lobelia dortmanna*
Water milfoil	Aquatic plant	North America and Europe	*Myriophyllum* spp.
Watershield	Aquatic plant	North America	*Brasenia schreberi*
Waterweed	Aquatic plant	North America and Europe	*Elodea* spp.
Wheat	Grass	North America and Europe	*Triticum* spp.
White water lily	Aquatic plant	Europe	*Nymphaea alba*
Wild cherry	Tree	Europe	*Prunus avium*
Willow	Trees	North America and Europe	*Salix* spp.
Willow-leafed cotoneaster	Flowering plant	North America and Europe	*Cotoneaster salicifolius*
Witch hazel	Tree	North America and Europe	*Hammamelis virginiana*
Wood anemone	Flowering plant	North America and Europe	*Anemone nemorosa*
Wood cranesbill	Flowering plant	North America and Europe	*Geranium sylvaticum*
Woolly butterbur	Flowering plant	North America and Europe	*Petasites spurius*
Yarrow	Flowering plant	North America and Europe	*Achillea millefolium*
Yellow dock	Flowering plant	North America and Europe	*Rumex crispus*
Yellow iris	Flowering plant	North America and Europe	*Iris pseudacorus*
Yellow loosestrife	Flowering plant	North America and Europe	*Lysimachia vulgaris*
Yellow water lily	Aquatic plant	North America and Europe	*Nuphar lutea*
Yellow wood anemone	Flowering plant	North America and Europe	*Anemone ranunculoides*

Source: After Campbell-Palmer et al. (2015), reprinted by permission of Pelagic Publishing.

period, 16 genera of herbaceous plants and foliage from 19 species of trees and woody shrubs were identified through stomach inspections of 165 beavers (Roberts and Arner, 1984). The proportion of woody food represented 86% of the total dry weight in the winter diets (December to March) but declined to 42% in May and dropped to 17% during August (Svendsen (1980)).

In most places, as non-woody vegetation senesces and dies during the colder months at northern lati-tudes, woody plant material dominates the diet. Beavers are fairly unique in their ability to fell large, mature trees (> 1 m in diameter) and utilize every part as either food or construction materials (see Chapter 4). They tend to favour smaller tree saplings (< 5 cm diameter) to obtain their bark, side branches, and leafy stems (Haarberg and Rosell, 2006, Margaletić et al., 2006) (Figure 5.2). However, they ingest only a proportion of the total biomass of trees harvested. Throughout the autumn, winter,

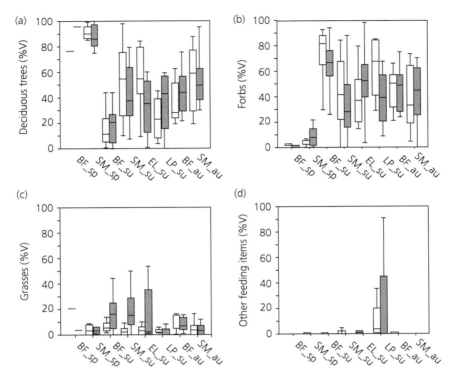

Figure 5.1 The proportion of main food categories (a) deciduous trees, (b) forbs, (c) grasses and (d) other feeding items (maize, fruits, ferns, mosses and unidentified fragments) in micro- (white box) and macro-fractions (grey box) of the beaver diet at particular study sites in the spring (sp), summer (su) and the autumn (au). The median, quartiles and the minimum and maximum values are given. (Krojerov.-Prokešov., J., Baranček ov., M., Hamš.kov., L., and Vorel, A. 2010. Feeding habits of reintroduced Eurasian beaver: spatial and seasonal variation in the use of food resources. Journal of Zoology, 281, 183–193. © 2010 The Zoological Society of London).

and until spring, the beaver's diet consists mainly of bark and branches from various broad-leaved trees and shrubs (Table 5.2), but they also feed on aquatic plants if present (Simonsen, 1973). It is in this period that the majority of trees and shrubs are felled; typically very few trees are felled during the summer months (Wilsson, 1971, Svendsen, 1980).

In the Czech Republic, beavers preferentially selected willows and poplars across all study sites (approximately 500 beavers in five populations in different environments) (Vorel et al., 2015). The total biomass of poplars and willows in the diet of beavers across the different sites varied between 78% and 94%, which indicated that these plants were key food items across a wide geographic range. Contrary to the view of beavers as opportunistic, such feeding behaviour could indicate much greater specialization with regard to these two genera.

Figure 5.2 A beaver kit cutting a sapling during winter (Photo supplied courtesy of Frank Rosell.)

Long-term foraging by beavers may decrease the diversity of woody plants and alter the composition to less palatable species (Fryxell, 2001, Raffel et al.,

Table 5.2 Preferred (cut) woody vegetation by beavers (*Cf* = Eurasian beaver; *Cc* = North American beaver) in some published studies across Europe and North America. All genera are given in descending order of preference.

Species	Country (region)/N°	Preferred forage	References
Cc	USA (Georgia) 32° 24' 57.24'' N	Sweetgum	Brzyski and Schulte (2009)
Cc	USA (New Mexico) 34° 31' 11.78'' N	Rio Grande cottonwood (*Populus* wislizeni), Russian olive cottonwood (*Elaeagnus angustifolia*), and narrowleaf willow (*Salix exigua*)	Barela and Frey (2016)
Cc	USA (North Carolina) 39° 24' 44.71'' N	Musclewood (*Carpinus carolinana*), sweetgum (*Liquidambar styraciflua*), and tag alder (*Alnus serrulata*)	Rossell et al. (2014)
Cc	USA (Ohio) 40° 19' 40.32'' N	Shingle oak (*Quercus imbricaria*), black oak (*Q. velutina*), red oak (*Q. rubra*), hackberry (*Celtis occidentalis*), sassafras (*Sassafras albidum*), and hop hornbeam (*Ostrya virginiana*)	Raffel et al. (2009)
Cc	USA (Ohio) 40° 25' 2.24'' N	Alder (*Alnus* spp.), soft maple (*Acer saccharinum*), and red elm (*U. rubra*)	Nixon and Ely (1969)
Cf	France (Rhone Valley) 45° 10' 22.44'' N	Dogwood (*Cornus* spp.), poplars, and willows[a]	Erome and Broyer (1984)
Cc	Canada (New Brunswick) 46° 33' 55.15'' N	Beaked hazelnut (*C. cornuta*), largetooth aspen (*P. grandidentata*), pin cherry (*Prunus pensylvanica*), trembling aspen, and willows	Gallant et al. (2004)
Cc	North America (Isle Royale, Michigan) 47° 07' 22.56'' N	Sugar maple and yellow birch (*Betula alleghaniensis*)	Belovsky (1984)
Cf	Hungary (Moson Danube) 47° 51' 25.56'' N	Hazel, bird cherry (*P. padus*), and willow	Varju and Jánoska (2015)
Cf	Austria (Vienna) 48° 12' 29.41'' N	Hazel and poplars	Krebs (1984)
Cf	Czech Republic (Chropynsky and Niva Dyje) 49° 18' 47.63'' N	Willows and poplars[b]	Vorel et al. (2015)
Cf	Czech Republic (Oderske vrchy) 49° 37' 13.944'' N	Willows, aspen, dogwood, birch, and alder	Dvořák (2013)
Cf	Czech Republic (Labe and Cesky les) 50° 38' 36.46'' N	Poplars and willows[b]	Vorel et al. (2015)
Cc	Canada (Ontario) 51° 15' 13.57'' N	Aspen, red maple (*A. rubrum*), sugar maple (*A. saccharum*), and beaked hazel (*C. cornuta*)	Donkor and Fryxell (1999); Donkor and Fryxell (2000)
Cf	Poland (Polesie National Park) 52° 01' 23.20'' N	Willow, aspen, and alders	Janiszewski et al. (2017)
Cf	Germany (Magdeburg) 52° 07' 54.70'' N	Poplars and willows[a]	Heidecke (1989)
Cf	Netherlands (Biesbosch) 52°07'57.48''N	Willows	Nolet et al. (1994)
Cc	Canada (Newfoundland) 53° 08' 07.84'' N	Aspen and alder	Northcott (1971)
Cc	Canada (British Columbia) 53° 43' 36.012'' N	Willows	Gerwing et al. (2013)
Cc	Canada (Alberta) 53° 55' 59.77'' N	Poplar and willow	Hood and Bayley (2008)

continued

Table 5.2 Continued

Species	Country (region)/N°	Preferred forage	References
Cf	Poland (Wigry National Park) 54° 04' 5.66'' N	Grey willow (*S. cinerea*) and common hazel (*C. avellana*)	Misiukiewicz et al. (2016)
Cf	Scotland (Bamff) 56° 37' 23.48'' N	Willows	Jones and Ardren (2003)
Cf	Denmark (Klosterheden State Forest District) 57° 12' 33.32'' N	Willows	Elmeros et al. (2003), Erome and Broyer (1984)
Cf	Norway (Telemark) 59° 23' 16.04'' N	Grey alder (*Alnus incana*), willows, rowan, and birches	Haarberg and Rosell (2006)
Cc	Finland (Evo) 61° 11' 33.83'' N	Aspen, downy birch (*B. pubescens*), and speckled alder (*A. incana*)	Salandre et al. (2017), Heidecke (1989)
Cc	Canada (Northwest Territories) 64° 49' 31.94'' N	Willows, poplar, and alder	Aleksiuk (1970)
Cf and Cc	Finland (Southern Lapland) 67° 55' 20.03'' N	Birch, aspen, and mountain ash	Lahti (1966)

[a] It is unclear if foraging comprised preference indices or only showed frequency of the vegetation selected.
[b] Same preference in two different areas.

2009). When beavers first recolonized Allegany State Park, New York, USA, in 1937 they foraged mostly on aspen (*Populus* spp.) (44%) and very few maple (*Acer* spp.), birches (*Betula* spp.), black cherry (*Prunus serotina*), beeches (*Fagus* spp.), and ash trees (*Fraxinus* spp.) (combined 19%) (Shadle and Austin, 1939). Approximately 45 years after recolonization, however, these less preferred species were a substantial part of the beaver's diet (combined 57%) (Müller-Schwarze and Schulte, 1999). Large variations in diet can also occur in areas with high species diversity compared to regions dominated by a few species of woody vegetation. In northern Karelia, Russia, where few alternative foods are available, 97.5% of the beaver's diet consists of birch, while in southern Karelia their diet is 53% willow (*Salix* spp.), 21% aspen, and 12% bird cherry (*P. padus*). Studies of beaver diets from different parts of the world report a preference for widely different woody species, which is expected since species composition varies across the beavers' range (see Table 5.2 for variation in preferred species among study sites).

In the spring and summer, the beaver will primarily eat herbs, forbs, grasses, and riparian and aquatic plants. These plants may account for up to 90% of their diet at this time of year in some areas

(Simonsen, 1973, Jenkins, 1980, Svendsen, 1980, Doucet and Fryxell, 1993). For example, in Patagonia, Chile, beavers can rely entirely on herbaceous vegetation (Pietrek et al., 2017). In some areas, bark, fresh shoots, and leaves of deciduous trees and shrubs form the most important food resources during summer months (Tevis, 1950, Histøl, 1989, Nolet et al., 1995) (Table 5.3). Buds and new leaves are especially attractive after beavers have spent the long winter eating bark from the branches stored in the food cache (see Section 4.3). Beavers tend to eat both the leaves and bark from species they fell, but there are some exceptions. For example, they readily eat the leaves but not the bark of blackthorn (*P. spinosa*) (Simonsen, 1973).

Variation in the beaver's diet is also dependent on nutrient content and digestibility of available forage. Individual beavers attempt to maximize energy intake over time (Doucet and Fryxell, 1993, Nolet et al., 1995). Being 'energy maximizers', beavers prefer to select foods with the greatest return of energy (Belovsky, 1984, Fryxell, 1992, Gallant et al., 2016). Doucet and Fryxell (1993) investigated the effect of nutritional content on beaver selection for five plant species in a cafeteria-style experiment. They ranked preferences in the following descending order: trembling aspen

Table 5.3 Summer diet of Eurasian beavers from different localities (a–f) and years in small lakes in Vennesla municipality, Vest-Agder county, Norway. It is indicated whether plants are seldom used as food (1), moderately used as food (2), or often used as food (3) (0 means no data). Food plants are grouped in the following categories: trees (A), shrubs (B), field layer plants (C), herbs and bog/marsh (D), and aquatic plants (E).

Plant category	Plant genus or species	1983 d e	1984 a b c d e f	1986 a b d e	1988 a b c d e f
A	Birch (*Betula* spp.)	3 2	3 3 3 2 2 3	3 3 3 3	3 3 3 3 0 1
A	Oak (*Quercus* spp.)	3 0	3 0 3 3 1 0	3 0 3 2	3 2 3 3 2 0
A	Rowan (*Sorbus aucuparia*)	0 0	3 2 1 0 1 0	2 2 1 0	2 2 1 1 1 0
A	Norway spruce (*Picea abies*)	1 0	1 2 0 2 0 0	0 0 0 1	0 0 0 1 0 0
A	Scots pine (*Pinus sylvestris*)	1 0	1 1 2 0 0 0	1 0 0 0	0 0 0 1 0 0
A	Quaking aspen (*Populus tremula*)	0 0	3 0 3 0 0 0	2 0 0 0	1 0 0 0 0 0
A	Hazel (*Corylus avellana*)	0 0	1 0 0 0 0 0	0 0 0 1	1 0 0 0 0 0
A	Grey alder (*Alnus incana*)	0 0	0 2 0 0 0 0	0 0 0 0	0 0 0 0 0 0
A	Alder (*Alnus glutinosa*)	0 0	0 0 0 0 0 0	0 0 0 0	0 0 0 1 0 0
A	Elder (*Sambucus* spp.)	0 0	0 2 0 0 0 0	0 0 0 0	0 1 0 0 0 0
B	Willow (*Salix* spp.)	1 2	1 3 2 1 1 2	2 1 2 0	1 1 2 0 1 0
B	Juniper (*Juniperus communis*)	1 1	2 1 1 1 1 0	0 0 0 1	1 0 1 1 0 0
B	Sweetgale (*Myrica gale*)	2 0	2 1 1 1 1 0	0 0 0 1	1 0 1 1 0 0
B	Raspberry (*Rubus idaeus*)	0 0	1 2 0 0 1 0	0 0 0 1	0 2 0 0 1 0
B	Buckthorn (*Rhamnus frangula*)	0 0	1 0 0 1 0 1	0 0 1 0	1 0 0 1 0 0
C	Bilberry (*Vaccinium myrtillus*)	2 0	1 1 1 1 0 0	1 0 1 0	1 1 0 1 0 0
C	Bracken (*Pteridium aquilinum*)	3 0	1 1 1 3 0 0	0 1 2 0	1 0 0 2 0 0
C	Heather (*Calluna vulgaris*)	1 0	1 1 1 0 0 1	0 0 0 0	1 0 0 0 0 0
C	Common lady fern (*Athyrium filix-femina*)	0 1	2 0 0 0 2 0	0 0 0 0	2 0 0 0 0 0
C	Bog bilberry (*Vaccinium uliginosum*)	1 0	0 1 0 0 0 0	0 2 0 0	0 0 0 0 0 0
C	Stone bramble (*Rubus saxatilis*)	0 0	0 1 0 0 0 0	0 0 0 0	0 0 0 0 0 0
C	Common cinquefoil (*Potentilla erecta*)	0 0	0 0 0 1 0 0	0 0 0 0	0 0 0 0 0 0
D	Swamp cinquefoil (*Potentilla palustris*)	1 3	0 1 0 1 3 0	0 0 0 3	0 0 0 0 3 0
D	Sedges (Cyperacea)[a]	0 0	0 0 0 1 1 3	0 1 0 1	0 0 [b]1 0 0
D	Common reed (*Phragmites australis*)	0 1	0 0 0 0 1 0	0 0 0 2	0 0 0 0 2 0
D	Milk parsley (*Peucedanum palustre*)	0 0	0 0 0 0 1 0	0 0 0 0	0 0 0 0 0 0
D	Bog bean (*Menyanthes trifoliata*)	0 0	0 1 0 0 0 0	0 0 0 0	0 0 0 0 0 0
E	Water lobelia (*Lobelia dortmanna*)	3 0	2 2 0 3 1 0	1 1 2 0	1 0 0 2 0 0
E	Water horsetail (*Equisetum fluviatile*)	3 3	0 3 0 3 3 0	0 0 1 2	0 1 0 2 0 0
E	Yellow water lily (*Nuphar lutea*)	2 1	0 3 0 2 0 0	3 3 1 1	0 0 0 0 0 0
E	Lake quillwort (*Isoetes lacustris*)	1 3	3 1 0 2 0 0	0 0 1 1	0 2 0 0 0 0
E	White water lily (*Nymphaea alba*)	1 0	0 0 0 0 0 1	0 0 0 0	0 0 0 0 0 0

Source: After Histøl (1989), reprinted by permission of Fauna.

[a] Mainly common cottongrass (*Eriophorum augustifolium*) and bottle sedge (*Carex rostrata*). The softer and less fibrous lower part of bottle sedge/beaked sedge was eaten and the tuber-like stem of water lobelia (*Lobelia dortmanna*).

[b] The beavers had probably left the locality and only one food item was registered.

(*P. tremuloides*), white water lily (*Nymphaea alba*), raspberry (*Rabus idaeus*), speckled alder (*Alnus incana*), and red maple (*A. rubrum*). Interestingly, this result correlates with the mean retention times. Aspen (16 hours), water lily (19 hours), and raspberry (13 hours) had a much shorter retention time than the less preferred alder (38 hours) and maple (42 hours). They concluded that the beavers selected a diet that maximized long-term energy intake.

The amount of time the beaver searches for food varies according to season. Beavers in a reintroduced population in the Netherlands spent on average 7.7 hours of their 12-hour active period (defined as time not in a lodge) foraging during ice-free winters and 5 hours of their 10 active hours foraging during spring and summer (Nolet and Rosell, 1994). This finding is proportionately similar to a Norwegian population in Telemark, where 2.4 hours were spent foraging out of the 4.6 hours when beavers were active and visible (Sharpe and Rosell, 2003). In Kouchibouguac National Park, Canada, beavers spent 18% of their time foraging. During the majority of the foraging time they fed on woody plants, such as trees (45%) and shrubs (31%), and only 15% on herbaceous plants and 7% on aquatic plants. Their foraging time increased as the summer progressed (Gallant et al., 2016). Thus, most beavers spend at least half their time budget searching for, gathering, and consuming food.

5.2 Foraging behaviours

Foraging behaviours are highly variable according to a wide range of factors that affect the distribution of the food items, including geographic location, topography, and associated hydrology (Rosell et al., 2005). Other factors that may influence beaver foraging choices are distance from the water's edge to the food item, forage density, composition and concentration of nutrient and secondary compounds in the plants, flood regime, disturbance (harvesting and human presence or constructions), and predator activity (Baker et al., 2005, Gerwing et al., 2013). Some plant species are avoided in some locations, while the same species can be preferred in other places. The preferred plant species can even vary from year to year, because the nutritional value of

the species can change. Individual beaver diet also varies among different families in the same population and even within families (Shelton, 1966, Vorel et al., 2015). However, some similarities in beaver foraging behaviour can be observed, such as increasing selectivity with increasing distance from the water.

5.2.1 Central-place foraging

Beavers adjust foraging intensity and preference with distance from water, following a central-place foraging strategy (Orians and Pearson, 1979, McGinley and Whitham, 1985, Haarberg and Rosell, 2006). For beavers, the water represents their central place and their main shelter. Cutting and transporting forage from land to the water is energetically costly, which affects the beaver's choice on where to forage as well as food preference (Fryxell and Doucet, 1991). It is easier to float woody material downstream, so beavers therefore forage at greater distances upstream from their main lodge than downstream (Boyce, 1981). With increasing distance from the water, beavers face increased transportation costs as well as increased risk of predation. Therefore, the number of cut stems declines sharply with increasing distance from the water (Donkor and Fryxell, 1999) (Figure 5.3). Foraging intensity declines while food selectivity and (within limits) tree diameter of cut trees increase with increasing distance from the water (Haarberg and Rosell, 2006, Salandre et al., 2017). Beavers chose larger trees and were more selective with the species taken as distance from shore increased in high-quality habitat (Pinkowski, 1983, Fryxell and Doucet, 1993, Gallant et al., 2004, Raffel et al., 2009). Other studies found that beavers selected smaller stems with increasing distance from the lodge (Jenkins, 1980, Belovsky, 1984, Donkor and Fryxell, 1999) or reported no change in species selectivity with distance from the water (Haarberg and Rosell, 2006). In low-quality habitats, beavers might need to forage further from the safety of water and, therefore, engage in riskier behaviour (Sih, 1980).

Most feeding activity occurs along or close to the water's edge, with most felling activity recorded within the first 20–40 m away (Haarberg and Rosell, 2006, Margaletić et al., 2006), and they rarely move

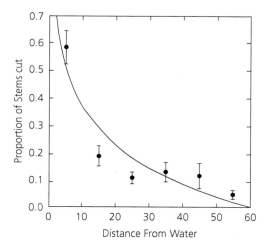

Figure 5.3 Proportion of stems cut by beavers, in relation to distance from water, at 15 ponds in Algonquin Park, Ontario. Over 50% of cut stems were located at 0–10 m from water, while less than 7% of stems were cut more than 40 m away from water (Reprinted from Forest Ecology and Management, 118, Donkor, N.T. & Fryxell, J.M., Impact of beaver foraging on structure of lowland boreal forests of Algonquin Provincial Park, Ontario, pages 83–92, © Elsevier Science (1999), with permission from Elsevier).

further than 60 m from the water to forage (Parker et al., 2001, Donkor and Fryxell, 1999, Barnes and Dibble, 1988). A large variation between study sites has been reported (Elmeros et al., 2003, Baskin and Sjöberg, 2003, Hood and Bayley, 2008) and the distance is also dependent on the presence of predators and beavers' innate response to predators (Rosell and Sanda, 2006). If there is low availability of food nearby, beavers can travel several hundred metres from the water's edge to find it. For example, they forage a few hundred metres away from water to obtain highly preferred forage species such as aspen and poplar (*Populus* spp.) (see Section 5.3). In Telemark, Norway, beavers usually foraged alone and they were accompanied by another beaver in only 6% of all foraging observations—in 51% of which the accompanying beaver was an off-spring and in 34% it was a partner (Lodberg-Holm et al., 2021).

In northern Alabama, USA, beavers moved faster with increasing distance from the lodge in wetland landcover than in terrestrial landcover, most likely to avoid predation risk from alligators (*Alligator mississippiensis*) (McClintic et al., 2014). Foraging

beavers have to balance energy requirements and predator avoidance. When facing predation risk, they trade off maximization of energy intake against minimization of predation risk. The selection of trees to cut based on their size and distance from the water represent such a trade-off between energy maximization and predation (Basey and Jenkins, 1995). The time it takes to either flee to or drag forage to water is important for their behavioural decisions (Gerwing et al., 2013). If predation risk is high, beavers are expected to become time minimizers instead of energy maximizers, but this response has never been verified (Gallant et al., 2016). Beavers most likely use their sense of smell to assess the risk of predation when going onto land to forage and alter their space use in response to an indirect cue of predation risk. Therefore, they might choose to forage in areas where the wind is stronger from the bank to the water, so that they can quickly detect whether there are any enemies nearby further inland. They should avoid leaving the water in places where the wind blows their scent ashore ahead of them, thus increasing their exposure to predators (see also Section 5.8).

Wolf (*Canis lupus*) urine reduced North American beavers' (*Castor canadensis*) use of foraging trails in Michigan, USA. Mean number of beavers and time spent decreased by 95% on urine-treated trails and was unchanged on untreated trails (Severud et al., 2011). When beavers have been exposed to various foraging trials where some food sources were tainted with predator scents, they fed more often on the unscented aspen sticks (experiment controls) than on sticks treated with predator scents (Engelhart and Müller-Schwarze, 1995, Rosell and Czech, 2000, Salandre et al., 2017). Smith et al. (1994) showed that on islands where the level of predation by black bears (*Ursus americanus*) was higher, beavers foraged closer (< 30 m) to the shore compared to islands with less predation pressure (here they travelled > 200 m inland to forage). Similar results are found in areas with large predators, such as wolf and lynx (*Lynx* spp.). In these areas, beavers limit the distance they travel while foraging. If predators were present, beavers foraged within 30–35 m from the bank, while the almost complete absence of predators allowed beavers to travel at a distance of 100–165 m from the water when foraging (Goryainova et al., 2014). Therefore, predator avoidance can also help explain the pattern of decreasing foraging activity with increasing distance from the water. Cut vegetation is usually brought to the water's edge and consumed there (Salandre et al., 2017). Water acts as a refuge for beavers, because their main predators are terrestrial mammals which cannot follow them in deep water (Gable et al., 2018), at least in areas without alligators (see Chapter 8). Therefore, aquatic vegetation can be consumed with minimal predation risk (Fryxell and Doucet, 1993). Aquatic resources also require less search and handling time than woody material on land since the cost of travel to foraging areas and in conveying selected items is reduced due to their buoyancy in water (Fryxell and Doucet, 1993, Severud et al., 2013).

5.2.2 Woody species processing and felling

While gathering and transporting branches, a beaver will often take a bundle of them, up to six or seven at a time, carrying them back to the water and to a favoured feeding place (Figure 5.4). Beavers are creatures of habit and develop preferred sites on the water's edge where distinctive piles of peeled, woody debris accumulate at feeding stations (Wilsson, 1971). Although these materials tend to gradually disappear, generally washing away with water movement over time, some have even been identified in archaeological sites (Coles, 2006). The beaver then sits near the water's edge and chews its food. Watching the beaver eat branches can be a special experience, as they display sophisticated techniques to prepare and feed on different food items. If the branches are thin (up to 5–6 mm in diameter), the beaver will fold them double and hold them with both fore paws while they quickly stuff them into their mouths and chew them into splinters with their front teeth. They often fold leaves into small bundles and eat these in the same manner. They only consume the bark on larger branches by clutching the branch with their front paws. One paw holds the branch in place between the outer toe and other toes, while the lower paw spins the branch quickly, while the lower incisors peel the bark away. A beaver will take short breaks during this process to chew the bark with its back

Figure 5.4 A beaver usually take branches back to a favoured feeding place (Photo supplied courtesy of Frank Rosell.)

teeth, swallow, and sharpen its incisors (see Chapter 3). Beavers cut plant material, including bark, with their incisors and then grind it up with their molars. Thereafter, a regular alternation of masticatory cycles from one to the other side has been observed (Druzinsky, 1995, Stefen et al., 2011), though grinding occurs on right and left cheek teeth rows simultaneously, thus increasing the effectiveness of chewing (Stefen et al., 2011).

The peeled sticks are discarded, often forming visible piles that makes it easy to identify a location as a feeding station (Campbell-Palmer et al., 2015). The manner of eating bark is similar to the way humans eat a cob of maize (*Zea mays*). Gnawing and rasping are distinctive noises beavers make when feeding on woody vegetation or crisp green vegetation, which can be quite audible, especially on still evenings. It is a useful method of locating beavers in dark conditions or thick vegetation. It is also possible that predators locate beavers this way. As a field sign, feeding on woody food is more easily detected than feeding on herbs or roots (Nolet et al., 1995).

A beaver uses olfactory sampling to assess which tree is worth felling (Jenkins, 1979) and can press its nose against the stem to take a scent test. It might also use olfaction to investigate the nutritional secondary substances in the potential food item. This is not surprising since beavers forage on land almost exclusively at night (Doucet et al., 1994, Fryxell, 1999). If researchers make 'hybrid plants', i.e. use aspen as the stem and combine it with branches from red maple or vice versa, beavers will cut more aspen stems with maple canopies than the other way around. Covering the stem with paper bags did not alter their response (Doucet et al., 1994). Beavers are also able to distinguish between aspen saplings with a full canopy and ones with no canopy, but they are unable to distinguish more subtle differences in canopy biomass. The use of olfaction might be how beavers distinguish between living and dead young trees in a natural setting (Doucet et al., 1994). There have been attempts to trick beavers into felling aspens without leaves, but the beavers avoided doing so. Once fresh leaves were hung from the top of the tree, the beavers felled the tree (Wilsson, 1971). After a tree has passed any initial selection test, the beaver removes a little piece of bark, possibly for the purpose of taking a taste test. This allows the beaver to judge the nutritional value of the tree and decide whether or not it is worth the effort of felling.

When the beaver starts to fell a tree, it stands upright and chews on the stem at around 50 cm above the ground (Wilsson, 1971). However, this behaviour can vary. A study in Norway identified an average height of 4.1 cm (range 1.4–30 cm) based on observations of 30 felled aspen trees with a diameter of 2 cm and a height of 1.5–3 m (Skøyen et al., 2018). It sets its upper teeth into the stem, while the lower teeth 'shave' the wood loose with the help of powerful jaw muscles. The placement of the upper teeth ensures stable anchoring to the tree, while the tail and legs hold the animal balanced and upright (Wilsson, 1971). Gnawing is accomplished by movement of the lower jaw against the skull by the muscles of mastication (see Section 3.1.3 for details on teeth morphology and biting force). Unilateral gnawing (cutting with a single incisor) concentrates the bite force into a small area compared to bilateral cutting (cutting with both incisors), thus making it easier for the animal to propagate a fracture and drive its lower incisor through the wood. Given these specialized behaviours, beavers have evolved an extremely effective wood-cutting and tree-felling apparatus (Cox and Baverstock, 2016). During unilateral gnawing, the tip of a single upper incisor is pressed against the stick, while the corresponding lower incisor that is sharper than the upper is used for cutting (Rybczynski, 2008; Figure 5.5a–c).

Trees with diameters under 20 cm can usually be felled by beavers chewing on just one side, but

larger trees are cut by gnawing from multiple sides (Wilsson, 1971, Raffel and Gatz, 2003). Younger trees or shrubs, which are still soft enough that the animal can bite them off with a short series of bites with the upper front teeth, can be cut down in just a few seconds. The head is positioned such that the edges of the front teeth form an approximately 135-degree angle against the stem while cutting perpendicularly across the fibres (Figure 5.6a–d). The cut surface is thin and angled, with distinctive tooth marks, which enables one to count the number of bites that the animal used (Wilsson, 1971). During stem harvesting a beaver's head is tilted so that one side rests near or against the stick being cut (Rybczynski, 2008).

In harder trees with coarser bark, beavers use another technique to cut them. First, they use a series of bites with the lower front teeth, with the head held in such a way that the angle of the teeth is parallel to the stem, in a way similar to when they fell smaller trees. Occasionally a splinter of wood will be removed from the bottom and upwards of the tree, but normally they turn their heads to about a 45-degree angle to the stem while they perform a similar series of bites a little further up the stem (Figure 5.6c). The remaining pieces of stem are torn apart from the powerful pressure. There are two types of wood chips (Shadle, 1957): those mainly composed of bark and those entirely or almost entirely composed of wood. The former are clearly longer than the latter type. Wood chips can be over 10 cm long and 3 cm thick on birch trees and even larger if the tree is an aspen or other species with softer wood (Wilsson, 1971).

With larger trees, the animal will first cut away smaller splinters lower down the trunk in an upwards movement, then later switch to cutting downwards. Before the actual felling, the beaver will debark the stem around the chewing site. Dry, hard bark is bitten loose and thrown aside, but if the bark is soft and juicy, the beaver will draw it out in long strips from the bottom to the top and eat them one at a time (Wilsson, 1971). During the felling process, beavers occasionally take a break to sharpen their teeth by grinding the enamel tip of their incisors against each other to self-sharpen (Stefen et al., 2016). If the beaver can manage to chew all around the tree, it sometimes moves around the stem as it works (Wilsson, 1971). The

(a)

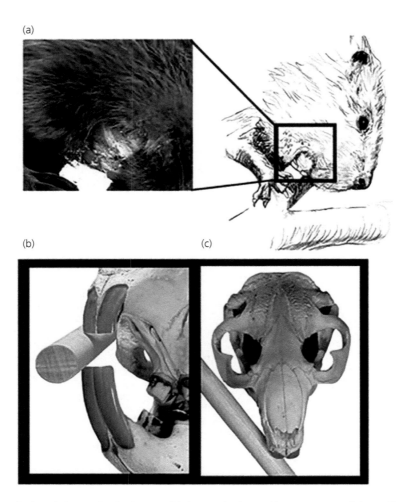

(b)   (c)

Figure 5.5 Position of incisors during woodcutting in *Castor*. (a) Photograph is from a video sequence recorded at Smithsonian Institution National Zoological Park. The lower incisor closest to the camera is clearly visible, indicating that only the incisor farthest from the camera is being used for cutting. *Castor* skull positioned as if during cutting through a stick, shown in anterolateral (b) and dorsal (c) views. Note the animal can maximize the cutting angle by leaning its cheek against the stick. (Republished with permission of Cambridge University Press from Woodcutting behaviour in beavers (Castoridae, Rodentia): estimating ecological performance in a modern and a fossil taxon, Paleobiology, by Rybczynski, N., volume 34, pages 389–402, copyright 2008; permission conveyed through Copyright Clearance Center, Inc.)

tree will then look like two sharpened pencils with the points together just before it falls (Figure 5.6d).

Beavers prefer harvesting trees with a stem diameter of less than 10 cm measured 30 cm above the ground, and the preferred diameter is between 1 cm and 5 cm (Haarberg and Rosell, 2006). However, beavers occasionally fell very large trees (Figure 5.7). In Aasterberg Limburg, the Netherlands, beavers felled a willow tree with a diameter of 80 cm (Willy de Koning-Bovenhoff, pers. comm.). A cottonwood tree with a diameter of 117 cm was reported as hav-

ing been felled by a beaver in North America (Warren, 1927), and they started to gnaw on a white willow (*S. alba*) with a diameter of 162 cm in Kamchatka, Russia (Alexander Saveljev, pers. comm.) and a 2-m-diameter poplar in Germany (Friedrich, 1907, Hinze, 1950). However, there can still be even larger, unreported beaver-felled trees out there.

Opportunistic observations of 67 instances of tree cutting by beavers indicated that cutting time increased sharply with sapling diameter (stem

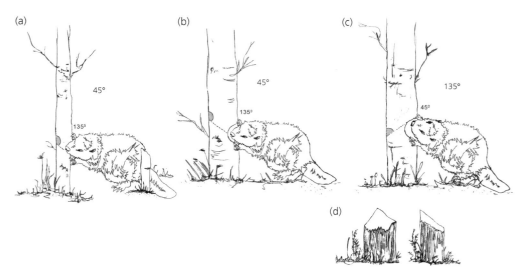

Figure 5.6 (a) Thin trunks are felled with a series of bites made with the 'edge line' of the incisors at an angle of about 45 degrees to the length of the trunk. (b) When felling thick trunks a series of chips are cut off, the 'edge line' of the incisors working at an angle of about 45 degrees to the length of the trunk, and then (c) the beaver works for a while with its head almost upside down, at an angle of approximately 135 degrees to the length of the trunk. (d) Equal work on all sides of the trunk results in conical cutting surfaces; if the beaver has to work on one side only the result is a slanting surface. (Redrawn from Wilsson 1971 by Rachael Campbell-Palmer.)

Figure 5.7 Beaver with a partially cut large tree. Distinctive bite marks are visible. (Photo supplied courtesy of Leopold Kanzler.)

diameter of 5 cm or less). The time required to cut down trees was related to tree size, with an allometric coefficient of 2.6 (Fryxell and Doucet, 1993). A poplar tree with a size similar to a finger can be bitten off in a single bite. A tree of approximately 8 cm in diameter can be cut down in less than 10 minutes, while an aspen with a diameter of 12 cm can be felled in less than half an hour, even if the animal does not work at full speed the entire time. An aspen with a diameter of 25 cm was felled by two animals working together efficiently in about 4 hours (Belovsky, 1984).

Beavers mostly work on trees alone, but on rare occasions several family members v = c an work together. Smaller trees can be felled quickly by one beaver and often brought immediately back to the water. Larger trees are usually felled in stages, and individuals can work in shifts or at different times. If the tree is so large that it cannot be dragged to the water by a single animal, beavers may cooperate and help each other with cutting branches and dragging them back to the water, where they share the food (Wilsson, 1971).

Stems and branches larger than about 12 cm in diameter will often be left behind at the harvest site. They will often be debarked on site but are as often left lying untouched. On some occasions, as many as five beavers can be observed chewing bark on the same stem. When trees are felled, their side branches or upper crowns are removed in manageable sections. Larger trees may need to be cut into smaller sections before they can be transported to water (Wilsson, 1971). These sections are then dragged back to the water to be consumed or incorporated into constructions such as lodges and dams (see Chapter 4).

It can appear as though the beaver knows when a tree is about to fall, because it will apply extra power to the last bite and quickly move aside as needed (Wilsson, 1971). The beaver is most likely reacting to the snapping sound made by the tree just before it falls. It nearly always moves to the correct side but not every time (Kile and Rosell, 1996; see Chapter 8). It seems like the beaver most of the time cannot control where the tree will fall (Raffel and Gatz, 2003).

It is advantageous for beavers to fell trees along banks, because these trees will usually fall into the water (they have a more developed crown and better-developed roots on that side), thus allowing beavers to remain in the safety of the water while foraging on the tree or dividing it into smaller pieces and to more easily transport the tree to the dam, lodge, or food cache. Beavers might have some insight into how to cut the trunk so it falls in the direction of the water but cannot prejudge about the canopy getting stuck on another tree, which hinders felling. Beavers, therefore, would only have full access to a fraction of the trees they harvest. However, beavers are tenacious and can cut the

stem up to 10–15 times to ensure that the tree falls down (Figure 5.8) (Warren, 1926). The expression 'busy as a beaver' is an appropriate description!

Sometimes very high or unusual beaver-gnawed shapes in tree trunks are found, which most likely is due to varying snow depths between each attempt. Spiral-chewed stems can result from several interrupted attempts to fell a tree. Stumps cut high above the ground have been the basis of rumours of giant beavers lurking about in the wilds. It is more likely that the tree was lying at an angle or was half fallen, which allows a beaver to 'climb' along the trunk. Young animals have been observed balancing along stems lying at angles of 35 degrees.

Beavers will often leave behind half-finished projects (Figure 5.9). It is possible that the tree is felled or processed later, but it is usually left to rot (Simonsen, 1973). The reason for this repeated felling of trees that are then not foraged on is not known. Perhaps the animals can meet a specific nutritional need through this chewing. It is possible that the beaver is trying to reduce the density of the forest. By cutting down large trees that have little value as food or building materials the beaver can stimulate new growth, thus securing both food and materials for the following generations. One study in Norway showed that 5.6% of felled trees were not used and that 2.3% of trees that beavers had begun to fell were abandoned and later only chewed

Figure 5.8 Aspen tree felling production line, with an eleventh in the process of being cut (note that one stem is not visible). (Photo supplied courtesy of Frank Rosell.)

Figure 5.9 Beavers will often leave behind half-finished projects like the aspen in this photo. Note the wood chips around the tree, and the gnawed branches and the two lodges. (Photo supplied courtesy of Frank Rosell.)

on occasionally (Rosell and Pedersen, 1999). Further, we cannot totally rule out that beavers are just maintaining their teeth through an innate need to chew (Glynnis Hood, pers. comm.).

There are also large tree stumps that can look as if they have been polished. The base of these stumps is commonly surrounded by very finely crushed sawdust that resembles coffee grounds, rather than the larger wood chips that are more common. These stumps have been given the name 'beaver grinding stones' because it was once believed that beavers needed to chew to keep their front teeth from growing too long, which we now know is not the case. The reason for this apparently 'unnecessary' chewing is not yet fully understood (but see Section 5.7).

5.2.3 Sex and age differences in foraging times and forage selection

Beaver foraging patterns are expected to vary according to sex and age of the beaver due to different physiological needs, but this is not always the case. In Telemark, Norway, adult males and females spent equal time foraging throughout the night from March to August (Sharpe and Rosell, 2003), but their selection for food items may vary by season. Body condition of females was positively influenced by the increase in forest cover and abundance of deciduous trees but had a negative effect on males. Grassland cover, on the other hand,

had a negative effect on body condition of females but males were unaffected (Arnberg, 2019). In Poland, females foraged more on deciduous shoots and tree leaves during summer and autumn than males (Bełżecki et al., 2018). In southeastern Ohio, USA, adults preferentially feed on terrestrial herbs on shore during the summer months, whereas subadults preferably fed on aquatic vegetation (Svendsen, 1980). Kits utilize food brought to the shoreline by adults or subadults, feed along the edge on emergent plants, or consume aquatic vegetation (Svendsen, 1980). The kits' hesitance to forage on land could reflect differential vulnerability to predation (Severud et al., 2013). Other studies found no difference between the sexes or among age groups in the dietary composition (Krojerová-Prokešová et al., 2010, Severud et al., 2013).

5.3 The importance of different woody species and genera

Beavers prefer woody species with softer wood and little resin, such as aspen and poplar (Hall, 1960, Aleksiuk, 1970, Basey et al., 1990, Fryxell and Doucet, 1993, Fryxell et al., 1994, Vorel et al., 2015) (Figure 5.10a,b). Other preferred woody plants are from the genera willow, birch, ash, cherry (*Prunus* spp.), hazel (*Corylus* spp.), alder (*Alnus glutinosa*), maple, beech, hornbeam (*Carpinus* spp.), and hawthorn (*Crataegus*), while the class conifers (*Pinopsida*) is generally avoided, but selection varies relative to composition of what is available (Curry-Lindahl, 1967, Johnston and Naiman, 1990, Nolet et al., 1994, Fustec et al., 2001, Haarberg and Rosell, 2006, Müller-Schwarze, 2011, Iason et al., 2014, Vorel et al., 2015, Shadle et al., 1943).

Aspen is one of the more preferred species (Johnston and Naiman, 1990, Fryxell and Doucet, 1993, Nolet et al., 1994). As seen along forage trails, beavers often ignore other species to find an aspen grove 300–400 m from the water (see Figure 4.14) and sometimes even further. The beaver's preference for aspen is not surprising, since the bark of the aspen contains more calcium than premium clover hay and as much phosphorus as the highest quality grass seeds. Aspen bark probably has the highest nutritional value per square centimetre than any other tree species (Doucet and Fryxell,

Figure 5.10 A beaver kit is cutting a branch to bring it back to the water (a) but can also gnaw at the branch on the spot (b). (Photo supplied courtesy of Frank Rosell.)

1993). The bark contains chlorophyll, which captures solar energy and converts it to carbohydrates. When beavers fell an aspen tree, they select the newer, grey-green bark at the top of the tree. Consumption of 1 kg of aspen bark equals three times as many calories as the same mass of birch bark (Danilov et al., 2011b). A study of beavers foraging on aspen trees in Utah, USA, suggests that there are differences between the nutritional value of mature trees and sprouts and that beavers differentiate between forage of various nutritional value (Masslich et al., 1988). Twigs from mature aspen and aspen sprouts are much higher in nitrogen, phosphorus, and iron than the stem bark. Zinc is much higher in the bark of mature aspen trees compared with twigs and particularly higher in the twigs of aspen sprout than bark. Calcium concentrations are much higher in mature aspen bark than twigs, and magnesium is much higher in mature aspen twigs than bark (Masslich et al., 1988).

Different species of the **willow family** (around 400 species) are also high on the list of the beaver's preferred foods. In Biesbosch, the Netherlands, willow bark comprised nearly 90% of the beavers' diet in winter. This preference might be due to Salicaceae containing much greater volumes of nutrients compared to other woody vegetation (Nolet et al., 1994). These species also regenerate quickly (as does aspen) and, in some locations, beavers thrive on this regrowth. Shrub willows are preferred to tree willows, most likely due to trees

having to be felled before branches become available (Nolet et al., 1994). However, beavers are able to differentiate within the genus. In British Columbia, Canada, Sitka willow (*S. sitchensis*), Drummond's willow (*S. drummondiana*), and Scouler's willow (*S. scouleriana*) were actively selected for, while long-beaked willow (*S. bebbiana*) was selected against, and there was no selection for or against shining willow (*S. lucida*), sandbar willow (*S. exigua*), and Mackenzie's willow (*S. prolix*) (Gerwing et al., 2013). Also, in the Netherlands differential selection across the willow genus was observed (Nolet et al., 1994).

Rowan (*Sorbus aucuparia*) is often a highly selected species. Beavers tend to eat the leaves, thin branches, and bast, but the bark can be ignored even though it is thin and smooth (Haarberg and Rosell, 2006).

Birch is one of the most often consumed species by beavers, partly due to its prevalence. All birch species accumulate resins in their bark and leaves (Nolet et al., 1995). Beavers ignore the bark but eat the soft, nutritious phloem or cambium layer under the bark. If one looks closely, one will see the discarded bark under a felled tree.

Alder and **blackthorn** are common in wet environments and are used primarily as building materials by beavers (see Chapter 4). Alder has a high resin content, which might deter the beavers (Nolet et al., 1995). The leaves of blackthorn will occasionally be eaten but never the bark. Stems and branches of this species, therefore, will be found with the bark intact among other debarked 'edible' building materials in dams and lodges. In some areas it is believed that the beaver eats primarily alder, most likely because the beaver does not have access to other more preferable species. Beavers consume the leaves and bark of grey alder (Haarberg and Rosell, 2006).

Oak (*Quercus* spp.) can also be used as food (Raffel et al., 2009) but usually not until the preferred species have diminished. It is only consumed in relatively small quantities compared to its availability. Beavers eat oak leaves in summer, which are preferred instead of the other parts of the tree, though bark of younger oak trees can be nutritionally important. If the bark is thin and smooth, it will usually be eaten by the beaver, but if it is thicker and more cork-like, the bark will be removed and the inner bast layer will be eaten. Larger oak trees ($\geq$ 60 cm diameter) can be felled in some areas (Margaletić et al., 2006).

Hazel is preferred by beavers along the Donau River in Austria (Krebs, 1984). In the highlands of western Poland, hazel, and beech are preferred (Ważna et al., 2018; see Table 5.2).

In September and October, once the grains are harvested, apple trees (*Malus* spp.) become popular for beavers in areas where available. People with a garden with apple trees should not be surprised to find a beaver on the lawn feasting on apples in the evening. When the beaver eats apples, it grunts happily; these seem to be special treats. The beaver will usually eat the fallen fruit, but they have been seen standing erect to try to reach apples hanging on the lower branches. If they cannot reach enough apples this way they can chew down an apple tree to get to the fruits. In captivity, they preferred to eat red apples instead of green apples.

Cherry, ash, basswood (*Tila* spp.), maple, and elm (*Ulmus* spp.) form a nutritional reserve and are generally felled only when other foraging options are limited. Leaves of bird cherry are commonly eaten in some areas. These species can also be used as building materials. Elm is high on the list for beavers along the Elbe River in Germany, and buckthorn (*Rhamus cathartica*) is used exclusively as a building material.

Bog myrtle (*Myrica gale*) can also be a regular part of the beaver's diet. The leaves and thin branches can be eaten in the summer, while the bark is used in the winter.

Juniper (*Juniperus* spp.), spruce (*Picea* spp.), pine (*Pinus* spp.), hemlock (*Tsuga* spp.) and fir (*Abies* spp.) are used early in the spring, as noted by branches from these species lying in feeding places along the water's edge. Beavers will often eat the needles of these species. Conifers have a high concentration of different types of sugar in their sap and inner bark in spring and early summer, and it is primarily these sugars that beavers seek. Sugar content is generally higher in trees growing near water. Beavers normally ignore conifers during the rest of the year (Jenkins, 1979, Svendsen, 1980). Presumably, lack of other vegetation, combined with renewed sugar flow in the pines render them attractive for a short period of time (Müller-Schwarze, 2011). Debarking of larger trees can be seen in some areas, especially in the springtime; debarking is done to gain access to the sap which they lick up (Svendsen, 1980). In several areas, beavers fell pines when they have reduced access to broad-leaved trees (Chabreck, 1958). Spruce is only rarely felled, but it is occasionally selected and can be used for dam building. Spruce is generally less attractive to beavers than pine, but some individuals take advantage of the smaller spruces in a stand. Beavers primarily fell pines and spruces with diameters less than 10 cm.

5.4 Important herbs and forbs

Herbs are generally good sources of protein, and beavers probably eat these to meet their protein requirements (Doucet and Fryxell, 1993) (Figure 5.11). Herbs and forbs are usually browsed in situ, although beavers can collect bundles of lighter plants in their mouths (Campbell, 2010). They can be quite selective when feeding, sweeping together clumps of lighter plants with their fore paws to form manageable sized mouthfuls. They tend to sever the stems of larger plants at a distinct angle and can create areas of closely cropped 'lawns', often in circular or semicircular patterns (Campbell-Palmer et al., 2015). In Wyoming, USA, beaver faeces were investigated and parts of 20 species of forbs and 24 species of graminoids (herbaceous plants with a grass-like morphology) were found (Collins, 1976).

Figure 5.11 Beaver swimming with herbaceous plants. (Photo supplied courtesy of Bjørnar Røstad Roheim.)

Figure 5.12 Beaver feeding on berry leaves. (Photo supplied courtesy of Kjell-Erik Moseid.)

Herbs have higher K (potassium) concentrations relative to Na (sodium) concentrations, which could disturb the Na:K balance. Herbaceous plants also tend to contain alkaloids (Nolet et al., 1995) and tend to be high in phosphorus in the spring and summer. Female beavers' need for both phosphorus and sodium is more than six times higher than normal when the young are nursing, thus making these plant sources important for reproduction (Nolet et al., 1995). A difference in preference between pregnant females and other individuals would, therefore, be expected. In mammals, pregnant females have an energy demand that is 17–37% higher than non-breeding females (Robbins, 2012). In northeastern Minnesota, USA, pregnant females spend 52% of their time foraging during May and June compared to only 29% by adult males (Buech, 1995). In Telemark in Norway, pregnant females did not forage more due to increased nutrient demands (Mayer et al., 2017). Whereas in Biesbosch, the Netherlands, beavers struggled with reproduction because there was a scarcity of herbs and aquatic plants in this area. Analysis of older beavers also revealed that their bones were porous in some areas (Piechocki, 1962), which is indicative of a lack of phosphorus (Nolet et al., 1994).

The most commonly eaten herbs and forbs include meadowsweet, wood cranesbill, purple marshlocks (*Comarum palustre*) (rhizomes only), rosebay willowherb (*Epilobium angustifolium*), and milk parsley (*Peucedanum palustre*). Rosebay willowherb is a particular favourite for much of the summer, as are raspberry bushes. Beavers eat the outermost shoots

and leaves of raspberry plants. Bracken fern (*Pteridium aquilinum*), broad buckler-fern (*Dryopteris dilatata*), swamp sawgrass (*Cladium mariscus*), devil's-bit scabious (*Succisa pratensis*), yellow loosestrife (*Lysimachia vulgaris*), blackcurrant (*Ribes nigrum*), sorrel (*Rumex acetosa*), clover (*Trifolium* spp.), dandelions (*Taraxacum* spp.), brambles, and stinging nettles (*Urtica dioica*) are also eaten. Berry bushes such as blueberry (*Cyancooccus* spp.) and bilberry (*Vaccinium myrtillus*) are grazed in some areas in southern Norway (Simonsen, 1973; Figure 5.12). In Ohio, USA, North American beavers extensively used fronds and rhizomes of the Christmas fern (*Polystichum acrostichoides*) during January and February and travelled up to 50 m from the water to obtain it (Svendsen, 1980). Nettles are eaten in some areas during summer, and in Russia beavers can even overgraze this plant (Djoshkin and Safonov, 1972). The most important herbaceous foods in east-central Mississippi, USA, were smartweeds (*Polygonum* spp.), marsh purslanes (*Ludwigia* spp.), grasses (Poaceae), soybean (*Cephalanthus occidentalis*), and Japanese honeysuckle (*Lonicera japonica*) (Roberts and Arner, 1984).

5.5 Important crop plants

Beavers consume a very wide range of vegetation, which can vary across the seasons and reflect what is available to them within their active territory. Some of these plants include crops farmed close to freshwater bodies, but beavers do not appear to be significant consumers of crops, compared to the scale of rabbits (Leporidae) and deer (Cervidae) for example.

A beaver living near grain fields that has developed a taste for **grain** cannot keep itself away from the field (Figure 5.13). The first sign of crop grazing by beavers is the large quantity of stems lying near the water's edge. On closer examination, one will notice that the head is missing and the root is still attached to the stem. It is easy to find the beaver trail used for nightly raids leading up to the grains. **Wheat** (*Triticum* spp.), barley (*Hordeum vulgare*) and **oats** (*Avena sativa*) are the beaver's favourite field grain in some areas. (Rosell and Pedersen, 1999, Mikulka et al., 2020). Unfortunately, very little information exists on the importance of crop grazing by beavers (but see Mikulka et al.,

2020), though from a management point of view no grain crop grown near the shoreline should be presumed as unattractive.

Many other crop plants may also be important for beavers. For example, **sugar beet** (*Beta vulgaris*) is an especially popular forage species in southern areas of Europe. In the Netherlands and Germany beavers will eat these crops adjacent to waterways. Some beavers can swim long distances through small dikes to find and consume sugar beet. Beavers also eat rape (*Brassica napus*), carrots (*Daucus carota*), cabbages (*B. oleracea*), swedes (*B. napobrassica*), potatoes (*Solanum tuberosum*), turnips (*B. rapa*), broccoli (*B. oleracea*), peas (*Pisum sativum*), maize (*Zea mays*), and sunflowers (*Helianthus* spp.) (Krojerová-Prokešová et al., 2010, Stefen et al., 2016, Bełżecki et al., 2018, Mikulka et al., 2020).

Figure 5.13 Foraging trail leading to a cleared area in a grain field. (Photo supplied courtesy of Hanna Lodberg-Holm.)

5.6 Aquatic vegetation

Recent stable isotope studies estimate that aquatic plants comprise ~60–80% of the winter and autumn diets of beavers in Quebec, Canada (Milligan and Humphries, 2010) and ~55% of annual diets in Minnesota (Severud et al., 2013). In Ohio, 50% of summer foraging time is spent feeding on aquatic plants (Svendsen, 1980). Beavers often dive repeatedly while feeding on submerged plants and, although these feeding dives can last for between 5 and 6 minutes, they are generally much shorter. Beavers feel for food in the water with their fore paws and carry the vegetation in their mouths (Wilsson, 1971; Figure 5.14).

Beavers incorporate aquatic plant rhizomes in winter food caches and also consume aquatic plants under the ice (Shelton, 1966, Ray et al., 2001). Calculations by Aleksiuk (1970) suggest that northern beavers balance their energy budget in winter by supplementing food resources in the cache with aquatic vegetation acquired away from the food cache. Seasonally, aquatic plant rhizomes often provide the greatest nutritional value in winter and spring when plants store nutrients in rhizomes in preparation for spring leaf growth (Harrison and Mann, 1975, Haraguchi, 2004). Aquatic plants seem

to be a more important component of the diet during summer and winter and in colder regions such as in boreal and subarctic areas (Vorel et al., 2015).

Aquatic plants might also offer better nutritional benefits than terrestrial plants, including increased digestibility, higher protein, sodium, and iron content, and lower amounts of cellulose, lignin, and secondary metabolites (Fraser et al., 1984, Fryxell and Doucet, 1993, Tischler, 2004). One study determined that beavers obtained 2.5 times more food per unit effort when foraging on aquatic vegetation (Belovsky, 1984). Beavers most likely eat aquatic plants to meet their nutritional needs for sodium (Doucet and Fryxell, 1993). Sodium is needed to maintain the Na:K balance in cells. Sodium levels in one species of American water lily (*Nymphaea odorata*) were ten times higher than in the nearby aspen and maple species. However, higher use of aquatic plants did not lead to improved body condition (Severud et al., 2013).

5.6.1 Important aquatic and wetland plants

Some of the most important aquatic and wetland plants for beavers include reeds, lakeshore bulrush (or common club-rush [*Schoenoplectus lactustris*]),

Figure 5.14 Beaver carrying a bundle of aquatic plants (*Equisetum* spp.) in its mouth. (Photo supplied courtesy of Lillian U. Gulliksen.)

water lobelia, broad-leaved pondweed (*Potamogeton natans*), yellow iris (*Iris pseudacorus*), water horsetail (*Equisetum fluviatile*), common (or lake) quillwort (*Isoetes lacustris*), common cottongrass (*Eriphorum angustifolium*), bottle sedge (*Carex rostrate*), bog bean (*Menyanthes trifoliata*), water-arum (*Calla palustris*), watershield (*Brasenia schreberi*), and white and yellow water lilies (*Nuphar lutea*) (Hinze, 1950, Brenner, 1962, Northcott, 1971, Simonsen, 1973, Svendsen, 1980, Histøl, 1989, Ganzhorn and Harthun, 2000); see Table 5.1). Important aquatic plant species published in studies across Europe and North America are presented in Table 5.4.

Beavers prefer specific parts of individual plants. With some plants, notably lakeshore bulrush, broad-leaved pondweed, and common cottongrass, only the roots are eaten, but with others, such as common quillwort, water horsetail, fenugreek (*Trignonella foenum-graecum*), and common reed, the entire plant is consumed (Simonsen, 1973). When beavers feed on the roots of aquatic plants, they first shake the plant in the water before they divide it up and start eating. This behaviour is probably to wash soil from the roots. Once the plants are cleaned, beavers turn them upside down and eat the tender lower parts. The upper parts do not appear to taste quite as good as beavers discard them (Simonsen, 1973).

In many areas, **water lily** often dominates the beaver's diet at midsummer. If water lily rhizomes are available, beavers prefer them compared to woody vegetation in all seasons (Jenkins, 1979). A beaver often lies at the water's surface, with its tail erect and with its fore legs gripping a water lily leaf. It looks almost as if the beaver is eating a pancake. Their ability to eat while swimming at the water's surface is impressive. While the beaver uses its back legs to swim, it will fold the leaf double with its fore legs and move the leaf to its mouth (Figure 5.15). It prefers leaves from the white water lily to those of the yellow in the summertime, while in the winter it will eat the rhizomes of the yellow water lily (Lahti, 1966). The rhizomes of the white water lily are of less importance to a beaver's nutrition. They are more difficult to acquire since they stretch downwards below the river- or lakebed, while those of the yellow water lily lie horizontally just below the

Table 5.4 Important aquatic plants (and waterside/wetland plants) species published in studies across Europe and North America (*Cf* = Eurasian beaver; *Cc* = North American beaver). See Table 5.1 for Latin names.

Species	Country (region)	Important species	References
Cf	Norway (Vest-Agder)	River horsetail, water lilies, common quillwort, common club-rush, and common reed	Simonsen (1973)
Cf	Norway (Vest-Agder)	Water lobelia, river horsetail, yellow water lily, and common quillwort	Histøl (1989)
Cf	Russia	Cattail, reed, water lilies, and pondweed	Gaevskaia (1969)
Cf	Finland	Bog bean, yellow water lily, and river horsetail	Lahti (1966)
Cf	Scotland (Perth)	Yellow iris and bog bean	Law et al. (2014b)
Cf	Scotland (Knapdale)	White water lily	Law et al. (2014a)
Cf	Russia (Karelia)	Water lilies, cattails, bulrush, reeds, sedges, meadowsweet, geum, stinging nettle, ground elder, marsh marigold, and buckbean	Danilov et al. (2011a)
Cc	USA (Ohio)	Waterweed and pondweed	Svendsen (1980)
Cc	Canada (Newfoundland)	Water lilies	Northcott (1972)
Cc	Canada (Saskatchewan)	Water sedges	Novakowski (1965)
Cc	Canada (Quebec)	Water lilies and water sedges	Milligan and Humphries (2010)
Cc	USA (Georgia)	Lizard's tail	Parker et al. (2007)

Figure 5.15 Beaver eating a folded bullhead lily (*Nuphar variegata*). (Photo supplied courtesy of Michael Runtz.)

Figure 5.16 Beaver feeding on watershield (*Brasenia schreberi*). (Photo supplied courtesy of Michael Runtz.)

sediments. Roots of yellow water lilies could lie between rocks and be difficult to reach (Northcott, 1972). Therefore, a mass of water lily leaves on the surface of a pond does not necessarily indicate a bountiful supply of food beneath.

In summer, beavers also eat small side roots and flowers of the white water lily. The flower from the yellow water lily is not that popular but is also eaten (Histøl, 1989). Law et al. (2014a) found no evidence of consumption of white water lily flowers by beavers. In their study in Scotland, beavers selected floating leaves that were much larger and heavier, which was most likely to avoid chemical defences associated with anthocyanin pigments that dominate in smaller leaves. Also, the leaves contained much more phosphorus and nitrogen per unit dry weight than rhizomes. When grazing on white water lily, beavers stayed close to the shore (3 m) and dived to shallow depths (56 cm) (Law et al., 2014a). Water lilies are important to North American beaver density in lakes and streams of Ontario over time (Fryxell, 2001) but not to Eurasian beavers in Swedish streams in a single year (Hartman, 1996).

Watershield (*Brasenia schreberi*) is particularly valuable to sustaining beavers in smaller lakes on Isle Royale. Several lakes with abundant watershield support high population densities and long-term occupancy for beavers (Bergman et al., 2018) (Figure 5.16).

In one study in Kristiansand, Norway, **quillwort** (*Isoetes lacustris*) was the primary food source for beavers in the early summer (Simonsen, 1973). Beavers consume the tuber-like stem of this plant

but not the leaves. Beavers eat quillwort until availability of other plant species improves, and then ignore it once other foods are available. Once a beaver starts eating quillwort it can keep eating it for an hour or more. When a beaver dives for quillwort, it can remain submerged for about 30 seconds and resurfaces with as many as 36 plants. Beavers have preferred foraging grounds for quillworts, with eating locations nearby (Simonsen, 1973). Beavers also eat the tuber-like stem of **water lobelia** (*Lobelia dortmanna*) but not the leaves (Histøl, 1989).

In cafeteria-style experiments, Eurasian beavers consumed greater proportions of **yellow iris** compared to all other grazed species, and there was no difference in the total biomass consumed between any of the other grazed aquatic plants (Law et al., 2014b; Figure 5.17). Beavers consumed the softer and less fibrous lower part and the leaf blade of **bottle sedge**/beaked sedge and yellow iris, and the stalks and rhizome but not the leaf blades of **bog bean** (Lahti, 1966, Histøl, 1989, Law et al., 2014b).

Later in summer, **water horsetail** becomes an increasingly important food item (Lahti, 1966, Simonsen, 1973). When beavers eat water horsetail, they will either bite the plant off at the water's surface or eat them 'on the root'. Most of the time they only eat the part of the plant that protrudes from the water. Beavers can float around in shallow water and collect 10–20 plants before taking them to their eating place. Beavers will also eat browned and wilted plants under the ice during winter (Simonsen, 1973).

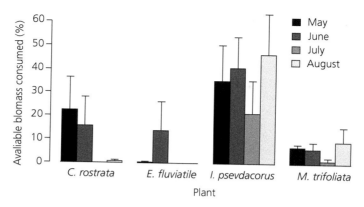

Figure 5.17 Mean proportion of available biomass consumed of grazed macrophyte species per month in 2012. Error bars indicate standard error. (Reprinted from *Aquatic Botany*, 116, Law, A., Jones, K.C. & Willby, N.J., Medium vs. short-term effects of herbivory by Eurasian beaver on aquatic vegetation, pages 27–34, © Elsevier Science (2014), with permission from Elsevier.)

Reed is a popular food in the middle of summer (Djoshkin and Safonov, 1972, Simonsen, 1973). Mature reeds are not especially attractive for beavers but the young shoots are, and beavers eat both the leaves and stems (Simonsen, 1973). Beavers are also fond of **algae** (Svendsen, 1980). Several species of algae contain large amounts of proteins and are very nutritious.

5.7 Variation in the diet

A beaver's fitness improves on varied diets rather than on single-species diet. A balanced diet contains proteins, carbohydrates, and fats, but vitamins and trace elements are also important. Deficiencies can lead to disease and death. For example, iodine deficiency may lead to goitre and ultimately thyroid cancer. Due to their high iodine content, aquatic plans might, therefore, be important for beavers (Müller-Schwarze, 2011). Tree species, such as alder, hazel, ash, aspen, and bird cherry, were not abundant in Biesbosch, the Netherlands, but were the preferred foods of beavers in the first year after reintroduction in this area. Beavers selected these species to supplement their diets (Nolet et al., 1994). They chose hazel and ash for sodium and cherry and poplar for phosphorus. Trees in the willow family are poor in phosphorus and nitrogen and very poor in sodium. A varied diet could be a strategy to minimize the risk of nutrient deficiencies (Nolet et al., 1994).

Beavers will not normally chew on dead or diseased trees, but it does happen. Beavers have also chewed on telephone poles, pallets, old roots, and half-rotted stumps. Dead branches are relatively rich in sodium. It is therefore possible that beavers use them to acquire this nutrient (Nolet et al., 1995). Beavers can obtain a variety of nutrients from different plant species, but they also encounter a wide variety of plant defences.

5.8 Plant defences and responses to beaver feeding

Plants have evolved various defences against herbivores. These defence mechanisms can be chemical, physiological, or mechanical. Chemical methods are most common. Woody vegetation reacts to intense grazing both by increasing regrowth and by chemical defences (toxins) to deter grazing (review by Bryant et al. 1991). These toxins affect grazing animals in a number of ways. Some affect the animal's digestion and cause illness. For example, tannins are a group of phenols that can reduce the animal's uptake of proteins and sodium.

Parts of plants easily reached by herbivores tend to have higher concentrations of harmful compounds than trees. These compounds are concentrated in the outermost part of the branches, for example in buds of birch trees and twigs of aspen. Plants can show their 'enemies' they have developed anti-herbivory compounds by changing their appearance, thus deterring predation by the animal. Through such defences plants can gain an advantage in the constant struggle between plant and beaver. Changes in plant chemistry or form as a response to predation might explain why a beaver family moves to another area even when there appears to be plentiful forage in the area.

5.8.1 Plant responses to beaver foraging

Plants respond to beaver foraging in different ways, but beavers eat many of them anyway. Fifty-seven toxic plants—22 woods and 35 herbs—used (for food and other utilization) by beavers have been identified (Nitsche, 2017). The beaver can reduce the effects of the plant's chemical defences by avoiding certain parts of the plant that contain higher concentrations of toxins. When a beaver comes to a new area not previously inhabited by other beavers, it prefers to consume smaller trees rather than larger ones. If the area has been occupied by beavers in the past, a newcomer cuts larger trees due to the chemical defences in the smaller trees. Beavers living near a pond periodically occupied over a period of more than 20 years chose aspens > 19.5 cm in diameter and avoided those < 4.5 cm in diameter (Basey et al., 1988). Beavers avoid juvenile-sized sprouts compared with adult-sized sprouts of uncut aspen clones. Aspen clones that regenerate after having been felled by beavers have higher concentration of secondary compounds that repel beavers (Basey and Jenkins, 1993). As such, aspen can deploy a chemical defence to reduce beaver grazing (Basey et al., 1988, Basey et al., 1990). Also, Fremont cottonwood (*P. fremontii*) growing furthest from water did not suffer from substantial harvesting, had an upright growth form, and likely reached maturity and reproduced sexually. In contrast, cottonwood trees close to water had dense, shrub-like growth because selective harvesting by beavers maintained them in perpetual juvenile condition, and reproduction was predominately vegetative. Narrowleaf cottonwood (*P. angustifolia*) and aspen reproduce clonally by producing adventitious buds by the roots when they are heavily harvested by beavers. These buds sprout around the base of the cut stumps (McGinley and Whitham, 1985).

Beavers strongly avoid red maple (Müller-Schwarze et al., 1994). Interestingly, when aspen stems were painted with red maple extracts containing secondary metabolites, there was diminished palatability of aspen stems for beavers (Müller-Schwarze et al., 1994). Similarly, beavers avoided aspen stems painted with extracts from bark and needles of Jeffrey pine (*P. jefferyi*) in feeding experiments (Basey, 1999). Avoidance of common elderberry (*Sambucus nigra*) by beavers in several studies is likely a consequence of the presence of a poisonous cyanogenic glycoside in the tree (Nolet et al., 1994).

Phenolic-laden tress, such as red maple and witch hazel, are avoided. Young trees often are more chemically defended than mature ones (Müller-Schwarze, 2011). However, beavers tend to leave sticks of these two species in the water for 1–3 days before they consume them, most likely to leach out undesirable compounds, which makes them more palatable (Müller-Schwarze et al., 2001). Beavers prefer tree species with linear growth (e.g. Fremont cottonwood) with low levels of condensed tannins (Bailey et al., 2004) rather than branched growth with condensed tannins (Hagerman and Robbins, 1993). Interestingly, beaver saliva contains proteins that bind only to the linear and not to condensed tannins and hydrolysable tannins (Müller-Schwarze, 2011).

Low concentrations of secondary compounds might make water lily more palatable than other food plants, as revealed in a cafeteria experiment (Doucet and Fryxell, 1993). In contrast, roots of yellow iris contain high concentrations of tannins (Nolet et al., 1995). Interestingly, the alkaloid desoxynupharidin of the yellow water lily was found in secretion from castor sacs in beavers along the River Elbe, which suggests that beavers can adapt to the intake of some of these compounds by expelling them through the castoreum (Luckner, 1969; see Section 7.2.2.1).

5.8.2 Physiological defence of plants

Early in the recovery phase, willows invest in regrowth rather than a chemical defence response. After heavy autumn and winter browsing by herbivores, cut shrubs have a better food quality (more water and nitrogen content) and fewer phenolics in the following spring (Veraart et al., 2006). These results are in accordance with a cafeteria experiment in the field from the same area (Veraart et al., 2006). In May, beavers preferred shoots from the cut treatment, but in November they showed no preference. Willows invest their carbon in regrowth rather than carbon-based defence (like phenolic glycosides). Therefore, increased plant secondary compounds do not affect the beaver's foraging choices when feeding on preferred plant genera, for example *Populus* spp. and *Salix* spp. In fact, the increased food quality after winter browsing of willows might have a positive effect on the reproduction of beavers (Veraart et al., 2006).

References

Aleksiuk, M. 1970. The seasonal food regime of Arctic beavers. *Ecology*, 51, 264–270.

Arnberg, M. P. 2019. The importance of vegetation availability in three scales on body condition, reproduction and territory size of the Eurasian beaver. MSc thesis, University of South-Eastern Norway.

Bailey, J. K., Schweitzer, J. A., Rehill, B. J., Lindroth, R. L., Martinsen, G. D., and Whitham, T. G. 2004. Beavers as molecular geneticists: a genetic basis to the foraging of an ecosystem engineer. *Ecology*, 85, 603–608.

Baker, B. W., Ducharme, H. C., Mitchell, D. C. S., Stanley, T. R., and Peinetti, H. R. 2005. Interaction of beaver and elk herbivory reduces standing crop of willow. *Ecological Applications*, 15, 110–118.

Barela, I. A. and Frey, J. K. 2016. Habitat and forage selection by the American beaver (*Castor canadensis*) on a regulated river in the Chihuahuan Desert. *The Southwestern Naturalist*, 61, 286–293.

Barnes, W. J. and Dibble, E. 1988. The effects of beaver in riverbank forest succession. *Canadian Journal of Botany*, 66, 40–44.

Basey, J. M. 1999. Foraging behavior of beavers (*Castor canadensis*), plant secondary compounds, and management concerns. In: Busher, P. E. and Dzięciołowski, R. M. (eds.) *Beaver Protection, Management, and Utilization in Europe and North America*. Boston: Springer.

Basey, J. M. and Jenkins, S. H. 1993. Production of chemical defenses in relation to plant growth rate. *Oikos*, 68, 323–328.

Basey, J. M. and Jenkins, S. H. 1995. Influences off predation risk and energy maximization on food selection by beavers (*Castor canadensis*). *Canadian Journal of Zoology*, 73, 2197–2208.

Basey, J. M., Jenkins, S. H., and Busher, P. E. 1988. Optimal central-place foraging by beavers: tree-size selection in relation to defensive chemicals of quaking aspen. *Oecologia*, 76, 278–282.

Basey, J. M., Jenkins, S. H., and Miller, G. C. 1990. Food selection by beavers in relation to inducible defenses of *Populus tremuloides*. *Oikos*, 59, 57–62.

Baskin, L. and Sjöberg, G. 2003. Planning, coordination and realization of northern European beaver management, based on the experience of 50 years of beaver restoration in Russia, Finland, and Scandinavia. *Lutra*, 46, 243–250.

Belovsky, G. E. 1984. Summer diet optimization by beaver. *American Midland Naturalist*, 111, 209–222.

Bełżecki, G., Miltko, R., Kowalik, B., Demiaszkiewicz, A. W., Lachowicz, J., Giżejewski, Z., Obidziński, A. and McEwan, N. R. 2018. Seasonal variations of the digestive tract of the Eurasian beaver *Castor fiber*. *Mammal Research*, 63, 21–31.

Bergman, B. G., Bump, J. K., and Romanski, M. C. 2018. Revisiting the role of aquatic plants in beaver habitat selection. *American Midland Naturalist*, 179, 222–246.

Boyce, M. S. 1981. Beaver life-history responses to exploitation. *Journal of Applied Ecology*, 18, 749–753.

Brenner, F. J. 1962. Foods consumed by beavers in Crawford county, Pennsylvania. *Journal of Wildlife Management*, 26, 104–107.

Bryant, J. P., Provenza, F. D., Pastor, J., Reichardt, P. B., Clausen, T. P., and du Toit, J. T. 1991. Interactions between woody plants and browsing mammals mediated by secondary metabolites. *Annual Review of Ecology and Systematics*, 22, 431–446.

Brzyski, J. R. and Schulte, B. A. 2009. Beaver (*Castor canadensis*) impacts on herbaceous and woody vegetation in southeastern Georgia. *American Midland Naturalist*, 162, 74–86.

Buech, R. R. 1995. Sex differences in behavior of beavers living in near-boreal lake habitat. *Canadian Journal of Zoology*, 73, 2133–2143.

Campbell, R. 2010. Demography and life history of the Eurasian beaver *Castor fiber*. PhD thesis, University of Oxford.

Campbell-Palmer, R., Gow, D., Needham, R., Jones, S., and Rosell, F. 2015. *The Eurasian Beaver*. Exeter, Pelagic Publishing.

Chabreck, R. H. 1958. Beaver–forest relationships in St. Tammany Parish, Louisiana. *Journal of Wildlife Management*, 22, 179–183.

Coles, B. 2006. *Beavers in Britain's Past*. Oxford, Oxbow Books.

Collins, T. C. 1976. Population characteristics and habitat relationships of beaver, *Castor canadensis*. PhD thesis, University of Wyoming.

Cox, P. G. and Baverstock, H. 2016. Masticatory muscle anatomy and feeding efficiency of the American beaver,

Castor canadensis (Rodentia, Castoridae). *Journal of Mammalian Evolution*, 23, 191–200.

Curry-Lindahl, K. 1967. The beaver, *Castor fiber* Linnaeus, 1758 in Sweden extermination and reappearance. *Acta Theriologica*, 12, 1–15.

Danilov, P., Kanshiev, V., and Fyodorov, F. 2011a. Characteristics of North American and Eurasian beaver ecology in Karelia. In: Sjöberg, G. and Ball, J. P. (eds.) *Restoring the European Beaver: 50 Years of Experience.* Sofia-Moscow: Pensoft.

Danilov, P., Kanshiev, V. Y., and Fedorov, F. 2011b. History of beavers in eastern Fennoscandia from the neolithic to the 21st century. In: Sjöberg, G. and Ball, J. P. (eds.) *Restoring the European Beaver: 50 Years of Experience.* Sofia-Moscow: Pensoft.

Dezhkin, V. V., D'yakov, Y. V., and Safonov, V. G. 1986. *Bobr* [*The Beaver*]. Moscow: Agropromizdat, pp, 45–46.

Djoshkin, W. W. and Safonov, V. G. 1972. *Die Biber der alten und neuen Welt.* Wittenberg: A. Ziemsen.

Donkor, N. T. and Fryxell, J. M. 1999. Impact of beaver foraging on structure of lowland boreal forests of Algonquin Provincial Park, Ontario. *Forest Ecology and Management*, 118, 83–92.

Donkor, N. T. and Fryxell, J. M. 2000. Lowland boreal forests characterization in Algonquin Provincial Park relative to beaver (*Castor canadensis*) foraging and edaphic factors. *Plant Ecology*, 148, 1–12.

Doucet, C. M. and Fryxell, J. M. 1993. The effect of nutritional quality on forage preference by beavers. *Oikos*, 67, 201–208.

Doucet, C. M., Walton, R. A., and Fryxell, J. M. 1994. Perceptual cues used by beavers foraging on woody plants. *Animal Behaviour*, 47, 1482–1484.

Druzinsky, R. E. 1995. Incisal biting in the mountain beaver (*Aplodontia rufa*) and woodchuck (*Marmota monax*). *Journal of Morphology*, 226, 79–101.

Dvořák, J. 2013. Diet preference of Eurasian beaver (*Castor fiber* L., 1758) in the environment of Oderské vrchy and its influence on the tree species composition of river bank stands. *Acta Universitatis Agriculturae et Silviculturae Mendelianae Brunensis*, 61, 1637–1643.

Elmeros, M., Madsen, A., and Berthelsen, J. P. 2003. Monitoring of reintroduced beavers (*Castor fiber*) in Denmark. *Lutra*, 46, 153–162.

Engelhart, A. and Müller-Schwarze, D. 1995. Responses of beaver (*Castor canadensis* Kuhl) to predator chemicals. *Journal of Chemical Ecology*, 21, 1349–1364.

Erome, G. and Broyer, J. 1984. Analyse des relations *Castor*–vegetation. *Bievre*, 5, 15–63.

Fraser, D., Chavez, E. R., and Palohelmo, J. E. 1984. Aquatic feeding by moose: selection of plant species and feeding areas in relation to plant chemical composition and characteristics of lakes. *Canadian Journal of Zoology*, 62, 80–87.

Friedrich, H. 1907. Biber in Wintersnot. *Sankt Hubertus*, 25, 29–31.

Fryxell, J. M. 1992. Space use by beavers in relation to resource abundance. *Oikos*, 64, 474–478.

Fryxell, J. M. 1999. Functional responses to resource complexity: an experimental analysis of foraging by beavers. In: Collen, P. (ed.) *Herbivores: Between Plants and Predators.* Oxford: Blackwell Scientific.

Fryxell, J. M. 2001. Habitat suitability and source–sink dynamics of beavers. *Journal of Animal Ecology*, 70, 310–316.

Fryxell, J. M. and Doucet, C. M. 1991. Provisioning time and central-place foraging in beavers. *Canadian Journal of Zoology*, 69, 1308–1313.

Fryxell, J. M. and Doucet, C. M. 1993. Diet choice and the functional response of beavers. *Ecology*, 74, 1297–1306.

Fryxell, J. M., Vamosi, S. M., Walton, R. A., and Doucet, C. M. 1994. Retention time and the functional response of beavers. *Oikos*, 71, 207–214.

Fustec, J., Lode, T., Le Jacques, D., and Cormier, J. P. 2001. Colonization, riparian habitat selection and home range size in a reintroduced population of European beavers in the Loire. *Freshwater Biology*, 46, 1361–1371.

Gable, T. D., Windels, S. K., Bruggink, J. G., and Barber-Meyer, S. M. 2018. Weekly summer diet of gray wolves (*Canis lupus*) in northeastern Minnesota. *American Midland Naturalist*, 179, 15–27.

Gaevskaia, N. S. 1969. *The Role of Higher Aquatic Plants in the Nutrition of the Animals of Fresh-Water Basins.* Boston Spa, Yorkshire: National Lending Library for Science and Technology.

Gallant, D., Bérubé, C. H., Tremblay, E., and Vasseur, L. 2004. An extensive study of the foraging ecology of beavers (*Castor canadensis*) in relation to habitat quality. *Canadian Journal of Zoology*, 82, 922–933.

Gallant, D., Léger, L., Tremblay, É., Berteaux, D., Lecomte, N., and Vasseur, L. 2016. Linking time budgets to habitat quality suggests that beavers (*Castor canadensis*) are energy maximizers. *Canadian Journal of Zoology*, 94, 671–676.

Ganzhorn, J. U. and Harthun, M. 2000. Food selection by beavers (*Castor fiber albicus*) in relation to plant chemicals and possible effects of flooding on food quality. *Journal of Zoology*, 251, 391–398.

Gerwing, T. G., Johnson, C. J., and Alström-Rapaport, C. 2013. Factors influencing forage selection by the North American beaver (*Castor canadensis*). *Mammalian Biology*, 78, 79–86.

Gleason, J. S., Hoffman, R. A., and Wendland, J. M. 2005. Beavers, *Castor canadensis*, feeding on salmon carcasses: opportunistic use of a seasonally superabundant food source. *Canadian Field-Naturalist*, 119, 591–593.

Gorshkov, D. and Gorshkov, Y. 2011. *Feeding Strategy of Beavers in Tatarstan Republic.* Sofia-Moscow: Pensoft.

Goryainova, Z. I., Katsman, E. A., Zavyalov, N. A., Khlyap, L. A., and Petrosyan, V. G. 2014. Evaluation of tree and shrub resources of the Eurasian beaver (*Castor fiber* L.) and changes in beaver foraging strategy after resources depletion. *Russian Journal of Biological Invasions*, 5, 242–254.

Haarberg, O. and Rosell, F. 2006. Selective foraging on woody plant species by the Eurasian beaver (*Castor fiber*) in Telemark, Norway. *Journal of Zoology*, 270, 201–208.

Hagerman, A. E. and Robbins, C. T. 1993. Specificity of tannin-binding salivary proteins relative to diet selection by mammals. *Canadian Journal of Zoology*, 71, 628–633.

Hall, J. G. 1960. Willow and aspen in the ecology of beaver on Sagehen Creek, California. *Ecology*, 41, 484–494.

Haraguchi, A. 2004. Seasonal changes in redox properties of peat, nutrition and phenology of *Menyanthes trifoliata* L. in a floating peat mat in Mizorogaike Pond, central Japan. *Aquatic Ecology*, 38, 351–357.

Harrison, P. G. and Mann, K. H. 1975. Detritus formation from eelgrass (*Zostera marina* L.): the relative effects of fragmentation, leaching, and decay. *Limnology and Oceanography*, 20, 924–934.

Hartman, G. 1996. Habitat selection by European beaver (*Castor fiber*) colonizing a boreal landscape. *Journal of Zoology*, 240, 317–325.

Heidecke, D. 1989. Ökologische bewertung von biberhabitaten. *Säugetierkundliche Informationen*, 3, 13–28.

Hilfiker, E. L. 1991. *Beavers, Water, Wildlife and History*. Interlaken, NY: Windswept Press.

Hinze, G. 1950. *Der Biber. Körperbau und Lebensweise Verbreitung und Geschichte*. Berlin: Akademie.

Histøl, T. 1989. Sommerdiett hos bever *Castor fiber* L. i et utvalg av skogsvann i Vennesla kommune, Vest Agder. *Fauna*, 42, 96–103.

Hood, G. A. and Bayley, S. E. 2008. The effects of high ungulate densities on foraging choices by beaver (*Castor canadensis*) in the mixed-wood boreal forest. *Canadian Journal of Zoology*, 86, 484–496.

Iason, G. R., Sim, D. A., Brewer, M. J., and Moore, B. D. 2014. *The Scottish Beaver Trial: Woodland Monitoring 2009–2013, Final Report*. Inverness: Scottish Natural Heritage.

Janiszewski, P., Kolasa, S., and Strychalski, J. 2017. The preferences of the European beaver *Castor fiber* for trees and shrubs in riparian zones. *Applied Ecology and Environmental Research*, 15, 313–327.

Jenkins, S. H. 1979. Seasonal and year-to-year differences in food selection by beavers. *Oecologia*, 44, 112–116.

Jenkins, S. H. 1980. A size–distance relation in food selection by beavers. *Ecology*, 61, 740–746.

Johnston, C. A. and Naiman, R. J. 1990. Browse selection by beaver: effects on riparian forest composition. *Canadian Journal of Forest Research*, 20, 1036–1043.

Jones, A. G. and Ardren, W. R. 2003. Methods of parentage analysis in natural populations. *Molecular Ecology*, 12, 2511–2523.

Kile, N. B. and Rosell, F. 1996. European beaver, *Castor fiber*, pinned by a felled tree. *Canadian Field-Naturalist* 110, 706–707.

Krebs, U. 1984. Analyse der monatlichen fällmengen einer isolierten gründerpopulation des bibers *Castor fiber* L. in den Donauauen bei Wien. *Säugetierkundliche Mitteilungen*, 31, 209–222.

Krojerová-Prokešová, J., Barančeková, M., Hamšíková, L., and Vorel, A. 2010. Feeding habits of reintroduced Eurasian beaver: spatial and seasonal variation in the use of food resources. *Journal of Zoology*, 281, 183–193.

Lahti, S. 1966. Majavan ravinnonvalinnasta ja ravinnon käytöstä [On the food habits of the beaver (*Castor* spp.) in northern Finland]. *Suomen Riista*, 18, 7–19.

Law, A., Bunnefeld, N., and Willby, N. J. 2014a. Beavers and lilies: selective herbivory and adaptive foraging behaviour. *Freshwater Biology*, 59, 224–232.

Law, A., Jones, K. C., and Willby, N. J. 2014b. Medium vs. short-term effects of herbivory by Eurasian beaver on aquatic vegetation. *Aquatic Botany*, 116, 27–34.

Lodberg-Holm, H. K., Steyaert, S. M. J. G., Reinhardt, S., and Rosell, F. 2021. Size is not everything: Differing activity and foraging patterns between the sexes in a monomorphic mammal. *Behavioral Ecology and Sociobiology, 75*, 76.

Luckner, M. 1969. *Der Sekundärstoffwechsel in Pflanze und Tier*. Jena: VEB Gustav Fischer.

Margaletić, J., Grubešić, M., Dušak, V., and Konjević, D. 2006. Activity of European beavers (*Castor fiber* L.) in young pedunculate oak (*Quercus robur* L.) forests. *Veterinarski Arhiv*, 76, S167–S175.

Masslich, W. J., Brotherson, J. D., and Cates, R. G. 1988. Relationships of aspen (*Populus tremuloides*) to foraging patterns of beaver (*Castor canadensis*) in the Strawberry Valley of central Utah. *Great Basin Naturalist*, 48, 250–262.

Mayer, M., Robstad, C., Serra, E. P., Hohwieler, K., Fuchs, B., Evans, A. L., Arnemo, J. M., Zedrosser, A., and Rosell, F. 2017. *Exposure of the Eurasian beaver (*Castor fiber*) towards hunters during the spring hunt*. Bø i Telemark: University College of Southeast Norway.

McClintic, L. F., Wang, G., Taylor, J. D., and Jones, J. C. 2014. Movement characteristics of American beavers (*Castor canadensis*). *Behaviour*, 151, 1249–1265.

McGinley, M. A. and Whitham, T. G. 1985. Central place foraging by beavers (*Castor canadensis*): a test of foraging predictions and the impact of selective feeding on the growth form of cottonwoods (*Populus fremontii*). *Oecologia*, 66, 558–562.

Mikulka, O., Homolka, M., Drimaj, J., and Kamler. J. 2020. European beaver (*Castor fiber*) in open agricultural landscapes: crop grazing and the potential for economic damage. European Journal of Wildlife Research 66: 101.

Milligan, H. E. and Humphries, M. M. 2010. The importance of aquatic vegetation in beaver diets and the seasonal and habitat specificity of aquatic–terrestrial ecosystem linkages in a subarctic environment. Oikos, 119, 1877–1886.

Misiukiewicz, W., Gruszczyńska, J., Grzegrzółka, B., and Januszewicz, M. 2016. Impact of the European beaver (Castor fiber L.) population on the woody vegetation of Wigry National Park. Scientific Annals of Polish Society of Animal Production, 12, 45–64.

Müller-Schwarze, D. 2011. The Beaver: Its Life and Impact. Ithaca: Cornell University Press.

Müller-Schwarze, D. and Schulte, B. A. 1999. Behavioral and ecological characteristics of a 'climax' population of beaver (Castor canadensis). In: Busher, P. E. and Dzięciołowski, R. M. (eds.) Beaver Protection, Management, and Utilization in Europe and North America. Boston: Springer.

Müller-Schwarze, D., Schulte, B. A., Sun, L., Müller-Schwarze, A., and Müller-Schwarze, C. 1994. Red maple (Acer rubrum) inhibits feeding by beaver (Castor canadensis). Journal of Chemical Ecology, 20, 2021–2034.

Müller-Schwarze, D., Brashear, H., Kinnel, R., Hintz, K. A., Lioubomirov, A., and Skibo, C. 2001. Food processing by animals: do beavers leach tree bark to improve palatability? Journal of Chemical Ecology, 27, 1011–1028.

Nitsche, K. A. 2017. Von bibern (Castor fiber L.) genutzte pflanzen mit toxischen wirkstoffen. Artenschutzreport, 36, 50–56.

Nixon, C. M. and Ely, J. 1969. Foods eaten by a beaver colony in southeast Ohio. Ohio Journal of Science, 65, 313–319.

Nolet, B. A. and Rosell, F. 1994. Territoriality and time budgets in beavers during sequential settlement. Canadian Journal of Zoology, 72, 1227–1237.

Nolet, B. A., Hoekstra, A., and Ottenheim, M. M. 1994. Selective foraging on woody species by the beaver Castor fiber, and its impact on a riparian willow forest. Biological Conservation, 70, 117–128.

Nolet, B. A., Van Der Veer, P. J., Evers, E. G. J., and Ottenheim, M. M. 1995. A linear programming model of diet choice of free-living beavers. Netherlands Journal of Zoology, 45, 315–337.

Northcott, T. H. 1971. Feeding habits of beaver in Newfoundland. Oikos, 22, 407–410.

Northcott, T. H. 1972. Water lilies as beaver food. Oikos, 23, 408–409.

Novakowski, N. S. 1965. Population dynamics of a beaver population in northern latitudes. PhD thesis, University of Saskatchewan.

Orians, G. H. and Pearson, N. E. 1979. On the theory of central place foraging. In: Orians, G. H. and Pearson, N. E. (eds.) Analysis of Ecological Systems. Columbus: Ohio State University Press, pp. 157–177.

Parker, H., Haugen, A., Kristensen, Ø., Myrum, E., Kolsing, R., and Rosell, F. 2001. Landscape use and economic value of Eurasian beaver (Castor fiber) on a large forest in southern Norway. Proceedings of the 1st European–American Beaver Congress, 2001 Kazan, Russia, pp. 77–95.

Parker, J. D., Caudill, C. C., and Hay, M. E. 2007. Beaver herbivory on aquatic plants. Oecologia, 151, 616–625.

Piechocki, R. 1962. Die Todesursachen der Elbe-Biber (Castor fiber albicus Matschie 1907) unter besonderer beriicksichtigung funktioneller Wirbelsäulen-Störungen. Nova Acta Leopoldina, 25, 1–75.

Pietrek, A. G., Escobar, J. M., Fasola, L., Roesler, I., and Schiavini, A. 2017. Why invasive Patagonian beavers thrive in unlikely habitats: a demographic perspective. Journal of Mammalogy, 98, 283–292.

Pinkowski, B. 1983. Foraging behavior of beavers (Castor canadensis) in North Dakota. Journal of Mammalogy, 64, 312–314.

Raffel, T. R. and Gatz, A. J. 2003. The orientation of beavers (Castor canadensis) when cutting trees. Ohio Journal of Science, 103, 143–146.

Raffel, T. R., Smith, N., Cortright, C., and Gatz, A. J. 2009. Central place foraging by beavers (Castor canadensis) in a complex lake habitat. American Midland Naturalist, 162, 62–73.

Ray, A. M., Rebertus, A. J., and Ray, H. L. 2001. Macrophyte succession in Minnesota beaver ponds. Canadian Journal of Botany, 79, 487–499.

Richards, D. 1983. Beaversprite. New York: Heart of the Lakes Publishing.

Robbins, C. 2012. Wildlife Feeding and Nutrition. London: Academic Press.

Roberts, T. H. and Arner, D. H. 1984. Food habits of beaver in east-central Mississippi. Journal of Wildlife Management, 48, 1414–1419.

Rosell, F. and Czech, A. 2000. Responses of foraging Eurasian beavers Castor fiber to predator odours. Wildlife Biology, 6, 13–21.

Rosell, F. and Pedersen, K. V. 1999. Bever [The Beaver]. Oslo: Landbruksforlaget.

Rosell, F. and Sanda, J. 2006. Potential risks of olfactory signaling: the effect of predators on scent marking by beavers. Behavioral Ecology, 17, 897–904.

Rosell, F., Bozér, O., Collen, P., and Parker, H. 2005. Ecological impact of beavers Castor fiber and Castor canadensis and their ability to modify ecosystems. Mammal Review, 35, 248–276.

Rossell, C. R., Arico, S., Clarke, H. D., Horton, J. L., Ward, J. R., and Patch, S. C. 2014. Forage selection of native and nonnative woody plants by beaver in a rare-shrub community in the Appalachian Mountains of North Carolina. Southeastern Naturalist, 13, 649–662.

Rybczynski, N. 2008. Woodcutting behavior in beavers (Castoridae, Rodentia): estimating ecological performance in a modern and a fossil taxon. *Paleobiology, 34,* 389–402.

Salandre, J. A., Beil, R., Loehr, J. A., and Sundell, J. 2017. Foraging decisions of North American beaver (*Castor canadensis*) are shaped by energy constraints and predation risk. *Mammal Research, 62,* 229–239.

Severud, W. J., Belant, J. L., Bruggink, J. G., and Windels, S. K. 2011. Predator cues reduce American beaver use of foraging trails. *Human–Wildlife Interactions, 5,* 17.

Severud, W. J., Windels, S. K., Belant, J. L., and Bruggink, J. G. 2013. The role of forage availability on diet choice and body condition in American beavers (*Castor canadensis*). *Mammalian Biology, 78,* 87–93.

Shadle, A. R. 1957. Sizes of beaver chips cut from aspen. *Journal of Mammalogy, 38,* 268–268.

Shadle, A. R. and Austin, T. S. 1939. Fifteen months of beaver work at Allegany State Park, N.Y. *Journal of Mammalogy, 20,* 299–303.

Shadle, A. R., Nauth, A. M., Gese, E. C., and Austin, T. S. 1943. Comparison of tree cuttings of six beaver colonies in Allegany State Park, New York. *Journal of Mammalogy, 24,* 32–39.

Sharpe, F. and Rosell, F. 2003. Time budgets and sex differences in the Eurasian beaver. *Animal Behaviour, 66,* 1059–1067.

Shelton, P. C. 1966. Ecological studies of beavers, wolves, and moose in Isle Royale National Park, Michigan. PhD thesis, Purdue University.

Sih, A. 1980. Optimal behavior: can foragers balance two conflicting demands? *Science, 210,* 1041–1043.

Simonsen, T. A. 1973. *Beverens næringsøkologi i Vest-Agder.* Trondheim: Direktoratet for jakt, viltstell og ferskvannsfiske.

Skøyen, K., Iversen, T. N. S., and Ihle, P. B. 2018. *Crown Cover Preferences by the Eurasian Beaver (*Castor fiber*) on Aspen (*Populus tremula*) in Telemark, Norway.* Notodden: University of South-Eastern Norway.

Smith, D. W., Trauba, D. R., Anderson, R. K., and Peterson, R. O. 1994. Black bear predation on beavers on an island in Lake Superior. *American Midland Naturalist, 132,* 248–255.

Stefen, C., Ibe, P., and Fischer, M. S. 2011. Biplanar X-ray motion analysis of the lower jaw movement during incisor interaction and mastication in the beaver (*Castor fiber* L. 1758). *Mammalian Biology, 76,* 534–539.

Stefen, C., Habersetzer, J., and Witzel, U. 2016. Biomechanical aspects of incisor action of beavers (*Castor fiber* L.). *Journal of Mammalogy, 97,* 619–630.

Svendsen, G. E. 1980. Seasonal change in feeding patterns of beaver in southeastern Ohio. *Journal of Wildlife Management, 44,* 285–289.

Tevis, L. 1950. Summer behavior of a family of beavers in New York State. *Journal of Mammalogy, 31,* 40–65.

Tischler, K. B. 2004. Aquatic plant nutritional quality and contribution to moose diet at Isle Royale National Park. PhD thesis, Michigan Technological University.

Varju, J. and Jánoska, F. 2015. Woody nutrient preferences and habitat use of the Eurasian beaver (*Castor fiber* Linnaeus, 1758) at the Moson Danube. *Erdészettudományi Közlemények, 5,* 129–144.

Veraart, A. J., Nolet, B. A., Rosell, F., and de Vries, P. P. 2006. Simulated winter browsing may lead to induced susceptibility of willows to beavers in spring. *Canadian Journal of Zoology, 84,* 1733–1742.

Vorel, A., Válková, L., Hamšíková, L., Maloň, J., and Korbelová, J. 2015. Beaver foraging behaviour: seasonal foraging specialization by a choosy generalist herbivore. *Behavioral Ecology and Sociobiology, 69,* 1221–1235.

Warren, E. R. 1926. *A Study of Beaver in the Yancey Region of Yellowstone National Park.* Logan: Utah State University.

Warren, E. R. 1927. *The Beaver, Its Work and Its Ways.* Baltimore: Williams & Wilkins Co.

Ważna, A., Cichocki, J., Bojarski, J., and Gabryś, G. 2018. Selective foraging on tree and shrub species by the European beaver *Castor fiber* in lowland and highland habitats in western Poland. *Polish Journal of Ecology, 66,* 286–300.

Wilsson, L. 1971. Observations and experiments on the ethology of the European beaver (*Castor fiber* L.). *Viltrevy, 8,* 160–203.

CHAPTER 6

Activity patterns and life history

6.1 Daily activity patterns

Beavers (*Castor* spp.) are crepuscular and nocturnal (Hodgdon and Lancia, 1983; Novak, 1987; Rosell et al., 2006), meaning they are most active at dusk and dawn as well as being active throughout the night. With variation, they typically emerge from lodges around 18:00–20:00 hours during the summer months and can be active for the next 11–13 hours (Lancia, 1979; Hodgdon and Lancia, 1983; Sharpe and Rosell, 2003; Mott et al., 2011; Swinnen et al., 2015; Table 6.1).

Numerous factors influence the length of their activity period such as predator pressure, competition, disturbance, and/or food availability (Swinnen et al., 2015). For example, in Russia, Eurasian beavers reduced the length of their activity period by 2.5 hours due to motorboat traffic disturbance (Tyurnin, 1984). Daylight hours tend to be spent resting and sleeping within their shelters (see Section 4.2.1) (Mott et al., 2011), though individuals can be sporadically active in late mornings and afternoons, but there are huge variations between sites (Tevis, 1950). In Telemark, Norway, activity was not affected by daylight length, with individuals being active during daylight periods before sunset and/or after sunrise during early autumn and late spring (Cabré et al., 2020). Summer days are very long at northern latitudes (almost 19 hours), so to be able to forage and defend the territory they need to be active also during bright hours. Adult beavers tend to spend their time patrolling and feeding during the first part of the night and they construct and maintain dams and lodges during the second part of the night (Wilsson, 1971; Müller-Schwarze, 2011). In Telemark, beavers dived more in the second half of the night (Graf et al., 2016; Box 6.1).

During the summer, beavers typically follow a 24-hour cycle, dividing their time between activity and rest (Lancia, 1979; Sharpe and Rosell, 2003). Resting is defined as the time a beaver spends in a shelter following its return after nocturnal activities until it exits the lodge or burrow for new activities. The amount of time beavers spend resting varies seasonally (see Table 6.1) but is stable across years (Nolet and Rosell, 1994). It can be a common misconception that beavers hibernate in winter, but this is not correct and they are active throughout (see Section 6.2). Beavers can be especially playful in the springtime, swimming around in circles, running in and out of the water, play fighting and wrestling (Figure 6.1). This increased activity is understandable following a long period of restricted movements in and around the shelter during winter; when thawing comes beavers can finally move more freely. In the summer beavers are fairly easy to observe outside of their shelters, especially on still, warm, light evenings at northern latitudes.

The members of a beaver family (see Section 6.3) usually become active at the same time every night, whilst other families in the same population may start earlier or later (Busher, 1975; Rosell et al., 2006). Beavers are influenced by the 'ghosts of predators past' and therefore react to predators or cues no longer present in the area or to which they have been previously exposed, even across multiple generations (Rosell and Sanda, 2006). Beavers in largely predator-free landscapes retain crepuscular and nocturnal behavioural patterns (Swinnen et al., 2015). Human hunting (where allowed) also affects beaver activity. Active human hunting (i.e. non-trap methods) occurs during daylight hours, which can influence their activity patterns. In Belgium, changes in day length had only a limited effect on

Beavers: Ecology, Behaviour, Conservation, and Management. Frank Rosell and Róisín Campbell-Palmer, Oxford University Press.
© Frank Rosell and Róisín Campbell-Palmer 2022. DOI: 10.1093/oso/9780198835042.003.0006

Table 6.1 Beaver emergence and return times to the lodge in different studies of both species (*Cf* = Eurasian beaver [*Castor fiber*]; *Cc* = North American beaver [*Castor canadensis*]). Latitude is given as degrees, minutes, seconds (DMS). All studies radio-tracked the beavers (unless other methods are specified).

Beaver species	Country/state and latitude (DMS)	Habitat type	Sample size	Time of year	Sex and age	Emergence time	Return time	Principal activity period (hours and minutes)	References
Cc	Prescott Peninsula, USA 34°32′24.00″ N	Varied	3 (one individual in each age/sex class)	Summer	Adult male Adult female Subadult female	16:46 17:27 17:23	06:02 05:41 06:20	13:16 11:14 12:57	Lancia (1979)[a]
Cc	Prescott Peninsula, USA 34°32′24.00″ N	Varied	4 (one individual in each age/sex class)	Summer	Adult male Subadult female Male (2 year old) Yearling male	16:55 16:45 16:51 17:17	03:52 04:56 03:11 04:21	10:57 12:11 10:20 11:04	Lancia (1979)[a]
Cc	Prescott Peninsula, USA 34°32′24.00″ N	Varied	2 (1 male, 1 female)	26 December–13 January (not free running) 18–24 January (free running)	Male (adult) Male (adult) Female (adult)	16:40 14:24 14:35	03:28 02:42 02:53	10:42 12:30 12:30	Lancia et al. (1982)[b]
Cc	Minnesota, USA 41°52′42.71″N	Lake	35 86	May–July August–November	Male/female (all age groups) Male/female (all age groups)	20:34 18:41	07:50 07:08	11:16 12:27	Buech (1995)[b]
Cf	Biesbosch, Netherlands 52°07′57.48″N	River estuary	31	Winter Early spring Late spring Summer	Male/female (>1 year old) Male/female (>1 year old) Male/female (>1 year old) Male/female (>1 year old)	17:57 18:46 19:25 19:46	06:05 06:21 05:19 05:46	12:08 11:35 09:54 10:00	Nolet and Rosell (1994)[c]

continued

Table 6.1 Continued.

Beaver species	Country/state and latitude (DMS)	Habitat type	Sample size	Time of year	Sex and age	Emergence time	Return time	Principal activity period (hours and minutes)	References
Cf	Telemark, Norway 59°23'16.35"N (Gvarv) 59°24'29.48"N (Saua)	River	12 (6 male, 6 female)	Early spring	Male (adult)	20:22	05:10	08:48	Sharpe and Rosell (2003)[d]
					Female (adult)	20:26	04:38	08:12	
				Late spring	Male (adult)	20:30	05:57	09:27	
					Female (adult)	20:29	04:41	08:12	
				Summer	Male (adult)	20:28	06:03	09:35	
					Female (adult)	20:24	05:59	09:35	
Cf	Telemark, Norway 59°23'16.35"N (Gvarv) 59°24'29.48"N (Saua) 63° 52' 15.10" N (Straumen)	River	35 (19 male, 16 female)	March–June 2010–2016	Male/female (all adults except for 3 subordinates [mean age of subordinates 2.7 years])	20:39	06:22	09:43	Mayer et al. (2017a)[b, e]

[a] Data are presented as mean emergence and return times, with principal activity period calculated from the difference between them.
[b] Data are presented as mean emergence time and return time is calculated by addition of mean length of principal activity period.
[c] Data are presented as mean emergence and return times and mean principal activity period.
[d] Data are presented as median emergence time and return time is calculated by addition of mean length of principal activity period.
[e] Data were collected using global positioning system (GPS).

Box 6.1 Diving behaviour

Only one study has investigated the diving behaviour of free-living beavers (Graf et al., 2017). Accelerometers combined with a depth sensor were deployed on 12 adult Eurasian beavers (six males, six females) living in rivers in Telemark, Norway, and a total of 2,596 dives that happened during April to November were analysed. Results demonstrated the following:

- Beavers dived an average of 39.6 times per night (range 16.3–102.3).
- Dives were generally aerobic, i.e. they were shallow and less than 1 m (range 0.30–3.6 m) and of short duration (< 30 seconds) (range 1–400 seconds).
- Duration increased with greater dive depth, and deeper dives tended to have a longer bottom phase.
- Diving activities increased during the night, and a peak in diving activity was observed between 03:00 and 07:00 hours.
- Water temperature (3.0–11.6°C) did not affect diving behaviour.
- Beavers dived for only 2.8% of their active time, probably due to low availability of aquatic plants in the river habitats.
- Dive profiles (depth and duration) were similar to other semi-aquatic freshwater divers such as, for example, American mink (*Neovison vison*).
- Beavers may dive to forage for aquatic plants, to enter the lodge, for submerged transport of sticks/building material, or to escape in the water. These dives may not necessitate particularly deep dives.

Figure 6.1 Beavers are highly territorial and will fight to defend their territory. Play wrestling can help them prepare for this. (Photo supplied courtesy of Bjørnar Røstad Roheim.)

the duration of beavers' activity (Swinnen et al., 2015). Beaver activity increased during bright moonlight, and they showed bimodal activity patterns during nights with bright moonlight but unimodal during dark nights. Beavers may have higher foraging success during bright moonlight, and this benefit outweighed a possible increased predation risk and human disturbance (Swinnen et al., 2015).

The first indications that beavers are active usually appear as rings on the water's surface in and around a lodge or burrow, silently arising, as animals sit near the lodge's 'indoor swimming pool', most likely eating or grooming (Patenaude and Bovet, 1984). Beavers self-groom by combing with their hind paw (grooming claw), nibbling with the incisors, and shaking. The main function of such grooming is to maintain an insulating air layer between hair and skin but also to remove foreign bodies (such as dirt and external parasites; see Section 8.5.3.2) from the hair or skin. Not all places on the body can be easily reached by self-grooming, but with the help from family members (social grooming) this can be taken care of. The role of social grooming in maintaining social bonds or as an appeasement gesture is probably a minimal secondary function (Patenaude and Bovet, 1984). Just before they emerge from the shelter, larger waves or air bubbles may be seen at the water's surface, soon followed by a beaver head. Often, when a dry beaver dives, a beaded chain of air bubbles comes up from the back of the beaver to the surface of the water. Air that is trapped in the fur after grooming is pressed backwards by the currents around the body and leaves the fur around the beaver's spine. An observer can follow the path of the beaver underwater by following the bubbles as they rise to the surface. Presence of these bubbles can also indicate where the beaver has been travelling under the ice in the winter if the ice is clear and free of snow.

On first emergence, beavers often linger quietly in the water with just their nose, ears, and eyes above the surface, taking in their environment. After confirmation that no immediate dangers are present, the beaver floats up, with more of its upper body and tail being visible, reminiscent of a miniature floating hippo (*Hippopotamus* spp.). They then tend to swim around at a leisurely pace to investigate their surroundings more closely (patrolling behaviour; see Section 7.2.1) before getting on with their nightly tasks. Any suspicions will generate curiosity and deployment of their senses (ideally smell) to investigate further. This is the time a human observer may also hear a tail slap, the beaver's way of letting you know you have been detected and other family members have been warned. The tail slap can also elicit a response from the suspected source of danger, allowing a beaver to better evaluate what the disturbance may be, making further decisions (see Section 7.4.2).

Some hunters believe beavers remain in their lodge or burrows during heavy rainfall and strong windy conditions, though this can vary between families and individuals. Of course, beavers do not mind getting wet, but such weather conditions may hinder their hearing and smell, making it difficult to sense approaching dangers. Radio-tracking in the Netherlands has shown that Eurasian beavers will stay in the lodge during powerful thunderstorms. Predation risk can also play an important role in keeping them inside during storms (Tevis, 1950) or at least make them more sceptical of going onto land (see chapter 8). The same argument may also apply to predators. In Telemark, Norway, Eurasian beavers were active for longer periods of time with increasing precipitation (Cabré et al., 2020). They also spent more time on land with increasing precipitation, which might indicate anti-predator behaviour. Furthermore, falling rain may decrease the acoustic and olfactory ability of predators to detect beavers. Beavers only partly adjusted their spatial movement patterns to moonlight, spending more time in water and moving greater distances when there was more moonlight. Individuals might face a higher predation risk during brighter nights and therefore avoid land activity, though factors such as energetic requirements and territory patrolling are also key (Cabré et al., 2020).

Interestingly, in a study in Telemark, Eurasian beavers demonstrated low cohesion with their partners, which may be a way to avoid predation on multiple family members or to reduce direct competition for resources (McClanahan, 2020). They moved independently from each other with relatively large distances between them (mean ± SD: 480 ± 427 m). In larger territories, they were farther apart, and they were the closest at the beginning and end of their nightly active periods when they were also closest to their lodge. During the autumn, they come closer together. A partner at the territory border zone did not affect if the other partner also was there. The partners also spend most of the time on the same side of the territory (McClanahan, 2020).

6.1.1 Time budgets and associated roles according to sex and age class

Whether males or females emerge or return from the main lodge or burrow first varies between studies. Typically, the first beaver to emerge is an adult beaver, male or female (Hodgdon and Lancia, 1983). Factors such as time of year, age, and reproductive status can have a strong effect. For example, in one study conducted from March to December in North America, the male adult was the first to emerge from the lodge in 89% of observations, preceding the youngest family members and adult female by about 20 minutes (Brady and Svendsen, 1981). Another study found the adult female emerged first in 93% of observations. New kits began to spend the night outside the lodge in August and were generally the last to emerge right up until December (Hodgdon and Larson, 1973; see Table 6.1).

Data from Norway (on large rivers) suggest that older Eurasian beavers have shorter activity periods than younger ones (Mayer et al., 2017e). The activity period was not affected by the sex of an individual in the spring (from March until May), which is the period leading up to birth. Pregnant females had similar spatial movement patterns and activity periods compared to other individuals. However, one female that had given birth had shorter activity periods, becoming active and making land forages later than males and non-nursing females in other families (Mayer et al., 2017e). Adult males returned to the daytime rest site later, had longer activity periods, and travelled more than their mate (Sharpe and Rosell, 2003). When kits were small, the adult male spent more time in the lodge than before parturition or after kits become more mobile (Lancia, 1979).

During the night, a beaver pair needs to carry out many important activities. These roles include territorial defence, construction work, parental care, foraging, and personal care. Timing and reported activities of adult males and females in the beaver family differ among studies (Hodgdon and Larson, 1973; Brady and Svendsen, 1981; Hodgdon and

Lancia, 1983; Busher and Jenkins, 1985; Woodard, 1994; Buech, 1995; Sharpe and Rosell, 2003; Rosell and Thomsen, 2006; Table 6.2). Unfortunately, studies on beaver differences between the sexes are either based on anecdotal observations, or have not focused on the breeding pair, or had a low sample size. There does seems to be general consensus that the adult pair share in parental care, construction, scent-marking, and other territorial behaviours (see Chapter 7), but there are reported differences on the relative effort contributed by adult males and females (Buech, 1995; Rosell and Thomsen, 2006; Busher, 2007). These discrepancies may be related to factors such as methods used (observations vs. radio-tracking), age differences, habitat (lake [Buech, 1995] or river [Sharpe and Rosell, 2003]), environmental factors (need for territory defence and family sizes), condition (food, predators, and weather), and energy requirements for growth, reproduction, and homeostasis (Samson and Raymond, 1995; Sharpe and Rosell, 2003; Busher, 2007). For this reason, we only report the specific results from two studies below: one on the North American beaver (*C. canadensis*) in a lake habitat and one on the Eurasian beaver in rivers.

In Birch Lake, northeastern Minnesota, USA, adults from three North American beaver families were observed and radio-tracked (Buech, 1995). The adult females and males spent respectively 91% and 86% of their time feeding, travelling, and being inside the lodge. Adult females spent the majority of their time in the late spring and early summer feeding, which coincides with parturition and lactation. In late summer and throughout the autumn, females with young spent their time foraging for food, working inside the lodge, and collecting branches for the food cache. The adult males, on the other hand, spent less time feeding and more time maintaining their territory (swimming around and scent marking), staying inside the lodge, and working on the lodge in the late spring and early summer (Buech, 1995). As the season progressed, adult males travelled less and spent more time feeding and working on the lodge.

Six mated pairs of the Eurasian beaver were radio-tracked on river systems in Telemark, Norway (Sharpe and Rosell, 2003). The time budgets of adult males and females did not differ for most activities, but males travelled (swimming/diving or walking when not actively foraging) further distances associated with territory defence. Large differences between the time budgets of each of the six pairs were found. In addition to travel, the amount of time beavers spent foraging, in the lodge, and undertaking 'other' activities varied significantly between pairs but not between males and females in the same family. In the 'other' category, scent marking and grooming varied significantly between pairs. The remaining behaviours such as social interactions, antagonistic interaction, and static (no movement) did not vary. The researchers concluded that mated pairs from the same territory behaved in a highly similar way but not identical.

An observational study of one marked family of beavers in Massachusetts, USA, found that the adult female was dominant to the adult male based on encounters and other behaviours (tail slapping twice as often, and she led the family in lodge maintenance, food cache building, and dam maintenance) (Hodgdon and Larson, 1973). Either adult within a pair may be more dominant, but it is just as common for both animals to be co-dominant (Busher, 2007). Older beavers always have a higher status than younger ones, and adults dominate yearlings and yearlings dominate kits (Hodgdon and Larson, 1973; Busher, 1983). Adult beavers spend more time defending their territory by scent marking (Hohwieler et al., 2018) and patrolling, in construction/maintenance of lodges, dams, and food caches (Busher and Jenkins, 1985), and bringing food to the younger animals (Brady and Svendsen, 1981). Kits spend their time feeding (41%), swimming (22%), and in social interactions (29%) with other beavers (Busher and Jenkins, 1985). Year-old males tend to behave in a similar way to kits. The behaviour of 1-year-old females tends to be more comparable to adult females, suggesting they develop more quickly (Busher and Jenkins, 1985). For example, the female yearlings spent over five times their time patrolling compared with yearling males. The contribution of the dominant pairs and subadults in scent marking and patrolling behaviour is discussed further in Chapter 7.

6.2 Winter: the secret time of year

The beaver's life during wintertime is mysterious for many, because most of us rarely, if ever, see any

Table 6.2 Comparison of behaviours performed by adult male and female beavers. Papers by Sharpe and Rosell, Rosell and Thomsen, and Mayer et al. studied the Eurasian beaver (*Castor fiber*) and the others the North American beaver (*Castor canadensis*).

Behaviour type	Sex	Hodgdon and Larson (1973)[a]	Hodgdon (1978)[a]	Busher (1980)[b]	Brady and Svendsen (1981)[a]	Busher and Jenkins (1985)[c]	Woodard (1994)[b]	Buech (1995)[b]	Sharpe and Rosell (2003)[d]	Rosell and Thomsen (2006)[e]	Mayer et al. (2017c)
Adding material to scent mounds	Male	9.00	–	–	–	–	–	–	–	–	–
	Female	11.00	–	–	–	–	–	–	–	–	–
Avoiding group feeding	Male	–	–	–	35.00	–	–	–	–	–	–
	Female	–	–	–	4.00	–	–	–	–	–	–
Canal work	Male	–	28.40	–	–	–	–	–	–	–	–
	Female	–	34.00	–	–	–	–	–	–	–	–
Dam inspection	Male	–	37.30	17.40	–	0.17	0.70	–	–	–	–
	Female	–	24.40	13.80	–	0.08	0.60	–	–	–	–
Dam work	Male	–	20.50	3.30	–	0.14	9.77	–	–	–	–
	Female	–	44.30	2.40	–	0.05	2.45	–	–	–	–
Deep-water circling[f]	Male	–	–	–	70.00	–	–	–	–	–	–
	Female	–	–	–	73.00	–	–	–	–	–	–
Feeding	Male	–	–	–	–	0.73	38.51	34.60	10.00	–	–
	Female	–	–	–	–	0.94	34.14	44.10	9.00	–	–
First reaction to experimental scent mounds (ESM)	Male	–	–	–	–	–	–	–	–	–	23.00
	Female	–	–	–	–	–	–	–	–	–	17.00
Food cache	Male	–	25.00	0.50	–	0.14	–	2.60	–	–	–
	Female	–	29.20	0.30	–	0.05	–	5.30	–	–	–
Grooming	Male	–	–	–	–	0.03	8.67	–	–	–	–
	Female	–	–	–	–	0.04	8.58	–	–	–	–
Initiating group feeding	Male	–	–	–	1.00	–	–	–	–	–	–
	Female	–	–	–	9.00	–	–	–	–	–	–
Initiating nasal to nasal contact	Male	–	–	–	2.00	–	–	–	–	–	–
	Female	–	–	–	3.00	–	–	–	–	–	–
Initiating nose to rump contact	Male	–	–	–	13.00	–	–	–	–	–	–
	Female	–	–	–	19.00	–	–	–	–	–	–
Inside lodge	Male	–	–	–	–	–	–	–	9.00	–	–
	Female	–	–	–	–	–	–	–	10.00	–	–

	Sex	1	2	3	4	5	6	7	8	9	10	11
Leading a caravan	Male	–	–	–	3.00	–	–	–	–	–	–	–
	Female	–	–	–	20.00	–	–	–	–	–	–	–
Lodge work	Male	–	22.40	0.50		0.19	0.04	8.50	–	–	–	–
	Female	–	31.30	2.00		0.14	0.05	1.10	–	–	–	–
Mutual grooming	Male	–	–	–	–	–	0.41	–	–	–	–	–
	Female	–	–	–	–	–	0.50	–	–	–	–	–
Number of contact with food and vocalization	Male	–	–	–	–	–	0.02	–	–	–	–	–
	Female	–	–	–	–	–	0.03	–	–	–	–	–
Number of vocalization incidents emitted	Male	–	13.00	–	–	–	–	–	–	–	–	–
	Female	–	21.00	–	–	–	–	–	–	–	–	–
Number of vocalization incidents received	Male	–	63.00	–	–	–	–	–	–	–	–	–
	Female	–	160.00	–	–	–	–	–	–	–	–	–
Patrolling	Male	–	–	–	–	1.92	2.15	–	–	–	–	–
	Female	–	–	–	–	1.60	2.12	–	–	–	–	–
Provisioning the lodge	Male	–	–	–	13.00	0.14	–	–	–	–	–	–
	Female	–	–	–	7.00	0.06	–	–	–	–	–	–
Receiving nose to rump contact	Male	–	–	–	8.00	–	–	–	–	–	–	–
	Female	–	–	–	1.00	–	–	–	–	–	–	–
Scent marking	Male	–	9.00	–	–	–	–	–	–	480.00	–	–
	Female	–	11.00	–	–	–	–	–	–	326.00	–	–
Tail slapping	Male	9.00	–	–	30.00	0.02	0.11	–	–	–	–	–
	Female	23.00	–	–	181.00	0.02	0.11	–	–	–	–	–
Touching with food and vocalization	Male	–	–	–	–	–	0.02	–	–	–	–	–
	Female	–	–	–	–	–	0.03	–	–	–	–	–

continued

Table 6.2 Continued.

Behaviour type	Sex	Hodgdon and Larson (1973)[a]	Hodgdon (1978)[a]	Busher (1980)[b]	Brady and Svendsen (1981)[a]	Busher and Jenkins (1985)[c]	Woodard (1994)[b]	Buech (1995)[b]	Sharpe and Rosell (2003)[d]	Rosell and Thomsen (2006)[e]	Mayer et al. (2017c)
Touching with vocalization	Male	–	–	–	–	–	0.60	–	–	–	–
	Female	–	–	–	–	–	0.70	–	–	–	–
Transporting a kit	Male	–	–	–	2.00	–	–	–	–	–	–
	Female	–	–	–	5.00	–	–	–	–	–	–
Travelling	Male	–	–	–	–	–	–	33.20	7.00	–	–
	Female	–	–	–	–	–	–	31.10	6.00	–	–

Dashes indicate no data available.

[a] Reported observations as the proportion of the behaviour performed by each age/sex class.

[b] Reported observations as the frequency of occurrence of behaviours in relation to all beavers.

[c] Reported data as relative frequencies according to case. The sum of each behaviour type was then calculated for both male and female adult beavers.

[d] Reported observations as mean number of 15-minute fixes recorded for each behaviour.

[e] Reported observations as frequency of occurrence from each sex class.

[f] Deep-water circling is a response to disturbance; adult beavers move to deep water, swimming in large circles with raised snouts in an attempt to localize the disturbance.

sign of them. Beavers adapt to wintertime conditions by becoming less active and live off their pre-prepared food cache and accumulated body fat. At high northern latitudes, beavers are forced to spend much of the winter inside the lodge due to the thick ice that can form on the water outside the lodge. At the extremes, they have been reported spending up to 9 or 10 months inside their shelters in some areas (Novak, 1987). Beavers cope well inside the lodge during winter, even when outside temperatures are far below freezing. Their fur becomes thicker in the winter for additional insulation. It is also vital that they are in good condition with significant fat reserves going into winter. Beavers thicken the lodge walls with additional mud and other materials during the autumn, and, along with a layer of snow, this maintains a milder internal temperature (Stephenson, 1969; Dyck and MacArthur, 1993). A beaver can also reduce its body temperature in the winter to save energy. Curling up into a ball shape has the smallest surface area to volume ratio, thus reducing heat loss. If the beavers curl up together in the lodge, press their tails against their bodies, and lie close to each other, they stay warm (Baker and Hill, 2003). Contrastingly, in the summer, the beavers tend to lie alone, stretched out on their backs, which allows them to lose more heat and cool themselves down (see Chapter 3 for more details).

Generally, a beaver's winter activity level is low when the weather is extremely cold, beavers only making routine trips out of the lodge to gather twigs from the food cache (Figure 6.2). During very

Figure 6.2 Food cache in a frozen water body. The hole next to the lodge will be maintained for as long as possible in the winter months to allow access to and from the lodge. (Photo supplied courtesy of Frank Rosell.)

cold conditions beavers only bring food from the cache into the lodge a few times every night, preferring to stay warm inside the lodge. Beavers are reported to rarely go on land when temperatures fall below −6 to −10°C, but this varies with location (Semyonoff, 1957; Wilsson, 1971; Lancia, 1979; Lancia et al., 1982). Activity increases in proportion to increases in temperatures. When milder (< −10 °C) temperatures occur in winter, beavers often go on land to cut down broad-leaved trees or forage on previously felled trees. This suggests that beavers prefer fresh food rather than food that has been stored in their food caches. Unfortunately, no nutritional data regarding standing trees in winter compared to sticks in the food cache exist. By foraging on land they can save their winter supplies for times when environmental conditions do not allow them to forage. Individual swimming events are usually shorter in winter and early spring (60 minutes) compared to late spring (165 minutes) and summer (135 minutes) (Nolet and Rosell, 1994). The daily time spent swimming increased from 2.9 hours in winter to 4.2 hours in late spring and summer. However, beavers may take short swimming trips even at air temperatures as low as −20°C.

Beavers can and do excavate snow tunnels, which start under the ice and end in foraging areas far away from the water. They have been observed to dig a 20-m-long snow tunnel almost directly to a stand of birch trees. They most likely find these trees under the snow using their sense of smell. It is also common to find beaver trails/paths in the snow or on the ice (Figure 6.3a–c). These can be several hundred metres long, equal to those found in autumn. In Norway, beavers living along smaller streams were observed to be active on the land for nearly the entire winter. These animals used burrows in the ice or on land to access fresh supplies of broad-leaved trees. They were only inactive when temperatures were extremely low. One reason for their high winter activity on land may be that beavers in streams did not have the opportunity to collect large food caches due to the shallow waters and thick ice. Eurasian beavers living at Hardangervidda in Norway had small food caches (5.2 m³), and ice was present for 7 months. Therefore, they had to forage on land during early spring, 6–8 weeks before the ice disappeared (Mossing, 2005). However, in North America even along small

(a)　　　　　　　　　　　(b)　　　　　　　　　　　(c)

Figure 6.3 It is common to find beaver trails/paths in the snow (a) and on the ice during winter (b and c). (Photo supplied courtesy of Frank Rosell.)

Figure 6.4 Beaver breaking through ice from below. (Photo supplied courtesy of Kjell-Erik Moseid.)

streams they are not active that much during winter (Peter Busher, pers. comm.).

Travelling up steep hillsides can be difficult in the winter since they are often covered with ice, but the beaver solves this problem easily. Gnawing such ice sheets to produce anti-slip grooves or 'steps' has been observed. More commonly, beavers will maintain small openings in the ice, especially around the lodge, for as long as they can, allowing easy access to land. Beavers can break up thin ice by pressing down on it with their fore legs, or if the ice is too thick for this approach they crawl out to the edge and use their body weight to break through (Figure 6.4). When under water in winter beavers can use their very strong backs to break through the ice from below by swimming under the ice and pressing upwards. Even small beaver kits can break

ice to access such holes, and they also use their incisors to break ice.

If the food cache runs out and the animals do not have an opportunity to go onto land to gather fresh food, they can subsist on aquatic vegetation even during winter. The tubers of water lilies and other aquatic vegetation can be valuable additions to the winter diet (see Section 5.6). If beavers become trapped in their lodge they can also begin feeding on the lodge's infrastructure (the branches that beavers used to create the lodge) and survive on that for a time. They do chew on the walls of the lodge regardless of bark or not on the branches, since they do get a little nutritive value from even barked branches. They can also survive on their body fat reserves for some time if necessary. It is therefore important for beavers to save their fat reserves for as long as possible by reducing all energy-demanding activities (Smith and Jenkins, 1997). There is one very exceptional case study of an ice-trapped beaver family displaying self-gnawing of their tails, presumably as a stress response to being physically trapped in their lodge with no access to food (Saveljev et al., 2016). This family was freed by local people and photographed with significant amounts of their tails gnawed and eaten.

In typical winter conditions beavers spend most of their time resting and feeding, and their metabolism slows in the darkest, coldest, most difficult months (see also Chapter 3). During winter, beavers in Canada (latitudes between 47°N and 51°N) that are confined to their lodges by heavy ice cover have been reported to adopt a free running circadian activity rhythm ranging from 26 to 29 hours (Bovet and Oertli, 1974; Potvin and Bovet, 1975; Lancia et al., 1982). This means that beavers have fewer total activity periods in the winter than in the summer. With a 26- or 29-hour activity cycle beavers will experience respectively 83 and 74 activity periods in a 3-month-long winter, while if they still maintained a 24-hour cycle they would have 90 activity periods. Though one study in central Massachusetts, USA, at lower latitudes (42°N) than the Canadian study found that during winter only the female lengthened the activity cycle to 26.6 hours. Males remained on a 24-hour cycle (Lancia, 1979). If they are again exposed to the light due to open water, beavers revert back to a regular 24-hour cycle (Potvin and Bovet, 1975). The differences reported may be due to local environmental conditions in Canada.

In Voyageurs National Park, Minnesota, USA, the behaviour of beavers living in lake habitats was studied to determine the effects of water decline during winter (Smith and Peterson, 1991). This resulted in altered behaviours, but beavers showed a high degree of adaptability to low water. During the winter, beavers living on drawdown reservoirs spent less time inside their lodges than beavers from environments with stable water levels, spending more time in dry areas beneath the ice, and constructed nests of wood chips. This illustrates that if sufficient food and no predators are present, beavers adapt to water loss in ponds during winter. It was noted that when the water retreated beneath, causing the ice to sag, beavers dug canals in the reservoir bottom to allow them to move between their new nest sites and the lodge (Smith and Peterson, 1991). This occurred in 24% of the families observed. It was also found that beavers in environments with unstable water levels foraged more above ice and were unable to fully use their food caches because they were either frozen in ice or surrounded by collapsed ice.

6.3 The family group

The term 'colony' is often used to describe the beaver's social group, although this is a term often used in connection with breeding birds or rodents such as prairie dogs (genus Cynomys) (Bradt, 1938; Payne, 1982). The term colony is not actually correct, as they do not meet the sociobiological definitions (Busher, 2007). It is thought the term colony emerged with European fur trappers hunting beavers in North America (Morgan, 1868). We will therefore use the more accurate term 'family group' (Busher, 2007). A family group typically consists of the adult breeding pair with their offspring from the current breeding season and the previous year's litter (Wilsson, 1971; Jenkins and Busher, 1979; Busher, 1983; McTaggart and Nelson, 2003; Busher, 2007). Family groups often show a wide range of relatedness, though outbreeding is the preferred and common strategy to avoid inbreeding depression (but see Section 6.5.2) (Crawford et al., 2008). Families typically include combinations of first-order relatives (an individual's parents, full siblings, or offspring), but second-order relatives (an individual's grandparents, grandchildren, aunts, uncles, nephews, nieces, or half-siblings) may also be present (Crawford et al., 2008). In dense and trapped populations in Illinois, USA, adult females within the same family were always identified as first-order relatives, while adult males always were unrelated mates of females in the family (Crawford et al., 2008). They also identified that 20% (3 of 15) of the families contained individuals ≥ 1 year old unrelated to others. If, for example, the dominant male dies, another unrelated male may take over as a stepfather for the remaining young. What happens to the young when a stepparent takes over warrants further investigation, but they may be tolerated in many situations.

Generally three basic types of beaver 'families' or groups are observed: a single animal, a mated pair, and a family (three or more animals) (Bradt, 1938; Bergerud and Miller, 1977; Hodgdon, 1978). However, beaver families can have many different structures. For example, a study of 230 Eurasian beaver settlements in Russia reported nine types (Zharkov, 1973):

1) Single animals (primarily territory-seeking 2-year-olds).
2) Pairs without offspring (usually newly established pairs of animals of the same age).
3) Pairs with offspring born in the spring of that year.
4) Pairs with offspring born that year and the year prior.
5) Pairs with 1- and 2-year-olds without offspring born in the current year.
6) Pairs with 1-year-olds without offspring born in the current year.
7) Pairs with offspring born in the current year and 2-year-olds.

8) Pairs with offspring from three age groups.
9) Groups with more than two adult animals.

Types 1 and 2 formed the reserves for future family establishment, whilst groups 3, 4, 7, and 8 included the population of reproducing families and were important contributors to the yearly increases in numbers of animals. Those that did not produce offspring that year are family groups 5 and 6, but these were in the minority. Type 9 was rare and was only found on eight occasions.

Beaver populations show a distinct variation in family group size. The average family group size was estimated to be 3.8 (mean range 2.4–5.5) in a review of 13 studies on the Eurasian beaver and 5.2 (mean range 2.7–9.2) for 51 studies of the North American beaver (Rosell and Parker, 1995). In two families in Russia, a maximum group size of 18 animals was reported (Goloduško and Fomenko, 1973), and in a long-term study in Telemark, Norway, it was 11 individuals. Density and habitat quality appear to be the proximate factors influencing the composition of a beaver family (Busher, 2007). For example, if mainly aspen (*Populus tremuloides*, a preferred species of beavers throughout their range) is present in the territory the family group size is larger (7.8) than families living on predominantly willow (*Salix* spp.) (5.1) (Hay, 1958).

Beaver populations in the establishment phase (a new population or reintroduced animals) have a lower than average family group size, because they are made up of a larger proportion of pairs and single animals (Payne, 1984a; Żurowski and Kasperczyk, 1986). It is also fair to assume that in populations exposed to hunting and trapping on a regular basis the average family group size is somewhat smaller and disrupted (Nordstrom, 1972; Payne, 1982). This may not be the case if the hunting is focused on animals living alone or in pairs, while the majority of families are left alone, though this may be difficult to determine. Hunting strategies that remove the most productive females will also lead to a large reduction in average family size. In poor-quality habitats or overgrazed territories there may be greater numbers of singles and pairs than usual because already as yearlings they leave their parental family to find better food sources (Semyonoff, 1951; Gunson, 1970), resulting in smaller

family group sizes. Extended occupation of habitats with the accompanying reduction in food resources will have the same effect (Bishir et al., 1983). Increasing population growth and density leads to increases in family group size (Molini et al., 1980; Bishir et al., 1983; Busher, 1983), since some of the 2-year olds remain in their natal territory or return to their parent families after attempting to disperse (Wilsson, 1971; Nordstrom, 1972; Bergerud and Miller, 1977; Payne, 1982; Hodgdon and Lancia, 1983; Payne, 1984a; Peterson and Payne, 1986; see also Section 6.9). The time of year family group size is measured is critical since group size varies between seasons. If the family size of North American beavers was measured before the younger animals dispersed, the family group size was larger (5.9) than if the measurement occurred between the time when the older offspring disperse and the birth of new kits (4.1). The family group size increased again after the current year's litter was born (6.0) (Svendsen, 1980a).

6.4 Mate choice and pair bonding

Beavers most often exhibit long-term monogamy which is rare in mammals and especially unique among rodents, as only mating occurs with a single individual each year (see also Section 6.5.1). Subdominant members of the family are not normally sexually active when there is a dominant animal of the same sex present. There are a few examples where multiple lactating or pregnant females have been documented within the same lodge or territory (Bergerud and Miller, 1977; Busher et al., 1983; Wheatley, 1993; Crawford et al., 2008; Fischer et al., 2010). Sterilization of the adult breeding male or female did not trigger reproduction by the rest of the family when both mates remained in the territory (see also Chapter 10) (Brooks et al., 1980). A captive family in England in which the breeding pair was sterilized also demonstrated that the family structure and behaviours were maintained afterwards (Campbell-Palmer et al., 2016).

In dense Norwegian populations, the average age for pair formation was 4.1 and 4.3 years for females and males, respectively, with an average age difference of 3.8 for mate change (one member of a pair

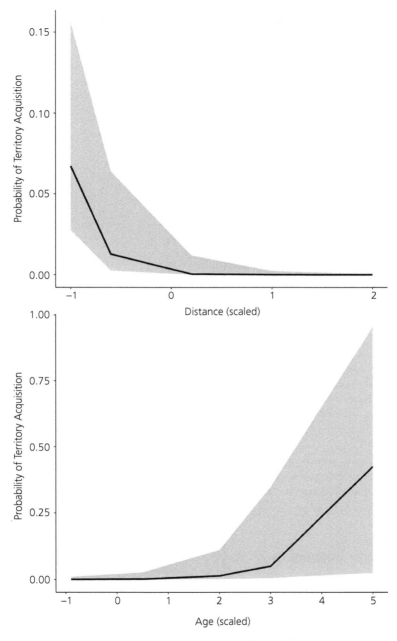

Figure 6.5 Predicted relationship between dispersal distance and age on the probability of territory acquisition in Eurasian beavers based on data collected between 1998 and 2017 in southeastern Norway. (Reprinted by permission of Priyank Sharad Nimje and the University of South-Eastern Norway under the CC BY-NC-SA 4.0 licence.)

Figure 6.6 Eurasian beaver pair Aegir and Rinda from the Netherlands, who have been together for at least 8 years. (Photo supplied courtesy of Frank Rosell.)

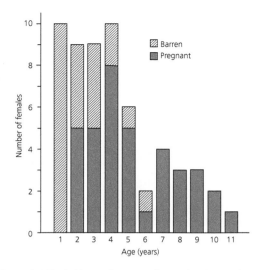

Figure 6.7 The incidence of pregnancy by age class for Eurasian beavers culled between 13 March and 15 May in Bø Township, Telemark County, Norway, 1997–1999 (© 2017 The Zoological Society of London, Parker et al., 2017.)

replaced) and 0.5 for pair replacement (both members replaced) (Nimje, 2018) (Figure 6.5) (see also Section 7.1.1). However, there are reports of beaver pairs staying together for much longer (Figure 6.6). A study in southeastern Ohio, USA, found that successful pair bonding did not occur between an older and larger female and a younger, smaller male (Svendsen, 1989). Fifty-six percent of all new pair relationships were formed during September to November. New pair relationships were usually formed between a resident beaver that either had not yet managed to acquire a mate or had recently lost its partner and a new arrival, with 90% of the newly formed pairs involving a 2-year disperser (Svendsen, 1989). Little evidence for active mate choice is apparent (Nimje, 2018).

6.5 Sexual maturity and mating

Both males and females generally become sexually mature during their second winter, at ~1.5 years (20 months) of age (Novak, 1987), though delayed primiparity has been recorded (Semyonoff, 1951; Doboszyńska and Żurowski, 1983). Beavers will normally have offspring for the first time during their third summer, providing they establish their own territory and find a mate (Parker et al., 2017) (Figure 6.7). In a dense population at carrying capacity in Norway the average age of first reproduction was much later: 5.2 years (62 months) for males and 4.8 years (57 months) for females, even when

they become sexually mature earlier (Nimje, 2018). Lean body mass, but not fat, seems to be an important determinant of age of primiparity in beavers (Parker et al., 2017). The proportion of 2-year-old Eurasian beaver females (i.e. approximately 21 months) reported breeding includes 8% in wild populations in Russia (Danilov and Kan'shiev, 1983), 29% among captive beavers in Poland (Doboszyńska and Żurowski, 1983), and 60% in wild beavers in Telemark, Norway (Parker et al., 2017). For North American beavers, the proportion of breeding 2-year-olds in a population also varies greatly among studies, with an average of 30% (range 11–44%) (Novak, 1987; see also Section 6.7).

Beavers exhibit reproductive seasonality. Rodents with highly variable sperm morphology generally have a relatively small testis mass (see Section 3.2.8), suggesting low levels of sperm competition between males (Breed, 2004). Beavers have in fact relatively small testes (Kenagy and Trombulak, 1986), like a walnut and sometimes flattened (Doboszyńska and Żurowski, 1983). This suggests that beavers exhibit low levels of sperm competition between males and therefore are preclusive to social monogamy (Bierla et al., 2007), which they are (Sun, 2003). The male's testicles begin to increase in size during the autumn in preparation for the mating season;

after mating, they shrink back to their normal size. Forty-two genes of the Eurasian beaver in Poland were differentially regulated in the testes during the breeding season (April) compared to the non-breeding season (November) (Bogacka et al., 2017). During the non-breeding season, a much larger group of genes (37) was significantly more highly expressed and a smaller group of genes (5) was less expressed than during the breeding season. These genes encode molecules that are involved in signal transduction, DNA repair, stress responses, inflammatory processes, metabolism, and steroidogenesis. These results could pave the way for further research into the processes that occur in beaver testes during various periods of reproductive activity.

At northern latitudes, mating peaks during mid-January and early February (Parker and Rosell, 2001) but with successful mating occurring as early as December and as late as May (Wilsson, 1971; Doboszyńska and Żurowski, 1983; Novak, 1987), emphasizing the considerable flexibility of this life history event in beavers. Though at southern latitudes peak mating is around July (Pietrek et al., 2017). Females are generally only receptive for a short time, usually from 10 to 14 hours after ovulation (Hinze, 1950; Wilsson, 1971). Ninety-three percent of copulations took place in the evening or at night (Wilsson, 1971). If mating does not occur during that period or conception fails, the female can become receptive again two to four more times within the following 7- to 15-day period. In a group of captive Eurasian beavers in Poland, only 16% (10 of 61) of the females conceived on their first attempts during a breeding season. The other 51 females became pregnant during their next heat (Doboszyńska and Żurowski, 1983).

Beavers normally mate in the lodge's swimming pool, but mating can also occur outside or within the main lodge or burrow (Hediger, 1970; Wilsson, 1971). The wide, flat tail and placement of the vaginal opening makes it impossible for beavers to mate in the same manner as the majority of other mammals. When the male is preparing his advances, the female stretches herself, hunches forward on her fore legs, and lifts her backside, breaking the water's surface. The male then pushes the female ahead of himself in the water. During the mating process, the female lies in the water with all four limbs outstretched while the male floats sideways and grabs

on to her. The male holds onto the female with his fore feet or teeth and then shoves his backside and tail under her so that both tails are at the correct angle and their genitalia come into contact. The penis, which has a bone and is about 3.0–3.5 cm in length, can now enter. The animals can be completely or partially submerged whilst mating, lasting from 30 seconds to 4 minutes, and this usually ends with the female pushing the male away with her fore legs or with a bite. They can mate a number of times during the course of the night and groom their pelts enthusiastically between each session, especially around the abdomen (Hediger, 1970; Wilsson, 1971).

If mating is successful, the female will develop a hard plug in the vagina 2 days later, which hinders any new successive attempts to mate (Doboszyńska and Żurowski, 1983). Gestation time is 105–107 days for the Eurasian beaver (Semyonoff, 1951; Doboszyńska and Żurowski, 1983) and 104–111 days for the North American beaver (Hediger, 1970). This is an unusually long time in rodents but results in very well-developed offspring (see Section 6.8).

6.5.1 Extra-pair copulations and paternity

Both sexes may engage in extra-pair copulations (EPC) that can lead to extra-pair paternity (EPP). In a North American beaver population with high beaver densities, exploited by trappers, and experiencing mild winters, EPC was recorded in 56% of litters (5 of 9), and > 50% had been sired by multiple males, often from neighbouring family groups (Crawford et al., 2008). In Eurasian beavers, one Russian study found little evidence for EPP (Syrůčková et al., 2015), though only a sample size of 10 kits was examined. Extensive snow and ice cover during the mating season restricted beaver movement in the study and may therefore have discouraged EPP (Syrůčková et al., 2015). In a dense beaver population at carrying capacity in Telemark, Norway, little snow and ice cover exist, and Nimje (2018) found that EPP was responsible for only 5.4% (9 of 166) of the kits (7 of 100 litters). The average age of pairs participating in EPP was 9.2 years, i.e. the individuals were quite old. The average age for males was 8.9 years (range 3–13) and 9.7 for females (range 4–15). In 78% of the cases (7/9 kits), EPP was due to a dominant male from the neighbouring

Table 6.3 Details from the seven Eurasian beaver pairs where evidence of EPP was found based on data collected from 1998 to 2016 in Telemark, Norway. Age of the male, female, and extra-pair mate (EPM) is age at time of conception of extra-pair young. Most ages are minimum ages, as individuals were first trapped as an adult (≥ 3 years).

Couple no.	Family	Length of pair bond$_{EPP}$	Length of pair bond$_{total}$	N$_{pair}$ offspring	N$_{EPP}$	Birth year of EPP kit	Age$_{Female}$	Age$_{Male}$	Age$_{EPM}$	Fate
3	Lunde 2a	> 2	> 6	1	1	2000	> 5	> 5	> 5	Male died in 2005, no more kits. Female moved together with father of EPP kit.
8	Lunde 4a	5	5*	0	2	2016	> 8	> 8	6	*Couple still together.
24	Gvarv lower	2	4*	4	1	2014	> 10	6	–	*Couple still together.
30	Patmos 1	6	8	4	1	2009	11	11	–	Both killed by hunter in 2011, no more kits.
32	Patmos 2a	9	9	6	2	2012	> 10	> 13	9	Both died in autumn/winter 2012, father of EPP kit took over territory with new female.
37	Patmos 3a	9	9	5	1	2012	> 13	> 15	9	Male died during autumn 2012.
46	Patmos 5	1	2*	1	1	2015	> 11	> 4	> 7	*Couple still together.

Source: After Nimje (2018), reprinted by permission of Priyank Sharad Nimje/University of South-Eastern Norway).

territory. None of the EPPs resulted in mate change or multiple litters. All pairs stayed together until one of the partners died (Table 6.3). Nimje (2018) suggested that EPP in beavers is caused by lapses in mate guarding (see Busher, 2007) by ageing male partners. Unfortunately, evidence for mate guarding in beavers is rare (Sun, 2003; Busher, 2007), and GPS studies of the pair at the same time during the mating period are needed to explore this. The Norwegian study also found evidence of multiple paternities. Two of the EPP kits were in two litters with kits fathered by another male (Nimje, 2018). These results suggest that beavers are generally both social and genetic monogamists (2 of 100 litters, i.e. 98%) but can exhibit plasticity in mating behaviour.

6.5.2 Inbreeding

Among strongly social animals, mating with siblings or parents can be fairly common. This is a survival strategy whereby isolated animals can produce offspring that are adapted to life in the environment to which they are born. It is not unusual for a Eurasian beaver to mate with its own offspring (Wilsson, 1971) and mother–son matings have been reported in North American beaver (Svendsen, 1980b). More recently, a known father–daughter breeding in the Knapdale trial in Scotland and a mother–son and sibling–sibling mating in Devon, England, were all confirmed by genetics (Campbell-Palmer et al., 2020). There are also observations that beaver sibling pairs can travel together from their territories and form pairs (Wilsson, 1971). Repeated mating with close relatives often results in disastrous genetic defects, a phenomenon called inbreeding depression (Müller-Schwarze, 2011). However, inbreeding might not cause a serious problem for beavers (Nolet and Baveco, 1996). When the average relatedness in the population is high and long-distance natal dispersal is not favoured, individuals tolerate inbreeding (Nimje, 2018). Results from a study in Telemark, Norway, suggested that the inbreeding coefficient was positively associated with mate choice, i.e. it was a selection for inbred individuals. However, the effect disappeared where the researchers had observations for dispersal distance. If the average related-

ness in a population is high and long-distance dispersal from the birth (natal) territory is not possible, beavers can tolerate inbreeding. Therefore, relatedness does not necessarily have to be associated with the fitness and reproductive success of beavers (Nimje, 2018).

It has been suggested that new populations founded by the reintroduction of a number of individuals from a single relict population are less successful than populations founded by individuals from several relict populations (Saveljev and Milishnikov, 2002; Frosch et al., 2014). The genetic diversity in mixed reintroduced populations is higher compared to relict populations (Senn et al., 2014), and this may account for the increased success of these reintroductions. The historical genetic variation in Eurasian beaver populations would have been much higher historically, indicated in western European populations even after severe bottleneck situations (Frosch et al., 2014). The genetic diversity in Swedish, Norwegian, and Russian beavers has been found to be quite low (Ellegren et al., 1993). However, the impact of original genetic diversity on successful reintroductions is not clear since the bulk of Finnish North American beaver populations originate from two pairs (seven individuals introduced in total) whose offspring were further reintroduced to different parts of the country (Lahti and Helminen, 1974; Ermala et al., 1989). The pregnancy rates and fetus numbers documented in the Finnish North American beaver populations do not differ from those reported for North American populations in North America. This suggests that the Finnish North American beaver populations are not suffering from inbreeding depression in reproductive success. Though these parameters do not indicate anything of the viability or fitness of the kits (Ruusila et al., 2000).

A translocation of 80 Eurasian beavers from Norway to Sweden between 1922 and 1938 led to a healthy population of 100,000 beavers by 1993 (Nolet and Baveco, 1996), although the genetic variability among these beavers was small (Ellegren et al., 1993). An inbred captive population originating from only 15 Eurasian beavers formed a successful source for the reintroduction of beavers in Poland (Żurowski, 1989). Reintroductions to date (from populations in Telemark in Norway, the Rhone valley in France, and the Elbe valley in Germany) have therefore demonstrated that successful population growth can be achieved from unmixed stocks, and reintroduction failure caused by inbreeding seems unlikely (Halley, 2011). Long-term genetic adaptability and robustness of these populations remains an important consideration in restorations projects (see Section 11.6).

Generally, the annual reproductive success was very low in Telemark, Norway, compared to other Eurasian beaver populations (Heidecke, 1984; Saveljev and Milishnikov, 2002; Halley, 2011). There may be several reasons for this. It might be due to the methods used to estimate litter size (see Section 6.6); resources having been depleted as beavers inhabited the area from at least 1920 (Olstad, 1937); population differences (Campbell et al., 2012; Steyaert et al., 2015); or inbreeding depression. The genetic diversity in this population was lower compared to other populations in Europe (Durka et al., 2005).

All of the evidence accumulated so far makes it difficult to determine if Eurasian beaver populations have suffered from inbreeding depression (Rosell et al., 2012). Only one Russian study indicates that inbreeding might negatively affect beavers, with incidence of dental anomalies far greater (> 30%) in introduced, inbred North American beavers in eastern Russia than in the endemic population of Eurasian beavers (< 1%) (Saveljev and Milishnikov, 2002). None of the typical indications of inbreeding such as cleft lips or palates and polydactyla has been observed (Rosell et al., 2012). Halley (2011) stated that data from Russia (Saveljev and Milishnikov, 2002) also strongly suggested inbreeding depression in unmixed stocks due to lower litter sizes. Again however, the methods used to estimate litter sizes are affected by a number of factors (see Section 6.6), and other confounding variables such as environmental pollution may be responsible for the differences observed (Rosell et al., 2012). Inbreeding in beavers may have therefore gone beyond the point where inbreeding depression would occur. Genes that would give rise to negative traits have most likely been removed through natural selection. Therefore, there is a need for more knowledge

on how much genetic diversity is still present within different populations. For example, a population in the Voronezh region in Russia appears to retain a high level of genetic diversity (Saveljev and Milishnikov, 2002).

6.6 Time of birth and litter size

After the 3.5-month gestation period, the female gives birth in the warm, dry sleeping chamber of the natal lodge or burrow (Wilsson, 1971). Beaver kits tend to be born from mid-April to mid-August in the northern latitudes, with parturition peaking mid-May to mid-June (Semyonoff, 1951; Doboszyńska and Żurowski, 1983; Parker and Rosell, 2001). The mean parturition date was estimated to be 12 May in a study in Telemark, Norway (Parker and Rosell, 2001; Figure 6.8), though kits have been reported to be born as early as February in Mississippi and Texas, USA (Miller, 1948, Wigley et al., 1983), and as late as November in Wyoming (Thomas, 1943). It is possible that beavers at higher altitudes or extremely high latitudes give birth later in the year because of later ice melting. Beaver kits are born around October in the southern hemisphere in the introduced North American populations in Tierra del Fuego, Argentina, in southern Patagonia (Pietrek et al., 2017). In captivity, 90% of births occurred during the daytime between 10:00 and 14:00 hours (see also Box 6.2), and the birthing process was from a few minutes to 2 hours in duration (Doboszyńska and Żurowski, 1983). However, the time it takes for a litter to be born can range from a few hours to up to 8 days (Doboszyńska and Żurowski, 1983).

There are large difficulties, as one can imagine, investigating parturition in the wild. A new promising method to determine birth dates is the use of temperature loggers. Preliminary data from a female beaver in Telemark, Norway, showed a clear increase in body core temperature on 29 May, probably indicating that she gave birth (see Figure 3.24). This was later confirmed when a kit was observed in the territory in September (their mother–daughter relationship was also confirmed by genetics). No change in body temperature of her mate during the same period was recorded (Mayer et al., 2017c).

The harvesting of adult males in families markedly delayed parturition in females in a study in

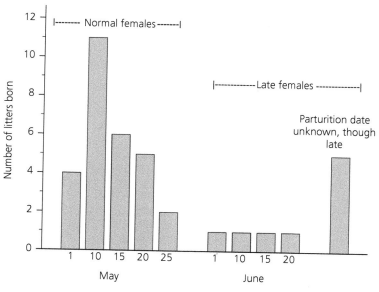

Figure 6.8 Distribution of predicted parturition dates based on fetus size for litters of Eurasian beavers from females culled between 13 March and 15 May in Bø Township, Telemark, Norway, 1997–1999. For five females, corpora lutea were present but fetuses were not visible due to particularly late breeding. For these females, parturition dates could not be predicted. (Parker, H., Zedrosser, A., and Rosell, F. 2017. Age-specific reproduction in relation to body size and condition in female Eurasian beavers. Journal of Zoology, 302, 236–243. © 2017 The Zoological Society of London.)

Box 6.2 Observed behaviours in a North American beaver family (Patenaude, 1983; Patenaude and Bovet, 1983; Patenaude, 1984)

The weeks leading up to birth

- Sleeping chamber of natal shelter was renovated. The beavers scraped the walls with their incisors and fore legs. This increased the size of the room from 130 to 160 cm, a nearly 50% increase in surface area. The renovation occurred from 8 to 52 days before birth in the observed lodge. The floor and ceiling were left untouched. The adults and yearlings contributed to this work.
- Floors of sleeping chamber were covered with fresh plant material 2 days before birth. This occurred in both of the years that observation took place. This behaviour was also observed in another family but never in lodges that contained non-reproducing or single animals.

The day before birth

- The female was very restless. She paced, groomed her pelt, or rested only for a few seconds between activities.
- The birth began with a series of contractions in the lower body, which lasted 2–3 seconds. Vertical tail movements were often observed at the same time.

During and immediately after birth

- In the first year four births were observed, which happened between 10:00 and 12:30 hours. In the second year, there were four births: two between 15:30 and 17:00 and two between 10:30 and 15:00 (though these were not observed).

The female's activities

- During the first birth the female made a sound similar to a bellow (which was heard only one time) and a series of high- and low-frequency sounds, which aroused the attention of the male and the 1-year-olds.
- The following typical positions were observed in all of the births: female sat upright with her tail in front, her head bent down towards her rear, and her back arched. The kits were born onto her tail within seconds of her assuming this position. She did not use her fore paws or mouth. In the two occasions where the kit could be observed emerging from the birth canal, they came out head-first. The umbilical cord was severed immediately

continued

Box 6.2 *Continued*

after they emerged. Only one of the kits retained a 3-cm piece of the cord, which was eaten by the female 4 hours later.

- The female ingested all of the afterbirth, and she drank water from the swimming pool a few minutes later.
- Females spent a lot of time grooming: licking the kits, her own anal area, different parts of her body, and the floor of the lodge. Licking was rarely observed on days other than birth days.

Activities of the male and 1-year-olds

- The male and yearlings were present at the birth in both of the observed years.
- The male and one of the 1-year-olds formed a triangle around the kits, which protected them and kept them warm.
- The 1-year-old licked the kits on two occasions.

Kits' behaviour immediately following birth

- They started to walk and climb immediately, albeit clumsily.
- They tried to reach their mother's nipples, but she was too restless just after the birth, and nursing happened later that evening.
- One kit fell into the swimming pool 3 hours after birth. It struggled and swam for a few seconds before climbing out and joining its siblings.

Movements in and out of the lodge

- The kits were left alone in the lodge within a few hours of having been born. They slept shortly after their parents left.
- All of the adult animals remained near the lodge after birth, where they could be heard whining and hissing as well as slapping their tails onto the water. This was most likely due to the researchers' presence in the observation chamber.

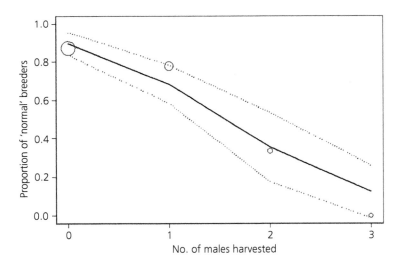

Figure 6.9 The relationship between the predicted parturition time ('normal' or 'late') in spring-shot Eurasian beaver (n=37) and the number of adult males harvested from each female's territory during the study (1997–1999) prior to her own death, southeast Norway. Dotted lines are 95% confidence intervals and circle size is directly proportional to the sample size (i.e. the number of pregnant females). (Reprinted with permission of Royal Society Publishing from Harvesting of males delays female breeding in a socially monogamous mammal; the beaver by Howard Parker, Frank Rosell and Atle Mysterud, volume 3, edition 1, copyright 2006; permission conveyed through Copyright Clearance Center, Inc.)

Telemark, Norway (Parker et al., 2007). The proportion of normal breeders declined from over 80% when no males had been shot in the territories of pregnant females to less than 20% when three males had been shot (Figure 6.9). The high population harvest rate of adult males might have hindered sufficient adult male recruitment to the families that lost the dominant male, leading to repeated ovulations and late breeding in some females (Parker et al., 2007). Alternatively, the death of a pregnant female, even late in pregnancy, might stimulate oestrus in a previously subordinate, resident female or a sexually mature transient (Wilsson, 1971; Svendsen, 1989), again leading to late oestrus. Delayed parturition may lead to reduced offspring survival and thus lower recruitment (Parker et al., 2007), but further investigations are needed.

Breeding pairs of both species have only one litter per year, and in many populations pairs produce a litter every year. However, intermittent years of non-breeding were common among sexually mature females in a population of Eurasian beavers in Telemark, Norway (Parker et al., 2017). This reproductive response is apparently linked to persistently high population densities and dwindling food supplies (Baker and Hill, 2003; Campbell, 2010). Irrespective of age of North American beaver females in a population in Finland, the number of fetuses between two consecutive years correlated negatively with each other. Females that did not have fetuses in the current year produced 2.4 times more fetuses the next year (Ruusila et al., 2000). Therefore, the lack of a litter the previous year had a positive effect on the following year's litter. Females that reproduced in two consecutive years had significantly more fetuses, stressing the importance of multiple reproduction attempts in maximizing lifetime reproductive success.

Typically, 2 to 4 kits are born in a litter, but this can vary from 1 to 6 for the Eurasian beaver (Żurowski

et al., 1974; Doboszyńska and Żurowski, 1983; Parker and Rosell, 2001; Campbell et al., 2005; Petrosyan et al., 2019; Figure 6.10) and 1 to 12 for the North American beaver (Hammond, 1943; Hill, 1982). As many as nine embryos have been recorded for the Eurasian beaver in Russia (Petrosyan et al., 2019). Beaver populations do show a distinct variation in litter size (Table 6.4). For example, the mean litter size for North American beavers was calculated to be 4.1 (range 3.1–6.1) based on only data collected by counting embryos from beavers in exploited populations (Table 6.4). The average litter size was estimated to be 2.6 (mean range 1.4–3.4) in a review of 10 studies on the Eurasian beaver in Russia (Saveljev and Milishnikov, 2002) and 3.5 (mean range 2.2–6.3) for 27 studies of the North American beaver in different states in the USA (Hill, 1982).

Reviews on litter size in North America suggest that beavers in the southeastern states tend to have smaller litters compared to northern and perhaps western state beavers (Baker and Hill, 2003). In a beaver research farm in Poland, one female gave birth to 49 kits in 13 litters during her 15-year lifetime (Doboszyńska and Żurowski, 1983). It should be noted that many different techniques are used to estimate litter sizes. For example, data from the same population showed a mean litter size of 0.7 when counting emerging kits (Rosell et al., 2006) but the number was 2.3 when counting fetuses in shot females 2–4 years later (Parker and Rosell, 2001). Studies that used a combination of corpora lutea, placental scars, and embryos may estimate a higher litter size than studies counting the number of offspring surviving until emergence when the kits are 2 months old, due to the difficulty of observing emerging kits or complete family removal. There is also the issue of fetus mortality before birth (see also Section 8.2). In some families, kits may not be observed in their first autumn but are first observed as yearlings the next year. Litter size comparisons should only be made between similar populations since so many factors can affect beaver reproductive success (see Section 6.7).

The sex ratio of fetuses for any age class did not deviate significantly from 50:50 but males were more prevalent among 2-year-olds (Parker et al., 2002). In a study of the North American beavers in eastern South Dakota, USA, males tended to

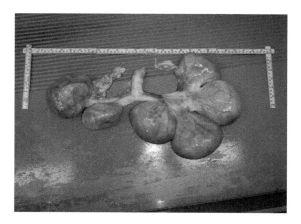

Figure 6.10 Ovaries from a dissected Eurasian beaver, with six kit fetuses. (Photo supplied courtesy of Frode Bergan.)

Table 6.4 Litter sizes of North American beaver. Latitude is given as degrees, minutes, seconds (DMS).

Latitude (DMS)	State, country	Habitat type	Approximate population age (years)	Exploited/ unexploited population	Method	Sample size	Litter size[a]	Expected timing of kit parturition	References
–50°									
54° 48' 6.88'' S	Tierra del Fuego, Argentina	Forest Steppe	71	Exploited	Kits	–	1.7 2.7	–	Pietrek et al. (2017)
30°									
30° 28' 26.58'' N	Texas, USA	River	–	Exploited	Fetuses	2	1.5	Late February (development of fetuses)	Miller (1948)
32° 19' 5.628'' N	Alabama, USA	Varied	–	–	Ovulation	–	2.8	–	Hill (1982)
32° 43' 26.31'' N	Mississippi, USA	Varied	–	Exploited	Fetuses	40	2.6	–	Wigley et al. (1983)
37°22'58.00"N	Illinois, USA	Varied	–	Unexploited	Kits and fetuses	17	3.6	April–June (neonatal period, data from remote videography)	Bloomquist and Nielsen (2010)
39° 21' 57.79'' N	Ohio, USA	Varied	–	Unexploited	Kits	17	2.7	May–June (sightings of females with hairless nipples and kits whining inside lodge)	Svendsen (1980a)
39° 25' 57.00'' N	California, USA	Varied	49	Exploited	Kits	15	2.1	–	Woodard (1994)
39° 27' 35.02'' N	Illinois, USA	Varied	–	Exploited	Embryos and corpora lutea	35	4.0	–	McTaggart and Nelson (2003)
40°									
40° 24' 46.98'' N	Ohio, USA	Varied	–	Exploited	Placental scars	105	3.9	–	Henny and Bookhout (1969)
40° 39' 59.22'' N	Colorado, USA	Varied	–	Exploited	Fetuses and placental scars	7	1.7	May–June (development of fetuses and post-partum females)	Harper (1968)
41° 39' 16.27'' N	Wyoming, USA	Varied	–	Exploited	Placental scars Fetuses Fetuses and placental scars	19 22 41	3.4 2.9 3.1	Late April–early June (development of fetuses)	Osborn (1953)
41° 39' 52.00'' N	Pennsylvania, USA	Varied	–	Exploited	Embryos	10	5.5	–	Brenner (1964)
42° 04' 28.92'' N	Iowa, USA	Varied	> 25	Exploited	Embryos	12	4.1	–	Sanderson (1953)

Latitude	Location	Habitat		Exploitation	Measure	n	Mean	Notes	Reference
44° 23' 31.99'' N	Illinois, USA	River	–	Exploited	Kits	12	2.5	–	Havens et al. (2013)
44°08' 02.80'' N	Michigan, USA	Varied	–	Exploited	Kits and embryos	65	3.7	–	Bradt (1938)
44°33' 31.69'' N	Vermont, USA	Varied	–	Exploited	Embryos	58	4.1	–	Bond (1956)
45° 18' 54.9'' N	Maine, USA	Varied	–	Exploited	Embryos and placental scars	11 / 30	3.6 / 4.0	–	Hodgdon (1949)
47° 10' 1.20'' N	Newfoundland, Canada	Varied	–	Unexploited	Embryos and placental scars	60 / 243	2.9 / 2.9	–	Payne (1984b)
48° 52' 25.824'' N	North Dakota, USA	Varied	> 8	Exploited	Embryos	18	4.1	–	Hammond (1943)
48°55' 59.66'' N	Newfoundland, Canada	Boreal forest	39	Exploited	Kits	198	2.7	Late May–early June (fetus observations and holes in uterine wall. Also parturition in 4 females)	Bergerud and Miller (1977)
50°									
53° 36' 26.36'' N	Alberta, Canada	Forest	< 13	Exploited	Embryos	8	6.1	–	Pearson (1960)
53° 57' 39.53'' N	Saskatchewan, Canada	Varied	–	Exploited	Embryos	43	3.8	–	Pearson (1960)
56° 00' 00.00'' N	Saskatchewan, Canada	Varied	–	Exploited	Fetuses / Ovulation	509 / 398	4.1 / 4.6	–	Gunson (1970)
58° 20' 00.00'' N	Alberta, Canada	Varied	–	Exploited	Embryos	18	3.1	–	Novakowski (1965)
60°									
64° 56' 20.51'' N 66° 15' 22.03'' N	Alaska, USA	River	–	Exploited and unexploited	Embryos and placental scars	305	3.2 (both sites)	–	Boyce (1981)
66° 15' 59.49'' N 64° 47' 44.00'' N	Alaska, USA	River	–	Exploited	Embryos	69	3.1	–	Boyce (1974)

Dashes indicate no data available.

[a] Note it has been found for beavers in Mississippi that estimates of corpora albicantia were significantly higher than any other method, but corpora lutea, placental scars, and fetuses did not differ (Wigley et al., 1983).

dominate among > 4.5-year-olds (Dieter, 1992). However, a review of the literature of the North American beaver found that the sex ratio heavily favoured males in the juvenile and 2-year-old classes and females among adults (Novak, 1987). The sex distribution of unborn kits in a population in Russia was influenced by the female's age. Females aged between 4 and 9 years old had a disproportionate number of male kits, while older dominant females had 1.5–2 times more male kits than female (Grevtsev, 1989). For details about the physiological knowledge of pregnancy and reproduction see Box 3.1.

6.7 Other factors affecting reproduction

Beavers' natural capability to produce offspring (fecundity) is affected by various factors such as hunting (Hendry, 1966; Nordstrom, 1972; Payne, 1984b), latitude (Gunson, 1970; Novak, 1987), altitude above sea level (Harper, 1968), mother's weight (Pearson, 1960; Wigley et al., 1983), mother's age (Osborn, 1953; Payne, 1984b), mother's physical condition (Parker et al., 2017), surrounding population density of other beavers (Boyce, 1974; Payne, 1984a; Payne, 1984b), habitat quality (Huey, 1956; Longley and Moyle, 1963; Gunson, 1970; Novak, 1987), climate change (Campbell et al., 2013), and duration of territory occupation (Mayer et al., 2017d).

6.7.1 Hunting

Beavers increase kit production as harvesting increases (Hendry, 1966; Nordstrom, 1972; Payne, 1984b). Heavy culling rates (approximately 70%) depressed embryo production. Female beavers residing in populations regularly subjected to hunting pressures begin to reproduce at a younger age, are physically smaller when sexually mature, and suffer a higher reproduction mortality compared to those in less-disturbed populations (Boyce, 1981). In Telemark, Norway, adults of both sexes and pregnant females were more likely to be the first individual shot within an active beaver territory (Parker et al., 2002). A 25% harvest

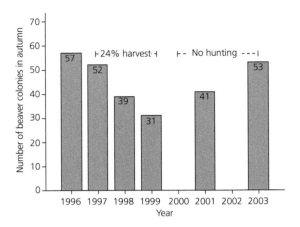

Figure 6.11 Change in number of *Castor fiber* family groups in autumn in an area 242 km² in Bø Township, Telemark, Norway, during a 3-year (1996–1999) experimental hunt with an estimated annual mean harvest rate of 24% (range 22–26%) and during the following 3 years with no hunting. Most beavers (94%) were shot with firearms during legal spring hunting from mid-March to mid-May. Data for the years 1996–1999 originally appeared in Parker et al. (2002). (© 2014 Howard Parker and Frank Rosell, 2014.)

each year for three consecutive years reduced the number of beaver families by 46%. However, the population recovered quickly when hunting or trapping pressure was stopped (Parker and Rosell, 2014; Figure 6.11).

6.7.2 Latitude

Litter size has been suggested to be higher in the south than the north as a result of better climatic conditions. Additionally, vegetation growing season is shorter in the northern hemisphere, which affects litter size (Novak, 1987), though more research is needed, given that a review did not find that geographical position correlated with litter size (or temperature) for the North American beaver (Hill, 1982). For example, the litter size (embryo count) was 2.2 in South Carolina, 2.5 in Alabama, 5.3 in Wisconsin, and 6.3 in Alberta. These litter sizes are the opposite of what was suggested (see also Table 6.4).

6.7.3 Altitude

North American beavers living at extreme elevations in Colorado, USA (> 2,740 m) had smaller litter sizes (1.7) even though they lived in a good-quality habitat with plenty of willows and aspens (Harper, 1968). Another study of North American beavers in Colorado showed that beavers living above 2,500 m had smaller litter sizes (2.3) than beavers at elevations less than 1,500 m (4.4) (Rutherford, 1964).

6.7.4 Weight and age of mother

Adult female body mass has been shown to correlate positively with reproduction in the North American beaver (Novakowski, 1965; Henry and Bookhout, 1969; Gunson, 1970; Boyce, 1974; Wigley et al., 1983). First-time mothers normally have just one kit, and reproductive success generally increases with age. For example, an age-related decline in annual maternal reproductive output in a population of the Eurasian beaver in Telemark, Norway, after a minimum age of 5 years has been found (Campbell et al., 2017). Dominant females reaching an older age were no more or less likely to breed in a given year and did not produce a greater number of or fewer offspring per year than those that died when younger. In Newfoundland, Canada, an increase in placental scars, indicating birth of beaver kits in North American beavers, until aged 9 was found (Payne, 1984b). However, there are exceptions. One beaver from Saskatchewan, Canada, which was at least 13 years old produced six embryos (Gunson, 1967). In central Illinois, USA, litter sizes were 3.0, 3.4, and 4.2 per female for yearlings, 2-year olds, and older adults, respectively (McTaggart and Nelson, 2003).

6.7.5 Physical condition of mother

In Telemark, Norway, the fecundity was significantly higher in females with greater tail fat, and this effect increased with age (Parker et al., 2017). For females 3 years of age or older, pregnant individuals had significantly more tail fat than non-reproducing individuals. Additionally, females conceiving at the typical peak time in late-January were significantly heavier than individuals that conceived from 1 to 3 months later. Late breeders had significantly more corpora lutea, possibly

because of the improved nutrient levels provided by early spring growth that led to higher ovulation rates. Fat accumulated during summer and autumn prior to winter breeding appears to be an important determinant of successful reproduction in female beavers.

6.7.6 Density of beavers

Fecundity in beavers is thought to be negatively correlated with population density (Boyce, 1974; Payne, 1984b), i.e. as density increases, reproductive rate decreases. Number of dispersing 2-year-old beavers is also known to decline with increasing density (Mayer et al., 2017d). Beavers that delay dispersal in high-density populations have been reported to have increased lifetime reproductive success compared with younger dispersers. By delaying dispersal beavers may gain habitat information via forays outside the natal area (e.g. on population density) as well as increasing their body mass (i.e. increasing their competitive ability) while remaining in their natal family group, which could influence later reproductive success (Mayer et al., 2017d; see also Box 7.1).

6.7.7 Habitat quality

When food availability is reduced, breeding is also reduced, but the beaver family can maintain its size by the inclusion of more adult animals (non-dispersers who remain at home) (Mayer et al., 2017f). In addition to breeding frequency, litter size may be reduced when competition for food is increased since some of the fetuses are reabsorbed into the uterus. In Saskatchewan, Canada, beavers in low-quality habitats displayed high frequency of resorption of embryos, resulting in fewer kits per female beaver. When food sources are scarce the remaining fetuses will be smaller at birth and more susceptible to diseases, increasing infant mortality (Gunson, 1970).

In Algonquin Provincial Park, Canada, more kits were produced per family group in areas with high woody cover and aquatic vegetation (Fryxell, 2001). Beavers in this area were best characterized as a population in which some territories ($n = 5$) were sources (offspring outnumbered immigrants) but most territories (*n* = 9) were sinks (immigrants out-numbered offspring). A smaller fraction of territories supported offspring production of sufficient magnitude to replace parents. In Biesbosch, the Netherlands, mean litter size of beavers increased significantly with the number of years since the first beavers were released (1.0 in 1989 to 3.3 in 1999) (Nolet et al., 2005). This slow population growth may be due to a climatic shift in food quality since animals were translocated from Germany to the Netherlands. The initially low reproduction rate was partly due to the relatively early sprouting of willow at the release site, along with other factors including increased exploration costs and disturbance from boat traffic (Nolet, 1992; Nolet et al., 2005). Regardless, the beavers seem to have adjusted well to both factors in one decade.

Adult females from low-quality territories are significantly less likely to reproduce as they age and also senesced at an earlier age (Campbell et al.,

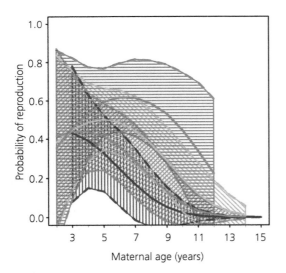

Figure 6.12 Relationship between probability of reproduction and maternal minimum age for four levels of territory quality (TQ), where blue = TQ1, green = TQ2, orange = TQ3, and red = TQ4. Lines present predictions and shading denotes areas within 95% prediction intervals averaged from the top set of generalized linear mixed models (GLMMs). Adult females from low-quality territories are significantly less likely to reproduce as they age. (Reprinted from Campbell, R. D., Rosell, F., Newman, C. and Macdonald, D. W. 2017. Age-related changes in somatic condition and reproduction in the Eurasian beaver: resource history influences onset of reproductive senescence. *PLoS ONE*, 12, e0187484 under CC BY 4.0 License.)

2017; Figure 6.12). In terms of resource effects, in poorer years, older mothers (≥ 6 years old) produced larger offspring than younger mothers (≤ 5 years old). Older mothers were less affected by short-term resource variation. Campbell et al. (2017) speculated that this response could arise if older mothers accumulate greater energy reserves during non-reproducing years in order to mitigate unexpected poor conditions during reproducing years. They proposed that such flexible life-history schedules could play a role in the dynamics of populations exhibiting reproductive skew, with earlier breeding opportunities leading to an earlier senescence schedule through resource-dependent mechanisms.

Interestingly, in a study of the introduced North American beaver in Tierra del Fuego Argentina, mean litter size per family per year was higher in open steppe (grassland and shrub) habitat (2.7 kits) compared to forested habitat (1.7 kits) (Pietrek et al., 2017). Pietrek et al. (2017) suggested that lower litter size in forest habitats may be evidence of density-dependent regulation. Across the range, beaver populations do occur in habitats with lower tree cover, and this may also be due to shrub availability and productivity in the steppe habitat.

6.7.8 Climatic factors

Higher rainfall resulted in fewer kits weaned per breeding female, for all ages (Campbell et al., 2017). A greater reproduction was linked to (influenced by) higher seasonal variability of temperature in Norway (Campbell et al., 2012). Greater temperature variability indicates a cooler spring/autumn and/or a warmer summer. Cooler spring temperatures slow plant development and therefore extend the period during which high-quality early plant forage is available (Nolet et al., 2005). These researchers suggested that increasing variability in temperature and rainfall reduced reproduction of kits in the breeding population. It is possible that if trends towards an extended growing season and reduced winter ice cover persist, these effects may in the long term outweigh the predicted negative effects of climate variability (Campbell et al., 2012). Also, a negative association was observed between higher rainfall and reproductive success in a population of Eurasian beavers in Telemark,

Norway (Campbell et al., 2013). Higher rainfall in the autumn prior to the spring, when females give birth, led to smaller litter sizes. The higher rainfall stressed important food plants by waterlogging their roots, resulting in lowered nutrient condition for beavers during the reproductive period. Alder (*Alnus* spp., the principal beaver food in the study area) growth rings exhibited a positive relationship with rainfall for trees growing at locations > 2 m above water level but a negative relationship for trees growing at < 0.5 m. They deduced that temperature influenced beavers at the landscape scale via effects on plant growth and winter thermoregulation, while rainfall limited beavers' ability to take advantage of better forage growth further from water during wetter years (Campbell et al., 2013).

6.7.9 Duration of territory occupation

The lifetime reproductive success increased with an increasing duration of territory occupancy. This result shows the importance of holding a territory (Mayer et al., 2017d; see more about this in Section 7.3).

6.8 Kit development and the life inside beaver lodges

Beaver kits are well developed at birth and show some characteristics of an extremely precocial animal, i.e. they are relatively mature and mobile from the moment of birth. They are born with a dense pelt, with their eyes partially open (complete opening occurs 2–4 hours after birth), and their incisors, which are sharp, pointed, and covered by a thin membrane, are partially erupted. Their tails are soft and covered with fine hair (Wilsson, 1971), which they retain well into the summer. The kits lack bony epiphyses (the rounded end of a long bone at its joint with adjacent bone(s)) in the arms and legs, a characteristic more typical of altricial animals (helpless at birth and dependent on parental care). The kits remain within the lodge for the first 2–5 weeks and in some cases do not venture from the lodge before they are 2 months old (Wilsson, 1971; Hodgdon and Lancia, 1983). Therefore, they lead a nidicolous life, a

status somewhat intermediate between being precocial and altricial (Müller-Schwarze, 2011).

At birth kits usually weigh between 500 and 550g but this can vary from about 300 to 700 g (Żurowski, 1977; Parker and Rosell, 2001). Large North American beaver mothers are more likely to give birth to heavier kits than smaller ones (Pearson, 1960; Boyce, 1974). However, in a population in Telemark, Norway, beaver mothers of all ages produced kits with similar body sizes (Campbell et al. 2017). It was also found that in terms of resource effects, in poorer years older mothers produced larger offspring than younger mothers. Generally, kits are about 31 cm long at birth, but both length and weight are dependent upon litter size (Wilsson, 1971).

Beavers display a high degree of biparental care (Wilsson, 1971; Sun, 2003; Busher, 2007) and hence are considered to be obligate monogamous species, which is rare among mammals (Kleiman, 1977). After birth kits are licked clean by their mother who continues to provide a high amount of direct care. The mother usually ingests the placenta. The female stays with her kits in the sleeping chamber for the first hours after birth but can leave the lodge for short times later the same day (Patenaude and Bovet, 1983). The female ingests the kit faeces in the first 2–3 days after birth, but thereafter kits can defecate in the water in and around the natal lodge or burrow (Owesen, 1979; Wilsson, 1995). This behaviour by the female might be because the stomach and anus of small mammals must be massaged in a manner similar to the way that the mother does when she licks her offspring to avoid constipation—an important consideration and requirement if hand-rearing beaver kits (Richards, 1983; see Section 10.3.1).

The mother has two pairs of nipples located in between and under the front legs (Franke-Radowiecka et al., 2016). The nipples become enlarged, 0.5–1.0 cm in length, during the second month of gestation and remain fairly prominent during lactation (Doboszyńska and Żurowski, 1983; Müller-Schwarze, 2011). The lactation period lasts between 45 and 90 days (Żurowski et al., 1974; Doboszyńska and Zurowski, 1975; Rusanov and Maksimenko, 2019). Kits suckle for so long even though they obtain little milk at the end. A reason for this may be that the kits want to maintain their bond with their mother (Baker and Hill, 2003). At the end of lactation, the nipples have many injuries, and that is why the kits are weaned at that time (Doboszyńska and Żurowski, 1983).

Beaver kits nurse up to nine times a day, usually for 5–10 minutes per session, but they can occasionally nurse for as long as 30 minutes (Patenaude and Bovet, 1983). The father may help shove the kits towards their mother's nipples. The kits press the milk out using rhythmic movements with their fore legs. The upper pair of teats has been reported to provide 50–75% less milk than the lower pair (Żurowski et al., 1974). The consistency of beaver milk leads to rapid growth of the kits (Figure 6.13). The teats begin to shrink 1 month after birth, but the kits can nurse for the two first months of their lives (Żurowski et al., 1974) (see also chapter 3).

Studies from cameras placed in natal lodges and captive families allow a glimpse into the early life of kits (see Section 10.4.2 for remote monitoring techniques). Wilsson (1971) in Sweden studied captive Eurasian beavers in artificial lodges, focusing particularly on which behaviours were inherited and which were learnt. He closely followed the

development of 14 kits, recording, amongst other things, the age at which important behaviours were first exhibited (Table 6.5). Life inside the lodge also has been studied in detail in North America (Patenaude, 1983; Patenaude and Bovet, 1983; Müller-Schwarze and Herr, 2003; Mott et al., 2011; Table 6.5). Patenaude (1984) built an observation chamber which gave direct visual access to wild North American beavers, giving us some of the greatest insight into the lodge life of a single family.

The newborn kits were observed to make their way to the swimming pool inside the natal lodge or burrow within 30 minutes of birth. After a short swim they climb out or older animals of both sexes may take them back to the sleeping chamber, typically by carrying them in their mouths (Shadle, 1930; Wilsson, 1971). The young kits have a swim every night, and their parents and older siblings struggle to constrain them indoors. At around 5 days old, their muscles are so well coordinated that they can begin to wrestle with each other. They are good swimmers after just 1 week and start to eat solid foods. After about 10 days they begin to take some shallow dives within the lodge pool, and it becomes increasingly difficult for the older animals to keep an eye on them. However, the young kits are not able to dive and

Figure 6.13 Three feeding Eurasian beaver kits. (Photo supplied courtesy of Kjell-Erik Moseid.)

Table 6.5 Ontogeny of behaviour patterns of North American and Eurasian beavers.

Behaviour pattern	Age of beaver at first observed performance			
	North American (studied in the wild)		Eurasian (studied in semi-liberty and captivity)	
	Patenaude (1984)	Other authors[a]	Other authors[a]	Wilsson (1971)
Locomotion				
Walking	< 1 minute	A few minutes [1]	A few minutes to hours [2,3,4,5]	A few hours
Opening eyes			A few hours [6,7]	
Swimming on surface	< 3 hours to 1 day	A few hours [1]	A few hours [4], < 3–8 days [2,5,8,9]	4 days
First attempt at diving underwater			6–7 days [5,10]	
Diving and swimming underwater	21–15 days		6–10 days [2,10], 20–26 days [11]	12 days
Galloping	8–10 days		22–27 days [2,11], 1 month [5]	30 days
Bipedal gait	45 days		1–2 months [2,5]	30 days
Hopping			1 month [5]	
Fully developed ability to stay underwater	60 days		53 days [2], 2 months [5,10]	60 days
Feeding				
First sampling of solid food			6–7 days [5]	
Gnawing on bark and leaves	11–12 days		5–6 days [2]	4 days
Gnawing on wood			11 days [5]	
Eating leaves	15–17 days		9 days [2]	11 days
Eating plants			16–18 days [5]	
Handling twigs and stalks	6–9 days		12–14 days [2,5]	14 days
Grasping twigs between finger V and other fingers	30 days		23 days [11]	29 days
Eating mainly solid food	30–35 days	30 days [1]	24 days [2], 1 month [5]	33 days
Caecotrophy	30–35 days		28–30 days [2], 41 days [11], 3–4 weeks [5]	33 days
Peak consumption of milk			1 month [5]	
Squashing leaves into a roll	30–35 days			38 days
Grasping leaves with one paw while biting off stalk				44 days
Peeling bark/making wood strippings	40 days	2 months [1]	21 days [12]	58 days
Cutting and felling small trees	< 12 months			2.5 months
Weaning			6–8 weeks [13,14], 10 weeks [15], 45–50 days [5]	
Milk consumption dropping to 0 ml			6–7 weeks [5]	
Collection of food stores underwater				6 months
Eating leaves in the characteristic way of adult beavers	30–35 days			12 months
Felling large trees	12 months		50 days [11]	16 months

continued

Table 6.5 Continued.

Behaviour pattern	Age of beaver at first observed performance			
	North American (studied in the wild)		Eurasian (studied in semi-liberty and captivity)	
	Patenaude (1984)	Other authors[a]	Other authors[a]	Wilsson (1971)
Sleeping, resting, and associated behaviours				
Stretching	14–30 days	2 months [1]		
Cleaning floor chamber	12 months			
Bringing litter from outside into lodge	12 months			
Care of fur				
Grooming with fore paws	< 1–5 hours		A few minutes [2,3,5,12]	4 days
Grooming with poor balance			7 days [5]	
Combing with double claw of hind paw	< 5 hours to 1 day	16 days [1]	16 days [2], 21 days [5]	4 days
Use of functional anal glands to grease fur	1–9 days		3–4 weeks [5,10]	
Shaking	1–9 days		26 days [2,11]	
Grooming with teeth	5–12 days		1 month [5]	6 days
Fully developed grooming with balance	30–35 days	28 days [1]	3 weeks [5]	19 days
Fully developed ability to keep fur water-repellent	30–35 days	45 days [1]	45 days [2], 6–8 weeks [10], 1.5–2 months [5]	60 days
Digging				
Scratching and shovelling after isolation from family group				6 days
Scratching, shovelling, and pushing earth within family group				14 days
Scratching, shovelling, and pushing earth	16–18 days		< 3 days [2], 20 days [11]	6–14 days
Shoving and packing	13–15 days			45 days
Digging a temporary nest				60 days
Digging a tunnel system				12 months
Building of lodges, dams, and winter stores				
Carrying and pushing sticks at random	2–9 days		11 days [2]	14 days
Dragging	20–26 days		16 days [2]	16 days
Pinning	30 days			23 days
Complete series of movements: shoving, lifting, pushing, and packing	60 days		48 days [2]	45 days
Carrying with front paws	12 months	12 months [16]		
Carrying with bipedal gait	12 months	> 12 months [16]	90 days [5]	
Lodge building	18 months	> 12 months [16]		4–5 months
Dam building	12 months	> 12 months [16]		6–7 months
Food caching	18 months	6 months [16]		

Behaviour pattern	Age of beaver at first observed performance			
	North American (studied in the wild)		Eurasian (studied in semi-liberty and captivity)	
	Patenaude (1984)	Other authors[a]	Other authors[a]	Wilsson (1971)
Anti-predator behaviours				
Gnashing teeth		A few hours [17]	A few hours [5]	
Hissing	4–16 days	A few days [1], a few hours [17]	A few hours [5]	1 day
Leaping aside				3 days
Escape dives			8–10 days [5]	
Seeking protection in water	6 days	10 days [1]	8–10 days [2,10], 3–4 weeks [5]	10 days
Tail slapping	13 days	30 days [1]	19 days [2], 3–4 weeks [5]	1 month
'Information posture'; standing on hind paws	19–20 days			
Fully developed escape behaviour	35 days			2 months
Social behaviour and reproduction				
Territorial behaviour				
Exploratory behaviour	< 1 day	A few days [1]	1 day [2]	1 day
Tail wagging				6 days
Movements of territorial marking behaviour				6 days
Aggressive tendencies (wrestling)			5 days [12], 20 days [2], 3 weeks [5]	30 days
Territorial (scent) marking	4–20 days		13–14 days [5]	5 months
Fully developed territorial behaviour				5 months
Socialization				
Vocalizations:				
Whines	< 1 minute	A few minutes [1]	A few minutes [3,4,8,12]	
Low-pitch whines	1–9 days	A few hours [17]	1 day [2]	1 day
Hisses	4–16 days	A few days [1]	1 day [2]	1 day
Whistles	15–35 days			
Growls	21 days			
Social play	< 2–5 hours			
Mutual grooming after isolation from family group				6 days
Grooming with teeth within family group				16 days
Fully developed mutual grooming	12–15 days		18 days [11], 3–4 weeks [5]	
Indicators of play behaviour: 'building', feeding, mutual grooming with young, rolling			8 weeks [10]	
Care of young	12 months			
Can be fully socialized				7 days
Can be partially socialized				8–30 days
Can be socialized to some small degree				31–60 days

Source: After Patenaude (1984), © 1984 Verlag Paul Parey, Berlin und Hamburg [a] (1) Patenaude (1984); (2) Lancia and Hodgdon (1982); (3) Hediger (1970); (4) Shadle (1930); (5) Lancia and Hodgdon (1982); (6) Bradt (1939); (7) Guenther (1948); (8) Hirai (1975); (9) Hediger (1966); (10) Müller-Schwarze and Herr (2003); (11) Kalas (1976); (12) Richards (1983); (13) Bailey (1927); (14) Jenkins and Busher (1979); (15) Patenaude and Bovet (1983); (16) Hodgdon (1978); (17) Patenaude (1983).

swim out of the lodge because they are too light and buoyant with so much air in their fur that they cannot execute a proper dive; instead they float like corks on the top of the water. If a spring flood should inundate the lodge or burrow the kits will need the older animals to help them safely escape. On these occasions, they can cling to the fur or tail of an adult or older sibling (Müller-Schwarze, 2011).

All family members help to look after the new kits in the lodge, taking it in turns to babysit for several hours at a time. The most important job of the babysitter is to fill the role of a lifeguard since the youngest kits cannot close their ears or nostrils, nor are their pelts water-repellent. Therefore, any young kit that gets wet needs to be dried off by an older animal to avoid getting cold. The babysitters also assist in cleaning the natal shelter. After the kits are born the lodge is cleaned every second or third day, indicating that cleanliness is more important when kits are present since they are not normally cleaned so often (Patenaude and Bovet, 1983).

Recent mothers may be quite elusive in the first 2 months following the birth of her kits. Some mothers rarely leave the lodge in the period shortly after birth and are very shy when outside the lodge. One good example of this was demonstrated by a radio-tracked nursing female in the Netherlands. This female was not observed visually during 10 tracking nights between May and August, and radio-tracking indicated that she remained in the lodge and took only short trips outside. When she did venture out, she swam long distances underwater. Interestingly, reproducing females were never caught in a long-term beaver study of 20 years in Telemark, Norway, during the last week in May or the first 2 weeks of June (Mayer et al., 2017d). This suggests that females remain mostly inside the lodge for approximately 2 weeks after parturition. Beaver parents and older siblings regularly bring leafy twigs and other vegetation for the kits to eat (Wilsson, 1971; Busher, 1980). Therefore, if beavers are observed diving towards the lodge with food in their mouths around this time of year, this may be an indication there are kits inside.

Kit development and growth is relatively rapid (Figure 6.14). When the kits first come out of the lodge they usually weigh around 2–3 kg. They are often led out by the older members of their family (Patenaude, 1983), and the majority of kits are not shy the first few times they emerge from the lodge. It is often possible to approach them as they swim and dive near the lodge, especially if you are in a boat or canoe, though they are much more wary if you approach on land. The kits are not developed enough to remain submerged before they are 2 months old.

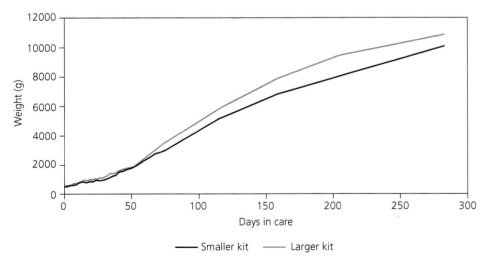

Figure 6.14 Growth rate of two captive-raised beaver kits in Belgium, 2018. (Source : Belgium Beaverworking group, Jorn Van Den Bogaert, beverwerkgroep@hotmail.com, www.beverwerkgroep.be.)

The largest kits are most often the first to dare to explore. At 2 months of age, their swimming, grooming, and feeding capabilities are fully developed. In the beginning, the older animals ensure that the kits stay near the lodge's feeding locations. The kits are very curious about their surroundings, and they swim around and explore the terrain adjacent to the lodge, though they rapidly develop anti-predator behaviours. When kits become tired, they can hitch a ride with one of the older animals by climbing onto their back or tail (Patenaude, 1983). Occasionally, two young animals will be carried together, with one on the adult's back and the other on its tail. When they are small they can also exhibit a rarely seen behaviour known as 'caravanning' where they swim in the wake of an older relative, often partially clinging onto their back (Müller-Schwarze, 2011). After a period of practice diving and tentative exploration around their lodge, they quickly start to independently forage further from the natal shelter. The youngsters can be quite comical in their explorations, as they have not quite developed the same bodily control as their older relatives. If a young beaver is observed sitting on its rump along the water's edge and grooming its pelt, it can suddenly lose its balance and topple over.

It is important for the kits to learn to swim in the pool inside the lodge as early as possible; this enables them to remember the routes into and out of the lodge, along with territory, before autumn sets in. When autumn arrives the older animals spend more time preparing for winter, while the kits are seemingly left to their own devices. One Californian study reported kits also engaging in food cache construction behaviour (Busher, 1980).

A remote videography system for monitoring behaviour and demography of North American beavers inside lodges has been developed (Bloomquist and Nielsen, 2009). This system was used to record 1,506 hours of within-lodge activity from 23 beaver family groups (17 lodges and 6 bank dens) and to characterize behavioural patterns based on sex and age over daily and monthly intervals (Mott et al., 2011). The time-budgets were similar among male and female adult beavers. Feeding, sleeping, allogrooming, and individual grooming accounted for more than 95% of all recorded behaviours. From spring onwards, sleeping time tends to decrease,

whilst grooming and feeding increase. Kits demonstrated significant diel behavioural trends. They exhibited sleep-recumbent behaviours during all hours, and periodic bouts in these activities were clearly evident. They had three separate sleep intervals, each lasting approximately 5 hours, so kits slept 15 hours per day. Adults, however, displayed a distinct sleep interval lasting from 05:00 to 21:00 hours, where sleeping/recumbent activities dominated all other behaviours. Kits displayed almost all adult behaviours within weeks of birth, and all behaviours recorded among adults had been detected among kits within 1 month post-birth. Agonistic interactions were seldom observed within these family groups. They also detected increased lodge maintenance immediately prior to parturition during early spring, which may reflect 'nesting' instincts. This lodge maintenance before birth is similar to that reported in other studies and may be a consistent behaviour in beavers (Patenaude, 1983; Patenaude, 1984).

In another behavioural study of beavers inside a lodge, adult beavers extruded their anal glands and dragged them over the substrate when leaving the sleeping platform and diving into the plunge hole to travel outside (Müller-Schwarze and Herr, 2003). In some scenes, a kit left behind could be seen sniffing the area just 'marked'. The researchers also observed an adult beaver pushing a kit repeatedly under the water surface in the plunge hole. Grooming, play behaviour, and grinding of teeth to sharpen them were also observed.

6.9 Dispersal of offspring

A beaver family's offspring tend to stay within their natal territory for two winters and leave it just before their second birthday or shortly thereafter, typically prior to the arrival of the next litter. Beavers usually become sexually mature at around 20 months of age (Novak, 1987). This extended stay occurs despite their ability to form pair bonds, establish territories, and having acquired vital skills such as dam and lodge building as 1-year-olds. Beavers maximize their fitness by dispersing from the natal territory, which increases the probability of finding a mate and reproducing. In other words, they must attempt to reproduce and pass their own genes on to new generations, which they cannot do if they remain in their natal territories since the presence of their parents (the adult pair) prevents this. By leaving they can also ensure that the family group does not grow too large, which helps to ensure sufficient food supplies for the rest of the family and for the newborn kits. Additionally, dispersal lowers the possibility of inbreeding and potential genetic abnormalities.

6.9.1 Age at dispersal

Many factors affect when the offspring decide to disperse. In an expanding beaver population in Sweden at low population density (0.13 families/km shoreline) the average age at dispersal was 1.4 years (17 months) (Hartman, 1997). In a study in New York state, USA, all beavers dispersed by the age of 3 years (Sun et al., 2000). The latter study showed that the majority of first-time dispersers were 2-year-olds (64%), 21% were 3-year-olds, and only 14% were 1-year-olds. However, in a population with a high density (0.64 families/km shoreline, with no unoccupied areas between territories) of beavers in Telemark, Norway, it was found that the mean age at dispersal of known aged beavers was 3.5 years (Mayer et al., 2017f). Individuals started dispersal at 1 year, and others remained within their natal family group up until the age of 7 (Figure 6.15). The high population density was possibly the reason why individuals delayed dispersal to increase their competitive ability. Beavers dispersed as solitary animals (not with other same-age siblings), and 59% dispersed when 1 to 3 years old, while 41% delayed dispersal until they were 4 to 7 years old. Only 9% of individuals remained philopatric (stayed within the family territory) and became dominant after both of their parents disappeared. It was also found that subordinates delayed dispersal with increasing age of the same-sex parent. This suggests that either the parents are more tolerant towards their offspring at an increased age or subordinates can perceive potential parental senescence, and thus 'queue' in the natal territory to take it over after the death of the parents (Mayer et al., 2017f).

A beaver's dispersal behaviours are intrinsic, rather than being chased away by their parents as reported by earlier researchers. It is unlikely that there are physical disputes between dominant

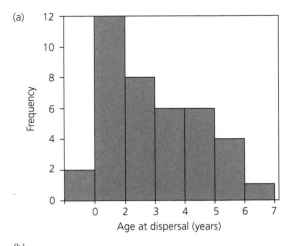

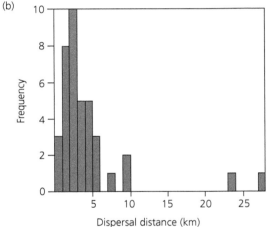

Figure 6.15 Frequency histogram of (a) dispersal age and (b) dispersal distance from data collected between 1998 and 2015 in southeast Norway (N = 39 Eurasian beavers). (Reprinted from *Animal Behaviour*, 123, Mayer, M., Zedrosser, A. and Rosell, F., When to leave: the timing of natal dispersal in a large, monogamous rodent, the Eurasian beaver, pages 375–382. © Elsevier Science (2017), with permission from Elsevier.)

parent and subordinate offspring (Crawford et al., 2015). In southwestern Illinois, USA, little aggression was observed prior to subadult dispersal, supporting the assertion that dispersal is instinctual and not mediated by agonistic interactions (Mott et al., 2011). The young adults instead chose to stay home to look after their younger siblings and to contribute to the family group. Helping behaviours (voluntary actions intended to help the others, with reward regarded or disregarded) can also be an effective way to ensure that their genes are spread through their siblings.

Helping behaviours can be found in a number of species and are almost always a case where individuals assist near-relatives in defending their territories or caring for the young. These behaviours are advantageous in a number of ways. Subordinates are shown to contribute towards the territorial defence (Tinnesand et al., 2013; Hohwieler et al., 2018), dam and lodge repairs, and the provisioning of the kits before they emerge from the lodge (Müller-Schwarze, 2011). This could be an important help for their senescing parents and an experience gained for the subordinates. They also increase their own fitness by remaining in the natal family group (McTaggart and Nelson, 2003; Mayer et al., 2017f). The number of reactions to experimental scent markings from intruding strangers by subordinates has been found to be positively related to the age of their parents (Hohwieler et al., 2018). This suggests that subordinates might take over territorial duties with increasing parental senescence, because they have increased chances of inheriting the territory. Increased experience in territorial activities such as scent marking possibly helps subordinates to successfully gain and defend a territory on their own. In addition, individuals are more likely to await a promising time to initiate dispersal, i.e. disperse when older and when population densities are lower (Mayer et al., 2017f; Mayer et al., 2020). In Telemark, Norway, individuals that dispersed at an older age had a greater body mass and were possibly more experienced compared to younger dispersers, which likely gave them a competitive advantage in order to successfully take over or establish a new territory (Mayer et al., 2017f). In western New York, USA, older dispersing North American beavers were more successful at establishing in neighbouring territories, suggesting that they have a competitive advantage (Sun et al., 2000). The onset of dispersal in Telemark was probably related to population density, which suggests that subordinates can perceive changes in the population density by conducting extra-territorial movements (Mayer et al., 2017e; see Section 6.9.2).

Secondary dispersal (a second dispersal after a successful natal dispersal) was found in 39% of all dispersal events in a study of North American beavers in western New York, USA (Sun et al., 2000). Males that left for the second time were most successful in taking over neighbouring families/territories. Also, on occasion, the adult male and 1-year-olds may leave the territory in the breeding season and 'wander' around the neighbourhood, especially if there is a good supply of food in the area. The adult female and the kits will then have all of the food within the territory for themselves (Sun et al., 2000). Mayer et al. (2017f) observed in Telemark, Norway, secondary dispersal in only one case (both members of a pair moved to a new territory).

6.9.2 Extra-territorial movements

Exploratory movements of subadults away from their natal sites have been reported in several studies (Aleksiuk, 1968; Hodgdon, 1978; Weaver, 1986; Van Deelen, 1991). Individuals will commonly make exploratory excursions into neighbouring areas to obtain predispersal information or to increase reproductive success via extra-pair copulations (Campbell et al., 2005; Mayer et al., 2017e). Through these trips, beavers detect suitable times for dispersal by gaining knowledge on neighbours, possibility for territory expansion, breeding opportunities, available territories, condition of territory owners, and population density fluctuations (Mayer et al., 2017e; Figure 6.16).

Three of eight subordinates (38%) made extra-territorial movements (forays) before dispersal in a North American beaver population in east-central Illinois, USA (Havens, 2006), and three of nine subordinates (33%) made predispersal forays ranging between 1.5 and 15 km in a Swedish beaver population (Hartman, 1997). Twenty of 46 (43.5%), dominant individuals and 6 of 10 (60%) subordinates conducted extra-territorial forays in a dense population of beavers in Telemark, Norway (Mayer et al., 2017e). Beavers spent up to 11% of their active time away from their territories during these forays. Beavers were documented to travel faster and spend more time in water when conducting forays compared to their intra-territorial movements, most likely to avoid detection by conspecifics that could end in a physical dispute (see Section 7.2.3;

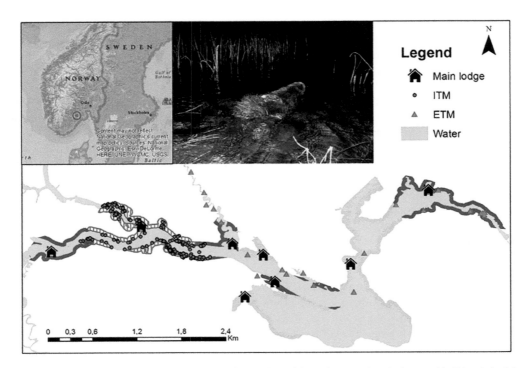

Figure 6.16 Location of study area in southeast Norway (top left), and photo of the study species, Eurasian beaver, with GPS on its back (top middle). Main map shows exemplary GPS data from a subordinate individual who conducted extra-territorial movements (ETM, red triangles) through five different territories (shown with grey shadings). Territory is shown in red hatching and intra-territorial movements (ITM, GPS positions) are shown as red dots. (Main map was created using ArcMap 10.1 [Esri, Redlands, CA, USA, http://www.esri. com/arcgis/about-arcgis]. Source of small map [top left]: National Geographic, Esrif DeLorme.) (Reprinted from Mayer, M., Zedrosser, A. and Rosell, F. 2017d. Extra-territorial movements differ between territory holders and subordinates in a large, monogamous rodent, *Scientific Reports*, 7, 15261, under CC BY 4.0 License.)

Figure 6.17). This suggests that forays are energetically costly and may be a trade-off between information and experience gained before dispersal and the costs of decreased health or even fitness due to stress. Individuals conducted fewer forays when occupying larger territories, possibly because they have sufficient resources compared with beavers in smaller territories, which more often assessed possibilities for territory expansion and foraging. The distance moved on individual forays ranged between 298 and 11,237 m, and beavers intruded into one to five different territories while conducting forays. Subordinate forays lasted longer (189 vs. 81 minutes), moved greater distances (4,756 vs. 1,919 m), and intruded into more territories compared to forays by territory holders (2.75 vs. 0.62 territories). The dominants mainly intruded only into adjacent territories (Mayer et al., 2017e; Figure 6.18).

Exploratory movements away from their native territory can sometimes lead to a true dispersal if the weather conditions permit and if they find available territories, but a change of weather and refreezing of the waterway can send the explorers home again (Molini et al., 1980).

6.9.3 Time of year

Several factors are known to affect the time of dispersal. Two-year-old beavers in captivity became very restless in the spring and early summer and attempted to leave their enclosure. Also, 1-year-olds were 'uneasy' (Wilsson, 1971). In the wild, the period between the spring thaw and early summer is the time when they usually leave their natal territory (Beer, 1955; Bergerud and Miller, 1977; Svendsen, 1980b; Sun et al., 2000), but this can vary considerably.

(a)

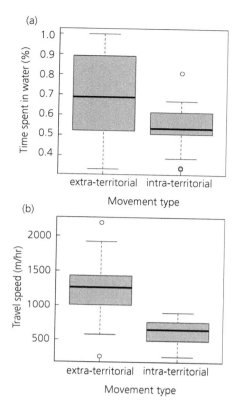

(b)

Figure 6.17 Box plots showing (a) proportion of time spent in water and (b) travel speed (both averaged per individual) for extra-territorial and intra-territorial movements of 25 GPS-tagged Eurasian beavers in southeast Norway (2009–2016). Plots show median, 25th and 75th percentile, range of data, and outliers (dots). (Reprinted from Mayer, M., Zedrosser, A. and Rosell, F. 2017d. Extra-territorial movements differ between territory holders and subordinates in a large, monogamous rodent, *Scientific Reports*, 7, 15261 under CC BY 4.0 License.)

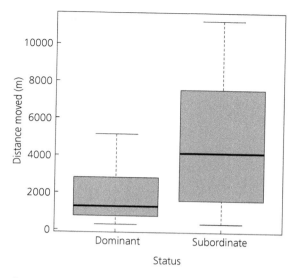

Figure 6.18 Box plot showing distance moved during extra-territorial movements of dominant (*n*=20 individuals, 43 forays) and subordinate (*n*=6 individuals, 8 forays) GPS-tagged Eurasian beavers in southeast Norway (2009–2016). Plots show median values, 25th and 75th percentile, and range of data. (Reprinted from Mayer, M., Zedrosser, A. and Rosell, F. 2017d. Extra-territorial movements differ between territory holders and subordinates in a large, monogamous rodent, *Scientific Reports*, 7, 15261 under CC BY 4.0 License.)

For example in the USA, most dispersal movements were made in spring (April–June) in Massachusetts (DeStefano et al., 2006). This is reasonably consistent with the average dispersal date in Montana that was 17 May but ranged between 7 April and 20 August (Van Deelen and Pletscher, 1996). However, in an unexploited population of beavers in the lower Mississippi River bottomlands of southern Illinois, USA, 13 of 38 (24%) juvenile (6 yearlings and 7 subadults) beavers dispersed during 1 October to 23 March (mean date was 28 December) (Bloomquist and Nielsen, 2010). Conversely, another study in southern Illinois reported that dispersal was initiated around 16 February (McNew and Woolf, 2005). In Telemark, Norway, dispersing beavers have also been observed during August to October. Geographical differences in dispersal dates are likely the result of regional climate differences in spring thaw and water flows and other variables. The young beavers usually leave home when the current is strongest and ample water is available so land predators can be avoided (Sun et al., 2000; McNew and Woolf, 2005).

At northern latitudes the timing of dispersal is of course dependent on the ice conditions (Sun et al., 2000). If the waterway in the natal territory is frozen, the animals cannot disperse until it is gone. The young dispersing animals usually find a place relatively near the waterway where they are born and will establish a neighbouring territory if there is room in the habitat (Crawford et al., 2008; Nimje, 2018). Hence, many waterways are usually populated with animals that are genetically related.

6.9.4 Obstacles, distance, and direction

During dispersal, subadult beavers are capable of travelling long distances along water bodies and will occasionally undertake trips overland. Beavers generally do not like to be far from the water or to travel over open land, though reports of individuals travelling on land up to 11.7 km away from freshwater exist (Saveljev et al., 2002). Watershed divides may act as dispersal barriers, but this varies depending on topography (Halley and Rosell, 2002; Halley et al., 2013). Beavers can disperse to new watersheds through brackish or saltwater, but establishment of permanent territories in these aquatic environments seems to be rare and may be dependent upon a local source of freshwater (see Section 4.1.3). Small waterfalls may not be dispersal obstacles since beavers can simply travel on land to go around them. The speed of the beaver's dispersal is affected by topography. Surveys in Sweden and Norway indicate that dispersal occurs more quickly within a watershed than between watersheds (Hartman, 1995; Halley et al., 2013). Urban areas and rough terrain can stop beavers from dispersing to new areas. Some beavers end up getting lost in the city and may not fare well.

Many studies have investigated the distance of dispersal, with inconsistent results concerning differences between males and females (Table 6.6), though not all of the studies separated natal dispersal by subadults from secondary dispersal of adults (Sun et al., 2000). The average dispersal distance for beavers in Telemark, Norway, was 4.5 km, and there was no sex difference in dispersal distance. Two individuals dispersed 23 and 27 km, respectively, making it likely that several beavers dispersed outside the study area (Mayer et al., 2017f). The average dispersal distance is similar to those in other studies on North American beavers (Table 6.6). In southern Illinois, USA, beavers (no sex differences) dispersed a mean distance of 5.9 km in streams (McNew and Woolf, 2005), and the mean dispersal distance was 4.5 km in Massachusetts (DeStefano et al., 2006). Beavers can disperse more than 50 km (31 miles) along streams (Müller-Schwarze, 2011), and the longest migration distance ever recorded was 500 km (310 miles) in Russia (Saveljev et al., 2002).

Although beavers have been reported to disperse downstream or upstream equally (Hodgdon, 1978; Van Deelen and Pletscher, 1996), the majority of beavers in a study in western New York, USA, dispersed downstream (Sun et al., 2000). Thirty-seven percent settled downstream, 20% settled upstream, 37% moved downstream and then come upstream, and 7% moved upstream and then headed downstream. This seems to make sense since downstream is the easiest direction due to the direction of water flow. If beavers cannot find an available territory, especially at high population densities, they often return to their natal territory.

6.9.5 Floaters

Some beavers can also live as temporary 'guests' within established territories or as floaters (animals without their own territories who move among established territories), and the number of animals living under these conditions increases as the population density increases. It has been estimated that in saturated habitats (high population densities) floaters make up as much as 20% of the population (Molini et al., 1980). These floaters are adult animals ready to take over territories if they become available or if a reproducing animal dies. Information regarding available territories appears to spread rapidly among beavers in the area. There is also the slight possibility of an unrelated beaver moving in with an established pair and helping them and therefore having the opportunity to take over the territory in the future (Crawford et al., 2008). It is also possible that these 'helpers' will attempt to mate with one of the territory's owners. Floating beavers have been found in late spring and early summer, and they often travel in pairs (Müller-Schwarze, 2011). Additionally, if a beaver lingers around established territories for an extended time, it may bump into a desirable mate who is also 'floating' as they both are looking for a future home. One beaver rehabilitator describes a new female adult being accepted by a resident breeding pair, sharing in the rearing of their subsequent kits and later going on to have a litter of her own in a nearby lodge. All three adults shared parental care, and the female stayed a few seasons before disappearing;

Table 6.6 Dispersal distances displayed by male and female beavers (*Cf* = Eurasian beaver; *Cc* = North American beaver).

Beaver species	State/country	Habitat type	Mean dispersal distance (km)		Longest dispersal distance recorded (km)	References
			Male	Female		
Cf	Grimsö Wildlife Research Area, Sweden	River	1.5	9.0	–	Hartman (1997)[a,b]
Cf	Telemark, Norway	River	4.5	4.5	27.0	Mayer et al. (2017f)[a,b]
Cf	Telemark, Norway	River	2.5	2.1	9.6	Nimje (2018)[b,d]
Cf	Todzhinskii kozhuun district, Tyvia Republic, Russia	River	11.1	2.1	85.0	Saveljev et al. (2002)[c,e]
Cc	Minnesota, USA	Stream in St. Cox State Park in Pine county	19.7	6.2	49.9	Beer (1955)[f,g,h]
Cc	New Mexico, USA	Waterbodies where they were expected to be beneficial	12.9	9.7	19.3	Berghofer (1961)[c,h,i]
Cc	Massachusetts, USA	Streams and rivers	4.6	4.6	–	DeStefano et al. (2006)[c,f,j]
Cc	Massachusetts, USA	Quban Reservoir tributary streams	13.1	10.2	14.9	Hodgdon (1978)[a,g]
Cc	Montana, USA	River	9.8	5.6	31.6	Jackson (1990)[a,c]
Cc	Idaho-Wyoming state line, USA	River	8.4	10.9	11.0	Leege (1968)[f,g,h]
Cc	Illinois, USA	Reclaimed surface mine and river	3.1	4.6	20.9	McNew and Woolf (2005)[a,g]
Cc	New York, USA	Allegany National Park—varied	3.5	10.2	–	Sun et al. (2000)[f,g]
Cc	Montana, USA	River	7.7	7.7	–	Van Deelen and Pletscher (1996)[a,c]

Dashes indicate no data available.
[a] Data on natal dispersal.
[b] Unclear if distance is determined by following the watercourse or straight line distance.
[c] Distances travelled by water.
[d] Data on mean dispersal distance for a selected mate.
[e] Records data on migration. Migration rates of adults are low, always less than 4 km from the first capture place, whereas 2-year-old and 3-year-old beavers travelled further, with means of 18 and 13 km, respectively.
[f] Data on natal and secondary dispersal.
[g] Distances by air (straight line distance).
[h] Converted from miles to kilometres by multiplying by 1.609.
[i] Data on beavers that were relocated.
[j] Results do not differentiate between male and female beavers.

the pair then occupied the area with some of her descendants for 24 years (Richards, 1983).

If the dispersing beaver is looking for unoccupied territories and a mate at the same time, it can continue to move away from the core population (Hartman, 1995). Many areas that are passed by in the early colonizing phase can later become densely populated. This suggests that the beaver passes by these areas because they cannot find partners there, even though there are enough resources. Beavers that have access to partners establish territories in turn. Several attempts at reintroduction have partially or completely failed because the released beavers have wandered so far from their placement areas and therefore struggle to find partners (Hartman, 1995; Saveljev et al., 2002; see also Chapter 11).

6.10 Mate change, length of pair bonds, and loss of family members

Once paired, beavers tend to remain together until one partner either dies or is displaced by another individual of the same sex in territorial disputes (Bradt, 1938; Mayer et al., 2017b). If one member of a pair dies, the survivor will try to find a new partner as quickly as possible. If the survivor is alone for an extended period of time it will often choose the first individual of the opposite sex that enters the area. If there are some young animals at home, they may also succeed their mother or father and take over their role in the family group (Wilsson, 1971).

The average pair bond length for beavers living along rivers in Telemark, Norway, was 5.0 years (Mayer et al., 2017b) as compared to 1.9 years (Woodard, 1994), 2.5 years (Svendsen, 1989), 3.0 years (Busher, 1980), 3.6 years (Taylor, 1970), and 4.3 years (Sun, 2003) in North American beaver studies. Mayer et al. (2017b) speculated that the differences between populations are due to population densities or environmental factors. In fact, beavers living in more stable conditions, such as in lakes, stayed together longer than those that lived in streams or more unstable conditions (Svendsen, 1989). Another study in Telemark investigated 86 different beavers and found that 65 beavers had only one partner

during their lifetime (n=32 females, n=33 males), 18 beavers had two partners (n=11 females, n=7 males), and three males had more than three partners in their lifetime (Nimje, 2018).

In southern Ohio, USA, in 39% (9 of 23) of occasions when the pairs were separated it was because of the death of one of the partners, and in 61% (14 of 23) of these it was the female that died (Svendsen, 1989). Desertion of a partner was recorded in only 4% of the occasions. Age differences within the bonded pair when one animal died before the other are usually the main reason for a beaver to find a new partner. The mate change was not initiated by a member of the mated couple but rather by the intrusion of a new individual in a study in Telemark, Norway (Mayer et al., 2017b). This mate change occurred in the seventh year of a partnership and the remaining individual re-paired with a younger, incoming individual. The fate of the replaced individual was mostly unknown. Additionally, mate change was observed to be caused by the accidental loss of a partner. In pairs that experience a mate change, the partnership with the first mate lasted on average significantly longer than with the second mate (Mayer et al., 2017b). Human-caused mortality (hunting and car accidents) accounted for 20% of the mate changes in the population. Of the 62 pairs observed, 25 pairs (40%) ended in a mate change (note that these

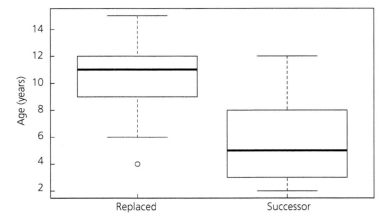

Figure 6.19 Box plot showing age of replaced and successor (n=38 individuals) Eurasian beavers in the year the mate change occurred in southeast Norway (1998–2014). Plots show median values, 25th and 75th percentile, range of data, and outlier (dot). (Reprinted by permission from Springer Nature Customer Service Centre GmbH: Springer Nature, Behavioral Ecology and Sociobiology, The 7-year itch: non-adaptive mate change in the Eurasian beaver, Mayer, M., Künzel, F., Zedrosser, A. and Rosell, F., © Springer-Verlag Berlin Heidelberg 2017.)

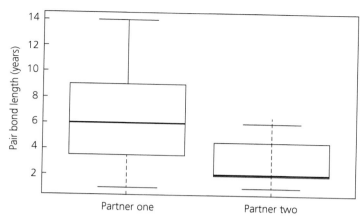

Figure 6.20 Box plot showing pair bond length with the first and second partner (*n*=19 pairs) of Eurasian beavers in southeast Norway (1998–2014). Plots show median values, 25th and 75th percentile, and range of data. (Reprinted by permission from Springer Nature Customer Service Centre GmbH: Springer Nature, Behavioral Ecology and Sociobiology, The 7-year itch: non-adaptive mate change in the Eurasian beaver, Mayer, M., Künzel, F., Zedrosser, A. and Rosell, F., © Springer-Verlag Berlin Heidelberg 2017.)

authors looked at pairs and not individual beavers), 19 pairs (30.6%) were still together at the end of the study, 11 cases (17.7%) had both members of the original pair being replaced by a new dominant pair (of these replaced pairs, seven pairs stayed together for life, i.e. had only one partner, and four had another partner previously), and seven cases (11.3%) were uncertain due to a lack of information on the end of the partnership or the new mate (Mayer et al., 2017b; Figures 6.19 and 6.20).

How do beavers react when family members die? After the death of his mate, a male beaver changed his behaviour throughout the winter (Wilsson, 1971). There have also been reports of a male and female which were kept together in captivity where they lived happily together until the female died. The male did not seem to notice that she was dead on the first day, and he even set food in front of her. After a while, and having received no response, he covered her body with branches and twigs. In Switzerland, a female that gave birth to a stillborn kit on the bank and away from the lodge bent over and examined the dead body very closely for 2 minutes. Thereafter, she dug a hole 2 m away (30 cm deep and 50 cm wide) for 20 minutes and put the stillborn kit into the hole and started to cover it with grass and branches. The reason for this behaviour was most likely to remove the dead body that will release strong odours and therefore could attract undesirable attention from predators or insects or even promote the spread of diseases because the kit could have died of an infection (Angst and Madlen, 2018).

6.11 Longevity

The average age at confirmed death (or disappearance) from a study population in Telemark, Norway, was 8.9 years for males and 9.4 for females (Nimje, 2018). In a reintroduced population in the Netherlands beavers lived an average of 14 years (Nolet and Baveco, 1996). In a sample of 2,400 North American beaver carcasses from New York state, USA, only eight beavers were older than 12.5–13 years (Brown, 1979), and of 108 Eurasian beaver carcasses from Telemark, Norway, only five individuals were older than 8 years (Rosell et al., 2010). However, some beavers in the wild have been reported to live up to 20–24 years (Larson, 1967; Brown, 1979; Gorbunova et al., 2008; see Chapter 8 for causes of death).

References

Aleksiuk, M. 1968. Scent-mound communication, territoriality, and population regulation in beaver (*Castor canadensis* Kuhl). *Journal of Mammalogy*, 49, 759–762.

Angst, C. and Madlen, P. 2018. Digging a hole: anti-preda-tor behaviour or funeral? *Proceedings of the 8th International Beaver Symposium*, Vemb, Denmark.

Bailey, V. 1927. Beaver habits and experiments in beaver culture. *USDA Technical Bulletins*, 21, 1–39.

Baker, B. W. and Hill, E. 2003. *Beaver (Castor canadensis)*, 2nd edn. Baltimore: Johns Hopkins University Press.

Beer, J. R. 1955. Movements of tagged beaver. *Journal of Wildlife Management*, 19, 492–493.

Bergerud, A. T. and Miller, D. R. 1977. Population dynam-ics of Newfoundland beaver. *Canadian Journal of Zoology*, 55, 1480–1492.

Berghofer, C. B. 1961. Movement of beaver. *Annual Con-ference, Western Association State Game and Fish Com-missioners*, pp. 181–184.

Bierla, J. B., Gizejewski, Z., Leigh, C. M., Ekwall, H., Söderquist, L., Rodriguez-Martinez, H., Zalewski, K., and Breed, W. G. 2007. Sperm morphology of the Eurasian beaver, *Castor fiber*: an example of a species of rodent with highly derived and pleiomorphic sperm populations. *Journal of Morphology*, 268, 683–689.

Bishir, J., Lancia, R. A., and Hodgdon, H. E. 1983. Beaver family organization: its implications for colony size. *Investigations on Beavers*, 4, 105–113.

Bloomquist, C. K. and Nielsen, C. K. 2009. A remote vide-ography system for monitoring beavers. *Journal of Wildlife Management*, 73, 605–608.

Bloomquist, C. K. and Nielsen, C. K. 2010. Demography of unexploited beavers in southern Illinois. *Journal of Wildlife Management*, 74, 228–235.

Bogacka, I., Paukszto, Ł., Jastrzębski, J. P., Czerwińska, J., Chojnowska, K., Kamińska, B., Kurzyńska, A., Smolińska, N., Giżejewski, Z., and Kamiński, T. 2017. Seasonal differences in the testicular transcriptome pro-file of free-living European beavers (*Castor fiber* L.) deter-mined by the RNA-Seq method. *PLoS ONE*, 12, e0180323.

Bond, C. F. 1956. Correlations between reproductive con-dition and skull characteristics of beaver. *Journal of Mammalogy*, 37, 506–512.

Bovet, J. and Oertli, E. F. 1974. Free-running circadian activity rhythms in free-living beaver (*Castor canaden-sis*). *Journal of Comparative Physiology*, 92, 1–10.

Boyce, M. S. 1974. Beaver population ecology in Interior Alaska. MSc thesis, University of Alaska.

Boyce, M. S. 1981. Beaver life-history responses to exploit-ation. *Journal of Applied Ecology*, 18, 749–753.

Bradt, G. W. 1938. A study of beaver colonies in Michigan. *Journal of Mammalogy*, 19, 139–162.

Bradt, G. W. 1939. Breeding habits of beaver. *Journal of Mammalogy*, 20, 486–489.

Brady, C. A. and Svendsen, G. E. 1981. Social-behaviour in a family of beaver, *Castor canadensis*. *Biology of Behaviour*, 6, 99–114.

Breed, W. G. 2004. The spermatozoon of Eurasian murine rodents: its morphological diversity and evolution. *Journal of Morphology*, 261, 52–69.

Brenner, F. J. 1964. Reproduction of the beaver in Crawford County, Pennsylvania. *Journal of Wildlife Management*, 28, 743–747.

Brooks, R. P., Fleming, M. W., and Kennelly, J. J. 1980. Beaver colony response to fertility control: evaluating a concept. *Journal of Wildlife Management*, 44, 568–575.

Brown, M. K. 1979. Two old beavers from Adirondacks. *New York Fish and Game Journal*, 26, 92.

Buech, R. R. 1995. Sex differences in behavior of beavers living in near-boreal lake habitat. *Canadian Journal of Zoology*, 73, 2133–2143.

Busher, P. E. 1975. Movements and activities of beavers, *Castor canadensis*, on Sagehen Creek, California. MSc thesis, San Francisco State University.

Busher, P. E. 1980. The population dynamics and behavior of beavers in the Sierra Nevada. PhD thesis, University of Nevada.

Busher, P. 1983. Interrelationships between behaviours in a beaver, *Castor canadensis*, population. *Journal of Tennessee Academic Science*, 58, 50–53.

Busher, P. 2007. Social organization and monogamy in the beaver. In: Wolff, J. O. and Sherman, P. W. (eds.) *Rodent Societies: An Ecological and Evolutionary Perspective.* Chicago: University of Chicago Press.

Busher, P. E. and Jenkins, S. H. 1985. Behavioral patterns of a beaver family in California. *Biology of Behaviour*, 10, 41–54.

Busher, P. E., Warner, R. J., and Jenkins, S. H. 1983. Population density, colony composition, and local movements in two Sierra Nevadan beaver populations. *Journal of Mammalogy*, 64, 314–318.

Cabré, L. B., Mayer, M., Steyaert, S., and Rosell, F. 2020. Beaver activity (*Castor fiber*) and spatial movement in response to light and weather conditions. *Mammalian Biology*, 100(3), 261–271.

Campbell, R. 2010. Demography and life history of the Eurasian beaver *Castor fiber*. PhD thesis, University of Oxford.

Campbell, R. D., Rosell, F., Nolet, B. A., and Dijkstra, V. A. A. 2005. Territory and group sizes in Eurasian bea-vers (*Castor fiber*): echoes of settlement and reproduc-tion? *Behavioral Ecology and Sociobiology*, 58, 597–607.

Campbell, R. D., Nouvellet, P., Newman, C., Macdonald, D. W., and Rosell, F. 2012. The influence of mean climate trends and climate variance on beaver survival and recruitment dynamics. *Global Change Biology*, 18, 2730–2742.

Campbell, R. D., Newman, C., Macdonald, D. W., and Rosell, F. 2013. Proximate weather patterns and spring green-up phenology effect Eurasian beaver (*Castor fiber*) body mass and reproductive success: the implications of climate change and topography. *Global Change Biology*, 19, 1311–1324.

Campbell-Palmer, R., Gow, D., Campbell, R., Dickinson, H., Girling, S., Gurnell, J., Halley, D., Jones, J., Lisle, S., Parker, H., Schwab, G., and Rosell, F. 2016. *The Eurasian Beaver Handbook: Ecology and Management of* Castor fiber. Exeter: Pelagic Publishing.

Campbell-Palmer, R., Senn, H., Girling, S., Pizzi, R., Elliott, M., Gaywood, M. and Rosell, F. 2020. Beaver genetic surveillance in Britain. *Global Ecology and Conservation*, 24, 1–10.

Campbell, R. D., Rosell, F., Newman, C., and Macdonald, D. W. 2017. Age-related changes in somatic condition and reproduction in the Eurasian beaver: resource history influences onset of reproductive senescence. *PLoS ONE*, 12, e0187484.

Crawford, J. C., Liu, Z., Nelson, T. A., Nielsen, C. K., and Bloomquist, C. K. 2008. Microsatellite analysis of mating and kinship in beavers (*Castor canadensis*). *Journal of Mammalogy*, 89, 575–581.

Crawford, J. C., Bluett, R. D., and Schauber, E. M. 2015. Conspecific aggression by beavers (*Castor canadensis*) in the Sangamon River Basin in central Illinois: correlates with habitat, age, sex and season. *American Midland Naturalist*, 173, 145–155.

Danilov, P. I. and Kan'shiev, V. Y. 1983. State of populations and ecological characteristics of European (*Castor fiber* L.) and Canadian (*Castor canadensis* Kuhl.) beavers in the northwestern USSR. *Acta Zoologica Fennica*, 174, 95–97.

DeStefano, S., Koenen, K. K., Henner, C., and Strules, J. 2006. Transition to independence by subadult beavers (*Castor canadensis*) in an unexploited, exponentially growing population. *Journal of Zoology*, 269, 434–441.

Dieter, C. D. 1992. Population characteristics of beavers in eastern South Dakota. *American Midland Naturalist*, 128, 191–196.

Doboszyńska, T. and Zurowski, W. 1975. Changes observed in the reproductive tract of a beaver female after high dosages of gonadotropic hormones. *Acta Theriologica*, 20, 105–111.

Doboszyńska, T. and Żurowski, W. 1983. Reproduction of the European beaver. *Acta Zoologica Fennica*, 174, 123–126.

Durka, W., Babik, W., Ducroz, J. F., Heidecke, D., Rosell, F., Samjaa, R., Saveljev, A., Stubbe, A., Ulevicius, A., and Stubbe, M. 2005. Mitochondrial phylogeography of the Eurasian beaver *Castor fiber* L. *Molecular Ecology*, 14, 3843–3856.

Dyck, A. P. and MacArthur, R. A. 1993. Seasonal variation in the microclimate and gas composition of beaver lodges in a boreal environment. *Journal of Mammalogy*, 74, 180–188.

Ellegren, H., Hartman, G., Johansson, M., and Andersson, L. 1993. Major histocompatibility complex monomorphism and low levels of DNA fingerprinting variability in a reintroduced and rapidly expanding population of beavers. *Proceedings of the National Academy of Sciences of the United States of America*, 90, 8150–8153.

Ermala, A., Helminen, M., and Lahti, S. 1989. Some aspects of the occurrence, abundance and future of the Finnish beaver population. *Suomen Riista*, 35, 108–118.

Fischer, J. W., Joos, R. E., Neubaum, M. A., Taylor, J. D., Bergman, D. L., Nolte, D. L., and Piaggio, A. J. 2010. Lactating North American beavers (*Castor canadensis*) sharing dens in the southwestern United States. *Southwestern Naturalist*, 55, 273–277.

Franke-Radowiecka, A., Giżejewski, Z., Klimczuk, M., Dudek, A., Zalecki, M., Jurczak, A., and Kaleczyc, J. 2016. Morphological and neuroanatomical study of the mammary gland in the immature and mature European beaver (*Castor fiber*). *Tissue and Cell*, 48, 552–557.

Frosch, C., Kraus, R. H. S., Angst, C., Allgöwer, R., Michaux, J., Teubner, J., and Nowak, C. 2014. The genetic legacy of multiple beaver reintroductions in central Europe. *PLoS ONE*, 9, e97619.

Fryxell, J. M. 2001. Habitat suitability and source–sink dynamics of beavers. *Journal of Animal Ecology*, 70, 310–316.

Goloduško, B. Z. and Fomenko, A. N. 1973. Jaktens inverkan på läget och dynamiken i bäverpopulationern. In: Den 5. allsovjetiska konferensen om Bäver. Svnesk översettning av Torsten Jansson. 3 pages.

Gorbunova, V., Bozzella, M. J., and Seluanov, A. 2008. Rodents for comparative aging studies: from mice to beavers. *Age*, 30, 111–119.

Graf, P. M., Hochreiter, J., Hackländer, K., Wilson, R. P., and Rosell, F. 2016. Short-term effects of tagging on activity and movement patterns of Eurasian beavers (*Castor fiber*). *European Journal of Wildlife Research*, 62, 725–736.

Graf, P. M., Wilson, R. P., Sanchez, L. C., Hackländer, K., and Rosell, F. 2017. Diving behavior in a free-living, semi-aquatic herbivore, the Eurasian beaver *Castor fiber*. *Ecology and Evolution*, 8, 997–1008.

Grevtsev, V. I. 1989. *Age and Sex Structure of Beaver* (Castor fiber *L.) Populations as Numbers Dynamics Factor*. Kirov: All-Union Research Institute of Game Management & Fur Farming, pp. 299–300.

Guenther, S. E. 1948. Young beavers. *Journal of Mammalogy*, 29, 419–420.

Gunson, J. R. 1967. Reproduction and productivity of beaver in Saskatchewan. *Proceedings of the 2nd Fur Workshop*, Quebec City.

Gunson, J. R. 1970. Dynamics of the beaver of Saskatchewan's northern forest. MSc thesis, University of Alberta.

Halley, D. J. 2011. Sourcing Eurasian beaver *Castor fiber* stock for reintroductions in Great Britain and Western Europe. *Mammal Review*, 41, 40–53.

Halley, D. and Rosell, F. 2002. The beaver's reconquest of Eurasia: status, population development and management of a conservation success. *Mammal Review*, 32, 153–178.

Halley, D., Teurlings, I., Welsh, H., and Taylor, C. 2013. Distribution and patterns of spread of recolonising Eurasian beavers (*Castor fiber* Linnaeus 1758) in fragmented habitat, Agdenes peninsula, Norway. *Fauna Norvegica*, 32, 1–12.

Hammond, M. C. 1943. Beaver on the Lower Souris Refuge. *Journal of Wildlife Management*, 7, 316–321.

Harper, W. R. 1968. Chemosterilant assessment for beaver. MSc thesis, Colorado State University.

Hartman, G. 1995. Patterns of spread of a reintroduced beaver *Castor fiber* population in Sweden. *Wildlife Biology*, 1, 97–104.

Hartman, G. 1997. Notes on age at dispersal of beaver (*Castor fiber*) in an expanding population. *Canadian Journal of Zoology*, 75, 959–962.

Havens, R. P. 2006. Beaver home ranges and movement patterns on the Embarras River watershed in east central Illinois. MSc thesis, Eastern Illinois University.

Havens, R. P., Crawford, J. C., and Nelson, T. A. 2013. Survival, home range, and colony reproduction of beavers in east-central Illinois, an agricultural landscape. *American Midland Naturalist*, 169, 17–30.

Hay, K. G. 1958. Beaver census methods in the Rocky Mountain Region. *Journal of Wildlife Management*, 22, 395–402.

Hediger, H. 1966. Rx: castoreum! *Animals*, 8:395-357.

Hediger, H. 1970. The breeding behavior of the Canadian beaver (*Castor canadensis*). *Forma et Functio*, 2, 336–351.

Heidecke, D. 1984. Untersuchungen zur Ökologie und Populationsentwicklung des Elbebibers (*Castor fiber albicus* Matschie, 1907). I: Biologische und populationsökologische Ergebnisse. *Zoologische Jahrbücher. Abteilung für Systematik, Ökologie und Geographie der Tiere*, 111, 1–41.

Hendry, G. 1966. *Kapuskasing Beaver Study*. Kapuskasing: Ontario Department of Lands and Forests.

Henry, D. B. and Bookhout, T. A. 1969. Productivity of beavers in northeastern Ohio. *Journal of Wildlife Management*, 33, 927–932.

Hill, E. P. 1982. Beaver (*Castor canadensis*). In: Chapman, J. A. and Feldhamer, G. A. (eds.) *Wild Mammals of North America—Biology, Management and Economics*. Baltimore: Johns Hopkins University Press.

Hinze, G. 1950. *Der Biber. Körperbau und Lebensweise Verbreitung und Geschichte*. Berlin: Akademie.

Hirai, N. 1975. Breeding of American beaver. *Animals and Zoos*, 27, 6–9.

Hodgdon, H. E. 1978. Social dynamics and behavior within an unexploited beaver (*Castor canadensis*) population. PhD thesis, University of Massachusetts.

Hodgdon, H. E. and Lancia, R. A. 1983. Behaviour of the North American beaver, *Castor canadensis*. *Acta Zoologica Fennica*, 174, 99–103.

Hodgdon, H. E. and Larson, J. S. 1973. Some sexual differences in behaviour within a colony of marked beavers (*Castor canadensis*). *Animal Behaviour*, 21, 147–152.

Hodgdon, K. W. 1949. Productivity data from placental scars in beavers. *Journal of Wildlife Management*, 13, 412–414.

Hohwieler, K., Rosell, F., and Mayer, M. 2018. Scent-marking behavior by subordinate Eurasian beavers. *Ethology*, 124, 591–599.

Huey, W. S. 1956. New Mexico beaver management. *New Mexico Department of Game and Fish Bulletin* 4, 1–49.

Jackson, M. D. 1990. Beaver dispersal in western Montana. MSc thesis, University of Montana.

Jenkins, S. H. and Busher, P. E. 1979. *Castor canadensis* Kuhl, 1820. *Mammalian Species*, 120, 1–8.

Kalas, K. 1976. Beobachtungen bie der Hanaufzucht eines kanadischen Bibers (*Castor canadensis* Kuhl, 1820). *Säugetierkundliche Informationen*, 24, 304–315.

Kenagy, G. J. and Trombulak, S. C. 1986. Size and function of mammalian testes in relation to body size. *Journal of Mammalogy*, 67, 1–22.

Kleiman, D. G. 1977. Monogamy in mammals. *Quarterly Review of Biology*, 52, 39–69.

Lahti, S. and Helminen, M. 1974. The beaver *Castor fiber* (L.) and *Castor canadensis* (Kuhl) in Finland. *Acta Theriologica*, 19, 177–189.

Lancia, R. A. 1979. Year-long activity patterns of radio-marked beaver (*Castor canadensis*). PhD thesis, University of Massachusetts.

Lancia, R. A. and Hodgdon, H. E. 1982. Observations on the ontogeny of behavior of hand-reared beavers (*Castor canadensis*). *Acta Zoologica Fennica*, 174, 117–119.

Lancia, R. A., Dodge, W. E., and Larson, J. S. 1982. Winter activity patterns of two radio-marked beaver colonies. *Journal of Mammalogy*, 63, 598–606.

Larson, J. S. 1967. Age structure and sexual maturity within a western Maryland beaver (*Castor canadensis*) population. *Journal of Mammalogy*, 48, 408–413.

Leege, T. A. 1968. Natural movements of beavers in southeastern Idaho. *Journal of Wildlife Management*, 32, 973–976.

Longley, W. and Moyle, J. B. 1963. *The Beaver in Minnesota.* St. Paul: Minnesota Department of Conservation, Division of Game and Fish, Section of Research and Planning.

Mayer, M., Robstad, C.A., Pi Serra, E., Hohwieler, K., Fuchs, B., Evans, A.L., Arnemo, J.M., Zedrosser, A. & Rosell, F. 2017a. Exposure of the Eurasian beaver (*Castor fiber*) towards hunters during the spring hunt. Rapport. Høgskolen i Sørøst-Norge. 50 sider.

Mayer, M., Künzel, F., Zedrosser, A., and Rosell, F. 2017a. The 7-year itch: non-adaptive mate change in the Eurasian beaver. *Behavioral Ecology and Sociobiology*, 71, 1–9.

Mayer, M., Robstad, C., Serra, E. P., Hohwieler, K., Fuchs, B., Evans, A. L., Arnemo, J. M., Zedrosser, A., and Rosell, F. 2017b. *Exposure of the Eurasian Beaver (Castor fiber) Towards Hunters During the Spring Hunt*. Bø i Telemark: University College of Southeast Norway.

Mayer, M., Zedrosser, A., and Rosell, F. 2017c. Couch potatoes do better: delayed dispersal and territory size affect the duration of territory occupancy in a monogamous mammal. *Ecology and Evolution*, 7, 4347–4356.

Mayer, M., Zedrosser, A., and Rosell, F. 2017d. Extra-territorial movements differ between territory holders and subordinates in a large, monogamous rodent. *Scientific Reports*, 7, 15261.

Mayer, M., Zedrosser, A., and Rosell, F. 2017e. When to leave: the timing of natal dispersal in a large, monogamous rodent, the Eurasian beaver. *Animal Behaviour*, 123, 375–382.

Mayer, M., Frank, S., Zedrosser, A., and Rosell, F. 2020. Causes and consequences of inverse density-dependent territorial behavior and aggression in a monogamous mammal. *Journal of Animal Ecology*, 89, 577–588.

Mcclanahan, K., Rosell, F. and Mayer, M. 2020. Minding your own business: low pair cohesion in a territorial, monogamous mammal. *Animal Behaviour*, 166: 1–29.

McNew, J., Lance B., and Woolf, A. 2005. Dispersal and survival of juvenile beavers (*Castor canadensis*) in southern Illinois. *American Midland Naturalist*, 154, 217–228.

McTaggart, S. T. and Nelson, T. A. 2003. Composition and demographics of beaver (*Castor canadensis*) colonies in central Illinois. *American Midland Naturalist*, 150, 139–150.

Miller, F. 1948. Early breeding of the Texas beaver. *Journal of Mammalogy*, 29, 419.

Molini, J. J., Lancia, R. A., Bishir, J., and Hodgdon, H. E. 1980. A stochastic model of beaver population growth. *Worldwide Furbearers Conference Proceedings*, Frostburg, pp. 1215–1245.

Morgan, L. H. 1868. *The American Beaver and his Works.* Philadelphia: J.B. Lippincott and Company.

Mossing, A. 2005. Kolonitetthet, landskapspåvirkning og vinterfurasjeringsstferd hos Eurasisk bever (*Castor fiber*)

i et subalpint habitat på Hardangervidda. MSc thesis, Høgskolen i Telemark.

Mott, C. L., Bloomquist, C. K., and Nielsen, C. K. 2011. Seasonal, diel, and ontogenetic patterns of within-den behavior in beavers (*Castor canadensis*). *Mammalian Biology*, 76, 436–444.

Müller-Schwarze, D. 2011. *The Beaver: Its Life and Impact.* New York: Cornell University Press.

Müller-Schwarze, D. and Herr, J. 2003. Are they 'listening in the dark'? Behaviour of beavers in their lodge. *Lutra*, 46, 255–257.

Nimje, P. S. 2018. The effect of social organization on genetic estimates of fitness in Eurasian beaver *Castor fiber*. PhD thesis, University of South-Eastern Norway.

Nolet, B. 1992. Reintroductionof beaver in the Rhine and Meuse estuary. *Semiaquatiche Saugetiere*, 130.

Nolet, B. A. and Baveco, J. M. 1996. Development and viability of a translocated beaver *Castor fiber* population in the Netherlands. *Biological Conservation*, 75, 125–137.

Nolet, B. A. and Rosell, F. 1994. Territoriality and time budgets in beavers during sequential settlement. *Canadian Journal of Zoology*, 72, 1227–1237.

Nolet, B. A., Broftová, L., Heitkönig, I. M., Vorel, A., and Kostkan, V. 2005. Slow growth of a translocated beaver population partly due to a climatic shift in food quality. *Oikos*, 111, 632–640.

Nordstrom, W. R. 1972. Comparison of trapped and untrapped beaver populations in New Brunswick. MSc thesis, University of New Brunswick.

Novak, M. 1987. Beaver. In: Novak, M., Baker, J. A., Obbard, M. E., and Mallock, B. (eds.) *Wild Furbearer Management and Conservation in North America.* Toronto: Queens Printer for Ontario, pp. 283–304.

Novakowski, N. S. 1965. Population dynamics of a beaver population in northern latitudes. PhD thesis, University of Saskatchewan.

Olstad, O. 1937. Beverens (*Castor fiber*) utbredelse i Norge. *Statens Viltundersøkelser. Nytt Magasin for Naturvidenskapen*, 77, 217–273.

Osborn, D. J. 1953. Age classes, reproduction, and sex ratios of Wyoming beaver. *Journal of Mammalogy*, 34, 27–44.

Owesen, A. 1979. *I beverskog.* Oslo: Gylendal Norsk Forlag.

Parker, H. and Rosell, F. 2001. Parturition dates for Eurasian beaver *Castor fiber*: when should spring hunting cease? *Wildlife Biology*, 7, 237–241.

Parker, H. and Rosell, F. 2014. Rapid rebound in colony number of an over-hunted population of Eurasian beaver *Castor fiber*. *Wildlife Biology*, 20, 267–269.

Parker, H., Rosell, F., Hermansen, T., Sørløkk, G., and Stærk, M. 2002. Sex and age composition of spring-hunted Eurasian beaver in Norway. *Journal of Wildlife Management*, 66, 1164–1170.

Parker, H., Rosell, F., and Mysterud, A. 2007. Harvesting of males delays female breeding in a socially monogamous mammal; the beaver. *Biology Letters*, 3, 107–109.

Parker, H., Zedrosser, A., and Rosell, F. 2017. Age-specific reproduction in relation to body size and condition in female Eurasian beavers. *Journal of Zoology*, 302, 236–243.

Patenaude, F. 1983. Care of the young in a family of wild beavers, *Castor canadensis*. *Acta Zoologica Fennica*, 174, 121–122.

Patenaude, F. 1984. The ontogeny of behavior of free-living beavers (*Castor canadensis*). *Zeitschrift Für Tierpsychologie*, 66, 33–44.

Patenaude, F. and Bovet, J. 1983. Parturition and related behavior in wild American beavers (*Castor canadensis*). *International Journal of Mammalian Biology*, 48, 136–145.

Patenaude, F. and Bovet, J. 1984. Self-grooming and social grooming in the North American beaver, *Castor canadensis*. *Canadian Journal of Zoology*, 62, 1872–1878.

Payne, N. F. 1982. Colony size, age, and sex structure of Newfoundland beaver. *Journal of Wildlife Management*, 46, 655–661.

Payne, N. F. 1984a. Mortality rates of beaver in Newfoundland. *Journal of Wildlife Management*, 48, 117–126.

Payne, N. F. 1984b. Reproductive rates of beaver in Newfoundland. *Journal of Wildlife Management*, 48, 912–917.

Pearson, A. M. 1960. A study of the growth and reproduction of the beaver (*Castor canadensis* Kuhl) correlated with the quality and quantity of some habitat factors. MSc thesis, University of British Columbia.

Petrosyan, V. G., Golubkov, V. V., Zavyalov, N. A., Khlyap, L. A., Dergunova, N. N., and Osipov, F. Peterson, R. P. and Payne, N. F. 1986. Productivity, size, age, and sex structure of nuisance beaver colonies in Wisconsin. *Journal of Wildlife Management*, 50, 265–268.

A. 2019. Modelling of competitive interactions between native Eurasian (*Castor fiber*) and alien North American (*Castor Canadensis*) beavers based on long-term monitoring data (1934–2015). *Ecological Modelling*, 409, 1–15.

Pietrek, A. G., Escobar, J. M., Fasola, L., Roesler, I., and Schiavini, A. 2017. Why invasive Patagonian beavers thrive in unlikely habitats: a demographic perspective. *Journal of Mammalogy*, 98, 283–292.

Potvin, C. L. and Bovet, J. 1975. Annual cycle of patterns of activity rhythms in beaver colonies (*Castor canadensis*). *Journal of Comparative Physiology*, 98, 243–256.

Richards, D. 1983. *Beaversprite: My Years Building an Animal Sanctuary*, Interlaken, NY: Heart of the Lakes Publishing.

Rosell, F. and Parker, H. 1995. *Beaver Management: Present Practice and Norway's Future Needs*. Bø i Telemark: Telemark University College.

Rosell, F. and Sanda, J. 2006. Potential risks of olfactory signaling: the effect of predators on scent marking by beavers. *Behavioral Ecology*, 17, 897–904.

Rosell, F. and Thomsen, L. R. 2006. Sexual dimorphism in territorial scent marking by adult Eurasian beavers (*Castor fiber*). *Journal of Chemical Ecology*, 32, 1301–1315.

Rosell, F., Parker, H., and Steifetten, Ø. 2006. Use of dawn and dusk sight observations to determine colony size and family composition in Eurasian beaver *Castor fiber*. *Acta Theriologica*, 51, 107–112.

Rosell, F., Zedrosser, A., and Parker, H. 2010. Correlates of body measurements and age in Eurasian beaver from Norway. *European Journal of Wildlife Research*, 56, 43–48.

Rosell, F., Campbell-Palmer, R., and Parker, H. 2012. More genetic data are needed before populations are mixed: response to 'Sourcing Eurasian beaver *Castor fiber* stock for reintroductions in Great Britain and Western Europe'. *Mammal Review*, 42, 319–324.

Rusanov, N. and Maksimenko, S. 2019. Features of population and distribution dynamics of the Eurasian beaver (*Castor fiber* L) in the Kursk region. *IOP Conference Series: Earth and Environmental Science*, 2019, 1–8.

Rutherford, W. H. 1964. *The Beaver in Colorado: Its Biology, Ecology, Management and Economics*. Colorado Game, Fish and Parks Technical Publication. Denver: Game Research Division, Colorado Game, Fish and Parks Department.

Ruusila, V., Ermala, A., and Hyvärinen, H. 2000. Costs of reproduction in introduced female Canadian beavers (*Castor canadensis*). *Journal of Zoology*, 252, 79–82.

Samson, C. and Raymond, M. 1995. Daily activity pattern and time budget of stoats (*Mustela erminea*) during summer in southern Québec. *Mammalia*, 59, 501–510.

Sanderson, G. C. 1953. Recent status of the beaver in Iowa. *Proceedings of the Iowa Academy of Science*, 1953, pp. 746–753.

Saveljev, A. and Milishnikov, A. 2002. Biological and genetic peculiarities of cross-composed and aboriginal beaver populations in Russia. *Acta Zoologica Lituanica*, 12, 397–402.

Saveljev, A. P., Stubbe, M., Stubbe, A., Unzhakov, V. V., and Kononov, S. V. 2002. Natural movements of tagged beavers in Tyva. *Russian Journal of Ecology*, 33, 434–439.

Saveljev, A. P., Batbayar, N., Boldbaatar, S., and Dashbiamba, B. 2016. Self-eating in beavers—trophic opportunism or reaction on stress? Extreme case from Mongolia. *Russian Journal of Theriology*, 15(1), 6815(1

Semyonoff, B. T. 1951. The river beaver in Archangel province. *Translation Russian Game Representatives*, 1, 5–45.

Semyonoff, B. T. 1957. Beaver biology in winter in Archangel province. *Translation Russian Game Representatives*, 1, 71–92.

Senn, H., Ogden, R., Frosch, C., Syrůčková, A., Campbell-Palmer, R., Munclinger, P., Durka, W., Kraus, R. H. S., Saveljev, A. P., Nowak, C., Stubbe, A., Stubbe, M., Michaux, J., Lavrov, V., Samiya, R., Ulevicius, A., and Rosell, F. 2014. Nuclear and mitochondrial genetic structure in the Eurasian beaver (*Castor fiber*)—implications for future reintroductions. *Evolutionary Applications*, 7, 645–662.

Shadle, A. R. 1930. An unusual case of parturition in a beaver. *Journal of Mammalogy*, 11, 483–485.

Sharpe, F. and Rosell, F. 2003. Time budgets and sex differences in the Eurasian beaver. *Animal Behaviour*, 66, 1059–1067.

Smith, D. W. and Jenkins, S. H. 1997. Seasonal change in body mass and size of tail of northern beavers. *Journal of Mammalogy*, 78, 869–876.

Smith, D. W. and Peterson, R. O. 1991. Behavior of beaver in lakes with varying water levels in northern Minnesota. *Environmental Management*, 15, 395.

Stephenson, A. B. 1969. Temperatures within a beaver lodge in winter. *Journal of Mammalogy*, 50, 134–136.

Steyaert, S., Zedrosser, A., and Rosell, F. 2015. Socio-ecological features other than sex affect habitat selection in the socially obligate monogamous Eurasian beaver. *Oecologia*, 179, 1023–1032.

Sun, L. 2003. Monogamy correlates, socioecological factors, and mating systems in beavers. In: Reichard, U. H. and Boesch, C. (eds.) *Monogamy: Mating Strategies and Partnerships in Birds, Humans and Other Mammals*. Cambridge: Cambridge University Press.

Sun, L., Müller-Schwarze, D., and Schulte, B. A. 2000. Dispersal pattern and effective population size of the beaver. *Canadian Journal of Zoology*, 78, 393–398.

Svendsen, G. E. 1980a. Population parameters and colony composition of beaver (*Castor canadensis*) in southeast Ohio. *American Midland Naturalist*, 104, 47–56.

Svendsen, G. E. 1980b. Seasonal change in feeding patterns of beaver in southeastern Ohio. *Journal of Wildlife Management*, 44, 285–289.

Svendsen, G. E. 1989. Pair formation, duration of pair-bonds, and mate replacement in a population of beavers (*Castor canadensis*). *Canadian Journal of Zoology*, 67, 336–340.

Swinnen, K. R., Hughes, N. K., and Leirs, H. 2015. Beaver (*Castor fiber*) activity patterns in a predator-free landscape. What is keeping them in the dark? *Mammalian Biology*, 80, 477–483.

Syrůčková, A., Saveljev, A. P., Frosch, C., Durka, W., Savelyev, A. A., and Munclinger, P. 2015. Genetic relationships within colonies suggest genetic monogamy in the Eurasian beaver (*Castor fiber*). *Mammal Research*, 60, 139–147.

Taylor, D. E. M. 1970. Growth, decline, and equilibrium in a beaver population at Sagehen Creek, California. PhD thesis, University of California.

Tevis, L. 1950. Summer behavior of a family of beavers in New York State. *Journal of Mammalogy*, 31, 40–65.

Thomas, G. 1943. Delayed parturition in beaver. *Wyoming Wildlife*, 8, 17–18.

Tinnesand, H. V., Jojola, S., Zedrosser, A., and Rosell, F. 2013. The smell of desperadoes? Beavers distinguish between dominant and subordinate intruders. *Behavioral Ecology and Sociobiology*, 67, 895–904.

Tyurnin, B. N. 1984. Factors determining numbers of the river beaver (*Castor fiber*) in the European North. *Soviet Journal of Ecology*, 14, 337–344.

Van Deelen, T. R. 1991. Dispersal patterns of juvenile beavers in western Montana. MSc thesis, University of Montana.

Van Deelen, T. and Pletscher, D. H. 1996. Dispersal characteristics of two-year-old beavers, *Castor canadensis*, in western Montana. *Canadian Field-Naturalist*, 110, 318–321.

Weaver, K. M. 1986. Dispersal patterns of subadult beavers in Mississippi as determined by implant radio-telemetry. Thesis, Mississippi State University.

Wheatley, M. 1993. Report of two pregnant beavers, *Castor canadensis*, at one beaver lodge. *Canadian Field-Naturalist*, 107, 103.

Wigley, T. B., Roberts, T. H., and Arner, D. H. 1983. Reproductive characteristics of beaver in Mississippi. *Journal of Wildlife Management*, 47, 1172–1177.

Wilsson, L. 1971. Observations and experiments on the ethology of the European beaver (*Castor fiber* L.). *Viltrevy*, 8, 160–203.

Wilsson, L. 1995. Bäver. Hundskolan i Sollefteå AB, Sollefteå Tryckeri AB,189 s.

Woodard, E. L. 1994. Behaviour, activity patterns and foraging strategies of beaver (*Castor canadensis*) on Sagehen Creek, California. PhD thesis, University of California.

Zharkov, I. V. 1973. Strukturen i et bäversamhälle. *Från 5. allsovjetiska konferensen om bävrar. Oversatt til svensk av Fil. Mag. Torsten Jansson*, 69–72.

Żurowski, W. 1977. Rozmnazanie sie bobrow europejskich w warunkach fermowych. *Rozprawy habilitacyjne. Polska Akademia Nauk. Instytut Genetyki i Hodowli Zwierząt*, 7, 31–42.

Żurowski, W. 1989. Dam building activity of beavers on the mountainous streams. *Abstracts: Fifth International Theriological Congress*, Rome, 1, pp. 316–317.

Żurowski, W. and Kasperczyk, B. 1986. Characteristics of a European beaver population in the Suwałki Lakeland. *Acta Theriologica*, 31, 311–325.

Żurowski, W., Kisza, J., Kruk, A., and Roskosz, A. 1974. Lactation and chemical composition of milk of the European beaver (*Castor fiber* L.). *Journal of Mammalogy*, 55, 847–850.

CHAPTER 7

Territoriality, communication, and populations

7.1 Territory establishment and size, and factors affecting territory size

Beavers (*Castor* spp.) are highly territorial and actively defend a territory that includes nutritional resources, one or more rest sites, a family over-wintering site, and a reproduction site. Beavers typically occupy a 'multipurpose/breeding territory' (Rosell and Bjørkøyli, 2002). Each family group occupies a defined, defended, and demarcated area. The specific defended area is termed a territory, while the entire area that beavers use is the home range. Home ranges tend to follow the irregular shoreline patterns of streams, rivers, pond, and lakes. Small ponds may contain only a family, but in other habitats the home ranges are typically larger and more linear (Novak, 1987; Wheatley, 1997c).

Usually, how beavers use the territory size changes throughout the season, i.e. the family has and maintains the territory but individuals may have different use patterns—home ranges—within the family territory. For example, in a study of pair cohesion in Telemark, Norway, the individual territory size ranged from 1.1 to 6.8 km bank length (average 3.7 ± 1.7 km) (McClanahan et al., 2020). However, we will use the term territory throughout this chapter, since for beavers, home range and territory are not too different (Müller-Schwarze, 2011). The overlap between neighbouring territories is usually minimal (0.5–2.2% overlap) (Bradt, 1938; Herr and Rosell, 2004), but up to 10% overlap has been recorded (Bloomquist et al., 2012). Overlap in adjacent territories is influenced by population density, habitat type, and resource availability. Since beavers rarely travel more than 40–60 m from

water (Barnes and Dibble, 1988; Donkor and Fryxell, 1999; Haarberg and Rosell, 2006), and most activity is within 20–30 m of the shoreline, they occupy landscapes along banksides in a linear fashion. The advantage of having a defended territory is that the animals are able to become familiar with their surroundings, making it easier for them to avoid and hide from predators. It also reduces competition for food resources and mates between families. In an evolutionary sense, beavers evolved territorial behaviour to increase reproductive success in a hostile, cold, and ice-covered environment by mate-guarding. Being close to the female when she went into oestrus was critical (Busher, 2007).

7.1.1 Establishment

New territories are normally established during early autumn, but this can vary (Svendsen, 1989), and they are usually established when dispersing beavers settle down in a new area of their own. These dispersing beavers are normally 2–3 years old (see Section 6.9). However, age of dispersal and establishment of a new territory has been reported to vary based on overall population density and habitat quality. For example, in a dense beaver population at carrying capacity in Telemark, Norway, new territories were acquired by dispersing beavers at a mean age of 5 years (Campbell et al., 2017). Older Eurasian beavers are reported to be more successful at acquiring a new territory than younger beavers due to increasing competitive ability and possibly experience (Mayer et al., 2017c). If the population is dense, beavers with a comparatively shorter dispersal distance have a higher chance of acquiring a territory

Beavers: Ecology, Behaviour, Conservation, and Management. Frank Rosell and Róisín Campbell-Palmer, Oxford University Press.
© Frank Rosell and Róisín Campbell-Palmer 2022. DOI: 10.1093/oso/9780198835042.003.0007

(Nimje, 2018) (see Figure 6.5a). Occasionally, beavers inherit a part of their parents' territory and convert it into their own (Müller-Schwarze, 2011).

7.1.2 Territory size

The territorial size is generally measured as the length of shoreline beavers use, but it can also be expressed as a conglomerate of terrestrial and aquatic patches used by family members in a two-dimensional space (Wheatley, 1997a; Bloomquist et al., 2012). The average size of a beaver territory is approximately 3 km of shoreline, but the size can vary greatly from 0.2 to 21.2 km of shore length (Busher, 1975, Busher; 1980; Campbell et al., 2005; Mayer et al., 2020b; Figure 7.1) depending on specific habitat type (Table 7.1). Reported territory size of beavers in Alabama, USA, was 11.9 ha (McClintic et al., 2014a) and 25.5 ha in southern Illinois but can range from 0.4 to 105.4 ha (Gillespie, 1976; Bloomquist et al., 2012).

The size of a beaver territory depends on many factors, with some of the most important being their settlement pattern, habitat quality and quantity, habitat type (stream, larger river, lake, or pond), density of beaver population, social factors (sex and age), and time of year (season) (Campbell et al., 2005). However, it is important to be aware that also the methodology (for example how territory size is estimated) used might influence the reported territory size (Campbell et al., 2005; Havens et al., 2013; Korbelová et al., 2016). Interestingly, the number of family members in a group was found to have no effect on territory size in Biesbosch in the Netherlands, Telemark in Norway (Campbell et al., 2005), or Newfoundland in Canada (Bergerud and Miller, 1977). However, Wheatley (1997b), in Canada, found that territories were larger for non-family beavers (solitary or couple) than for those living in a family during summer. It is important to remember that the territory size is multifactorial and that local environmental conditions determine territory size. It makes sense energetically that territory size is relatively small, since beavers invest in their territory and defend it. Energetic costs to defend a large territory would be prohibitive (Mayer et al., 2017b; see also Section 7.3).

7.1.3 Factors affecting territory size

7.1.3.1 Settlement pattern

It has been shown that Eurasian beavers usually first settle down in the most suitable habitats, then utilize the poorer-quality sites, before being 'floaters' without their own territories. In the Netherlands, where beavers were introduced into unoccupied habitat, it was found that animals released initially still had the largest and best territories (plentiful food supplies) after 5 years, while those that were released later had smaller, poorer territories (little food) (Nolet and Rosell, 1994; Campbell et al., 2005; Figure 7.2). This pattern of initial colonizing beavers utilizing the highest-quality habitats suggests that the beavers' territorial behaviour can reduce population densities in local habitats. The sequence of arrival of pairs in unoccupied areas seems to play an important long-term role in determining the size of the territory (Campbell et al., 2005).

7.1.3.2 Habitat quality and quantity

Resource availability appears to be the main determinant of territory size in both beaver species. In northern Alabama, USA, territory size was positively related to the diversity of landcover within the territory (McClintic et al., 2014a). Territory size also decreased with increasing temporal variability in green plant material. North American beavers (*Castor canadensis*) increased territory size in relation to the relative amount of woody plants to include more green plant material within their territories. Therefore, as a central-place forager, beavers most likely increase territory size to increase the availability of woody plants for food consumption and avoid food resource depletion. For instance, along a stream beavers may move their activity centre depending on resource use (see Section 4.1.2 for more information about 'crop rotation').

Resource availability also appears to be the main determinant of territory size of Eurasian beavers. Territory size and proportion of deciduous species in the habitat were positively correlated in both Biesbosch in the Netherlands and Telemark in Norway but was negatively correlated in the River Loire region in France (Fustec et al., 2001). Larger

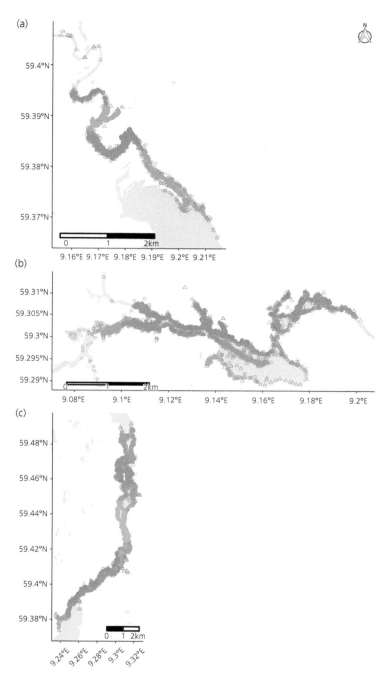

Figure 7.1 Territory sizes of Eurasian beavers across three connected rivers—Gvarv (a), Straumen (b), and Saua (c)—in Telemark, southeastern Norway, calculated using GPS data. Different colours are used for different territories, and triangles for males and circles for females. (Figure provided courtesy of Rasmus Mohr Mortensen.)

Table 7.1 Territory/home range sizes of male and female beavers of different ages across varying habitats (*Cf* = Eurasian beaver [*Castor fiber*]; *Cc* = North American beaver [*Castor canadensis*]). Latitude is given as degrees, minutes, seconds (DMS). For family densities see Table 7.3.

Species	State/country and latitude (DMS)	Habitat type	Age group	Sex	Relative patrolling time (%)	Territory/ home range	Territory/home range size (km or ha)	Season	References
Cc	Alabama, USA 34° 41′ 3.00′′ N	Varied	Yearling/ subadult/adult	Male/ female	–	Home range	20.89 ha (11.86 when 3 beavers with unusually large home ranges are excluded)	Taken from data collected over a year	McClintic et al. (2014b)
Cc	Illinois, USA 37°16′20.16′′N	Varied	Juvenile/ subadult/adult	Male/ female	– – – – – –	Home range	26.80 ha 28.90 ha 25.10 ha 35.30 ha 27.50 ha 11.30 ha	Spring 2005 Summer 2005 Autumn 2005 Winter 2006 Spring 2006 Summer 2006	Bloomquist et al. (2012)
Cc	Illinois, USA 39° 27′ 35.03′′ N	River	Subadult/adult	Male/ female	– – – –	Home range	2.10 km 2.20 km 4.00 km 1.50 km	Autumn Winter Spring Summer	Havens et al. (2013)
		Stream			–		1.80 km	Over 2 years	
		River			–		3.60 km	Over 2 years	
Cf	Czech Republic 49° 04′ 48.83′′ N (South Moravia)	Streams	Subadult/adult	Male/ female	–	Home range	3.96 ha 2.14 ha 6.00 ha	Autumn Winter Spring	Korbelová et al. (2016)[a]
	49° 05′ 11.00′′ N (North Bohemia)	Rivers			–		6.65 ha 4.15 ha –	Autumn Winter Spring	
	49° 59′ 53.59′′ N (West Bohemia)	Varied			–		4.79 ha 4.66 ha 7.45 ha	Autumn Winter Spring	
Cc	Winnipeg, Manitoba 51° 02′ 40.00′′ N	Varied	All age ranges	Male/ female	– – –	Home range	10.34 ha 3.07 ha 0.25 ha	Summer Autumn Winter	Wheatley (1997a)

	Location	Habitat	Age/category	Sex		Value	Measure	Season	Reference
Cc	Winnipeg, Manitoba 51° 02' 40.00'' N	Varied	All age ranges (family)	Male/female	–	7.74 ha		Summer	Wheatley (1997b)
			All age ranges (non-family)	Male/female	–	17.56 ha			
			Adult (family)	Male	–	10.89 ha			
			Adult (family)	female	–	4.39 ha			
			Yearling	Male	–	8.25 ha			
			Yearling	Female	–	7.69 ha			
			Juvenile	Female	–	7.25 ha			
			Unspecified (non-family)	Male	–	13.31 ha			
			Unspecified (non-family)	Female	–	20.95 ha			
			All age ranges (family)	Male/female	–	2.87 ha		Autumn	
			All age ranges (non-family)	Male/female	–	4.25 ha			
			Adult (family)	Male	–	3.96 ha			
			Adult (family)	female	–	1.90 ha			
			Yearling	Male	–	3.13 ha			
			Yearling	Female	–	2.25 ha			
			Juvenile	Female	–	5.25 ha			
			Kits	–	–	1.42 ha			
			Unspecified (non-family)	Male	–	2.75 ha			
			Unspecified (non-family)	Female	–	4.75 ha			
Cc	Winnipeg, Manitoba 51° 02' 40.00'' N	Pond	All beavers	–	–	6.54 ha	Home range	Summer	Wheatley (1997c)
		Lake	Family only	–	–	6.19 ha			
		River	All beavers	–	–	9.77 ha			
			Family only	–	–	8.78 ha			
			All beavers	–	–	16.94 ha			
			Family only	–	–	10.01 ha			
		Pond	All beavers	–	–	2.69 ha		Autumn	
		Lake	Family only	–	–	2.69 ha			
		River	All beavers	–	–	2.65 ha			
			Family only	–	–	2.46 ha			
			All beavers	–	–	4.50 ha			
			Family only	–	–	5.00 ha			

continued

Table 7.1 Continued.

Species	State/country and latitude (DMS)	Habitat type	Age group	Sex	Relative patrolling time (%)	Territory/ home range	Territory/home range size (km or ha)	Season	References
Cf	Biesbosch, Netherlands 52°7'57.48"N	Freshwater estuary	Over 1 year old	Male/ female	45.60[b] 56.28[b] 45.24[b]	Territory	7.53 km 16.88 km 16.68 km	Winter Late spring Summer	Nolet and Rosell (1994)[c,d]
Cf	Biesbosch, Netherlands 52°7'57.48"N	Freshwater estuary	Subadult/adult	Male/ female	–	Territory	12.80 km	Spring and summer	Campbell et al. (2005)
	Telemark, Norway 59°23'16.35"N (Gvarv) 59°24'29.48"N (Saua)	Rivers			–		4.00 km		
Cf	Telemark, Norway 59°23'16.35"N (Gvarv) 59°24'29.48"N (Saua)	Rivers	Adult	Male	28.70 (own border) 22.80 (joint border) 51.50 (total)	Territory	3.90 km (during entire study) 4.12 km (after parturition)	Spring and summer	Herr and Rosell (2004)
				Female	27.60 (own border) 19.60 (joint border) 47.20 (total)		3.48 km (during entire study) 3.36 km (after parturition)		
Cf	Telemark, Norway 59°23'16.35"N (Gvarv) 59°24'29.48"N (Saua) 63° 52' 15.10'' N (Straumen)	Rivers	Adult	Male/ female Male Female	13.75[e] 13.93[e,f] 13.57[e,f]	Territory	4.07 km 4.07 km[f] 3.96 km[f]	Spring	Graf et al. (2016)
				Male/ female Male Female	13.40[e] 17.45[e,f] 9.35[e,f]		3.07 km 4.83 km[f] 4.52 km[f]	Autumn	

Dashes indicate no data available.

[a] Sex and age had no effect on results so were omitted.

[b] Values given in hours per day swimming and converted to percentage from mean principal activity period for each season.

[c] Did not differentiate between male and female beavers or different age classes.

[d] Values calculated as mean for each season across a 4-year period (all banks).

[e] Values given are absolute patrolling.

[f] Values calculated from territories where data are recorded for both sexes to enable comparisons between sexes.

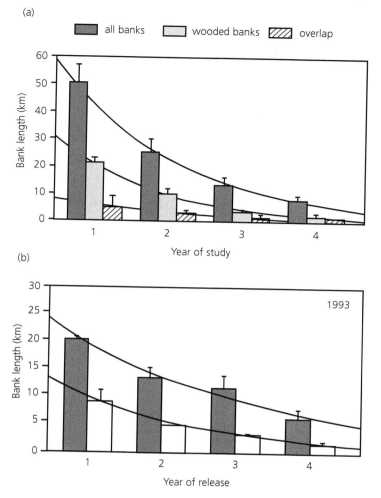

Figure 7.2 (a) Year-round territory sizes, length of wooded banks within these territories, and overlap with neighbouring territories (mean + SE) for radio-tagged beavers in the course of the study (*n*=3, 5, 4, 2). (b) Territory sizes and lengths of wooded banks (mean ± SE) for all beavers in 1993 plotted against year of release (of that member of a pair which was released first) (*n*=3, 4, 2, 3). Exponential curves were fitted to the data. (Republished with permission of Canadian Science Publishing Territoriality and time budgets in beavers during sequential settlement by Bart A. Nolet and Frank Rosell, volume 72, copyright 1994; permission conveyed through Copyright Clearance Center, Inc.)

territories were proportionately richer in terms of resource abundance (Campbell et al., 2005; Figure 7.3; see also Section 4.1.4). Beavers hold larger territories with plants with increasing average stem diameter due to resource depletion close to their central place. By occupying larger territories they can incorporate preferred stem diameters of deciduous trees (Arnberg, 2019). Beavers living at more northerly latitudes should have larger territory sizes than those at more southerly latitudes, because primary production decreases with increasing latitudes.

Habitat fragmentation by human-developed areas likely also results in larger territories (McClintic et al., 2014a).

7.1.3.3 Habitat type (stream, river, or pond)

The size of territories can vary between different habitat types. Territories normally follow the convoluted shorelines of lakes, the meandering routes of rivers, or the natural, irregular outlines of ponds (Wheatley, 1997b). In the Czech Republic, the territory size observed in lowland floodplain forests

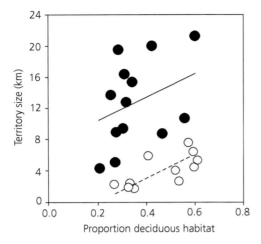

Figure 7.3 Beaver territory sizes (km) against proportion of deciduous habitat of these territories, with the selected general linear model, in Biesbosch (solid symbols and lines) and Telemark (open symbols and dashed lines). (Reprinted by permission from Springer Nature Customer Service Centre GmbH: Springer Nature, Behavioral Ecology and Sociobiology, Territory and group size in Eurasian beaver: echoes of settlement and reproduction? Campbell, R.D., Rosell, F., Nolet, B.A. and Vilmar, A.A. © Springer-Verlag 2005.)

was largest, and the smallest was found in sub-mountain hilly streams (Korbelová et al., 2016). The submountain hilly stream territories may be smaller because dam maintenance requires more time, and consequently smaller territory sizes would be more preferable (Figure 7.4). Also, beavers living on rivers have larger territory sizes than those creating dam systems on smaller streams or those living on small ponds (Wheatley, 1997c; Havens, 2006; Korbelová et al., 2016).

7.1.3.4 Density of beaver populations

A reintroduced Eurasian beaver population in Biesbosch, the Netherlands, with low population density had a larger territory size when compared with a more established population with higher density in Telemark, Norway (Campbell et al., 2005). In densely populated areas along larger rivers, territories are usually aligned adjacent to each other. Although it is rare to find two lodges from different family groups closer together than 400 m, though they may be as close as 100 m in dense populations. If the beaver family has a territory on a larger lake (ca. > 100 m wide), they usually do not

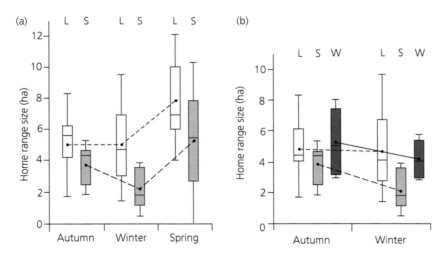

Figure 7.4 Seasonal pattern of home range size a) within two different habitat types in three consecutive seasons, and b) within three different habitat types in two consecutive seasons. Horizontal lines inside boxes are medians, points represent mean home range sizes, and line connecting these points shows trend in home range size between consecutive seasons (each habitat type is distinguished by a different line style: lowland floodplain forests (L)—short dashed; small submountain streams (S)—long dashed; wide unregulated river (W)—full). Habitat types are differentiated by colours: (L)—white; (S)—light grey; (W)—dark grey. (Reprinted by permission from Springer Nature Customer Service Centre GmbH: Springer Nature, Mammal Research, Seasonal variation in the home range size of the Eurasian beaver: do patterns vary across habitats? Korbelová, J., Hamšíková, L., Maloň, J., Válková, L. and Vorel, A. ©Mammal Research Institute, Polish Academy of Sciences, Białowieża, Poland 2016.)

share the lake with another family. However, if the density is high, each family may occupy their own shoreline (see Figure 7.1b). If the lake is narrower, then one family will most likely control the entire shoreline on both sides of the lake. Most landscapes include lengths of unsuitable beaver habitat. At the landscape scale, the distribution of beaver territories is often highly discontinuous (Parker et al., 2001). In these situations, territories will be located at greater distances from each other, with relatively large 'no man's land' between them (Hodgdon, 1978; Busher, 1980; Schulte, 1998). For example, in a study area in Grimsö, Sweden, which was not densely populated, the average distance between territories was 2.1 km (range 1.5–4.5 km) (Hartman, 1996).

7.1.3.5 Social factors (sex and age)

Similar territory sizes for different sexes and age classes have been recorded (Wheatley, 1997b; Bloomquist et al., 2012; Korbelová et al., 2016). However, females decreased their territory size when taking care of kits (Wheatley, 1997b; Herr and Rosell, 2004). Beaver kits need food and care during the early summer, which keeps the adults nearer to the lodge area. The adult female by necessity is in the lodge/natal burrow more than the other family members, since she needs to nurse, which gives her a smaller activity area in the summer. In the Tennessee River, Alabama, USA, subadults (11–16 kg and 25–36 months old) had larger territories than yearlings, but this was most likely due to the inclusion of forays (McClintic et al., 2014b; see Section 6.9.2). If a dispersing subadult is suddenly forced to settle down due to harsh winter conditions and find a place to overwinter, it can use a site (pond) that is as small as 5–10 m². However, in most cases it will move on as soon as it can during spring to find a mate.

7.1.3.6 Time of year (season)

The beaver's territory size varies according to season, with seasonal patterns being similar among different habitats (Korbelová et al., 2016). In the Czech Republic, territory sizes were largest in spring, followed by autumn. Smallest territory sizes were observed during winter (Korbelová et al., 2016). However, in South Carolina, USA, largest territory sizes were found during autumn and winter (Davis et al., 1994), while in Manitoba, Canada,

largest territory sizes were found in summer, intermediate in autumn, and smallest in winter (Wheatley, 1997a). In a temperate region in Illinois, USA, variation occurred, with largest territory sizes during winter and summer, but this seasonal difference did not occur every year (Bloomquist et al., 2012). Bloomquist et al. (2012) suggested that beavers foraged within newly flooded areas during winter when high water-level fluctuations occurred. These differences might be explained by water temperature and length of ice cover—and therefore movement restrictions under colder conditions at northern latitudes compared to southern latitudes (Nolet and Rosell, 1994; Wheatley, 1997a). Heterogeneously distributed food resources and the collecting of food for the cache and building up fat reserves before the winter could also explain the increased territory sizes during autumn (Svendsen, 1980b; Nolet and Rosell, 1994). Larger territories during spring might be due to intense patrolling of the territory to protect its borders from dispersing subadults (Aleksiuk, 1968; Rosell and Nolet, 1997; Rosell et al., 1998; see Section 7.2.1). In winter, the territories shrink down to the size that a beaver can patrol daily without losing condition (7.9 km in the Netherlands) (Nolet and Rosell, 1994).

7.2 Territorial defence

Beavers vigorously defend their territories from intruders, year-round, with both sexes and all family members, except kits, participating in defensive behaviours (Aleksiuk, 1968; Nolet and Rosell, 1994). The dominant adult pair is especially active in this behaviour, patrolling their territory to look out for intruders, scent marking and over-marking scent mounds from non-family members (Campbell-Palmer and Rosell, 2010; Hohwieler et al., 2018; Mayer et al., 2020b). The territorial defence can also include direct fighting (Nolet and Rosell, 1994; Crawford et al., 2015; Mayer et al. 2020a) and/or boundary stick displays (Thomsen et al., 2007).

7.2.1 Patrolling and travelling

Patrolling to look out for intruders and travelling to foraging areas are two very different behaviours that may be hard to differentiate. Also, these two behaviours are not mutually exclusive. Patrolling often occurs when beavers first emerge from the lodge in the evening and in the morning again. A beaver can be travelling to a foraging site but use this activity to patrol its territory at the same time.

Beavers living under higher population densities patrol their territories daily to look for intruders, and the distance moved per night increases with territory size (Busher, 1980; Nolet and Rosell, 1994; Herr and Rosell, 2004; Graf et al., 2016). Unfortunately direct encounters between residents and invaders are rare, and no quantitative data exist (Sun, 2003). Time spent travelling by both sexes increases with increasing air and water temperatures (Nolet and Rosell, 1994; Sharpe and Rosell, 2003; Figure 7.5).

The dominant pair exhibit similar space use and movement behaviour within their territories outside the reproductive season, i.e. there are no sexual dimorphic behavioural differences, and they do not reduce their patrolling activity in relation to the number of subordinate helpers in the family (Herr and Rosell, 2004). In Telemark, Norway, it was found that the dominant adult pair were located on the same side of the river on average 67% of the time (McClanahan et al., 2020). After the kits are born, it seems that males invest more time in territorial defence than females because males cannot initially nurse the kits (Busher and Jenkins, 1985; Buech, 1995; Herr and Rosell, 2004). The increase in travel time by males, which is most likely associated with territorial patrol (Nolet and Rosell, 1994), would also explain the increase in activity period and later return time to the main lodge/burrow of males during this time (Sharpe and Rosell, 2003). In the Tennessee River, Alabama, USA, the hourly distances moved from lodges by beavers (adults, subadults, and yearlings) were distributed bimodally during the breeding season (McClintic et al., 2014b). This was probably due to a trade-off between long foraging trips to self-feed and short foraging trips close to the lodge to nurse and protect the kits. Territory owners stay closer to the lodge with increasing number of kits (Steyaert et al., 2015). This may indicate a trade-off between patrolling/travelling effort and parental care.

To own and defend a territory comes with a cost, and these defence costs must be smaller than the benefits of maintaining a territory. The costs of defending a territory increase with its size due to the increase in patrolling distances. Since animals

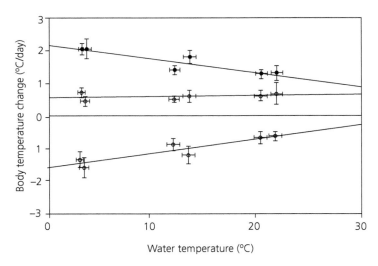

Figure 7.5 Daily body temperature change (mean + SE) in relation to water temperature over 6 months (*n*=9, 14, 11, 14, 9, 6). Black dot indicates difference between daily maximum and minimum body temperatures; white dot, difference between daily maximum (upper data points) or minimum (lower data points) body temperature, respectively, and value at resting site. (Republished with permission of Canadian Science Publishing, Territoriality and time budgets in beavers during sequential settlement by Bart A. Nolet and Frank Rosell, volume 72, copyright 1994; permission conveyed through Copyright Clearance Center, Inc.)

will not use more energy for defence than they gain in benefits, the size of the territory will vary according to the access it gives to the resources it is protecting. Territory size seems to act as a counterbalancing factor for patrolling and foraging, making both owning larger and smaller territories viable strategies in beaver populations at high densities. Beaver territories are not configured to a minimum economically defensible area (Campbell et al., 2005). Rather, they seem to occupy larger territories to reduce the rate of resource depletion during initial settlements in an area, whereas in populations at or near carrying-capacity, territories that become vacant are conquered independent of size. In Telemark, Norway, there was a territory size-dependent trade-off between patrolling and foraging (Graf et al., 2016). Beavers in larger territories moved greater distances each night, thereby spending more time patrolling, and stayed closer to the shoreline when being on land, i.e. when they were foraging. Inversely, in smaller territories beavers patrolled less and foraged further inland from the shoreline. However, the short distance between the up- and downstream border reduced patrolling costs and resulted in greater efficiency in territorial defence, thus compensating for increased foraging costs in smaller territories. These findings suggest that individual's trade off the costs of patrolling larger territories against the benefits of foraging closer to the shoreline. Smaller territories might be more prone to resource depletion (Goryainova et al., 2014), thus making foraging further from the shoreline a strategy to ensure sustainable resource use. As beavers in larger territories have to cover greater distances to reach the borders, they face higher patrolling costs for two reasons: (1) swimming has been shown to decrease the body temperature compared to being on land, especially during winter and early spring at northern latitudes (Nolet and Rosell, 1994); and (2) an increased patrolling effort constrains the time that beavers can spend foraging (Graf et al., 2016).

Older beavers have a greater absolute patrolling effort since they spent more time on land and close to territory borders than younger ones. This suggest a behavioural change with age, possibly due to increased experience and boldness (Graf et al., 2016). Beavers may gain experience related to predator and hunting pressure (both were low in their study area) over the years as a territory holder, leading to increased boldness and dominance. However, the distance moved in water did not differ between age groups, suggesting that older beavers spent more time per visit at a border. Spending more time at territory borders may allow beavers to spend more time on land instead of swimming between up- and downstream borders (Figure 7.6; see also Section 7.5.2).

7.2.2 Using scent communication

A major way beavers communicate is by using odours (see also Section 7.4). With scent communication,

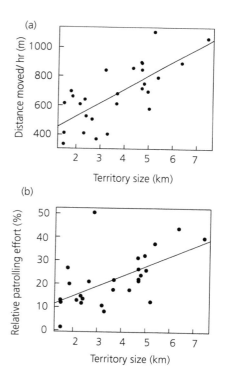

Figure 7.6 Predicted relationship between territory size (given as bank length in kilometres) and (a) average distance moved in water (metres per hour) and (b) relative patrolling effort (defined as proportion of GPS positions close to territory borders, within lower and upper 5% of territory) for 25 Eurasian beavers in southeast Norway. (Reprinted by permission from Springer Nature Customer Service Centre GmbH: Springer Nature, Mammalian Biology, Territory size and age explain movement patterns in the Eurasian beaver, Graf, P. M., Mayer, M., Zedrosser, A., Hackländer, K. and Rosell, F. © Deutsche Gesellschaft für Säugetierkunde 2016.)

neither the donor nor recipient needs to be present when the signal is sent. The receiver can, by using its odour receptors, identify, interpret, and give a physical response to the signal. Scent signals are quick and allow for large amounts of information to be sent across long distances in a short time (Müller-Schwarze, 2006).

7.2.2.1 Castoreum

Castoreum is produced in the castors sacs (see Chapter 3 for anatomy and development). Castoreum is special because of its ability to bind and slowly release aromatic scent compounds with low molecular weight (94% of the compounds have a molecular weight below 300 Dalton), which means they evaporate quickly and are therefore carried long distances (Rosell, 2003). The chemical composition in the castor sacs is very complicated, and more than 75 different compounds have been identified (Tang et al., 1993; Tang et al., 1995; Rosell and Sundsdal, 2001; Burdock, 2007; Stoye et al., 2009). Castoreum is mainly composed of dietary by-products (Müller-Schwarze, 1992; Sun and Müller-Schwarze, 1999). The information in castoreum is dependent on the physiological condition of the animal, i.e. what it eats, and maybe also its health and stress level. Beavers sequester potentially adverse plant secondary compounds and recycle them as signals for territorial advertisement (Müller-Schwarze, 2001). Chemical groups such as alcohols, phenols, aldehydes, amines (nitrogen bases), ketones, aromatic acids, and esters are present in castoreum (Lederer, 1946; Lederer, 1949; Müller-Schwarze and Houlihan, 1991; Tang et al., 1993; Tang et al., 1995).

Chemical analyses of castoreum using a gas chromatograph with a mass spectrometer (GC-MS) found no differences between the sexes (Rosell and Sundsdal, 2001). There is aroma variation, as dogs (*Canis lupus familiaris*) are able to distinguish between castoreum from males and females in controlled laboratory experiments (Rosell, 2018). In this study 92% of the samples were correctly identified as male or female by four dogs.

Interestingly, a female beaver will accept a foreign newborn if it has been rubbed with castoreum from her mate (Lavrova, 1957, cited in Wilsson, 1971). If it has an unknown scent she will attack the kit. A difference between the chemical profiles of the two beaver species' castoreum has also been found (Pedersen, 1999; Rosell, 2002b) and between beavers from Norway and Germany (Rosell and Steifetten, 2004). Since castoreum is dependent on habitat and animal condition it is not surprising that geographic differences and species differences exist. See Table 7.2 for other information coded in the castoreum.

7.2.2.2 The anal glands

The anal glands are separate structures from the castor sacs (see Chapter 3 for anatomy and development), surrounded by a pale pink membrane, and have two openings on either side of the anal opening. The anal glands are true glands. They are thick-walled bladders with groups of glandular tissue attached to the epithelial walls of the bladders (Svendsen, 1978; Walro and Svendsen, 1982; Valeur, 1988). The glands produce a secretion that fills the bladders. The glands have a restricted pathway that ends in one papilla on each gland and they are normally turned inward towards the wall of the cloaca but are exposed during scent marking. The papillae (small, pale-pink/red tips) are covered with a few brush-like hairs. When expelling the secretion from the anal glands, the papillae are forced outwards, and the secretion is emitted when the hairs are in contact with a solid object. This has been observed both in captive (Wilsson, 1971) and in free-living beavers both inside (Müller-Schwarze and Herr, 2003) and outside their lodges (Rosell and Bergan, 1998).

Anal gland secretion (AGS) supplements the sebaceous glands of the skin and maintains the water-repellent quality of the fur (Walro and Svendsen, 1982). By protruding the anal papillae through the cloaca, beavers can transfer AGS to their fore feet and spread it into their fur. AGS is a better waterproofing agent than castoreum. AGS is insoluble in water (castoreum is not), and smearing AGS into the fur increase its water-resistance. This coating of oily secretion will also help the beaver trap air within its underfur while it is underwater. The water-resistant properties of fur are lost if beavers are prevented from using their fore legs to groom or if their anal glands are surgically removed even though the grooming movement patterns are

Table 7.2 Information coded in the two scent types in the two beaver species. Scent types are classed as '✓' (does code for information), 'x' (does not code for information), or '–' (not investigated).

Information coded for	Castoreum			Anal gland secretion		
	Eurasian	North American	References	Eurasian	North American	References
Individual recognition	–	–		✓	✓	Sun (1996); Sun and Müller-Schwarze (1999); Tinnesand and Rosell (2012)
Sex	✓	x	Müller-Schwarze (1992); Sun and Müller-Schwarze (1999); Rosell (2018)	✓	✓	Schulte et al. (1995b); Rosell and Sun (1999); Rosell and Sundsdal (2001)
Age/dominance	x	x	Schulte (1998); Herr et al. (2006); Tinnesand et al. (2013)	✓	–	Tinnesand et al. (2013); Benten et al. (2020)
Family members	✓	✓	Schulte (1998); Sun and Müller-Schwarze (1998a)	✓	✓	Sun and Müller-Schwarze (1998a)
Neighbour	✓	✓	Schulte (1998); Rosell and Bjørkøyli (2002)	✓	–	Rosell and Bjørkøyli (2002)
Distant neighbour	–	–		✓	–	Benten et al. (2020)
Strangers	✓	✓	Schulte (1998); Rosell et al. (2000); Rosell and Bjørkøyli (2002)	✓	–	Rosell and Bjørkøyli (2002)
Siblings	–	x	Sun and Müller-Schwarze (1997)	–	✓	Sun and Müller-Schwarze (1997)
Relatedness	–	x	Sun and Müller-Schwarze (1997)	✓	✓	Sun and Müller-Schwarze (1997); Sun and Müller-Schwarze (1998b and c); Tinnesand et al. (2013)
Subspecies	✓	x	Rosell and Steifetten (2004); Peterson et al. (2006)	x	x	Rosell and Steifetten (2004); Peterson et al. (2006)
Species	✓	✓	Rosell (2002b); Rosell et al. (2019)	✓	✓	Rosell and Sun (1999); Rosell (2002b)

still in place (Wilsson, 1971; Walro and Svendsen, 1982). Younger animals' fur is not water-resistant, and a youngster that has become wet must be groomed by an older beaver to avoid getting cold. Adult beavers have been observed grooming their AGS into the fur of other beavers (Ryden, 1990). Clumps of castoreum have also been found in the beaver's fur, which indicates that the animal has had its paws near the anal opening and likely taken castoreum along with the AGS. However, it is also possible that the castoreum has come onto the fur when the animal has pulled itself over a scent mark (Rosell and Sundsdal, 2001). Despite these observations, it is still possible to read in the literature that the beaver does not transfer its AGS to its fur to make it waterproof. Some people believe they do not need these compounds, since their fur is so dense already. Also, the chemicals from AGS are not found regularly in the pelage (Rosell, 2002a).

There is another advantage to smearing AGS into your fur: it is an effective way of telling other beavers who you are. Beavers can swim around like huge 'scent mounds', and other individuals will quickly find out whether it is a family member, neighbour, or a stranger that is approaching before making direct contact (Table 7.2). The secretions in the fur may tell other animals living downstream (neighbours) or animals wandering upstream who is living nearby, as the secretions may be spread on the water surface (Rosell, 2002a). It is also a possibility that their faeces are covered with the secretions when they drop them in the water (Rosell, 2003). However, no studies have been carried out.

At least 163 compounds have been found in the anal glands—many more in males (126) than females (56) (Rosell and Sundsdal, 2001). Only 12.5% of AGS compounds in males and 32.5% of AGS compounds in females have a molecular weight below 300 Dalton. Therefore, most compounds in the AGS do not easily spread through the air (Rosell, 2003), since the upper size limit for airborne pheromones is a molecular weight of about 300 (Bradbury and Vehrencamp, 1998). Chemical analyses of AGS in the North American beaver have shown that AGS contains information about individuality, sex, relatedness, and family membership (Sun and Müller-Schwarze, 1998a; Sun and Müller-Schwarze, 1998c; Sun and Müller-Schwarze, 1999). Chemical analyses have also shown that the compounds in the anal glands are very different between individuals (Tinnesand and Rosell, 2012) and sexes of the Eurasian beaver (Grønneberg, 1979; Grønneberg and Lie, 1984; Rosell and Sundsdal, 2001). Some of the chemical compounds occur in both male and female beavers, while others are specific to a single sex. For example, wax esters and their corresponding free fatty acids and fatty alcohols are found in male secretions but not in female secretions. The colour, smell, and consistency of AGS is also different between the sexes (Kolbin 1962, cited in Saveljev, 2003) as well as the two species (Rosell and Sun, 1999; see Chapter 10 for details). In addition, AGS contains chemical information about geographic variations, i.e. beavers from Norway and Germany (Rosell and Steifetten, 2004) and species (Rosell, 2002b). Dogs are able to distinguish between castoreum and AGS in controlled laboratory experiments (Rosell, 2018). Three dogs were 100% correct and one dog made one mistake. Four dogs managed also to distinguish between AGS from males and females in 88% of the trials; three dogs were 100% correct (the dog that was unsuccessful did not receive sufficient training). See Table 7.2 for other information coded in AGS.

7.2.2.3 Scent marking behaviour

Due to the importance of scent communication among beavers, understanding how they ensure transmission of the scent messages is critical. Members of established beaver families or adult pairs scent mark their territory to tell other beavers that the area is occupied. The primary purpose of scent marking (depositing scent compounds) therefore appears to be territorial defence. While scent marking the beaver places castoreum from the castor sacs or secretion from the anal glands, or both at the same time (Rosell and Bergan, 1998; Baker and Hill, 2003). Scent marks are used by family groups to identify their territory and to warn off potential challengers (Aleksiuk, 1968; Hodgdon, 1978; Rosell and Nolet, 1997; Rosell et al., 1998).

Usually beavers build scent mounds, but they may also scent mark directly on a stone or a log. Scent compounds are mainly deposited on specially constructed piles that are called scent mounds. These mounds are small piles of mud, sand, vegetation (grass, leaves, and mosses), sticks, and other detritus that beavers gather from the surrounding environment. When depositing castoreum the Eurasian beaver bends its back, scrapes together a scent mound with its forelegs, and kicks with its back legs while the castoreum is sprayed/urinated against the substrate, which is sometimes audible. This deposition process is not dissimilar to urination except that urine flushes through the contents of the castor sacs. When depositing secretions from the anal glands, the two papillae are pushed out (Figure 7.7) and AGS is released when the hairs contact a solid substrate (Wilsson, 1971; Svendsen, 1978). It has been reported that North American beavers often dive to the bottom of the pond, dredge up a batch of mud with its forefeet, and holding it against its chest. Thereafter it walks on its hind feet onto land and creates a scent mound (Müller-Schwarze, 2011). This way of marking has never been described or observed for the Eurasian beaver. However, North American beavers can construct mounds in a similar manner as the Eurasian beavers (P. Busher, pers. comm.). More than 100 scent mounds can be found within a single territory, which the animals will visit on a regular basis. In Ohio, one adult female made more than 70 mounds during one week. In warmer, ice-free regions, scent marking can occur all year, and is more intense during the spring dispersal period. In Alabama, scent marking has been observed in December and January (Baker and Hill, 2003).

The size of these mounds can vary considerably, but they are usually around 15 cm in diameter and about 10 cm tall (Figure 7.8), though they can be as large as 80 cm in diameter and 50 cm high (Müller-Schwarze, 2011). Beavers can scent mark directly on clumps of grass (tussocks), stones, and tree trunks, or right on the ground. Although some scent marks can be hard to spot, the camphor-like smell associated with these structures is quite distinctive. These freshly marked scent marks can often be smelled by humans 10 m away (depending on the weather conditions and the human nose), and it is the castoreum that is smelled.

Scent marks can function as both 'no trespassing' signs and 'advertising' signs at the same time. If a single, newly established animal moves into an area it will scent mark to tell other beavers what sex it is. The scent mark may then attract the opposite sex (advertising) and let them know that they are welcome. Individuals of the same sex receive the message that this area is taken (no trespassing) and that they are not welcome. Research has also indicated that scent marking is important in established pairs in connection with the mating period, especially in habitats where there is no or little ice that can restrict movements in the winter. The number of scent marks is found to increase in the breeding season compared to the non-breeding portion of the winter (Rosell and Bergan, 2000).

The scent-matching hypothesis proposes that the scent marks provide an olfactory link between resident animals and their territory, which enables

Figure 7.7 Scent marking by pushing papillae outwards, depositing anal gland secretion (AGS) when the hairs come into contact with a solid surface. (Photo supplied courtesy of Frank Rosell.)

Figure 7.8 Scent mound, characterized most by its distinctive smell. (Photo supplied courtesy of Frank Rosell.)

intruders to recognize the probability of conflict escalation, i.e. both the signaller (territory owner) and the signal (scent mark) need to be present (Gosling, 1982). The intensity of response (over-marking of experimental scent mounds) should be the same or decrease because, without the presence of the intruding signaller coupled with the chemical signal, the presence of the scent itself does not advertise the ownership of a territory. Support for the scent-matching hypothesis was found for the North American beaver (Sun and Müller-Schwarze, 1998b). The family response was stable to strangers' AGS and decreased to strangers' castoreum during a period of 6 days (Sun and Müller-Schwarze, 1998b).

Beavers appear to invest more in scent marking in good-quality territories and when a territory has been occupied for a relatively long time (Hodgdon, 1978; Rosell and Nolet, 1997). The number of scent marks also is dependent on population density (Houlihan, 1989; Rosell and Nolet, 1997; Ulevičius and Balciauskas, 2000). Beaver families with close neighbours scent mark more than families that are more isolated. The number of scent marks increases dramatically with increasing number of neighbour territories and the number of individuals in the neighbouring territory (Rosell and Nolet, 1997). Experimental scent mounds have been constructed to manipulate colonization of beavers (Welsh and Muller-Schwarze, 1989). These researchers showed that only one (4%) of the scented sites had been colonized at the end of the experiment compared to four (16%) of the control sites without scent (Figure 7.9). However, application of the scent must re-occur regularly and the effect is reduced in populations with a high density (Müller-Schwarze and Heckman, 1980; Welsh and Muller-Schwarze, 1989).

All Eurasian beavers older than 5 months old mark the family territory through scent marking (Wilsson, 1971). Rudimentary scent marking begins as early as 13–14 days of age (Baker and Hill, 2003). Beavers that are 2–3 months old will perform the movements associated with scent marking, but it is uncertain whether any liquid/secretions are deposited, since the compounds within the castor sacs and anal glands are not fully developed at this age. Both sexes deposit scent marks within their territories. Most studies have focused on the adult males

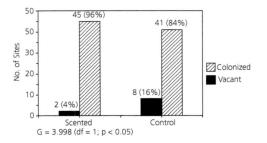

Figure 7.9 Number of vacant sites colonized (black bars) and not colonized (cross-hatched bars) after experimental scenting in Fulton County, Georgia, USA. (Redrawn by Rachel Hinds from Müller-Schwarze, D. 1990. Leading them by their noses: animal and plant odours for managing vertebrates from Chemical Signals in Vertebrates, © Oxford University Press. Reproduced with permission of the Licensor through PLSclear.)

and females, and few studies have investigated the younger beavers' scent-marking activities. Dominant beavers scent mark more frequently compared to other family members (Buech, 1995; Hohwieler et al., 2018). Female kits scent mark less than male kits (Hodgdon, 1978). Territorial behaviours by both adult sexes are suggested to have evolved from a mate-guarding strategy and/or a resource defence strategy (for both food and the physical family area) (Busher, 2007). Male beavers spend significantly more time in territorial defence at territory borders and deposit more scent marks and over-marks (see below) during the summer than females (Hodgdon and Lancia, 1983; Rosell and Thomsen, 2006).

Scent marks are constructed throughout the year (Aleksiuk, 1968; Rosell and Nolet, 1997; Rosell et al., 1998; Baker and Hill, 2003). Though April to June are the most active months for scent mound creation, while dispersing subadults search for new territories (Svendsen, 1980a; Rosell and Nolet, 1997; Rosell et al., 1998; Figure 7.10). This represents the time of greatest threat to territory holders as dispersers seek feeding resources and opportunities to take over a territory (see Chapter 6). Additionally, in the Eurasian species, we have recorded a secondary peak in scent marking around February when females are in oestrus (Rosell et al., 1998; Rosell and Bergan, 2000). This may be a way of males mate-guarding their females (Busher, 2007) or a way for females to efficiently advertise their reproductive

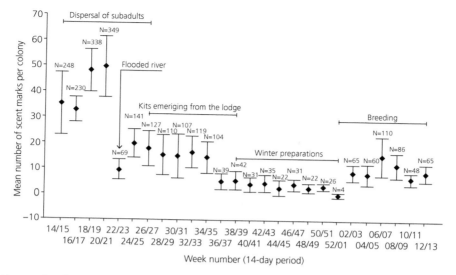

Figure 7.10 Mean number of scent marks in seven adjacent territories along the Bø River observed at 14-day intervals throughout one year (1 April 1995 to 31 March 1966). Vertical bars show 95% confidence intervals. N is the total number of scent marks for each 14-day period in all seven territories combined. A flood during weeks 22/23 hampered scent-marking activity. (Reprinted by permission from Springer Nature Customer Service Centre GmbH: Springer Nature, *Journal of Chemical Ecology*, Scent-marking in the Eurasian beaver (*Castor fiber*) as a means of territorial defence, Rosell, F., Bergan, F. and Parker, H. © Plenum Publishing Corporation 1998.)

status. In South Carolina, USA, peak scent marking occurred during autumn and winter due to increased food competition among families, and very little scent marking happened during summer (Davis et al., 1994). Therefore, scent marking activity may differ between southern and northern beavers.

Scent marks tend to be found along territorial boundaries and near feeding and resting sites close to the water's edge (Rosell and Nolet, 1997; Baker and Hill, 2003). They are more prevalent near up- and downstream borders (Hodgdon, 1978; Rosell et al., 1998; Sun and Müller-Schwarze, 1999). From October to December, when beavers mark at lower frequencies at northern latitudes, almost all scent marks were found at the territorial borders in a study in the Bø River region in Telemark, Norway (Rosell et al., 1998). In the Voronezh Nature Reserve in Russia, which has been occupied by beavers since 1950, beavers scent marked few times at the borders but frequently at the centres of the territory (Zavyalov et al., 2016). After moving to an area, beavers also scent marked more in the centre of their new territory previously used by another family (Aleksiuk, 1968; Bau, 2001; Zavyalov et al., 2016).

Along some rivers beavers mark more upstream of the lodge than downstream, even if they have adjacent neighbours in both places. This is most likely because the normal dispersal direction in this river is downstream, and by marking the upstream borders animals that are on the move will know that the territory is occupied immediately after they cross the border. Those that are coming from downstream may also be able to smell that the upstream territory is taken. As AGS is insoluble in water (Svendsen, 1978) it is not impossible that the chemical compounds from the anal glands float on the water surface downstream (Rosell et al., 1998). In this instance, the beaver's vomeronasal organ may be important in receiving scent at the air–water interface (Rosell and Pedersen, 1999).

If a scent mark from a neighbour or an unknown individual (stranger) appears in an established territory, the resident beavers will be attracted to that scent mark. They will first sniff the scent marking, then destroy it and over-mark it with their own scents (Müller-Schwarze et al., 1983; Müller-Schwarze and Houlihan, 1991; Schulte et al., 1994; Schulte et al., 1995a; Schulte, 1998; Rosell and Bjørkøyli, 2002). They may partly or fully over-mark it or make their own scent mark next to the intruder's scent mark. After destroying and over-marking the intruder's mound, the territory holder will go into the water to search for the intruder. Distinguishing among multiple scent marks is essential for beavers if they are to identify potential mates, competitors, and territory owners (Rosell, 2003). Scent marking has multiple functions and beavers must be able to identify what information the scent marks are conveying (see Table 7.2).

Castoreum functions as the main scent used in territorial defence and was always deposited (96 of 96 times) while AGS was only deposited in 4 of 96 times, at least during the wintertime at northern latitudes (Rosell and Sundsdal, 2001). Castoreum has not been reported to contain detailed information about the individual beaver and seems to state 'I live here; this is my territory; this territory is occupied'. However, AGS contains many compounds that convey different types of information (see Table 7.2). It is likely that the lighter, more volatile castoreum compounds direct beavers toward the less-volatile and more informative compounds in the anal glands. AGS is probably more energetically costly to produce compared to castoreum (Rosell, 2003).

There have been many scent play-back experiments performed to find out what kind of information a beaver will sniff out by smelling a scent mark with castoreum and/or AGS from another beaver. These studies have discovered that beavers can gain a great amount of information from these marks (see Table 7.2 for more studies and other information coded in castoreum and AGS). The main findings from some of these studies are summarized as follows:

- *Male versus female strangers*: Both sexes of the Eurasian beaver reacted more strongly to AGS from male strangers compared to AGS from female strangers. Intruding males may pose more of a threat to resident males via intrasexual competition and resident females due to the long-term cost, such as infanticide and abortion, of a new dominant male (Cross et al., 2014). To date there is no empirical evidence for infanticide in beavers (Sun, 2003).
- *Age and social status*: Resident Eurasian beavers reacted more strongly to AGS from a stranger

son (i.e. a non-territory-holding floater; ≥ 2 years old) than his father (i.e. a territory owner), suggesting the non-territory holder is viewed as a bigger territorial threat. Stranger sons might be more motivated to take over a territory and engage in physical conflict than their fathers that already have their own territory. Therefore, it appears that territory owners can be identified by their AGS by other territory owners (Tinnesand et al., 2013).

- *Family, neighbours, and strangers*: Eurasian beavers manage to differentiate family members from neighbours and strangers. They have a stronger reaction to both castoreum and AGS from a stranger than from a neighbour (Rosell and Bjørkøyli, 2002). A territorial neighbour poses less threat to an individual's territory (referred to as the 'dear enemy phenomenon'; Temeles, 1994). Strangers could be non-territorial floaters that would pose a higher threat. Little support for the dear enemy phenomenon has been found for the North American beaver, but they do discriminate castoreum from family and non-family adult males (Schulte, 1998). This was most likely due to the large distance between neighbours—there was always an unoccupied stretch of stream between territories and therefore no real neighbours (only strangers). Eurasian beavers also discriminate between distant neighbours (i.e. two territories away) and strangers (i.e. further away than the known dispersal distance) (Benten et al., 2020). Males responded stronger to subdominant distant neighbours, especially males, than subdominant strangers. No such discrimination was found by females or between dominant distant neighbours and strangers. Benten et al. (2020) suggested that the 'nasty neighbour' response by male resident beavers towards subdominant distant neighbours relates to the relative threat levels due to repeated intrusions during dispersal attempts.
- *Kinship*: North American beavers (dispersers or the siblings of dispersers) have been reported to react more strongly to AGS from non-relatives compared to related individuals. Beavers responded less strongly to their siblings' AGS than to non-relatives' AGS (Sun and Müller-Schwarze, 1997). The mates of dispersers also

showed a lower response to AGS from dispersers' siblings than to those from dispersers' non-relatives. Therefore, information about kinship is coded in AGS, and these results provide evidence that cues used for discrimination can be acquired indirectly via a third individual as the reference.

- *Geographic variation (subspecies) and species*: Eurasian beavers react stronger to their own subspecies' castoreum, but there is no difference in their reaction to AGS (Rosell and Steifetten, 2004). Geographical isolation has most likely developed these discriminatory abilities. Eurasian beavers perceived conspecifics to be a greater threat to territory than North American beavers (Rosell, 2002b). A study of the North American beaver failed to show a difference in response to their own species and the Eurasian beaver and also between their own subspecies and a different subspecies by AGS and castoreum (Peterson et al., 2006). However, in this study the researchers only looked at the family response overnight and not at the response after the first beaver's visit. Therefore, the discrimination may have been overlooked.

7.2.3 Fighting, aggression, and tail scars

Intruders are treated aggressively and territorial combat can result in serious or even fatal injuries (Piechocki, 1977; Nolet and Rosell, 1994; Crawford et al., 2015; Figure 7.11). Fighting accounted for

27.5% of beaver pelt damage in Mississippi, USA (Arner and Roberts, 1981). A beaver that is wandering through a foreign territory that contains many scent markings will normally swim through as quickly as possible, spending very little if any time on land. The resident beavers usually do not bother to chase away the intruder when it hurries through. However, the passage is not always peaceful, and the animals can occasionally fight.

In an aggressive conflict, one beaver moves towards the other, stares, slaps its tail, moves its head towards the other's rump, moves its body, snaps, and bites. Fighting usually starts with the territorial beaver swimming in circles to most likely demonstrate ownership and the will to fight. The antagonists will then lift the fronts of their bodies out of the water, grab each other in a 'wrestler's grip', and try to push each other backwards while grunting. If one of them loses its balance it will hit the water with its tail to try to recover its position. The wrestling match will continue while they both strike at each other with their front teeth (see Figure 6.1). These aggressive bouts can last for 10 minutes or longer. The closer to the lodge or rest site the animals are, the more intense the battle. The fight usually ends with the one beaver being forced underwater. Bites normally occur near the shoulders, the nearest part of the back, and on the flanks, but bites lower on the back can also occur (Arner and Roberts, 1981). If an animal is bitten, it might die from it or of infections later. Blood is usually seen in the water after a fight (Figure 7.12).

Figure 7.11 As beavers are territorial, fights can result in injuries. They can sometimes recover from injuries such as these in 3.5 weeks. This beaver made a full recovery. (Photo supplied courtesy of Frank Rosell.)

Figure 7.12 Two beavers engaged in a territorial fight. Note the blood. (Photo supplied courtesy of Michael Runtz.)

Scars are also common on the tail (Mayer et al. 2020a), and the tails of older animals can have a very deformed appearance (Figure 7.13). Numerous tail scars on an adult animal have been considered as a sign that the population is at or near carrying capacity and that there are limited opportunities for the animals to pass uninjured through to new available territories. In Müller-Schwarze and Schulte (1999a) study areas in New York state, USA, more tail scars (most likely cumulative injuries throughout the life of the beaver) were found in males (63%) compared with females (37%). Also, adults had more tail scars than other age classes, but no tail scars were found on kits (Müller-Schwarze and Schulte, 1999a).

Crawford et al. (2015) trapped and examined recent injuries on the pelt from North American beavers in central Illinois, USA, to investigate conspecific aggression. However, no difference in pelt scars between yearlings, subadults, and adults were found, but kits had ten times fewer recent injuries than adults. Reports from a North American population in Massachusetts, in the exponential phase of growth, have indicated that in 14 of 19 fights the adult male was involved (Hodgdon, 1978). These results suggested that males were more often involved in territorial defence battles, but another study found no difference between the sexes, indicating that both sexes participate in territorial fights (Crawford et al., 2015). Beavers on small streams had fewer recent injuries than beavers on rivers. Only 16% had injuries on streams compared to 44% on rivers. This is most likely due to fewer beavers passing through in streams compared to rivers. No differ-

ences were found between seasons (predispersal or dispersal) (Crawford et al., 2015). Only one study has looked into tail scars of the Eurasian beaver. In a dense population in Telemark, Norway, 1,158 beaver tails from 368 beavers were investigated for tail scars (Estalella, 2018). Males had more tail scars than females, suggesting they participated more in territorial fights. Tails scars accumulate during the years, and dominant and subordinate beavers had more tails scars than kits and yearlings. No difference in tails scars was found between spring and autumn.

7.2.4 Stick display

In a wild population of Eurasian beavers in Telemark, Norway, an additional defensive behaviour, stick display, was observed 131 times (Thomsen et al., 2007). At least six adult beavers, of both sexes, from three families have been observed carrying out this unique behaviour on two different rivers that are connected. The behaviour was not sex-specific, and both sexes displayed the stick display to the opposite sex. Stick display was often carried out after scent marking (35%) or another stick display (34%). This behaviour involves a beaver picking up an object (a stick if available, but occasionally other objects such as weeds are used), rising up on its hind legs, and moving its upper body rapidly up and down while holding the object in its mouth and fore paws. The behaviour was usually witnessed at disputed territory boundaries and may be wholly confined to individuals within this population, though similar (beavers bobbed their front body at human observers but without holding anything in their mouths, except for one male who was lodge-building and held a branch in his mouth) behaviour has been witnessed three times in two different sites in Kouchibouguac National Park, New Brunswick, Canada, with North American beavers (Thomsen et al., 2007). The two most active stick display families had their main lodges within 100 m of the shared border and the authors speculated that the stick display might only be triggered under such high-pressure situations. The stick display may be a better or at least additional indicator of strength and aggressiveness than the information in a scent mark and could act to reduce physical conflict with neighbours (Thomsen et al., 2007; Figure 7.14). Recently

Figure 7.13 Tail scars can be a result of territorial disputes. (Photo supplied courtesy of Rachel Hinds.)

Figure 7.14 Stick display is another, less common territorial behaviour. (Photo supplied courtesy of Bjørnar Røstad Roheim.)

(in 2018), several other beavers in both of the rivers mentioned above as well as in an additional river have been observed doing the stick display.

7.3 Duration of territory occupation

The duration of territory occupation can vary greatly, and it is important for the fitness of the territory holder. The duration of territory occupancy ranged from 1 to 11 years (mean 6.2 years) in a population in Telemark, Norway, at carrying capacity (Figure 7.15) and was a predictor for the lifetime reproductive success of an individual (Mayer et al., 2017b; Figure 7.16). The duration of territory

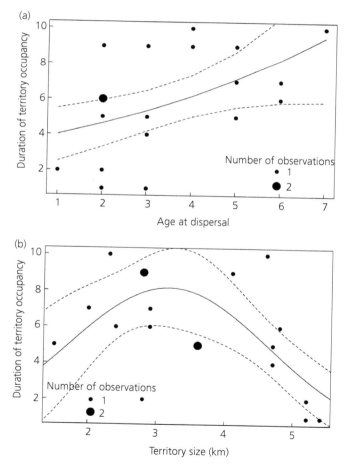

Figure 7.15 Predicted relationship (solid line) between (a) age at dispersal (in years) and duration of territory occupancy (DTO, in years) and (b) territory size and DTO for 19 Eurasian beavers in southeast Norway (1998–2015). Dashed lines present upper and lower 95% confidence intervals. (Reprinted from Mayer, M., Zedrosser, A. and Rosell, F. 2017. Couch potatoes do better: Delayed dispersal and territory size affect the duration of territory occupancy in a monogamous mammal, *Ecology and Evolution*, 7, 4347–4356 under CC BY 4.0 License.)

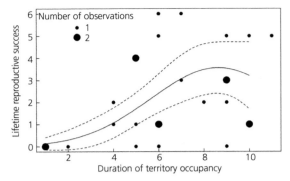

Figure 7.16 Predicted relationship (solid line) between duration of territory occupancy (in years) and lifetime reproductive success (measured as total number of kits produced in a lifetime) for 25 Eurasian beavers in southeast Norway (1998–2015). Dashed lines present upper and lower 95% confidence intervals. (Reprinted from Mayer, M., Zedrosser, A. and Rosell, F. 2017. Couch potatoes do better: Delayed dispersal and territory size affect the duration of territory occupancy in a monogamous mammal, *Ecology and Evolution*, 7, 4347–4356 under CC BY 4.0 License.)

occupancy was positively related to dispersal age, suggesting that individuals that delayed dispersal had a competitive advantage due to larger body mass (22.1 ± 3.0 kg vs. 16.6 ± 4.5 kg for normal dispersers). Beavers that dispersed at an older age and established intermediate-sized territories occupied them for longer than younger dispersers and individuals establishing smaller or larger territories. This suggests that an individual is better off waiting until it is physically and behaviourally mature before acquiring a territory, and it demonstrates that intermediate-sized territories follow the optimization criterion, ensuring sufficient resource availability and decreased costs of territorial defence at the same time. These data also indicate that large territories are more costly to defend due to increased patrolling effort and overall territorial defence effort, and small territories might not have sufficient resources (see also Section 7.2.1). Beavers holding an intermediate-sized territory and delaying dispersal had important consequences for the evolutionary fitness of beavers, since both the duration of territory occupancy and lifetime reproductive success increase (Mayer et al., 2017b).

7.4 Communication

In addition to communication with odours (see Section 7.2.2), beavers use different vocalizations,

tail slapping, poses, and movements to communicate with other beavers and other species. One limitation of these signalling methods compared to scent communication is that the intended recipient of the signal must be available and paying attention at the same instant that the signal is sent.

7.4.1 Vocalization

Seven different vocalizations (hiss, whine, cry, nasal sound, summons, soft churr, and guttural sound) have been identified in North American beavers living in Nova Scotia, Canada (Leighton, 1933), but only four (hiss, whine, whistling, and clacking) of these have been identified for the Eurasian beaver in Sweden (Wilsson, 1971). Four sounds were also described from a Eurasian beaver family in captivity in Germany: grunts, sighs, tooth chattering, and hissing (Frommolt and Heidecke, 1992). Five vocalization types were recorded for the North American beaver living in Quebec, Canada—whine, low-pitch whine, hiss, whistle, and growl (Patenaude, 1984)—and only three sounds were recognized outside the lodge for the North American beaver in Massachusetts, USA—hiss, whine, and guttural growl (Hodgdon and Larson, 1985).

A high-frequency whistling sound, difficult for humans to hear, was reported mostly outside the lodge by newly caught Eurasian beaver kits (Wilsson, 1971) and for North American beaver kits inside the lodge (Patenaude, 1984). Researchers categorize sounds differently and it is currently impossible to say whether beavers produce four, seven, or even more sounds. This is not surprising since different beaver groups are studied by different authors in various contexts and by using different methods. Therefore, different authors most likely use different names for very similar sounds. Also, some sounds mentioned above are probably whine variations. The majority of researchers, however, describe three main types of vocalization: hissing, whining (whistling/peeping), and growling. The first two types of sounds are primarily produced outside of the lodge, while the last is mostly heard inside the lodge. Inside the lodge you can also hear burps (Müller-Schwarze, 2011). Vocalizations occur throughout the year but most commonly during late summer and early autumn (Hodgdon, 1978).

Beavers have defensive vocalizations, mainly in the form of hissing and growling (Hodgdon and Lancia, 1983). Hissing can be heard in response to capture, handling, and unfamiliar scents, or towards other animals. The hiss is normally used against unknown beavers or enemies, but it can also be used against family members. When researchers approach a beaver that has been caught in a live trap, net, or sack, they most often hear the beaver hiss. Beavers can also show aggression by grunting and tooth chattering and/or grinding. Whining by kits is important in initiating grooming and play (Hodgdon and Larson, 1985), when kits beg for food, and to reduce aggressive tendencies from older beavers (Hodgdon, 1978; Brady and Svendsen, 1981). When standing outside a lodge you can often hear whining, which may be a show of satisfaction, or growling, which could mean that a beaver needs more space in their sleeping chamber or is not happy with other family members in the lodge.

The young cohorts vocalize more often than the older animals and have a noticeably louder, higher-pitched whine call than the adults (Tevis, 1950; Novakowski, 1969; Hodgdon and Larson, 1985; Pilleri et al., 1985). The whine is the most commonly reported vocalization and is considered the typical beaver sound (Hodgdon and Larson, 1985). Kits made 64.8% of all whine calls, yearlings 23.2%, 2- and 3-years-olds 5.3%, and adults 6.7%. Females of all age classes were more vocal than males of the same age class. Beaver voices become coarser with age, and age classes have been reported to be distinguished based on sound. Most (94.1%) of the whine calls occurred when two or more animals met or were together (Hodgdon and Larson, 1985).

Immediately after birth, the kits will make sounds such as sharp cries and whines, and they continue to be noisy the rest of the day (Patenaude, 1984). If you are standing near a beaver home and listening you can hear the kits in the lodge's sleeping chamber and occasionally from a good distance from the lodge. The kits' sounds are reminiscent of the whining of small children (Wilsson, 1971). The whine call that is produced by the kits just after birth is different from that produced some hours later (Hodgdon, 1978). They will 'cry' day and night until they have been weaned. If you record the sounds coming from inside the lodge you may be able to record the time of birth, and if you continue recording

you will be able to individually identify up to four of the newborn kits from each other (Żurowski, 1979; Żurowski and Kasperczyk, 1986). The loudest sounds are made when their parents or older siblings come in with food. When family members cuddle with each other to sleep they will often make a special low-frequency version of the whining just before they fall asleep (Wilsson, 1971).

The kits have a fine, peeping, barely audible squeak in danger and conflict situations. The parents can detect this danger signal from a long distance and react quickly by coming to their rescue. Begging sounds from the kits to the adults can often be heard outside of the lodge in the summer and these are a form of whining. A kit that wants a twig that an adult is peeling will approach while whining, and then make contact nose to nose, before taking the twig, usually without protest from the older animal. The youngsters do not always get what they want and if the adult wants to eat its stick in peace it will turn away from the youngster or emit a deep grunt. Kits primarily use contact sounds outside the lodge to secure food at the bank edge (Hodgdon and Larson, 1985).

When animals meet in the water in the evening, they will often make a high-pitched whining sound. When the beaver is eating something especially good it can make a humming sound. If another animal approaches at this time, the feeding beaver may make a hissing sound, warning the approaching beaver to stay away. If the other animal does not react to the warning and tries to steal the food, the first animal will make a growling sound. This sound warns the 'thief' that they will be attacked if they do not leave immediately, or it may illustrate dominance.

7.4.2 The tail slap

Beavers are well known for using their tails to generate alarm signals to warn other family members of approaching danger, to frighten or drive potential predators away, and to elicit more information from the stimulus source (Hodgdon, 1978; Busher, 1980, Hodgdon and Lancia, 1983). An undisturbed beaver will dive without making a sound by performing a quiet dive and disappearing quickly under the surface without making any strong movements in the water (Wilsson, 1971) (Figure 7.17a,b). This is the

Figure 7.17 (a) North American beaver raising its tail for tail slapping. (Photo supplied courtesy of Michael Runtz.); (b) An Eurasian beaver has tail slapped. (Photo supplied courtesy of Frank Rosell.)

quickest way to dive and can also be performed when a beaver is surprised with an unexpected danger. However, because beavers can detect scents and movements a good distance away, the response to an unexpected situation often leads to a dive with a tail slap. Normally the beaver dives after a tail slap, but it may occasionally continue to swim and/or slap its tail several times in a row, with only a few seconds' rest between each slap without diving. The beaver can also slap its tail against the ground if it gets annoyed while on land. The tail slap in water can be heard from a distance of several hundred metres under quiet conditions and usually before the beaver is spotted. Predator and human presence, and unfamiliar noises and odours (including castoreum from unknown beavers) can all stimulate tail slapping (Hodgdon, 1978; Müller-Schwarze, 2011).

When threatened, one beaver's loud tail slap on the water surface may be a means of communicating the threat to family members. In the water, the beaver will first lift its tail in an arc over its back while simultaneously kicking both back legs backwards, sending the water up in a powerful cascade, and then powerfully slapping the underside of its tail against the water surface. (Figure 7.17 a, b) This creates a large 'hole' in the water, and the beaver disappears into the hole in a strong dive. Since this slows the beaver down rather than speeding it up, the tail slap may serve multiple purposes (Wilsson, 1971).

The tail slap is most likely a warning signal to invaders, telling them that they have been discovered and do not stand a chance, since predators are often dependent upon the element of surprise. The invader can be an unknown beaver, a human,

or another predator. A powerful tail slap may also tell the invader that the beaver is not easy prey but is an animal in top condition and ready to defend itself. It is well known that predators will often decide not to attack prey that display physical strength. Beavers have also been observed slapping with only the top side of their tail or with both sides in turn. Which tail slap is performed depends upon the position of the body in the water when the beaver dives.

Other family members do not always react to the tail slap but will continue with their activities. An animal already deep underwater will normally show little response to a tail slap, since it is already in a place of safety. Beavers that are eating or resting in shallow water will usually swim immediately into deeper water (Hodgdon, 1978). Those on land will usually gallop to the water and dive in as quickly as possible.

Like the young of other mammals, beaver kits are alarmed easily (Brady and Svendsen, 1981). They often slap their tail on the water when alarmed (Busher, 1980), but other age classes do not respond to kits' 'false alarms' as often (29.7%) (Hodgdon, 1978). They are not as strong as the adults' and they cannot achieve the same 'clap' against the water. Yearlings and 2- and 3-year-olds trigger an escape response to tail slaps in 44.0% to 46.4% of cases, respectively. The adults' tail slap triggers an escape response in 62.8% of cases (Hodgdon, 1978). Young animals will usually react quickly to their parents' tail slaps by going inside the lodge, but the reaction to the tail slap is not instinctive. The tails differ in size and shape and it is likely that beavers discriminate tail slaps from different individuals (Müller-Schwarze, 2011) or at least an adult from a kit and maybe even a yearling. Reports state that females use the tail slap more often than males and that other animals in the family will more often respond to the female's tail slap by swimming into deep water than they will for the male's (Hodgdon and Larson, 1973).

An interesting observation was made in south-eastern Norway when an adult beaver tried to frighten off a red fox (*Vulpes vulpes*) that had taken a beaver kit 5 minutes earlier (Kile et al., 1996). The beaver swam to the fox's location, circled in the water a few times, and then went on land to where the fox had begun eating the kit. The fox stopped and walked a few steps towards where the beaver was lying in wait. The beaver began to slap with its tail, in a likely attempt to frighten the fox away. The fox, however, ignored the slapping and continued to eat the kit. This time the adult beaver was too late to save the kit. Other observations have shown beavers using tail slaps to scare away Canada geese (*Branta canadensis*), great crested grebes (*Podiceps cristatus*), loons (*Gavia* spp.), otters (Lutrinae), elk (*Cervus canadensis*), and moose (*Alces alces*) (Belant, 1985). Another possible purpose of the slap is to elicit a response from an unknown in the environment. When an animal (including human) approaches a pond but stands still, the beaver knows it is there but may not be sure if it is a threat or not. One way to know animate from inanimate is to tail slap. If it startles the unknown and it moves, the beaver knows it may be a threat.

7.4.3 Body posture and movements

Beavers can also communicate using poses (body position) and movements. Examples of this include grooming each other's furs, nose-to-nose contact, and play. (Grooming is described in Section 6.1.)

Defining play in animal behaviour is not an easy task. Some have simply labelled behaviours that do not have any other explanation as play. Play is not as violent as a fight; it is a pleasurable activity that 'trains' beavers for real-world fighting. It is therefore rare for play to result in injuries to the animals. Adult beavers can dive under a younger animal and lift it up on their head such that it falls back into

the water with a splash. The youngsters seem to enjoy it, since they immediately swim back for more. Young beavers engage in these activities most often, since the adults are not especially playful. Wilsson (1971) describes wrestling as dancing, or a show, or body language that one beaver uses to demonstrate its dominance over another. According to Wilsson (1971) the 'loser' will often react by grooming its 'dance' partner to please it. These wrestling matches can begin when the animals are just 5 days old. The kits push, embrace, and roll around with each other.

When beavers encounter one another while swimming they can often take a moment for nose-to-nose contact in passing. This often leads to them 'standing up' in the water and waving their arms at each other. This can look a lot like a boxing match. After a few blows are exchanged the boxing match can become a wrestling match. They can grab each other's skin with their fore legs and shove as hard as they can until the strongest makes the other fall backwards into the water. They can continue in this manner for several minutes until one of them finally gives up. These wrestling matches may determine the youngsters' dominance rankings. Teeth are not used in these 'friendly' wrestling matches, as they are used in fighting matches between adult animals (see Section 7.2.3). There are many indications that the youngest animals engage in these activities to prepare for real situations they can encounter later in life. Their muscles are strengthened and they learn the right combination of movements that will help them to win a battle with an unknown beaver once they mature. The young often imitate the adults when they play. It is very likely that they will eventually have to engage in a real fight with an unknown animal when they disperse to find their own territory.

7.5 Populations

7.5.1 Pattern of population development

Population growth in beavers can be divided into several phases (Figure 7.18). In general, growth rates in newly established beaver populations are initially slow due to sparsely distributed individuals (mainly subadults) with low probability for

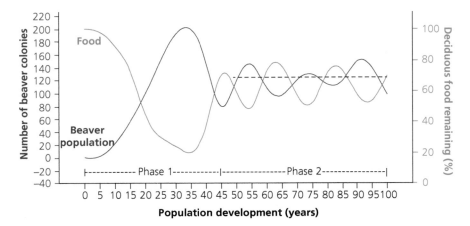

Figure 7.18 Graphic presentation of predicted relationship between a developing population of beaver (*Castor fiber*) and its prime winter food source of deciduous trees and bushes. Assumptions for the model are an area of 500 km² of typical beaver habitat in Fennoscandian boreal forest with a peak in colony density occurring at 0.40 colonies/km² after 35 years and mean colony density of 0.25/ km² (dotted line) following population recovery. Harvest pressure is light or non-existent. The basic shape of curve for colony number in 'phase 1' is relatively well founded in field studies. The shape in 'phase 2' is based on limited anecdotal observation. (© Howard Parker and Frank Rosell/University of South-Eastern Norway.)

mating. As individuals find suitable territories, form pair bonds, and produce enough offspring, dispersers are able to establish on their own. In release projects, if large (~40+) numbers of animals are released, migrating offspring will meet each other sooner and therefore population growth will be higher. As dispersers may travel dozens of kilometres from their family territories, the process of population establishment creates a 'patchwork' pattern of beaver territories. The length of time required for rapid population expansion varies depending on the characteristics of the river system and may take 15–20 years on larger river systems (Hartman, 1995). Beavers spread rapidly within water drainage systems, resulting initially in scattered colonization and very low population densities in relation to range (Hartman, 1995). When partners are found and new territories are established, the population will accelerate rapidly (Hartman, 1994; Halley and Rosell, 2002). Thereafter, beaver populations exhibit a classic 'irruptive' pattern, with a slowdown in population increase after the rapid expansion phase, occupation of marginal habitat not cable of sustaining beavers permanently, and a consequent decline in population as the 'capital' of these marginal areas is depleted (Hartman, 1994; Busher and Lyons, 1999; Halley and Rosell, 2002; Petrosyan et al., 2016).

The growth of beaver populations follows a general logistic pattern, even in populations that are not exploited (Hartman, 1994) and even with predation pressure (Balodis, 1990). The population reaches a level where the growth flattens out, i.e. a saturation point or the population's carrying capacity. The highly territorial behaviour of beaver families has a regulatory effect on beaver populations. This aspect of beaver ecology exerts therefore a considerable influence on the growth of beaver populations (see Section 7.2). In established populations, growth will decrease over time and eventually reach zero, but periods of increase and decrease can occur. Population increase can be small or non-existent due to factors such as mortality, migration connected to reintroduction, or a lack of partners. Equilibrium state with quasi-periodic oscillations was reached between 14 and 26 years in Russia (Petrosyan et al., 2013). This means that a population's *per capita* growth rate gets smaller and smaller as population size approaches a maximum imposed by limited resources in the environment. Food availability decreases over time, and the utilization of food resources is faster than they are renewed. The overbrowsing by beavers of deciduous trees and bushes for food and building material followed by territory abandonment seems to experience a time-lag due to the ensuing regeneration time for deciduous food sources (Hartman, 1994; Fryxell, 2001). This is the most common explanation for population irruptions (Hartman, 2003).

In Sweden, for example, abrupt population growth was replaced by decline 34 and 25 years after the reintroduction, and population growth rates became negative at densities of 0.25 and 0.20 families/km^2, respectively (Hartman, 2003). A similar pattern of population development has been observed in North America (Johnston and Naiman, 1990; Fryxell, 2001). In Prescott Peninsula in Massachusetts beavers returned after 200 years. The population had been unexploited for 45 years, and periods of population change were recorded. The population showed slow growth for the first 15 years, then 15 years of very rapid growth, and then a rapid decline in numbers, until it stabilized at 23% of its peak. Changes in productivity, family composition, dispersal, and interaction with whitetail deer (*Odocoileus virginianus*) were the reasons for these periods of population change. The same pattern of growth and decline was found in a population in Sagehen Creek in California, USA (Busher and Lyons, 1999). Šimůnková and Vorel (2015) found that beavers established dense populations faster in smaller catchments than larger. Therefore, the growth of the populations was not similar among differentially sized catchments. The researchers also concluded that the growth was probably based on other factors than habitat quality, such as size of catchment, progress of colonization front, and orographic dispositions. Populations that were established in the 1940s in Poland increased at mean annual rates of 0–15% between 1980 and 1988 (Żurowski and Kasperczyk, 1988), in Norway it increased 5–6% annually between 1880 and 1965 (Myrberget, 1967), and in the River Elbe catchment, Germany, there was a 7% annual increase between 1948 and 1981 (Heidecke, 1977).

Petrosyan et al. (2016) developed a single-species model and revealed four growth patterns of beaver population dynamics: (1) irruptive, (2) single-step with quasi-periodic oscillations, (3) multistep with quasi-periodic oscillations, and (4) periodic oscillations around a logistic growth curve. They concluded that their single-species model could predict the main patterns of Eurasian beaver population dynamics in different habitats.

7.5.2 Densities

Beavers can only have one litter per year, with a small number of offspring, and therefore their populations are relatively low compared to other typical rodent species. Population densities can vary greatly both spatially and temporally and are dependent upon the quality of the inhabited area (water quality, landscape type, and access to and quality of the food), the amount of time beavers have been present in the area, mortality (primarily due to hunting, winter conditions, predation, and disease), management strategies, and behaviour (territorial activity and aggression) ((Baker and Hill, 2003; Table 7.3; see Section 7.2 and Chapter 8). Population density can be measured along linear transects (per kilometre of shoreline) or within a total area (square kilometres or hectares). Since beavers stay along waterways, a linear transect can give a good enough impression of population density, but if the population will be compared to other species it is better to measure the total area (Table 7.3).

In some populations, it has taken several decades for beavers to attain typical population densities (Hartman, 1994; Halley et al., 2013). A survey in a province in southwestern Sweden found that the population density increased from 0.10 families/km^2 in 1976, to 0.19 in 1987, to 0.21 in 1999, and peaked at 0.34 families/km^2 25 years later (Hartman, 2003). In 58 different areas in Värmland, Sweden, the population increased quickly the first 25 years after reintroduction, before stabilizing or shrinking. If the populations were registered along linear transects, a completely different picture emerges. In one area in Sweden the density was 1 family per 7.6 km in 1987, which had increase to 1 family per 1.7 km in 1995. In Voyageurs National Park, Minnesota, USA, no exploitation has been allowed since 1975 and the density increased from 0.33 families/km^2 in 1958 to 1.1 families/km^2 in the mid-1980s through the mid-1990s, then decreased to 0.9 families/km^2 thereafter (Johnston and Windels, 2015). In southern Scandinavian landscapes ≥ 100 km^2 in area and encompassing beaver habitats of varying quality, the mean density of occupied sites in four studies was 0.26 families/km^2 (range = 0.32–0.20 families/km^2). On a smaller scale, the highest densities tended to

Table 7.3 Density of beaver families in varying habitat types (*Cf* = Eurasian beaver; *Cc* = North American beaver). Latitude is given as degrees, minutes, seconds (DMS).

Species	State/country and latitude (DMS)	Habitat type (families per km or km²)			Habitat description (other)	Census method	Exploited/unexploited	Approximate population age	References
		Forest	River	Other					
Cc	Navarino Island, Chile 55° 03′ 24.26″ S	–	–	1.10 km	Forest and peatland mosaic	Total count	Exploited	50	Skewes et al. (2006)
Cc	Navarino Island, Chile 55° 03′ 24.26″ S	–	–	0.79 km	Forest and peatland mosaic	Dam count	Exploited	51	Davis et al. (2016)
Cc	Tierra del Fuego Islands, Argentina 54° 50′ 35.91″ S	–	–	0.70 km²	Forest and wetland	Aerial surveys and ground surveys	Exploited	47	Lizarralde (1993)
Cc	Tierra del Fuego Islands, Chile 54° 12′ 41.80″ S	–	–	1.03 km	Forest and peatland mosaic	Total count	Exploited	50	Skewes et al. (2006)
Cc	Brunswick Peninsula, Chile 53° 30′ 0.00″ S	–	–	0.52 km	Forest and peatland mosaic	Dam counts	Exploited	22	Davis et al. (2016)
Cc	Illinois, USA 37° 22′ 58.00″ N	–	–	3.27 km²	Wetlands, hardwood forest, and agriculture	Spatial organization data and ground searches	–	–	Bloomquist and Nielsen (2010)
Cc	California, USA 39° 25′ 55.45″ N	–	1.25–1.71 km	–	–	Total count	–	34	Busher (1987)
Cc	California, USA 39° 25′ 57.00″ N	–	0.62 km	–	–	Total count	Exploited	49	Woodard (1994)
Cc	Illinois, USA 39° 27′ 35.02″ N	–	0.40 km	–	–	Total count	–	–	Cox (2005)
Cc	Colorado, USA 40° 28′ 54.65″ N	–	0.35 km	–	–	Total count	Unexploited (3 years prior to and during study)	–	Breck et al. (2001)[a]
Cc	Colorado, USA 40° 46′ 40.11″ N	–	0.50–0.60 km	–	–	Total count	Unexploited (3 years prior to and during study)	–	Breck et al. (2001)[a]
Cc	New York state, USA 42° 05′ 58.26″ N (main study site)	–	–	0.24–1.14 km	Main study site is dominated by hardwood forest and includes rivers and ponds	–	Unexploited (in 3 study areas) Exploited (in 1 site)	62 (main study site)	Müller-Schwarze and Schulte (1999b)
Cc	Massachusetts, USA 42° 21′ 37.08″ N	0.44 km	–	0.55 km	Shoreline of reservoir	Aerial photographs, ground surveys, and interviews		26	Hodgdon (1978)
Cc	Northern Wisconsin, USA 46° 07′ 46.44″ N	0.20 km	–	–	–	Total count	–	–	Ribic et al. (2017)[a]

Cc	Quebec, Canada 46° 48' 11.82'' N	–	–	0.00–1.36 km²	Province-wide	Aerial surveys	–	–	Jarema et al. (2009)
Cf	Loire Valley, France 47° 33' 11.66'' N	–	< 0.13 km	–	–	Total records	–	25	Fustec et al. (2001)
Cc	Quebec, Canada 48°06'0.65'' N	0.55 km²	–	–	–	Estimates from dam counts	Exploited	–	Tremblay et al. (2017)
Cc	Minnesota, USA 48° 26' 44.89'' N	1.40 km²	–	–	–	Aerial surveys	Unexploited	–	Johnston and Windels (2015)
Cc	Newfoundland, Canada 48° 36' 23.24'' N	–	–	<2.60 km²	Province-wide	Aerial surveys	Exploited	–	Bergerud and Miller (1977)
Cf	North Bohemia, Czech Republic 49° 05' 11.00'' N	–	0.20–0.5 km	–	–	Total records	Unexploited[b]	20	Barták et al. (2013)
Cf	Czech Republic 49° 04' 48.83'' N (forest) 49° 05' 11.00'' N (river) 49° 59' 53.59'' N (varied)	0.28km	0.29 km	0.21 km	Production forest and mosaic pastures	–	Unexploited[b]	>20	Korbelová et al. (2016)
Cf	West Bohemia, Czech Republic 49° 59' 53.59'' N	–	0.10–0.30 km	–	–	Total records	Unexploited[b]	34	Barták et al. (2013)
Cf	Moravia, Czech Republic 50° 05' 45.17'' N	–	0.11–0.15 km	–	–	Total records	Unexploited[b]	34 (central) 20 (south)	Barták et al. (2013)
Cf	Mordovia, Russia 54° 14' 13.00'' N	–	0.04–0.46 km	–	–	Ecological statistic method	Exploited	33 (oldest population)	Brozdnyakov et al. (1997)
Cf	Mordovia, Russia 54° 18' 46.00'' N	–	0.52 km	–	–	Total count	Unexploited since 2005	80	Andreychev (2017)
Cf	Poland 54° 47' 30.00'' N	–	–	0.15 km²	Almost 3.4% of area covered by water—this includes rivers, lakes, and marshes	Use of dogs, nets, and traps; trapped and fleeing beavers were counted	–	> 30	Żurowski and Kasperczyk (1986)

continued

Table 7.3 Continued.

Species	State/country and latitude (DMS)	Habitat type (families per km or km²)				Census method	Exploited/unexploited	Approximate population age	References
		Forest	River	Other	Habitat description (other)				
Cc	Molėtai district, Lithuania 55° 13' 59.988'' N	–	–	1.90 km²	Forests, meadows, lakes, and swamps (hilly moraine upland)	All beaver sites were registered and put on a topographical map with their activity status	Exploited	–	Bluzma (2003)
Cc	Alberta, Canada 55° 54' 0.00'' N[c]	–	–	3.90 km²	–	Total count	–	–	Novak (1987); Bromley and Hood (2013)
Cc	Alberta, Canada 57° 04' 0.63'' N	–	0.42 km	–	–	Aerial surveys	Exploited	–	Penner (1976)
Cf	Vestfold, Norway 59° 10' 14.83'' N	–	0.40 km	–	–	Total count	Exploited	47	Parker and Rønning (2007)
Cf	Telemark, Norway 59° 27' 36.59'' N	0.11 km², 0.24 km²[d]	–	0.25 km², 0.50 km²	Urban, Agricultural	Total count and moose hunting teams	Exploited	70	Parker et al. (2002)
Cf	Värmland, Sweden 60° 50' 8.71'' N	–	–	0.21 km²	Province-wide, around 75% boreal forest	Moose hunting teams	Exploited (however, there is only low hunting pressure)	80	Hartman (2003)
Cc	Alaska, USA 64° 56' 20.51'' N	–	0.42 km	–	–	Aerial surveys (verified by visiting locations by foot or canoe)	Exploited	–	Boyce (1974)
Cc	Northwest Territories, Canada 65° 08' 10.01'' N	–	0.39 km²	–	–	Aerial surveys	–	–	Aleksiuk (1968)[e]
Cc	Alaska, USA 66° 15' 22.03'' N	–	0.41 km	–	–	Aerial surveys (verified by visiting locations by foot or canoe)	Exploited (lightly, < 10 harvested each year for 7 years previous)	–	Boyce (1974)

Dashes indicate no data available.

[a] Mean values.

[b] Although hunting is banned, a few instances of illegal hunting may still occur in South Moravia.

[c] Latitude taken from the centre of Alberta; specific location not provided.

[d] Values are from within the moose hunting zone.

[e] Mean values from 1962 to 1965; however, 1963 was omitted due to likely being an underestimate (food cache counts were completed on 6 September before all caches were constructed). Values were converted from square miles to square kilometres by dividing by 2.59.

occur in agricultural landscapes (0.50 families/km^2) with a prevalence of low-gradient streams (Parker et al., 2013). These densities are similar to those found in other study areas worldwide with established populations, with some exceptions (Table 7.3). The maximum recorded density was 3.90 families/km^2 (Novak, 1987). The mean density of beaver families was 0.52 families/km with a range of 0.07–1.48 families/km. For densities given in square kilometres, the mean is 1.03 families/km^2 with a range of 0.11–3.9 families/km^2 (values calculated from data across all habitat types). For data given as a range, the median was calculated (see Table 7.3).

Jarema et al. (2009) examined the spatial variation in abundance of North American beavers across 1 million km^2 in the northeastern range in Quebec, Canada. They evaluated which climate (temperature and precipitation) and non-climate (nature and extent of waterways, shorelines, vegetation cover, soil composition, slope, beaver harvest intensity, and predator abundance) variables explained this variation. The density of beaver families followed a roughly logistic pattern, with high but variable density across the southern portion of the province, a sharp decline in density at about 49°N, and a long tail of low density extending as far as 58°N (Figure 7.19). Climate variables explained 97.1% of the variation that could be explained by a combination of climate and non-climate variables. Climate variables alone explained 17.4% of the variation in beaver density, non-climate variables alone explained 1.5%, and climate and non-climate variables jointly explained 33.3% (47.7% was unexplained). Climate sensitivity (change in density per unit change in climate) was greatest in the interior and lowest at the edge of the range. Jarema et al. (2009) concluded that relatively warm climates appear necessary but not sufficient for beavers to attain high densities in Quebec. Beavers survive and reproduce in the extreme climatic conditions in northern Quebec, where annual temperature is –5°C, lakes are frozen for 8 months of the year, and the only trees present are riparian shrubs, but appear to be unable to attain high densities in these regions. They also concluded that some of the most dramatic responses to climate change may be happening in the core of species' range, far away from the edge-of-the-range focus of most current modelling and monitoring efforts.

7.5.3 Abandonment of sites

An aspect of beaver ecology that exerts considerable influence on the spatial and temporal distribution of beavers and density of occupied territories is the alternating pattern of site occupation and abandonment. As the number of beavers in an environment increases, the gaps between territories and areas of poorer habitat are gradually infilled by dispersers, a process that in turn creates more territories. Later, a decrease in territorial sizes may also become evident (Campbell et al., 2005). At this point, the availability of habitat becomes a limiting factor, and territorial disputes become more common. Mortality rates increase directly through fighting and indirectly through the subsequent stresses of living in smaller territories that need to be defended more vigorously. This whole process can become physically evident by a decrease in breeding rates, delayed dispersal, and lighter individual body weights (Busher et al., 1983). While a developing beaver population with abundant habitat can display growth rates of 15–20% per annum (Hartman, 1995), this process will stagnate with no demonstrable growth once their whole environment is occupied. Therefore, although environmental factors will influence the fluctuation of beaver populations, on the whole they are self-regulating. As long as there is suitable habitat available, beaver populations will grow and expand. It is worth noting that a lack of genetic diversity may result in inbreeding depression that could limit the rate of population growth and geographic expansion (Saveljev and Safonov, 1999).

Following extensive foraging, sites may be abandoned for a varying number of years if food resources become limited. Once preferred food species have grown back in sufficient quantity, a new period of colonization will eventually occur. Thus a dynamic source–sink pattern of site occupation and abandonment becomes established, with rotation times varying depending on habitat quality and harvest levels (Fryxell, 2001). In these populations, and at a landscape scale including both good and poor beaver habitats (e.g. > ~100 km^2), the proportion of sites occupied at any time, i.e. the site occupation rate, tends to vary between 0.33 and 0.50 km^2. Thus, in populations that have somewhat stabilized following an initial population peak, between a third and half of the potential beaver habitat within a

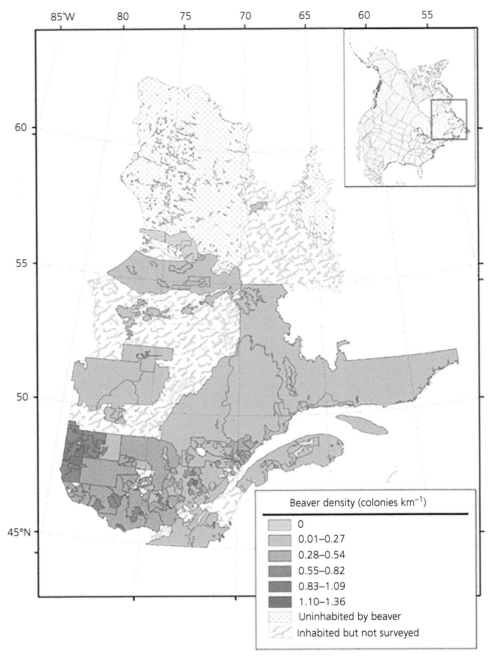

Figure 7.19 Local abundance of North American beaver across the province of Quebec. Densities were derived from 161 helicopter surveys conducted between 1976 and 2004. Average number of active beaver colonies (family groups) per square kilometre for each survey area was calculated by locating active colonies along watercourses and dividing this number by the total area. Inset: approximate North American range of *C. canadensis*. (© Stacey I. Jarema, Jason Samson, Brian J. McGill and Murray M. Humphries/Blackwell Publishing Ltd. 2008.)

Box 7.1 Inverse density-dependent territorial behaviour and aggression

Mayer et al. (2020b) investigated the spatial movement patterns and territorial behaviours by territory owners using data from marked individuals, GPS data, scent experiments (experimental scent marks were created; Box Figure 7.1), and camera trap observations for a population living on large rivers in Norway. Their results suggested that efforts to defend territories were inversely density-dependent. Territory owners spent more time patrolling at lower population densities (but still high) and therefore detected more simulated intrusions (i.e. the experimental scent marks). Also, conspecific aggression, as evidenced by number of tail scars, increased at lower population densities (Box Figure 7.2) (see also Mayer et al. 2020a). They suggested that lower population densities caused an increased mate change rate (the replacement of one pair member) (Mayer et al., 2017a) and an increased dispersal probability of subordinates (Mayer et al., 2017c). The increased mate change rate most likely increased dispersal rates due to intolerance of family group members by the new territory owner (Mayer et al., 2017c). This demonstrated an inverse density-dependent feedback mechanism between dispersal and mate change, which in turn affected the population dynamics.

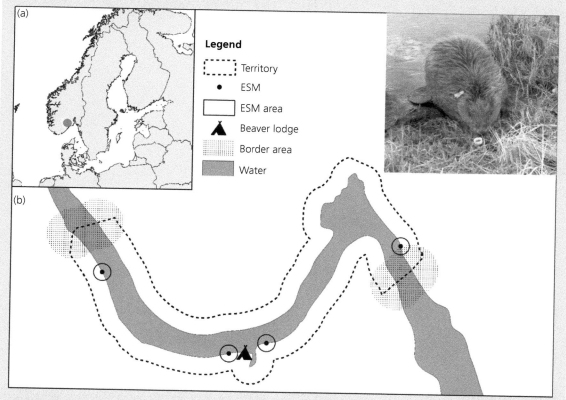

Box Figure 7.1 (a) Study area in southeast Norway (red dot) and (b) schematic representation of beaver territory with main lodge, location of experimental scent mounds (ESM), ESM areas, and border areas. (c) Eurasian beaver sniffing an ESM. The figure was created using ArcMap10.1 (Esri: http://www.esri.com/arcgis/). Source of small map (a): http://www.naturalearthdata.com/downloads/10m-cultural-vectors/ (© 2019 Martin Mayer, Shane C. Frank, Andreas Zedrosser and Frank Rosell/*Journal of Animal Ecology* © 2019 British Ecological Society.)

continued

Box 7.1 *Continued*

(a)

(b)

(c)

Box Figure 7.2 (a) Effect of population density on proportion of GPS positions in water. (b) Proportion of GPS positions inside experimental scent mound (ESM) areas before (control period) and after (treatment period) ESMs were constructed. ESMs that were not detected are shown in grey and ESMs that were detected are shown in black. (c) Proportion of detected ESM by 37 territory owners in relation to population density. GPS positions were obtained from 46 territory-holding Eurasian beavers in southeast Norway (2008–2016). Covariate effects are shown with dots or black lines and lower and upper 95% confidence intervals with bars or shaded bands. (© 2019 Martin Mayer, Shane C. Frank, Andreas Zedrosser and Frank Rosell/*Journal of Animal Ecology* © 2019 British Ecological Society.)

larger area will be in use at any particular time (Parker and Rosell, 2012).

It is the area's nutritional resources, and therefore the possibility for surviving the winter by assembling a sufficient food cache, that determines the amount of time a beaver will occupy an area. A beaver population that is not put under pressure by humans or predators has a tendency to eat itself 'out of house and home'. In other places, beavers are observed moving on occasion. They will often move up and down a waterway if possible. In Russia, previously abandoned habitats are repopulated after 3–50 years (Zavyalov et al., 2016).

When animals are forced to travel long distances on land to gather enough food, the risk of predation increases, and the animal will usually relocate. The exception to this occurs in areas where willow (*Salix* spp.) is the main food source. Willow is a nearly inexhaustible food source for beavers due to its rapid vegetative regrowth. Along rivers in Telemark,

Norway, beavers have primarily stored willow in their food caches, and many of the lodges in these areas have been regularly occupied since the 1930s. Studies in the Netherlands have shown that beavers will only remove about 10% of the regrowth of willow trees. Overexploitation is therefore not likely in these productive areas (Nolet et al., 1994).

Beavers can occasionally leave an area even when there are still sufficient food resources and a low population density. This can be explained by other factors, such as an inner need to move, disturbances from humans, death of one or both of the adult pair, low water levels, high sedimentation, or that the plants have developed chemical defences (see Chapters 4–8 and 11). The remaining food can also be too monotonous, as beavers appreciate a varied diet (see Section 5.7).

Movement of the family to another site also occurs (Fryxell, 2001; Cunningham et al., 2006). Family migrations occur primarily in the spring before the new litter is born or late in the summer when their offspring are at least 2 months old. It appears as if beaver parents wait until their young are strong enough to travel before they set out on a long move.

References

Aleksiuk, M. 1968. Scent-mound communication, territoriality, and population regulation in beaver (*Castor canadensis* Kuhl). *Journal of Mammalogy*, 49, 759–762.

Andreychev, A. 2017. Population density of the Eurasian beaver (*Castor fiber* L.) (Castoridae, Rodentia) in the Middle Volga of Russia. *Forestry Studies*, 67, 109–115.

Arnberg, M. P. 2019. The importance of vegetation availability in three scales on body condition, reproduction and territory size of the Eurasian beaver. MSc thesis, University of South-Eastern Norway.

Arner, D. H. and Roberts, T. H. 1981. Beaver pelt damage in Mississippi. *Annual Conference of South-Eastern Association Fish & Wildlife Agencies*, 35, 80–83.

Baker, B. W. and Hill, E. 2003. Beaver (*Castor canadensis*), 2nd edn. Baltimore: Johns Hopkins University Press.

Balodis, M. 1990. The Beaver. Biology and Management in Latvia. Riga: Zinátne.

Barnes, W. J. and Dibble, E. 1988. The effects of beaver in riverbank forest succession. *Canadian Journal of Botany*, 66, 40–44.

Barták, V., Vorel, A., Šímová, P., and Puš, V. 2013. Spatial spread of Eurasian beavers in river networks: a comparison of range expansion rates. *Journal of Animal Ecology*, 82, 587–597.

Bau, L. 2001. Behavioural ecology of reintroduced beavers (*Castor fiber*) in Klosterheden State Forest, Denmark. Masters thesis, University of Copenhagen, pp. 67–82.

Belant, J. L. 1985. Common loon aggression toward river otters and a beaver. *Passenger Pigeon*, 54, 233–234.

Benten, A., Cross, H. B., Zedrosser, A., and Rosell, F. 2020. Distant neighbours: friends or foes? Eurasian beavers show context-dependent responses to simulated intruders. *Behavioural Ecology and Sociobiology*, 74(2).

Bergerud, A. T. and Miller, D. R. 1977. Population dynamics of Newfoundland beaver. *Canadian Journal of Zoology*, 55, 1480–1492.

Bloomquist, C. K. and Nielsen, C. K. 2010. Demography of unexploited beavers in southern Illinois. *Journal of Wildlife Management*, 74, 228–235.

Bloomquist, C. K., Nielsen, C. K., and Shew, J. J. 2012. Spatial organization of unexploited beavers (*Castor canadensis*) in southern Illinois. *American Midland Naturalist*, 167, 188–198.

Bluzma, P. 2003. Beaver abundance and beaver site use in a hilly landscape (eastern Lithuania). *Acta Zoologica Lituanica*, 13, 8–14.

Boyce, M. S. 1974. Beaver population ecology in interior Alaska. MSc thesis, University of Alaska.

Bradbury, J. W. and Vehrencamp, S. L. 1998. *Principles of Animal Communication*. Sunderland: Sinauer Associates.

Bradt, G. W. 1938. A study of beaver colonies in Michigan. *Journal of Mammalogy*, 19, 139–162.

Brady, C. A. and Svendsen, G. E. 1981. Social-behaviour in a family of beaver, *Castor canadensis*. *Biology of Behaviour*, 6, 99–114.

Breck, S. W., Wilson, K. R., and Andersen, D. C. 2001. The demographic response of bank-dwelling beavers to flow regulation: a comparison on the Green and Yampa rivers. *Canadian Journal of Zoology*, 79, 1957–1964.

Bromley, C. K. and Hood, G. A. 2013. Beavers (*Castor canadensis*) facilitate early access by Canada geese (*Branta canadensis*) to nesting habitat and areas of open water in Canada's boreal wetlands. *Mammalian Biology—Zeitschrift für Säugetierkunde*, 78, 73–77.

Brozdnyakov, V., Skobelev, A., and Shestun, K. 1997. Population dynamics of the beaver in Samarskaya Oblast. *Russian Journal of Ecology*, 28, 245–249.

Buech, R. R. 1995. Sex differences in behavior of beavers living in near-boreal lake habitat. *Canadian Journal of Zoology*, 73, 2133–2143.

Burdock, G. A. 2007. Safety assessment of castoreum extract as a food ingredient. *International Journal of Toxicology*, 26, 51–55.

Busher, P. 2007. Social organization and monogamy in the beaver. In: Wolff, J. O. and Sherman, P. W. (eds.) *Rodent Societies: An Ecological and Evolutionary Perspective*. Chicago: University of Chicago Press.

Busher, P. E. 1975. Movements and activities of beavers, *Castor canadensis*, on Sagehen Creek, California. MSc thesis, San Francisco State University.

Busher, P. E. 1980. The population dynamics and behavior of beavers in the Sierra Nevada. PhD thesis, University of Nevada.

Busher, P. E. 1987. Population parameters and family composition of beaver in California. *Journal of Mammalogy*, 68, 860–864.

Busher, P. E. and Jenkins, S. H. 1985. Behavioral patterns of a beaver family in California. *Biology of Behaviour*, 10, 41–54.

Busher, P. E. and Lyons, P. J. 1999. Long-term population dynamics of the North American beaver *Castor canadensis* on Quabbin Reservation, Massachusetts, and Sagehen Creek, California. In: Busher, P. E. and Dzieciolowski, R. (eds.) *Beaver Protection, Management, and Utilization in Europe and North America*. New York: Kluwer Academic/Plenum.

Busher, P. E., Warner, R. J., and Jenkins, S. H. 1983. Population density, colony composition, and local movements in two Sierra Nevadan beaver populations. *Journal of Mammalogy*, 64, 314–318.

Campbell-Palmer, R. and Rosell, F. 2010. Conservation of the Eurasian beaver *Castor fiber*: an olfactory perspective. *Mammal Review*, 40, 293–312.

Campbell, R. D., Rosell, F., Nolet, B. A., and Dijkstra, V. A. A. 2005. Territory and group sizes in Eurasian beavers (*Castor fiber*): echoes of settlement and reproduction? *Behavioral Ecology and Sociobiology*, 58, 597–607.

Campbell, R. D., Rosell, F., Newman, C., and Macdonald, D. W. 2017. Age-related changes in somatic condition and reproduction in the Eurasian beaver: resource history influences onset of reproductive senescence. *PLoS ONE*, 12, e0187484.

Cox, D. R. 2005. An evaluation of beaver habitat models for Illinois rivers. MSc thesis, Eastern Illinois University.

Crawford, J. C., Bluett, R. D., and Schauber, E. M. 2015. Conspecific aggression by beavers (*Castor canadensis*) in the Sangamon river basin in central Illinois: correlates with habitat, age, sex and season. *American Midland Naturalist*, 173, 145–155.

Cross, H. B., Zedrosser, A., Nevin, O., and Rosell, F. 2014. Sex discrimination via anal gland secretion in a territorial monogamous mammal. *Ethology*, 120, 1044–1052.

Cunningham, J. M., Calhoun, A. J. K., and Glanz, W. E. 2006. Patterns of beaver colonisation and wetland change in Acadia National Park. *Northeastern Naturalist*, 13, 583–596.

Davis, E. F., Valenzuela, A. E. J., Murcia, S., and Anderson, C. B. 2016. Habitat use by invasive North American beaver during intermediate and long-term colonization periods in southern Patagonia. *Mastozoología Neotropical*, 23, 51–61.

Davis, J. R., Guynn Jr, D. C., and Gatlin, G. W. 1994. Territorial behavior of beaver in the Piedmont of South Carolina. *Proceedings of the Annual Conference of the Southeastern Association of Fish and Wildlife Agencies*, Biloxi, Mississippi, pp. 152–161.

Donkor, N. T. and Fryxell, J. M. 1999. Impact of beaver foraging on structure of lowland boreal forests of Algonquin Provincial Park, Ontario. *Forest Ecology and Management*, 118, 83–92.

Estalella, C. A. 2018. Intraspecific competition in the Eurasian beaver (*Castor fiber*) and variation between sex, age, social status and season. BSc thesis, University of Barcelona.

Frommolt, K. H. and Heidecke, D. 1992. Lautäußerungen des bibers *Castor fiber* L., 1758. *Semiaquatiche Saugetiere*, 169–174.

Fryxell, J. M. 2001. Habitat suitability and source–sink dynamics of beavers. *Journal of Animal Ecology*, 70, 310–316.

Fustec, J., Lodé, T., Le Jacques, D., and Cormier, J. 2001. Colonization, riparian habitat selection and home range size in a reintroduced population of European beavers in the Loire. *Freshwater Biology*, 46, 1361–1371.

Gillespie, P. S. 1976. Summer activities, home range and habitat use of beavers. MSc thesis, University of Toronto.

Gosling, L. 1982. A reassessment of the function of scent marking in territories. *Zeitschrift für Tierpsychologie*, 60, 89–118.

Goryainova, Z. I., Katsman, E. A., Zavyalov, N. A., Khlyap, L. A., and Petrosyan, V. G. 2014. Evaluation of tree and shrub resources of the Eurasian beaver (*Castor fiber* L.) and changes in beaver foraging strategy after resources depletion. *Russian Journal of Biological Invasions*, 5, 242–254.

Graf, P. M., Mayer, M., Zedrosser, A., Hackländer, K., and Rosell, F. 2016. Territory size and age explain movement patterns in the Eurasian beaver. *Mammalian Biology*, 81, 587–594.

Grønneberg, T. Ø. 1979. Analysis of a wax ester fraction from anal gland secretion of beaver (*Castor fiber*) by chemical ionization mass-spectrometry. *Chemica Scripta*, 13, 56–58.

Grønneberg, T. Ø. and Lie, T. 1984. Lipids of the anal gland secretion of beaver (*Castor fiber*). *Chemica Scripta*, 24, 100–103.

Haarberg, O. and Rosell, F. 2006. Selective foraging on woody plant species by the Eurasian beaver (*Castor fiber*) in Telemark, Norway. *Journal of Zoology*, 270, 201–208.

Halley, D. and Rosell, F. 2002. The beaver's reconquest of Eurasia: status, population development and management of a conservation success. *Mammal Review*, 32, 153–178.

Halley, D., Teurlings, I., Welsh, H., and Taylor, C. 2013. Distribution and patterns of spread of recolonising Eurasian beavers (*Castor fiber* Linnaeus 1758) in fragmented habitat, Agdenes peninsula, Norway. *Fauna Norvegica*, 32, 1–12.

Hartman, G. 1994. Long-term population development of a reintroduced beaver (*Castor fiber*) population in Sweden. *Conservation Biology*, 8, 713–717.

Hartman, G. 1995. Patterns of spread of a reintroduced beaver *Castor fiber* population in Sweden. *Wildlife Biology*, 1, 97–104.

Hartman, G. 1996. Habitat selection by European beaver (*Castor fiber*) colonizing a boreal landscape. *Journal of Zoology*, 240, 317–325.

Hartman, G. 2003. Irruptive population development of European beaver (*Castor fiber* L.) in southwest Sweden. *Lutra*, 46, 103–108.

Havens, R. P. 2006. Beaver home ranges and movement patterns on the Embarras River watershed in east central Illinois. MSc thesis, Eastern Illinois University.

Havens, R. P., Crawford, J. C., and Nelson, T. A. 2013. Survival, home range, and colony reproduction of beavers in east-central Illinois, an agricultural landscape. *American Midland Naturalist*, 169, 17–30.

Heidecke, D. 1977. Untersuchungen zur ökologie und populationsentwicklung des elbebibers, *Castor fiber*. MSc, Martin-Luther-University.

Herr, J. and Rosell, F. 2004. Use of space and movement patterns in monogamous adult Eurasian beavers (*Castor fiber*). *Journal of Zoology*, 262, 257–264.

Herr, J., Müller-Schwarze, D., and Rosell, F. 2006. Resident beavers (*Castor canadensis*) do not discriminate between castoreum scent marks from simulated adult and sub-adult male intruders. *Canadian Journal of Zoology*, 84, 615–622.

Hodgdon, H. E. 1978. Social dynamics and behavior within an unexploited beaver (*Castor canadensis*) population. PhD thesis, University of Massachusetts.

Hodgdon, H. E. and Larson, J. S. 1973. Some sexual differences in behaviour within a colony of marked beavers (*Castor canadensis*). *Animal Behaviour*, 21, 147–152.

Hodgdon, H. E. and Lancia, R. A. 1983. Behaviour of the North American beaver, *Castor canadensis*. *Acta Zoologica Fennica*, 174, 99–103.

Hodgdon, H. E. and Larson, J. S. 1985. Vocal communication outside the lodge by Canadian beavers (*Castor canadensis*). In: Pilleri, G. (ed.) *Investigations on Beavers*. Berne: Brain Anatomy Institute.

Hohwieler, K., Rosell, F., and Mayer, M. 2018. Scent-marking behavior by subordinate Eurasian beavers. *Ethology*, 124, 591–599.

Houlihan, P. W. 1989. Scent mounding by beaver *Castor canadensis*: functional and semiochemical aspects. MSc thesis, State University of New York.

Jarema, S. I., Samson, J., McGill, B. J., and Humphries, M. M. 2009. Variation in abundance across a species' range predicts climate change responses in the range interior will exceed those at the edge: a case study with North American beaver. *Global Change Biology*, 15, 508–522.

Johnston, C. A. and Naiman, R. J. 1990. Browse selection by beaver: effects on riparian forest composition. *Canadian Journal of Forest Research*, 20, 1036–1043.

Johnston, C. A. and Windels, S. K. 2015. Using beaver works to estimate colony activity in boreal landscapes *Journal of Wildlife Management*, 79, 1072–1080.

Kile, N. B., Nakken, P. J., Rosell, F., and Espeland, S. 1996. Red fox, *Vulpes vulpes*, kills a European beaver, *Castor fiber*, kit. *Canadian Field-Naturalist*, 110, 338–339.

Korbelová, J., Hamšíková, L., Maloň, J., Válková, L., and Vorel, A. 2016. Seasonal variation in the home range size of the Eurasian beaver: do patterns vary across habitats? *Mammal Research*, 61, 243–253.

Lederer, E. 1946. Chemistry and biochemistry of the scent glands of the beaver (*Castor fiber*). *Nature*, 157, 231–232.

Lederer, E. 1949. Castoreum and ambergris: chemical and biochemical constituents. *Perfumery and Essential Oil Record*, 40, 353–359.

Leighton, A. H. 1933. Notes on the relations of beavers to one another and to the muskrat. *Journal of Mammalogy*, 14, 27–35.

Lizarralde, M. S. 1993. Current status of the introduced beaver (*Castor canadensis*) population in Tierra del Fuego, Argentina. *Ambio*, 22, 351–358.

Mayer, M., Künzel, F., Zedrosser, A., and Rosell, F. 2017a. The 7-year itch: non-adaptive mate change in the Eurasian beaver. *Behavioral Ecology and Sociobiology*, 71, 1–9.

Mayer, M., Zedrosser, A., and Rosell, F. 2017b. Couch potatoes do better: delayed dispersal and territory size affect the duration of territory occupancy in a monogamous mammal. *Ecology and Evolution*, 7, 4347–4356.

Mayer, M., Aparicio Estalella, C., Windels, S.K., Rosell, F. 2020a. Landscape structure and population density affect intraspecific aggression in beavers. *Ecology and Evolution*, 10, 13883–13894.

Mayer, M., Frank, S.C., Zedrosser, A. & Rosell, F. 2020b. Causes and consequences of inverse density-dependent territorial behavior and aggression in a monogamous mammal. Journal of Animal Ecology 89:577-588.

Mayer, M., Zedrosser, A., and Rosell, F. 2017c. When to leave: the timing of natal dispersal in a large, monogamous rodent, the Eurasian beaver. *Animal Behaviour*, 123, 375–382.

McClanahan, K., Rosell, F., and Mayer, M. 2020. Minding your own business: low pair cohesion in a territorial, monogamous mammal. *Animal Behaviour*, 166: 119–128.

McClintic, L., Taylor, J., Jones, J., Singleton, R., and Wang, G. 2014a. Effects of spatiotemporal resource heterogeneity on home range size of American beaver. *Journal of Zoology*, 293, 134–141.

McClintic, L. F., Wang, G., Taylor, J. D., and Jones, J. C. 2014b. Movement characteristics of American beavers (*Castor canadensis*). *Behaviour*, 151, 1249–1265.

Müller-Schwarze, D. 1992. Castoreum of beaver (*Castor canadensis*): function, chemistry and biological activity of its components. In: Doty, R. L. and Müller-Schwarze, D. (eds.) *Chemical Signals in Vertebrates*. New York: Plenum.

Müller-Schwarze, D. 2001. From individuals to populations: field studies as proving grounds for the role of chemical signals. In: Marchlewska-Koj, A., Lepri, J. J., and Müller-Schwarze, D. (eds.) *Chemical Signals in Vertebrates 9*. Boston: Springer.

Müller-Schwarze, D. 2006. *Chemical Ecology of Vertebrates*. Cambridge: Cambridge University Press.

Müller-Schwarze, D. 2011. *The Beaver: Its Life and Impact*. New York: Cornell University Press.

Müller-Schwarze, D. and Heckman, S. 1980. The social role of scent marking in beaver (*Castor canadensis*). *Journal of Chemical Ecology*, 6, 81–95.

Müller-Schwarze, D. and Herr, J. 2003. Are they 'listening in the dark'? Behaviour of beavers in their lodge. *Lutra*, 46, 255–257.

Müller-Schwarze, D. and Houlihan, P. W. 1991. Pheromonal activity of single castoreum constituents in beaver, *Castor canadensis*. *Journal of Chemical Ecology*, 17, 715–734.

Müller-Schwarze, D. and Schulte, B. A. 1999a. Behavioral and ecological characteristics of a 'climax' population of beaver (*Castor canadensis*). In: Busher, P. E. and Dzięciołowski, R. M. (eds.) *Beaver Protection, Management, and Utilization in Europe and North America*. New York: Kluwer Academic/Plenum.

Müller-Schwarze, D. and Schulte, B. A. 1999b. Characteristics of a 'climax' population of beaver (*Castor canadensis*). In: Busher, P. E. and Dzięciolowski, R. M. (eds.) *Beaver Protection, Management and Utilisation in Europe and North America*. New York: Kluwer Academic/Plenum.

Müller-Schwarze, D., Heckman, S., and Stagge, B. 1983. Behavior of a free-ranging beaver (*Castor canadensis*) at scent marks. *Acta Zoologica Fennica*, 174, 111–113.

Myrberget, S. 1967. The Norwegian population of beaver *Castor fiber*. *Papers of the Norwegian State Game Research Institute*, 2, 1–40.

Nimje, P. S. 2018. The effect of social organization on genetic estimates of fitness in Eurasian beaver *Castor fiber*. PhD thesis, University of South-Eastern Norway.

Nolet, B. A. and Rosell, F. 1994. Territoriality and time budgets in beavers during sequential settlement. *Canadian Journal of Zoology*, 72, 1227–1237.

Nolet, B. A., Hoekstra, A., and Ottenheim, M. M. 1994. Selective foraging on woody species by the beaver *Castor fiber*, and its impact on a riparian willow forest. *Biological Conservation*, 70, 117–128.

Novak, M. 1987. Beaver. In: Novak, M., Baker, J. A., Obbard, M. E. and Mallock, B. (eds.) *Wild Furbearer Management and Conservation in North America*. Toronto: Queens Printer for Ontario.

Novakowski, N. S. 1969. The influence of vocalization on the behavior of beaver, *Castor canadensis* Kuhl. *American Midland Naturalist*, 81, 198–204.

Parker, H. and Rønning, Ø. C. 2007. Low potential for restraint of anadromous salmonid reproduction by beaver *Castor fiber* in the Numedalslågen River catchment, Norway. *River Research and Applications*, 23, 752–762.

Parker, H. and Rosell, F. 2012. *Beaver Management in Norway—A Review of Recent Literature and Current Problems*. Porsgrunn: Telemark University College.

Parker, H., Haugen, A., Kristensen, Ø., Myrum, E., Kolsing, R., and Rosell, F. 2001. Landscape use and economic value of Eurasian beaver (*Castor fiber*) on a large forest in southern Norway. *Proceedings of the 1st European–American Beaver Congress*, Kazan, Russia, pp. 77–95.

Parker, H., Rosell, F., and Gustavsen, Ø. 2002. Errors associated with moose-hunter counts of occupied beaver *Castor fiber* lodges in Norway. *Fauna Norvegica*, 22, 23–31.

Parker, H., Steifetten, Ø., Uren, G., and Rosell, F. 2013. Use of linear and areal habitat models to establish and distribute beaver *Castor fiber* harvest quotas in Norway. *Fauna Norvegica*, 33, 29–34.

Patenaude, F. 1984. The ontogeny of behavior of free-living beavers (*Castor canadensis*). *Zeitschrift für Tierpsychologie*, 66, 33–44.

Pedersen, B. 1999. Kjemisk analyse av flyktige stoffer i castoreum fra europeisk og amerikansk bever. Trondheim: Norwegian University of Science and Technology.

Penner, D. F. 1976. Preliminary Baseline Investigations of Furbearing and Ungulate Mammals Using Lease No. 17. Environmental Research Monograph. Edmonton: Syncrude Canada Ltd.

Peterson, A. M., Sun, L., and Rosell, F. 2006. Species and sub-species recognition in the North American beaver. In: Doty, R. L. and Müller-Schwarze, D. (eds.) *Chemical Signals in Vertebrates*. New York: Plenum.

Petrosyan, V. G., Golubkov, V. V., Goryainova, Z. I., Zav'yalov, N. A., Al'bov, S. A., Khlyap, L. A., and Dgebuadze, Y. Y. 2013. Modeling of the Eurasian beaver (*Castor fiber* L.) population dynamics in the basin of a small Oka river tributary, the Tadenka River (Prioksko-Terrasnyi Nature Reserve). *Russian Journal of Biological Invasions*, 4, 45–53.

Petrosyan, V., Golubkov, V., Zavyalov, N., Goryainova, Z., Dergunova, N., Omelchenko, A., Bessonov, S., Albov, S., Marchenko, N., and Khlyap, L. 2016. Patterns of population dynamics of Eurasian beaver (*Castor fiber* L.) after reintroduction into nature reserves of the European part of Russia. *Russian Journal of Biological Invasions*, 7, 355–373.

Piechocki, R. 1977. Ökologische Todesursachenforschung am Elbe-biber (*Castor fiber albicus*). *Beiträge zur Jagd und Wildforschung*, 10, 332–341.

Pilleri, G., Gihr, M., and Krause, C. 1985. Vocalization and behaviour in the young beaver (*Castor canadensis*). In: Pilleri, G. (ed.) *Investigations on Beavers*. Berne: Brain Anatomy Institute.

Ribic, C. A., Donner, D. M., Beck, A. J., Rugg, D. J., Reinecke, S., and Eklund, D. 2017. Beaver colony density trends on the Chequamegon-Nicolet national forest, 1987–2013. *PLoS ONE*, 12, e0170099.

Rosell, F. 2002a. Do Eurasian beavers smear their pelage with castoreum and anal gland secretion? *Journal of Chemical Ecology*, 28, 1697–1701.

Rosell, F. 2002b. The function of scent marking in beaver (*Castor fiber*) territorial defence. PhD thesis, Norwegian University of Science and Technology.

Rosell, F. 2003. Territorial scent marking behaviour in the Eurasian beaver (*Castor fiber* L.). *Denisia 9, zugleich Kataloge der OÖ. Landesmuseen Neue Serie 2*, 147–161.

Rosell, R. 2018. *Secrets of the Snout: The Dog's Incredible Nose*. Chicago: University of Chicago Press.

Rosell, F. and Bergan, F. 1998. Free-ranging Eurasian beavers, *Castor fiber*, deposit anal gland secretion when scent marking. *Canadian Field-Naturalist*, 112, 532–535.

Rosell, F. and Bergan, F. 2000. Scent marking in Eurasian beaver *Castor fiber* during winter. *Acta Theriologica*, 45, 281–287.

Rosell, F. and Bjørkøyli, T. 2002. A test of the dear enemy phenomenon in the Eurasian beaver. *Animal Behaviour*, 63, 1073–1078.

Rosell, F. and Nolet, B. A. 1997. Factors affecting scent-marking behavior in Eurasian beaver (*Castor fiber*). *Journal of Chemical Ecology*, 23, 673–689.

Rosell, F. and Pedersen, K. V. 1999. *Bever [The Beaver]*. Oslo: Landbruksforlaget.

Rosell, F. and Steifetten, Ø. 2004. Subspecies discrimination in the Scandinavian beaver (*Castor fiber*): combining behavioral and chemical evidence. *Canadian Journal of Zoology*, 82, 902–909.

Rosell, F. and Sun, L. X. 1999. Use of anal gland secretion to distinguish the two beaver species *Castor canadensis* and *Castor fiber*. *Wildlife Biology*, 5, 119–123.

Rosell, F. and Sundsdal, L. J. 2001. Odorant source used in Eurasian beaver territory marking. *Journal of Chemical Ecology*, 27, 2471–2491.

Rosell, F. and Thomsen, L. R. 2006. Sexual dimorphism in territorial scent marking by adult Eurasian beavers (*Castor fiber*). *Journal of Chemical Ecology*, 32, 1301–1315.

Rosell, F., Bergan, F., and Parker, H. 1998. Scent-marking in the Eurasian beaver (*Castor fiber*) as a means of territory defense. *Journal of Chemical Ecology*, 24, 207–219.

Rosell, F., Johansen, G., and Parker, H. 2000. Eurasian beavers (*Castor fiber*) behavioral response to simulated territorial intruders. *Canadian Journal of Zoology*, 78, 931–935.

Rosell, F., Cross, H. B., Johnsen, C. B., Sundell, J., and Zedrosser, A. 2019. Scent-sniffing dogs can discriminate between native Eurasian and invasive North American beaver. *Scientific Reports*, 9, 1–9.

Ryden, H. 1990. *Lily Pond: Four Years with a Family of Beavers*. New York: HarperCollins.

Saveljev, A. P. 2003. *Beaver's Anal Glands: View from the East*. Kirov: Russian Research Institute of Game Management and Fur Farming.

Saveljev, A. P. and Safonov, V. G. 1999. The beaver in Russia and adjoining countries. Recent trends in resource changes and management problems. In: Busher, P. and Dzieciolowski, R. (eds.) *Beaver Protection, Management and Utilization in Europe and North America*. New York: Kluwer Academic/Plenum.

Schulte, B. A. 1998. Scent marking and responses to male castor fluid by beavers. *Journal of Mammalogy*, 79, 191–203.

Schulte, B. A., Müller-Schwarze, D., Tang, R., and Webster, F. X. 1994. Beaver (*Castor canadensis*) responses to major phenolic and neutral compounds in castoreum. *Journal of Chemical Ecology*, 20, 3063–3081.

Schulte, B. A., Mullerschwarze, D., Tang, R., and Webster, F. X. 1995a. Bioactivity of beaver castoreum constituents using principal component analysis. *Journal of Chemical Ecology*, 21, 941–957.

Schulte, B. A., Müller-Schwarze, D., and Sun, L. 1995b. Using anal gland secretion to determine sex in beaver. *Journal of Wildlife Management*, 59, 614–618.

Sharpe, F. and Rosell, F. 2003. Time budgets and sex differences in the Eurasian beaver. *Animal Behaviour*, 66, 1059–1067.

Šimůnková, K. and Vorel, A. 2015. Spatial and temporal circumstances affecting the population growth of beavers. *Mammalian Biology*, 80, 468–476.

Skewes, O., Gonzalez, F., Olave, R., Ávila, A., Vargas, V., Paulsen, P., and König, H. E. 2006. Abundance and distribution of American beaver, *Castor canadensis* (Kuhl 1820), in Tierra del Fuego and Navarino islands, Chile. *European Journal of Wildlife Research*, 52, 292–296.

Sokolov, V. Y. and Shchennikov, G. N. 1990. Identification of the species *Castor fiber* and *C. canadensis* (Rodentia, Castoridae) by the structure of the anal glands. *Zoologiceskij Zurnal*, 69, 158–159.

Steyaert, S., Zedrosser, A., and Rosell, F. 2015. Socio-ecological features other than sex affect habitat selection in the socially obligate monogamous Eurasian beaver. *Oecologia*, 179, 1023–1032.

Stoye, A., Quandt, G., Brunnhöfer, B., Kapatsina, E., Baron, J., Fischer, A., Weymann, M., and Kunz, H. 2009. Stereoselective synthesis of enantiomerically pure nupharamine alkaloids from castoreum. *Angewandte Chemie International Edition*, 48, 2228–2230.

Sun, L. 1996. Chemical kin recognition in the beaver (*Castor canadensis*): behavior, relatedness and information coding. PhD thesis, State University New York.

Sun, L. 2003. Monogamy correlates, socioecological factors, and mating systems in beavers. In: Reichard, U. H.

and Boesch, C. (eds.) *Monogamy: Mating Strategies and Partnerships in Birds, Humans and Other Mammals.* Cambridge: Cambridge University Press.

Sun, L. and Müller-Schwarze, D. 1997. Sibling recognition in the beaver: a field test for phenotype matching. *Animal Behaviour*, 54, 493–502.

Sun, L. and Müller-Schwarze, D. 1998a. Anal gland secretion codes for family membership in the beaver. *Behavioral Ecology and Sociobiology*, 44, 199–208.

Sun, L. and Müller-Schwarze, D. 1998b. Beaver response to recurrent alien scents: scent fence or scent match? *Animal Behaviour*, 55, 1529–1536.

Sun, L. and Müller-Schwarze, D. 1998c. Anal gland secretion codes for relatedness in the beaver, *Castor canadensis*. *Ethology*, 104, 917–927.

Sun, L. and Müller-Schwarze, D. 1999. Chemical signals in the beaver. One species, two secretions, many functions? In: Johnston, R. E., Müller-Schwarze, D. and Sorensen, P. W. (eds.) *Advances in Chemical Signals in Vertebrates.* New York: Kluwer Academic/Plenum.

Svendsen, G. E. 1978. Castor and anal glands of the beaver (*Castor canadensis*). *Journal of Mammalogy*, 59, 618–620.

Svendsen, G. E. 1980a. Patterns of scent-mounding in a population of beaver (*Castor canadensis*). *Journal of Chemical Ecology*, 6, 133–148.

Svendsen, G. E. 1980b. Population parameters and colony composition of beaver (*Castor canadensis*) in southeast Ohio. *American Midland Naturalist*, 104, 47–56.

Svendsen, G. E. 1989. Pair formation, duration of pair-bonds, and mate replacement in a population of beavers (*Castor canadensis*). *Canadian Journal of Zoology*, 67, 336–340.

Tang, R., Webster, F. X., and Müller-Schwarze, D. 1993. Phenolic compounds from male castoreum of the North American beaver, *Castor canadensis*. *Journal of Chemical Ecology*, 19, 1491–1500.

Tang, R., Webster, F. X., and Müller-Schwarze, D. 1995. Neutral compounds from male castoreum of North American beaver, *Castor canadensis*. *Journal of Chemical Ecology*, 21, 1745–1762.

Temeles, E. J. 1994. The role of neighbours in territorial systems: when are they 'dear enemies'? *Animal Behaviour*, 47, 339–350.

Tevis, L. 1950. Summer behavior of a family of beavers in New York State. *Journal of Mammalogy*, 31, 40–65.

Thomsen, L. R., Campbell, R. D., and Rosell, F. 2007. Tool-use in a display behaviour by Eurasian beavers (*Castor fiber*). *Animal Cognition*, 10, 477–482.

Tinnesand, H. V. and Rosell, F. 2012. The potential for long-term individual recognition in the anal gland secretion of Eurasian beavers (*Castor fiber*). *Proceedings of the 16th International Symposium on Olfaction and Taste*, Stockholm, Sweden.

Tinnesand, H. V., Jojola, S., Zedrosser, A., and Rosell, F. 2013. The smell of desperadoes? Beavers distinguish between dominant and subordinate intruders. *Behavioral Ecology and Sociobiology*, 67, 895–904.

Tremblay, G., Valeria, O., and Imbeau, L. 2017. Characterisation of beaver habitat parameters that promote the use of culverts as dam construction sites: can we limit the damage to forest roads? *Forests*, 8, 494–507.

Ulevičius, A. and Balciauskas, I. 2000. Scent marking intensity of beaver (*Castor fiber*) along rivers of different sizes. *Mammalian Biology-Zeitschrift fur Saugetierkunde*, 65, 286–292.

Valeur, P. 1988. Beverens territorial-adferd som populasjons-regulerende faktor. *Fauna*, 41, 20–34.

Walro, J. M. and Svendsen, G. E. 1982. Castor sacs and anal glands of the North American beaver (*Castor canadensis*): their histology, development, and relationship to scent communication. *Journal of Chemical Ecology*, 8, 809–819.

Welsh, G. and Muller-Schwarze, D. 1989. Experimental habitat scenting inhibits colonization by beaver, *Castor canadensis*. *Journal of Chemical Ecology*, 15, 887–893.

Wheatley, M. 1997a. Beaver, *Castor canadensis*, home range size and patterns of use in the taiga of southeastern Manitoba: I. Seasonal variation. *Canadian Field-Naturalist*, 111, 204–210.

Wheatley, M. 1997b. Beaver, *Castor canadensis*, home range size and patterns of use in the taiga of southeastern Manitoba: II. Sex, age, and family status. *Canadian Field-Naturalist*, 111, 211–216.

Wheatley, M. 1997c. Beaver, *Castor canadensis*, home range size and patterns of use in the taiga of southeastern Manitoba: III. Habitat variation. *Canadian Field-Naturalist*, 111, 217–222.

Wilsson, L. 1971. Observations and experiments on the ethology of the European beaver (*Castor fiber* L.). *Viltrevy*, 8, 160–203.

Woodard, E. L. 1994. Behaviour, activity patterns and foraging strategies of beaver (*Castor canadensis*) on Sagehen Creek, California. PhD thesis, University of California.

Zavyalov, N., Albov, S., and Khlyap, L. 2016. Mobility of settlements and elements of the biological signaling field of beavers (*Castor fiber*) in the basin of the Tadenka River (Priosko-Terrasny Nature Reserve). *Biology Bulletin*, 43, 1099–1109.

Żurowski, W. 1979. Preliminary results of European beaver reintroduction in the tributary streams of the Vistula River. *Acta Theriologica*, 24, 85–91.

Żurowski, W. and Kasperczyk, B. 1986. Characteristics of a European beaver population in the Suwałki Lakeland. *Acta Theriologica*, 31, 311–325.

Żurowski, W. and Kasperczyk, B. 1988. Effects of reintroduction of European beaver in the lowlands of the Vistula basin. *Acta Theriologica*, 33, 325–338.

CHAPTER 8

Mortality and morbidity

8.1 Mortality

People commonly ask: How long do beavers (*Castor* spp.) live? With any wild animal this question is less than straightforward as we know relatively little about the various causes of mortality, and these can be complex to study. Only a few detailed studies on beaver mortality and morbidity have been undertaken. As both beaver species recover across their formal native range and are subject to various reintroduction projects, details on their health status and mortality are increasing.

Mortality rates in beavers vary greatly depending on a range of factors such as habitat quality, population density, climatic conditions, weather extremes (particularly spring flooding), hunting and trapping, predators (especially wolves [*Canis lupus*]), and diseases (Novak, 1987; Rosell et al., 2005; Girling et al., 2019c). These and other known causes of mortality, i.e. human and biological factors, in beavers are summarized in Table 8.1.

In Eurasia, territorial fights, infections and inflammations, predator attacks, and hunting are major mortality sources. Several causes of mortality were summarized in a reintroduced Eurasian beaver populations in the Netherlands and compared with two other studies from the Elbe River in Germany (source of these reintroduced individuals) (Nolet et al., 1997). Nolet et al. (1997) categorized the different causes of death as human factors, biological factors, or diseases/infections, which they found to be the most common, a similar finding to a German study (Piechocki, 1977). In dense beaver populations, increased territorial fighting often generates increased infections following wounds, whilst in another German study avian tuberculosis was found to be the major mortality factor, followed by hunting (Hinze, 1950). More recently,

a study in eastern Germany investigated causes of death of Eurasian beavers (analyzing 1,282 carcasses collected since 1957) and found that from the 1980s onwards road traffic accidents were the most significant (Stefen, 2019). These traffic accidents may indeed be a large source of mortality, but sampling bias no doubt played a part in these findings. Infections and inflammations were the predominant natural cause. In Voronzeh Reserve, Koper Nature Reserve, and Oka Nature Reserve in Russia, the main causes of beaver mortality were beaver bites during fights, predator attacks, natural diseases, natural disasters, poaching, asphyxiation under ice, and starvation (Petrosyan et al., 2019).

The mortality sources in North America are very similar, though tularaemia appears to be a much more significant mortality risk. Many North American mortality studies are usually limited to a few individual observations of causes of death. In southern Illinois, USA, 62 beavers were monitored in an unexploited population (Bloomquist and Nielsen, 2010). All of the 15 beavers found dead occurred during autumn and winter. Predation was responsible for 13% of the mortalities, and 20% of deaths were attributed to tularaemia. Disease and hunting were responsible for 74% of mortalities during the first year of monitoring in another study from an agricultural landscape in east-central Illinois, USA (Havens et al., 2013). Again, tularaemia was the main cause of death, but no cases were recorded the following year. A study on relocated North American beavers in Wyoming, USA, found that coyotes (*Canis latrans*) were responsible for 27% of all mortalities, black bears (*Ursus americanus*) and grizzly bears (*U. arctos*) 10% each, cougars (*Puma concolor*) 2%, and humans 5% (McKinstry and Anderson, 2002).

Beavers: Ecology, Behaviour, Conservation, and Management. Frank Rosell and Róisín Campbell-Palmer, Oxford University Press.
© Frank Rosell and Róisín Campbell-Palmer 2022. DOI: 10.1093/oso/9780198835042.003.0008

Table 8.1 Causes of mortality (biological and human factors) of beavers and their occurrence (shown as a percentage; excluding viral, fungal, bacterial, and parasitic diseases/infections; see Table 8.4).

Eurasian beaver

Country	Germany	Germany	The Netherlands	Sweden	Austria	Germany	Germany	Serbia	Croatia	Germany
References	Hinze (1950)[a]	Piechocki (1977)[a]	Nolet et al. (1997)[a]	Mörner (1999)	Steineck and Sieber (2003)	Weber and Weber (2013)	Zinke et al. (2013)	Grubešić et al. (2015)	Grubešić et al. (2015)	Stefen (2019)
Biological causes										
Abscesses	–	–	–	14.0	–	–	–	–	–	–
Bite trauma	0.0	17.9	0.0	–	10.0	–	–	–	–	4.9
Dog/other carnivore	9.6	3.0	5.0	–	–	1.51	–	3.0	1.0	1.8
Emaciation	–	–	–	21.0	–	–	–	–	–	–
Felled tree	0.0	1.2	–	–	–	–	–	–	–	0.4
Inundated/drowned	–	–	–	26.0	–	–	–	–	–	1.7
Old age	0.0	6.0	10.0	–	–	–	–	–	–	1.8
Perforation of the stomach	–	–	–	–	–	–	–	–	–	0.4
Problems giving birth	–	–	–	–	–	–	–	–	–	0.8
Shock (cause unknown)	–	–	–	–	–	–	–	–	–	0.2
Starvation due to tooth abnormality	–	–	–	–	–	–	–	–	–	0.9
Trauma (unspecified)	–	–	–	10.0	–	–	–	–	–	–
Winter victim	0.0	4.1	5.0	–	–	–	–	–	–	2.3
Human causes										
Accidentally caught	5.5	5.4	0.0	–	–	–	–	11.0	16.0	1.7
Club/strike	–	–	–	–	–	–	–	–	–	6.2
Hunting (gun)	–	–	–	–	–	–	–	–	–	3.5
Illegal poaching	–	–	–	–	–	–	–	8.0	6.0	–
Incised wound	0.0	0.0	5.0	–	–	–	–	–	–	–
Metal trap	–	–	–	–	–	–	–	–	–	0.9
Poisoned	0.0	2.0	0.0	–	–	–	–	–	–	1.8

Stress during capture	–	–	–	–	–	–	–	–	–	0.4
Traffic victim (road and train)	4.1	6.0	20.0	–	46.7	83.3	44.1	16.7	40.0	27.0
Transmitter (VHF)	0.0	0.0	5.0	–	–	–	–	–	–	–

North American beaver

State/country	Illinois, USA	Massachusetts, USA	Illinois, USA	Wyoming, USA	Illinois, USA	Illinois, USA	Illinois, USA	Newfoundland, Canada
References	**Bloomquist and Nielsen (2010)**	**DeStefano et al. (2006)**	**Havens et al. (2013)**	**McKinstry and Anderson (2002)**	**McNew and Woolf (2005)[b]**	**McNew and Woolf (2005)[c]**	**McTaggart and Nelson (2003)**	**Payne (1984)[d]**
Biological causes								
Age-specific mortality[e]	–	–	–	–	–	–	25.0	–
Disease	4.8	1.5	37.0	–	18.2	6.3	–	–
Felled tree	–	1.5	–	–	–	–	–	–
Predation								
Black bear (*Ursus americanus*)	–	–	–	10.0	–	–	–	–
Bobcat (*Lynx rufus*)	–	1.5	–	–	–	–	–	–
Coyote (*Canis latrans*)	–	1.5	–	27.0	–	–	–	–
Domestic dog (*Canis lupus familiaris*)	1.5	–	–	–	–	–	–	–
Grizzly bear (*Ursus arctos*)	–	–	–	10.0	–	–	–	–
Cougar (*Puma concolor*)	–	–	–	2.0	–	–	–	–
Unknown	1.5	–	–	–	20.3	9.5	–	–
Trauma	1.0	–	–	–	–	–	–	–
Unspecified	–	–	–	–	–	–	–	4.9

continued

Table 8.1 Continued.

North American beaver

State/country	Illinois, USA	Massachusetts, USA	Illinois, USA	Wyoming, USA	Illinois, USA	Illinois, USA	Illinois, USA	Newfoundland, Canada
References	Bloomquist and Nielsen (2010)	DeStefano et al. (2006)	Havens et al. (2013)	McKinstry and Anderson (2002)	McNew and Woolf (2005)[b]	McNew and Woolf (2005)[c]	McTaggart and Nelson (2003)	Payne (1984)[d]
Human causes								
Poisoned	1.6	–	–	–	–	–	–	–
Shot	–	3.0	–	–	–	–	–	–
Traffic victim (road)	–	–	5.0	–	–	–	–	–
Trapping/hunting (unspecified)	3.0	–	37.0	5.0	–	–	–	26.0
Unspecified	–	4.0–24.0	–	–	–	–	–	–
Unknown cause	9.7	–	21.0	24.0	–	15.0	–	–

Source: After Nolet et al. (1997) and Stefen (2019), reprinted by permission from John Wiley and Sons (© The Zoological Society of London) and Springer Nature Customer Service Centre GmbH: Springer Nature, Mammal Research. Causes of death of beavers (*Castor fiber*) from eastern Germany and observations on parasites, skeletal diseases and tooth anomalies—a long-term analysis, Stefan, C. © Mammal Research Institute, Polish Academy of Sciences, Białowieża, Poland (2018).

Dashes indicate no data available.

[a] Percentages were calculated by dividing the number of beavers for each mortality type by the number of beavers that the cause of death was determined for.

[b] AMAX deltaArea; values are means of adults, juveniles, and kits.

[c] Union County Conservation Area; values are means of adults and juveniles.

[d] Mean values for adult beavers between 1960 and 1970.

[e] Mean values from birth, 18 months, and 1–15 years old. Values were highest during the first 8 months of life, ranging from 36.0 to 57%. Values at 10 years old were also high at 50%.

8.2 Pre- and postnatal mortality

Pre- and postnatal mortality rates vary between studies, most likely due to whether a population has been exploited or not and according to methods used to estimate the mortality rate. Prenatal mortality can be caused by implantation failure of fertilized ova or some embryos being restricted during development (Provost, 1958; Henry and Bookhout, 1969). A review of the North American literature found that pre-implantation losses can vary from 3.8 to 38.2%, and post-implantation mortality of embryos can be an additional 2.7–17.2% (Novak, 1987). In central Illinois, USA, trapping analysis kit mortality was reported to be as high as 57% during the perinatal period (immediately before and after birth) and the first 8 months of life (McTaggart and Nelson, 2003).

Most studies suggest that kit mortality is quite high, especially compared to adult mortality rates. At one captive breeding facility, mortality rates were highest at birth and the first 3 days afterward. Mortality rates of newborn kits is higher when the mother is > 9 years old (Doboszyńska and Żurowski, 1983). Kit mortality during the first year has in some areas been estimated to be below 50% (McTaggart and Nelson, 2003; Bloomquist and Nielsen, 2010), but Payne (1984) reported as many as 52% dying during the first 6 months of life in a trapped area in Newfoundland, Canada. Twenty-eight percent of the kits survived to 11 months in a radio-tracking study in southern Illinois (Bloomquist and Nielsen, 2010). Kit survival rates over winter are still relatively unknown for those losing one or both of the parents. If food caches exist and older siblings remain, they have a much greater chance of survival. Similarly, kit survival rates when one parent is replaced by a new-comer during their first autumn are generally unknown.

Many studies have demonstrated that mortality rates are also high in the first 2–3 years of life, often during the dispersal phase, though survival of beavers > 1 year old can be very high, e.g. 92% vs. 87% for breeding adults in a dense population in Telemark, Norway (Campbell et al., 2012). In Newfoundland, Canada, mortality rates of 40% were recorded for beavers between 1.5 and 2.5 years (Payne, 1984), while in central Illinois, USA, mortality was highest

for beavers between 8 months and 2 years (McTaggart and Nelson, 2003). Animals seeking new territories can often end up fighting with territory-owning individuals. This can result in serious wounds which may become infected and potentially result in death. Examination of any tanned beaver pelt often reveals many healed scars from such fights, especially if the pelt came from an older individual (Nordstrom, 1972). Such fighting is most common in dense populations with few available territories. It is assumed that this competition between individual beavers can help to regulate the population and limit it in some areas. Fighting with others was the cause of death for 16% of beavers found in a study in Switzerland (Stocker, 1985) and 25% in a study in New Brunswick in North America (Nordstrom, 1972) (see also Section 7.2.3 and Table 8.1).

Mortality rates usually decrease after dispersal when beavers settle down within their own territory, though in the following 2 years after this, mortality was still ~40% (Petrosyan et al., 2019). In Newfoundland, Canada, mortality rates were reported to be 32% for adults (2.5 years and older) and around 20% for beavers 4.5–10.5 years old (varying between 9 and 26%) (Payne, 1984). When beavers reach 5–9 years, very few died in unexploited populations in Saskatchewan's northern forest in Canada (Gunson, 1970). Both sexes have reported similar mortality rates (Petrosyan et al., 2019).

8.3 Natural mortality factors

8.3.1 Predation

The semi-aquatic lifestyle offers an effective anti-predatory strategy. By spending most of their time in and around water or taking shelter in underground shelters with underwater entrances, beavers can avoid many land-based predators. In the winter, outer lodge material can be frozen solid, which creates an impenetrable 'armour'. An additional protection against terrestrial predators is free-standing lodges surrounded by water, i.e. a 'moat'. However, beavers do not always build a free-standing lodge, instead often building into a bank where it is not surrounded by water (see Section

4.2.1). Beaver dam systems also function as anti-predation devices due to the increased amount of water and escape routes (Fryxell and Doucet, 1991; see Section 4.4).

Beavers are most vulnerable to predation when they are away from the water. It has been suggested that beavers feed at the water's edge because it gives them better protection against predators (Boyce, 1981; Hill, 1982). Shorelines without much vegetation, for example, can force beavers to forage farther inland, thus increasing their exposure. Curiously, beavers often feed alone and avoid group foraging, particularly when outside of the main lodge/burrow (Lodberg-Holm et al., 2021; see Section 5.2.1). This has been interpreted as an adaptation to minimize predation risk (Brady and Svendsen, 1981; McClanahan, 2019). Kits face higher predation risk when they first start to explore leaving the natal lodge or burrow after birth (Bloomquist and Nielsen, 2010). Havens et al. (2013) postulated that logjams and other physical structures, including aquatic vegetation, may provide vulnerable kits escape cover while also slowing the water currents, making it easier for kits to retreat to the lodge. In east-central Illinois, there was a weak positive correlation between number of kits per family and number of logjams near the lodge, and Havens et al. (2013) found that the estimated number of kits per family had a weak positive correlation to the number of logjams near the shelter. Beavers relocated to large rivers (> 2.83 m/s) in Wyoming had greater aquatic escape cover from predators than their counterparts released in small streams (< 0.283 m/s) (McKinstry and Anderson, 2002).

Despite their seemingly adaptive defences, the beaver has its natural enemies. If water is not immediately accessible, beavers may 'freeze' (remain motionless) when threatened (this has been reported for both wild and captive beavers). Freezing is accompanied by fear bradycardia (slowing of the heart rate). This behaviour may help conceal the beaver on land and reduce the likelihood of its detection by predators (Swain et al., 1988). Long dive times (~15 minutes) have also been recorded in extremely frightened beavers, which have been observed pressing themselves against the bottom of a river or lake, where they remain motionless, or they can swim considerable distances (800 m)

underwater to escape the threat (Irving and Orr, 1935; Wilsson, 1971).

Wolves travelling along shorelines have been known to kill beavers by trapping a foraging beaver on land (Gable et al., 2016). In the water, beavers have a better chance of escaping. Beavers have been shown to select water streams deeper than 40 cm to be able to more easily escape predation (Baskin and Novoselova, 2008). A good example of this was shown in an American nature documentary where a cougar surprised a beaver that was working on its dam. The beaver went straight into the water, where after an intense fight it got away with no visible injuries. If a predator attempts to enter the lodge, an adult beaver may respond by making loud noises like hissing (see section 7.4.1), grinding its teeth, or charging and attacking. There are very few mammals that risk and even manage to enter a beaver lodge, although otters (*Lontra canadensis* and *Lutra lutra*) and bears have been known to do so through either stealth or full-blown brute strength (Table 8.2).

The majority of predation events on beavers are based on finding and examining carcasses or investigating predator stomach contents or scats; therefore cause of death is not always easy to determine. Unfortunately, from the latter data it is not possible to know if beavers were already dead and scavenged upon or predated. Only a few studies have reported a direct attack or provided evidence from video-footage and/or camera traps (Table 8.2).

8.3.2 Common predators

Not many predators regularly take on adult beavers, given their size, strength, and razor-sharp teeth powered by strong jaw muscles (Gable et al., 2018a). As such, there are only a handful of large predators (for example bears, cougars, and wolves) that commonly predate on adult beavers. However, there is little evidence that beaver populations are actually regulated through predation.

In most parts of the world, wolves are the main predator of beavers and some evidence suggests that individual wolves can specialize in beaver hunting (Gable et al., 2018b; Figure 8.1). Wolves primarily prey on beaver during the ice-free season, when wolf predation may be significant on beavers of all age classes. Beavers make up between 5 and

Table 8.2 Possible predators of beavers.

Common name	Scientific name	Evidence	References
American alligator	*Alligator mississippiensis*	Carcass and attack observed	Wilkinson (1962); Arner et al. (1981)
American mink	*Mustela vison*	In lodge and scats	Recker (1997); Valenzuela et al. (2013)
Black bear	*Ursus americanus*	From live traps, dug into lodges, and scats	Gunson (1970); Jackson (1990); Smith et al. (1994)
Bobcat	*Lynx rufus*	Scats and digestive tract	Litvaitis et al. (1986); Toweill and Anthony (1988)
Brown bear	*Ursus arctos*	Killed in lodge/dens, hairs and scats from kill site and carcass	McKinstry and Anderson (2002); Rosell et al. (2005); Baskin (2011a); Safonov (2016)
Canada lynx	*Lynx canadensis*	Stomach and scats	Saunders (1963)
Cougar/mountain lion	*Puma concolor*	Stomach and scats	Spalding and Lesowski (1971); Ackerman et al. (1984); Lowrey et al. (2016)
Coyote	*Canis latrans*	Observed attacks and scats	Packard (1940); Basey and Jenkins (1995)
Culpeo fox[a]	*Pseudalopex culpaeus*	Video footage and camera traps	Tadich et al. (2018)
Dog	*Canis lupus familiaris*	Dug out dens and observed attacks	Rosell et al. (2005); Baskin (2011a)
Eagle owl	*Bubo bubo*	Scats	Baskin (2011a)
Eurasian lynx	*Lynx lynx*	Carcass	Rosell et al. (2005)
Eurasian otter	*Lutra lutra*	Scats	Baskin (2011a)
European badger	*Meles meles*	Puncture wounds	
European marten	*Martes martes*	In lodge and scats	Rosell and Hovde (1998); Janiszewski et al. (2014)
Fisher	*Martes pennanti*	?	Payne (1984)
Grey wolf	*Canis lupus*	Video footage of attacks, kill sites, scats, and dug out lodges	Gable et al. (2016); Gable et al. (2018b)
Northern pike	*Esox lucius*	Observed attacks	Baskin (2011a); Janiszewski et al. (2014)
Raccoon dog	*Nyctereutes procyonoides*	?	Janiszewski et al. (2014)
Red fox	*Vulpes vulpes*	Attacked, scats, carcass, and dug out dens	Payne and Finlay (1975); Jones and Theberge (1983); Kile et al. (1996); Sidorovich and Sidorovich (2011)
Red wolf	*Canis rufus*	?	Müller-Schwarze (2011)
River otter	*Lontra canadensis*	Scats	Green (1932); Greer (1955); (Reid, 1984); LeBlanc et al. (2007)
Sea eagle	*Haliaeetus albicilla*	Carcass	Nolet et al. (1997)
Snapping turtle	*Chelydra* spp. and *Macrochelys* spp.	?	Bailey (1927)
Taimen	*Hugo taimen*	Observed attacks	Baskin (2011a); Janiszewski et al. (2014)
Wolverine	*Gulo gulo*	Attacks and stomach, activity sites with a carcass	Rausch and Pearson (1972); Tyurnin (1984); Scrafford and Boyce (2018)

Source: After Gable et al. (2018b), reprinted by permission of John Wiley and Sons (© 2018 The Mammal Society and John Wiley & Sons Ltd).

[a] Only document attempted predation event on beavers, and there is no evidence that a Culpeo fox has been successful.

Figure 8.1 Progression of a wolf hunting a beaver: (a) wolf waiting for and then ambushing the beaver; (b) dragging the beaver out of a small stream and into the road; (c) chasing the escaping beaver; (d) briefly releasing the beaver after being bitten; (e) attacking the beaver again shortly after releasing it; (f) continuing to attack and subsequently kill the beaver. (Reprinted from Gable, T. D., Stanger, T., Windels, S. K. and Bump, J. K. 2018. Do wolves ambush beavers? Video evidence for higher-order hunting strategies, *Ecosphere*, 9, e02159 under CC BY 4.0 License.)

60% of the wolves' diet in different areas in North America and between 6 and 36% in different areas in Eurasia (Gable et al., 2018b). Beavers are relatively vulnerable to predation when on land, but not all encounters are fatal for the beavers (Gable et al., 2018b). In Minnesota, USA, some beavers display old canine puncture tail wounds most likely from previous wolf attacks. Beaver kits and yearlings might be pretty easy to kill, but large sub-adults and adults are quite difficult for wolves to kill (Gable et al., 2018a; Gable et al., 2018b). Beavers may also be attacked in water and killed onshore (Gable et al., 2016), though in the water adult beavers might be able to drown smaller wolves (Gorbunova et al., 2008). A typical wolf hunting strategy has three components (Gable et al., 2016): '1) waiting near areas of high beaver use (e.g. feeding trails) until the beaver comes near shore or ashore, 2) using vegetation, the dam, or other habitat features for concealment, and 3) attacking the

beaver by cutting off access to water, or immediately attacking the beaver (e.g. ambush).' Wolves have also been found to dig into active lodges, though there are no accounts of wolves managing to kill beavers this way (Gable et al., 2018b). During spring (April–May), wolves primarily hunt and kill beavers below or on shorelines (58% of kills), whereas during autumn (September–October) wolves primarily hunt and kill beavers around feeding canals and trails (80% of kills) (Gable et al., 2016).

If any predator is going to influence beaver population dynamics, it would be wolves, given their propensity to hunt and kill beavers. A study that analysed and synthesized all available literature on wolf–beaver interactions concluded that there is little evidence to suggest wolves influence beaver population dynamics in North America or Eurasia in any meaningful way (Gable et al., 2018b). For example, in northern Minnesota, one adult male wolf killed 22 beavers during a single ice-free season, and the entire pack was estimated to remove 38–42% of the beaver population in the pack's territory, though beavers still increased by 43% the following year (Gable and Windels, 2018). Wolf–beaver interactions have been relatively understudied and more research is needed to fully understand the effect of wolf predation on North American beaver populations (Gable et al., 2016).

In Europe, beavers are only considered to be a minor food source for wolves, with the exceptions of Belarus (up to 30% of their diet) (Sidorovich et al., 2017) and Latvia (36% of the diet) (Andersone and Ozoliņš, 2004; Gable et al., 2018b). This could change in future as wolves and beavers continue expanding across their historical ranges. In northwestern Poland, the adult wolf diet consisted of only 5.6% beavers; interestingly, wolf pups' diets were 19.8%, indicating beavers might actually be quite an important food source in some areas (Mysłajek et al., 2019).

Most bear predation on beavers appears to be largely opportunistic, and bears do not appear to target beavers, except in rare instances where other food sources are limited. In the Stockton Islands in Wisconsin, USA, black bears managed to dig their way into 18 of 26 (69%) beaver lodges, even reaching the sleeping chamber of some (Smith et al., 1994).

It is suggested that bears are attracted by the sounds of newborn kits (Swank, 1949; Nash, 1951). Bears have also eaten beavers trapped in Hancock live-traps during autumn, and six of 90 bear scats (7%) contained beaver hair (Smith et al., 1994). Heavy predation pressures from the bears combined with reduced food access contributed to a decrease in beaver population on Stockton Island—from 13 active families in 1987 to only one in 1992 (Smith et al., 1994). In Montana, USA, evidence was provided that suggested that black bears may be a significant mortality factor, especially on dispersal-aged beavers (Jackson, 1990).

8.3.3 Infrequent predators

Although much less reported and with limited scientific information, a range of additional predators of adult beavers have been sporadically documented, including alligators (*Alligator mississippiensis*), lynxes (*Lynx canadensis*, *L. rufus*, and *L. lynx*), cougars, coyotes, domestic dogs (*Canis lupus familiaris*), and wolverines (*Gulo gulo*) (see Tables 8.1 and 8.2). Beaver kits appear to be particularly vulnerable to small-to-medium-sized predators such as foxes (*Pseudalopex culpaeus* and *Vulpes vulpes*), otters, martens and fishers (*Martes* spp.), European badgers (*Meles meles*), raccoon dogs (*Nyctereutes procyonoides*), American mink (*Mustela vison*), ferret (*M. putorius furo*), snapping turtles (*Chelydra* spp. and *Macrochelys*), and even large fish such as northern pike (*Esox lucius*) and taimen (*Hugo taimen*) (Bailey, 1927; Zharkov and Sokolov, 1967; Hilfiker, 1991; Baskin, 2011a). Only one study from Russia has reported beaver remains in any bird scats, though it was found that 47% of 23 eagle owl (*Bubo bubo*) scats contained hair and teeth from beaver kits (Baskin, 2011a). However, it has been speculated that other large owls, hawks, buzzards, and falcons may also kill beaver kits (Hilfiker, 1991; Baskin, 2011b).

In one slightly misguided event, four large alligators (1.2–3 m) and six beavers (4.5–15.8 kg) were released into a pond in Mississippi, USA. Perhaps unsurprisingly, the alligators killed and ate all except one of the beavers (Arner et al., 1981). At another site, the release of 33 alligators was later demonstrated to have had a major reduction impact on local beaver populations. A 2-m-long alligator may kill beavers as heavy as 13 kg (Arner et al., 1981). A cougar in Colorado, USA, became specialized in beaver hunting, spending up to six times more time in beaver habitat and consuming ten times more individual beavers than expected (Lowrey et al., 2016). Coyotes have hunted and killed beavers in the Rocky Mountain National Park, Colorado, USA (Packard, 1940). Coyote scats also contained 10.1% beavers during spring and 5.6% during winter in the Adirondack Mountains of New York (Müller-Schwarze, 2011). Coyotes killed 27% of transplanted beavers in Wyoming, USA (McKinstry and Anderson, 2002). Interestingly, it is estimated that a beaver may escape a coyote attack on land if not further than 7.1 m from water (Basey and Jenkins, 1995). Another predator is the lynx, which were responsible for 15% of beaver deaths in some regions of Russia (Glushkov et al. 2001, referred to in Janiszewski et al., 2014). Stomach and colon contents from 193 wolverines contained remains from only two beavers (Rausch and Pearson, 1972). In northern Alberta, Canada, activity sites were located from GPS-collared wolverines in winter, and the remains of 21 beavers were found. Also, documentation of wolverines residing in beaver lodges or under ice after killing them were provided (Scrafford and Boyce, 2018). The few reported studies of beavers killed by wolverines may be due to wolverines living in more mountainous areas where fewer beavers live.

There have been occasions where domestic dogs have mauled beavers to death, but in most cases the dogs are on the losing side of such encounters. In Scotland, where beavers have been more recently reintroduced, a number of instances of dogs bringing newly emerged kits into the houses of their human owners have been reported. In Russia, stray dogs were responsible for 27% of beaver deaths in some areas (see references in Janiszewski et al., 2014)). Numerous beaver-caused injuries in dogs have been reported. It is therefore good practice to keep dogs on a leash or out of the water especially during the breeding season in areas where there is a possibility of confrontation with a beaver (Salvesen, 1927).

Several researchers have claimed that red foxes can kill beavers (Figure 8.2), but few direct observations are documented. In southeast Norway, a red fox predation of a female Eurasian beaver kit (ca. 2.5 kg) was recorded while it was feeding 2 m from the shoreline near the lodge (Kile et al., 1996). The fox killed the beaver with a bite to the spine. Beavers from 1 to 2 months of age are the most at risk from attacks by foxes while they feed on land. Their flight instinct has not yet fully developed at such a young age. Red foxes attacked eight young beavers and killed them and remains of 30 other beavers were found near the exits of fox dens during July and August in Belarus (Sidorovich and Sidorovich, 2011). In Tierra del Fuego Island, Argentina, beaver remains were recorded in American mink scats (1.3%). Mink scavenging on dead beavers were observed several times, but direct predation on beaver kits may also be a cause of death (Valenzuela et al., 2013).

Otters are excellent swimmers and active hunters. They can easily outmanoeuvre beavers in the water and manage to enter the lodge at any time of the year. It is primarily the kits that are most at risk, especially in the first month(s) after birth when they are still inside the lodge. There have been observations of otters being chased away by both species of beavers, especially during the breeding season and if they enter the lodge. There have even been observations of beavers working together to attack an otter. Remains of North American beavers were

found in only 0.4% of 1,140 otter scats in Alberta, Canada (Reid, 1984). In Russia, only 1.8% of 165 otter scats contained beaver hair, and only two cases of otter predation of beavers are known there (Baskin, 2011a). Many researchers believe that the beaver and the otter live in a begrudged truce throughout the year (Janiszewski et al., 2014). This has been confirmed both via direct observations and through analysis of the food choices of the otter (Tyurnin, 1984). However, beaver-killed otters have been found inside lodges and near dams. The beaver has one advantage in such conflicts in that its incisors are long and sharp, as opposed to those of the otter which are much shorter (Rosell and Pedersen, 1999).

8.3.4 Drowning, water regulations, and floods

Although beavers are very capable swimmers and spend their lives in and around water, drowning can be a mortality factor (see Table 8.1 for rates). Numerous beavers are recovered, drowned or near death, along sea coastlines, especially when dispersing from areas of dense populations (see also Section 4.1.3). Juvenile beavers searching for new territories are often at the greatest risk of drowning. Other less common causes of drowning are entanglement in fishing nets and trout-traps (Mörner, 1992; Stefen, 2019). Beavers may also drown under the ice during sudden snow melts in midwinter or violent spring breakups that raise water levels in streams and destroy lodges (Hill, 1982).

Beavers can experience drastic changes in water levels even throughout a single growing season in a pond. For example, a dam may be destroyed in spring, the pond drains, and the beavers build the dam back up. The evolutionary forces have most likely selected for beavers that are able to survive during changing water conditions. However, rapid midwinter or early spring snowmelt can lead to dramatic increases in water levels, which can destroy lodges and food caches, or sometimes even drown the beavers trapped under the ice. Families in larger rivers and regulated waterways are especially at risk. Because of regular floods, these beavers tend to expend more energy and can lose weight, in some instances drowning or starving to

Figure 8.2 Camera trap footage of a red fox predating on a beaver kit. (Photo supplied courtesy of Volker Zahner.)

Figure 8.3 Beavers on top of a lodge during a flood. (Photo supplied courtesy of Willy de Koning in the Netherlands.)

death (Hill, 1982; Nitsche, 2001). Beavers may find temporary shelter elsewhere, such as other lodges/dens, trees, platforms for game, or dikes, during flood periods, and young beavers can be brought to these safe places by the adults (Nitsche, 2001; Figure 8.3).

In a river in Colorado, USA, North American beavers were very adaptable to flooding (Breck et al., 2001). The beavers moved to a burrow located farther from the centre of the river during floods. Flooding events also gave them some advantages by minimizing the predation risk because beavers had increased access to preferred foods with the increased water level. In Oslo, Norway, two Eurasian beaver families survived a 5.5-m decrease in water level during a dam repair that happened in January and November 2016. The beaver families were helped by humans with extra food (apples and wood) and isolation mats to keep the water open, and the direction of a stream was changed to increase the water level in front of the lodge. However, no kits were reportedly produced that year (Bergan and Rosell, 2017).

In Sweden, several beavers in the Fax River died as a result of waterway dam regulations in the 1950s. In regulated waterways the food caches were washed away or left lying above the water's surface during dramatic changes in water level. Beavers can also be carried away by currents and over waterfalls, especially in the springtime.

A snow melt in only 24 hours created a flood in January 1997 in Sagehen Creek, California, and the beavers either perished or were washed downstream (Busher, 1999). There have also been recordings of large numbers of dead kits during severe floods overlapping with the birthing period (Wilsson, 1971; Wilsson, 1995). The kits are very vulnerable to floods in their first month because they are not able to dive properly out of the lodge's flooded entrances. The parents will often attempt to save their offspring by carrying them out in their mouths and assembling them in temporary lairs above the waterline (see Section 6.8).

8.3.5 Dental issues

Beaver's incisors grow continuously throughout their lives, and overgrowth of the incisors can occasionally occur if not worn down appropriately (see also Section 10.3.2 for captive issues). The reasons for incisor overgrowth can include deformities such as twists, shortened or lengthened jawbones, injuries such as gun wounds leading to the loss or damage of incisors, or a change in the positioning of the incisors (Wiggs, 1990; Kim et al., 2005) (Figure 8.4). The incisors can become very long and grow in a circular form (Cave, 1984). This makes foraging and cutting wood a problem, but despite this, captured beavers with overgrown incisors have occasionally been otherwise healthy. In many cases, the animal may lose its ability to feed or close its mouth fully. A 3-year-old Eurasian beaver in southern Norway, otherwise healthy and weighing 20 kg, was observed with unusually long lower incisors (Rosell and Kile, 1998). The beaver was observed noisily feeding on birch branches. The lower incisor in the left part of the jaw had grown into an arc that nearly reached the animal's eye. The incisor was 13.5 cm long and was pointed at the end. As beavers are born with pointed incisors and the lower jaw was damaged, it was concluded that the incisor had grown so long due to a birth defect or injury shortly after birth. Another example was a North American beaver with a lower incisor that had grown into a nearly perfect circle through the skin of the lower jaw, stopping behind the last molar. Even with this handicap the animal survived to reproductive age (Cave, 1984).

8.3.6 Harsh winters

Unusually severe winters can lead to beavers dying of hunger and malnutrition, especially if the ice is extra thick (Figure 8.5a), and if the entrance to the burrow or lodge is located above the water level (Figure 8.5b). If the entire food cache is frozen in ice and the lodge's exits are frozen solid, the beaver must dig through the walls of the lodge to come out and look for food on land. This was observed in two lodges in Telemark, Norway, in the winter of 1995–1996, which was unusually cold. These types of observations are fairly common during colder winters. Beavers can be trapped inside a lodge during persistent cold snaps, when the exits become frozen and the lodge itself covered with a thick layer of ice. The animals can then die of hunger inside. In some cases, a family may not manage to store enough food, and the caches can be emptied during the winter. This can happen to younger or inexperienced animals or during particularly long winters, leading to the early exhaustion of resources. The beaver may then attempt to change the depth of the waterway by making a benthic canal in front of the lodge and adjacent to the food cached, to prevent the water freezing (Grudzinski et al., 2019; see Figure 4.15; point number 8). In Alberta, Canada, it was determined that increasing water levels by 10 cm was sufficient to prevent wetlands from freezing solid to their bottom. In adjacent localities without canals, 10% of the families did not survive the winter due to water freezing beneath their lodges (Hood and Larson, 2015). It was also found that the pond water near lodges defrosted 11 days earlier than other pond water, because the beaver lodges and food caches had a warming effect on the water and the surrounding land (Bromley and Hood, 2013). Its estimated that beaver populations can experience mortality rates up to 30% during severe winters in Newfoundland, Canada, with kits being especially vulnerable (46% mortality) (Bergerud and Miller, 1977).

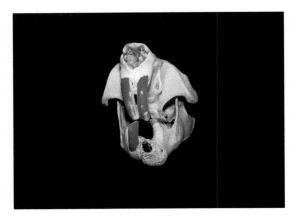

Figure 8.4 Beaver skull displaying a significant dental abscess that has permeated the lower jawbone. (Photo supplied courtesy of Roisin Campbell-Palmer.)

Figure 8.5 An Eurasian beaver kit in harsh winter conditions, with a frozen waterbody. Its burrow entrance is above the water level and some few meters behind the animal (a). A beaver kit emerging from its burrow, with the entrance exposed due to a drop in water levels (b). (Photo supplied courtesy of Frank Rosell.)

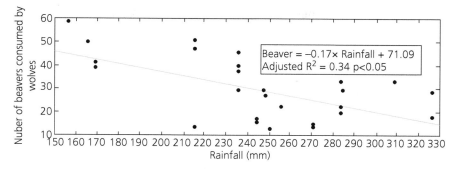

Figure 8.6 Negative relationship between beaver consumption by wolves and rainfall and, analysed by linear regression. The lower the rainfall, the higher the consumption (This article was published in Sidorovich, V., Schnitzler, A., Schnitzler, C., Rotenko, I., and Holikava, Y., (2017). Responses of wolf feeding habits after adverse climatic events in central-western Belarus, *Mammalian Biology*, 83, 44–50, Copyright Elsevier 2017.

8.3.7 Drought

Severe drought situations have been linked to increased beaver loss. For example, a significant drought period in 2007 was the likely reason for a 32% decline in the northwestern Wisconsin beaver population between 2005 and 2008 (Ribic et al., 2017) and in beavers in western Nebraska (Novak, 1987). In Belarus, eastern Europe, droughts greatly decreased the water level in rivers and canals, and therefore there was increased wolf predation on beavers (Sidorovich et al., 2017; Figure 8.6).

8.3.8 More unusual endings

Beavers have also reportedly died in a variety of unusual and surprising ways. These have included being killed by charging male moose (*Alces alces*) in Norway and in the USA (Salvesen, 1936) and bites from poisonous vipers (Viperidae) (Baskin, 2011a). In Russia, young animals are known to be crushed by floating logs (Tyurnin, 1984).

To fell a tree is not entirely without risk for beavers, and occasionally an animal pays with its life (Kile and Rosell, 1996; Nitsche, 1996; Figure 8.7). The beaver can be killed by the sharp point of the tree if it hits the animal as the tree falls, or the tree can itself fall on the beaver. The tree can occasionally fall on the beaver's tail or back leg, which may trap it, leading to a slow and painful death, evident by obvious digging tracks around the trapped animal and consumed vegetation nearby. The beaver will finally die of hunger or thirst. Since the majority of tree felling

Figure 8.7 Although they usually avoid it, beavers can be crushed by trees they are felling. (Photo supplied courtesy of Beate Strøm Johansen.)

occurs during the autumn, this is also the time more of these types of accidents are recorded. Six of eight beavers that have been found dead as a result of tree felling were found during the period from August to November, and those that were measured for age were all adults (Kile and Rosell, 1996).

8.4 Human factors

8.4.1 Pollutants

Humans have released many pollutants into beaver habitats, though how pollutants affect beavers is largely unknown as few studies have examined the toxicological impact of environmental contaminants (Peterson and Schulte, 2016). Deaths resulting from direct poisoning are rare, but sublethal effects (low-level to high-level, not fatal exposure) can cause organ failure and disease, as well as affect reproductive success and population dynamics (Reich et al., 2001). It is also possible that pollutants may influence the endo-parasite communities in the gut of beavers, but more research is needed (Borchert et al., 2019).

Generally, beavers appear unaffected by water pollution. Agricultural runoff may contribute to the death of some beavers, but it doesn't seem to be a major cause of mortality. It is potentially more likely such runoff, full of fertilizers and herbicides, could affect the food resources beavers rely on in agricultural areas. Fertilizers and herbicides may reduce the diversity of aquatic plants and herbs, some of which can be essential supplementary food sources for beavers. A reduction of these plants can influence

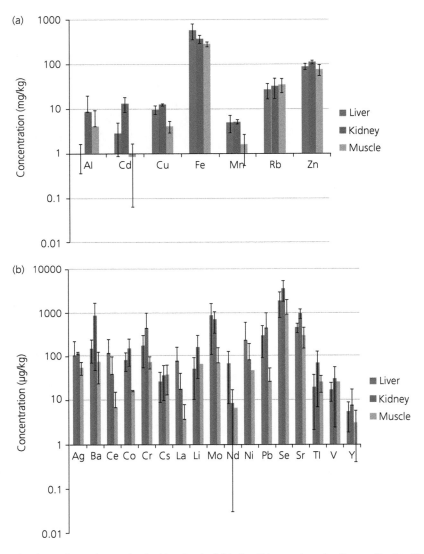

Figure 8.8 Average abundance of trace elements (mg/kg (a) and μg/kg (b)) in liver, kidney, and muscle of beaver. (Reprinted from Shotyk, W., Bicalho, B., Dergousoff, M., Grant-Weaver, I., Hood, G., Lund, K., and Noernberg, T., (2019). A geochemical perspective on the natural abundance of trace elements in beaver (*Castor canadensis*) from a rural region of southern Ontario, Canada, *Science of the Total Environment*, 672, 40–50, © Elsevier Science (2019), with permission from Elsevier.)

reproduction and survival (Rosell and Pedersen, 1999), though in some places runoff from agriculture and urban areas can be advantageous to beavers, since it contributes to plant growth via increased fertilizer levels. Toxic substances produced by humans released into the watercourse can commonly damage cells and tissues at critical levels. Some common pollutants are cadmium, copper, mercury, nickel, lead, selenium, and zinc (Figure 8.8a,b). These pollutants are of particular interest, and their concentrations (and some other ones) within beaver areas are presented in Table 8.3. Concentrations of pollutants can vary greatly among individual beavers, beaver species, and ecosystems, driven by a variety of biotic and abiotic factors (Borchert et al., 2019).

Acid rain can have a harmful effect on the vegetation beavers rely on and on the beavers themselves. Precipitation that is unusually acidic with low pH is called acid rain. One problem with the effects of acid rain is that substances that were previously bound to soils or sediments can be released and taken up first by vegetation and later by beavers. One such harmful substance in acid rain is cadmium. Cadmium accumulates primarily in the liver and kidneys of beavers, so these organs should not be ingested by humans (Fimreite et al., 2001). Liver cadmium concentrations in North American beavers from South Carolina and Georgia were on average 0.9 and 0.11 µg/g wet weight, respectively (Borchert et al., 2019) and less than reported from Canada (0.19 µg/g wet weight) (Wren, 1984) and Poland (0.22–0.88 µg/g wet weight) (Zalewski et al., 2012; Giżejewska et al., 2015). Higher liver and kidney cadmium concentrations have also been found in adult beavers compared to juveniles (Giżejewska et al., 2014). Mean cadmium in the kidney of Eurasian beavers in Poland was nearly eightfold higher than recommended values (Giżejewska et al., 2014). Elevated levels of cadmium and lead were found in the kidneys and liver of Eurasian beavers from a former military airport area (Zalewski et al., 2012). Analyses of cadmium concentrations in Eurasian beavers' livers and kidneys have been performed in Telemark, Norway, and average values of 1.03 µg/g wet weight for liver and 10.25 µg/g wet weight for kidneys were found (Fimreite et al., 2001). These elevated values (the recommended maximum is 0.5 µg/g wet weight; Commission Regulation [European Commission], 2006) were found in beavers whose diet consisted primarily of willow (*Salix* spp.), a shrub that readily accumulates cadmium. Cadmium concentrations in beavers have also been found to increase with age, i.e. cadmium is stored in the body once ingested. Fimreite et al. (2001) suggested that higher than normal cadmium concentrations may reduce fat content in beavers and therefore increase not only their susceptibility to diseases but also the risk of overwintering starvation. High cadmium and lead content in the tissue of one beaver in Poland changed its sperm morphology, and cadmium can also induce changes in the concentration of sex hormones (Zalewski et al., 2012). High concentrations of cadmium in beaver meat can have health implications for humans eating them.

The Mulde River beaver population in Germany seems to be thriving despite their high cadmium intake (Nolet et al., 1994). A beaver living here had the highest cadmium concentration ever measured in the kidneys of a free-living herbivore, i.e. 93.4 µg/g wet weight. Cadmium levels in both hair and kidney were correlated with concentrations of cadmium in both the bark and leaves of willow and popular, and the accumulation increased with time. Interestingly, a threefold increase in cadmium concentration in hair in less than 6 months was found. Negative effects cannot be ruled out. Beavers in this area were often afflicted with porous bones from a young age, and displayed symptoms of this from between 2 and 5 years old (Nolet et al., 1994).

Levels of radio-caesium (^{137}Cs), copper, and zinc in the tissue of Eurasian beavers in Poland were low and according to Zalewski et al. (2012) did not pose a threat to their health and reproduction, but Nolet et al. (1997) and Fimreite et al. (2001) considered copper and zinc as risk factors in certain Eurasian beaver populations. Liver and kidney copper and zinc concentrations have been fairly similar among studies (copper: 1.60–9.2 µg/g wet weight; zinc: 16.00–35.70 µg/g wet weight) (Wren, 1984; Hillis and Parker, 1993; Fimreite et al., 2001; Zalewski et al., 2012; Giżejewska et al., 2014). Zinc concentration in the muscular tissue of Eurasian beavers in Poland ranged from 27.13 to 50.59 µg/g wet weight (Zalewski et al., 2012).

Table 8.3 Different compounds found in the liver and kidney of beavers (*Cf* = Eurasian beaver; *Cc* = North American beaver). All values are given in micrograms per gram of wet weight unless otherwise specified. Maximum levels for cadmium and lead for human consumption in the liver set by the European Commission of Farm Animals are 0.5 µg/g wet weight for both and in the kidney 1.0 and 0.5 µg/g wet weight, respectively (values above maximum are indicated in bold) (Commission Regulation (European Commission), 2006).

Liver[a]

Species	Cadmium	Copper	Mercury	Nickel	Lead	Selenium	Zinc	Type of site	State/country	References
Cf	**1.03**	2.80	–	–	–	–	27.66	Unpolluted	Telemark, Norway	Fimreite et al. (2001)
Cf	0.21	9.20	–	–	0.08	–	35.70	Unpolluted	Poland	Giżejewska et al. (2015)
Cf	**0.88**	–	0.02	–	0.11	–	–	Unpolluted	Poland	Giżejewska et al. (2014)
Cf	**0.58**	4.04	–	–	0.19	–	23.73	Former military airport	Poland	Zalewski et al. (2012)
Cf	0.22	4.40	–	–	0.14	–	27.52	Unpolluted	Poland	Zalewski et al. (2012)
Cc	0.11	5.91	0.01	–	–	–	33.92	Varied	Georgia, USA	Borchert et al. (2019)
Cc	0.09	2.71	0.01	–	–	–	31.90	Varied	South Carolina, USA	Borchert et al. (2019)
Cc	0.11	3.16	0.01	0.03	0.02	0.31	27.31	Savannah River site (mix of industrial and forested areas)	South Carolina, USA	Borchert et al. (2019)
Cc	**2.00**	–	–	–	–	–	–	Mackenzie River delta	Northwest Territories, Canada	Gamberg et al. (2005)[b]
Cc	0.46	1.60	–	0.30	**0.54**	–	16.00	15 km from ore smelter	Ontario, Canada	Hillis and Parker (1993)[b,c]
Cc	0.46	1.60	–	0.24	0.28	–	16.00	90 km from ore smelter	Ontario, Canada	Hillis and Parker (1993)[b,c]
Cc	**0.58**	1.94	–	0.05	0.06	0.36	17.40	Wye River, 1.5 km north of Emvale village. Used as reference of antimony (Sb) in bottled water due to leaching	Ontario, Canada	Shotyk et al. (2019)[b]
Cc	0.19	2.90	0.03	0.50	–	0.20	29.60	175 km from ore smelter	Ontario, Canada	Wren (1984)[d]

Kidney[a]

Species	Cadmium	Copper	Mercury	Nickel	Lead	Selenium	Zinc	Type of site	State/country	References
Cf	**10.25**	2.17	–	–	–	–	23.11	Unpolluted	Telemark, Norway	Fimreite et al. (2001)
Cf	**2.81**	3.70	–	–	0.08	–	21.50	Unpolluted	Poland	Giżejewska et al. (2015)
Cf	**7.93**	–	0.03	–	0.06	–	–	Unpolluted	Poland	Giżejewska et al. (2014)
Cf	**93.40**	–	–	–	–	–	–	Pollution by heavy metals—peaked in 1970s	The Netherlands	Nolet et al. (1994)[b]
Cf	**7.10**	2.73	–	–	0.13	–	25.23	Former military airport	Poland	Zalewski et al. (2012)
Cf	**2.44**	3.15	–	–	0.09	–	27.05	Unpolluted	Poland	Zalewski et al. (2012)
Cc	**11.00**	–	–	–	–	–	–	Mackenzie River delta	Northwest Territories, Canada	Gamberg et al. (2005)[b]

Cc	3.76	1.86	–	0.52	**0.68**	–	19.86	15 km from ore smelter	Ontario, Canada	Hillis and Parker (1993)[b]
Cc	3.76	1.86	–	0.42	0.40	–	19.86	90 km from ore smelter	Ontario, Canada	Hillis and Parker (1993)[b]
Cc	2.66	2.46	–	0.02	0.08	0.70	22.20	Wye River, 1.5 km north of Emvale village. Used as reference of Sb in bottled water due to leaching	Ontario, Canada	Shotyk et al. (2019)[b]
Cc	1.44	3.00	0.03	0.40	–	0.90	25.40	175 km from ore smelter	Ontario, Canada	Wren (1984)[d]

Source: After Giżejewska et al. (2015) and Shotyk et al. (2019). Reprinted from Giżejewska, A., Spodniewska, A., Barski, D. and Fattebert, J. 2015. Beavers indicate metal pollution away from industrial centers in northeastern Poland, Environmental Science and Pollution Research, 22, 3969–3975 under CC BY license and Science of the Total Environment. Shotyk, W., Bicalho, B., Dergousoff, M., Grant-Weaver, I., Hood, G., Lund, K. and Noernberg, T. A geochemical perspective on the natural abundance of trace elements in beaver (Castor canadensis) from a rural region of southern Ontario, Canada, 40–50 © 2019 Elsevier B.V. with permission from Elsevier. Dashes indicate no data available.

a Values for ^{226}Ra are included in Clulow et al. (1991) across three sites: Elliot Lake (<1 km from mine site)—kidney = 9.00 mBq/g, liver = 2.30 mBq/g; Mid-serpent (~20 km downstream)—kidney = 4.50 mBq/g, liver = 2.80 mBq/g; Low-serpent (~30 km downstream)—kidney = 4.00 mBq/g, liver = 2.90 mBq/g. Total dioxin-like polychlorinated biphenyl (dl-PCB) concentration in liver varied between 221 and 1,590 pg/g wet weight in a Polish study (Zalewski et al., 2012). In Poland 1.6–39 ng/g wet weight was found for perfluorooctanesulfonate, 0.06–0.28 ng/g wet weight for perfluorooctanoic acid, and 0.05–1.4 ng/g wet weight for perfluoroonoonanoic acid in liver samples (Falandysz et al., 2007). In northeastern Poland perfluorooctanoic acid was only found in both sexes' livers in a range from 0.48 to 0.72 ng/g wet weight and perfluorooctanesulfonate at levels of 1.66 to 8.45 ng/g wet weight, while perfluoroonoonanoic acid was only found in females, with the highest content of 6.61 ng/g wet weight (Surma et al., 2015).

b Results were converted to wet weight by calculating 20% of the dry weight value.

c No significant difference found in levels of cadmium, copper, and zinc across sites, so values given are a mean taken across data from both sites. Results were converted to wet weight by calculating 20% of dry weight value (Nolet et al., 1994).

d Values for liver and kidney were also included: liver—Al=3.70, Ca=109.00, Fe=251.00, K=2087.00, Mg=148.00, Mn=3.70, Na=1258.00, Mn=3.70; kidney—Al=2.90, Ca=128.00, Fe=95.00, K=2282.00, Mg=154.00, Mn=3.40, Na=1555.00, P=2048.00.

North American beavers living in close proximity to an ore smelter in Sudbury, Ontario, Canada, had higher levels of nickel and lead in their livers and kidneys than beavers living further away (Hillis and Parker, 1993). Radioactive substances (^{226}Ra) in bones, muscles, and kidneys were highest in North American beavers living close to uranium mining. However, intake of radionuclides by people eating beavers from the area were below the maximum recommended values set by the Canadian regulatory authorities (Clulow et al., 1991).

Environmental contamination with mercury is a particular concern for beavers as it has an ability to accumulate in food webs and is lethal (Borchert et al., 2019). Liver tissue mercury concentrations from North American beavers have been reported to vary from 0.01 µg/g wet weight in South Carolina to 0.04 µg/g wet weight in Wisconsin (Sheffy and Amant, 1982), and 0.02 µg/g wet weight was reported from Eurasian beavers in Poland (Giżejewska et al., 2014). Selenium plays a key role in the prevention of the toxic effects of mercury by modulating the activity of several selenium-dependent enzymes, and experimental and environmental data have indicated a protective role for selenium against mercury toxicity (Pinheiro et al., 2009). In Ontario, Canada, a positive relationship was found between mercury and selenium concentrations in liver tissue of North American beavers. Thus selenium modifies the toxic effect of mercury in these beavers (Wren, 1984).

In Alberta, Canada, researchers investigated whether one adult and two juvenile North American beavers could survive exposure to bitumen (asphalt) contamination following rehabilitation at the Edmonton Wildlife Hospital (Hood, 2015). They were released into a fenced area and monitored for 14 months. One juvenile died of unknown cause during the 14-month study. Two kits were born, and normal behaviours such as lodge building and food caching were displayed by the adult. However, before their release and during the spill, two of the five beavers in the family had to be euthanized due to organ failure and decreased mentation. This suggests that beavers may be vulnerable to contamination from asphalt roads close to watercourses.

8.4.2 Hunting and trapping

The history of beavers and humans has been a bloody one, causing an almost worldwide extirpation of both beaver species until hunting became more carefully regulated (Nolet and Rosell, 1998; see also Chapters 2 and 11). Humans are the most important mortality factor in many areas and can regulate beaver densities by trapping and shooting (Müller-Schwarze, 2011). More modern mortality rates from harvest vary between 11 and 36% in Newfoundland, Canada (Payne, 1984), 6% in Sweden, 10% in Norway, and 20% in Finland (Hartman, 1999), though it should be noted that the Eurasian beaver is a protected species across much of Europe. Testing the sustainability of different harvesting regimes of Eurasian beavers in Telemark, Norway, revealed that a harvest of 22–26% (mean = 24%) of the estimated population reduced the number of beaver families by 46% during the next 3 years. Four years with no harvesting caused a rebound in the number of families by 93% (Parker and Rosell, 2014). Adult beavers, and especially pregnant females, are highly susceptible to being shot during spring hunts, and this can reduce the population's reproduction significantly (Parker et al., 2002; see Figure 6.11).

Beavers may still be killed illegally in areas where hunting is prohibited or out of hunting season (Payne, 1984; Grubešić et al., 2015). Moreover, shot pellets have been found under the skin of living beavers, and in some individuals these seem not to have hindered the animal in its daily activities. Shot pellets have become encapsulated in the body and not led to lead poisoning. However, there are many stories of less-fortunate individuals. A trapped beaver was found living with shot pellets in its head, which had certainly caused the animal a great deal of suffering. One of its front teeth had been forced out of position and was growing in sideways and was completely black at the root. A pellet had also blinded the animal in one eye, which caused infection, and several other pellets were stuck in its head (Rosell and Pedersen, 1999). Beavers are sometimes found with shots of several different calibers in their bodies, indicating that they were shot by more than one hunter at some point. In North America,

7–18% of the beavers found dead had shot wounds from poachers (Payne, 1984), and in Poland 16% of beaver deaths were due to illegal hunting (Żurowski, 1987; see also Table 8.1).

Beavers have been reported losing a foot in illegal traps such as scissor traps. When they have been captured again later, the sores are occasionally healed, but they can become infected (Rosell and Pedersen, 1999). On rare occasions beaver lodges have been illegally destroyed by dynamite in Norway (Rosell and Pedersen, 1999), and evidence of burnt lodges have been found in parts of Scotland (Campbell-Palmer et al., 2018).

8.4.3 Road traffic and motorboat accidents

Increases in beaver populations combined with increases in traffic mean that beavers are sometimes hit by vehicles (Grubešić et al., 2015). When dispersers are looking for a new territory they will often follow or cross roads. Beavers with established territories may also cross roads, especially when they are gathering food in the autumn. Being hit by a car nearly always results in death of the beaver. The ultimate cause of death is most often internal bleeding or direct skull impact. There have been reports linking 37% of beaver mortality events in France (Estève, 1988) and 16% in Switzerland to traffic accidents (Stocker, 1985). Encounters with motorboats and trains also kill beavers (Stefen, 2019; see also Table 8.1).

8.5 Diseases, pathogens, and parasites

In more recent times, diseases associated with beavers have been well investigated, especially in light of reintroduction projects, to ensure not only that individuals are fit for release but also that they do not spread diseases or parasites at the release site to other wildlife, livestock, or humans. Like any other mammal beavers may spread certain diseases that may impact human health, but there are also a wide range of diseases that spread within beaver communities. Infectious diseases caused by viruses and bacteria are the most common, since beavers live in family groups with regular direct contact between individuals. This creates a favourable environment for spreading infections, as does their aquatic environment. Changing climate is an additional important factor, with warmer and more humid weather increasing the survival of microorganisms (virus, fungi, bacteria, and protozoa) (Alberts et al., 2002).

Beavers may regularly harbour several pathogens (Table 8.4). From a health and biosecurity perspective, beavers are currently considered to present no greater risk to human, livestock, or other wildlife health than any other mammal. Infectious diseases have so far not contributed significantly to mortality among Eurasian beavers. However, several large disease outbreaks have occurred in North American beaver populations (Knudsen, 1953). Infection of diseases from beavers usually occurs through contamination of the environment through faeces, but it can also occur via bites and direct contact. Some people are skeptical of handling beavers and even more hesitant of eating them. As long as standard precautions are taken, there is no reason to fear contamination or infection from handling beavers. For example, the hantavirus common in many other rodents has not been found in Eurasian beavers (Girling et al., 2019b). There is also little danger involved in eating fresh prepared beaver meat, where permitted. It has been suggested that the consumption of Eurasian beaver meat in Latvia can be a risk to human health if it is consumed improperly cooked due to the disease trichinellosis and the nematode *Trichinella britovi* (Segliņa et al., 2015). However, the nematode was present in only one of 182 (0.5%) beavers, which contained 148 larvae.

The highest-risk pathogens for Eurasian beavers reintroduced to Britain were identified to be the parasites *Cryptosporidium parvum*, *Echinococcus multilocularis*, *Eimeria* spp., *Fasciola hepatica*, *Giardia* spp., and *T. britovi* (Girling et al., 2019c). Other high-risk pathogens were the bacteria *Escherichia coli*, *Franciscella tularensis*, *Mycobacterium avium*, *Salmonella* spp., *Yersinia* spp., the fungus *Chrysosporium parvum*, and the terrestrial rabies virus *Lyssa virus*. Girling et al.'s review also identified many other potential pathogens that are not currently found in beavers. *Campylobacter* is considered one of the most common causes of bacterial diseases worldwide (Rosef et al., 1983), yet faecal samples from neither North American beavers in Washington, USA (Pacha et al., 1987) nor Eurasian beavers in Telemark, Norway (Rosell et al., 2001) showed the

Table 8.4 Possible pathogens of beavers: viral, fungal, bacterial, and parasitic (haemo- and endoparasites). The pathogen/organism was classified as 'rare', 'moderate', or 'often' based on how often it had been identified in beavers (%), number of studies, and/or number of samples investigated.

Pathogen/organism	Cause	Eurasian beaver	North American beaver	References
Viral				
Cowpox	*Orthopoxvirus* spp.	–	Rare	Hentschke et al. (1999)
Papillomatosis	*Papillomavirus*	–	Rare	Carlson et al. (1983)
Rabies	*Lyssavirus*	–	Rare	Brakhage and Sampson (1952); Belcuore et al. (2002)
Fungal				
Adiaspiromycosis	*Chrysosporium parvum*	Rare	Rare	Erickson (1949); Mörner et al. (1999)
Bacterial				
Aeromonas	*Aeromonas hydrophila*	–	Rare	Cullen (2003)
Avian tuberculosis	*Mycobacterium avium*	Moderate	–	Hinze (1950); Nolet et al. (1997)
Bronchopneumonia	*Haemophilus* spp.	–	Rare	Havens et al. (2013)
	Klebsiella pneumonia	–	Rare	Havens et al. (2013)
Clostridium	*Clostridium sordellii*	–	Rare	Cullen (2003)
Corynebacterium	*Corynebacterium* spp.	Rare	–	Mörner (1999)
Enterobacter	*Enterobacter* spp.	–	Rare	Cullen (2003)
Escherichia	*Escherichia coli*	Rare	Rare	Goodman et al. (2012); Wong et al. (2016)
Leptospirosis	*Leptospira* spp.	Moderate	Moderate	Stuart et al. (1978); López-Pérez et al. (2017); Marreros et al. (2018); Girling et al. (2019a)
Micrococcus	*Micrococcus* spp.	–	Rare	Cullen (2003)
Peptostreptococcus	*Peptostreptococcus* spp.	–	Rare	Cullen (2003)
	Yersinia enterocolitica	Moderate	–	Nolet et al. (1997); Platt-Samoraj et al. (2015)
Pseudotuberculosis	*Yersinia pseudotuberculosis*	Moderate	Moderate	Langford (1972); Nolet et al. (1997); Gaydos et al. (2009)

Pseudomonas	*Pseudomonas* spp.	–	Rare	Cullen (2003)
Salmonellosis	*Salmonella* spp.	Rare	–	Romasov (1992)
Staphylococcus	*Staphylococcus* spp.	–	Rare	Cullen (2003)
Streptococcus	*Streptococcus castoreus* spp.	Rare	Rare	Cullen (2003); Lawson et al. (2005)
Tularaemia	*Francisella tularensis*	Moderate (but resistant)	Often	Stenlund (1953); Mörner and Sandstedt (1983); Schulze et al. (2016)
Parasitic				
Acanthocephalan	*Polymorphus paradoxus*	–	Rare	Bush and Samuel (1981)
Cestoda	*Echinococcus granulosus*	Rare	–	Romaschov (1969); Demiaszkiewicz et al. (2014)
	Echinococcus multilocularis	Moderate	–	Janovsky et al. (2002); Barlow et al. (2011); Ćirović et al. (2013); Campbell-Palmer et al. (2015a)
	Hymenolepis spp.	Rare	–	Romaschov (1969)
	Paranoplocephala omphalodes	Rare	–	Romaschov (1969)
	Rodentolepis straminea	Rare	–	Romaschov (1969)
	Taeniidae spp.	–	Rare	Romaschov (1969)
Coccidiosis	*Eimeria causeyi*	Rare	Rare	Ernst et al. (1970)
	Eimeria sprehni	–	Rare	Yakimoff (1934); Ernst et al. (1970)
	Sarcocystis neurona	Rare	Rare	Jordan et al. (2005)
Cryptosporidiosis	*Cryptosporidium parvum* spp.	Rare	Rare	Bajer et al. (1997); Fayer et al. (2006); Goodman et al. (2012); Sroka et al. (2015)
Giardiasis	*Giardia* spp.	Moderate	Often	Fayer et al. (2006); Bajer et al. (2008)
Hymenolepis	*Hymenolepididae* spp.	Rare	–	Romaschov (1969)

continued

Table 8.4 Continued.

Pathogen/organism	Cause	Eurasian beaver	North American beaver	References
Nematoda	*Ascaris castoris/A. lumbricoides*	Rare	–	Joszt (1964); Romaschov (1969)
	Baylisascaris spp.	–	Rare	McKown et al. (1995); Havens et al. (2013)
	Capillaria hepatica	Rare	Rare	Chitwood (1934); Romaschov (1969); Fuehrer (2014)
	Castorstrongylus castoris	Rare	Rare	Romaschov (1969); Bush and Samuel (1981)
	Cysticerosis coenuriasis	Rare	–	Campbell-Palmer et al. (2015b)
	Dipetalonema sprenti spp.	–	Rare	Price (1934); Fedynich et al. (1986)
	Dirofilaria immitis	–	Rare	Foil and Orihel (1975)
	Dracunculus spp.	–	Rare	McKown et al. (1995)
	Gongylonema pulchrum	–	Rare	Ogburn-Cahoon and Nettles (1978)
	Heterobilharzia americana	–	Rare	Fedynich et al. (1986)
	Oxyuris spp.	–	Rare	Canavan (1929)
	Psilotrema castoris	Moderate	–	Demiaszkiewicz et al. (2014)
	Rhabditis spp.	Rare	–	Romaschov (1969)
	Taenia martis	Rare	–	Schuster and Heidecke (1992); Campbell-Palmer et al. (2015a)
	Travassosius americanus	Rare	Often	Romaschov (1969); Bush and Samuel (1981)
	Travassosius rufus	Often	–	Romaschov (1969); Bush and Samuel (1981)
	Trichinella britovi	Rare	–	Segliņa et al. (2015)
	Trichinella native	–	Rare	Dick and Pozio (2001)
	Trichinella spiralis	–	Rare	Romaschov (1969)
	Trichocephalus castoris	Rare	–	Romašov and Stubbe (1992)
	Trichostrongylus axei	Rare	–	Romaschov (1969)
	Trichostrongylus capricola	Rare	–	Joszt (1964); Demiaszkiewicz et al. (2014)
Toxoplasmosis	*Toxoplasma gondii*	Rare	Moderate	Forzán and Frasca (2004); Jordan et al. (2005); Stefen (2019)

Trematoda				
	Dicrocoelium lanceatum	Rare	–	Romaschov (1969)
	Echinoparyphium spp.	Rare	–	Romaschov (1969)
	Echinostoma armigerum	–	Rare	Smith and Archibald (1967)
	Echinostoma orlovi	Rare	–	Romaschov (1969)
	Echinostoma revolutum	Rare	Rare	Bush and Samuel (1981)
	Fasciola hepatica	Rare	Rare	Joszt (1964); Romaschov (1969); Lang (1977); Shimalov and Shimalov (2000)
	Opisthorchis felineus	Rare	–	Bush and Samuel (1981)
	Plagiorchis castoris	Rare	–	Romaschov (1969)
	Plagiorchis massino	Rare	–	Romaschov (1969)
	Psilotrema castoris	Rare	–	Romaschov (1969)
	Renifer ellipticus	–	Rare	Canavan (1934)
	Stephanoproraoides lawi	–	Rare	Smith and Archibald (1967); Addison (1973)
	Stichorchis subtriquetrus	Often	Often	Bush and Samuel (1981); Demiaszkiewicz et al. (2014)

Dashes indicate no data available.

presence of this bacteria, though it has been found in a very small number of beavers in Britain (Campbell-Palmer et al. 2021). Additionally, haemoparasites (blood parasites) have not been found in Eurasian beavers (Cross et al., 2012).

Beavers have been shown to die of other diseases such as pneumonia, intestinal disorders, and liver diseases. For example, if the fur is not maintained properly, water can permeate the undercoat and come into contact with the skin, causing the animal to become sick with pneumonia. Infections following bites from other beavers or pneumonia caused by poor water quality have also been documented (Nolet et al., 1997). A yearling Eurasian beaver from Switzerland with apathy and a reduced appetite was diagnosed with possible bronchopneumonia and Fallot's tetralogy (congenital heart defect) (Wenger et al., 2010). However, heart flow murmurs seem to be common (in 26 of 27 beavers; 96%) with no evidence of heart disease and may actually be related to their semi-aquatic lifestyle (Devine et al., 2012).

8.5.1 Viruses

Viruses are ultramicroscopic, infectious organisms. Rabies is an example of a greatly feared viral disease with a widespread occurrence globally and across various species. One of the first reported beaver cases with rabies attacked a dog in Missouri, USA (Brakhage and Sampson, 1952). Rabies was reported in three North American beavers by Krebs et al. (2002). Another beaver in Florida, USA, that exhibited aggressive behaviour toward canoes and kayaks was diagnosed with rabies in the brain tissue (Belcuore et al., 2002). Rabies has never been recorded in the Eurasian beaver.

Unfortunately, no data are available on cancer rates in beavers (Gorbunova et al., 2008). Neoplasm, an uncontrolled growth of cells, sometimes self-limiting as in benign tumours, sometimes malignant, has been reported in North American beavers, e.g. a mandibular periodontal fibroma (Hamerton, 1934) and a rhabdomyoma (a benign tumor of striated muscle) (Dieterich, 1981). An 11-year-old female North American beaver died after a 3.5-month course of intermittent diarrhea, lethargy, and anorexia due to thyroid follicular carcinoma with

pulmonary metastases (Anderson et al., 1989). This was the first documented observation of this. A North American beaver's hind foot with multiple papillomas (small wart-like growths on the skin) was investigated in Woodstock, Connecticut, USA, and virus-like particles were found in the nuclei of cells. It was suggested that these tumours were caused by a papillomavirus (Carlson et al., 1983).

8.5.2 Bacteria

Bacteria are single-celled microscopic organisms lacking nuclei (prokaryotes). A wide range of diseases in beavers are caused by bacteria, and below we briefly discuss the most important pathogens (see also Table 8.4).

Only two studies have found Eurasian beavers that had *avian tuberculosis*, caused by *Mycobacterium avium*. The most common infection routes are through food, water, and direct contact with infected animals (Nishiuchi et al., 2017). It is closely related to the bacteria that causes human tuberculosis. Infection with avian tuberculosis in humans is extremely rare and immunocompromised individuals are most at risk (Davis et al., 2015). Twenty-five (34.2%) Eurasian beavers in Germany (Hinze, 1950) and one reintroduced beaver (5%) in the Netherlands died due to this bacteria (Nolet et al., 1997).

Leptospirosis is caused by the bacterium *Leptospira* and is particularly associated with rodents and mammals living in aquatic and semi-aquatic habitats (Gelling et al., 2015; Ayral et al., 2016). It is considered to be the most widespread zoonotic disease in the world (Adler and De la Peña Moctezuma, 2010). The bacteria reside in the host's kidney and spread via urine, and they can enter the human body via for example the mouth, eyes, or wounds (Kurucz et al., 2018). Eight out of 16 (50%) Scottish beavers tested positive for *Leptospira* antibodies, indicating they had been exposed but not necessarily were carrying this disease (Goodman et al., 2016) and many later went on to test negative. In another study in Scotland, 4.6% (7 of 151) tested positive for *Leptospira* (Girling et al., 2019c). None had abnormal physical, biochemical, or haematological changes. However, 13 beavers reportedly died of leptospirosis in Switzerland (Marreros et al., 2018). A reintroduction study in the Netherlands revealed that leptospirosis was a contributing cause of death, and three individuals (15%) were found dead between 24 and 31 days after their release (Nolet et al., 1997). Leptospirosis was detected in 13 of 25 (52%) of the North American beavers in Louisiana, USA (Stuart et al., 1978) and in two of three (67%) North America beavers in northwestern Mexico (López-Pérez et al., 2017).

Pseudotuberculosis (*rodentiosis* or *yersiniosis*) has been found in beavers in Europe and North America. This disease in beavers can be acute, with fever and death within 24 hours, or chronic, with internal abscessation, chronic weight loss, and shedding via the faecal route of the bacterium. *Yersinia pseudotuberculosis* and the closely related *Y. enterocolitica* both spread via the faecal–oral route from faeces and urine of sick animals. Infected animals stop eating, become malnourished, and eventually die. The spleen becomes enlarged, and small white infection granules/nodules can be found in the liver. These granules/nodules can also be found in the kidneys, often larger than those in the liver. Infections in the intestines and lungs can also occur. This disease can be transmitted to humans, most commonly via contaminated food or vegetation (Girling et al., 2019c). Three beavers died of *Y. pseudotuberculosis* (Far East scarlet-like fever) in the Netherlands, and one died of the closely related bacteria *Y. enterocolitica* (in total 20%). The beavers were found between 32 and 85 days after they were released into the country (Nolet et al., 1997). Captive beavers in Norway have tested positive for *Y. pseudotuberculosis* (Olstad, 1937). *Y. enterocolitica* was isolated from the oropharyngeal and cloacal area in 16.7% (4 out of 24) of wild Polish beavers (Platt-Samoraj et al., 2015). *Y. pseudotuberculosis* has also been reported as the cause of death of one North American beaver juvenile in Washington state, USA (Gaydos et al., 2009). In southern Illinois, USA, 3 of 8 (37.5%) dispersing beavers' (yearlings and sub-adults) deaths were caused by *Y. pseudotuberculosis* (McNew and Woolf, 2005).

Bacteria in the genera *Staphylococcus* and *Streptococcus* can cause illness in both people and animals. These two genera have been found in Eurasian beavers living in Voronezh, Russia (Romasov, 1992), and *Streptococcus* was possibly found in one Eurasian beaver in Britain (Lawson et al., 2005). *Streptococcus* was found in 2 of 20 (10%) North American beavers' eyes (Cullen, 2003). The usual infection route is through animals with latent or active infections or via puncture wounds. Therefore, it is a good practice to use gloves when handling beavers, especially if you have a cut or sores on your hands.

The composition of the bacterial microbiome in the North American beaver gut system consists primarily of Bacteroidetes (49%) and Firmicutes (48%) (Gruninger et al., 2016), while for the Eurasian beaver also Actinobacteria dominated and Bacteroidetes was only found in male juveniles (Pratama et al., 2019). Anaerobic bacteria in the castor sacs of the North American beaver have been found (Walro and Svendsen, 1982) in the form of Gram-positive, facultatively anaerobic, pleomorphic cocci. It is unknown if the bacterium is involved in chemical signalling (see Section 7.2.2.1). High numbers of aerobe *Escherichia coli* and anaerobe *Bacteroides fragilis* were isolated from the anal glands of both sexes and different age classes of the North American beaver (Svendsen and Jollick, 1978). Twenty-four cultures of bacteria have been found in the beaver (unfortunately it is not clear from the paper from which beaver species) (Saveljev, 2003).

Both beaver species are reported as a host for *Franciscella tularensis*, but only North American beavers seem to regularly become impacted and have a high risk of dying from the disease it causes, *tularaemia*. Its scientific name comes from the place where it was first detected, Tulare. This happened in 1906 after survivors of the great earthquake in San Francisco were afflicted with a plague-like illness. It transmits by blood-sucking insects such as ticks and biting flies predominantly but can under some conditions have a waterborne life cycle (De Barba et al., 2010). Aquatic animals such as beavers may then be an important part of such cycles. There is also a chance that humans can be infected with tularaemia via contact with beavers. The tularaemia bacteria cause white spots on the beaver's liver or spleen, and these can also appear on the kidneys, lymph nodes, the back of the abdomen, lungs, or stomach. If you are unfortunate to be infected with tularaemia, you will experience a strong flu-like illness with high fever, aches, and a long convalescence (Spickler, 2017). The Eurasian beaver is only occasionally reported to be a host (Mörner, 1992; Schulze et al., 2016). Analyses of blood samples from beavers collected in Sweden in the 1970s revealed that they were relatively often infected with tularaemia, but they were rather resistant and the disease did not cause death of the animals (Mörner, 1992). Up to 40% of the beavers had antibodies against the disease. In North America, beavers have been reported to be a significant reservoir. In a farm in Oregon in 1960, 116 of 150 (77%) beavers died due to this disease (Bell et al., 1962), and in Voronezh, Russia, 50% of the North American beavers died but all Eurasian beavers survived even when they lived side by side with sick North American beavers (Avrorov and Borisov, 1947, cited in Lavrov, 1983). Large populations of beavers in North America have been exterminated because of tularaemia. In Minnesota, USA, as many as 10,000–15,000 beavers probably died of tularaemia during the winter of 1951–1952 (Knudsen, 1953), and 407 dead beavers were found in the same area most likely affected by the disease (Stenlund, 1953). In 1981 and 1982 a 'substantial' number of beavers died of tularaemia at the Necedah Wildlife Refuge and adjacent states in central Wisconsin (US Department of Interior Fish and Wildlife Service, 1982).

8.5.3 Parasites

Parasitism is a relationship between organisms where one organism (the parasite) lives at the expense of the other (host organism). Parasites are usually smaller than the host and do not necessarily kill them directly but may weaken the host and lead to indirect population reductions. Parasites live on the host with the life motto: As much food, warmth, and protection as possible, but not so much that the host immediately dies. If excessive parasitic burdens do kill the host, it will take the parasite with it in death (Anderson, 2000).

Parasites can be divided into two groups: internal parasites (endoparasites), which live inside the host, and external parasites (ectoparasites), which live outside the host's body. Beavers can be infected by several parasites at the same time, both endo- and ectoparasites. A major attack by one parasite reduces the beaver's resistance against infection by other parasites and diseases, and the combined effects of these stressors can lead to the animal's collapse. Note that many parasitic infections are not necessarily detrimental because many hosts and parasites have coevolved. Parasites do not usually exhibit clinical signs in the host, unless at very high parasite burdens where progressive weight loss, anaemia, intoxication, and even death can occur (Skrjabin, 1949; McKown et al., 1995).

8.5.3.1 Endoparasites

The majority of beavers have internal parasites such as intestinal worms or flukes (Figures 8.9 and 8.10). However, when the beaver host loses condition, becomes ill, or is excessively stressed, for example, the number of intestinal worms can become so high that they can contribute to mortality. At least 49 endoparasites have been reported in beavers, 34 in the Eurasian beaver and 26 in the North American beaver (see Table 8.4). Only the most important endoparasites are described below.

Beavers can act as intermediate hosts, those in which an intermediate (usually non-sexually mature or reproducing) stage of the parasite develops and is passed on, for example if an aquatic snail ingests an egg or if a fox ingests an infected rodent. The fox tapeworm (*Echinococcus multilocularis*) is a pathogenic parasitic zoonosis present in central Europe (Campbell-Palmer et al., 2015). This tapeworm only becomes sexually mature and sheds eggs once

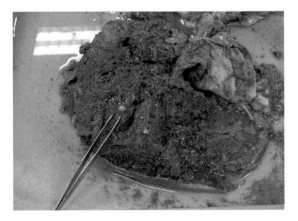

Figure 8.9 Beaver intestinal fluke (*Stichorchis subtriquetrus*), more typically found in the caecum and large intestine. (Photo supplied courtesy of Inger Sofie Hamnes.)

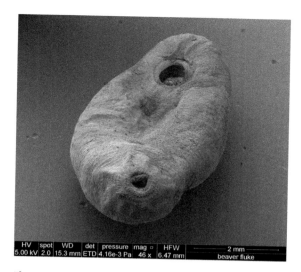

Figure 8.10 Electron microscope image of beaver intestinal fluke. (Photo supplied courtesy of Inger Sofie Hamnes.)

inside a red fox, but under certain circumstances cats (*Felis catus*) and dogs can also act as the final host (Eckert and Deplazes, 1999). Infection can be determined through post-mortem inspection of the beaver's liver and/or via minimally invasive exploratory laparoscopy in live beavers (Campbell-Palmer et al., 2015). Infection by this tapeworm was observed in an imported Bavarian beaver from a captive population in England (Barlow et al., 2011) and suspected in 2 of 11 (18.2%) wild Bavarian beavers (Campbell-Palmer et al., 2015). In Serbia and Switzerland, the parasite has been found in one beaver from each country (Janovsky et al., 2002, Ćirović et al., 2013). In Poland, *E. granulosus* was found in 1 out of 48 beavers (2.1%) (Demiaszkiewicz et al., 2014).

Cryptosporidium spp. is an intestinal protozoan parasite which is a source of water-borne diarrhoea in humans and is responsible for the disease *crypto-sporidiosis*. It has a complicated life cycle, one stage of which is the formation of oocysts that are secreted with the faeces and which causes infection if we drink contaminated water. This parasite causes disease in the small intestine, particularly in immunocompromised and immunologically naïve animals. The transmission route is faecal–oral. It has been found in 2 of 19 (10.5%) Eurasian beavers tested in Poland (Bajer et al., 1997) and 2 of 4 (50%) North

American beavers from Massachusetts, USA (Fayer et al., 2006). One kit born in Scotland tested positive for *Cryptosporidium* oocysts in a faecal sample obtained post mortem (Goodman et al., 2012). In northeastern Poland, water close to Eurasian beaver lodges was significantly contaminated with cysts of *Cryptosporidium* spp. (32.9%) (Sroka et al., 2015).

Although many are familiar with the term '*beaver fever*', they are probably unaware that it is caused by a tiny intestinal flagellate protozoan parasite—*Giardia* spp.—common in many other mammals including humans (Isaac-Renton et al., 1987). *Giardia* lives in the small intestine and causes the disease *giardiasis* in humans, the primary symptom of which is diarrhoea. Other symptoms include cramps, vomiting, listlessness, low-grade fevers, chills, and headache. These symptoms can lead to dehydration, which is problematic when one is travelling in the backcountry. The protozoan is found in the beaver's faeces, which is why we can contract '*beaver fever*' by drinking contaminated water (Tsui et al., 2018). In the USA, *Giardia* is the most common parasitic infection in humans, and the North American beaver is seen as an important factor in the spread of the disease. Newer research, however, indicates that other species are more important vectors than beavers. The beaver must be infected for at least 3 months before it can begin to spread cysts of *Giardia* into the water. The cysts, which are resistant to chlorination, can survive for up to 2 months in water. Untreated water should be boiled for at least 1 minute before it is used, and very cloudy or contaminated water should be boiled for at least 5 minutes prior to use (Erlandsen, 1994; Erlandsen et al., 1996; Tsui et al., 2018). Infection rates of *Giardia* spp. varied from 0 to 8% for the Eurasian beaver in Poland (Bajer et al., 2008). Fayer et al. (2006), reviewing the prevalence of *Giardia* in 27 studies in North America, found that the mean prevalence was 21.4% (range 0–100%). Sex and age of beavers are not related to *Giardia* infection (Dunlap and Thies, 2002).

The nematode *Travassosius rufus* is a roundworm with a round cross-section, and its body is pointed at both ends (Figures 8.11a and b). It only has muscles along the length of its body which gives it a characteristic stiff wiggling movement. It is pink/red in coloration, and the female is slightly larger

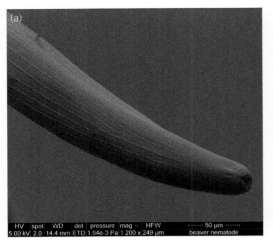 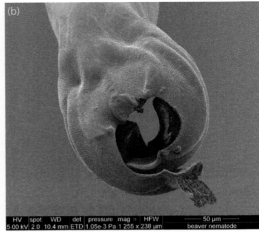

Figure 8.11 (a) Electron microscope images of the anterior end with mouth, and (b) of the tail region of the beaver stomach nematode (*Travassosius rufus*). (Photos supplied courtesy of Inger Sofie Hamnes.)

than the male. It grows to about 1.5 cm long and at the front of the worm are two distinct papillae. This species-specific nematode is found in the mucosal layer of beaver stomachs and within its contents (Anderson, 2000; Demiaszkiewicz et al., 2014). It has a direct life cycle, with eggs expelled with beaver faeces. The infectious larval stage is ingested by the beavers while feeding on aquatic and/or emergent vegetation (Khalil, 1922; Åhlén, 2001). All of the 30 beavers tested from Sweden were infected with *T. rufus* (Åhlén, 2001). This species was present in 68.7% of Eurasian beavers in Poland, and one individual had as many as 4,336 specimens (Demiaszkiewicz et al., 2014). In the Czech Republic, 63% had the nematode (Drózdz et al., 2004). Eurasian beavers living in streams and small lakes had significantly more *T. rufus* than beavers living in rivers (Campbell-Palmer, 2017), and older beavers had significantly more of this nematode than younger beavers: 80% vs. 39%, respectively (Drózdz et al., 2004). A similar species, *T. americanus* from North America, has been found in 93% of beavers from Alberta, Canada (Bush and Samuel, 1978), and the number of individuals ranged from 10 to 10,290 (average 1,352) per infected beaver in a study in Nova Scotia, Canada (Smith and Archibald, 1967).

The disease *toxoplasmosis* is caused by a tiny single-celled parasite known as *Toxoplasma gondii* and occurs in many animal species, including humans. It can cause miscarriage, and birth defects of the brain and eyes in offspring. Toxoplasmosis is spread via feline faeces to herbivores, and then from them to carnivores. The life cycle is completed when a cat eats an infected prey animal. The beaver can therefore serve as an intermediate host, and the parasite has so far been recorded in North American beavers in Massachusetts (6 of 62 animals; 10%) (Jordan et al., 2005), Connecticut (5-month-old orphaned male) (Forzán and Frasca, 2004), and Kansas (1 of 14; 7.1%) (Smith and Frenkel, 1995) and in only 1 of 1,137 (0.1%) beavers in eastern Germany (Stefen, 2019). Beavers that are infected with this disease can be diagnosed by the yellow-white bumps on an otherwise dark-brown liver. If you find these spots on the liver of a beaver you have harvested, you should not eat it (Maldonado, et al., 2017).

The trematode most often found in beavers is the beaver fluke (*Stichorchis subtriquetrus*) (Orlov, 1941). It can be up to 3.5 cm in length, circular in transversal section, pale-brown to orange in colour, with a non-segmented body and a spineless, 5- to 20-µm-thick tegument (Campbell-Palmer et al., 2013). The typical life cycle of the beaver fluke involves an intermediate host, such as, for example, the common bithynia (*Bithynia tentaculata*). This aquatic snail is widespread throughout Europe and in North America and found in a range of aquatic habitats, being less common in smaller ponds, preferring shallow lakes and stream with heavily vegetated banks and silty substrate (Jokinen, 1992; Seddon, 2014). During spring and summer the snails live and lay eggs on a range of aquatic macrophytes (Pennak, 1989; Vincent et al., 1991). The free-living larva encapsulates itself on the leaves and stems of aquatic plants. If the beaver feeds on these, the larva passes into the beaver's intestine where it grows to maturity.

This parasite causes a number of stomach and intestinal infections, and prefers to live in the beaver's caecum and to a lesser extent in the colon (Demiaszkiewicz et al., 2014). Beavers living in places with an abundant snail community will most likely have a higher occurrence of the beaver fluke (Borchert et al., 2019). Every beaver (30) in Sweden that was checked for the beaver fluke was infected (Åhlén, 2001), whereas the prevalence was 77.4% in Norway (Campbell-Palmer, 2017), 82% in the Czech Republic (Máca et al., 2015), and > 90% in northeastern Poland (Drózdz et al., 2004; Demiaszkiewicz et al., 2014). As many as 661 flukes were found in one beaver from Nova Scotia, Canada (Smith and Archibald, 1967), and 893 flukes were found in a beaver in Poland (Demiaszkiewicz et al., 2014). The beaver fluke caused illness in beavers in Voronezh, Russia, where it was responsible for high mortality rates within the population (Orlov, 1941; Orlov, 1948). Eurasian beavers living in rivers and small lakes had significantly more flukes than beavers living in streams (Campbell-Palmer, 2017). In North America, the prevalence of the beaver fluke was 72% in Alberta (Bush and Samuel, 1981), 83% in Texas (Fedynich et al., 1986), and 89% in Kansas (McKown et al., 1995).

The beaver can also be a vector for the liver fluke (*Fasciola hepatica*) (Romaschov, 1969; Lang, 1977). In Belarus, 2 of 20 (10%) beavers had the liver fluke (Shimalov and Shimalov, 2000). The sexually mature fluke is 3 cm long, 1 cm wide, and resides in the beaver's liver and bile ducts, where it nourishes itself on the tissue and blood. It has a leaf-like shape and is a member of the flatworm family (Platyhelminthes). The egg is spread through beaver faeces and needs to come into contact with water or damp ground, where it hatches 2 or 3 weeks after incubation in the water. The larva then infects a pond snail, where it forms a free-living larva that forces its way out of

the snail and swims around until it can form a cyst on aquatic vegetation. Beavers become infected after ingesting vegetation, on which the flukes encyst, or by drinking water contaminated with the flukes. The fluke migrates from the stomach to the liver or the bile duct, where it reaches sexual maturity approximately 3 months after being ingested by the beaver. The complete life cycle of the liver fluke is between 4 and 6 months. Infected animals may display anaemia and reduced growth. In cases of serious infection, the beaver can be killed by the parasite. Liver fluke infection causes thicker bile ducts than normal, and the surface of the liver is bumpy and uneven (Romaschov, 1969).

8.5.3.2 Ectoparasites

External parasites are usually visible on the animal, especially if the fur is inspected, and the majority of hunters have become acquainted with one or more of these. If you look closely at the fur on any recently dead beaver you will almost certainly find either a beaver parasite beetle (*Platypsyllus castoris*) or a beaver nest beetle (*Leptinillus validus*). You may also find a fur mite from the genus *Schizocarpus*, but they are much harder to spot due to their small size. Ticks (*Ixodes* spp.) are not especially common on beavers (due to their regular grooming habits and specialized grooming claw being so effective in parasite removal) but can be found in larger numbers in some locations, especially in and around the ears. In North America, leeches (Hirudinea) have been observed on open sores and on the hind feet of beavers, but this is not a common occurrence (Novak, 1987). At least 65 ectoparasites have been reported on beavers, and these are listed in Table 8.5. The most common ones are described below.

The specialist beaver parasite beetle looks almost flea-like and was first classified as a louse, until further researched. Beaver beetles are specifically adapted to live in beaver fur and are especially easy to see as they crawl or 'swim' around the fur, being rusty-orange in coloration. There is no evidence that beavers are adversely affected by the presence of the beetles (Peck, 2006). The beetle neither damages the fur nor causes any sign of pathology. Therefore, the relationship may more precisely be termed as a commensal one (Behrendt, 2003).

Table 8.5 Ectoparasites on beavers. Note that we have just listed *Schizocarpus* spp., and not all species in this genus (see text for more info). The pathogen/organism was classified as either 'rare', 'moderate', or 'often' based on how often they have been identified in beavers (%), number of studies, and number of samples investigated.

Ectoparasites	Eurasian beaver	North American beaver	References
Beetles			
Leptinillus validus	–	Moderate	Peck (2007)
Platypsyllus castoris	Often	Often	Peck (2006); Duff et al. (2013)
Fleas			
Corrodopsylla curvata	–	Rare	Lawrence et al. (1965)
Megabothris quirini	–	Rare	Lawrence et al. (1965)
Monopsyllus quirini	–	Rare	Lawrence et al. (1965)
Leeches			
Hirudinea	–	Rare	Novak (1987)
Mites			
Androlaelaps fahrenholzi	–	Rare	Nelder and Reeves (2005)
Acarus farris	Rare	–	Haitlinger (1991)
Cunaxoides kielczewskii	Rare	–	Haitlinger (1991)
Demodex castoris	Moderate	–	Izdebska et al. (2016)
Neotrombicula whartoni	–	Rare	Nelder and Reeves (2005)
Ondatralaelaps multispinosus	–	Rare	Lawrence et al. (1965)
Phytoseiidae spp.	Rare	–	Haitlinger (1991)
Psorobia castoris	–	Rare	Kok et al. (1970)
Schizocarpus spp.	Often	Often	Izdebska et al. (2016)
Ticks			
Dermacentor albipictus	–	Rare	Stiles (1910)
Ixodes banksi	–	Rare	Lawrence et al. (1961)
Ixodes cookie	–	Rare	Judd (1954)
Ixodes hexagonus	Rare	–	Haitlinger (1991); Kadulski (1998)
Ixodes ricinus	Moderate	–	Goodman et al. (2012)
Screw worms			
Cochliomyia americana	–	Rare	Cook (1940)

Dashes indicate no data available.

The adult beaver parasite beetle is about 2.2–3.0 mm long, hind wings are absent, and its antennae are club-like. The body is broadly ovate, it is strongly dorsoventrally flattened, and the cuticle is thin and translucent (Duff et al., 2013). It feeds on epidermal tissue, skin secretion, wound exudates, and fur mites in the beaver's undercoat. Some have suggested that it also feeds on blood (Wood, 1965), but investigations of the insect's mouth show that this is not possible. It lacks eyes, and the dark spots at the front of its head are just spots. The head has a hair comb which other beetle species lack (Peck, 2006; Figure 8.12).

Beavers are the primary host for the beaver parasite beetle, with only one record of an accidental host switch to a North American river otter (*L. canadensis*) (Belfiore, 2006). This host switch possibly occurred when the otter entered a beaver

Figure 8.12 Ventral and dorsal views of the beaver parasite beetle *Platypsyllus castoris* Ritsema. (Photo supplied courtesy of Richard Naylor.)

lodge and was not long term. There is no evidence that the beaver parasite beetle can reproduce on other mammalian species. Adults can be found throughout the year and appear to spend their entire lives on the host animal, except in the case of pregnant females that temporarily leave to lay eggs in beaver nesting material in a lodge or burrow (Peck, 2006). Adult beetles lay about 10–25 eggs at a time, in crevices typically at the ends of sticks within the lodge. Newly emerged larvae are very active and probably immediately seek out live beavers and enter their fur. Larvae are regularly found on the host animal, sometimes in large numbers and especially around the shoulders and spine, but they may also disperse to other host individuals scavenging inside the beaver lodge or burrow. The larvae have sharp mandibles and can cause small insignificant lesions on the skin of their hosts, feeding on skin particles and fatty secretions from the skin. The mature larva then leaves the host and constructs a smooth-walled pupation cell in the earth within a lodge or burrow. Emergence from the pupal stage into the adult beetle takes place after about 3 weeks (Janzen 1963; Wood 1965; Peck 2006).

Beavers do not appear to be bothered by the feeding of this beetle (Wood, 1965). Beetles are often found around the neck and head region and can be collected by combing the fur with a fine comb (Peck, 2006). If the beaver has lain dead and cold for a while, and you place our warm hand for example on the animal's snout, beetles will immediately crawl onto your hand. They can live for more than 16 days at temperatures of 4°C, which is not surprising since they must survive long periods of cold when beavers are swimming or sitting on the ice during winter (Janzen, 1963). The adult beetles float when placed in water but drown when immersed (Duff et al., 2013). They do not survive long off the beaver, owing to desiccation or extremes in temperature (Janzen, 1963). Infestation rates in North American beavers have been recorded to be between 0 and 92 adult beetles per beaver (Janzen, 1963) and 1 and 742 beetles per beaver in Sweden (Åhlén, 2001).

The beaver parasite beetle is found on both species of beaver (Sarwar and Rauf, 2018), which is interesting considering the beaver's evolution. This suggests that both species of beavers have a shared ancestor and have since evolved separately. Another species of beetle, the beaver nest beetle (*Leptinillus validus*), has been found on beavers throughout the more northern parts of North America, i.e. from Minnesota and further north (Peck, 2007). Both beaver beetle species have also been found on the same beaver (Wood, 1965). The beaver nest beetle has a body length of about 4.5–5.0 mm and is larger and morphologically different from the beaver parasite beetle (Peck, 2007; Figure 8.13). Both larvae and adult beaver nest beetles feed on the same food as the beaver parasite beetle. Adult beaver nest beetles cluster on twigs and sticks hanging from the ceiling of lodges, and 1,000 beetles per lodge have been recorded. Interestingly, both larvae and adult beaver nest beetles spend more time off the host than on it (Peck, 2007). The beaver nest beetle also occurs on introduced and captive populations of North American beavers in Europe and on Eurasian beavers where the two species have come into direct contact (Peck, 2006; Peck, 2007).

Sixty-two species of fur mites belonging to the genus *Schizocarpus* have been identified on beavers: 45 species on the Eurasian beaver and 24 on the North American beaver (Bochkov et al., 2012; Izdebska et al., 2016). Seven fur mites have also been found on both species when held at the same fur farms in Russia (Izdebska et al., 2016). Mites are usually spread through direct contact. Infected animals often shake, scratch, or rub their ears and

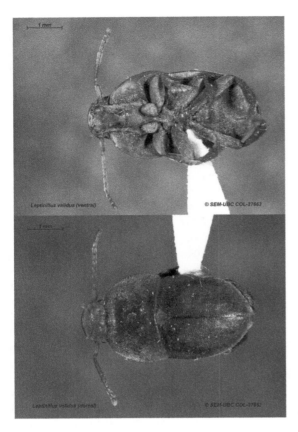

Figure 8.13 Ventral and dorsal view of the beaver nest beetle *Leptinillus validus*. (Photo courtesy of the Spencer Entomological Collection, Beaty Biodiversity Museum, UBC.)

head. Mites that cause itching can lead to injuries to the skin and secondary infections (Stocker, 2000). More than 10 species may live on an individual beaver at any one time, and they are often restricted to particular areas of the body. The beaver mite attaches itself to the root of the hair shaft and feeds on sebaceous secretions that build up in this area. In addition to the ears, mites can live on the head, neck, chest, sides, base of the tail, fore legs, and back legs of the beaver (Bochkov et al., 2012). All 15 Eurasian beavers examined from various parts of Sweden had ears infected with mites. On average, 140 individual mites were found per beaver, with the number of mites varying from 59 to 618 (Åhlén, 2001).

Various tick species are found throughout the ranges of both beaver species. Ticks have been reported on some beavers in Scotland (Goodman et al., 2012; Jones and Campbell-Palmer, 2014), largely in the fur underneath tracking tags and around the ears (Lawrence et al., 1956), indicating lack of grooming access. Ticks (*I. banksi*) have been found in 34% of North American beaver lodge bedding material in Michigan, USA. Of 68 beavers captured, 30% had ticks on their bodies, and one of the beavers had 18 ticks (Lawrence et al., 1956). In another investigation, 29 of 53 (55%) beavers had ticks, and one of these had 106 ticks on its body (Novak, 1987). No pathological disease has been associated with ticks, but they can clearly act as a vector for a number of blood-borne diseases (e.g. borreliosis), and heavy infection may lead to anemia (Brown et al., 2005). More research should be carried out on the beaver–tick relationship.

References

Ackerman, B. B., Lindzey, F. G., and Hemker, T. P. 1984. Cougar food habits in southern Utah. *Journal of Wildlife Management*, 48, 147–155.

Addison, E. M. 1973. Life cycle of *Dipetalonema sprenti* Anderson (Nematoda: Filaroidea) of beaver (*Castor canadensis*). *Canadian Journal Zoology*, 51, 403–416.

Adler, B. and de la Peña Moctezuma, A. 2010. *Leptospira* and leptospirosis. *Veterinary Microbiology*, 140, 287–296.

Åhlén, P. A. 2001. *The Parasitic and Commensal Fauna of the European Beaver (Castor fiber) in Sweden*. Umeå: Sveriges Lantbruksuniversitet.

Alberts, B., Johnson, A., Lewis, J., Raff, M., Roberts, K., and Walter, P. 2002. *Molecular Biology of the Cell*. New York: Gartland Science.

Anderson, R. C. 2000. *Nematode Parasites of Vertebrates: Their Development and Transmission*. Wallingford, Oxon: CABI.

Anderson, W. I., Schlafer, D. H., and Vesely, K. R. 1989. Thyroid follicular carcinoma with pulmonary metastases in a beaver (*Castor canadensis*). *Journal of Wildlife Diseases*, 25, 599–600.

Andersone, Ž. and Ozoliņš, J. 2004. Food habits of wolves *Canis lupus* in Latvia. *Acta Theriologica*, 49, 357–367.

Arner, D. H., Mason, C., and Perkins, C. J. 1981. Practicality of reducing a beaver population through the release of alligators. *Worldwide Furbearer Conference*, Frostburg, MD.

Ayral, F., Djelouadji, Z., Raton, V., Zilber, A.-L., Gasqui, P., Faure, E., Baurier, F., Vourc'h, G., Kodjo, A., and Combes, B. 2016. Hedgehogs and mustelid species: major carriers of pathogenic *Leptospira*, a survey in 28 animal species in France (20122015). *PLoS ONE*, 11, e0162549.

Bailey, V. 1927. Beaver habits and experiments in beaver culture. *USDA Technical Bulletins*, 21, 1–39.

Bajer, A., Bednarska, M., and Sinski, E. 1997. Wildlife rodents from different habitats as a reservoir for *Cryptosporidium parvum*. *Acta Parasitologica*, 42, 192–194.

Bajer, A., Bednarska, M., Paziewska, A., Romanowski, J., and Siński, E. 2008. Semi-aquatic animals as a source of water contamination with *Cryptosporidium* and *Giardia*. *Wiad Parazytologii*, 54, 315–318.

Barlow, A. M., Gottstein, B., and Mueller, N. 2011. *Echinococcus multilocularis* in an imported captive European beaver (*Castor fiber*) in Great Britain. *Veterinary Record*, 169, 339.

Basey, J. M. and Jenkins, S. H. 1995. Influences off predation risk and energy maximization on food selection by beavers (*Castor canadensis*). *Canadian Journal of Zoology*, 73, 2197–2208.

Baskin, L. 2011a. Predators as determinants of beaver alertness and shelter-making behavior. *Restoring the European Beaver*, 50, 271–280.

Baskin, L. M. 2011b. Predators as determinants of beaver alertedness and shelter-making behaviour. In: Sjöberg, G. and Ball, J. P. (eds.) *Restoring the Eurasian Beaver: 50 Years of Experience*. Sofia: Pensoft Publishers.

Baskin, L. M. and Novoselova, N. S. 2008. A fear of predators as a factor determining the length of feeding routes in beavers. *Zoologichesky Zhurnal*, 87, 226–230.

Behrendt, M. 2003. *Beaver Beetles. Article from Buckeye Trapper*. Ohio: Ohio State Trappers.

Belcuore, T., Conti, I., Childs, J., Hanlon, C., Krebs, J., and Guerra, M. 2002. Rabies in a beaver—Florida, 2001. *Journal of the American Medical Association*, 287, 3202–3203.

Belfiore, N. M. 2006. Observation of a beaver beetle (*Platypsyllus castoris* Ritsema) on a North American river otter (*Lontra canadensis* Schreber) (Carnivora: Mustelidae: Lutrinae) in Sacramento county, California (Coleoptera: Leiodidae: Platypsyllinae). *The Coleopterists Bulletin*, 60, 312–314.

Bell, J. F., Owen, C. R., Jellison, W. L., Moore, G. J., and Bukker, E. O. 1962. Epizootic tularemia in pen-raised beavers and field trials of vaccines. *American Journal of Veterinary Research*, 23, 884–888.

Bergan, F. and Rosell, F. 2017. *Konsekvensene av betydelig redusert vannstand for to beverfamilier i Nøklevann*. Bø i Telemark: Høgskolen i Sørøst-Norge.

Bergerud, A. T. and Miller, D. R. 1977. Population dynamics of Newfoundland beaver. *Canadian Journal of Zoology*, 55, 1480–1492.

Bloomquist, C. K. and Nielsen, C. K. 2010. Demography of unexploited beavers in southern Illinois. *Journal of Wildlife Management*, 74, 228–235.

Bochkov, A. V., Labrzycka, A., Skoracki, M., and Saveljev, A. P. 2012. Fur mites of the genus *Schizocarpus trouessart*

(Acari: Chirodiscidae) parasitizing the Eurasian beaver *Castor fiber belorussicus* Lavrov (Rodentia: Castoridae) in NE Poland (Suwałki). *Zootaxa*, 3162, 39–59.

Borchert, E. J., Leaphart, J. C., Bryan Jr, A. L., and Beasley, J. C. 2019. Ecotoxicoparasitology of mercury and trace elements in semi-aquatic mammals and their endoparasite communities. *Science of the Total Environment*, 679, 307–316.

Boyce, M. S. 1981. Beaver life-history responses to exploitation. *Journal of Applied Ecology*, 18, 749–753.

Brady, C. A. and Svendsen, G. E. 1981. Social-behaviour in a family of beaver, *Castor canadensis*. *Biology of Behaviour*, 6, 99–114.

Brakhage, G. K. and Sampson, F. W. 1952. Rabies in beaver. *Journal of Wildlife Management*, 16, 226.

Breck, S. W., Wilson, K. R., and Andersen, D. C. 2001. The demographic response of bank-dwelling beavers to flow regulation: a comparison on the Green and Yampa rivers. *Canadian Journal of Zoology*, 79, 1957–1964.

Bromley, C. K., and Hood, G. A. 2013. Beaver (*Castor canadensis*) facilitate early access by Canada geese (*Branta canadensis*) to nesting habitat and open water in Canada's boreal wetlands. *Mammalian Biology*, 78, 73–77. https://doi.org/10.1016/j.mambio.2012.02.009.

Brown, R. N., Lane, R. S., and Dennis, D. T. 2005. Geographic distributions of tick-borne diseases and their vectors In: Goodman, J., Dennis, D. T., and Sonenshine, D. (eds.) *Tick-Borne Diseases of Humans*. Washington, DC: ASM Press.

Bush, A. O. and Samuel, W. M. 1978. The genus *Travassosius* Khalil, 1922 (Nematoda, Trichostrongyloidea) in beaver, *Castor* spp.: a review and suggestion for speciation. *Canadian Journal of Zoology*, 56, 1471–1474.

Bush, A. O. and Samuel, W. M. 1981. A review of helminth communities in beaver (*Castor* spp.) with a survey of *Castor canadensis* in Alberta, Canada. *Worldwide Furbearer Conference*, pp. 3–11.

Busher, P. E. 1999. The long-term demographic patterns of unexploited beaver populations in the United States. *Proceedings of the 1st Euro-American Beaver Congress*, 24–27 August, pp. 39–50.

Campbell-Palmer, R. 2017. Importance of founder screening in beaver restoration programmes. PhD thesis, University College of Southeast Norway.

Campbell, R. D., Nouvellet, P., Newman, C., Macdonald, D. W., and Rosell, F. 2012. The influence of mean climate trends and climate variance on beaver survival and recruitment dynamics. *Global Change Biology*, 18, 2730–2742.

Campbell-Palmer, R., Girling, S., Pizzi, R., Hamnes, I. S., Øines, Ø., and Del Pozo, J. 2013. *Stichorchis subtriquetrus* in a free-living beaver in Scotland. *Veterinary Record*, 173, 72.

Campbell-Palmer, R., Del Pozo, J., Gottstein, B., Girling, S., Cracknell, J., Schwab, G., Rosell, F., and Pizzi, R. 2015. *Echinococcus multilocularis* detection in live Eurasian beaver (*Castor fiber*) using a combination of laparoscopy and abdominal ultrasound under field conditions. *PLoS ONE*, 10, e0130842.

Campbell-Palmer R,, Rosell F, Naylor A, Cole, G., Mota, S., Brown, D., Fraser, M., Pizzi, R., Elliott, M., Wilson, K., Gaywood, M. and Girling, S. 2021. Eurasian beaver (*Castor fiber*) health surveillance in Britain: Assessing a disjunctive reintroduced population. *Vet Rec.* e84.

Campbell-Palmer, R., Puttock, A., Graham, H., Wilson, K., Schwab, G., Gaywood, M., and Brazier, R. 2018. Survey of the Tayside area beaver population 2017–2018. *Scottish Natural Heritage Commissioned Report*. Scottish Natural Heritage.

Canavan, W. P. 1929. Nematode parasites of vertebrates in the Philadelphia Zoological Garden and vicinity. *Parasitology*, 26, 63–102.

Canavan, W. P. 1934. On trematode *Allopyge indulatus* n. sp. parasitic in Lilfords crane (*Mega grus lilfordi*). *Journal of Parasitology* 26, 117–120.

Carlson, B. L., Hill, D., and Nielsen, S. W. 1983. Cutaneous papillomatosis in a beaver. *Journal of the American Veterinary Medical Association*, 183, 1283–1284.

Cave, A. J. E. 1984. *Dentitional Anomalies in the Beaver and Some Other Mammals*. Berne: Brain Anatomy Institute.

Chitwood, B. G. 1934. *Capillaria hepatica* from the liver of *Castor canadensis canadensis*. *Proceedings of the Helminthological Society of Washington*, 1, 10.

ćirović, D., Pavlović, I., Kulišić, Z., Ivetić, V., Penezić, A., and Cosić, N. 2013. *Echinococcus multilocularis* in the European beaver (*Castor fibre* L.) from Serbia: first report. *Veterinary Record*, 1, e100879.

Clulow, F., Mirka, M., Dave, N., and Lim, T. 1991. ^{226}Ra and other radionuclides in water, vegetation, and tissues of beavers (*Castor canadensis*) from a watershed containing U tailings near Elliot Lake, Canada. *Environmental Pollution*, 69, 277–310.

Commission Regulation (European Commission) 2006. Setting maximum levels of certain contaminations in foodstuff. *Official Journal of the European Union*, L354/5, 1–20.

Cook, A. H. 1940. Screwworms infest beaver in Texas. *Journal of Mammalogy*, 21, 93.

Cross, H. B., Campbell-Palmer, R., Girling, S., and Rosell, F. 2012. The Eurasian beaver (*Castor fiber*) is apparently not a host to blood parasites in Norway. *Veterinary Parasitology*, 190, 246–248.

Cullen, C. 2003. Normal ocular features, conjunctival microflora and intraocular pressure in the Canadian beaver (*Castor canadensis*). *Veterinary Ophthalmology*, 6, 279–284.

Davis, M. A., Butcher, G. D., and Mather, F. B. 2015. *Avian Diseases Transmissible to Humans*. Institute of Food and Agricultural Sciences Extension. Gainesville: University of Florida.

de Barba, M., Waits, L. P., Garton, E. O., Genovesi, P., Randi, E., Mustoni, A., and Groff, C. 2010. The power of genetic monitoring for studying demography, ecology and genetics of a reintroduced brown bear population. *Molecular Ecology*, 19, 3938–3951.

Demiaszkiewicz, A. W., Lachowicz, J., Kuligowska, I., Pyziel, A. M., Bełżecki, G., Miltko, R., Kowalik, B., Gogola, W., and Giżejewski, Z. 2014. Endoparasites of the European beaver (*Castor fiber* L. 1758) in north-eastern Poland. *Bulletin of the Veterinary Institute in Pulawy*, 58, 223–227.

Destefano, S., Koenen, K. K., Henner, C., and Strules, J. 2006. Transition to independence by subadult beavers (*Castor canadensis*) in an unexploited, exponentially growing population. *Journal of Zoology*, 269, 434–441.

Devine, C., Girling, S. J., Martinez-Pereira, Y., Pizzi, R., Brown, D., Campbell-Palmer, R., and Rosell, F. 2012. Physiological heart murmurs in isoflurane anaesthetized Eurasian beavers. *Proceedings of the 6th International Beaver Symposium*, Ivanić Grad, Croatia.

Dick, T. A. and Pozio, E. 2001. *Trichinella* spp. and *Trichinellosis*. In: Samuel, W. M., Pybus, M. J., and Kocan, A. A. (eds.) *Parasitic Diseases of Wild Mammals*, 2nd edn. Ames: Iowa State University Press.

Dieterich, R. A. 1981. *Tumors. Alaskan Wildlife Diseases*. Fairbanks: Institute of Arctic Biology, University of Alaska Fairbanks.

Doboszyńska, T. and żurowski, W. 1983. Reproduction of the European beaver. *Acta Zoologica Fennica*, 174, 123–126.

Drózdz, J., Demiaszkiewicz, A. W., and Lachowicz, J. 2004. Endoparasites of the beaver *Castor fiber* (L.) in northeast Poland. *Helminthologia*, 41, 99–102.

Duff, A., Campbell-Palmer, R., and Needham, R. 2013. The beaver beetle *Platypsyllus castoris* Ritsema (Leiodidae: Platypsyllinae) apparently established on reintroduced beavers in Scotland, new to Britain. *Coleopterist*, 22, 9–19.

Dunlap, B. G. and Thies, M. L. 2002. *Giardia* in beaver (*Castor canadensis*) and nutria (*Myocastor coypus*) from east Texas. *Journal of Parasitology*, 88, 1254–1258.

Eckert, J. and Deplazes, P. 1999. *Alveolar echinococcosis* in humans: the current situation in central Europe and the need for countermeasures. *Parasitology Today*, 15, 315–319.

Erickson, A. B. 1949. The fungus (*Haplosporangium parvum*) in the lungs of the beaver (*Castor canadensis*). *Journal of Wildlife Management*, 13, 419–420.

Erlandsen, S. L. 1994. *Biotic Transmission—Is Giardiasis a Zoonosis? Giardia: From Molecules to Disease*. Wallingford, Oxon: CAB International, pp. 83–97.

Erlandsen, S. L., Macechko, P. T., van Keulen, H., and Jaroll, E. L. 1996. Beaver and *Giardia* in the environment: a current perspective on the existence of 'beaver fever'.

In: Rowan, A. N. and Weer, J. C. (eds.) *Living with Wildlife: The Biology and Sociology of Suburban Deer and Beaver*. North Grafton, MA: Tufts Center for Animals and Public Policy, p. 210

Ernst, J. V., Cooper, W. L., and Frydendall, M. J. 1970. *Eimeria sprehni* Yakimoff, 1934, and *E. causeyi* sp. n. (Protozoa: Eimeridae) from the Canadian beaver, *Castor canadensis*. *Journal of Parasitology*, 56, 30–31.

Estève, R. 1988. An analysis of beaver mortality in Haute-Savoie. *Biévre*, 9, 171–176.

Falandysz, J., Taniyasu, S., Yamashita, N., Rostkowski, P., Zalewski, K., and Kannan, K. 2007. Perfluorinated compounds in some terrestrial and aquatic wildlife species from Poland. *Journal of Environmental Science Health Part A*, 42, 715–719.

Fayer, R., Santín, M., Trout, J. M., Destefano, S., Koenen, K., and Kaur, T. 2006. Prevalence of *Microsporidia*, *Cryptosporidium* spp., and *Giardia* spp. in beavers (*Castor canadensis*) in Massachusetts. *Journal of Zoo and Wildlife Medicine*, 37, 492–498.

Fedynich, A. M., Pence, D. B., and Urubek, R. L. 1986. Helminth fauna of beaver from central Texas. *Journal of Wildlife Diseases*, 22, 579–582.

Fimreite, N., Parker, H., Rosell, F., Hosen, D. A., Hovden, A., and Solheim, A. 2001. Cadmium, copper, and zinc in Eurasian beaver (*Castor fiber*) from Bø, Telemark, Norway. *Bulletin of Environmental Contamination and Toxicology*, 67, 503–509.

Foil, L. and Orihel, T. C. 1975. *Dirofilaria immitis* (Leidy, 1856) in the beaver, *Castor canadensis*. *Journal of Parasitology*, 61, 433–433.

Forzán, M. J. and Frasca, S. 2004. Systemic toxoplasmosis in a five-month-old beaver (*Castor canadensis*). *Journal of Zoo and Wildlife Medicine*, 35, 113–115.

Fryxell, J. M. and Doucet, C. M. 1991. Provisioning time and central-place foraging in beavers. *Canadian Journal of Zoology*, 69, 1308–1313.

Fuehrer, H.-P. 2014. An overview of the host spectrum and distribution of *Calodium hepaticum* (syn. *Capillaria hepatica*): part 1—Muroidea. *Parasitology Research*, 113, 619–640.

Gable, T. D. and Windels, S. K. 2018. Kill rates and predation rates of wolves on beavers. *Journal of Wildlife Management*, 82, 466–472.

Gable, T. D., Windels, S. K., Bruggink, J. G., and Homkes, A. T. 2016. Where and how wolves (*Canis lupus*) kill beavers (*Castor canadensis*). *PLoS ONE*, 11, e0165537.

Gable, T. D., Stanger, T., Windels, S. K., and Bump, J. K. 2018a. Do wolves ambush beavers? Video evidence for higher-order hunting strategies. *Ecosphere*, 9, e02159.

Gable, T. D., Windels, S. K., Romanski, M. C., and Rosell, F. 2018b. The forgotten prey of an iconic predator: a review of interactions between grey wolves *Canis lupus* and beavers *Castor* spp. *Mammal Review*, 48, 123–138.

Gamberg, M., Braune, B., Davey, E., Elkin, B., Hoekstra, P. F., Kennedy, D., Macdonald, C., Muir, D., Nirwal, A., and Wayland, M. 2005. Spatial and temporal trends of contaminants in terrestrial biota from the Canadian Arctic. *Science of the Total Environment*, 351, 148–164.

Gaydos, J. K., Zabek, E., and Raverty, S. 2009. *Yersinia pseudotuberculosis* septicemia in a beaver from Washington state. *Journal of Wildlife Diseases*, 45, 1182–1186.

Gelling, M., Zochowski, W., Macdonald, D., Johnson, A., Palmer, M., and Mathews, F. 2015. Leptospirosis acquisition following the reintroduction of wildlife. *Veterinary Record*, 177, 440.

Girling, S. J., Goodman, G., Burr, P., Pizzi, R., Naylor, A., Cole, G., Brown, D., Fraser, M., Rosell, F. N., and Schwab, G. 2019a. Evidence of *Leptospira* species and their significance during reintroduction of Eurasian beavers (*Castor fiber*) to Great Britain. *Veterinary Record*, 185, 482.

Girling, S. J., McElhinney, L. M., Fraser, M. A., Gow, D., Pizzi, R., Naylor, A., Cole, G., Brown, D., Rosell, F., Schwab, G., and Campbell-Palmer, R. 2019b. Absence of hantavirus in water voles and Eurasian beavers in Britain. *Veterinary Record*, 184, 253.

Girling, S. J., Naylor, A., Fraser, M., and Campbell-Palmer, R. 2019c. Reintroducing beavers *Castor fiber* to Britain: a disease risk analysis. *Mammal Review*, 49, 30–323.

Giżejewska, A., Spodniewska, A., and Barski, D. 2014. Concentration of lead, cadmium, and mercury in tissues of European beaver (*Castor fiber*) from the north-eastern Poland. *Bulletin of the Veterinary Institute in Pulawy*, 58, 77–80.

Giżejewska, A., Spodniewska, A., Barski, D., and Fattebert, J. 2015. Beavers indicate metal pollution away from industrial centers in northeastern Poland. *Environmental Science and Pollution Research*, 22, 3969–3975.

Goodman, G., Girling, S., Pizzi, R., Meredith, A., Rosell, F., and Campbell-Palmer, R. 2012. Establishment of a health surveillance program for reintroduction of the Eurasian beaver (*Castor fiber*) into Scotland. *Journal of Wildlife Diseases*, 48, 971–978.

Goodman, G., Meredith, A., Girling, S., Rosell, F., and Campbell-Palmer, R. 2016. Outcomes of a 'One Health' monitoring approach to a five-year beaver (*Castor fiber*) reintroduction trial in Scotland. *EcoHealth*, 14, 139–143.

Gorbunova, V., Bozzella, M. J., and Seluanov, A. 2008. Rodents for comparative aging studies: from mice to beavers. *Age*, 30, 111–119.

Green, H. U. 1932. Observations on the occurrence of otter in the Riding Mountain National Park, Manitoba, in relation to beaver life. *Canadian Field Naturalist*, 46, 204–206.

Greer, K. R. 1955. Yearly food habits of the river otter in the Thompson Lakes region, northwestern Montana, as indicated by scat analyses. *American Midland Naturalist*, 54, 299–313.

Grubešić, M., Margeletić, J., Čirović, D., Vucelja, M., Bjedov, L., Burazerović, J. and Tomljanović, K. 2015. Analysis of beaver (*Castor fiber* L.) mortality in Croatia and Serbia. *Journal of Forestry Society Croatia*, 139, 137–143.

Grudzinski, B. P., Cummins, H., and Vang, T. K. 2019. Beaver canals and their environmental effects. *Progress in Physical Geography: Earth and Environment*, 1–23.

Gruninger, R. J., McAllister, T. A., and Forster, R. J. 2016. Bacterial and archaeal diversity in the gastrointestinal tract of the North American beaver (*Castor canadensis*). *PLoS ONE*, 11, e0156457.

Gunson, J. R. 1970. Dynamics of the beaver of Saskatchewan's northern forest. MSc thesis, University of Alberta.

Haitlinger, R. 1991. Stawonogi występujące na bobrze europejskim (*Castor fiber* L.) w Polsce. *Wiadomości Parazytologiczne*, 37, 107–109.

Hamerton, B. C. A. E. 1934. Report on deaths occurring in the society's gardens during the year 1933. *Proceedings of the Zoological Society of London*, 104, 389–422.

Hartman, G. 1999. Beaver management and utilization in Scandinavia. In: Busher, P. and Dzieciolowski, R. (eds.) *Beaver Protection, Management, and Utilization in Europe and North America*. New York: Springer.

Havens, R. P., Crawford, J. C., and Nelson, T. A. 2013. Survival, home range, and colony reproduction of beavers in east-central Illinois, an agricultural landscape. *American Midland Naturalist*, 169, 17–30.

Henry, D. B. and Bookhout, T. A. 1969. Productivity of beavers in northeastern Ohio. *Journal of Wildlife Management*, 33, 927–932.

Hentschke, J., Meyer, H., Wittstatt, U., Ochs, A., Burkhardt, S., and Aue, A. 1999. Kuhpocken bei kanadischen bibern (*Castor fiber canadensis*) und katzenbaren (*Ailurus fulgens*). *Tierärztliche Umschau*, 54, 311–317.

Hilfiker, E. L. 1991. *Beavers, Water, Wildlife and History*. Interlaken, NY: Windswept Press.

Hill, E. P. 1982. Beaver (*Castor canadensis*). In: Chapman, J. A. and Feldhamer, G. A. (eds.) *Wild Mammals of North America—Biology, Management and Economics*. Baltimore: Johns Hopkins University Press.

Hillis, T. and Parker, G. 1993. Age and proximity to local ore-smelters as determinants of tissue metal levels in beaver (*Castor canadensis*) of the Sudbury (Ontario) area. *Environmental Pollution*, 80, 67–72.

Hinze, G. 1950. *Der Biber. Körperbau und Lebensweise Verbreitung und Geschichte*. Berlin: Akademie.

Hood, G. A. 2015. *Post-Release Survival of Beavers Exposed to Bitumen*. Technical Report, Prepared for Canadian Natural Resources Limited. Camrose: University of Alberta.

Hood, G. A. and Larson, D. G. 2015. Ecological engineering and aquatic connectivity: a new perspective from beaver-modified wetlands. *Freshwater Biology*, 60, 198–208.

Irving, L. and Orr, M. 1935. The diving habits of the beaver. *Science*, 82, 569.

Isaac-Renton, J. L., Moricz, M. M., and Proctor, E. M. 1987. A *Giardia* survey of fur-bearing water mammals in British Columbia, Canada. *Journal of Environmental Health*, 50, 80–83.

Izdebska, J. N., Fryderyk, S., and Rolbiecki, L. 2016. *Demodex castoris* sp. nov. (Acari: Demodecidae) parasitizing *Castor fiber* (Rodentia), and other parasitic arthropods associated with *Castor* spp. *Diseases of Aquatic Organisms*, 118, 1–10.

Jackson, M. D. 1990. Beaver dispersal in western Montana. MSc thesis, University of Montana.

Janiszewski, P., Hanzal, V., and Misiukiewicz, W. 2014. The Eurasian beaver (*Castor fiber*) as a keystone species—a literature review. *Baltic Forestry*, 20, 277–286.

Janovsky, M., Bacciarini, L., Sager, H., Gröne, A., and Gottstein, B. 2002. *Echinococcus multilocularis* in a European beaver from Switzerland. *Journal of Wildlife Diseases*, 38, 618–620.

Janzen, D. H. 1963. Observations on populations of adult beaver beetles, *Platypsyllus castoris* (Platypsyllidae: Coleoptera). *Pan-Pacific Entomologist*, 39, 215–228.

Jokinen, E. 1992. *The Freshwater Snails (Mollusca: Gastropoda) of New York State*. Albany: University of the State of New York, State Education Department, New York State Museum.

Jones, D. M. and Theberge, J. B. 1983. Variation in red fox, *Vulpes vulpes*, summer diets in northwest British Columbia and southwest Yukon. *Canadian Field-Naturalist*, 97, 311–314.

Jones, S. and Campbell-Palmer, R. 2014. The Scottish Beaver Trial: The Story of Britain's First Licensed Release into the Wild. Final report. Argyll: Royal Zoological Society of Scotland.

Jordan, C. N., Kaur, T., Koenen, K., Destefano, S., Zajac, A. M., and Lindsay, D. S. 2005. Prevalence of agglutinating antibodies to *Toxoplasma gondii* and *Sarcocystis neurona* in beavers (*Castor canadensis*) from Massachusetts. *Journal of Parasitology*, 91, 1228–1230.

Joszt, L. 1964. The helminth parasites of the European beaver, *Castor fiber* L., in Poland. *Acta Parasitologica Polonica*, 12, 85–88.

Judd, W. W. 1954. Some records of ectoparasitic *Acarina* and *insecta* from mammals in Ontario. *Journal of Parasitology*, 40, 483–484.

Kadulski, S. 1998. Ectoparasites of the beaver *Castor fiber* L. from Popielno. *Wiadomosci Parazytologizne*, 44, 729–736.

Khalil, M. 1922. XXXV. *Travassosius rufus*, gen. et sp. n.: a nematode (Trichostrongylidæ) parasitic in the stomach of the Norwegian beaver. *Annals and Magazine of Natural History*, 10, 281–289.

Kile, N. B. and Rosell, F. 1996. European beaver, *Castor fiber*, pinned by a felled tree. *Canadian Field-Naturalist* 110, 706–707.

Kile, N. B., Nakken, P. J., Rosell, F., and Espeland, S. 1996. Red fox, *Vulpes vulpes*, kills a European beaver, *Castor fiber*, kit. *Canadian Field-Naturalist*, 110, 338–339.

Kim, J.-H., Lee, J. Y., Han, T.-S., Han, K.-B., Kang, S. S., Bae, C. S., and Choi, S. H. 2005. A case of malaloccluded incisor teeth in a beaver (*Castor canadensis*). *Journal of Veterinary Science*, 6, 173–175.

Knudsen, G. J. 1953. Beaver die-off. *Wisconsin Conservation Bulletin*, 18, 20–23.

Kok, N. J., Lukoschus, F. S., and Clulow, F. V. 1970. *Psorobia castoris* spec. nov. (Acarina: Psorergatidae), a new itch mite from the beaver, *Castor canadensis*. *Canadian Journal of Zoology*, 48, 1419–1423.

Krebs, J. W., Noll, H. R., Rupprecht, C. E., and Childs, J. E. 2002. Rabies surveillance in the United States during 2001. *Journal of the American Veterinary Medical Association*, 221, 1690–1701.

Kurucz, K., Madai, M., Bali, D., Hederics, D., Horváth, G., Kemenesi, G., and Jakab, F. 2018. Parallel survey of two widespread renal syndrome-causing zoonoses: *Leptospira* spp. and *Hantavirus* in urban environment, Hungary. *Vector-Borne and Zoonotic Diseases*, 18, 200–205.

Lang, B. Z. 1977. Snail and mammalian hosts for *Fasciola hepatica* in eastern Washington. *Journal of Parasitology*, 63, 938–939.

Langford, E. 1972. Pasteurella pseudotuberculosis infections in western Canada. *Canadian Veterinary Journal*, 13, 85–87.

Lavrov, L. S. 1983. Evolutionary development of the genus *Castor* and taxonomy of the contemporary beavers of Eurasia. *Acta Zoologica Fennica*, 174, 87–90.

Lawrence, W. H., Fay, L., and Graham, S. 1956. A report on the beaver die-off in Michigan. *Journal of Wildlife Management*, 20, 184–187.

Lawrence, W. H., Hays, K. L., and Graham, S. A. 1961. Ectoparasites of the beaver (*Castor canadensis* Kuhl). *Wildlife Diseases*, 12, 14.

Lawrence, W. H., Hays, K. L., and Graham, S. A. 1965. Arthropodous ectoparasites from some northern Michigan mammals. *Occasional Papers of the Museum of Zoology, University of Michigan*, 639, 1–7.

Lawson, P. A., Foster, G., Falsen, E., Markopoulos, S. J., and Collins, M. D. 2005. *Streptococcus castoreus* sp. nov., isolated from a beaver (*Castor fiber*). *International Journal of Systematic and Evolutionary Microbiology*, 55, 843–846.

Leblanc, F. A., Gallant, D., Vasseur, L., and Léger, L. 2007. Unequal summer use of beaver ponds by river otters: influence of beaver activity, pond size, and vegetation cover. *Canadian Journal of Zoology*, 85, 774–782.

Litvaitis, J. A., Clark, A. G., and Hunt, J. H. 1986. Prey selection and fat deposits of bobcats (*Felis rufus*) during autumn and winter in Maine. *Journal of Mammalogy*, 67, 389–392.

Lodberg-Holm, H. K., Steyaert, S. M. J. G., Reinhardt, S., and Rosell, F. 2021. Size is not everything: Differing activity and foraging patterns between the sexes in a monomorphic mammal. *Behavioral Ecology and Sociobiology*, 75, 80.

López-Pérez, A. M., Carreón-Arroyo, G., Atilano, D., Vigueras-Galván, A. L., Valdez, C., Toyos, D., Mendizabal, D., López-Islas, J., and Suzán, G. 2017. Presence of antibodies to *Leptospira* spp. in black-tailed prairie dogs (*Cynomys ludovicianus*) and beavers (*Castor canadensis*) in northwestern Mexico. *Journal of Wildlife Diseases*, 53, 880–884.

Lowrey, B., Elbroch, L. M., and Broberg, L. 2016. Is individual prey selection driven by chance or choice? A case study in cougars (*Puma concolor*). *Mammal Research*, 61, 353–359.

Máca, O., Pavlásek, I., and Vorel, A. 2015. *Stichorchis subtriquetrus* (Digenea: Paramphistomatidae) from Eurasian beaver (*Castor fiber*) in the Czech Republic. *Parasitology Research*, 114, 2933–2939.

Maldonado, Y. A., Read, J. S., and AAP Committee on Infectious Disease 2017. Diagnosis, treatment, and prevention of congenital toxoplasmosis in the United States. *Pediatrics*, 139, e20163860.

Marreros, N., Zürcher-Giovannini, S., Origgi, F., Djelouadji, Z., Wimmershoff, J., Pewsner, M., Akdesir, E., Batista Linhares, M., Kodjo, A., and Ryser-Degiorgis, M. P. 2018. Fatal leptospirosis in free-ranging Eurasian beavers (*Castor fiber* L.), Switzerland. *Transboundary and Emerging Diseases*, 65, 1297–1306.

McClanahan, K. 2019. Pair movement of a monogamous mammal, the Eurasian beaver. MSc thesis, University of South-Eastern Norway.

McKinstry, M. C. and Anderson, S. H. 2002. Survival, fates, and success of transplanted beavers, *Castor canadensis*, in Wyoming. *Canadian Field-Naturalist*, 116, 60–68.

McKown, R., Veatch, J., Robel, R., and Upton, S. 1995. Endoparasites of beaver (*Castor canadensis*) from Kansas. *Journal of the Helminthological Society of Washington*, 62, 89–93.

McNew, J., Lance, B., and Woolf, A. 2005. Dispersal and survival of juvenile beavers (*Castor canadensis*) in southern Illinois. *American Midland Naturalist*, 154, 217–228.

McTaggart, S. T. and Nelson, T. A. 2003. Composition and demographics of beaver (*Castor canadensis*) colonies in central Illinois. *American Midland Naturalist*, 150, 139–150.

Mörner, T. 1992. *Liv och död bland vilda djur*. Stockholm: Sellin & Partner.

Mörner, T. 1999. Monitoring disease in wildlife—a review of diseases in the orders Lagomorpha and Rodentia in Sweden. *Verhber Erkg Zootiere*, 39, 255–262.

Mörner, T. and Sandstedt, K. 1983. A serological survey of antibodies against *Francisella tularensis* in some Swedish mammals. *Nordisk Veterinaermedicin*, 35, 82–85.

Mörner, T., Avenäs, A., and Mattsson, R. 1999. Adiaspiromycosis in a European beaver from Sweden. *Journal of Wildlife Diseases*, 35, 367–370.

Müller-Schwarze, D. 2011. *The Beaver: Its Life and Impact.* New York: Cornell University Press.

Mysłajek, R. W., Tomczak, P., Tołkacz, K., Tracz, M., Tracz, M., and Nowak, S. 2019. The best snacks for kids: the importance of beavers *Castor fiber* in the diet of wolf *Canis lupus* pups in north-western Poland. *Ethology Ecology & Evolution*, 31, 1–8.

Nash, J. R. 1951. An investigationof some problems of ecology of the beaver *Castor canadensis canadensis* Kuhl, in northern Manitoba. MSc thesis, University of Manitoba.

Nelder, M. P. and Reeves, W. K. 2005. Ectoparasites of road-killed vertebrates in northwestern South Carolina, USA. *Veterinary Parasitology*, 44, 729–736.

Nishiuchi, Y., Iwamoto, T., and Maruyama, F. 2017. Infection sources of a common non-tuberculous mycobacterial pathogen, *Mycobacterium avium* complex. *Frontiers in Medicine*, 4, 27.

Nitsche, K. A. 1996. Über unfälle des bibers beim baumfällen. *Säugetierkundliche Mitteilungen*, 37, 157–159.

Nitsche, K.-A. 2001. Behaviour of beavers (*Castor fiber albicus* Matschie, 1907) during the flood periods. In: Czech, A. and Schwab, G. (eds.) The European beaver in a new millenium. *Proceedings of the 2nd European Beaver Symposium*, 27–30 September 2000, Bialowieza. Krakow: Carpathian Heritage Society, pp. 85–90.

Nolet, B. A. and Rosell, F. 1998. Comeback of the beaver *Castor fiber*: an overview of old and new conservation problems. *Biological Conservation*, 83, 165–173.

Nolet, B., Dijkstra, V. A., and Heidecke, D. 1994. Cadmium in beavers translocated from the Elbe River to the Rhine/Meuse estuary, and the possible effect on population growth rate. *Archives of Environmental Contamination and Toxicology*, 27, 154–161.

Nolet, B., Broekhuizen, S., Dorrestein, G., and Rienks, K. 1997. Infectious diseases as main causes of mortality to beavers *Castor fiber* after translocation to the Netherlands. *Journal of Zoology*, 241, 35–42.

Nordstrom, W. R. 1972. *Comparison of trapped and untrapped beaver populations in New Brunswick.* MSc thesis, University of New Brunswick.

Novak, M. 1987. Beaver. In: Novak, M., Baker, J. A., Obbard, M. E., and Mallock, B. (eds.) *Wild Furbearer Management and Conservation in North America.* Toronto: Queens Printer for Ontario.

Ogburn-Cahoon, H. and Nettles, V. F. 1978. *Gongylonema pulchrum* Molin 1857, (Nematoda: Spiruridae) in a beaver. *Journal of Parasitology*, 64, 812.

Olstad, O. 1937. Beverens (*Castor fiber*) utbredelse i Norge. *Statens Viltundersøkelser. Nytt Magasin for Naturvidenskapen*, 77, 217–273.

Orlov, I. 1941. Investigation of the cycle of development of the trematode *Stichorchis subtriquetrus* Rud., parasitic in beavers. *Comptes Rendus (Doklady) de l'Académie des Sciences de l'URSS*, 31, 641–643.

Orlov, I. V. 1948. Studies on the helminth fauna of beavers. *Parazitofauna i Zabolevaniya*. Moscow: Glav Upr po Zapovednikam.

Pacha, R. E., Clark, G. W., Williams, E. A., Carter, A. M., Scheffelmaier, J. J., and Debusschere, P. 1987. Small rodents and other mammals associated with mountain meadows as reservoirs of *Giardia* spp. and *Campylobacter* spp. *Applied Environmental Microbiology*, 53, 1574–1579.

Packard, F. M. 1940. Beaver killed by coyotes. *Journal of Mammalogy*, 21, 359–360.

Parker, H. and Rosell, F. 2014. Rapid rebound in colony number of an over-hunted population of Eurasian beaver *Castor fiber*. *Wildlife Biology*, 20, 267–270.

Parker, H., Rosell, F., Hermansen, T., Sørløkk, G., and Stærk, M. 2002. Sex and age composition of spring-hunted Eurasian beaver in Norway. *Journal of Wildlife Management*, 66, 1164–1170.

Payne, N. F. 1984. Mortality rates of beaver in Newfoundland. *Journal of Wildlife Management*, 48, 117–126.

Payne, N. and Finlay, C. 1975. Red fox attack on beaver. *Canadian Field-Naturalist*, 89, 450–451.

Peck, S. B. 2006. Distribution and biology of the ectoparasitic beaver beetle *Platypsyllus castoris* Ritsema in North America (Coleoptera: Leiodidae: Platypsyllinae). *Insecta Mundi*, 20, 85–94.

Peck, S. B. 2007. Distribution and biology of the ectoparasitic beetles *Leptinillus validus* (Horn) and *L. aplodontiae* Ferris of North America (Coleoptera: Leiodidae: Platypsyllinae). *Insecta Mundi*, 3, 1–7.

Pennak, R. 1989. *Fresh-Water Invertebrates of the United States.* New York: John Wiley & Sons.

Peterson, E. K. and Schulte, B. A. 2016. Impacts of pollutants on beavers and otters with implications for ecosystem ramifications. *Journal of Contemporary Water Research & Education*, 157, 33–45.

Petrosyan, V. G., Golubkov, V. V., Zavyalov, N. A., Khlyap, L. A., Dergunova, N. N., and Osipov, F. A. 2019. Modelling of competitive interactions between native Eurasian (*Castor fiber*) and alien North American (*Castor canadensis*) beavers based on long-term monitoring data (1934–2015). *Ecological Modelling*, 409, 1–15.

Piechocki, R. 1977. Ökologische Todesursachenforschung am Elbe-biber (*Castor fiber albicus*). *Beiträge zur Jagd und Wildforschung*, 10, 332–341.

Pinheiro, M. D. C. N., Luiz Martins do Nascimento, J., Carlos de Lima Silveira, L., Batista Teixeira da Rocha, J., and Aschner, M. 2009. Mercury and selenium—a review on aspects related to the health of human populations in the Amazon. *Environmental Bioindicators*, 4, 222–245.

Platt-Samoraj, A., Syczyło, K., Bancerz-Kisiel, A., Szczerba-Turek, A., Giżejewska, A., and Szweda, W. 2015. *Yersinia enterocolitica* strains isolated from beavers (*Castor fiber*). *Polish Journal of Veterinary Sciences*, 18, 449–451.

Pratama, R., Schneider, D., Böer, T., and Daniel, R. 2019. First insights into bacterial gastrointestinal tract communities of the Eurasian beaver (*Castor fiber*). *Frontiers in Microbiology*, 10, 1–13.

Price, E. W. 1934. A new trematode from a beaver. *Proceedings of the Helminthological Society of Washington*, 1, 1–2.

Provost, E. E. 1958. Sudies on reproduction and population dynamics in beaver. PhD thesis, Washington State University.

Rausch, R. A. and Pearson, A. 1972. Notes on the wolverine in Alaska and the Yukon Territory. *Journal of Wildlife Management*, 36, 249–268.

Recker, W. 1997. Seltene todesursache des bibers, *Castor fiber*. Der mink, *Mustela (Lutreola) vision*, als prädator des bibers im Bau. *Säugetierkundliche Mittenlungen*, 39, 87.

Reich, P. B., Peterson, D. W., Wedin, D. A., and Wrage, K. 2001. Fire and vegetation effects on productivity and nitrogen cycling across a forest–grassland continuum. *Ecology*, 82, 1703–1719.

Reid, D. G. 1984. Ecological interactions of river otter and beavers in boreal ecosystems. MSc thesis, University of Calgary.

Ribic, C. A., Donner, D. M., Beck, A. J., Rugg, D. J., Reinecke, S., and Eklund, D. 2017. Beaver colony density trends on the Chequamegon-Nicolet national forest, 1987–2013. *PLoS ONE*, 12, e0170099.

Romasov, B. V. 1992. Krankheiten der biber (beaver diseases). *Semiaquatische Säugetiere, Wissenschaftliche Beitrage der Martin-Luther-Universitat Halle-Wittenberg*, 199–203.

Romaschov, V. A. 1969. Helminth fauna of European beaver in its aboriginal colonies of Eurasia. *Acta Parasitologica Polonica*, 17, 55–64.

Romašov, V. A. and Stubbe, M. 1992. Stabilität der helminthen-fauna beim elbebiber (*Castor fiber* L.) in Deutschland. *Semiaquatische Säugetiere, Wissenschaftliche Beitrage der Martin-Luther-Universitat Halle-Wittenberg*, 204–206.

Rosef, O., Gondrosen, B., Kapperud, G., and Underdal, B. 1983. Isolation and characterisation of *Campylobacter jejuni* and *Campylobacter coli* from domesic and wild animals in Norway. *Applied Environmental Microbiology*, 46, 855–859.

Rosell, F. and Hovde, B. 1998. Pine marten, *Martes martes*, as a Eurasian beaver, *Castor fiber*, lodge occupant and possible predator. *Canadian Field Naturalist*, 112, 535–536.

Rosell, F. and Kile, N. 1998. Abnormal incisor growth in Eurasian beaver. *Acta Theriologica*, 43, 329–332.

Rosell, F. and Pedersen, K. V. 1999. *Bever [The Beaver]*. Oslo: Landbruksforlaget.

Rosell, F., Rosef, O., and Parker, H. 2001. Investigations of waterborne pathogens in Eurasian beaver (*Castor fiber*) from Telemark County, southeast Norway. *Acta Veterinaria Scandinavica*, 42, 479–482.

Rosell, F., Bozser, O., Collen, P., and Parker, H. 2005. Ecological impact of beavers *Castor fiber* and *Castor canadensis* and their ability to modify ecosystems. *Mammal Review*, 35, 248–276.

Safonov, V. G. 2016. Beaver trapping in Russia and Belarus and problems of resources management. *Russian Journal of Theriology*, 15, 2–7.

Salvesen, S. 1927. Om beveren i Norge. *Naturen*, 51, 193–210.

Salvesen, S. 1936. A fight between a beaver and a bull moose. *Journal of Mammalogy*, 17, 290.

Sarwar, M. and Rauf, A. 2018. Ectoparasitic insects genera of veterinary importance and some aspects of their control. *American Journal of Economics, Finance and Management*, 4, 116–123.

Saunders, J. K. 1963. Food habits of the lynx in New foundland. *Journal of Wildlife Management*, 27, 384–390.

Saveljev, A. P. 2003. *Beaver's Anal Glands: View from the East*. Warsaw: Russian Research Institute of Game Management and Fur Farming.

Schulze, C., Heuner, K., Myrtennäs, K., Karlsson, E., Jacob, D., Kutzer, P., Große, K., Forsman, M., and Grunow, R. 2016. High and novel genetic diversity of *Francisella tularensis* in Germany and indication of environmental persistence. *Epidemiology and Infection*, 144, 3025–3036.

Schuster, R. and Heidecke, D. 1992. *Helminthenfunde bei Castor fiber albicus Matschie, 1907. Semiaquatische Säugetiere, Wissenschaftliche Beitrage*. Halle: Martin-Luther-Universitat Halle-Wittenberg, pp. 207–213.

Scrafford, M. A. and Boyce, M. S. 2018. Temporal patterns of wolverine (*Gulo gulo luscus*) foraging in the boreal forest. *Journal of Mammalogy*, 99, 693–701.

Seddon, M. B. 2014. Bithynia tentaculata. *The IUCN Red List of Threatened Species*. http://dx.doi.org/10.2305/iucn.uk.20141.rlts.t15796a42436927.en [Accessed 12 March 2016].

Seglina, Z., Bakasejevs, E., Deksne, G., Spuņģis, V., and Kurjušina, M. 2015. New finding of *Trichinella britovi* in a European beaver (*Castor fiber*) in Latvia. *Parasitology Research*, 114, 3171–3173.

Sheffy, T. B. and Amant, J. R. S. 1982. Mercury burdens in furbearers in Wisconsin. *Journal of Wildlife Management*, 46, 1117–1120.

Shimalov, V. and Shimalov, V. 2000. Findings of *Fasciola hepatica* Linnaeus, 1758 in wild animals in Belorussian Polesie. *Parasitology Research*, 86, 527.

Shotyk, W., Bicalho, B., Dergousoff, M., Grant-Weaver, I., Hood, G., Lund, K., and Noernberg, T. 2019. A geochemical perspective on the natural abundance of trace elements in beaver (*Castor canadensis*) from a rural region of southern Ontario, Canada. *Science of the Total Environment*, 672, 40–50.

Sidorovich, A. A. and Sidorovich, V. E. 2011. The role of the Eurasian beaver *Castor fiber* in feeding of red fox *Vulpes vulpes* during warm season and effect of fox predatory to beaver population under Belarus conditions. In: Saveljev, A. (ed.) *Investigations of Beavers in Eurasia*. Kirov: Russian Res. Inst. of Game Management and Fur Farming, Issue 1, 95–104.

Sidorovich, V., Schnitzler, A., Schnitzler, C., Rotenko, I., and Holikava, Y. 2017. Responses of wolf feeding habits after adverse climatic events in central-western Belarus. *Mammalian Biology*, 83, 44–50.

Skrjabin, K. I. 1949. *Trematodes of Animals and Man*. Moscow: Academy of Sciences of the USSR.

Smith, D. D. and Frenkel, J. 1995. Prevalence of antibodies to *Toxoplasma gondii* in wild mammals of Missouri and east central Kansas: biologic and ecologic considerations of transmission. *Journal of Wildlife Diseases*, 31, 15–21.

Smith, D. W., Trauba, D. R., Anderson, R. K., and Peterson, R. O. 1994. Black bear predation on beavers on an island in Lake Superior. *American Midland Naturalist*, 132, 248–255.

Smith, H. J. and Archibald, R. M. 1967. On the incidence of gastrointestinal parasites in Nova Scotia beaver. *Canadian Journal of Zoology*, 45, 659–661.

Spalding, D. J. and Lesowski, J. 1971. Winter food of the cougar in south-central British Columbia. *Journal of Wildlife Management*, 35, 378–381.

Spickler, A. R. 2017. Tularemia. http://www.cfsph.iastate.edu/diseaseinfo/factsheets.php.

Sroka, J., Giżejewski, Z., Wójcik-Fatla, A., Stojecki, K., Bilska-Zając, E., Dutkiewicz, J., Cencek, T., Karamon, J., Zając, V., and Kusyk, P. 2015. Potential role of beavers (*Castor fiber*) in contamination of water in the Masurian Lake District (north-eastern Poland) with protozoan parasites *Cryptosporidium* spp. and *Giardia duodenalis*. *Bulletin of the Veterinary Institute in Pulawy*, 59, 219–228.

Stefen, C. 2019. Causes of death of beavers (*Castor fiber*) from eastern Germany and observations on parasites, skeletal diseases and tooth anomalies—a long-term analysis. *Mammal Research*, 64, 279–288.

Steineck, T. and Sieber, J. 2003. Ergebnisse pathologischer Untersuchungen bei Bibern (*Castor fiber* L.). *Denisia*, 9, 131–133.

Stenlund, M. H. 1953. Report of Minnesota beaver die-off, 1951–1952. *Journal of Wildlife Management*, 17, 376–377.

Stiles, C. W. 1910. *The Taxonomic Value of the Microscopic Structure of the Stigmal Plates in the Tick Genus Dermacentor*. Washington, DC: US Government Printing Office.

Stocker, G. 1985. The beaver (*Castor fiber* L.) in Switzerland. Biological and ecological problems of re-establishment. *Swiss Federal Institute of Forestry Research Reports*, 242, 1–49.

Stocker, L. 2000. *Practical Wildlife Care*. Oxford: Blackwell Publishing.

Stuart, B. P., Crowell, W. A., Adams, W. V., and Morrow, D. T. 1978. Spontaneous renal disease in beaver in Louisiana. *Journal of Wildlife Diseases*, 14, 250–253.

Surma, M., Giżejewski, Z., and Zieliński, H. 2015. Determination of perfluorinated sulfonate and perfluorinated acids in tissues of free-living European beaver (*Castor fiber* L.) by d-SPE/micro-UHPLC-MS/MS. *Ecotoxicology and Environmental Safety*, 120, 436–444.

Svendsen, G. E. and Jollick, J. D. 1978. Bacterial contents of the anal and castor glands of beaver (*Castor canadensis*). *Journal of Chemical Ecology*, 4, 563–569.

Swain, U. G., Gilbert, F., and Robinette, J. D. 1988. Heart rates in the captive, free-ranging beaver. *Comparative Biochemistry and Physiology Part A: Physiology*, 91, 431–435.

Swank, W. G. 1949. Beaver ecology and management in West Virginia. *West Virginia Conservation Committee, Game Management*, 1, 65.

Tadich, T. A., Novaro, A. J., Kunzle, P., Chacón, M., Barrientos, M., and Briceño, C. 2018. Agonistic behavior between introduced beaver (*Castor canadensis*) and endemic culpeo fox (*Pseudalopex culpaeus lycoides*) in Tierra del Fuego Island and implications. *Acta Ethologica*, 21, 29–34.

Toweill, D. E. and Anthony, R. G. 1988. Annual diet of bobcats in Oregon's cascade range. *North-West Science*, 62, 99–103.

Tsui, C. K.-M., Miller, R., Uyaguari-Diaz, M., Tang, P., Chauve, C., Hsiao, W., Isaac-Renton, J., and Prystajecky, N. 2018. Beaver fever: whole-genome characterization of waterborne outbreak and sporadic isolates to study the zoonotic transmission of giardiasis. *MSphere*, 3, e00090–18.

Tyurnin, B. N. 1984. Factors determining numbers of the river beaver (*Castor fiber*) in the European North. *Soviet Journal of Ecology*, 14, 337–344.

US Department of Interior Fish and Wildlife Service 1982. *Reports on Fisheries and Wildlife Research.* Denver: Fish and Wildlife Service.

Valenzuela, A. E., Rey, A. R., Fasola, L., Samaniego, R. A. S., and Schiavini, A. 2013. Trophic ecology of a top predator colonizing the southern extreme of South America: feeding habits of invasive American mink (*Neovison vison*) in Tierra del Fuego. *Mammalian Biology*, 78, 104–110.

Vincent, B., Rioux, H., and Harvey, M. 1991. Factors affecting the structure of epiphytic gastropod communities in the St. Lawrence River (Quebec, Canada). *Hydrobiologia*, 220, 57–71.

Walro, J. M. and Svendsen, G. E. 1982. Castor sacs and anal glands of the North American beaver (*Castor canadensis*): their histology, development, and relationship to scent communication. *Journal of Chemical Ecology*, 8, 809–819.

Weber, A. and Weber, J. 2013. Biber—Totfundanalytik im Naturpark Drömling, Sachsen-Anhalt. *Säugetierkundliche Informationen 9H*, 47, 131–138.

Wenger, S., Gull, J., Glaus, T., Blumer, S., Wimmershoff, J., Kranjc, A., Steinmetz, H., and Hatt, J.-M. 2010. Fallot's tetralogy in a European beaver (*Castor fiber*). *Journal of Zoo and Wildlife Medicine*, 41, 359–362.

Wiggs, R. B. 1990. Fractured maxillary incisors in a beaver. *Journal of Veterinary Dentistry*, 7, 21–22.

Wilkinson, P. M. 1962. A life history study of the beaver in east-central Alabama. MSc thesis, Auburn University.

Wilsson, L. 1971. Observations and experiments on the ethology of the European beaver (*Castor fiber* L.). *Viltrevy*, 8, 160–203.

Wilsson, L. 1995. *Bäver*. Sollefteå: Hundskolan, 189 s.

Wong, M. T., Wang, W., Lacourt, M., Couturier, M., Edwards, E. A., and Master, E. R. 2016. Substrate-driven convergence of the microbial community in lignocellulose-amended enrichments of gut microflora from the Canadian beaver (*Castor canadensis*) and North American moose (*Alces americanus*). *Frontiers in Microbiology*, 7, 1–13.

Wood, D. M. 1965. Studies on the beetles *Leptinillus validus* (Horn) and *Platypsyllus castoris* Ritsema (Coleoptera: Leptinidae) from beaver. *Proceedings of the Entomological Society of Ontario*, 1965, 33–63.

Wren, C. D. 1984. Distribution of metals in tissues of beaver, raccoon and otter from Ontario, Canada. *Science of the Total Environment*, 34, 177–184.

Yakimoff, W. L. 1934. Die Biberkokzidiose. *Berliner Tierarztliche Wochenschrift*, 50, 294.

Zalewski, K., Falandysz, J., Jadacka, M., Martysiak-żurowska, D., Nitkiewicz, B., and Giżejewski, Z. 2012. Concentrations of heavy metals and PCBs in the tissues of European beavers (*Castor fiber*) captured in northeastern Poland. *European Journal of Wildlife Research*, 58, 655–660.

Zharkov, I. V. and Sokolov, V. E. 1967. The European beaver (*Castor fiber* Linnaeus, 1758) in the Soviet Union. *Acta Theriologica*, 12, 27–46.

Zinke, O., Reusse, P., Kramp, P., Peters, T., and Ricklinkat, A. 2013. Todesursachen von Elbebiber (*Castor fiber albicus* Matschie, 1907) aus Ostsachsen und Südbrandenburg im Zietraum 1990 bis 2011. *Säugetierkundliche Informationen 9H*, 47, 153–158.

Żurowski, W. 1987. Difference in effects of the European beaver re-introduction into the lowland and mountainous tributaries of the Vistula River. *Proceedings of the 18th Beaver Congress*, Krakow.

CHAPTER 9

The ecological engineer

9.1 Shapers of landscapes

Beavers (*Castor* spp.) are one of the most significant modifiers of landscapes. As recognized ecological engineers, beavers and their activities have increased the complexity and biodiversity of freshwater ecosystems for millennia (Rosell et al., 2005; Stringer and Gaywood, 2016). Although beavers will readily impact and live along rivers and the shorelines of lakes (Figure 9.1), it is really when they modify smaller watercourses that exciting and dramatic changes can occur (Figure 9.2). As fittingly described by Eric Dolin in his fascinating history of the beaver fur trade, damming of a single stream can see it transformed into 'successive ponds dotting the terrain like a strand of liquid pearls' (Dolin, 2010), bursting with life. Numerous testimonies and texts, from Native American belief systems to modern restoration programmes, time and time again describe lands that have lost their beavers as dry, eroded, canalized, ecological, and biological deserts. The Blackfeet who refused to hunt beavers for the European fur trade wars, largely based this rejection on religious beliefs that water-beings should be respected. They also recognized the ecological role beavers played in keeping surface water on the plains where this tribe lived (Goldfarb, 2018). To keep beavers and the wetlands they create and maintain is to keep life and the continual provisioning of resources. These potential ecosystem benefits will only occur if beavers and their activities are allowed to function on a landscape scale, with management processes agreed that are flexible and pragmatic (Campbell-Palmer et al., 2016; see Chapter 11).

The two key traits for the beaver's success and survival have been attributed to their ability to utilize practically any plant species available (including the bark, perhaps the most unpalatable and fairly un-nutritious of materials, along with a number of species poisonous to other animals) and the ability to transform their environments to suit their needs (Hilfiker, 1991; see Chapter 5). Ancient relationships between beavers and other species are again revealing themselves as we often begin to reunderstand their ecology and the implications for other creatures and the habitats they reform (Dittbrenner et al., 2018). The reaction of many native tree species to even significant beaver harvesting is growth rather than death, suggesting a long evolutionary arms race. Until the end of the Quaternary, Eurasian river systems functioned naturally, with beavers being a significant factor in the generation of dead wood and channel complications (Francis et al., 2008; Table 9.1; Figure 9.3). Historical literature identifies (Wohl, 2014; Wohl, 2015), vast natural wood rafts in some of the larger North America rivers; one of which for example on the Red River, Louisiana, spanned nearly 500 km² of river channel and persisted for a century (Triska, 1984). These structures would have undoubtedly occurred in Eurasia too with any absence of reference to their existence simply reflecting their much earlier removal. Early explorers describe some North American rivers as being 'very much infested with logs' (Oconee River, 1818; Morrison, 2003). All this serves to highlight how wood was much more prolific and an integrated feature of our riverine systems but has been systematically removed for so long that we are experiencing a perception gap when it comes to its inclusion in river management and restoration (Wohl, 2015). Modern river

Beavers: Ecology, Behaviour, Conservation, and Management. Frank Rosell and Róisín Campbell-Palmer, Oxford University Press.
© Frank Rosell and Róisín Campbell-Palmer 2022. DOI: 10.1093/oso/9780198835042.003.0009

Figure 9.1 Beaver foraging and burrowing activities can increase the complexity of riparian systems and species diversity, without damming. (Photo supplied courtesy of Uli Messlinger.)

Figure 9.3 Beaver are significant producers of dead wood, both fallen and standing, through both foraging and damming behaviours (Photo supplied courtesy of Frank Rosell.)

Figure 9.2 Beaver activities can have landscape-wide impacts. Aerial surveys can provide data on beaver habitat use and extent of distribution, especially in remote areas. (Photo supplied courtesy of Tom Gable.)

management as well as the removal of beavers is a double blow to rivers' natural functioning, nutrient and sediment cycles, habitat, and species diversity.

An extensive body of scientific research indicates that beaver-generated landscapes contain higher levels of biodiversity and biomass on a landscape scale than environments from which they are absent (Rosell et al., 2005; Bashinskiy, 2020). Wetlands in themselves represent a habitat covering < 1% of the earth's surface but include nearly 12% of all its biodiversity (Collen et al., 2014). The activity of beavers creates habitats that are dynamic in nature and change over type according to activity and age,

going through a series of successional stages. These can result in the formation and maintenance of wetland habitats, with a positive effect on plant and animal diversity; however, there can be 'winners and losers', new habitat and opportunities for some, whilst others may be negatively impacted, more usually at a local scale.

Many northern hemisphere riparian landscapes have historically seen widespread and prolonged decline due to intensive drainage and grazing practices. While this process has in large part accelerated between the seventeenth to nineteenth centuries, profound changes to wetland drainage also occurred in earlier times in a Eurasian context (Purseglove, 2017). As technology and transport infrastructure grew, especially in the nineteenth century, the US Congress actively encouraged states to 'reclaim all swamps and overflow lands' to make way for roads, railways, and farming, thought to constitute > 65 million acres of wetland (Jakobi, 2017). For example, in Alberta, Canada, alone > 60% of wetlands have been lost since the Europeans first arrived, with an associated significant decrease in fish and waterfowl (Gibbs, 2000; Schindler and Donahue, 2006). By 1884 in Ohio an estimated 20,000 miles of ditches existed draining 11 million acres of wetlands (Wooten and Jones, 1955). Mastering and controlling powerful water systems such as the 'mighty Mississippi'

Table 9.1 Patterns and processes associated with wood and beavers in river corridors.

Influence of wood	Example references
Wood within the channel	Curran and Wohl (2003)
Increases hydraulic roughness	Buffington et al., (2002)
Deflects flow, creating local scour of bed and banks, including increased pool volume	
Creates low velocity zones with enhanced deposition of mineral sediment and particulate organic matter	Brooks et al. (2004)
	MacFarlane and Wohl (2003)
Alters the type and dimensions of bedforms	Montgomery et al. (1996)
Creates forced-alluvial reaches	Collins and Montgomery (2002); Wohl (2011)
Logjams that force channel migration, overbank flow, and multiple channel planforms	Nagayama et al. (2012)
Enhances habitat diversity with respect to substrate, flow depth, velocity, and overhead cover	Sawyer et al. (2012)
Enhances hyporheic exchange	Beckman and Wohl (2014)
	Benke and Wallace (2003)
Enhances nutrient retention and biological uptake	Nagayama et al. (2012)
Increases biodiversity and biomass of aquatic communities	
Beaver dams within the channel	Pollock et al. (2014)
Reduce the downstream velocity of peak flows	Green and Westbrook (1994)
Reduce bed and bank erosion	Butler and Malanson (1995); Pollock et al. (2007)
Enhance storage of sediment	Correll et al. (2000)
Enhance retention of organic matter and nutrients	Briggs et al. (2012)
Enhance hyporheic exchange	Levine and Meyer (2014)
Increase habitat diversity	Pollock et al (2003)
Increase biodiversity and biomass of aquatic communities	Westbrook et al. (2006)
Increase magnitude, duration, and frequency of overbank flows	
Wood on the floodplain	
Increases hydraulic roughness for overbank flows	Jeffries et al. (2003)
Creates erosionally resistant 'hard points' that influence the rate of floodplain turnover	Collins et al. (2012)
Increases habitat diversity and abundance for amphibians, reptiles, birds, and small mammals and, during overbank flows, for aquatic macro invertebrates and fish	Sedell et al. (1998)
Provides germination sites for plants	Harmon et al. (1986)
Beaver dams on the floodplain	Westbrook et al. (2006)
Reduce the velocity of overbank flows and increase floodplain sediment storage	Rosell et al. (2005)
Increase habitat abundance and diversity	Naiman et al. (1986)
Increase retention and biological uptake of nutrients	Wright et al. (2002)
Increase biodiversity and biomass of floodplain communities	

Source: After Wohl (2015), reprinted by permission of Wiley Periodicals, Inc.

and draining of swamps like the Everglades then became winning political strategies (Jakobi, 2017). It has been estimated that beavers historically were responsible for generating and maintaining >200,000 km² of wetlands in the upper Mississippi and Missouri river catchments before the arrival of the Europeans (Hey and Philippi, 1995). The loss of both beavers and these wetlands has resulted in significant changes in biochemical and hydrological regimes, including leading to the eutrophication of Chesapeake Bay (Brush, 2009). The impacts of their loss of beavers from riverine systems can dry out floodplains, result in desertification, and impact woodland and forest development (Wolf et al., 2007; Green and Westbrook, 2009; Westbrook et al., 2011). In landscapes throughout Eurasia this impact must also have occurred and very likely have been understood by the people of that time. Our modern patterns of habitat fragmentation as a result of competing land-use pressures and the drive for the intensification of meeting agricultural demand poses a major challenge to more natural habitats and their dependent wildlife.

9.1.1 A keystone species

Beavers are widely known as *'ecological engineers'* (Wright et al., 2002; Hartman and Törnlöv, 2006; Brazier et al., 2021), but what does this mean in reality? Only a few species are afforded such a grand term, which refers to those species that can create, significantly modify, or maintain habitats (Jones et al., 1994). As keystone species—those that have a disproportionate effect on the ecosystems within which they live (McKinstry et al., 2001)—beavers are renowned for their ability to create, modify, and maintain riverine habitats and influence floodplain connection processes on a landscape scale (Westbrook et al., 2011; Figure 9.4). Beaver activity types have been discussed in previous chapters (see Chapters 4 and 5), but generally these are grouped into foraging (including the felling of mature trees), burrowing, and damming. Through these activities

beavers have the ability to change both abiotic features, such as hydrology (water availability and storage) (Westbrook et al., 2006; Hood and Bayley, 2008), geomorphology, water chemistry and quality (Naiman et al., 1986; Maret et al., 1987), and temperatures, and biotic characteristics, including plant and animal assemblages and diversity (Pollock et al., 1995; Gurnell, 1998). Beaver habitat alterations can provide key ecosystem services, especially in degraded environments, such as agricultural landscapes, including trapping of sediment and pollutant runoffs; nutrient assimilation; regulation of water flow (Puttock et al., 2017); and enhanced aquatic biodiversity (Law et al., 2016).

Beaver-altered habitats create an array of opportunities for other species (Table 9.2) and can increase local biodiversity (Wright et al., 2002). The creation of surface water brings life, promoting riparian communities and water provision for other wildlife

Figure 9.4 Beavers are renowned for being a keystone species of wetlands of the northern hemisphere, bringing water and life. (www.ark.eu/bever. © Jeroen Helmer/ARK Nature.)

Table 9.2 Meta-analysis of published studies investigating beaver impacts on biodiversity.

Species group	Total	Positive	Neutral	Negative
Plants	10	7	3	0
Aquatic invertebrates	See Hering et al. (2001)			
Terrestrial invertebrates	5	5	0	0
Fish	See Kemp et al. (2012)			
Frogs and toads	10	8	1	1
Salamanders and newts	8	4	2	2
Reptiles	2	1	1	0
Birds	17	15	0	2
Mammals	11	6	4	1
Total	63	46	11	6
Percentage		73%	17%	10%

Source: After Stringer and Gaywood (2016), reprinted by permission of The Mammal Society and John Wiley & Sons Ltd.

and livestock (Demmer and Beschta, 2008), and can even act as breakers for forest wildfires (Baldwin, 2015; Stewart, 2020). The flooding of trees by beavers to increase their forage range creates standing and fallen dead wood to be perfectly exploited by a host of other species, for example woodpeckers, which in turn provides homes for other species such as wood ducks incapable of carving such features themselves (Hilfiker, 1991). While some species are able to exploit these to their benefit, others may be reduced, displaced, or even lost, especially less-mobile species of conservation value (Willby et al., 2018). Despite this, a recent meta-analysis determined that, overall, beavers have a positive effect on biodiversity (Stringer and Gaywood, 2016). As we live alongside beavers once more in our modern landscapes, our perceptions of traditional woodland and freshwater management will be challenged, away from uniform and tidy forests and canalized waterways. A range of practical management techniques have been developed to enable beaver presence and activity to be tolerated within accepted limits (see Chapter 11).

9.2 Wetland creation and floodplain reconnection

The goal of many reintroduction projects is the conservation value of the species being restored.

Another vital function of reintroductions, and one by which the beaver is famous for, is as a 'tool' for providing or improving habitat quality, improving ecosystem function, or increasing biodiversity. Beavers are often key characters in habitat restoration and 'rewilding' projects aimed at improving degraded landscapes (Foreman, 2004; Pereira and Navarro, 2015; Law et al., 2017). An estimated 85% of total stream length in the USA compromise headwater streams (Leopold et al., 1964). Left to their own devices beavers can influence up to 30–50% of the total length of fifth-order and smaller streams (Naiman and Melillo, 1984). Extensive evidence from American projects demonstrate that the use of dam analogues, to encourage beaver damming activity, a very practical and applicable technique, can create wetlands, restore incised streams, reconnect floodplains, and improve fish habitat (Pollock et al., 2014; Bouwes et al., 2016). Scottish studies have implied that wider beaver restoration can be used as a positive tool to combat wetland loss (Law et al., 2017), demonstrating that plant species richness increased by 46% and heterogeneity increased by 71% in measured plots after 12 years of beaver presence. Wetland area increase by beavers has been impressive, accounting for an 80% increase in open water in boreal forest as beavers returned to Alberta (Hood and Bayley, 2008), while in Wyoming, width of streams without beavers averaged 10.5 m, rising to 33.9 m in streams with beaver ponds (McKinstry et al., 2001).

The beaver's ability to influence and change landscapes is impressive, with damming often being the most significant altering activity (Butler and Malanson, 2005). Beaver dams influence ecosystem processes such as vegetation composition and succession (Johnston and Naiman, 1987; Terwilliger and Pastor, 1999; Hood and Bayley, 2008), sediment deposition and erosion (Butler and Malanson, 1995), nutrient cycling (Naiman et al., 1994), organic matter retention (Naiman and Melillo, 1984; Devito and Dillon, 1993), and water flow regimes and water quality (Puttock et al., 2017), often changing terrestrial habitat to wetland and lotic to lentic ecosystems as dams function as water and sediment traps (Bigler et al., 2001; Westbrook et al., 2011; Figure 9.5).

The ecological value of wetlands created by beaver activities is considered to be high but undervalued

Figure 9.5 Where beavers dam, water is stored and quality is improved (Photo supplied courtesy of Frank Rosell.)

in terms of water storage, water purification, and their ability to regenerate degraded aquatic ecosystems (Czech and Lisle, 2003). Generally, beavers prefer wooded stream networks, with gradients of < 6%, composed of sediment rather than stone or gravel (Gurnell, 1998). By constructing dams beavers can drastically alter the hydrology regime of a system, including increased structural heterogeneity (Rolauffs et al., 2001), creation of new and diversified habitats (Rosell et al., 2005), increased water storage capacity and reconnection of channels on floodplains (Collen and Gibson, 2001; Macfarlane et al., 2017), and reduced flow velocity (Burchsted and Daniels, 2014). This can make beavers an important factor in landscape-restoration projects directly through their ability to regenerate rivers, create new wetland habitats, and improve riparian forest health (Law et al., 2017) but also indirectly by acting as a charismatic flagship species for conservation projects (see Chapter 11).

9.2.1 Water storage

Beaver dams slow water runoff and reduce stream velocity; they also constantly leak and release water which frees up flood storage capacity. Dams have a dampening effect on low- and high-flow discharge peaks in watercourses (Gurnell, 1998; Puttock et al., 2017). The structural complexity of the channels created, with multiple dams, burrows, and fallen trees, serves to generate surface friction, which increases lag time and reduces peak flows. Water travelling speed was measured in a North American study; in a dam-free watercourse it was calculated

Figure 9.6 Wild boar utilizing a beaver pond for water and feeding. (Photo supplied courtesy of Derek Gow.)

to take 3–4 hours to travel ~2.6 km, whereas in the equivalent stretch with dams it took 11 days to travel the same distance (Müller-Schwarze, 2011), with this flow attenuation impact shown to persist even in large storms (Westbrook et al., 2020). Beaver ponds provide standing water, especially during periods of drought (Westbrook et al., 2006), affording foraging and shelter opportunities for a whole host of plant and animal species (Figure 9.6). During dry summers, beaver dams and canals have been shown to hold 60% more water (including ground water table) than comparable environments without beaver activity (Hood, 2012). Besides damming activity maintaining perennial surface water, ground water is increased (and stored) along with aquifer recharge (Lowry, 1993).

9.2.2 Erosion and sedimentation

Beavers have various geomorphic influences on the environments they inhabit. The phrase *busy as a beaver* was certainly not based on fiction. In any active beaver territory, you see numerous evidence of digging, such as canals, burrows, and feeding scrapes where they forage for roots, along with well-trodden forage trails and bankslides, all of which modify this riparian strip they make their home. The accumulative effect of all this sediment erosion and modification of the banking often alters the watercourse, which may then be trapped behind beaver-built dams, and this is an impressive cycle which nowadays scientists are starting to attribute figures too, especially as modern land-use practices drastically increase the generation of local surface runoff, sending water, pollutants, and hugely increased sediment loads into our watercourses (O'Connell et al., 2007). Such trapped sediment can remain in the general pond area, even after dams are abandoned and break down (Butler and Malanson, 2005).

Erosion generated by beaver activity is often quoted as a negative consequence of their presence. A section (> 800 m) of Bolin Creek in Orange county, North Carolina, was investigated to attempt to provide volumetric estimations for bank- and tunnel-slide erosion rates associated with typical beaver activities after 5 years of beaver occupation. The beavers were determined to be important agents of geomorphic change, with an estimated 21.84 m³ of sediment released (Meentemeyer et al., 1998). The trampling effect of repeatedly used forage trails is often an underestimated form of erosion, with many authors and landowners generally more concerned with burrowing impacts and geomorphic consequences (Meentemeyer et al., 1998) such as bank slumping (Butler and Malanson, 1994). However, beavers also trap sediment and diffuse pollutants behind dams (Cirmo and Driscoll, 1993). The trapping and storage of sediment can positively improve water quality downstream (Naiman et al., 1994). The loss of soils, especially in intensively managed landscapes, is a major environmental concern, as these rates vastly exceed typical soil formation rates of 0.1 ton/ha/year under intensive land-use practices (Verheijen et al., 2009), compared to typical soil erosion rates of 0.5–1.2 ton/ha/year in UK intensively managed grasslands (Bilotta et al., 2010; Gregory et al., 2015). Sediment accumulation rates behind dams vary greatly and are influenced by multiple factors but have been documented as 2–28 cm/year (Butler and Malanson, 1995) and 4–39 cm/year (Meentemeyer and Butler, 1999). One study in a semi-arid watershed in Oregon demonstrated aggradation rates of 0.47 m/year behind 13 new dams, though this rate had fallen to 0.075 m/year when the dams had reached 6 years (Pollock et al., 2007). All beaver ponds at 13 study sites across the coastal plain of rural North Carolina acted as sediment traps, reducing suspended sediment concentrations by at least 40%, with older ponds having a greater impact (Bason et al., 2017). The intensive monitoring of a study site in Devon saw a pair of released beavers construct 13 dams and by doing so hold back up to 100 tonnes of sediment, almost 16 tonnes of carbon, and 1 ton of nitrogen, all stored in only 1.8 ha of land (Puttock et al., 2018).

Beaver dams can also cause sediment deposition as plumes escaping from dams during overbank flooding events. Average sediment depths in such plumes have been found to average 6 cm SE ± 1.6 cm but ranged up to 120 cm, with the percentage of fine sediments increasing exponentially as distance from the river increases (Westbrook et al., 2011). Over 6.5 years a dam complex was found to reconnect the Colorado River to its riparian zone, with a deposition of approximately 750 m³ of sediment in a series of terrace formations (Westbrook et al., 2006).

9.2.3 Water quality and nutrient cycling

As a consequence of their activity, particularly the impact of beaver dams in changing river systems from free-flowing lotic to ponded lentic systems, beavers can have major effects on water quality and nutrient levels. Beaver activity can generate a shift towards sediment, detritus, and nutrient trapping and hence anaerobic biogeochemical cycling (Naiman et al., 1986). Water quality is generally improved by reducing nitrate concentrations and suspended sediment (Maret et al., 1987; Naiman et al., 1988; Cirmo and Driscoll, 1993; Klotz, 2010; Kroes and Bason, 2015). Nitrogen levels have been widely measured in beaver ponds, which have been determined to offer significant storage through nitrogen fixation associated with microbes in the trapped sediment and uptake by aquatic vegetation

(Songster-Alpin and Klotz, 1995). Beaver damming can increase the total nitrogen accumulation in the associated sediment of a section of stream by 9–44 times (Francis et al., 1985). Beaver ponds at study sites across the rural coastal plain of North Carolina were responsible for reducing nitrate concentrations by up to 19% (Bason et al., 2017), though drainage area and channel gradient were determined as the most influential site characteristics when it comes to nitrate concentrations in beaver ponds, with beaver ponds on smaller streams having more significant roles in modifying nitrogen concentrations (Bason et al., 2017).

Beaver wetland creation affects carbon biogeochemistry, with significant potential for carbon sequestration, which can be economically important by adding large quantities of organic matter to freshwater environments (Naiman et al., 1986; Wohl, 2013). However, beaver dams also create complex alterations to the carbon cycle (Figure 9.7).

Multiple studies have found that the concentration of dissolved organic carbon increases due to beaver activity (Puttock et al., 2017; Gatti et al., 2018). This increase is attributed to enhanced sediment and nutrient storage in addition to the overall increase in wetland extent creating an environment rich in organic matter, as previously shown by Vecherskiy et al. (2011), meaning stores but also losses of carbon may be greater. Beaver dams increase sediment and organic matter accumulation, creating anoxic conditions for methanogenesis; therefore some stored carbon in beaver ponds is released as methane (Ford and Naiman, 1988; Puttock et al., 2017; Gatti et al., 2018). Even though beavers can be a source if atmospheric methane, their ponds also sequester carbon (both in sediments and vegetation), thus acting as both a sink and source for carbon (Grosse et al., 2011; Johnston, 2014). For example, beaver meadows can account for up to 23% of total carbon stored in some landscapes (Wohl, 2013).

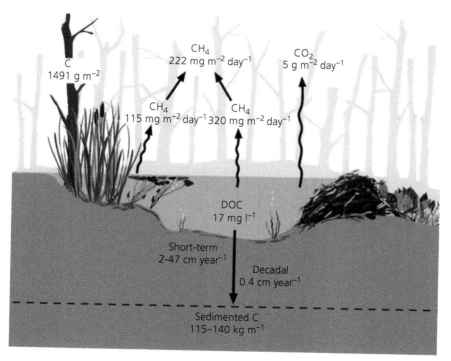

Figure 9.7 Carbon dynamics in a beaver wetland, with the main states shown. Carbon is emitted in gaseous form as methane and carbon dioxide, while stored in sediment and deadwood. During early impoundment the amount of dissolved organic carbon is high (From Nummi et al. (2018), reprinted by permission of The Mammal Society and John Wiley & Sons Ltd.)

The changes in dissolved organic matter (DOM) can be complex and variable according to catchment-specific features and age of beaver dams (Catalán et al., 2017). In a study of beaver damming in Swedish streams, new dams tended to leach humic-like DOM from the surrounding soils compared to older ponds, so that beaver ponds provide a boost of organic matter, as dams across a system will be transient in nature (Catalán et al., 2017). Beaver ponds have indeed been measured as having higher methane evasion and dissolved carbon dioxide rates (Ford and Naiman, 1988)—up to 23 times higher in western Siberian beaver ponds compared to flowing sections of the River Ob (Gatti et al., 2018). Methane emissions from permanently flooded areas with beavers ranged from 8 to 11 gcm^{-2}, whereas in areas infrequently inundated this ranged from 0.2 to 0.4 gcm^{-2} (Gatti et al., 2018). It has been suggested that up to 1% of the recent rise in atmospheric methane could be attributed to pond creation by beavers in North America (Naiman et al., 1991; Bubier et al., 1993). However, given the complex interaction of both press and pulse disturbance of carbon, beavers have been strongly proposed as agents in mitigating global climate change events, i.e. greenhouse gas emissions (Hood and Bayley, 2008).

The story with phosphorus seems less clear, with no general consensus on the role of beaver ponds as sinks or sources of this nutrient (Maret et al., 1987; Correll et al., 2000; Harthun, 2000). A recent review determined that in general phosphorus is not retained by new dams, as young beaver ponds act as a source, though as beaver ponds age, their phosphorus retention capacities increase (Ecke et al., 2017). This retention potential is associated with changes in sediment properties, organic matter input, and beaver activities such as digging (Naiman et al., 1986; Ecke et al., 2017). Naturally occurring mercury concentrations appear to be higher, on average increasing 1.3–1.7 times, associated with beaver damming, compared to concentrations downstream or in non-impounded streams, including a significant increase in predatory invertebrates (Painter et al., 2015; Ecke et al., 2017). This needs further research on nutrient cycling and any effects on basal food webs as beaver populations expand (Painter et al., 2015).

9.2.4 Water temperature

The impacts of beaver dams on water temperature appear to be conflicting. Studies have documented both rises in stream temperatures, downstream of beaver dams, and little or no changes (Błędzki et al., 2011; Majerova et al., 2015), suggesting instead beaver ponds have a regulating effect, preventing water from overheating during periods of high summer temperatures (McRae and Edwards, 1994; Ham et al., 2006).

To try to reconcile this, Weber et al. (2017) undertook an 8-year monitoring study of beaver dam distribution and water temperatures associated with them over ~40 km of stream length of Bridge Creek, Oregon. They determined two main ways in which beavers may alter stream temperatures: a buffering effect of diel stream temperature cycles during summer base flow periods downstream of dams; and reduction of shading effects of riparian vegetation through foraging, especially soon after initial damming when the surface area of water increases—through a combination of a water cooling effect when impoundments hold back cooler ground water and reduced shading enabling water to heat up during the day. As surface water volume increases it takes longer to reach extremes in temperature—hence there is a buffer effect. Monitoring longitudinal temperatures, they determined a reduction in maximum summer temperatures, arguing that this is an important factor in the provision of salmonid habitat (Weber et al., 2017), as juvenile steelhead trout (*Oncorhynchus mykiss*) experience stress, reduced condition, and reduced metabolic efficiency in water temperatures of 25°C and above (Feldhaus et al., 2010; Kammerer and Heppell, 2012), with 29°C becoming lethal (Rodnick et al., 2004). Elliott et al. (1995) found that optimal growth for UK brown trout (*Salmo trutta*) was reached at 13.11°C, but growth ceased when water reached 19.48°C, emphasizing the importance of lower water temperatures for salmonids. On Weber et al.'s section, Bridge Creek water temperatures rarely exceeded 25°C, whereas they commonly did so during August prior to beaver damming (2007–2010), thereby providing more constant habitat conditions and refuges. This is especially important in areas where thermal regimes may already be pushing fish populations to the edge of their thermal tolerance; thus this could be demonstrating an existing commensalistic relationship between these species.

Ponds in Alberta typically freeze over winter, though those occupied by beavers melt on average 10 days earlier than beaver-less ponds come the spring, with thawing first evident around lodges and food caches (Bromley and Hood, 2013). Noting beavers also physically break ice with their bodies. Open water in frozen landscapes provides valuable resources for other animals such as migrating birds. Therefore, it can be concluded that beaver dams impact stream temperature regimes at differing spatial and temporal scales, and often these are more apparent during low-flow conditions and depend on a range of interacting factors, including ground-water-surface water exchanges (Janzen and Westbrook, 2011; Majerova et al., 2015).

9.3 Positive and negative effects on plants and animals

Beaver activities can create spatial heterogeneity and consequently complex microhabitats, including deep open pools, shallow marshes, shrub swaps, and flooded woodlands (Bush and Wissinger, 2016),

the cumulative effects of which can generate incredible diversity in plants and animals (Wissinger and Gallagher, 1999; Wissinger et al., 2001). The effects are split into different taxonomic groups and discussed below.

9.3.1 Riparian and aquatic plants

The habitat modification abilities of beavers impacts on their feeding resources, and therefore plant communities and compositions, both directly through selective foraging and felling activities and indirectly via flooding, which has been determined to be the more significant factor (Hyvönen and Nummi, 2008; Figure 9.8a and b). Active beaver ponds when compared with other wetlands habitats typically have increased plant richness, largely generated by the higher disturbance factor of beaver activities such as water-level changes, foraging, and building behaviours (Willby et al., 2018). Damming, which raises and stabilizes the surrounding water table, increases soil moisture levels, and can drown out certain plants, in addition to beaver grazing act to increase light penetration (Naiman et al., 1994; Johnston et al., 1995; Wright et al., 2002). Changes in nutrient cycling and soil properties in impounded beaver ponds also

Figure 9.8a and b Beavers impact on plants through direct herbivory and habitat modification. Vegetation diversity increases as a result of water level changes and foraging activity (a). (Photo supplied courtesy of Uli Messlinger.) Plant species diversity also increases as woodland canopy is opened up through tree felling (b). (Photo supplied courtesy of Derek Gow.)

impact on vegetation (Naiman et al., 1994). All of this serves to increase the within-habitat heterogeneity of beaver ponds. An increase in soil waterlogging alone caused a reduction in the dominant tall understorey of common nettle (*Urtica dioica*) from 30 to 5%, thus enabling the growth of light-loving herbs and grasses (Law et al., 2017), whereas moss, for example, grows abundantly in beaver meadows in Alaska (Pollock et al., 1998).

Aquatic plant studies near Lake Superior in North America found that beaver-dammed ponds and lakes produced significant biomass compared to beaver-free riverine ponds or glacial lakes in the same area, though catchment variations in response to herbivory by beavers and moose (*Alces alces*) were apparent (Bergman and Bump, 2015). Along with changes in elk (*Cervus canadensis*) browsing, the restoration of willow communities at Yellowstone National Park has been dependent on the hydrological conditions generated by beaver damming activities (Marshall et al., 2013). Algae species richness and diversity are high in beaver-inhabited small lakes in Belarus (Makarevich et al., 2016). Algae biomass also increases in beaver ponds due to eutrophication, in part due to addition of beaver excretory products (Krylov et al., 2016).

Even in wetlands not created by beavers, their presence and herbivory alone generated positive increases in plant species richness. Beavers can feed on a long list of plant species, being generalists, and take advantage of what is available to them (see Chapter 5). For example, in their non-native range, introduced North American beavers in Argentina are reported as feeding on a range of bryophytes and algae, which is not typical (Briones et al., 2001). Food preference has a significant impact on plant communities, succession, and species composition and diversity (Rosell et al., 2005). Patches of beaver feeding are thought to generate up to 25% of the total species richness of herbaceous plants in riparian zones (Wright et al., 2002).

Beavers produce distinctive riparian habitats which are structurally complex, dynamic, mosaic habitats (Stringer and Gaywood, 2016). These can be typically characterised from aquatic plant communities, to flooded emergent vegetation, with a grass–forb–shrub composition, to open coppiced woodland (Ray et al., 2001; Martell et al., 2006). The diversity of herbaceous plants was found to be highest in beaver ponds between 11 and 40 years of age (Ray et al., 2001). Vegetation composition changes over time, for example as beaver ponds become meadows (Bonner et al., 2009). Beaver meadows can persist for decades as patches of distinct vegetation dominated by herbaceous or shrubby plants, which in turn increases the heterogeneity of a landscape (Neff, 1957; Remillard et al., 1987; Wright et al., 2002). Newly exposed sediments can be caused by several beaver activities including digging, areas of formally drowned vegetation, and sediment plumes from dams, providing new habitat for plant colonization. New sediment deposition has been found colonized by 81 vascular and 1 nonvascular plant species at a site on the Colorado River (Westbrook et al., 2011).

Along with factors such as wind, insects, and large herbivores, beavers are key natural agents of forest disturbance, key in generating variability and increased biodiversity (Kuuluvainen and Aakala, 2011; Nummi and Kuuluvainen, 2013). Beavers act as disturbance agents to forest in two main ways: as an ecosystem engineer (damming and flooding) and as a herbivore (especially of mature trees, which is rare) (Nummi and Kuuluvainen, 2013). Through herbivory they impact on tree stand structure and composition, opening up patches of the canopy and influencing plant colonization. Next to humans and most likely elephants (Elephantidae) (perhaps the only other animal today capable of felling mature trees), they typically select deciduous trees along the riparian shoreline, leading to changes in species composition (Barnes and Dibble, 1988; Johnston and Naiman, 1990; Nolet et al., 1994).

Many bryophytes and lichens are specialized in growing on deadwood, some of which are particularly rare and sensitive species (Stokland et al., 2012). Through flooding and foraging, beavers can significantly increase the amount of standing and coarse woody debris in a system (Thompson et al., 2016). However, if these changes happen too rapidly or impact on isolated tree stands, this may be detrimental to any species that are slow to colonize and/or have short colonization capabilities. Calicioids or pin lichens are good bio-indicators of forest health (Selva, 2003). Their species richness was found to be greater where beaver activity was present and around beaver ponds due to the amount and diverse forms of deadwood produced (Vehkaoja et al., 2017; Figure 9.3). Deadwood amount and heterogeneity is higher at beaver sites, varying in size, species, stage of decay, and location (e.g. on land or submerged in water), all of which can be lacking in modern managed forests (Thompson et al., 2016). This deadwood also acts as a form of carbon storage which then gets released slowly during decomposition (Naiman et al., 1988). The building up of sediment and organic matter provides nutrients and substrate for aquatic plants, which can also colonize due to changes in hydrology (Derwich and Mróz, 2008).

Beaver presence and activities encourage dynamic ecosystems including varying plant and tree structure and densities, the creation of messy forests including standing and fallen dead wood, and irregular waterways including pools, flooded forests, and swamp-type habitats. Selective tree felling for both food and building materials by beavers produces gaps in the tree canopy, opening up the woodland floor so that understorey plants get an opportunity to grow and compete amongst each other for light, space, and resources. Beaver herbivory alone can significantly influence the vegetation of wetlands in terms of both biomass and species diversity (Parker et al., 2007; Law et al., 2014), and plant responses to beaver activities will vary according to their life history traits.

Many native tree species across the northern hemisphere have a long evolutionary history with beavers and therefore have defences and reproductive strategies to readily protect against and deal with beaver activities and impacts (Basey et al., 1988; Johnston and Naiman, 1990). Willow (*Salix* spp.) for example responds to the selective foraging by beavers by increased productivity or can rapidly grow from cut stems and from wind-borne seeds in wet environments, thereby resulting in a feedback mechanism creating structural heterogeneity communities (Peinetti et al., 2009). For example, when cut by beavers red willow (*S. lasiandra*) stem elongation rates increase from 0.4 to 3.3 cm/diameter, with stem production proportional to the number of stems cut (Kindschy, 1989). When felled, aspen

(*Populus* spp.) will rapidly respond to beaver feeding by sending up multiple suckers to regenerate. Beaver browsing also stimulates sprouting in alders (*Alnus* spp.) and cottonwoods (*Populus* spp.) (Northcott, 1971). In the Scottish Beaver Trial, the regeneration of beaver-felled mature tree stumps of Atlantic hazel (*Corylus avellana*) rose from 26 to 52% following increased beaver foraging (Iason et al., 2014). Cottonwood trees (*Populus fremontii*) exposed to repeated beaver foraging change their structure completely into shrub-like forms 25% the height of unfelled trees with up to three times more basal stems (McGinley and Whitham, 1985), whereas other tree species such as southern beeches (*Nothofagus* spp.) native to South America, where beavers have been introduced, lack any evolved defensive mechanisms (Anderson et al., 2006). Conifer trees are highly susceptible to anoxia following flooding by beavers (Gill, 1970), but certain deciduous trees display varying levels of tolerance; therefore short-term flooding of an area may give a competitive advantage to deciduous species (Nummi, 1989; Hyvönen and Nummi, 2008). Willow possesses adventitious roots (those that can form into roots from any non-root tissue), resulting in some trees surviving at least a decade of flooding (Hyvönen and Nummi, 2011), whereas downy birch (*Betula pubescens*) and grey alder (*Alnus incana*) typically survive 2–3 years of flooding before dying off (Nummi, 1989).

Concern regarding the potential of negative impacts of beaver reintroduction on Eurasian aspen (*P. tremula*) conservation have been expressed in Britain, particularly in relation to important mature stands of high biodiversity value and to sensitive, dependent species such as the aspen hoverfly (*Hammerschmidtia ferruginea*) and a range of rare Diptera, lichens, and fungi. The felling of mature trees creating even small gaps in tree cover may prove to prevent the dispersal and colonization of such species, resulting in local extinction (Genney, 2015). Mature aspen stands can be rare in some areas and especially fragmented due to land management and overgrazing (for example in the UK), which makes them vulnerable, and some are concerned beavers could have significant impacts on specific stands (Puplett, 2008). For example, in Eagle Creek, Yellowstone National

Table 9.3 Summary of main beaver activities and the impacts these can generate for ecological features, plant and animal groups.

Area	Burrowing/building activities	Foraging activities	Damming activities	References
Hydrology and biochemistry	Add sediment and nutrients to watercourses. Provide debris accumulation/infilling.	Messy feeders—alter organic material. Increase leaf litter/detritus in watercourses. Fallen wood influences nutrient cycles. Can create anaerobic zones in ponds.	Ponds recharge groundwater. Dams trap sediments, contaminants, and pollutants, improving water quality. Dams provide more sustained flows year-round, increasing water storage and attenuating flow. Supplement low stream flows during dry periods. Provides ground water increase and storage and cooler ground water released downstream of dams. Provide open-water source for other wildlife and livestock.	Lowry (1993); Rosell et al. (2005); Pollock et al. (2007); Demmer and Beschta (2008); Gangloff (2013); Gibson and Olden (2014); Puttock et al. (2017); Westbrook et al. (2020)
Wetlands	Create, enhance, and maintain wetland habitats—increasing complexity and biodiversity. Provide habitat for rare species.			Gibbs (2000); Fustec et al. (2001); Syphard and Garcia (2001); Pollock et al. (2003); Cunningham et al. (2007)
Plants	Dredged and newly exposed sediment provides new ground for colonization.	Promote growth of riparian vegetation. Create mosaic microhabitat structure.	Promote growth of aquatic, semi-emergent, and riparian vegetation.	Ray et al. (2001); Pollock et al. (2003); Demmer and Beschta (2008); Gibson and Olden (2014); Celewicz-Gołdyn and Kuczynska-Kippen (2017); Clemente et al. (2019)
Invertebrates	Additional woody debris creates shelter and feeding opportunities and increases microhabitats for higher biodiversity.	Create habitats. Create dead wood for feeding.	Ponds provide more stable habitats. Create drought refuges.	Wissinger and Gallagher (1999); Gossner et al. (2016); Law et al. (2016)
Fish	Create new habitat and connectivity between water bodies.	Addition of plant material into watercourse provides biodiversity and microhabitats, including shelter and foraging opportunities.	Improve habitat quality. Increase invertebrate and other prey abundance. Dams not a barrier to native fish passage. Ponds provide overwintering habitat. Downstream of dams provide sediment and free spawning habitat.	Barabash-Nikiforov (1950); Ray et al. (2001); Kemp et al. (2012); Bouwes et al. (2016); Malison and Halley (2020)

Taxon				References
Amphibians	Burrowing creates spawning habitats. Lodges, dams, and burrows provide habitat for larval food production, shelter, predator avoidance, and hibernation. Beaver-dug channels aid dispersal and connectivity.	Submerged vegetation and increased woody debris provide shelter and predator avoidance. Tree felling creates sunlight corridors for dispersal, reduces habitat fragmentation, and increases habitat for open-land species.	Create new ponds in previously dry areas. Increase size, number, and diversity of lentic zones. Increase pond connectivity.	Cunningham et al. (2007); Dalbeck et al. (2007); Stevens et al. (2007); Karraker and Gibbs (2009); Browne and Paszkowski (2010); Alvarez et al. (2013); Anderson et al. (2014); Dalbeck et al. (2020)
Reptiles	Lodges, dams, and burrows provide habitat for feeding, basking, shelter, predator avoidance, and hibernation.	Felled trees provide habitat for feeding, basking, shelter, predator avoidance, and hibernation. Dead wood provides habitat for prey.	Habitat changes provide foraging opportunities.	Russell et al. (1999); Metts et al. (2001); Reddoch and Reddoch (2005); Graham (2013)
Birds	Disturbance can expose foraging opportunities	Increase plant diversity and promote riparian communities. Increase density of breeding opportunities. Felled trees provide habitat for feeding, shelter, predator avoidance, and hibernation. Dead wood provides habitat for prey.	Habitat for prey. Ponds provide more stable habitats. Create drought refuges.	Aznar and Desrochers, (2008); Cooke and Zack (2008); Nummi and Hahtola (2008); Johnston (2011); Nummi and Holopainen (2014)
Mammals	Lodges, dams, and burrows provide habitat for feeding, basking, shelter, predator avoidance, and hibernation.	Increase plant diversity and promote riparian communities. Increase plant biomass. Felled trees provide habitat for feeding, shelter, predator avoidance, and hibernation. Dead wood provides habitat for prey.	Ponds provide more stable habitats. Create drought refuges.	Medin and Clary (1991); Frey and Malaney (2009)

Park, mature aspen decreased by 62% due to beaver activities, though aspen sapling cover more than tripled; however, ungulate browsing prevented aspen suckers from reaching recruitment height (McColley et al., 2012). Experiences from Europe and North America generally conclude that, on balance, beavers have a beneficial effect on aspen, creating fallen and dead wood breeding resources for Diptera species and facilitating aspen clone regeneration (Batty, 2002; Runyon et al., 2014). Sensitive mature specimens and stands near watercourses should be protected, but more importantly conservation efforts should focus on increasing the quality (range of sizes and ages) and quantity of aspen. Beavers can indeed be used to facilitate aspen clone regeneration as long as grazing pressure from ungulates and livestock is low or managed (Runyon et al., 2014).

Left to their own devices beavers can fell a lot of trees, exceeding their dietary and building requirements. Various estimates to quantify this exist in the literature: 215–225 trees annually per beaver (Bradt, 1938) or 1 metric ton cut annually per beaver (Naiman et al., 1988). One beaver family (six or seven individuals) in British Colombia, Canada, was found to consume up to 1 ha of deciduous woodland per year (McKenna et al., 2000). Selective feeding can also influence forest composition, similar to the impacts of moose herbivory, and riparian shorelines with small numbers of deciduous trees can see a shift to greater conifer dominance (Nummi and Kuuluvainen, 2013), whereas beaver-generated flooding may see a shift in tree stand composition favouring deciduous species (Hyvönen and Nummi, 2011). An often-ignored component of beaver activities in riparian plant recruitment is their messy feeding, with beaver woody cuttings often transported downstream (Cottrell, 1995), which can accumulate on stream point bars, thereby facilitating plant colonization (Levine and Meyer, 2019). Such regeneration can in turn both add roughness to the stream channel and trap sediment; radio carbon dating demonstrates that this natural process has been occurring over thousands of years in floodplains of Montana (Levine and Meyer, 2019) and has been lost in other places where beavers have disappeared.

Beaver also impact trees through they damming activity, thereby drowning tree roots, with most flooded-out trees dying within a couple of years (Nummi, 1989). A greater abundance of both standing and submerged dead wood habitat further enhances this process, and invertebrate densities can rise in beaver-generated wetlands by up to 80% (Gurnell et al., 2008).

Beavers have been linked to the spread of non-native riparian plants. Studies in Oregon have found that the invasive reed canary grass (*Phalaris arundinacea*) can reduce wetland plant richness where it establishes, and in areas in which beavers repeatedly dam and abandon ponds this may provide a window for this plant to establish (Perkins and Wilson, 2005). By reducing cottonwood canopy cover, beavers have been implicated in encouraging invasive species such as Russian olive (*Elaeagnus angustifolia*) and tamarisk or salt cedar (*Tamarix* spp.) growth rates in eastern Montana (Lesica and Miles, 2004). Beaver tree felling, with the associated opening up of the woodland floor, has result in greater understorey plant species richness and productivity in South American beech forests. This is not to say of course that this is always a positive change, as here ultimately primary forest has been lost and introduced plant species, such as Kentucky bluegrass (*Poa pratensis*) and common dandelion (*Taraxacum officinale*), have benefited in the succession process (Anderson et al., 2006; Martínez Pastur et al., 2006).

9.3.2 Invertebrates

Beaver-generated wetlands are high in invertebrate biodiversity, largely due to the high habitat heterogeneity they create (Bush and Wissinger, 2016), providing both a greater expanse and increased variety of feeding and breeding opportunities for invertebrates (Figure 9.9). This is achieved via a range of interacting and dynamic processes (Naiman et al., 1988), including the variation of open water; edge habitat vegetation diversity and wetted edges; trapped sediment and beaver digging for canals; woody debris and dead wood from beaver feeding, flooding, and constructions such as lodges and dams; and aquatic vegetation (Benke

Figure 9.9 Dragonfly sunning itself on a beaver-cut stick. (Photo supplied courtesy of Gerhard Schwab.)

et al., 1999; Rolauffs et al., 2001; Hood and Larson, 2014). The diverse vegetated edges of beaver habitats have been determined as important drivers of invertebrate abundance and species diversity, becoming increasingly significant as new beaver-dug channels mature (Hood and Larson, 2014). Invertebrate biomass is higher (1.3–11.1 g/m²) in beaver ponds than in stream sections without dams (0.01–0.6 g/m²) in steeper-gradient and middle-order streams (McDowell and Naiman, 1986; Naiman et al., 1986). In a study of beaver-modified and -unmodified headwater streams draining pastureland in Scotland, Law et al. (2016) found that species richness, abundance, and diversity of invertebrates were greater in beaver-modified streams. This seems to be constant, with Sjöberg et al. (2000) finding a greater range of invertebrates in beaver habitats than those without beavers in southern Finland, whilst German studies demonstrate that beaver territories have significantly greater numbers of dragonflies and damselflies (order Odonta) (Harthun, 1999; Figure 9.9).

As flooded vegetation dies and organic matter including beaver excretion is broken down by a whole range of bacteria and protozoa, these in turn provide food for hosts of freshwater zooplankton, such as daphnia (Shevchenko et al., 2017). Leaf litter input can remain high up to the first 2 years in a flooded area (~100 g/m²/year) (Nummi, 1989); therefore inputs of carbon are higher in new rather

than older ponds (Naiman et al., 1986). This decomposing matter forms the basis of nutrient release for subsequent food chains. The slowing of water and its associated warming in beaver ponds increases plankton production, with the highest species diversity typically found in the centre (Janiszewski et al., 2014). In a study of Polish streams impounded by beavers, the greatest increase of taxa number and abundance occurred around and upstream of the dams, particularly by rotifers and cladocerans, mainly due to changes in hydrology but also due to increases in chlorophyll and macrophytes (Czerniawski et al., 2017). Beaver dams can increase the habitat available for sessile, benthic invertebrates upstream of dams by slowing the water and increasing the accumulation of finer substratum (Johnston and Naiman, 1987). Benthic macroinvertebrates are often used as a measure of stream ecosystem health (Rosenberg and Resh, 1993).

Russian studies have determined that beaver presence, even without constructions, still resulted in an increase in zooplankton population abundance, biomass, and species diversity through increased eutrophication stimulating growth (Krylov, 2008). More food is available here for filter feeders such as river mussels (species of Palaeoheterodonta), though at some sites colonization and dispersal of mussels may be impacted by beaver dams, depending on distance from source populations, number of fish hosts, and ease of fish passage (Bylak et al., 2019). Studies on western pearlshell (*Magaritifera falcate*) concluded that the colonization of a stream in western Washington state, by beavers, and their resultant damming led to significant mortality through altered water regime, sediment accumulation, and potential fish impediment (Hastie and Toy, 2008). However, it should also be noted that significant declines had also occurred on the beaver-free control stream studied, indicating other contributing factors, as demonstrated by the closely related freshwater pearl mussel (*M. margaritifera*), now a protected endangered species across Europe (Gibson, 1998). The thick-shelled river mussel (*Unio crassus*), for example, relies on fish hosts for dispersal at the glochidia life stage to a more significant degree than physical adult mussel movement (Barnhart et al., 2008). However, in Poland, the common minnow (*Phoxinus phoxinus*) is one of the most suitable hosts, which regularly migrates to utilize beaver ponds and is in fact probably most responsible for the dispersal of mussel glochidia to smaller tributaries where beavers are more likely to dam, thus in turn creating suitable mussel habitat in what were once unsuitable faster-flowing streams (Bylak et al., 2019).

Beaver impacts on aquatic, semi-aquatic, and terrestrial invertebrates are well documented (Table 9.4). Increased organic matter accumulating behind dams and in this slower water benefits numerous aquatic invertebrates, such as true flies (Diptera), mayflies (Ephemeroptera), and caddisflies (Trichoptera), all of which are crucial food supplies to birds, fish, amphibians, and reptiles (Rosell et al., 2005; Janiszewski et al., 2014). Regularly found invertebrates in upland beaver wetlands in Colorado include the caddisfly larva (*Limnephilus externus*), water boatman (*Hesperocorixa*), dragonfly and damselfly, meniscus midge larva (*Dixella* sp.), and dytiscid diving beetle (*Agabus tristus*) (Bush and Wissinger, 2016). Fruit fly (*Drosophila virilis*) numbers often dramatically increase in beaver-flooded areas, laying their eggs in decaying bark (Spieth, 1979). In one stand of beaver-flooded Norway spruce (*Picea abies*), up to 20 species of beetles alone have been recorded, including the bark beetle (*Hylurgops palliates*) and

striped ambrosia beetle (*Trypodendron lineatum*), as a result of the significant amount of dead wood generated (Saarenmaa, 1978). Wolf spider (Lycosidae), a dominant terrestrial invertebrate, abundance is increased by up to 60% around beaver ponds compared to non-beaver ponds in Montana (McCaffery and Eby, 2016). This was directly related to increased aquatic invertebrate prey emergence rates, which were > 200 higher around beaver ponds, with amount of aquatic-derived carbon decreasing linearly with distance from them. Some quite subtle relationships have been determined. For example, some popular species such as Fremont and narrowleaf cottonwood (*Populus fremontii* and *P. angustifolia*) react to beaver felling by producing regenerating sprouts higher in nitrogen phenol glycoside as an anti-herbivore deterrent, but this substance is attractive to the leaf beetle (*Chrysomela confluence*) (Martinsen et al., 1998). Predatory species were found to dominate beaver wetland habitat edges and channels, Gerridae, Gyrinidae, and Chaoboridae larvae being the most abundant species found (Hood and Larson, 2014).

Dead wood is an important biodiversity indicator (Stokland et al., 2004), with over 7,500 saproxylic species (typically invertebrates and fungi), compromising 20–25% of forest species, recorded in Nordic countries alone (Siitonen, 2001; Stokland et al., 2012). Saproxylic insects such as the redbay ambrosia beetle (*Xyleborus cryptophagus*) specialize on newly dead aspen trees with coarse bark, while the four-banded longhorn beetle (*Leptura quadrifasciata*) instead utilizes partially decomposed deciduous trees (Ehnström, 2002). Other terrestrial insects have been recorded as utilizing beaver-cut woody plants to forage on leaking sap, especially on aspen and willow stumps. These include various species of Calliphoridae flies, solitary bees (family Andrenidae), and mourning cloak butterflies (*Nymphalis antiopa* L.) (Rea and Lindgren, 2009). Abandoned beaver ponds typically contain higher amounts of organic matter, providing habitat and nutrients for a wide range of Tipulidae larvae (up to 26 species identified), along with surface-dwelling Ephemeroptera, Plecoptera, Megaloptera, and Trichoptera (Hodkinson, 1975a; Hodkinson, 1975b).

Table 9.4 Most abundant large-bodied taxa in new, mature, and abandoned beaver wetlands in Georgia (excluding Ostracoda, Branchiopoda, Copepoda, Chironomidae, and Ceratopogonidae).

Newly created	Mature	Abandoned
Oligochaeta	Caenidae	Entomobryidae
Dogielinotidae	Dogielinotidae	Coccoidea
Coenagrionidae	Coenagrionidae	Oligochaeta
Non-oribatid Acarina	Non-oribatid Acarina	Non-oribatid Acarina
Libellulidae	Libellulidae	Araneae
Spaeriidae	Spaeriidae	Coenagrionidae
Caenidae	Baetidae	Spaeriidae
Baetidae	Veliidae	Oribatidae
Dytiscidae	Oribatidae	Corethrellidae
Araneae	Scirtidae	Delphacidae

Source. Reproduced with permission from Bush and Wissinger (2015).

9.3.3 Fish

Beaver and fish interaction is perhaps one of the more controversial points of contention, with strong views held on the subject by both the fishing and pro-beaver communities. This may be reinforced by the fact that it is still policy in some US states to remove beavers and their dams as part of salmonid population management (USDA, 2011).

Certain freshwater species are of conservation, sport, and economic importance; for example, the Atlantic salmon (*Salmo salar*) and brown/sea trout (*S. trutta*) have declined across most of their range over the last few decades, and the European eel (*Anguilla anguilla*) and river lamprey (*Lampetra fluviatilis*) are listed on Annex II of the EC Habitats Directive making them subject to protection against exploitation (Youngston and Hay, 1996; Windsor et al., 2012). Both beaver species are sympatric with salmonid species across their entire ranges. The presence of beavers tends to have a positive and beneficial impact on numerous fish species populations through creation of foraging and refuge opportunities for a wide variety of species (Collen and Gibson, 2001; Pollock et al., 2004; Rosell et al., 2005). However, concerns with migratory fish, especially salmonid species, have been raised as it is believed beavers and salmon have a complex relationship, expected to vary according to catchment type and beaver densities, and over time (Beaver Salmonid Working Group, 2015). Beavers and salmonid species coexist in large populations in North America and a number of European countries, including Denmark, France, Norway, Sweden, and Finland, with limited intervention management, though in parts of Lithuania, for example, beaver dam removal is implemented to maintain brown trout fisheries (Kesminas et al., 2013).

In parts of Canada, the Miramichi Salmon Association in New Brunswick, with government funding, implemented a regular dam notching or removal programme, with some lethal beaver control to facilitate salmon spawning throughout the catchment (Mitchell and Cunjak, 2007; Reid, 2012; Ginson, 2013). For example, 88 dams were removed or notched in 2011 and 15 beavers culled out, with 112 dams managed in 2013 over 22 tributaries (Mitchell and Cunjak, 2007; Reid, 2012). This obviously involved significant resource and time investments; however, surveys also revealed salmon redds upstream of several beaver dams on the Catamaran Brook, where human intervention had occurred, leading to reviews of such management techniques (Parker, 2013).

Kemp et al. (2012) undertook one of the most thorough reviews of beaver and fish interactions in low-order streams. These positive interactions for fish include higher survival rates (Quinn and Peterson, 1996), faster growth rates (Sigourney et al., 2006; Malison et al., 2015), increased fish densities (Bouwes et al., 2016), and improved habitat heterogeneity with increased food resources (McDowell and Naiman, 1986; Keast and Fox, 1990; Rolauffs et al., 2001). Beaver ponds, formed behind dams, can increase density, abundance, and species richness of trout and coarse fish (Schlosser, 1995; Snodgrass and Meffe, 1998). These ponds can also support larger trout compared to unimpounded sections (Hägglund and Sjöberg, 1999; Malison et al., 2015) as these deeper pools provide more suitable habitat for larger fish (Armstrong et al., 2003).

Beaver dams can lead to significant changes in local stream morphology, hydrology, and chemistry, directly impacting on fish communities and generally shifting from lotic to lentic species (Collen and Gibson, 2001). In Fish Brook, River Ipswich, Massachusetts, for example, a series of dam complexes increased habitat heterogeneity—upstream of dams, deeper, wider, and slower moving waters are created, whilst downstream of dams, faster flows occur over coarser sediment, cumulatively resulting in increased fish species richness (Smith et al., 2003). Such changes may facilitate improved growth due to provision of more refuge from adverse flows (Sigourney et al., 2006). Atlantic salmon parr, for example, remaining in beaver ponds grew faster (length and weight) during the summer than those living either above or below beaver impoundments (Sigourney et al., 2006). Tree-felling and dam, food cache, and lodge building activities result in a greater abundance of submerged woody debris, providing refugia for multiple species (Collen and Gibson, 2001).

Beaver ponds can provide overwintering habitat for numerous fish species; indeed, densities of bull trout (*Salvelinus confluentus*) and cutthroat trout

(*Oncorhynchus clarkii*) were lower in all habitats except beaver ponds during low temperatures (Cunjak, 1996), where both species aggregated overwinter (Jakober et al., 1998). Fish diversity was found to be highest in beaver ponds that were between 9 and 17 years old (Snodgrass and Meffe, 1998). Additionally, beaver digging activities have been demonstrated to create new habitats for fish including the weather fish (*Misgurnus fossilis*), burbot (*Lota lota*), and wels catfish (*Silurus glanis*), which were found in beaver burrows at the Voronezh Nature Reserve, Russia (Barabash-Nikiforov, 1950).

Negative reported impacts include lowering dissolved oxygen concentrations in older ponds and associated reduced fish species richness (Snodgrass and Meffe, 1998); siltation of fish spawning grounds; and seasonal restriction of fish passage (Schlosser, 1995; Kemp et al., 2012). The impediment, or more precisely the perceived impediment, of beaver dams on the passage of migratory fish is a real point of contention in beaver reintroduction to Britain for example, grabbing many headlines, circulating numerous rumours, and resulting in the destruction of many dams. Both beaver species are sympatric with salmonid species, but solid evidence of dams being an active barrier to fish migration is lacking in both the USA and Europe, with conflicting studies existing (Thorstad et al., 2008; Figure 9.10). Many of our modern watercourses have seen centuries of control, straightening, and clearance, for the objectives of land drainage and facilitating fish runs, with salmonid species being of most concern, given their high monetary value.

Throughout western North America, habitat loss has been implicated in the decline of cutthroat trout (*Oncorhynchus clarkia*) (Budy et al., 2007), resulting in much restoration effort (Behnke, 2002). Certainly, there is no doubt some dams, at some points in time, can impede or partially restrict salmonid passage, especially at low water flow rates (Collen and Gibson, 2001). In North America some instances of dams temporarily limiting upstream migration of adult anadromous salmonids and juvenile downstream movements during low water flow have been documented (Gibson et al., 1996; Murphy et al., 1989; Alexander, 1998; Cunjak and Therrien, 1998). This was also determined by the monitored

passage of Bonneville cutthroat trout (*O. clarkia Utah*) and brook trout (*S. fontinalis*) through 21 beaver dams in two Utah streams (Lokteff et al., 2013). Upstream migration of Arctic grayling (*Thymallus arcticus*) displayed an average unbreached beaver dam passage probability of 88%, though this fell to < 50% at specific dams (Cutting et al., 2018). The researchers suggested upstream passage was affected by three main factors—temperature, breach status, and hydrologic leakage around the dam—so by connecting above- and below-stream sections, they concluded that upstream fish passage is strongly correlated with hydrological conditions and moderately controlled by beaver dams. They demonstrated that even one hydrological link (breach) in a dam increased average fish passage probability at least twofold compared to a completely unbreached dam. However, another study monitoring Atlantic salmon passage at a large dam in New Brunswick reported their passage was impossible except during flood events and thereby over half the stream could not be occupied (Mitchell and Cunjak, 2007). This information could certainly inform management plans, especially in regard to timing dam breaching, so that the benefits of damming to fish, including increased habitat diversity, fish shelter and feeding opportunities, and restoration of stream connectivity, could be retained

Figure 9.10 Beaver dams are not permanent barriers to fish migration, some dams especially during periods of low water flow may restrict immediate movement, though additional time in beaver ponds is also associated with greater fish growth, numbers and species diversity. Here a sea trout navigates a beaver dam in Devon, England, being especially attracted to any breaches with running water. (Photo supplied courtesy of Roger Auster.)

(Gurnell, 1998; Levine and Meyer, 2014; Weber et al., 2017).

It should be noted that beaver dams will typically force water out to each side of a dam (and even under), thereby offering continual water leakage, though this will vary with surrounding topography. The formation of scour pools below a dam can also impact on fish passage (Cutting et al., 2018). Shallow scour pools (< 10 cm) limited grayling and brook trout passage for individuals generally over 10 cm in length (Kondratieff and Myrick, 2006; Cutting et al., 2018), whereas if scour ponds were deepened to > 40 cm, all fish sizes passed, including vertical leaps from 63.5 to 73.5 cm heights (Kondratieff and Myrick, 2006; Cutting et al., 2018).

What is evident then is that the hydrological changes within a system will have significant influences on fish migration and spawning, and beaver dams will not act as barriers provided enough water passes over or around a dam, they can swim either up or down, or there is a pool at the foot of the dam from which they can leap (Mitchell and Cunjak, 2007; Thorstad et al., 2008). Some beaver dams can become impassable to fish species when rainfall (or water input) is low; this could result in some dams presenting significant obstacles to migratory fish species (Taylor et al., 2010).

Other studies find almost the reverse, with salmon reproduction viewed as unhindered by beaver presence in studies on productive salmon rivers in Norway (Parker and Rønning, 2007; Malison and Halley, 2020). In order to improve salmon habitat in low-order streams, beavers have been actively released in parts of western USA (Pollock et al., 2014; Petro et al., 2015; Bouwes et al., 2016). Native fish diversity and abundance has been found to be higher in and around dams due to increased habitat heterogeneity, including downstream through generating more fluvial habitat by revealing hard substrata and flowing water (Smith et al., 2003). In other projects, humans have attempted to encourage beavers to dam in preferred locations through the installation of beaver dam analogues, to provide foundation structures against which beavers can maintain dams (Bouwes et al., 2016). Although Paiute cutthroat trout spawning beds have reportedly been covered in sediment trapped by beaver impoundments in Inyo National Forest, California

(Hunter, 1976), at a catchment scale beavers and fish seem to have positive cycles. An interesting comparison study was undertaken on salmon rivers on two expansive floodplains: the Kol River, Kamchatka, Russia, on which beavers have never existed, and the Kwethluk River, Alaska, with abundant beavers following the collapse of the fur trade in this region (Malison et al., 2016). This study suggested that habitat loss for juvenile salmon did occur in beaver systems through converting spring brooks into beaver ponds, with densities significantly lower in beaver ponds in the orthofluvial zone with numerous dams, but salmon were abundant throughout the beaver-free floodplain (Malison et al., 2016). A trade-off was apparent though, with salmon achieving higher growth rates in beaver ponds located in the parafluvial zone in which regular inundation facilitates salmon movement (Malison et al., 2015). In conclusion, the effects of beavers on stream fish can be variable and site specific, positive especially in incised and degraded streams (see Section 11.4.4) but potentially reducing productivity of juvenile habitat in others (Kemp et al., 2012; Pollock et al., 2012; Malison et al., 2016).

9.3.4 Amphibians

Amphibians are often one of the first vertebrates to arrive in new beaver ponds and utilize these new and dynamic habitats. Where there was once dry forest floor, for example, can rapidly see amphibian spawning grounds appear out of seemingly nowhere (Figure 9.11). In fact, beaver-created habitats and ponds, especially through altering small, low-order streams, often in upland areas provide essential amphibian habitat, which is often lacking (Dalbeck et al., 2007). In comparison, beaver ponds in larger rivers typically support lower amphibian species numbers (Günther, 1996; Weddeling and Willigalla, 2011). As ponds mature, the amphibian species assemblage often changes. Questions have been raised whether positive correlation between increased beaver activity and fish numbers should not have a deleterious impact on amphibians. However, the demonstrated increases in habitat heterogeneity significantly aid in predator avoidance, along with increased larval food provision

Figure 9.11 Beaver-generated wetlands can have significant positive impact on amphibians, providing breeding and foraging grounds, along with numerous shelter and habitat connectivity opportunities. (Photo supplied courtesy of Mark Elliott.)

and hibernation opportunities (Dalbeck et al., 2007; Karraker and Gibbs, 2009; Alvarez et al., 2013).

All 19 European amphibian species are found in association with beaver ponds, despite the differing habitat requirements between all these species (Dalbeck et al., 2020). Beaver ponds in Europe have greater amphibian species richness than comparable artificial ponds, even in the presence of fish predators (Dalbeck and Weinberg, 2009). Russian studies have demonstrated that amphibian reproduction rates increase in sites with beaver activity; even partially drained ponds still provide spawning habitats, seeing maximal tadpole numbers from common frogs (*Rana temporaria*), moor frogs (*R. arvalis*), and common toads (*Bufo bufo*), though these may also be exposed to greater mortality (Bashinskiy, 2008). Overall, it has been estimated that beavers are crucial determinants of amphibian species richness in headwater streams (which

constitute 60–80% of temperate European water bodies) (Dalbeck et al., 2020).

In Europe beaver ponds are hugely important for common frog abundance; for example, two beaver families between them produced 20 ponds in a valley length of 2,325 m containing 2,500 frogspawn clumps (Dalbeck et al., 2020). At a study site in central Europe in the Rhenish Massif, beaver ponds made up around half of all lentic water bodies in the area but contained 82.5% of all common frogspawn clumps by increasing habitat availability, heterogeneity, and connectivity (Dalbeck et al., 2014).

Frog and toad abundance has been shown to be significantly higher in beaver ponds in western South Carolina and Alberta compared to similar stream reaches without dams (Metts et al., 2001; Stevens et al., 2007). However, a previous study suggested that amphibian species diversity and abundance in new and mature beaver ponds compared with beaver-free watercourses did not demonstrate a significant difference (Russell et al., 1999). Newly created beaver ponds in North America are often rapidly colonized by swamp cricket frogs (*Pseudacris triseriata*) and spring peepers (*Hyla crucifer*), often followed by wood frogs (*R. sylvatica*) and leopard frogs (*Lithobates pipiens*) which can take advantage of the flooded backwaters and the temporary pools created by beavers (Hilfiker, 1991). Beaver-dug channels actually aid the dispersal of wood frogs (Anderson et al., 2014).

The northern leopard frog has declined significantly after being a once common species across western USA (Clarkson and Rorabaugh, 1989), with multiple factors blamed, including habitat destruction (Smith and Keinath, 2007), introduced predators, and disease (Voordouw et al., 2010). However, colonization rates by this frog are higher in beaver compared to non-beaver ponds, indicating beavers provide not only increasing habitat but also one that is of higher quality and has been especially linked with appropriate breeding sites and hydroperiod conditions (Zero and Murphy, 2016). Beaver-generated habits have been demonstrated to actually affect the development and physiology of wood frog (*R. sylvatica*) larvae (Skelly and Freidenburg, 2000). In beaver-modified open wetlands, the average and peak water temperature were recorded as 2°C and 15°C, respectively, higher than comparable shaded wetlands. As a result of this, the critical thermal maxima or the upper thermal temperature this species can tolerate have increased in a relatively short period of time (< 36 years) in beaver wetlands. More mature beaver ponds are often required to accommodate the breeding cycle of other amphibian species that cannot tolerate flash or temporary water bodies, such as bullfrogs and green frogs which require more permanent water (Hilfiker, 1991). Similarly, predicted climate changes may see the reduced natural range of mink frogs (*R. septentrionalis*); mitigation actions have been proposed that could include the future range expansion of beavers in certain areas, New York state for example, given their positive association with beaver ponds (Popescu and Gibbs, 2009).

Further research is required for certain species—for example, native to New Mexico and Arizona, the threatened Chiricahua leopard frog (*L. chiricahuensis*) requires perennial water for breeding, so it is highly likely to benefit from beaver damming activity in these dryland habitats (Gibson and Olden, 2014). However, these perennial wetlands also act as perfect spawning areas for invasive American bullfrog (*L. catesbeianus*), which impacts significantly on native fishes and frogs (Maret et al., 2006). Beaver ponds and damming have been questioned as potential threats to the endangered Arroyo toad (*Anaxyrus californicus*), through the inundation of breeding habitats and provision of more habitat for non-native crayfish (*Procambarus clarkii*) and African clawed frogs (*Xenopus laevis*) which prey on these toads (United States Fish and Wildlife Service (USFWS), 2009).

As many species of salamanders prefer flowing water, their ability to utilize and therefore benefit from beaver ponds may be variable (Metts et al., 2001; Dalbeck et al., 2007). However, spotted (*Ambystoma maculatum*) and Jefferson salamanders (*A. jeffersonianum*) are often found in the quieter backwaters of beaver ponds during breeding seasons, afterwards leaving for winter hibernation sites in the surrounding landscape (Hilfiker, 1991). Newts especially benefit. In North America, red-spotted newts (*Notophthalmus viridescens*) rapidly colonize newly formed beaver ponds, which can appear and disappear over shorter time frames than newt population saturation rates, thus creating dynamic habitats which the newts must react to promptly (Müller-Schwarze, 2011).

At Stoneman Creek, southwestern Idaho, beavers were removed around 1992 by local hunters and the remaining dam was artificially plugged with T-posts and plastic tarps to reinforce the dam structure following changes to the upstream wetlands (Pilliod et al., 2018), causing declines in rare Columbian spotted frog (*R. luteiventris*) numbers (Lingo, 2013). Five beavers were released, quickly securing this dam and building several more, resulting in significant increases in frog abundance (Lingo, 2013). In fact, the persistence of this species of conservation concern along its southern range is credited to beaver activity, enabling this frog to be more resistance to the impacts of climate changes and buffering these effects through habitat stability, quality improvement, and connectivity activities (Arkle and Pilliod, 2015). Not only do wood frogs in beaver-generated ponds have higher survival rates, but also they are on average 1.3 times larger as metamorphic juveniles, in turn influencing survival and fitness as adults (Karraker and Gibbs, 2009).

In Europe, amphibians are highly endangered and all species are protected by the EU Habitats Directive 92/43/EEC (Temple and Cox, 2009). The link between beaver restoration and amphibian abundance and diversity, including endangered species, is so important, and the contributing factors to long-term amphibian conservation are recognized (Dalbeck et al., 2020), as many amphibians are generally highly endangered (Hoffman et al., 2010; Lannoo, 2012). Beaver activities and created habitats provide vital opportunities for amphibian conservation.

9.3.5 Reptiles

Not many studies specifically addressing beaver–reptile interactions exist, but similar to beaver influences on amphibians, this positive relationship between beaver ponds and species richness and diversity tends also to be true for reptiles (Metts et al., 2001). A range of reptile species have been recorded directly utilizing beaver ponds, such as the European pond turtle (*Emys orbicularis*) and water snakes (*Natrix* spp.) (Hilfiker, 1991; Reddoch and Reddoch, 2005; Janiszewski et al., 2014). Beaver ponds in western South Carolina, especially compared to unimpounded streams, have greater reptile abundance and diversity, and this increases with pond age (Russell et al., 1999; Metts et al., 2001). Up to 18 reptile species were recorded at South Carolina beaver sites, with the highest degree of species diversity at older beaver ponds, fewer at newly created beaver ponds, and even less associated with unimpounded streams (Russell et al., 1999). Lizard species, like fence lizards (*Sceloporus undulatus*), green anole (*Anolis carolinensis*), and little brown skinks (*Scincella lateralis*), were common, along with turtles, including common box turtles (*Terrapene carolina*), painted turtles (*Chrysemys picta*), and common musk turtles (*Sternotherus odoratus*), and a range of snake species, such as worm snakes (*Carphophis amoenus*) and redbelly snakes (*Storeria occipitomaculata*). Other beaver structures such as felled trees and lodges are used as basking sites by species such as cottonmouth snakes (*Agkistrodon piscivorus*) (Graham, 2013).

9.3.6 Birds

It has been well documented that beaver-generated landscapes have a positive impact on the richness, diversity, density, and abundance of bird species, particularly through the creation of new wetland habitats and altered riparian structure, which enhance foraging, nesting, and breeding opportunities (Grover and Baldassarre, 1995; Brown et al., 1996; Rosell et al., 2005; Figure 9.12). Beaver ponds create vastly productive waterfowl-rearing habitats through increased numbers of invertebrates, along with safety of nesting, and refuge sites from predators (Nummi and Hahtola, 2008). Wetland utilization by waterfowl and their densities increased with the expansion of beaver-created wetlands, though generally with a lag of one to several years as they mature (McCall et al., 1996). Increased bird species diversity, biomass, and density have been recorded at beaver ponds (106 species recorded), compared to unimpounded stream sections in

Figure 9.12 Numerous bird species utilize beaver-generated habitats for foraging. Here a heron uses a beaver lodge for the perfect fishing advantage point. (Photo supplied courtesy of Gerhard Schwab.)

semi-arid Wyoming watercourses (Medin and Clary, 1991). Furthermore, riparian bird abundance and diversity was positively related to beaver dam density (Cooke and Zack, 2008). The structural diversity of beaver ponds is often cited as the reason why they are so attractive to many birds, with 92 species alone identified at four beaver ponds in South Carolina (Reese and Hair, 1976). Beaver damming in semi-arid areas, such as parts of western North America, in particular experience significant positive relationships between the arrival of beavers, dam density, and bird life, given the expansion of both water areas and the riparian zone (Cooke and Zack, 2008).

Species such as woodpeckers (*Picidae* spp.), wood ducks (*Aix sponsa*), tree swallows (*Tachycineta bicolor*), mergansers (*Mergus merganser*), and nuthatch (*Sitta europea*) exploit the standing dead timber, and nocturnal avian predators such as owls (*Strigiformes* spp.) are attracted by increased prey and will utilize abandoned woodpecker nests (Hilfiker, 1991). Teal breeding demonstrably increased in beaver ponds compared to non-beaver ponds, with brood densities per 1 km of shoreline of nearly 0.6 compared to > 0.1, respectively, and mortality rates of ducklings were three times higher here (Nummi and Hahtola, 2008). Diurnal avian predators such as various kestrel (falcon) species, redtailed hawks (*Buteo jamaicensis*), and ospreys (*Pandion haliaetus*) use flooded dead trees as vantage points to hunt (Grover and Baldassarre, 1995; Longcore et al., 2006).

A literature review on the ecological impacts of beavers found that 88% of studies on beaver and bird interactions recorded a positive influence, whilst 12% suggest a negative one (out of 47 studies) (Stringer and Gaywood, 2016). Others require further clarification. For example, the endangered south-western willow flycatcher (*Empidonax traillii*) breeds in dense riparian vegetation (Finch and Stoleson, 2000), and therefore beaver restoration should create suitable habitat, such as vegetated backwaters. However, beaver removal has been undertaken due to concerns over beavers removing vegetation and therefore the habitat utilized by these birds (Finch and Stoleson, 2000).

Beaver meadows are sometimes a forgotten diverse bird habitat, especially for grassland and shrub species, with their wide range of plant species readily created by beaver tree felling and damming activities (Askins et al., 2007). A survey of 37 beaver meadows in western Massachusetts found numerous regional bird species of conservation concern and general abundance was positively correlated with vegetation complexity (Chandler et al., 2009). The authors argue that beaver meadows should in fact be given conservation priority in grassland bird recovery, with wetland species including red-winged blackbirds (*Agelaius phoeniceus*) and swamp sparrows (*Melospiza georgiana*) present whilst absent in neighbouring upland areas. These species, along with yellow warbler (*Setophaga petechia*) and eastern kingbird (*Tyrannus tyrannus*), are positively associated with beaver meadow areas. In parts of eastern Germany beaver restoration was associated with the return of cranes, black storks are increasing and nesting in beaver wetlands in Belgium, while Finland has seen significant duck populations rises (Müller-Schwarze, 2011).

Negative beaver impacts on particular birds have only been recorded in two studies as part of their meta-analysis (Stringer and Gaywood, 2016). A lack of Slavonian grebes (*Podiceps auritus*) was attributed to beaver presence and their reduction of forest, especially willow, close to the water's edge (Kuczynski et al., 2012), whilst a similar local decline of whitethroat (*Sylvia communis*) has been reported

in Denmark where beavers had been reintroduced (Elmeros et al., 2003). However, without doubt, non-native species such as Mandarin duck (*Aix galericulata*) benefit from the increased habitat generated by beaver-created wetlands throughout parts of Europe.

9.3.7 Mammals

Numerous mammalian species utilize the wide variety of niche habitats, vegetation generation, and prey abundance provided by beaver-generated landscapes (Suzuki and Mccomb, 2004; Ulevičius and Janulaitis, 2007). Mammals are associated with beaver ponds, lodges, burrows, and meadows, and are found living amongst felled trees, with various species facilitated during the varying phases of beaver colonization and succession—from increased numbers of bats during the flooding habitat phase (e.g. Nummi et al., 2011) to food provision for large herbivores during the terrestrial succession phase (e.g. Wolfe, 1974).

As messy feeders, beavers leave food remains at multiple feeding stations, and by felling taller vegetation they can provide leftovers for muskrats (*Ondatra zibethicus*) and water voles (*Arvicola amphibious* and *Microtus richardsoni*) to feed on (Grasse, 1951) and utilize air holes and food caches created by beavers during the winter (Tyurnin, 1984). Other small mammal communities benefit greatly from the insect abundances, dense vegetation for food, and increased shelter provisioned by more complex riparian zones. Non-native species, such as the introduced American mink (*Neovison vison*) and Asian racoon dog (*Nyctereuts procyonoides*), along with the semi-aquatic coypu/nutria (*Myocastor coypus*), and muskrat in many parts of Europe, also benefit from beaver-created wetlands (Ruys et al., 2011). Smaller rodents, especially muskrat, water voles, water shrews (*Neomys fodiens*), white-footed mice, and bank voles (*Myodes glareolus*), are readily found living in both active and abandoned beaver structures and feeding along beaver ponds and canals (Hilfiker, 1991, Müller-Schwarze, 2011). Even dried-out beaver ponds, following long periods of drought, in Russia had a greater range of mammalian species and increased visits compared to similar beaver-free areas. Camera trap images revealed wild boar (*Sus scrofa*), elk (*Alces alces*), roe deer (*Capreolus capreolus*), red foxes (*Vulpes vulpes*), badgers (*Meles meles*), pine martens (*Martes martes*), polecats (*Mustela putorius*), even stray dogs (*Canis lupus familiaris*) all utilizing the dried pond. The predators seemed to be attracted by the numerous amphibians congregating in the small remaining amounts of residual water around the lodge, with herbivores drawn to the coppiced vegetation (Mishin and Trenkov, 2016).

Biomass of small mammals, particularly voles (*Microtus* spp.) and shrews (*Sorex* spp.), was 2.7 times higher in riparian areas in Idaho associated with beaver pond complexes compared to those without (Medin and Clary, 1991). Another study in Oregon again found greater numbers of shrew species, voles, deer mice (*Peromyscus* spp.), and Pacific jumping mice in beaver-occupied streams (Suzuki and Mccomb, 2004), while the abundance of voles and shrews increased 2–2.5 times in southern Karelian beaver ponds, based on a 10-year study and 44 beaver-impounded sites (Fyodorov and Yakimova, 2012). Similarly, a study in Montana, using stable isotopes to trace aquatic-derived carbon, determined deer mice abundance rates up to 75% higher in beaver compared to non-beaver sites (McCaffery and Eby, 2016). Aquatic-derived carbon was higher in these mice than at non-beaver sites, which declined linearly with distance from the beaver streams. This nicely demonstrates the provisioning of nutrient subsidies by beaver activity to terrestrial mammals. Combined camera trapping and snow track surveying techniques used to monitor beaver-modified sites compared with non-beaver-modified or -occupied sites favoured detection of larger mammals (Nummi et al., 2019). Overall, these researchers determined that mammalian species richness and occurrence were 83% and 12% higher, respectively, in beaver-modified areas, with 14 species identified (excluding small mammals) including hare (*Lepus* spp.), deer (roe and white-tailed, *Odocoileus virginianus*), moose, squirrel (*Sciurus vulgaris*), weasel (*Mustela* spp.), badger, lynx (*Lynx lynx*), otter (*Lutra lutra*), racoon dog, fox, and wolf (*Canis lupus*) (Figure 9.13).

Large herbivores may compete for food resources with beavers and exploit the regrowth

Figure 9.13 Wolf cubs in North America utilizing an abandoned beaver lodge for shelter. (Photo supplied courtesy of Tom Gable.)

their coppicing generates (Brookshire et al., 2002; Hood and Bayley, 2008). Beaver meadows and foraged shorelines provide perfect foraging opportunities for various deer (Cervidae) including roe deer, elk, and red deer (*Cervus elaphus*), white-tailed deer, moose, livestock, and bears (*Ursus* spp.) (Windels, 2017; Nummi et al., 2019). This grazing can cause loss especially of willow and woody shrubs and in time can even limit beaver colonization (Baker et al., 2005; Ripple and Beschta, 2012). Moose will regularly feed on aquatic plants, such as water lilies and algae growing in beaver ponds (Grasse, 1951; Hilfiker, 1991). There is a demonstrated greater activity of moose in beaver-modified areas compared to absent controls in Finland; these represented areas of increased sapling growth (Johnston and Naiman, 1990; Nummi et al., 2019), whilst felled trees provide bark stripping opportunities for moose and deer species, with regenerating coppice often being preferred by grazing browsers. Additionally, mature beaver ponds provide habitat for sought-after aquatic plants such as water lilies (*Nymphaeales* spp.) and have even been suggested as provisioning submergence opportunities for relief from biting insects (Stoffyn-Egli and Willison, 2011; Nummi et al., 2019).

'Moose willow' refers to small tree forms created through heavy grazing by these large herbivores who, like beavers, are one of the few animals that can efficiently digest this browse. These are often associated with beaver meadows and along the fringes of beaver ponds to such an extent that in Yellowstone National Park beaver population growth rates are thought to be kept in check by such grazing, along with herds of elk that will selectively feed on fresh sprouting growth (Hilfiker, 1991). Similarly, livestock and deer browsing on beaver-coppiced trees across Europe may have a deleterious impact on tree regeneration, especially in heavily modified systems. This can drastically reduce or even eliminate more deeper-rooted plants, in turn increasing bankside erosion and soil compaction and decreasing riparian plant and animal communities (Kauffman and Krueger, 1984; Giullano and Homyack, 2004; Frey, 2018). Riparian habitats can be particularly sensitive to livestock grazing and especially prone to excessive grazing, which in turn can have negative impacts on the structure and function of this ecosystem (Trimble and Mendel, 1995; Fleischner, 2010). Heavy grazing by livestock and ungulates can negatively influence both woody species, especially willow, and beaver populations, especially where top predators have been removed. Herbivory often induces woody plants to send more resources from their roots to their shoots, increasing growth rates, photosynthesis, secondary compounds, and branching (Hobbs, 1996; Strauss and Agrawal, 1999). Larger mammals will selectively and repeatedly browse on any newer riparian shrub shoot growth, therefore exploiting the very resources utilized by beavers ((Hood and Bayley, 2008; see Section 11.6). In simulated beaver-cut experiments, unbrowsed plants recovered 84% of their precut biomass within two growing seasons, whereas those browsed by ungulates only recovered 6% (Baker et al., 2005). Simulated modelling has determined that beaver persistence in an area decreases exponentially as elk numbers rise from 30 to 60 elk/km^2, with the most pronounced change in willow structure (i.e. tall trees to short shrubs) occurring at ≥30 elk/km^2, around the apparent tipping point for loss of winter food and rapid beaver declines (Baker et al., 2012). This effect can be so pronounced that the natural mutualism between beavers and willow will decline or 'not thrive' in heavily grazed riparian environments, largely due to the lack of winter food resources as a result of this foraging competition (Baker and Hill, 2003; Baker et al., 2005).

Another interesting relationship between declining bat populations and beavers has been documented. Trees surrounding beaver ponds, particularly mature specimens, provide numerous roosting opportunities (Menzel et al., 2001). Riparian zones in general, along with beaver meadows and ponds, provide excellent feeding opportunities for bats as they are hotspots for emerging insects (Brooks and Ford, 2005; Fukui et al., 2006; Nummi et al., 2011). The shallow waters and natural breakdown of flooded and foraged vegetation generated by beavers provide the release of nutrients upon which invertebrates feed (McDowell and Naiman, 1986). In such beaver habitats, up to five times more numbers of emerging insects can be generated (Nummi and Pöysä, 1995). Vespertilionid bat activity at newly occupied Polish beaver sites was higher than control sites due to newly created wetland and forest canopy gaps, both increasing insect biomass and facilitating predation by bats (Ciechanowski et al., 2011). The northern bat (*Eptesicus nilssoni*) especially prefers to utilize beaver-generated ponds over non-beaver ponds (Nummi et al., 2011). Thus, such wetlands may be important in bat conservation, particularly more endangered species which are dependent on prey emerging from aquatic environments, such as pond bat (*Myotis dasycneme*) and lesser horseshoe bat (*Rhinolophus hipposideros*) (Nummi et al., 2011).

Mink, skunk (*Mephitis mephitis*), and raccoon (*Procyon lotor*) commonly hunt aquatic insects, fish, crayfish, reptiles, amphibians, birds, and small rodents in beaver-generated wetlands (Knudsen, 1962; Danilov, 1992). Reports exist of raccoons raiding the nests of ducks in both banks and trees, and turtles associated with beaver ponds (Hilfiker, 1991), while mink and racoon are reported to hunt snakes and crayfish amongst beaver structures in Wisconsin (Knudsen, 1962). Videographic studies of within-lodge beaver behaviours revealed that muskrats were the most common species present and in fact cohabiting, with lodge maintenance, locomotion, and grooming being the most frequent behaviours observed (Mott et al., 2013). In turn the beaver homebuilders largely ignored these alternative occupants, and when both species where present at the same time, beavers were most commonly sleeping/recumbent (80.2% of all observations) or grooming themselves (17.3%). Beaver dams are recognized as influential for short-tailed weasel occurrence, of conservation concern in southwest USA (Frey and Calkins, 2013), with their ponds also linked to increased habitat for the endangered New Mexico meadow jumping mouse (*Zapus hudsonius luteus*) (Frey and Malaney, 2009).

Small carnivores, particularly otter but also badger, red fox, stoat (*Mustela erminea*), and pine marten, can benefit through enhanced prey populations (such as fish, amphibians, and birds) and the presence of beaver lodges and burrows which they utilize for shelter and breeding (Snodgrass and Meffe, 1998; LeBlanc et al., 2007; Brugger et al., 2020). Red foxes especially have been documented as being more active around beaver ponds compared to control sites, most likely directly linked to increased prey abundance (Mishin and Trenkov, 2016; Nummi et al., 2019).

The relationship between beavers and otters is a complex one (see Section 8.3.3; Brugger et al., 2020). Both are large semi-aquatic mammals that share the same riparian habitat across North America and throughout Eurasia. They are often mistaken for each other and their ecological roles are sometimes confused. Their relationship has been described as a commensal one (Tumlison et al., 1982; Polecha, 1989), though it seems likely otters may sometimes opportunistically feed on beaver kits (Green, 1932; Greer, 1955). The North American otter (*Lontra canadensis*) has especially declined in the drylands of southwest USA, such as the upper Colorado basin, where they have been especially associated with beaver ponds (Depue and Ben-David, 2010). Greater otter activity has been determined in beaver-occupied ponds, especially through increasing the availability of shelter and prey, although they did not necessarily avoid inactive beaver ponds (LeBlanc et al., 2007). Beaver-generated landscapes benefit otters, providing foraging and shelter opportunities in terms of habitat for higher concentrations of fish (including larger and slower-moving fish) and as otters tend to depend on other animals to dig the shelters they utilize (Melquist and Hornocker, 1983). Otters have been observed in within-lodge beaver investigations (Mott et al.,

2013). Whilst within beaver lodges they were most commonly seen sleeping/in recumbence (86.7%) followed by grooming (6.3%), exploration (4.3%), and locomotion (1%) (see section 6.1, chapters 2 and 8). Otter population recovery in parts of North America has even been linked to beaver re-establishment (Vogt, 1981; Tumlison et al., 1982). For example, in Idaho, otter den and resting sites (32% and 6%, respectively) consisted of beaver burrows and lodges (Melquist and Hornocker, 1983). Otters will regularly use beaver ponds, dams, and lodges for hunting opportunities (Danilov, 1992) and will utilize air holes cut into ice by beavers during the winter (Tyurnin, 1984). By keeping areas of ponds open from the ice, along with opening up ponds from ice earlier in the season compared to unoccupied ponds, beavers provide otters with access to water that may otherwise be unavailable in winter periods, increasing their foraging opportunities and survival (Reid et al., 1994; Bromley and Hood, 2013).

Beavers at all life stages can of course act as prey to a number of predators, though only larger carnivores such as wolves, larger cat species, coyotes, and bears may take on adults (see Section 8.3.2). Predation on beaver populations by wolves is actually thought to be relatively minor across most of their range (Gable et al., 2018), thought they can be important sources of prey for some populations which may even specialize in beaver hunting (Newsome et al., 2016). One study in Belarus demonstrated that beavers were the main, all-year-round prey of local wolf populations (Sidorovich et al., 2017), whereas another study in Manitoba found for one individual wolf 83% of its summer diet consisted of beaver, while 78 other wolves examined revealed only 3–42% (Moayeri, 2013; see section 8.3.2). Therefore, the beaver–wolf relationship can be complex. Recently it has been determined that that by killing dispersing beavers, wolves can alter wetland development (Gable et al., 2020). Such dynamics require further research as these are undoubtedly further linked to wolf–ungulate systems (see Box 11.3). Probably more important are the vast range of prey species that utilize beaver-generated habitats; as such, it is these wetlands that are functionally more significant to such predators (Rosell and Czech, 2000; Rosell and Sanda, 2006).

References

Alexander, M. D. 1998. Effects of beaver (*Castor canadensis*) impoundments on stream temperature and fish community species composition and growth in selected tributaries of Miramichi River, New Brunswick. *Canadian Technical Report of Fisheries and Aquatic Sciences*, 2227, 53.

Alvarez, J. A., Shea, M. A., and Foster, S. M. 2013. *Rana draytonii* (California red-legged frog). Association with beaver. *Herpetological Review*, 44, 127–128.

Anderson, C. B., Griffith, C. R., Rosemond, A. D., Rozzi, R., and Dollenz, O. 2006. The effects of invasive North American beavers on riparian plant communities in Cape Horn, Chile: Do exotic beavers engineer differently in sub-Antarctic ecosystems? *Biological Conservation*, 128, 467–474.

Anderson, C. B., L.M., V., Wallem, P. K., Valenzuela, A. E. J., Simanonok, M. P., and Martínez Pastur, G. 2014. Engineering by an invasive species alters landscape-level ecosystem function, but does not affect biodiversity in freshwater systems. *Diversity and Distributions*, 20, 214–222.

Arkle, R. S. and Pilliod, D. S. 2015. Persistence at distributional edges: Columbia spotted frog habitat in the arid Great Basin, USA. *Ecology and Evolution*, 5, 3704–3724.

Armstrong, J. D., Kemp, P. S., Kennedy, G. J. A., Ladle, M., and Milner, N. J. 2003. Habitat requirements of Atlantic salmon and brown trout in rivers and streams. *Fisheries Research*, 62, 143–170.

Askins, R. A., Chávez-Ramírez, F., Dale, B. C., Haas, C. A., Herkert, J. R., Knopf, F. L., and Vickery, P. D. 2007. Conservation of grassland birds in North America: understanding ecological processes in different regions—report of the AOU Committee on Conservation. *Auk*, 124, 1–46.

Aznar, J.-C. and Desrochers, A. 2008. Building for the future: abandoned beaver ponds promote bird diversity. *Ecoscience*, 15, 250–257.

Baker, B. and Hill, E. 2003. Beaver (*Castor canadensis*). In: Feldhamer, G. A., Thompson, B. C., and Chapman, J. A. (eds.) *Wild Mammals of North America: Biology, Management and Conservation*. Baltimore: Johns Hopkins University Press.

Baker, B. W., Ducharme, H. C., Mitchell, D. C., Stanley, T. R., and Peinetti, H. R. 2005. Interaction of beaver and elk herbivory reduces standing crop of willow. *Ecological Applications*, 15, 110–118.

Baker, B. W., Peinetti, H. R., Coughenour, M. B., and Johnson, T. L. 2012. Competition favors elk over beaver in a riparian willow ecosystem. *Ecosphere*, 3, 1–15.

Baldwin, J. 2015. Potential mitigation of and adaptation to climate-driven changes in California's highlands through increased beaver populations. *California Fish and Game*, 101, 218–240.

Barabash-Nikiforov, I. I. 1950. Materials to study of biotic links of beaver (burrow cohabitants). *Doklady AN USSR*, 70, 1057–1059.

Barnes, W. J. and Dibble, E. 1988. The effects of beaver in riverbank forest succession. *Canadian Journal of Botany*, 66, 40–44.

Barnhart, M. C., Haag, W. R., and Roston, W. N. 2008. Adaptations to host infection and larval parasitism in Unionoida. *Journal of the North American Benthological Society*, 27, 370–394.

Basey, J. M., Jenkins, S. H., and Busher, P. E. 1988. Optimal central-place foraging by beavers: tree-size selection in relation to defensive chemicals of quaking aspen. *Oecologia*, 76, 278–282.

Bashinskiy, I. V. 2008. The effect of beaver (*Castor fiber* Linnaeus, 1758) activity on amphibian reproduction. *Inland Water Biology*, 1, 326–331.

Bashinskiy, I.V. 2020. Beavers in lakes: a review of their ecosystem impact. *Aquatic Ecology*, 54(3).

Bason, C. W., Kroes, D. E., and Brinson, M. M. 2017. The effect of beaver ponds on water quality in rural coastal plain streams. *Southeastern Naturalist*, 16, 584–602.

Batty, D. 2002. Beavers: aspen heaven or hell? In: Cosgrove, P. and Amphlett, A. (eds.) *The Biodiversity and Management of Aspen Woodlands. Proceedings of a One-Day Conference*, 25 May 2001, Kingussie, Scotland. Granton-on-Spey: The Cairngorms Biodiversity Action Plan, pp. 41–44.

Beaver Salmonid Working Group 2015. *Final report of the Beaver Salmonid Working Group. National Species Reintroduction Forum*, Inverness, Scotland, pp. 1–80.

Beckman, N. D., and Wohl, E. 2014. Carbon storage in mountainous headwater streams: The role of old-growth forest and logjams. *Water Resources Research*, 50(3), 2376–2393.

Behnke, R. J. 2002. *Trout and Salmon of North America*. New York: Free Press.

Benke, A., Ward, G., and Richardson, T. 1999. Beaver impounded wetlands of the southeastern coastal plain. In: Batzer, D. P., Rader, R. B., and Wissinger, S. A. (eds.) *Invertebrates in Freshwater Wetlands of North America: Ecology and Management*. New York: Wiley.

Benke, A. C., and Bruce Wallace, J. 2015. High secondary production in a Coastal Plain river is dominated by snag invertebrates and fuelled mainly by amorphous detritus. *Freshwater Biology*, 60, 236–255.

Bergman, B. G. and Bump, J. K. 2015. Experimental evidence that the ecosystem effects of aquatic herbivory by moose and beaver may be contingent on water body type. *Freshwater Biology*, 60, 1635–1646.

Bigler, W., Butler, D. R., and Dixon, R. W. 2001. Beaver-pond sequence morphology and sedimentation in northwestern Montana. *Physical Geography*, 22, 531–540.

Bilotta, G. S., Krueger, T., Brazier, R. E., Butler, P., Freer, J., Hawkins, J. M. B., Haygarth, P. M., Macleod, C. J. A., and Quinton, J. N. 2010. Assessing catchment-scale erosion and yields of suspended solids from improved temperate grassland. *Journal of Environmental Monitoring*, 12, 731–739.

Błędzki, L. A., Bubier, J., Moulton, L., and Kyker-Snowman, T. 2011. Downstream effects of beaver ponds on the water quality of New England first- and second-order streams. *Ecohydrology*, 4, 698–707.

Bonner, J. L., Anderson, J. T., Rentch, J. S., and Grafton, W. N. 2009. Vegetative composition and community structure associated with beaver ponds in Canaan valley, West Virginia, USA. *Wetlands Ecology and Management*, 17, 543–554.

Bouwes, N., Weber, N., Jordan, C. E., Saunders, W. C., Tattam, I. A., Volk, C., Wheaton, J. M., and Pollock, M. M. 2016. Ecosystem experiment reveals benefits of natural and simulated beaver dams to a threatened population of steelhead (*Oncorhynchus mykiss*). *Scientific Reports*, 6, 28581.

Bradt, G. W. 1938. A study of beaver colonies in Michigan. *Journal of Mammalogy*, 19, 139–162.

Briggs MA, Lautz LK, McKenzie JM, Gordon RP, Hare DK. Using high-resolution distributed temperature sensing to quantify spatial and temporal variability in vertical hyporheic flux. *Water Resource Research*, 48:W02527.

Briones, M., Schlatter, R., Wolodarsky, A., and Venegas, C. 2001. Clasificación ambiental para habitat de *Castor canadensis* (Kuhl 1820, Rodentia), de acuerdo a características de cuencas en un sector de Tierra del Fuego. *Anales del Instituto de la Patagonia*, 29, 75–93.

Brazier, R.E., Puttock, A., Graham, H.A., Auster, R.E., Davies, K.H. and Brown, C.M., 2021. Beaver: Nature's ecosystem engineers. *Wiley Interdisciplinary Reviews: Water*, 8, p.e1494.

Bromley, C. K. and Hood, G. A. 2013. Beavers (*Castor canadensis*) facilitate early access by Canada geese (*Branta canadensis*) to nesting habitat and areas of open water in Canada's boreal wetlands. *Mammalian Biology—Zeitschrift für Säugetierkunde*, 78, 73–77.

Brooks, R. T. and Ford, W. M. 2005. Bat activity in a forest landscape of central Massachusetts. *Northeastern Naturalist*, 12, 447–462.

Brooks, A.P., Gehrke, P.C., Jansen, J.D., and Abbe, T.B. 2004. Experimental reintroduction of woody debris on the Williams River, NSW: geomorphic and ecological responses. *River Research. Application*, 20, 513–536.

Brookshire, J. E., Kauffman, B. J., Lytjen, D., and Otting, N. 2002. Cumulative effects of wild ungulate and livestock herbivory on riparian willows. *Oecologia*, 132, 559–566.

Brown, D. J., Hubert, W. A., and Anderson, S. H. 1996. Beaver ponds create wetland habitat for birds in mountains of southeastern Wyoming. *Wetlands*, 16, 127–133.

Browne, C. L. and Paszkowski, C. A. 2010. Hibernation sites of western toads (*Anaxyrus boreas*): characterization and management implications. *Herpetological Conservation and Biology*, 5, 49–63.

Brugger, M., Jährig, M., Peper, J, Nowak, C, Cocchiararo, B, and Ansorge, H (2020). Influence of Eurasian Beaver (*Castor fiber*) on Eurasian Otter (*Lutra lutra*) evaluated by Activity Density Estimates in Anthropogenic Habitats in Eastern Germany. *IUCN Otter Spec. Group Bull.* **37**, 98–119.

Brush, G. S. 2009. Historical land use, nitrogen, and coastal eutrophication: a paleoecological perspective. *Estuaries and Coasts*, 32, 18–28.

Bubier, J. L., Moore, T. R., and Roulet, N. T. 1993. Methane emissions from wetlands in the midboreal region of northern Ontario, Canada. *Ecology*, 74, 2240–2254.

Budy, P., Thiede, G. P., and McHugh, P. 2007. Quantification of the vital rates, abundance, and status of a critical, endemic population of Bonneville cutthroat trout. *North American Journal of Fisheries Management*, 27, 593–604.

Burchsted, D. and Daniels, M. D. 2014. Classification of the alterations of beaver dams to headwater streams in northeastern Connecticut, USA. *Geomorphology*, 205, 36–50.

Bush, B. M. and Wissinger, S. A. 2016. Invertebrates in beaver-created wetlands and ponds. Invertebrates in freshwater wetlands. Cham: Springer.

Butler, D. R. and Malanson, G. P. 1994. Beaver landforms. *Canadian Geographer/Le Géographe Canadien*, 38, 76–79.

Butler, D. R. and Malanson, G. P. 1995. Sedimentation rates and patterns in beaver ponds in a mountain environment. In: Hupp, C. R., Osterkamp, W. R., and Howard, A. D. (eds.) *Biogeomorphology, Terrestrial and Freshwater Systems*. Amsterdam: Elsevier.

Butler, D. R. and Malanson, G. P. 2005. The geomorphic influences of beaver dams and failures of beaver dams. *Geomorphology*, 71, 48–60.

Bylak, A., Szmuc, J., Kukuła, E., and Kukuła, K. 2019. Potential use of beaver *Castor fiber* L., 1758 dams by the thick-shelled river mussel *Unio crassus* Philipsson, 1788. *Molluscan Research*, 40, 44–51.

Campbell-Palmer, R., Gow, D., Campbell, R., Dickinson, H., Girling, S., Gurnell, J., Halley, D., Jones, S., Lisle, S., Parker, H., Schwab, G., and Rosell, F. 2016. *The Eurasian Beaver Handbook: Ecology and Management of* Castor fiber. Exeter: Pelagic Publishing.

Catalán, N., Ortega, S. H., Gröntoft, H., Hilmarsson, T. G., Bertilsson, S., Wu, P., Levanoni, O., Bishop, K. and

Bravo, A. G. 2017. Effects of beaver impoundments on dissolved organic matter quality and biodegradability in boreal riverine systems. *Hydrobiologia*, 793, 135–148.

Celewicz-Gołdyn, S. and Kuczynska-Kippen, N. 2017. Ecological value of macrophyte cover in creating habitat for microalgae (diatoms) and zooplankton (rotifers and crustaceans) in small field and forest water bodies. *PLoS ONE*, 12(5), e0177317.

Chandler, R. B., King, D. I., and DeStefano, S. 2009. Scrub-shrub bird habitat associations at multiple spatial scales in beaver meadows in Massachusetts. *The Auk*, 126, 186–197.

Ciechanowski, M., Kubic, W., Rynkiewicz, A., and Zwolicki, A. 2011. Reintroduction of beavers *Castor fiber* may improve habitat quality for vespertilionid bats foraging in small river valleys. *European Journal of Wildlife Research*, 57, 737–747.

Cirmo, C. P. and Driscoll, C. T. 1993. Beaver pond biogeochemistry: acid neutralizing capacity generation in a headwater wetland. *Wetlands*, 13, 277–292.

Clarkson, R. W. and Rorabaugh, J. C. 1989. Status of leopard frogs (*Rana pipiens complex*: Ranidae) in Arizona and southeastern California. *Southwestern Naturalist*, 34(4), 531–538.

Clemente, J.M., Boll, T., Teixeira-de Mello, F., Iglesias, C., Roer, A., Pedersen, E.J., and Meerhoff, M. 2019. Role of plant architecture on littoral macroinvertebrates in temperate and subtropical shallow lakes: a comparative manipulative field experiment. *Limnetica*, 38(2), 759–772.

Collen, B., Whitton, F., Dyer, E. E., Baillie, J. E., Cumberlidge, N., Darwall, W. R., Pollock, C., Richman, N. I., Soulsby, A. M., and Böhm, M. 2014. Global patterns of freshwater species diversity, threat and endemism. *Global Ecology and Biogeography*, 23, 40–51.

Collen, P. and Gibson, R. J. 2001. The general ecology of beavers (*Castor* spp.), as related to their influence on stream ecosystems and riparian habitats, and the subsequent effects on fish—a review. *Reviews in Fish Biology and Fisheries*, 10, 439–461.

Collins, B.D., Montgomery, D.R., Fetherston, K.L., Abbe, T.B. 2012. The floodplain large wood cycle hypothesis: a mechanism for the physical and biotic structuring of temper- ate forested alluvial valleys in the North Pacific coastal eco-region. *Geomorphology*, 139–140:460–470.

Cooke, H. A. and Zack, S. 2008. Influence of beaver dam density on riparian areas and riparian birds in shrub steppe of Wyoming. *Western North American Naturalist*, 68, 365–374.

Correll, D. L., Jordan, T. E., and Weller, D. E. 2000. Beaver pond biogeochemical effects in the Maryland coastal plain. *Biogeochemistry*, 49, 217–239.

Cottrell, T. R. 1995. Willow colonization of Rocky Mountain mires. *Canadian Journal of Forest Research*, 25, 215–222.

Cunjak, R. A. 1996. Winter habitat of selected stream fishes and potential impacts from land-use activity. *Canadian Journal of Fisheries and Aquatic Sciences*, 53, 267–282.

Cunjak, R. A. and Therrien, J. 1998. Inter-stage survival of wild juvenile Atlantic salmon, *Salmo salar* L. *Fisheries Management and Ecology*, 5, 209–223.

Cunningham, J. M., Calhoun, A. J., and Glanz, W. E. 2007. Pond-breeding amphibian species richness and habitat selection in a beaver-modified landscape. *Journal of Wildlife Management*, 71, 2517–2526.

Curran, J,H, Wohl, E.E. 2003. Large woody debris and flow resistance in step-pool channels, Cascade Range, Washington. Geomorphology, 51:141–157.

Cutting, K. A., Ferguson, J. M., Anderson, M. L., Cook, K., Davis, S. C., and Levine, R. 2018. Linking beaver dam affected flow dynamics to upstream passage of Arctic grayling. *Ecology and Evolution*, 8, 12905–12917.

Czech, A. and Lisle, S. 2003. Understanding and solving the beaver (*Castor fiber* L.)–human conflict: an opportunity to improve the environment and economy of Poland. *Denisia*, 9, 91–98.

Czerniawski, R., Sługocki, Ł., and Kowalska-Góralska, M. 2017. Effects of beaver dams on the zooplankton assemblages in four temperate lowland streams (NW Poland). *Biologia*, 72, 417–430.

Dalbeck, L. and Weinberg, K. 2009. Artificial ponds: a substitute for natural beaver ponds in a central European highland (Eifel, Germany)? *Hydrobiologia*, 630, 49–62.

Dalbeck, L., Lüscher, B., and Ohlhoff, D. 2007. Beaver ponds as habitat of amphibian communities in a central European highland. *Amphibia-Reptilia*, 28, 493–501.

Dalbeck, L., Janssen, J., and Völsgen, S. L. 2014. Beavers (*Castor fiber*) increase habitat availability, heterogeneity and connectivity for common frogs (*Rana temporaria*). *Amphibia-Reptilia*, 35, 321–329.

Dalbeck, L., Hachtel, M., and Campbell-Palmer, R. 2020. A review of the influence of beaver *Castor fiber* on amphibian assemblages in the floodplains of European temperate streams and rivers. *Herpetological Journal*, 30, 135–146.

Danilov, P. I. 1992. *Introduction of North-American Semiaquatic Mammals in Karelia and its Consequences for Aboriginal Species. Semiaquatische Säugetiere.* Halle: Wissenschaftliche Beiträge, Martin-Luther Universität Halle-Wittenberg, pp. 267–276.

Demmer, R. and Beschta, R. L. 2008. Recent history (1988–2004) of beaver dams along Bridge Creek in central Oregon. *Northwest Science*, 82, 309–318.

Depue, J. E. and Ben-David, M. 2010. River otter latrine site selection in arid habitats of western Colorado, USA. *Journal of Wildlife Management*, 74, 1763–1767.

Derwich, A. and Mróz, I. 2008. Bóbr europejski *Castor fiber*. 1758 jako czynnik wspomagajacy renaturyzacje siedlisk nad górnym Sanem. *Studia i Materialy Centrum Edukacji Przyrodniczo-Lesnej*, 2, 173–183.

Devito, K. J. and Dillon, P. J. 1993. Importance of runoff and winter anoxia to the P and N dynamics of a beaver pond. *Canadian Journal of Fisheries and Aquatic Sciences*, 50, 2222–2234.

Dittbrenner, B. J., Pollock, M. M., Schilling, J. W., Olden, J. D., Lawler, J. J., and Torgersen, C. E. 2018. Modeling intrinsic potential for beaver (*Castor canadensis*) habitat to inform restoration and climate change adaptation. *PLoS ONE*, 13, e0192538.

Dolin, E. J. 2010. *Fur, Fortune and Empire*. New York, Norton and Company.

Ecke, F., Levanoni, O., Audet, J., Carlson, P., Eklöf, K., Hartman, G., McKie, B., Ledesma, J., Segersten, J., and Truchy, A. 2017. Meta-analysis of environmental effects of beaver in relation to artificial dams. *Environmental Research Letters*, 12, 113002.

Ehnström, B. 2002. *Insektsgnag i bark och ved*. Uppsala: Artdatabanken.

Elliott, J. M., Hurley, M. A., and Fryer, R. J. 1995. A new, improved growth model for brown trout, *Salmo trutta*. *Functional Ecology*, 9, 290–298.

Elmeros, M., Madsen, A., and Berthelsen, J. P. 2003. Monitoring of reintroduced beavers (*Castor fiber*) in Denmark. *Lutra*, 46, 153–162.

Feldhaus, J. W., Heppell, S. A., Li, H., and Mesa, M. G. 2010. A physiological approach to quantifying thermal habitat quality for redband rainbow trout (*Oncorhynchus mykiss gairdneri*) in the south Fork John Day River, Oregon. *Environmental Biology of Fishes*, 87, 277–290.

Finch, D. M. and Stoleson, S. H. 2000. *Status, Ecology, and Conservation of the Southwestern Willow Flycatcher*. Odgen, UT: USDA Forest Service, Rocky Mountain Research Station.

Fleischner, T. L. 2010. Livestock grazing and wildlife conservation in the American West: historical, policy and conservation biology perspectives. In: Du Toit, J. T., Kock, R., and Deutsch, J. C. (eds.) *Wild Rangelands: Conserving Wildlife While Maintaining Livestock in Semi-Arid Ecosystems*. Oxford: Blackwell Publishing, pp. 235–265.

Ford, T. E. and Naiman, R. J. 1988. Alteration of carbon cycling by beaver: methane evasion rates from boreal forest streams and rivers. *Canadian Journal of Zoology*, 66, 529–533.

Foreman, D. 2004. *Rewilding North America: A Vision for Conservation in the 21st Century*. Washington, DC: Island Press.

Francis, M. M., Naiman, R. J., and Melillo, J. M. 1985. Nitrogen fixation in subarctic streams influenced by beaver (*Castor canadensis*). *Hydrobiologia*, 121, 193–202.

Francis, R. A., Petts, G. E., and Gurnell, A. M. 2008. Wood as a driver of past landscape change along river corridors.

Earth Surface Processes and Landforms: Journal of the British Geomorphological Research Group, 33, 1622–1626.

Frey, J. K. 2018. Beavers, livestock, and riparian synergies: bringing small mammals into the picture (Chapter 6). In: Johnson, R. R., Carothers, S. W., Finch, D. M., Kingsley, K. J., and Stanley, J. T. (eds.) *Riparian Research and Management: Past, Present, Future*. Fort Collins, CO: Rocky Mountain Research Station.

Frey, J. K. and Calkins, M. T. 2013. Snow cover and riparian habitat determines distribution of the short-tailed weasel (*Mustela ermine*) at its southern range limits in western North America. *Mammalia*, 78, 45–56.

Frey, J. K. and Malaney, J. L. 2009. Decline of the meadow jumping mouse (*Zapus hudsonius luteus*) in two mountain ranges in New Mexico. *Southwestern Naturalist*, 54, 31–44.

Fukui, D., Murakami, M., Nakano, S., and Aoi, T. 2006. Effects of emergent aquatic insects on bat foraging in a riparian forest. *Journal of Animal Ecology*, 75, 1252–1258.

Fustec, J., Lodé, T., Le Jacques, D., and Cormier, J. 2001. Colonization, riparian habitat selection and home range size in a reintroduced population of European beavers in the Loire. *Freshwater Biology*, 46, 1361–1371.

Fyodorov, F. V. and Yakimova, A. E. 2012. Changes in ecosystems of the Middle Taiga due to the impact of beaver activities, Karelia, Russia. *Baltic Forestry*, 18, 278–287.

Gable, T. D., Johnson-Bice, S. M., Homkes, A. T., Windels, S. K., and Bump, J. K. 2020. Outsized effect of predation: Wolves alter wetland creation and recolonization by killing ecosystem engineers. *Science advances*, 6(46), eabc5439.

Gable, T. D., Windels, S. K., Bruggink, J. G., and Homkes, A. T. 2016. Where and how wolves (*Canis lupus*) kill beavers (*Castor canadensis*). *PLoS ONE*, 11, e0165537.

Gable, T. D., Windels, S. K., Romanski, M. C., and Rosell, F. 2018. The forgotten prey of an iconic predator: a review of interactions between grey wolves *Canis lupus* and beavers *Castor* spp. *Mammal Review*, 48, 123–138.

Gangloff, M. M. 2013. Taxonomic and ecological tradeoffs associated with small dam removals. *Aquatic Conservation: Maine and Freshwater Ecosystems*, 23, 475–480.

Gatti, R. C., Callaghan, T. V., Rozhkova-Timina, I., Dudko, A., Lim, A., Vorobyev, S. N., Kirpotin, S. N., and Pokrovsky, O. S. 2018. The role of Eurasian beaver (*Castor fiber*) in the storage, emission and deposition of carbon in lakes and rivers of the River Ob flood plain, western Siberia. *Science of the Total Environment*, 644, 1371–1379.

Genney, D. R. 2015. *The Scottish Beaver Trial: Lichen Impact Assessment 2010–2014, Final Report*. Inverness: Scottish Natural Heritage.

Gibbs, J. P. 2000. Wetland loss and biodiversity conservation. *Conservation Biology*, 14, 314–317.

Gibson, M. 1998. Issues involved in protecting important populations of *Margaritifera margaritifera*. *Journal of Conchology Special Publication*, 2, 243–244.

Gibson, P. P. and Olden, J. D. 2014. Ecology, management, and conservation implications of North American beaver (*Castor canadensis*) in dryland streams. *Aquatic Conservation: Marine and Freshwater Ecosystems*, 24, 391–409.

Gibson, R. J., Hillier, K. G., and Whalen, R. R. 1996. Status of Atlantic salmon (*Salmo salar*) in the Highlands River, St. George's Bay (SFA 13), Newfoundland, 1995. *Canadian Journal of Fisheries and Aquatic Sciences*, 96/39, 35.

Gill, C. J. 1970. The flooding tolerance of woody species—a review. *Forest Abstracts*, 31, 671–688.

Ginson, R. 2013. *Miramichi Salmon Association Conservation Field Programme Report 2012*. South Esk, New Brunswick: Miramichi Salmon Association.

Giullano, W. M. and Homyack, J. D. 2004. Short-term grazing exclusion effects on riparian small mammal communities. *Journal of Range Management*, 57, 346–350.

Goldfarb, B. 2018. Why two countries want to kill 100,000 beavers. https://www.washingtonpost.com/news/animalia/wp/2018/08/09/why-two-countries-want-to-kill-100000-beavers/ [Accessed 10/09/2018].

Gossner, M. M., Wende, B., Levick, S., Schall, P., Floren, A., Linsenmair, K. E., Steffan-Dewenter, I., Schulze, E. D., and Weisser, W. W. 2016. Deadwood enrichment in European forests—which tree species should be used to promote saproxylic beetle diversity? *Biological Conservation*, 201, 92–102.

Graham, S. P. 2013. How frequently do cottonmouths (*Agkistrodon piscivorus*) bask in trees? *Journal of Herpetology*, 47, 428–431.

Grasse, J. E. 1951. Beaver ecology and management in the rockies. *Journal of Forestry & Estates Management*, 49, 3–6.

Green, H. U. 1932. Observations on the occurrence of otter in the Riding Mountain National Park, Manitoba, in relation to beaver life. *Canadian Field Naturalist*, 46, 204–206.

Green, K. C., and Westbrook, C. J. 2009. Changes in riparian area structure, channel hydraulics, and sediment yield following loss of beaver dams. *Journal of Ecosystems and Management*, 10(1).

Green, K. C. and Westbrook, C. J. 2009. Changes in riparian area structure, channel hydraulics, and sediment yield following loss of beaver dams. *Journal of Ecosystems and Management*, 10, 68–79.

Greer, K. R. 1955. Yearly food habits of the river otter in the Thompson Lakes region, northwestern Montana, as indicated by scat analyses. *American Midland Naturalist*, 54, 299–313.

Gregory, A. S., Ritz, K., McGrath, S. P., Quinton, J. N., Goulding, K. W. T., Jones, R. J. A., Harris, J. A., Bol, R., Wallace, P., Pilgrim, E. S., and Whitmore, A. P. 2015. A review of the impacts of degradation threats on soil properties in the UK. *Soil Use and Management*, 31, 1–15.

Grosse, G., Harden, J., Turetsky, M., McGuire, A. D., Camill, P., Tarnocai, C., Frolking, S., Schuur, E. A. G., Jorgenson, T., Marchenko, S., Romanovsky, V., Kimberly P. Wickland, K. P., French, N., Waldrop, M., Bourgeau-Chavez, L., and Striegl, R. G. 2011. Vulnerability of high-latitude soil organic carbon in North America to disturbance. *Journal of Geophysical Research*, 116, G00K06.

Grover, A. M. and Baldassarre, G. A. 1995. Bird species richness within beaver ponds in south-central New York. *Wetlands*, 15, 108–118.

Günther, R. 1996. *Die Amphibien und Reptilien Deutschlands*. Jena: Gustav Fischer.

Gurnell, A. M. 1998. The hydrogeomorphological effects of beaver dam-building activity. *Progress in Physical Geography*, 22, 167–189.

Gurnell, J., Demeritt, D., Lurz, P. W. W., Shirley, M. D. F., Rushton, S. P., Faulkes, C. G., Nobert, S., and Hare, E. J. 2008. *The feasibility and acceptability of reintroducing the European beaver to England. Report prepared for Natural England and the People's Trust for Endangered Species*. York and London: Natural England and People's Trust for Endangered Species.

Hägglund, A. and Sjöberg, G. 1999. Effects of beaver dams on the fish fauna of forest streams. *Forest Ecology and Management*, 115, 259–266.

Ham, J., Toran, L., and Cruz, J. 2006. Effect of upstream ponds on stream temperatures. *Environmental Geology*, 50, 55–61.

Harmon, M.E., Franklin, J.F., Swanson, F.J., Sollins, P., Gregory, S.V., Lattin, J.D., Anderson, N.H., Cline, S.P., Aumen, N.G., Sedell, J.R., et al. 1986. Ecology of coarse woody debris in temperate ecosystems. *Adv Ecol Res*, 15,133–302.

Harthun, M. 1999. The influence of the European beaver (*Castor fiber albicus*) on the biodiversity (Odonata, Mollusca, Trichoptera, Ephemeroptera, Diptera) of brooks in Hesse (Germany). *Limnologica*, 29, 449–464.

Harthun, M. 2000. Influence of the damming-up by beavers (*Castor fiber albicus*) on physical and chemical parameters of highland brooks (Hesse, Germany). *Limnologica*, 30, 21–35.

Hartman, G. and Törnlöv, S. 2006. Influence of watercourse depth and width on dam-building behaviour by Eurasian beaver (*Castor fiber*). *Journal of Zoology*, 268, 127–131.

Hastie, L. C. and Toy, K. A. 2008. Changes in density, age structure and age-specific mortality in two western pearlshell (*Margaritifera falcata*) populations in Washington

(1995–2006). *Aquatic Conservation: Marine and Freshwater Ecosystems*, 18, 671–678.

Hering, D., Gerhard, M., Kiel, E., Ehlert, T. and Pottgiesser, T. 2001. Review study on near-natural conditions of Central European mountain streams, with particular reference to debris and beaver dams: results of the "REG Meeting" 2000. *Limnologica*, 31, 81–92.

Hey, D. L. and Philippi, N. S. 1995. Flood reduction through wetland restoration: the upper Mississippi River basin as a case history. *Restoration Ecology*, 3, 4–17.

Hilfiker, E. L. 1991. *Beavers, Water, Wildlife and History*. Interlaken, NY: Windswept Press.

Hobbs, N. T. 1996. Modification of ecosystems by ungulates. *Journal of Wildlife Management*, 60, 695–713.

Hodkinson, I. D. 1975a. A community analysis of the benthic insect fauna of an abandoned beaver pond. *Journal of Animal Ecology*, 44, 533–551.

Hodkinson, I. D. 1975b. Energy flow and organic matter decomposition in an abandoned beaver pond ecosystem. *Oecologia*, 21, 131–139.

Hoffman, M., Hilton-Taylor, C., Angulo, A., Böhm, M., and Brooks, T. M. 2010. The impacts of conservation on the status of the world's vertebrates. *Science*, 330, 1503–1509.

Hood, G. and Bayley, S. 2008. The effects of high ungulate densities on foraging choices by beaver (*Castor canadensis*) in the mixed-wood boreal forest. *Canadian Journal of Zoology*, 86, 484–496.

Hood, G. A. and Larson, D. G. 2014. Beaver-created habitat heterogeneity influences aquatic invertebrate assemblages in boreal Canada. *Wetlands*, 34, 19–29.

Hood, W. G. 2012. Beaver in tidal marshes: dam effects on low-tide channel pools and fish use of estuarine habitat. *Wetlands*, 32, 401–410.

Hunter, H. C. 1976. *Proposed beaver removal in Cottonwood Creek. Environmental Analysis Report*. US Forest Service, Inyo National Forest, White Mountain Ranger District, California, 7 May.

Hyvönen, T. and Nummi, P. 2008. Habitat dynamics of beaver *Castor canadensis* at two spatial scales. *Wildlife Biology*, 14, 302–308.

Hyvönen, T. and Nummi, P. 2011. Plant succession in beaver patches during and after flooding. In: Sjöberg, G. and Ball, J. P. (eds.) *Restoring the European Beaver: 50 Years of Experience*. Sofia: Pensoft Publishers.

Iason, G. R., Sim, D. A., Brewer, M. J., and Moore, B. D. 2014. *The Scottish Beaver Trial: Woodland Monitoring 2009–2013, Final Report*. Inverness: Scottish Natural Heritage.

Jakobi, S. R. 2017. *Birds, Bats, Bugs, Beavers, Bacteria: Lessons from Nature*. North Charleston, SC: CreateSpace.

Jakober, M. J., McMahon, T. E., Thurow, R. F., and Clancy, C. G. 1998. Role of stream ice on fall and winter movements and habitat use by bull trout and cutthroat trout in Montana headwater streams. *Transactions of the American Fisheries Society*, 127, 223–235.

Janiszewski, P. A. W. E. Ù., Hanzal, V., and Misiukiewicz, W. 2014. The Eurasian beaver (*Castor fiber*) as a keystone species—a literature review. *Baltic Forestry*, 20, 277–286.

Janzen, K. and Westbrook, C. J. 2011. Hyporheic flows along a channelled peatland: influence of beaver dams. *Canadian Water Resources Journal*, 36, 331–347.

Jeffries, R., Darby, S.E., and Sear, D.A. 2003. The influence of vegeta- tion and organic debris on flood-plain sediment dynam- ics: case study of a low-order stream in the New Forest, England. Geomorphology, 51,61–80.

Johnston, C. A. 2014. Beaver pond effects on carbon storage in soils. *Geoderma*, 213, 317–378.

Johnston, C. A. and Naiman, R. J. 1987. Boundary dynamics at the aquatic–terrestrial interface: the influence of beaver and geomorphology. *Landscape Ecology*, 1, 47–57.

Johnston, C. A. and Naiman, R. J. 1990. Browse selection by beaver: effects on riparian forest composition. *Canadian Journal of Forest Research*, 20, 1036–1043.

Johnston, C. A., Pinay, G., Arens, C., and Naiman, R. J. 1995. Influence of soil properties on the biogeochemistry of a beaver meadow hydrosequence. *Soil Science Society of America Journal*, 59, 1789–1799.

Jones, C. G., Lawton, J. H., and Shachak, M. 1994. Organisms as ecosystem engineers. *Oikos*, 69, 373–386.

Kammerer, B. D. and Heppell, S. A. 2012. Individual condition indicators of thermal habitat quality in field populations of redband trout (*Oncorhynchus mykiss gairdnei*). *Environmental Biology Fish*, 96, 823–835.

Karraker, N. E. and Gibbs, J. P. 2009. Amphibian production in forested landscapes in relation to wetland hydroperiod: a case study of vernal pools and beaver ponds. *Biological Conservation*, 142, 2293–2302.

Kauffman, J. B. and Krueger, W. C. 1984. Livestock impacts on riparian ecosystems and streamside management implications: a review. *Journal of Range Management*, 37, 430–438.

Keast, A. and Fox, M. G. 1990. Fish community structure, spatial distribution and feeding ecology in a beaver pond. *Environmental Biology of Fishes*, 27, 201–214.

Kemp, P. S., Worthington, T. A., Langford, T. E., Tree, A. R., and Gaywood, M. J. 2012. Qualitative and quantitative effects of reintroduced beavers on stream fish. *Fish and Fisheries*, 13, 158–181.

Kesminas, V., Steponenas, A., Piluraite, V., and Virbickas, T. 2013. Ecological impact of Eurasian beaver (*Castor fiber*) activity on fish communities in Lithuanian trout streams. *Rocznik Ochrona Srodowika*, 15, 59–80.

Kindschy, R. R. 1989. Regrowth of willow following simulated beaver cutting. *Wildlife Society Bulletin*, 17, 290–294.

Klotz, R. L. 2010. Reduction of high nitrate concentrations in a central New York state stream impounded by beaver. *Northeastern Naturalist*, 17, 349–356.

Knudsen, G. J. 1962. *Relationship of beavers to forests, trout and wildlife in Wisconsin*. Technical Bulletin. Madison: Wisconsin Conservation Department.

Kondratieff, M. C. and Myrick, C. A. 2006. How high can brook trout jump? A laboratory evaluation of brook trout jumping performance. *Transactions of the American Fisheries Society*, 135, 361–370.

Kroes, D. E. and Bason, C. W. 2015. Sediment-trapping by beaver ponds in streams of the Mid-Atlantic Piedmont and Coastal Plain, USA. *Southeastern Naturalist*, 14, 577–595.

Krylov, A. 2008. Impact of the activities of beaver on the zooplankton of a Piedmont river (Mongolia). *Inland Water Biology*, 1, 73–75.

Krylov, A. V., Chalova, I. V., Lapeeva, N. S., Tselmovich, O. L., Roma-nenko, A. V., and Lavrov, V. L. (2016) Experimental studies of the effect of beaver (*Castor fiber* L.) vital activity products on the formation of zooplankton structure (by the example of growth of two cladoceran species of different sizes). *Contemporary Problems of Ecology*, 9(4), 494–502.

Kuczynski, E. C., Paszkowski, C. A., and Gingras, B. A. 2012. Horned grebe habitat use of constructed wetlands in Alberta, Canada. *Journal of Wildlife Management*, 76, 1694–1702.

Kuuluvainen, T. and Aakala, T. 2011. Natural forest dynamics in boreal Fennoscandia: a review and classification. *Silva Fennica*, 45, 823–841.

Lannoo, M. J. 2012. A perspective on amphibian conservation in the United States. *Alytes*, 29, 133–144.

Law, A., Jones, K. C., and Willby, N. J. 2014. Medium vs. short-term effects of herbivory by Eurasian beaver on aquatic vegetation. *Aquatic Botany*, 116, 27–34.

Law, A., McLean, F., and Willby, N. J. 2016. Habitat engineering by beaver benefits aquatic biodiversity and ecosystem processes in agricultural streams. *Freshwater Biology*, 61, 486–499.

Law, A., Gaywood, M. J., Jones, K. C., Ramsay, P., and Willby, N. J. 2017. Using ecosystem engineers as tools in habitat restoration and rewilding: beaver and wetlands. *Science of the Total Environment*, 605, 1021–1030.

LeBlanc, F. A., Gallant, D., Vasseur, L., and Leger, L. 2007. Unequal summer use of beaver ponds by river otters: influence of beaver activity, pond size, and vegetation cover. *Canadian Journal of Zoology*, 85, 774–782.

Leopold, L. B., Wolman, M. G., and Miller, J. P. 1964. *Fluvial Processes in Geomorphology*. San Francisco: W.H. Freeman and Company.

Lesica, P. and Miles, S. 2004. Beavers indirectly enhance the growth of Russian olive and tamarisk along eastern Montana rivers. *Western North American Naturalist*, 64, 12.

Levine, R. and Meyer, G. A. 2014. Beaver dams and channel sediment dynamics on Odell Creek, Centennial Valley, Montana, USA. *Geomorphology*, 205, 51–64.

Levine, R. and Meyer, G. A. 2019. Beaver-generated disturbance extends beyond active dam sites to enhance stream morphodynamics and riparian plant recruitment. *Scientific Reports*, 9, 8124.

Lingo, H. A. 2013. Beaver reintroduction correlates with spotted frog population restoration and terrestrial patterns of the newly metamorphosed Columbia spotted frog in the Owyhee uplands of southwestern Idaho. MSc thesis, Boise State University.

Lokteff, R. L., Roper, B. B., and Wheaton, J. M. 2013. Do beaver dams impede the movement of trout? *Transactions of the American Fisheries Society*, 142, 1114–1125.

Longcore, J. R., McAuley, D. G., Pendelton, G. W., Bennatti, C. R., Mingo, T. M., and Stromborg, K. L. 2006. Macroinvertebrate abundance, water chemistry, and wetland characteristics affect use of wetland by avian species in Maine. *Hydrobiologia*, 567, 143–167.

Lowry, M. M. 1993. Groundwater elevations and temperature adjacent to a beaver pond in central Oregon. MSc thesis, Oregon State University.

Macfarlane, W. W., Wheaton, J. M., Bouwes, N., Jensen, M. L., Gilbert, J. T., Hough-Snee, N., and Shivik, J. A. 2017. Modeling the capacity of riverscapes to support beaver dams. *Geomorphology*, 277, 72–99.

MacFarlane, W. A., and Wohl, E. 2003. Influence of step composition on step geometry and flow resistance in step-pool streams of the Washington Cascades. Water Resources Research, 39(2).

Majerova, M., Neilson, B. T., Schmadel, N. M., Wheaton, J. M., and Snow, C. J. 2015. Impacts of beaver dams on hydrologic and temperature regimes in a mountain stream. *Hydrology and Earth System Sciences*, 12, 839–878.

Makarevich, T.A., Belous, V.V., and Gurchunova, T.A. 2016. Algoflora of oxbow ponds transformed with beavers' activity. Lake ecosystems: biological processes, anthropogenic transformation, water quality. *Proceedings of Materials of the 5th International Scientific Conference*, 12–17 September, Minsk, Naroch, pp. 160–162.

Malison, R. L. and Halley, D. J. 2020. Ecology and movement of juvenile salmonids in beaver-influenced and beaver-free tributaries in the Trøndelag province of Norway. *Ecology of Freshwater Fish*, 29(4), 623–639.

Malison, R. L., Eby, L. A., and Stanford, J. A. 2015. Juvenile salmonid growth, survival, and production in a large river floodplain modified by beavers (*Castor canadensis*). *Canadian Journal of Fisheries and Aquatic Sciences*, 72, 1639–1651.

Malison, R. L., Kuzishchin, K. V., and Stanford, J. A. 2016. Do beaver dams reduce habitat connectivity and salmon productivity in expansive river floodplains? *PeerJ*, 4, e2403.

Maret, T. J., Parker, M., and Fannin, T. E. 1987. The effect of beaver ponds on the nonpoint source water-quality of a stream in southwestern Wyoming. *Water Research*, 21, 263–268.

Maret, T. J., Snyder, J. D., and Collins, J. P. 2006. Altered drying regime controls distribution of endangered salamanders and introduced predators. *Biological Conservation*, 127, 129–138.

Marshall, K. N., Hobbs, N. T., and Cooper, D. J. 2013. Stream hydrology limits recovery of riparian ecosystems after wolf reintroduction. *Proceedings of the Royal Society B: Biological Sciences*, 280, 20122977.

Martell, K. A., Foote, A. L., and Cumming, S. G. 2006. Riparian disturbance due to beavers (*Castor canadensis*) in Alberta's boreal mixed wood forests: implications for forest management. *Ecoscience*, 13, 164–171.

Martínez Pastur, G., Lencinas, M. V., Escobar, J., Quiroga, P., Malmierca, L., and Lizarralde, M. 2006. Understorey succession in Nothofagus forests in Tierra del Fuego (Argentina) affected by *Castor canadensis*. *Applied Vegetation Science*, 9, 143–154.

Martinsen, G. D., Driebe, E. M., and Whitham, T. G. 1998. Indirect interactions mediated by changing plant chemistry: beaver browsing benefits beetles. *Ecology*, 79, 192–200.

McCaffery, M. and Eby, L. 2016. Beaver activity increases aquatic subsidies to terrestrial consumers. *Freshwater Biology*, 61, 518–532.

McCall, T. C., Hodgman, T. P., Diefenbach, D. R., and Owen, R. B. 1996. Beaver populations and their relation to wetland habitat and breeding waterfowl in Maine. *Wetlands*, 16, 163–172.

McColley, S. D., Tyers, D. B., and Sowell, B. F. 2012. Aspen and willow restoration using beaver on the northern Yellowstone winter range. *Restoration Ecology*, 20, 450–455.

McDowell, D. M. and Naiman, R. J. 1986. Structure and function of a benthic invertebrate stream community as influenced by beaver (*Castor canadensis*). *Oecologia*, 68, 481–489.

McGinley, M. A. and Whitham, T. G. 1985. Central place foraging by beavers (*Castor canadensis*): a test of foraging predictions and the impact of selective feeding on the growth form of cottonwoods (*Populus fremontii*). *Oecologia*, 66, 558–562.

McKenna, G., Keys, M., van Meer, T., Roy, L., and Sawatsky, L. 2000. The impact of beaver dams on the design and construction of reclaimed mine sites. *Proceedings of the 24th Annual British Columbia Mine Reclamation Symposium*, Williams Lake, British Columbia.

McKinstry, M. C., Caffrey, P., and Anderson, S. H. 2001. The importance of beaver to wetland habitats and waterfowl in Wyoming. *Journal of the American Resources Association*, 37, 1571–1577.

McRae, G. and Edwards, C. J. 1994. Thermal characteristic of Wisconsin headwater streams occupied by beaver: implications for brook trout habitat. *Transactions of the American Fisheries Society*, 123, 641–656.

Medin, D. E. and Clary, W. P. 1991. *Small Mammals of Beaver Pond Ecosystem and Adjacent Riparian Habitat in Idaho*. Research Paper INT-445. Ogden, UT: US Department of Agriculture, Forest Service, Intermountain Research Station, 4.

Meentemeyer, R. K. and Butler, D. R. 1999. Hydrogeomorphic effects of beaver dams in Glacier National Park, Montana. *Physical Geography*, 20, 436–446.

Meentemeyer, R. K., Vogler, J. B., Hill, C., and Butler, D. 1998. The geomorphic influences of burrowing beavers on streambanks, Bolin Creek, North Carolina. *Zeitschrift fur Geomorphologie*, 42, 453–468.

Melquist, W. E. and Hornocker, M. G. 1983. Ecology of river otters in west-central Idaho. *Wildlife Monographs*, 83, 60.

Menzel, M. A., Carter, T. C., Ford, W. M., and Chapman, B. R. 2001. Tree-roost characteristics of subadult and female adult evening bats (*Nycticeius humeralis*) in the upper coastal plain of South Carolina. *American Midland Naturalist*, 145, 112–119.

Metts, B. S., Lanham, J. D., and Russell, K. R. 2001. Evaluation of herpetofaunal communities on upland streams in the Upper Piedmont of South Carolina. *American Midland Naturalist*, 145, 54–65.

Mishin, A. S. and Trenkov, I. P. 2016. Dry beaver ponds as habitats attracting large mammals. *Russian Journal of Theriology*, 15, 75–77.

Mitchell, S. C. and Cunjak, R. A. 2007. Streamflow, salmon and beaver dams: roles in the structuring of stream fish communities within an anadromous salmon dominated stream. *Journal of Animal Ecology*, 76, 1062–1074.

Moayeri, M. 2013. Reconstructing the summer diet of wolves in a complex multi-ungulate system in northern Manitoba, Canada. MSc thesis, University of Manitoba.

Montgomery, D. R., Abbe, T. B., Buffington, J. M., Peterson, N. P., Schmidt, K. M., & Stock, J. D. 1996. Distribution of bedrock and alluvial channels in forested mountain drainage basins. *Nature*, 381(6583), 587–589.

Morrison, C. A. 2003. *Running the River: Poleboats, Steamboats and Timber Rafts on the Altamaha, Ocmulgee, Oconee and Ohoopee*. St Simons Island, GA: Saltmarsh Press.

Mott, C. L., Bloomquist, C. K., and Nielsen, C. K. 2013. Within-lodge interactions between two ecosystem engineers, beavers (*Castor canadensis*) and muskrats (*Ondatra zibethicus*). *Behaviour*, 150, 1325–1344.

Müller-Schwarze, D. 2011. *The Beaver: Its Life and Impact*, 2nd edn. New York: Cornell University Press.

Murphy, M. L., Heifetz, J., Thedinga, J. F., Johnson, S. W., and Koski, K. V. 1989. Habitat utilization by juvenile pacific salmon (*Onchorynchus*) in the glacial Taku River, southeast Alaska. *Canadian Journal of Fisheries and Aquatic Sciences*, 46, 1677–1685.

Nagayama, S., Nakamura, F., Kawaguchi, Y., Nakano, D. 2012. Effects of configuration of instream wood on autumn and winter habitat use by fish in a large remeandering reach. *Hydrobiologia*, 680, 159–170.

Naiman, R. J. and Melillo, J. M. 1984. Nitrogen budget of a subarctic stream altered by beaver (*Castor canadensis*). *Oecologia*, 62, 150–155.

Naiman, R. J., Melillo, J. M., and Hobbie, J. E. 1986. Ecosystem alteration of boreal forest streams by beaver (*Castor canadensis*). *Ecology*, 67, 1254–1269.

Naiman, R. J., Johnston, C. A., and Kelley, J. C. 1988. Alteration of North American streams by beaver. *BioScience*, 38, 753–762.

Naiman, R. J., Manning, T., and Johnston, C. A. 1991. Beaver population fluctuations and tropospheric methane emissions in boreal wetlands. *Biogeochemistry*, 12, 1–15.

Naiman, R. J., Pinay, G., Johnston, C. A., and Pastor, J. 1994. Beaver influences on the long-term biogeochemical characteristics of boreal forest drainage networks. *Ecology*, 75, 905–921.

Neff, D. J. 1957. Ecological effects of beaver habitat abandonment in the Colorado Rockies. *Journal of Wildlife Management*, 21, 80–84.

Newsome, T. M., Boitani, L., Chapron, G., Ciucci, P., Dickman, C. R., Dellinger, J. A., López-Bao, J. V., Peterson, R. O., Shores, C. R., and Wirsing, A. J. 2016. Food habits of the world's grey wolves. *Mammal Review*, 46, 255–269.

Nolet, B. A., Hoekstra, A., and Ottenheim, M. M. 1994. Selective foraging on woody species by the beaver *Castor fiber*, and its impact on a riparian willow forest. *Biological Conservation*, 70, 117–128.

Northcott, T. H. 1971. Feeding habits of beaver in Newfoundland. *Oikos*, 22, 407–410.

Nummi, P. 1989. Simulated effects of the beaver on vegetation, invertebrates and ducks. *Annales Zoologici Fennici*, 26, 43–52.

Nummi, P. and Hahtola, A. 2008. The beaver as an ecosystem engineer facilitates teal breeding. *Ecography*, 31, 519–524.

Nummi, P. and Holopainen, S. 2014. Whole-community facilitation by beaver: ecosystem engineer increases waterbird diversity. *Aquatic Conservation: Marine and Freshwater Ecosystems*, 24, 623–633.

Nummi, P. and Kuuluvainen, T. 2013. Forest disturbance by an ecosystem engineer: beaver in boreal forest landscapes. *Boreal Environmental Research Supplement A*, 13–24.

Nummi, P. and Pöysä, H. 1995. Habitat use by different-aged duck broods and juvenile ducks. *Wildlife Biology*, 1, 181–187.

Nummi, P., Kattainen, S., Ulander, P., and Hahtola, A. 2011. Bats benefit from beavers: a facilitative link between aquatic and terrestrial food webs. *Biodiversity and Conservation*, 20, 851–859.

Nummi, P., Vehkaoja, M., Pumpanen, J., and Ojala, A. 2018. Beavers affect carbon biogeochemistry: both short-term and long-term processes are involved. *Mammal Review*, 48, 298–311.

Nummi, P., Liao, W., Huet, O., Scarpulla, E., and Sundell, J. 2019. The beaver facilitates species richness and abundance of terrestrial and semi-aquatic mammals. *Global Ecology and Conservation*, 20, e00701.

O'Connell, P. E., Ewen, J., O'Donnell, G., and Quinn, P. 2007. Is there a link between agricultural land-use management and flooding? Hydrology and Earth System Sciences Discussions. *European Geosciences Union*, 11, 96–107.

Painter, K. J., Westbrook, C. J., Hall, B. D., O'Driscoll, N. J., and Jardine, T. D. 2015. Effects of in-channel beaver impoundments on mercury bioaccumulation in Rocky Mountain stream food webs. *Ecosphere*, 6, 1–17.

Parker, A. 2013. *Miramichi Salmon Association Conservation Field Program Report 2013*. South Esk, New Brunswick: Miramichi Salmon Association.

Parker, H. and Rønning, Ø. C. 2007. Low potential for restraint of anadromous salmonid reproduction by beaver *Castor fiber* in the Numedalslågen River catchment, Norway. *River Research and Applications*, 23, 752–762.

Parker, J. D., Caudill, C. C., and Hay, M. E. 2007. Beaver herbivory on aquatic plants. *Oecologia*, 151, 616–625.

Peinetti, H. R., Baker, B. W., and Coughenour, M. B. 2009. Simulation modeling to understand how selective foraging by beaver can drive the structure and function of a willow community. *Ecological Modelling*, 220, 998–1012.

Pereira, H. M. and Navarro, L. M. 2015. *Rewilding European Landscapes*. Cham: Springer.

Perkins, T. E. and Wilson, M. V. 2005. The impacts of *Phalaris arundinacea* (reed canarygrass) invasion on wetland plant richness in the Oregon Coast Range, USA depend on beavers. *Biological Conservation*, 124, 291–295.

Petro, V. M., Taylor, J. D., and Sanchez, D. M. 2015. Evaluating landowner-based beaver relocation as a tool to restore salmon habitat. *Global Ecology and Conservation*, 3, 477–486.

Pilliod, D. S., Rohde, A. T., Charnley, S., Davee, R. R., Dunham, J. B., Gosnell, H., Grant, G. E., Hausner, M. B., Huntington, J. L., and Nash, C. 2018. Survey of beaver-related restoration practices in rangeland streams of the western USA. *Environmental Management*, 61, 58–68.

Polecha, P. J. 1989. More evidence of a commensal relationship between the river otter and the beaver. In: Amlaner, C. J. (ed.) *Proceedings of the 10th International Symposium on Biotelemetry*. Fayetteville: University of Arkansas Press, pp. 217–236.

Pollock, M. M., Naiman, R. J., Erickson, H. E., Johnston, C. A., Pastor, J., and Pinay, G. 1995. *Beaver as Engineers: Influences on Biotic and Abiotic Characteristics of Drainage Basins. Linking Species & Ecosystems*. Boston: Springer.

Pollock, M. M., Naiman, R. J., and Hanley, T. A. 1998. Plant species richness in riparian wetlands—a test of biodiversity theory. *Ecology*, 79, 94–105.

Pollock, M. M., Heim, M., and Werner, D. 2003. Hydrologic and geomorphic effects of beaver dams and their influence on fishes. *Proceedings of the American Fisheries Society Symposium 37*, Bethesda, Maryland, pp. 213–233.

Pollock, M. M., Pess, G. R., Beechie, T. J., and Montgomery, D. R. 2004. The importance of beaver ponds to coho salmon production in the Stillaguamish river basin, Washington, USA. *North American Journal of Fisheries Management*, 24, 749–760.

Pollock, M. M., Beechie, T. J., and Jordan, C. E. 2007. Geomorphic changes upstream of beaver dams in Bridge Creek, an incised stream channel in the interior Columbia River basin, eastern Oregon. *Earth Surface Processes and Landforms*, 32, 1174–1185.

Pollock, M. M., Wheaton, J. M., Bouwes, N., Volk, C., Weber, N., and Jordan, C. E. 2012. *Working with beaver to restore salmon habitat in the Bridge Creek intensively monitored watershed: design rationale and hypotheses.* NOAA Technical Memorandum NMFS-NWFSC-120. Washington, DC: NOAA.

Pollock, M. M., Beechie, T. J., Wheaton, J. M., Jordan, C. E., Bouwes, N., Weber, N., and Volk, C. 2014. Using beaver dams to restore incised stream ecosystems. *BioScience*, 64, 279–290.

Popescu, V. D. and Gibbs, J. P. 2009. Interactions between climate, beaver activity, and pond occupancy by the cold-adapted mink frog in New York state, USA. *Biological Conservation*, 142, 2059–2068.

Puplett, D. 2008. Beavers and aspen: looking to the future. In: Parrott, J. and MacKenzie, N. (eds.) *Aspen in Scotland: Biodiversity and Management Highland Aspen Group. Proceedings of Conference, 3–4 October*, Boat of Garten, Scotland, pp. 27–30.

Purseglove, J. 2017. *Taming the Flood: Rivers, Wetlands and the Centuries-old Battle Against Flooding.* William Collins, Oxford.

Puttock, A., Graham, H. A., Cunliffe, A. M., Elliott, M., and Brazier, R. E. 2017. Eurasian beaver activity increases water storage, attenuates flow and mitigates diffuse pollution from intensively managed grasslands. *Science of the Total Environment*, 576, 430–443.

Puttock, A., Graham, H. A., Carless, D., and Brazier, R. E. 2018. Sediment and nutrient storage in a beaver engineered wetland. *Earth Surface Processes and Landforms*, 43, 2358–2370.

Quinn, T. P. and Peterson, N. P. 1996. The influence of habitat complexity and fish size on over-winter survival and growth of individually marked juvenile coho salmon (*Oncorhynchus kisutch*) in Big Beef Creek, Washington.

Canadian Journal of Fisheries and Aquatic Sciences, 53, 1555–1564.

Ray, A. M., Rebertus, A. J., and Ray, H. L. 2001. Macrophyte succession in Minnesota beaver ponds. *Canadian Journal of Botany*, 79, 487–499.

Ray, H. L., Ray, A. M., and Rebertus, A. J. 2004. Rapid establishment of fish in isolated peatland beaver ponds. *Wetlands*, 24, 399–405.

Rea, R. and Lindgren, B. 2009. Beaver labor serves up spring sweets for butterflies. *Wildlife Afield*, 6, 170–172.

Reddoch, J. M. and Reddoch, A. H. 2005. Consequences of beaver, *Castor canadensis*, flooding on a small shore fen in southwestern Quebec. *Canadian Field-Naturalist*, 119, 385–394.

Reese, K. P. and Hair, J. D. 1976. Avian species diversity in relation to beaver pond habitats in the Piedmont region of South Carolina. *Proceedings of the Annual Conference of the Southeastern Association of Fish and Wildlife Agencies*, Jackson, Mississippi, pp. 437–447.

Reid, D. G., Code, T. E., Reid, A. C. H., and Herrero, S. M. 1994. Food habits of the river otter in a boreal ecosystem. *Canadian Journal of Zoology*, 72, 1306–1313.

Reid, J. 2012. *Miramichi Salmon Association Conservation Field Programme Report 2011.* South Esk, New Brunswick: Miramichi Salmon Association.

Remillard, M. M., Gruendling, G. K., and Bogucki, D. J. 1987. Disturbance by beaver (*Castor canadensis* Kuhl) and increased landscape heterogeneity. In: Turner, M. G. (ed.) *Landscape Heterogeneity and Disturbance*. New York: Springer.

Ripple, W. J. and Beschta, R. L. 2012. Trophic cascades in Yellowstone: the first 15 years after wolf reintroduction. *Biological Conservation*, 145, 205–213.

Rodnick, K., Gamperl, A., Lizars, K., Bennett, M., Rausch, R., and Keeley, E. 2004. Thermal tolerance and metabolic physiology among redband trout populations in southeastern Oregon. *Journal of Fish Biology*, 64, 310–335.

Rolauffs, P., Hering, D., and Lohse, S. 2001. Composition, invertebrate community and productivity of a beaver dam in comparison to other stream habitat types. *Hydrobiologia*, 459, 201–212.

Rosell, F. and Czech, A. 2000. Responses of foraging Eurasian beavers *Castor fiber* to predator odours. *Wildlife Biology*, 6, 13–21.

Rosell, F. and Hovde, B. 1998. Pine marten, *Martes martes*, as a Eurasian beaver, *Castor fiber*, lodge occupant and possible predator. *Canadian Field-Naturalist*, 112, 535–536.

Rosell, F. and Sanda, J. 2006. Potential risks of olfactory signaling: the effect of predators on scent marking by beavers. *Behavioral Ecology*, 17, 897–904.

Rosell, F., Bozser, O., Collen, P., and Parker, H. 2005. Ecological impact of beavers *Castor fiber* and *Castor canadensis* and their ability to modify ecosystems. *Mammal Review*, 35, 248–276.

Rosenberg, D. M. and Resh, V. H. 1993. *Freshwater Biomonitoring and Benthic Macroinvertebrates.* New York: Chapman and Hall.

Runyon, M. J., Tyers, D. B., Sowell, B. F., and Gower, C. N. 2014. Aspen restoration using beaver on the northern Yellowstone winter range under reduced ungulate herbivory. *Restoration Ecology*, 22, 555–561.

Russell, K. R., Moorman, C. E., Edward, J. K., Metts, B. S., and Guynn, D. C. J. 1999. Amphibian and reptile communities associated with beaver (*Castor canadensis*) ponds and unimpounded streams in the Piedmont of South Carolina. *Journal of Freshwater Ecology*, 14, 149–158.

Ruys, T., Lorvelec, O., Marre, A., and Bernez, I. 2011. River management and habitat characteristics of three sympatric aquatic rodents: common muskrat, coypu and European beaver. *European Journal of Wildlife Research*, 57, 851–864.

Saarenmaa, H. 1978. The occurrence of bark beetles (*Col. scolytidae*) in a dead spruce stand flooded by beavers (*Castor canadensis* Kuhl). *Silva Fennica*, 12, 201–216.

Schindler, D. W. and Donahue, W. F. 2006. An impending water crisis in Canada's western prairie provinces. *Proceedings of the National Academy of Sciences*, 103, 7210–7216.

Schlosser, I. J. 1995. Dispersal, boundary processes, and trophic-level interactions in streams adjacent to beaver ponds. *Ecology*, 76, 908–925.

Sedell, J.R., Bisson, P.A., Swanson, F.J., and Gregory, S.V. 1998. What we know about large trees that fall into streams and rivers. In Maser C, Tarrant RF, Trappe JM, Franklin JF, eds. *From the Forest to the Sea: A Story of Fallen Trees.* USDA Forest Service General Technical Report PNW-GTR-229.

Selva, S. B. 2003. Using calicioid lichens and fungi to assess ecological continuity in the Acadian forest ecoregion of the Canadian Maritimes. *Forestry Chronicle*, 79, 550–558.

Shevchenko, N. S., Chalova, I. V., Tselmovich, O. L., Romanenko, A. V., Sakharova, E. G., and Krylov, A. V. 2017. Effect of the vital activity products of beaver (*Castor fiber* L.) on the formation of zooplankton structure: changes in the quantitative parameters of two cladocera species of different sizes in a beaver pound (in situ experiment). *Inland Water Biology*, 10, 375–383.

Sidorovich, V., Schnitzler, A., Schnitzler, C., Rotenko, I., and Holikava, Y. 2017. Responses of wolf feeding habits after adverse climatic events in central-western Belarus. *Mammalian Biology*, 83, 44–50.

Sigourney, D. B., Letcher, B. H., and Cunjak, R. A. 2006. Influence of beaver activity on summer growth and condition of age-2 Atlantic salmon parr. *Transactions of the American Fisheries Society*, 135, 1068–1075.

Siitonen, J. 2001. Forest management, coarse woody debris and saproxylic organisms: Fennoscandian boreal forests as an example. *Ecological Bulletins*, 49, 11–41.

Sjöberg, K., Pöysä, H., Elmberg, J., and Nummi, P. 2000. Response of mallard ducklings to variation in habitat quality: an experiment of food limitation. *Ecology*, 81, 329–335.

Skelly, D. K. and Freidenburg, L. K. 2000. Effects of beaver on the thermal biology of an amphibian. *Ecology Letters*, 3, 483–486.

Smith, B. E. and Keinath, D. A. 2007. Northern leopard frog (*Rana pipiens*): a technical conservation assessment. Colorado: US Department of Agriculture, Rocky Mountain Region: Forest Service.

Smith, D. W., Peterson, R. O., and Houston, D. B. 2003. Yellowstone after wolves. *BioScience*, 53, 330–340.

Snodgrass, J. W. and Meffe, G. K. 1998. Influence of beavers on stream fish assemblages: effects of pond age and watershed position. *Ecology*, 79, 928–942.

Songster-Alpin, M. and Klotz, R. 1995. A comparison of electron transport system activity in stream and beaver pond sediments. *Canadian Journal of Fisheries and Aquatic Sciences*, 52, 1318–1326.

Spieth, H. T. 1979. The virilis group of *Drosophila* and the beaver *Castor. American Naturalist*, 114, 312–316.

Stevens, C. E., Paszkowski, C. A., and Scrimgeour, G. J. 2006. Older is better: beaver ponds on boreal streams as breeding habitat for the wood frog. *Journal of Wildlife Management*, 70, 1360–1371.

Stevens, C. E., Paszkowski, C. A., and Foote, A. L. 2007. Beaver (*Castor canadensis*) as a surrogate species for conserving anuran amphibians on boreal streams in Alberta, Canada. *Biological Conservation*, 134, 1–13.

Stewart, E. 2020. Beavers Buffering Blazes: The Potential Role of *Castor canadensis* in Mitigating Wildfire Impacts on Stream Ecosystems. https://www.pugetsound.edu/files/resources/stewart-erin.pdf.

Stoffyn-Egli, P. and Willison, J. M. 2011. Including wildlife habitat in the definition of riparian areas: the beaver (*Castor canadensis*) as an umbrella species for riparian obligate animals. *Environmental Reviews*, 19, 479–494.

Stokland, J. N., Tomter, S. M., and Söderberg, U. 2004. Development of dead wood indicators for biodiversity monitoring: experiences from Scandinavia. *Proceedings of Conference, Monitoring and Indicators of Forest Biodiversity in Europe—From Ideas to Operationality*, EFI proceedings no. 51, pp. 207–226.

Stokland, J. N., Siitonen, J., and Jonsson, B. G. 2012. *Biodiversity in Dead Wood.* Cambridge: Cambridge University Press.

Strauss, S. Y. and Agrawal, A. A. 1999. The ecology and evolution of plant tolerance to herbivory. *Trends in Ecology & Evolution*, 14, 179–185.

Stringer, A. P. and Gaywood, M. J. 2016. The impacts of beavers *Castor* spp. on biodiversity and the ecological basis for their reintroduction to Scotland, UK. *Mammal Review*, 46, 270–283.

Suzuki, N. and Mccomb, B. C. 2004. Associations of small mammals and amphibians with beaver-occupied streams in the Oregon Coast Range. *Northwest Science*, 78, 286–293.

Syphard, A. D. and Garcia, M. W. 2001. Human-and beaver-induced wetland changes in the Chickahominy River watershed from 1953 to 1994. *Wetlands*, 21, 342–353.

Taylor, B. R., MacInnis, C., and Floyd, T. A. 2010. Influence of rainfall and beaver dams on upstream movement of spawning Atlantic salmon in a restored brook in Nova Scotia, Canada. *River Research and Applications*, 26, 183–193.

Temple, H. J. and Cox, N. A. 2009. *European Red List of Amphibians*. Luxembourg: Publications Office of the European Union.

Terwilliger, J. and Pastor, J. 1999. Small mammals, ectomycorrhizae, and conifer succession in beaver meadows. *Oikos*, 85, 83–94.

Thompson, S., Vehkaoja, M., and Nummi, P. 2016. Beaver-created deadwood dynamics in the boreal forest. *Forest Ecology and Management*, 360, 1–8.

Thorstad, E. B., Økland, F., Aarestrup, K., and Heggberget, T. G. 2008. Factors affecting the within-river spawning migration of Atlantic salmon, with emphasis on human impacts. *Reviews in Fish Biology and Fisheries*, 18, 345–371.

Trimble, S. W. and Mendel, A. C. 1995. The cow as a geomorphic agent—a critical review. *Geomorphology*, 13, 233–253.

Triska, F. J. 1984. Role of wood debris in modifying channel geomorphology and riparian areas of a large lowland river under pristine conditions: a historical case study. *Internationale Vereinigung für theoretische und angewandte Limnologie: Verhandlungen*, 22, 1876–1892.

Tumlison, C. R., Karnes, M. R., and King, A. W. 1982. River Otter in Arkansas: II. Indications of a beaver-facilitated commensal relationship. *Journal of the Arkansas Academy of Science*, 36, 73–75.

Tyurnin, B. N. 1984. Factors determining numbers of the river beaver (*Castor fiber*) in the European North. *Soviet Journal of Ecology*, 14, 337–344.

Ulevičius, A. and Janulaitis, M. 2007. Abundance and species diversity of small mammals on beaver lodges. *Ekologija*, 53, 38–43.

USDA. 2011. Cooperative Beaver Damage Management Program-Protecting Valued Resources in Wisconsin. USDA-APHIS Wildlife Services.

United States Fish and Wildlife Service (USFWS). 2009. *Arroyo Toad (Bufo californicus), 5-Year Review: Summary and Evaluation*. Ventura, CA: US Fish and Wildlife Service.

Vecherskiy, M., Korotaeva, V., Kostina, N., Dobrovol'skaya, T., and Umarov, M. 2011. Biological activities of 'beaver landscape' soils. *Moscow University Soil Science Bulletin*, 66, 175–179.

Vehkaoja, M., Nummi, P., and Rikkinen, J. 2017. Beavers promote calicioid diversity in boreal forest landscapes. *Biodiversity and Conservation*, 26, 579–591.

Verheijen, F. G. A., Jones, R. A., Rickson, R. J., and Smith, C. J. 2009. Tolerable versus actual soil erosion rates in Europe. *Earth-Science Reviews*, 94, 23–38.

Vogt, B. 1981. What ails the river otter? *National Wildlife*, 22, 8–15.

Voordouw, M. J., Adama, D., Houston, B., Govindarajulu, P., and Robinson, J. 2010. Prevalence of the pathogenic chytrid fungus, *Batrachochytrium dendrobatidis*, in an endangered population of northern leopard frogs, *Rana pipiens*. *BioMed Central Ecology*, 10, 6.

Weber, N., Bouwes, N., Pollock, M. M., Volk, C., Wheaton, J. M., Wathen, G., Wirtz, J., and Jordan, C. E. 2017. Alteration of stream temperature by natural and artificial beaver dams. *PLoS ONE*, 12, e0176313.

Weddeling, K. and Willigalla, C. 2011. Artenzahlen, Vergesell- schaftung und Einfluss der Landnutzung. In: Hachtel, M., Schlüpmann, M., Weddeling, K., Thiesmeier, B., Geiger, A., and Willigalla, C. (eds.) *Handbuch der Amphibien und Reptilien Nordrhein-Westfalens*. Laurenti. www.laurenti.de.

Westbrook, C. J., Cooper, D. J., and Baker, B. W. 2006. Beaver dams and overbank floods influence groundwater–surface water interactions of a Rocky Mountain riparian area. *Water Resources Research*, 42, 1–12.

Westbrook, C. J., Cooper, D., and Baker, B. 2011. Beaver assisted river valley formation. *River Research and Applications*, 27, 247–256.

Westbrook, C. J., Ronnquist, A., and Bedard-Haughn, A. 2020. Hydrological functioning of a beaver dam sequence and regional dam persistence during an extreme rainstorm. *Hydrological Processes*, 34(18), 3726–3737.

Willby, N. J., Law, A., Levanoni, O., Foster, G., and Ecke, F. 2018. Rewilding wetlands: beaver as agents of within-habitat heterogeneity and the responses of contrasting biota. *Philosophical Transactions of the Royal Society B: Biological Sciences*, 373, 20170444.

Windels, S. K. 2017. *Beavers as Engineers of Wildlife Habitat. Beavers: Boreal Ecosystem Engineers*. Cham: Springer.

Windsor, M. L., Hutchinson, P., Hansen, L. P., and Reddin, D. G. 2012. *Atlantic salmon at sea: findings from recent research and their implications for management*. NASCO document CNL (12). Edinburgh: NASCO.

Wissinger, S. A. and Gallagher, L. J. 1999. Beaver pond wetlands in northwest Pennsylvania. Modes of colonization and succession after drought. In: Batzerr, D., Rader, R., and Wissinger, S. (eds.) *Invertebrates in Freshwater Wetlands of North America: Ecology and Management*. New York: John Wiley & Sons.

Wissinger, S. A., Ingmire, S., and Bogo, J. 2001. Invertebrate and plant communities in restored wetlands in Pennsylvania. In: Rader, R. R., Batzer, D. P., and Wissinger, S. A. (eds.) *Biomonitoring and Management of North American Wetlands*. New York: John Wiley & Sons.

Wohl, E. 2013. Landscape-scale carbon storage associated with beaver dams. *Geophysical Research Letters*, 40, 3631–3636.

Wohl, E. 2014. A legacy of absence: wood removal in US rivers. *Progress in Physical Geography*, 38, 637–663.

Wohl, E. 2015. Of wood and rivers: bridging the perception gap. *Wiley Interdisciplinary Reviews: Water*, 2, 167–176.

Wolf, E. C., Cooper, D. J., and Hobbs, N. T. 2007. Hydrologic regime and herbivory stabilize an alternative state in Yellowstone National Park. *Ecological Applications*, 17, 1572–1587.

Wolfe, M. L. 1974. An overview of moose coactions with other animals. *Naturaliste Canadienne*, 101, 437–456.

Wooten, H. H. and Jones, L. A. 1955. *The History of our Drainage Enterprises*. Yearbook of Agriculture. US Department of Agriculture, 84th Congress, 1st Session. Washington, DC: US Department of Agriculture.

Wright, J. P., Jones, C. G., and Flecker, A. S. 2002. An ecosystem engineer, the beaver, increases species richness at the landscape scale. *Oecologia*, 132, 96–101.

Youngston, A. and Hay, D. 1996. *The Lives of Salmon*. Shrewsbury, UK: Swan Hill Press.

Zero, V. H. and Murphy, M. A. 2016. An amphibian species of concern prefers breeding in active beaver ponds. *Ecosphere*, 7, e01330.

CHAPTER 10

Animal management and population monitoring

10.1 Not a typical zoo animal

Some of the first documented holdings of beavers (*Castor spp.*) do not paint a happy existence. Similar to stories of the first gorilla held on show to relentless public inspection at London Zoo in 1887, living on a diet of sausages, beer, and cheese sandwiches (Kawata, 2008), the early captive lives of beavers were no doubt just as miserable, with numerous tales of diets consisting of bread, frequent dog attacks, and 'strange convulsions'. Early captive beavers are often depicted with short lifespans and were occasionally kept as curiosities (Poliquin, 2015). A pair of beavers briefly held captive in parts of the battlements in Cardiff Castle around 1874 made such an impression that they were commemorated by decorative panels in several rooms and feature in the garden fountains, but their reign was short (Coles, 2019). There are also historic records of captive beavers in zoological settings, such as Basel Zoo in 1876 (Dollinger et al., 1999). Transylvanian medieval transcripts (1211–1452) mention how young beavers were caught and bred in captivity for their meat and fur and record specialized beaver hunters and '*hodos*' meaning 'places with a lot of beavers' (Bejenaru et al., 2015). Early writings record beaver kits being easy to tame and being made pets (Martin, 1892), and numerous social media posts humorously document their mischiefs, testing the nutritional value of furniture and determined attempts to dam up doorways with household objects; however, the actual domestication of beavers has never been achieved. In the times of Frederick the Great of Prussia (1712–1788) records of failed attempted beaver domestication exist and, though details are lacking,

early attempts in Russia, Germany, and France are described as failures (Martin, 1892). In North America, the Hudson's Bay Company is said to have learnt from experiences of Native Americans of leaving certain wild families in situ as 'beaver reserves' which were then harvested every 3 years, whilst attempts at using artificial enclosures were unsuccessful (Charlevoix, 1744; Martin, 1892).

Since these times, beavers have been kept in captivity for a variety of reasons—from entertainment, to fur production, as habitat management tools, and for breeding for species restoration projects. Today, much of the knowledge on captive beavers and their requirements comes from wildlife rehabilitators and to a lesser extent zoological collections, so as a result information is often scattered across various websites, reports, and the personal accounts of individuals. Much of our beaver captive and monitoring knowledge has originated and been developed through restoration projects (see Chapter 11), with temporary captive holding facilities often assisting in the restoration process, enabling animals to be sexed, and the formation of compatible pairs/family units for release and health screen, for example the Methow Beaver Project, Washington, USA (Brick and Woodruff, 2019). Common field techniques, such as restraint without sedation, sexing, and sample collection, can be applied in captivity. Both North America and Europe have a long history of reintroducing and translocating beavers to regions where they have been extirpated. Although early projects focused little on genetics, health, sex, animal welfare, and even species origin in some cases, it is a very different story today.

Beavers: Ecology, Behaviour, Conservation, and Management. Frank Rosell and Róisín Campbell-Palmer, Oxford University Press.
© Frank Rosell and Róisín Campbell-Palmer 2022. DOI: 10.1093/oso/9780198835042.003.0010

Currently there are calls for greater consideration of population genetics, health screening, and animal welfare (Halley, 2011; Campbell-Palmer and Rosell, 2015; Campbell-Palmer et al., 2020; 2021; Pollock et al., 2015), as mortality during translocation can still be considerable (McKinstry and Anderson, 2002).

10.2 Captive husbandry requirements

The housing and husbandry requirements for any species should at the very least ensure their 'five freedoms' are met to maintain a minimal welfare standard, and in most countries it is also a legal requirement (Webster, 1994). Freedom from thirst and hunger, freedom from pain, injury, and disease, and freedom from discomfort tend to be the most obvious and easiest to ensure. However, it is worth noting that in larger, more naturalistic enclosures, individual beaver welfare is likely to necessitate additional monitoring requirements, due to their daily activity patterns, semi-aquatic and burrowing lifestyle. The remaining two of the five freedoms, freedom from fear and distress, and freedom to express natural behaviours, tend to be more difficult to assess and provide for (Webster, 1994). As prey species, beavers may not display obvious fear-associated behaviours—freezing, inactivity, lethargy, and weight loss are some typical stress responses which can be subtle to assess. Previous chapters describe the range and extent of behaviours beavers will display, including territoriality (Chapter 7), building constructions, and hydrological modifications (Chapter 4), which are important considerations in captivity. Whilst no captive setting can fully replicate the wild, beavers have high drives to build, chew, and modify; therefore preventing or significantly reducing this behavioural expression can be viewed as impacting directly on their welfare. Experience would show that detailed studies on the captive needs of beavers and improving their welfare are lacking, with the extent of behavioural enrichment often requiring the provision of browse and differing water regimes. At the other end of the captivity spectrum, they are kept in enclosed natural habitats, where there is no doubt they can express a wider range of behaviours and wild diets, though more often with the loss of an ability to monitor individual welfare, for example administering veterinary care. Given this range in enclosure type, key enclosure requirements are discussed in Section 10.2.1.

Beavers are highly social species within their family groups, exhibiting strong social bonds (Figure 10.1; and see Section 6.3). Therefore, it is not surprising that they want and should be housed together.

Single-housed and/or long-term separation, for example of an injured individual separated from the family for treatment (see Figure 7.15), potentially leads to increased welfare challenges. Single-housed animals have been found dead after spending time alone, and captive facilities have recorded beavers breaking into the pens of others to attack and even fatally wound their neighbours (Wilsson, 1968). Their highly aggressive nature with non-related/non-compatible beavers is notorious and can make animal grouping challenging, though some experts suggest North American beavers are generally less aggressive. Unless establishing a new pair, which should be a careful and supervised process with adequate room for escape, families should be kept separately and not share joint boundaries. Typically, humans will judge an area large enough or a level of habitat quality high enough to support multiple family groups within an enclosed area, where often beavers will perceive this differently and fighting with significant injuries can result. Records of beavers fighting to the death are quoted, but often it is the infection of wounds in captivity that can lead to fatalities, though they can be fairly tolerant to minor wounds and abrasions.

Figure 10.1 Sometimes beavers can be relatively relaxed in captivity! As social rodents, beavers flourish when kept in pairs or family groups, though monitoring of resources and older offspring needs particular management. (Photo supplied courtesy of Christof Angst.)

Beavers do not always thrive in captivity and the reasons may not be obvious. There are numerous stories of them losing weight and their fur becoming dull and lustreless if conditions are not suitable.

Obviously, the aquatic environment is key in beaver husbandry; as potential prey species they feel most secure here and will often enter available water to avoid capture. Beavers defecate, urinate, feed, and drink in water. Therefore, hygiene is crucial, without overcleaning or working in an enclosure, moving quietly and calmly to reduce stress (Campbell-Palmer and Rosell, 2015; Pollock et al., 2015).

10.2.1 Key enclosure requirements

10.2.1.1 Animal grouping

Members of the same family should be housed together; this may include animals from a range of sexes, ages, and litters, as long as they are related and have lived together without periods of separation. Ideally, if trapping/capturing and relocating a family group, the trapping interval between all family members should be kept to a minimum, there have been instances in captivity of infighting or non-acceptance of offspring. Individuals from the same family have been accepted even with trapping intervals of several weeks (Gow, 2002), but this may not always be the case. Additionally, a thorough trapping period should ensure all members of the same family have been trapped over as short a time frame as possible, so that family members are not separated too long, leading to conflicts when they are reunited, but also for wild populations, so that dispersing beavers do not act as mate replacements (Svendsen, 1989; Mayer et al., 2017a). This can occur more readily and even in a matter of days, especially in wild populations at carrying capacity in which vacant suitable habitat is a limited resource. Animals from different families must not be housed together, unless trying to establish a breeding pair. This requires a carefully controlled and staged introduction protocol to ensure that there is only one female and only one male involved, and that there is adequate room for escape should aggression occur.

On the whole, a single male and female of similar size should readily accept each other; usually individuals of dispersing age tend to more readily accept

each other. Wilsson (1968) tells an intriguing story of introducing two unrelated females which did go on to live happily together though their introduction was fraught with concerns. He humorously describes how the resident female reacted to aggressive rejection of her advances of friendship by sealing the new female in the sleeping den with sticks and mud; however, we can conclude that this was a special case rather than the rule. Perhaps the younger female, having been in captivity longer, was craving social companionship and in this case their behavioural characteristics made them more accepting of each other than is typical of two adult females.

Regardless of such stories, if attempting to build up beaver numbers within a single enclosure, a pair should be allowed to breed or a whole family translocated rather than trying to create a group from unrelated animals or trying to add multiple unrelated pairs, as they have shown a strong ability to find and attack each other. Severe fighting between two females previously housed together but then separated for some time has been described (Wilsson, 1968). The morning after, the 'known' but in this case 'new-comer' was found dead, caused by a bite through the spine just above the tail. Fighting wounds may become infected by bacteria (see Section 7.2.3 and Chapter 8), leading to diseases including pneumonia, septicaemia, and bacterial endocarditis. These may result in cardiovascular compromise, dramatic weight loss, and condition loss, which in turn may prove fatal (Saunders, 2016).

Beavers have not commonly been recorded as part of mixed species exhibits in zoos, though instances of sharing with European otter (*Lutra lutra*) and capybaras (*Hydrochoerus hydrochaeris*) have had their concerns (Campbell-Palmer and Rosell, 2015). In larger and more naturalistic enclosures there seems no reason why other wetland species could not be included, and there are numerous reports of enclosed beaver ponds with multiple species utilizing the same ponds, such as breeding storks and large fish species.

10.2.1.2 Shelter

Any enclosure should ensure beavers can live within suitable climatic conditions; therefore shelter provisions should not only ensure appropriate climate control but also function to allow them to retreat from public view (Hosey et al., 2009). Giving animals access to 'privacy' is a more common trend in zoological setting, with use of appropriate screening and within-lodge cameras employed to increase public viewing opportunities (Hosey et al., 2009). In more naturalistic enclosures and with the correct resources (friable substrate and building materials) beavers will quickly construct their own shelters. Artificial lodges or sleeping dens should be provided for beavers when first entering an enclosure if shelter is a concern or especially in more artificial enclosures (Campbell-Palmer and Rosell, 2015). These should be situated near the water's edge, with the opening facing the pool/deepest area of water. Often the easiest form is created by using bales of straw with a weighted-down plywood or metal-sheet roof, providing the beavers with additional browse, and letting them chew, build, and modify. Other facilities have used concrete blocks with a metal-sheet roof as shelter. Breaking open a lodge should be avoided unless necessary. If this has to occur, beavers should return to the same lodge but may be unsettled, so additional lodge-building materials should be provided to allow them to modify and restore their lodge. If a permanent artificial lodge is provided (perhaps with viewing facilities as in some zoos), ideally these should have two to three interconnecting chambers, with each being about 1 m³ to encourage use and allow for breeding. Cameras may be placed in these areas to enable public viewing during visiting times.

10.2.1.3 Substrate

Any substrate should allow for digging and manipulation, to provide exercise and facilitate natural behaviours. This may make effective display to the public difficult as beavers will readily manipulate any movable materials to adapt their environment. Beavers will split any light timber into shreds to form their own bedding materials. Straw or wood chips can also be provided for bedding material. Solid concrete flooring should be avoided as abrasion causes damage to foot pads and tails. Less-naturalistic enclosures, in which planting is not an option, should provide branches and logs regularly to enable beavers to gnaw, feed, and engage in a range of building activities (e.g. Henderson et al., 2015). Beaver constructions should be left in place as long as they do not cause a risk to any animals or keeping staff, or provide means of escape, or present a hygiene risk.

10.2.1.4 Water

Being semi-aquatic beavers should always be provided with suitable access to water (beyond drinking facilities) for swimming, submerging, feeding, and grooming. As a potential prey species beavers will often naturally enter and remain in the water to avoid capture, which should be taken into consideration before planning any catch-up procedure. In captive settings, areas of both deep and shallow water provisions have been recommended, so that a range of behaviours can be achieved and also to maximize public viewing of the animals. For example, at Minnesota Zoological Gardens, the Minnesota Trial presents its captive beaver family in various dry and wet areas, with over- and underwater viewing (Henderson et al., 2015). Depending on enclosure type and water source (e.g. naturally fed, free-flowing systems), filtration systems may be required, with the accumulation of beaver-generated debris and potential attempts to dam up outlets common husbandry experiences. It should be noted that without flowing water and regular cleaning, pools foul quickly.

It is possible to keep beavers out of water for short periods of time, for example, after surgery or whilst in temporary holdings; however, behavioural and ethical consequences must be taken into consideration. Under such circumstances a heavy metal, non-tippable water dish must be provided, as beavers will often attempt to bathe and defecate in any available water and therefore tip water dishes or move them around.

10.2.1.5 Retention

Beavers escapes have generally been linked to poor perimeter fencing, flood events, or a failure to cater for burrowing alongside water inflows or outflows. Collections have also reported beavers climbing over fences after methodically gathering materials and building up against fences when left unchecked. Signs of digging near the fence should be filled in and covered with weld mesh. Any trees within falling distance of the fence line should be protected with tree guards or anti-game paint (see Chapter 11.4.2), removed, or coppiced to ensure beavers do not fell them on to the fence line. Any watercourses entering or leaving an enclosed area, including drainage, should be especially reinforced as beavers tend to

Figure 10.2 Captive conditions for beavers vary greatly, from relatively sterile concrete enclosures with little opportunities for wide behavioural expression, to highly naturalistic ones which enclose existing wetland and woodland areas. (Photo supplied courtesy of Róisín Campbell-Palmer.)

disperse along water and these areas will particularly be tested in escape attempts.

Standard methods of beaver retention have evolved, given the range of resources and budgets available, but mainly as the escape abilities of beavers become ever more documented. Britain is perhaps one country in which beaver fence development has become more of a controversial issue as one of the last states in Europe to begin its beaver reintroduction journey, or where, perhaps more accurately, beavers seem determined to restore themselves. Of course, numerous, often comical stories from American wildlife rehabilitators of beavers in their care decidedly concluding themselves when it is time for them to leave their captive carers (Tournay, 2003); such leaks are not so much of a concern when dealing with a native beaver species, with surrounding permitted populations, though any risk of pathogen aquirement in captivity should be acknowledged, along with the legality for that country/ state and any risk of beaver species introduction.

Beavers have hatched escape plots, or more likely decided it is time for territory expansion and/or mate seeking, after fence failures involving electric power breaks, trees falling onto structures, unnoticed beaver buildings or excavations, biting or digging through unmaintained fence lines, and unexpected flooding events aptly allowing them to swim out,

Figure 10.3 Beaver retention fencing needs to be robust; a wire-mesh skirt or buried fencing is key to prevent burrowing and escape. (Photo supplied courtesy of Róisín Campbell-Palmer.)

all providing breakout opportunities. Fencing for more zoological-type settings includes concrete, brick, or reinforced glass walls and weld-mesh fencing sometimes with 'hot wires' (Campbell-Palmer and Rosell, 2013; Figure 10.2). In more naturalistic enclosures, often high-tensile wire-mesh fencing is used, run between fence posts buried into the ground at around 60 cm (timber posts are cheaper, noting that posts should be on the outside of the enclosure or protected with weld-mesh), with a mesh skirt at the fence base extending ~90 cm into the enclosure so beavers cannot dig under (Campbell-Palmer et al., 2016; Figure 10.3). The mesh skirt seems to be a key retention feature, often cheaper and more practical than building a buried fence line, and enables vegetation to grow through the mesh panels, providing extra grounding. Fence heights of around 120 cm minimum give good scope for ensuring any felled material or larger animal cannot gain a potential escape route by pulling themselves over, though an overhang in the fence design will fully prevent this. Overhangs and/or electrical fences are also used if potential predator entry is a concern. Mesh square dimensions and wire gauge are key, with 10 × 10 cm panels recommended to prevent the smallest of offspring going through. Slight variations around this, especially in line with commercially available products, make sense, though it should be noted that beavers have been witnessed squeezing through remarkably small gaps given their size and general body bulk; for example, a 17-kg individual pulling itself through a 13-cm vertical gap was caught on remote cameras.

10.2.2 Dietary requirements

In the wild, beavers exhibit seasonal and spatial variations in their diet, eating a wide range of plant species (see Section 5.1) which cannot typically be represented in captivity. In one beaver stomach alone, 42 trees species, 36 green plant genera, four types of woody vines, and various grasses were identified (Harper, 1969). Grasses, sedges, and forbs are significant parts of beaver summer diets, difficult to make adequate provisions for in many captive circumstances. Aquatic plants are an important source of iodine and sodium in wild beaver diets (Müller-Schwarze, 2011), and zoo-captive diets are known to be deficient in iodine, so dietary supplementation may be required. Older animals have displayed highly porous bones, suggesting that phosphorus deficiency may need monitoring (Piechocki, 1962; Nolet et al., 1994).

Whilst diet variation is recommended in captivity, sudden and significant changes in diet should be avoided to prevent the disruption of gut microflora. Commercial rodent pellets (such as Mazuri Rodent Chow) are commonly given in wildlife /zoo-type facilities to provide complete nutrition. Browse is a vital component of beaver diets and long-term provisions should be planned in advance, whether in the form of daily cut material, planting strategy within the enclosure, and/or enclosures large enough to function as a natural habitat. Providing a varied diet high enough in fibre is a challenge for some captive environments. The nutritional value of browse will vary with species and time of year, but it is also vital to provide the high-fibrous diets beavers are evolved to cope with, and allow captive beavers to express a fuller behavioural repertoire. Captive North American beavers fed only a diet of one to two tree species lost body weight at rates of 0.1–0.6% per day (O'Brien, 1938). Captive diets of 40% carrots, 30% apples and pears, 20% potatoes, and 10% green salads are described (Aeschbacher and Pilleri, 1983; Figure 10.4), with daily quantities of 14–15 kg for seven animals, along with three to five branches (~2.5 m) of browse such as beech (*Fagus sylvaticus*), ash (*Fraxinus excelsior*), sycamore (*Acer pseudoplatanus*), various fruit trees, and willow (*Salix acutifolia*). Many captive beaver diets are predominantly composed of apples and root vegetables, cheap and readily sourced. These are strongly advised against, with experience of animals displaying significant condition and weight loss on these carbohydrate-heavy diets (S. Girling and K. Woodruff, pers. comm.).

Beavers need to eat large amounts of food daily, because their digestive efficiency can be low especially for woody plant material (Currier et al., 1960). Quantities of approximately 2 kg of woody vegetation per day (Van der Ouderaa, 1985) or 0.6–0.1 kg body mass (willow bark, twigs, and leaves), which is equivalent to 1.2–1.9 kg of willow per day for a 20-kg beaver in captivity, have been recommended (Nolet et al., 1994). Another study determined that

Figure 10.4 Sweet fruits and vegetables such as apples, carrots and sweet potatoes are readily taken by beavers, but should only form a small part of any captive daily diet. (Photo supplied courtesy of Frank Rosell.)

wild beavers of 15 kg body weight required on average 1,213 kcal/day and spent on average 269 minutes/day foraging to achieve this (Belovsky, 1984). Required amounts of 0.7–1 kg of digestible energy per day or 850 kcal (203.07 KJ) per day to maintain themselves during the summer have been stated, with 2,040 kcal (487.36 KJ) for maximal growth and 3,340 kcal (797.94 KJ) in winter for captive North American beavers (Brenner, 1962). Other published figures, based on metabolic data for North American beavers, determined the daily energy expenditure of free-living beavers to be 6,755 KJ in an ice-free winter, 5,823 KJ in spring, and 5,397 KJ in summer, but only 4,141 KJ in winter when movement was restricted by ice (Nolet and Rosell, 1994).

Other authors have determined that adult beavers (>2 years) require 1,700, yearlings 1,400, and kits 900 calories/day (Dezhkin et al., 1986). Aspen trees with a diameter at breast height (dbh) of 2.5 cm provide 1.3 kg of beaver food on average, whilst those with a dbh of 7.5 cm equate to 21.3 kg, and trees with a dbh of 25 cm can amount to 101.2 kg (Stegeman, 1954). A single beaver is said to meet its daily nutritional requirements by eating about 1,500 cm² of bark (0.5 m length of 10-cm-diameter trunk) or 6,000 cm² (2 m of 10-cm-diameter trunk) for a family of four. An estimated 1,400 kg/ha/year is proposed as the amount of woody biomass harvested per beaver (Johnston and Naiman, 1990). Such stated volumes should be taken with slight caution as beavers will often forage more biomass than they consume, for behavioural and ecological function reasons, so strict quantities should be avoided and ad libitum opportunities provided, as diversity in diets is important. Beavers in captivity that are fed on a monotonous diet of aspen only for several weeks lost approximately 0.1% of body mass per day. When only red maple, birch, and hoary alder were offered, weight loss was even more severe (0.3–0.6% of body mass per day). One beaver reportedly died when fed this monospecific diet (O'Brien, 1938). They lost more nitrogen than they took in, and their calcium and phosphorus levels were unbalanced (Jenkins, 1981). If captive beavers are offered aspen, rowan, poplar, and birch, they eat aspen first, although if fed solely on aspen over a longer period of time and then offered birch, they will take the birch first (Owesen, 1979). It is important to always provide a range of browse in long-term captivity.

10.3 Captive issues

10.3.1 Behavioural concerns

Several authors have spent years detailing the behaviours of beavers in their care or observations of them in the wild, from personal anecdotal accounts to more rigorous detailing of ethology (Wilsson, 1971; Richard, 1975; Pilleri et al., 1985; Ryden, 1990; see Section 6.1.1 on behavioural time budgets). As social mammals with complex chemical communication systems (see Section 7.2.2) and with such an ability to modify their environments, studies of wild counterparts suggest the captive requirements of beavers may actually be more sophisticated than generally perceived. A lack of stressful events in captivity may result in negative behaviours such as boredom, depression, and stereotypic behaviours. In the 1960s Lars Wilsson, a Swedish ethologist, studied the development of beaver behaviours in a number of animals he kept. Over a range of experiments, he described cases of redirected aggression and/or displacement behaviour apparently triggered by unfamiliar beaver scent, and repeated and long-lasting digging behaviours against solid objects such as walls (Wilsson, 1971). Today, we can perhaps look back at the questionable nature of the welfare of young beavers deprived of specific environmental aspects, e.g. substrate for digging, but through this he determined that most behaviours were hard-wired and displayed if and when the correct provisions were provided (Wilsson, 1971).

Evidence of negative behaviours resulting from unsuitable captive conditions determines if there is a need to identify and evaluate welfare indicators for this species. Activity time budgets for captive beavers have yet to be fully investigated or compared to wild counterparts, which could also be used to indicate areas of concern (Veasey, 2006). No matter how naturalistic an enclosure, it may still be viewed as a restrictive and monotonous environment for a captive animal. Behavioural enrichment is a science in itself but a routine animal husbandry practice in many zoos, and it is considered an essential part of improving the welfare of captive animals

(Shepherdson et al., 1998). Periods of repetitive but functionless digging and/or scratching against a wall while standing on their back legs in captivity have been documented (Wilsson, 1968). Captive individuals may experience a range of stressful events including capture, transportation, restraint and examination, and close proximity to humans and other animals (including scent). Some observations of beavers held in captive facilities at Voronezh State Nature Reserve describe up to 10–15% of the adult animals trapped in order to restock the captive breeding population as being 'unfit for living in cages' due to their extreme aggression, and immediate release is recommended (Lavrov, 1969; Milishnikov et al., 1994).

More recent experience at captive facilities associated with the Methow Beaver Project and beaver translocations in Scotland describe individualistic variations, with some animals readily accepting short captive periods and engaging in eating, gaining weight, and resting behaviours, whilst others can be unsettled and lose weight (K. Woodruff, pers. comm.). Threatened beavers tend to retreat to the water, tail slap, grind their teeth, or 'freeze' if water is not immediately accessible. This beaver research facility also described that a similar amount of adults are 'discarded' after a month of captivity and almost one-fifth of trapped animals are removed when establishing a captive breeding stock (Lavrov, 1969; Milishnikov et al., 1994). Although welfare standards have undoubtedly improved since some of these early beaver breeding farms, these accounts provide insight into the reactions of beavers to highly restricted captive conditions. Such projects have resulted in large numbers of beavers released for reintroduction projects, e.g. 71% of all beavers released to Wielkopolska, Poland, were sourced from Popielno, Polish Academy of Sciences, and the Wiartel Fur Animal Farm (Graczyk, 1985).

Touching accounts of hand-rearing and bottle-feeding orphaned beavers have been written by numerous authors, with varying success stories (Wilsson, 1968; Richards, 1983; Tournay, 2003; Figure 10.5). There are antidotal accounts of wild orphans being adopted by neighbouring lactating females following death of their parents though this is difficult to confirm (Mills, 1913), and difficult to achieve successfully in captivity so any cross-fostering

is advised against, being better to match offspring in size for company and warm. Individual babies have both thrived and gone on to breed and live long lives, though, unfortunately, other tales of listlessness, appetite loss, diarrhoea, and eventually decline and death are also common (Wilsson, 1968). Some of the first properly documented cases of hand-rearing of kits come from the beaver farm at Popielno, Poland, where hand-rearing began in 1958, with the process later described as 'troublesome and painstaking', particularly due to the frequency of feeding, bathing, and contact with the keeper (Zbigniew et al., 1995). Therefore, the costs and benefits of hand-rearing should be carefully evaluated, including any long-term implications, resource availability, social factors, and individual welfare.

At Popielno kits were removed from their mothers at 5, 8, and 14 days of age (Zbigniew et al., 1995). The composition of beaver milk was not fully known or understood, so these early diets were composed of herbal teas (such as oak bark, fennel fruits, chamomile flowers, and rhizomes of calamus) and condensed cow's milk; kits were fed up to six times a day. As documented and acknowledged by the authors, this milk was too high in sugars (lactose) and too low in fat and protein, and diarrhoea was common at the start (Zbigniew et al., 1995). Milk feeds continued up until 15 weeks of age, by then only really consisting of one feed a day, with most of the diet then composing green fodder, first taken in small amounts at 4 weeks, with increased feeding on this diet at 8 weeks. The largest intake of milk occurred during weeks 8 and 12. Similarly, there are accounts of North American rehabilitators at first attempting to feed cow's and goat's milk to orphaned kits and this disagreeing with them after a few days (Richards, 1983). A range of dried, evaporated, raw, and boiled milk were all tried, but this still lead to intestinal issues. Richards (1983) found that if poplar leaves were ground into a pulp and this along with their resultant juices were added to milk, kits seemed to show improvement. 'Pablum', a commercially available, processed cereal designed for human infants, along with similar brands in Germany have been reported as aiding young kit digestion during weaning (Richards, 1983), though sugar content and amounts fed should be carefully assessed.

Beaver milk is very rich and thick, yellowish in coloration, with a consistency more similar to toothpaste. It is higher in fat, protein, and energy but lower in sugar content compared to that of many other mammals (Żurowski et al., 1974). The actual composition of beaver milk has been broken down into 67% water, 19% fat, 11.2% protein, 1.7% sugars, and 1.1% ash (Müller-Schwarze, 2011). The calcium to phosphorus ratio is 1.19:1.80, with overall energy content of 92.46 kcal/g (Żurowski et al., 1974). Kits in the wild will typically nurse for 6–10 weeks and start to take solid foods as well at around 3–4 weeks. Any hand-reared kits should be fed substitute milk, such as Esbilac® (Saunders, 2016) or Fox Valley, with a low sugar content to avoid diarrhoea (Wilsson, 1971). Oral rehydration solution can also be given for the first feeds, before gradually changing to a milk substitute over several feeds (Saunders, 2016).

Food should be presented approximately every 2 hours until the kit starts to take food for itself. Whilst feeding on formula milk, the anogenital region should be stimulated after each feed to induce urination and defecation (Sainsbury, 2003), though beavers really do need access to water in order to pass waste effectively. Around 3 days after birth kits should have access to water at all times for drinking themselves and swimming; this will also encourage urination and defecation. During the 'milk period' the stomach and small intestine develop, especially the duodenum (Nikulin, 1954; Dezhkin and Romashov, 1976). During the transition to plant-based nutrition, the large intestine develops more significantly; in beavers, varying parts of the digestive system (see section 3.2.2) mature according to the dietary requirements, and therefore providing vegetation too early can be damaging and prevent appropriate nutrient uptake. Solids should be offered from around 2 weeks onwards, such as wild herbaceous plants and dark-green leafy vegetables. Kits have reportedly taken a few, new-growth leaves from around a week of age (Richards, 1983), though these should not be used as a milk substitute and could upset the developing digestive system. Weaning foods can include small pieces of yam, sweet potato, apple, carrot, dandelion, trefoil, romaine lettuce, broccoli, plantain, water lily pads and roots, new-growth willow, and aspen or raspberry leaves (Tournay, 2003). As a general rule for hand-rearing

Figure 10.5 Orphan beaver kits can be handed into wildlife rescue facilities if recovered from lodges following parent death or have even been retrieved by dogs bringing their owners an unusual gift. In this photo two orphans were recovered from a known lodge in Belgium after mother was killed on the road. (Photo supplied courtesy of Jorn Van Den Bogaert).

rodents, up to 35–40% of body weight can be fed daily, though only 25–50 ml/kg bodyweight should be given per individual feed (Saunders, 2016).

Beavers are social animals and young individuals do not tend to fare well alone for long periods in isolation. At Popielno, they describe the constant need for a kit to be with its human keeper, to the point they conclude such demands would prevent large numbers of kits being reared in this way (Zbigniew et al., 1995). Wilsson (1968) also notes in his ethological studies how he found it difficult to keep wild beavers alone for any length of time in captivity. Some wildlife rehabilitation staff (Anne G. Miller, Alabama Wildlife Center) talk of dedicated volunteers making surrogate fake-fur 'beaver mothers' for orphans in their care in which milk bottles were inserted, also providing the kits with a source of warmth and comfort. Heat retention in young kits without other family members is a big concern, though overheating of kits in summer months should also be monitored closely. Access to cool water so animals can regulate their temperature is vital.

10.3.2 Health and hygiene

Beavers can be challenging animals to view in captivity, being semi-aquatic and crepuscular in nature, commonly held in more naturalistic enclosures, and often left to their own devices. With regular effort, they may become habituated to consistent feeding times and areas, allowing assessment of health including locomotion and movement, behaviours such as feeding, and physical qualities such as tail condition, wounds or injuries, presence of discharge, or erratic breathing, for example (Goodman et al., 2012). Beavers are diligent groomers, investing much time keeping fur in good quality to ensure effective insulation and buoyancy (Fish et al., 2002). A lack of grooming, observed as matted or waterlogged fur, and a 'scruffy' or 'unkempt' appearance may actually be evidence of an underplaying health or behavioural issue.

Body condition and weight changes are valuable measures that may indicate underlying health or welfare issues. Weights vary depending on the time of year, mainly due to food quality and quantity (Hartman, 1992; Parker et al., 2001), with all age classes losing weight and body condition over the winter period, apart from kits born that year which will generally increase in weight (Campbell et al., 2005). As captive individuals tend to receive a more constant and unvarying feeding regime, they may not display significant seasonal variation. Therefore, sudden or progressive weight loss is likely to be indicative of an underlying medical condition. Beavers have large digestive systems to assist the processing of woody material, so that even in poor body condition an unfamiliar observer may still describe them as 'fat'. Good-condition animals will have plump tails, as opposed to thin and sunken. Beaver tails function as fat storage and so tail dimensions vary depending on deposition and mobilization of fat (Aleksiuk, 1970; Smith and Jenkins, 1997). The ratio of tail dimensions (length, width, and thickness) varies seasonally and with age, being positively correlated with body mass (Smith and Jenkins, 1997). Using the ratio of tail size (length × width), which is seasonally variable, to body length (measured from the nose to base of tail, following curvature of the spine), which is stable over each season, gives a tail fat index; the larger this ratio, the higher the tail fat content and therefore a better body condition can be reliably assumed (Parker et al., 2007; Parker et al., 2017; Figure 10.6). A reliable indicator of body condition, as opposed to visual observation, is to palpate the spine and pelvis for fat coverage.

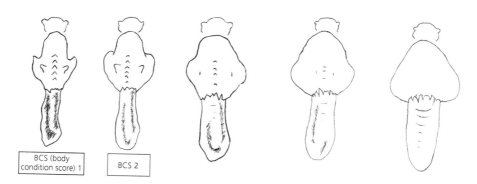

BCS (body condition score) 1   BCS 2

Figure 10.6 General body and tail condition scoring from 1 to 5 (left to right). Standard rodent body scoring can be used to determine condition, with particular attention paid to pelvic region, backbone and tail plumpness.

BCS 1 Beaver is emaciated: skeletal structure extremely prominent, little flesh cover. Vertebrae distinctly segmented. Tail arch very prominent, with tail sunken on either side of midline, owing to low fat reserves.

BCS 2 Beaver is in poor condition: segmentation of vertebral column evident. Dorsal pelvic bones are readily palpated. Tail arch prominent, tail sunken, low fat reserves.

BCS 3 Beaver is in normal condition: vertebrae and dorsal pelvis not prominent, but palpated with slight pressure. Tail arch is visible, but tail is thick with good healthy fat reserves.

BCS 4 Beaver is overweight: spine is a continuous column. Vertebrae palpated only with firm pressure. Tail arch not really visible, tail thick and more rounded.

BCS 5 Beaver is obese: body is bulky. Bones disappear under flesh and subcutaneous fat layer. Tail is thick and rounded. Note obese animals in the wild are incredibly rare.

(Illustrations provided courtesy of Rachael Campbell-Palmer.)

Tail fat index (I) = Tail length × Tail width / Body length

Cleaning regime and hygiene requirements will vary greatly depending on the degree of artificial nature and water sourcing of the enclosure. For example, in more naturalistic enclosures with stream-fed ponds, water changes and cleaning requirements will be reduced. Pool size and number of animals will determine the regularity of water changes, especially with non-filtered systems. Emptying and refilling pools with large volumes of water may not be viable to maintain water quality, so alternative water treatment methods should be employed. Static water bodies will quickly become fouled; ideally larger pools should employ a constant input/output water system. Debris such as uneaten food and faeces should be scooped from the water every day. Water in smaller pools/tubs should also be changed on a daily basis. It should be noted that such cleaning and water changes can be stressful to beavers. Any water treatment methods should address chemistry and visual quality, and remove suspended and dissolved solids and pathogens.

Beavers should be allowed to build their own shelters without regular disturbance. These must be monitored to ensure they are not continually damp throughout, with fresh bedding provided as required. Although lodges should also be checked for any carcasses, if individuals are unaccounted for, this is not always easily accomplished. Beaver-built structures can be extremely robust and the crepuscular nature of the species ensures that they are not easy to monitor or observe. Other techniques such as monitoring of food taken and camera traps should also be employed.

10.3.2.1 Dental issues

Buccal and dental issues (see Section 8.3.5 for dental issues in wild beavers) may be difficult to ascertain unless an animal is anaesthetized. Behavioural cues such as rubbing at the mouth or problems chewing food or a reluctance to eat may be indicative of underlying issues. Constant excessive drooling, with wet and matted fur around the mouth, chest, and even fore legs, progressive weight loss, bacterial infections, and even hair loss are further clinical signs (Kim et al., 2005). Mandibular incisor dentoalveolar abscesses following infections have been recorded and are evident in some skulls; this can

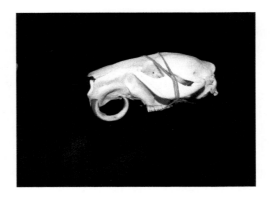

Figure 10.7 Beaver skull depicting exaggerated overgrown teeth due to lack of opposing incisor. Lack of appropriate browse provision in captivity can lead to dental and behavioural issues. (Photo supplied courtesy of Róisin Campbell-Palmer.)

lead to necrosis of the teeth and damage and even destruction of the supporting jawbone (Cave, 1984).

Abnormal incisor growth, in particular malocclusion (irregular contact between teeth of the lower and upper jaw), excessive growth, unusual curvature, and atypical occlusal wear patterns have been recorded in both wild and captive individuals (Figure 10.7). Dental abnormalities in captive Eurasian beavers, with some eventually resulting in deaths, have been described (Żurowski and Krzywinski, 1974). Again, at the beaver breeding farm at Popielno abnormal incisor wear was reported in old and long-term ill individuals, often causing interrupted feeding on browse to the point that jaw movements were impeded, gaping was evident when animals could no longer fully close their mouths, and in some circumstances this lead to drowning. Improper feeding and poor husbandry have been cited as causes of malocclusion in zoo beavers, including a lack of appropriate gnawing materials and nutritional deficiencies (Kim et al., 2005). Tooth loss, whether a birth defect or caused by disease or injury, can also lead to significant dental and feeding issues.

10.4 Animal husbandry and field monitoring techniques

10.4.1 Capturing and handling

Beaver live-capture/trapping and handling is required in a range of situations including not only the captive setting for routine tagging and health screening but also in the wild for ongoing research, restoration, and animal management programmes (Figures 10.8a and b). Any trapping programme should assess the trapping requirements, resource availability, animal welfare requirements, and trapping site. Capture and handling should only be undertaken if necessary, as these can induce stress responses (Wingfield, 2005) and lead to physical, physiological, and behavioural changes (Arnemo et al., 2006; Harcourt et al., 2010; Becciolini et al., 2019). Beavers can appear quite passive but can turn aggressive in the capturing process. Being strong animals equipped with sharp teeth and claws, risks should not be underestimated. It has been repeatedly documented that beavers can learn from experience and can readily become trap-wary and shun trapping attempts, though this can vary between individual animals. Both authors of this book regularly trap beavers, and whilst some individuals will regularly come to a trap or repeatedly put up little resistance on handling, others will be especially trap wise, evasive, and resistant to handling procedures.

Trapping regimes are highly determined by legal regulations and permitted protocols, which vary between countries. In live trapping for any mammal there can be individualistic differences, so animals may become trap shy and completely avoid traps as they may be suspicious of new objects (Grasse and Putnam, 1950), whilst others may happily return nightly for a free feed. Trap rate can also decline after the first few family members have been trapped, with subsequent family members more alert and suspicious and have even been recorded leaving a territory (Rosell and Kvinlaug, 1998). Giving the trapping site a rest, relocating traps, and then returning to attempt trapping at a later date have proven successful (Hammerson, 1994).

When approached, a beaver's main defence behaviour will be to head towards water; therefore these routes should be blocked off (e.g. captive pools drained) in advance. In captive situations, beavers may be walked into transport crates, for example, using guiding or 'pig' boards. Nets may also be used, especially if veterinary or tagging procedures are required. Nets may require custom-made specifications, must be of appropriate size to accommodate large adults, and must be robust enough to support their weight.

10.4.1.1 Welfare during trapping and handling

Historically, more inhumane trapping techniques have been employed and are often described in older texts. Specially bred beaver hunting dogs known as bibracco, a type of Laconian dog, were used in Europe in the Middle Ages (Marcuzzi, 1986). Even today dogs are used by hunters in parts of Lithuania, western Manitoba and around Lake Superior (Backhouse, 2015). Early methods of catching young beavers for fur farms included driving out kits by shaking the lodge and/or thrusting a stick into the nest cavity to encourage them to leave the lodge and then be picked out of the water by the tail by people in canoes. This should most definitely be avoided in such a body-heavy animal (Bailey, 1927).

Numerous trapping guidelines and regulations exist to promote welfare standards of live traps, encourage a humane death where appropriate and avoid non-target species (see Section 11.4.9). Typical non-target species reported by beaver trapping programmes include raccoons (*Procyon lotor*), both river otter species (*Lutra lutra* and *Lontra canadensis*), fresh water turtles, muskrat (*Ondatra zibethicus*), mink (*Neovison vision*), pine marten (*Martes martes*), badger (*Meles meles*), red fox (*Vulpes vulpes*), various game birds and even pets (McNew et al., 2007). Live traps should also avoid non-target species and in the main seek to minimize the time in the trap, ensuring a trapped individual is not suffering unnecessarily: for example, not setting or ensuring additional trap checks in poor weather conditions or periods of extended temperatures below freezing, so that any one animal is not left sitting for extended times at risk of hypothermia. Similarly, long exposures to periods in the sun are potentially harmful for beavers and should always be avoided (Hill, 1982). The degree of submergence (depending on type of trap used) and weather conditions during trapping will drastically impact on rate of exposure of any trapped animal. These are successful traps but have their drawbacks, and they risk holding an animal for a long time partially submerged in cold water, where they are at risk of hypothermia or drowning if water levels fluctuate (MacArthur and Dyck, 1990). Kits in particular may die after 4–5 hours in the water, as their fur does not yet have the same waterproof capabilities as adults (Wilsson, 1968). Any spring or cage traps risk animals injuring themselves if escape attempts are made; these commonly include bruised or cut noses, chipped incisors, and damaged claws. Although these tend to be relatively minor they are worth examination before re-release. Multiple trap checks within a 24-hour period are also recommended, noting that beavers seem most likely to be caught in early evening and through the night before dusk. Trapping seasons exist in many regions to avoid the trapping of heavily pregnant and/or during the early kit dependency period.

Capture and handling effects have been investigated in both beaver species (e.g. Graf et al., 2016; Smith et al., 2016). Short-term impacts following the application of telemetry devices have been found, including more time spent in the lodge following release and less activity in general including foraging (Graf et al., 2016). This clearly suggests a behavioural response to trapping and handling and a reluctance to leave protective shelters. Monitoring long-term impacts, particularly over several variables, is an important consideration, especially if these involve repeated capture and handling events, though these may often be overlooked, with priority commonly given to more immediate responses to capture and procedures such as tagging (Wilson and McMahon, 2006). North American beavers tagged with transmitters experienced greater weight and tail area loss over winter than untagged individuals (Smith et al., 2016). A detailed analysis of capture and handling effects on a Eurasian beaver population exposed to long-term, repeated trapping over 20 years has recently been undertaken in Norway. The dataset includes capture rates of up to 25 events for some individuals, with an average of 4.7 ± 4.2 events per adult (Mortensen and Rosell, 2020). A number of key variables were compared over time, including body condition (measured as body mass, size, and tail fat index), annual reproduction, litter size, and survival. Overall, none of these variables demonstrated significant differences with repeated capture and handling; however, during the initial years of this 20-year project negative impacts on litter size and annual reproduction were evident. Reproductive success later recovered to normal, showing that this negative trend decreased over time considerably with habituation to the capture and handling process (Mortensen and Rosell, 2020).

10.4.1.2 Live capture methods

Several beaver live trap types exist (Figure 10.8a), though permitted types and trapping periods vary between countries and required permissions should be secured in advance. Methods for live trapping beavers have been reviewed by Rosell and Kvinlaug (1998), which we summarize here along with more recent updates, though effectiveness and welfare standards are often only as good as the trapper employing them. See Table 10.1 for advantages and disadvantages of the various traps, together with trapping and mortality rates.

The use of snares is subject to varying legality, public accepatbility and legality in some countries. However, there are questionable welfare impacts and higher mortality rates are reported (McKinstry and Anderson, 1998; McKinstry and Anderson, 2002). This method is typically used in fur trapping and is typically coupled with an immersion set (to ensure drowning) for lethal control (see Section 11.4.9).

In small enclosures or restricted spaces, access to water should be restricted and then beavers can either be walked into transport crates or netted by hand. In larger or more complex enclosures, especially those in which water cannot be drained or covered, the easier way to catch beavers is to use a cage trap. Beavers can be routinely fed in the same location to get them used to coming to the bait. A disbanded trap (doors securely tied open or removed for example) can provide a regular feeding area which the beavers become habituated to using, and this can be monitored with trial cameras. When ready to catch the animals, the traps can then be set to facilitate a swift capture. Any live trapping in the field, for beaver restoration or sample collection for example, will require careful reconnaissance of the trapping site in advance. Information should be gathered on recent beaver activity and habitat use, landownership boundaries, general hydrology of the site (e.g. water level fluctuations, currents), land-use type, and accessibility by the public and pets.

Bailey and Hancock traps are the most commonly used beaver live traps (Buech, 1983). They both operate on the principle of a large spring-loaded trap, the plates of which snap shut together when the middle treadle plate is triggered. They can be

Table 10.1 Beaver trapping methods, with general advantages and disadvantages have been summarized by various authors (Taber and McCowan, 1969; Hodgdon, 1978; Buech, 1983; Rosell and Kvinlaug, 1998; McNew et al., 2007), and trapping success and mortality rates from various studies.

Trapping method	Advantages	Disadvantages	Trapping rates	Mortality rates
Snares	Light-weight, cheap, easy to carry, and many can be set in a short space of time.	Mortality due to a range of factors including entanglement and suffocation, predation whilst trapped, drowning, and bites inflicted by other beavers (McNew et al., 2007). Capture of non-target species can be reduced using a stop to restrict closing dimensions and reduce by-catch of otters for example. Even with such measures non-target species may still be trapped (McNew et al., 2007).	4% combined capture rate for Bailey traps and snares. 48 beavers in 1,099 trap nights (7% success with Bailey traps and 2% with snares) (Davis et al., 1984). 231 (166 different individuals) trapped in 4,316 trap nights, giving a 5% success rate and 1.5 beavers per night effort (McNew et al., 2007). Capture success of 7% (Weaver, 1986). Capture success of 5.2% (Mason et al., 1983).	4% (*n*=277) (McKinstry and Anderson, 2002). 7.5% (McKinstry and Anderson, 1998). 2 out of 48 captured beavers died (Davis et al., 1984). 12.2% (Weaver, 1986).
Bailey traps	Not easily seen by beaver. Useful for trap-shy/-wary animals. Set underwater so can be used in blind sets (without bait) but limited by water depth.	Risk to trap setter. Heavy (~11 kg), can be difficult to handle. Accidental trapping of non-target species. High incidence of misfiring, up to 50%. Beaver injury and risk of heat loss and exposure to other animals whilst in water. Risk of beaver drowning.	4.3 trap nights per capture (Busher, 1975). 7.8 trap nights per capture (25 individuals during 196 trap nights) (Flemming, 1977). 3 beavers trapped in 100 trap nights, and 1 beaver trapped in 41 trap nights (Lavsund, 1977). 16% capture rate (111 captures in 703 trap nights) (Hodgdon, 1978). 10 beavers trapped in 203 trap nights, using 8 traps (Rosell and Parker, 1997).	2 out of 48 captured beavers died (Davis et al., 1984). No mortality (Collins, 1976).
Hancock traps	Tend to be less prone to setting off falsely. Not limited by water depth. Useful for trap-naive animals. Useful for when hypothermia may be an issue. Take less time to set than Hancock trap.	Hold animal out of water so less subject to heat loss in water. Heavy (~13 kg), can be difficult to handle. Trapped animal may be exposed to other animals. Risk of beaver drowning with sudden water level changes. Predation whilst trapped has been reported (McKinstry and Anderson, 2002).	3.6 trap nights per capture (73 individuals during 266 trap nights) (Flemming, 1977). 11% capture rate (125 beavers in 1,165 trap nights). 14.1 trap nights per capture of untrapped individuals, 18.6 trap nights per capture of previously trapped individuals (Hodgdon, 1978). Capture success of 10% (Jackson, 1990). Capture success of 12.2% (Van Deelen, 1991).	1% (*n*=277) (McKinstry and Anderson, 2002). 3.1% (Jackson, 1990). Other studies reported no mortality.
Cage/box traps (many variations but based on Bavarian beaver traps)	Robust and safely hold and shelter a trapped animal from weather and other animals.	Can be heavy and awkward to move in the field. Trap placement is key, though trap-shy or trap-happy individuals may emerge, making it harder to trap remaining animals.	Capture rates of 21% (21 out of 99 trap nights) (Koenen et al., 2005).	No mortalities reported.
Aquatic netting	Can target individuals for capture. No trapping of non-target species. No injuries reported.	Requires specialized equipment and experience. Can only be used in specific hydrological conditions.	84 beavers trapped in 22 nights (129.8 hours), 1.9 hours of effort/ beaver capture (Rosell and Hovde, 2001).	No mortalities reported.

used in shallow water so that a beaver can swim over the set trap unnoticed and then once triggered is held within the trap folds. The main differences with Bailey traps is that both trap jaws move and lie below the water line, so that a trapped beaver is generally held partially in water, whereas in Hancock traps one jaw lies in the water, the other out, and only one jaw moves, so that the beaver is scooped upwards and out of the water (Vantassel, 2006). Animal welfare is of growing public concern; therefore alternative trapping methods, particularly in Europe, have been developed and tend to be increasingly used as beaver restoration projects develop. Others have used fishing-net-type setups to try to trap animals leaving lodges or burrows; however, this may have several drowning risks for the animals, be generally unsuccessful, and present questionable health and safety issues for the human trappers involved (Müller-Schwarze and Haggart, 2005).

Aquatic netting from a boat is one such method, and in principle it applies to netting beavers on land as well. Regularly undertaken by the Norwegian Beaver Project (Rosell and Hovde, 2001) and used in Scottish and Mongolian projects, this highly effective method allows the specific targeting of individuals which are captured by either a scoop or landing net (Figure 10.8a). Although this technique requires experience and can be labour intensive, overall it

excludes heavy trap deployment, days of lack of trap interest, can target trap-shy animals, and avoids the trapping of non-target individuals or the potential of many hours in a trap. By driving a lightweight metal powerboat and using handheld spotlights along the shoreline of a river or lake with clear water and suitable banking, beavers will often dive into the water on approach, where they can be netted by hand using specially designed landing nets (Figure 10.8a). Most animals tend to be trapped between 02:00 and 03:00 hours and an equal balance of sexes is achieved (Rosell and Hovde, 2001). Trapping success using this method is hindered slightly according to inaccessible banking, when beavers escape into bank burrows, turbid water, and weather conditions such as heavy rain and wind.

Cage or box traps are also commonly used, with a range of designs and evolving from relatively basic beginnings. The 'Bavarian' beaver trap (Figure 108.b) has been specially developed to improve welfare, including a tail-sized gap on closing should any oversized individuals leave some hanging out! All the traps work on the principle of encouraging a beaver to enter for food or scent, and in doing so trigger a treadle plate, dropping a trap door either behind them or both in front and behind them. Another positive of the Bavarian design is the use of sheet as opposed to mesh metal, so that any trapped

Figure 10.8 Various live traps exist. A boat netting method has been successfully developed by beaver researchers in Norway. This involves the use of a specially designed diving net, using the boat to guide the beaver towards a suitable shoreline so that the trapper can net the animal in shallow water. The net is long so the beaver has space to rise up to the surface, the net safely secured behind it, and taken to shore for processing (a). Live traps, that safely enclose and keep it protected from the elements until the beaver is removal, have been developed by beaver managers in Bavaria (b) (Photos supplied courtesy of USN and Róisin Campbell-Palmer.)

animal is protected from the elements and kept dark apart from one mesh viewing panel. Using little mesh and ensuring mesh diameter is small enough to prevent beavers from gripping with their teeth seems to significantly reduce trap injuries. Box traps allow trapped animals to move around, eat, and groom (and keep dry), as opposed to 'suitcase'-type designs which may hold animals in uncomfortable or painful positions for long periods of time, potentially exposing them to hypothermia (Grasse and Putnam, 1950), vulnerability to predation (e.g. mountain lions [*Puma concolor*] have been recorded predating on beavers in Hancock traps), or drowning (Buech, 1983; Davis et al., 1984; K. Woodruff, pers. comm.).

Double-door traps may encourage beavers to enter as they may feel less inclined to be entering an enclosed box. Koenen et al. (2005) discuss some of the early studies on the use of box traps for American beaver trapping, achieving higher capture rates as the design was improved including features such as door locking mechanisms and trap placement knowledge. Accidental triggering of traps, especially when animals walk past rather than through these types of traps, is reported from several projects, along with a failure of traps to latch properly on occasions. Of course, as for any trap, there can be occasions where animals somehow remove bait without triggering the trap or non-target species are caught, but these traps tend to exhibit minor injuries for species across the board. Minor bruising and cuts, particularly to the nose, mouth, and front claws, have been reported across various projects (Koenen et al., 2005).

10.4.1.3 Trap placement and bait

Two main principles in beaver trap placement are generally applied: either aim to lure beavers to a trap (active trapping) or place the trap in an area regularly used already by the beavers (passive trapping and 'traffic traps') (Rosell and Kvinlaug, 1998). Of course, a combination of both may be used to improve trapping success, i.e. careful placement of a baited trap on top of an actively used forage trail, typically on the worn tracks leading from the water to an active felling site. Similarly, certain traps working on the same principle can be set into a destroyed section of a dam, so that a beaver coming to repair the broken dam is likely to be trapped. Several studies record improved trapping success, especially for

terrestrial-placed traps, during the autumn and winter months, most likely as beavers increase their land foraging requirements, their diet shifts to more woody vegetation, and they engage in increased lodge maintenance behaviours (e.g. McNew et al., 2007). The major welfare risk with any trap placement in water is sudden water level rises (heavy rains, snow melt, tidal surges, etc.) that can lead to drowning and should be completely avoided. Whole beaver families with mature dams have been successfully removed by using this method based on their strong instincts to repair leaks in the dam (Bradt, 1938). However, one must be extremely careful that a subsequent family member doesn't repair the dam breach below a trapped individual, thereby returning water levels to normal and potentially drowning the trapped individual (Rosell and Kvinlaug, 1998).

A range of baits can be used for trapping, though variations and preferences may be displayed by different families. Natural browse baits include leafy branches of aspen (*Populus* spp.) (Bailey, 1927; Simonsen, 1973), rowan (*Sorbus* spp.), birch (*Betula* spp.), willow (*Salix* spp.) (Simonsen, 1973; Nolet et al., 1994), and fruits trees such as apple (Hodgdon, 1978; Smith and Peterson, 1988). Other successful foodstuffs regularly used are apples, parsnips carrots, sweetcorn, sweet potato, and grain (Hodgdon, 1978; De Almeida, 1987). Scent lures can be very successful in the form of castoreum from other beavers, though artificial versions are commercially available. These lures are especially productive as they

Figure 10.9 Castoreum based scent lure products are available. (Photo supplied courtesy of Frank Rosell.)

encourage the territorial behaviour of beavers to react to the scent of stranger beavers in their territory, and they may have the added benefit of masking any human scent associated with the trap (Rosell and Nolet, 1997; see Section 7.2.2; Figure 10.9).

10.4.1.4 Chemical immobilization

Light anaesthesia can facilitate examination for health checks safely, though full immobilization is a requirement for more invasive health checks and veterinary procedures. One of the key considerations for the use of any anaesthesia regime is how recovery will occur, and there should be restricted access to water until an individual is deemed safe for release. Chemical immobilization with long or variable recovery times is not ideal in field conditions unless animals can be retained in travel crates or captivity until any risk of drowning is removed (Girling et al., 2015; Saunders, 2016). Extensive beaver health screening following their reintroduction to Scotland and parts of England has determined that beaver restraint in a hessian/cloth sac with face mask induction with isoflurane as the sole induction agent enables physical examination, sample collection, and tagging of beavers with quick recovery and significantly reduces the risk of post-anaesthetic drowning, and can be used under field conditions (Goodman et al., 2012; Girling et al., 2015). Numerous

American studies discuss the administration of injectable anaesthetics, especially associated with population monitoring studies involving radio-tagging (McNew and Woolf, 2005; Havens et al., 2013). Beavers have been crated or placed in plastic bins until fully recovered from anaesthetic before release (Havens et al., 2013). See Table 10.2 for drugs used and Figures 10.10 and 10.11 illustrating two methods of administering anaesthesia (via face mask and intubation).

10.4.1.5 Handling without anaesthetic

Short-term handling can occur without anaesthesia, with experienced handlers, if a beaver is to be physically examined or tagged and/or samples are to be collected (Rosell and Hovde, 2001; Sharpe and Rosell, 2003; Goodman et al., 2012). This is greatly facilitated by the placing of a beaver in a cloth or hessian sack or funnel-shaped 'Cordura' bag, making both handling, sample collection, and tagging easier, but this also improves animal welfare, so it is strongly encouraged. No aggressive interactions are reported with this method, with beavers tending to remain calm (Sharpe and Rosell, 2003). The key to this technique is that the eyes remain covered and a beaver can be handled with less resistance and increased safety for the handler (Grasse and Putnam, 1950).

Figure 10.10 Chemical immobilization is possible in the field with careful recovery times, facilitating more invasive sample collection and tagging if required. Here a beaver is being given anaesthetic through a face mask. (Photo supplied courtesy of Romain Pizzi.)

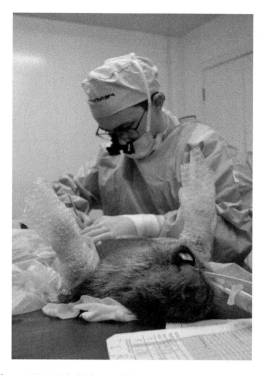

Figure 10.11 Tubal induction for longer procedures, such as laparoscopic surgery. Bubble wrap is a ready technique to mitigate heat loss. (Photo supplied courtesy of Jon Cracknell.)

Some older fur farm accounts and trapper manuals state beavers can be caught or lifted by their tails, but this should absolutely be avoided. Despite the large flat appearance of the tail, the vertebral bones inside are actually very small. The beaver tail vertebrae joints are much more inflexible than most other long-tailed mammals, with a more limited range of motion possible. The beaver tail, while flattened to allow slapping of the water, cannot resist the forces applied when the animal is lifted by the tail, and this will be painful to them. It is easy to luxate (dislocate) tail vertebrae or damage the ligaments connecting tail vertebrae, and this could cause long-term adverse effects.

10.4.1.6 Sample collection

Sample collection varies with capture project objectives, monitoring techniques employed, and individual animal requirements. Included here are the typical ranges of biological samples collected from a restrained or anaesthetized animal. Faecal samples, largely for parasite and bacteriology analysis, can be easily collected by putting gentle pressure on either side and behind the cloaca. Faecal matter if present can typically be felt in the rectum and gently encouraged to be expelled (Campbell-Palmer

Table 10.2 Chemical immobilization dosage rates used in beavers across both species and various studies.

Anaesthesia drugs	Dosage	Comments	References
Isoflurane	Via face mask, 3–4% for induction and 1–2% for maintenance with oxygen flow rate of 2 L	Light anaesthesia	Goodman et al. (2012)
Isofluorine and sevofluorine	5% isoflurane and 7% sevoflurane	Gas anaesthesia	Breck and Gaynor (2003)
Diazepam and ketamine	0.2 mg/kg diazepam and 10–12.5 mg/kg ketamine	Intramuscular injection	Goodman et al. (2012)
Medetomidine and ketamine	0.06–0.08 mg/kg medetomidine and 6.6–9.5 mg/kg ketamine 0.1 mg/kg medetomidine and 7 mg/kg ketamine	Intramuscular injection	Goodman et al. (2012) Smith et al. (2016)
Ketamine hydrochloride and xylazine hydrochloride	6–12 mg/kg or 10 mg/kg ketamine hydrochloride and 0–1.25 mg/kg or 1.0 mg/kg xylazine hydrochloride, respectively	Intramuscular injection	McNew and Woolf (2005)
Ketamine, butorphanol, and medetomidine	5 mg/kg ketamine, 0.1 mg/kg butorphanol, and 0.05 mg/kg medetomidine		Ranheim et al. (2004)
Ketamine, butorphanol, medetomidine, and midazolam	5 mg/kg ketamine, 0.1 mg/kg butorphanol, 0.05 mg/kg medetomidine, and 0.25 mg/kg midazolam	Reported as a better muscle relaxation compared to above regime	Ranheim et al. (2004)

and Rosell, 2013). Note often castoreum may also be expelled, but removal of faecal matter from the rectum should be undertaken first, and then castoreum can be 'milked' by feeling for two sizeable internal lumps behind the cloaca. By applying a gentle rolling, downward pressure towards the bladder and over the cloaca, castoreum should be released if present (Schulte, 1998; Rosell and Bjørkøyli, 2002). Castoreum collection is more often performed for scent lures for trapping. Anal gland secretion (AGS) is mainly collected in order to sex an individual (see Section 10.4.3). AGS is specifically excreted from the anal papillae within the body cavity. To expose these, press either side of the cloaca and in front of the castor sacs; once exposed, one papilla should be gently but firmly squeezed and pressure applied until secretions start to emerge (Rosell and Bjørkøyli, 2002); note this takes practice! The papillae may be difficult to grasp as they will tend to contract back inside the cloaca (Campbell-Palmer and Rosell, 2013).

Blood is most typically and easily taken from the ventral mid-tail blood vessels (Greene et al., 1991; Girling et al., 2015; Figure 10.12). Hair samples, often collected for genetic analysis, can be taken from the lower back region, taking a few guard-hairs at a time and tugging upwards in a swift movement, ensuring root bulbs are collected (Pollock et al., 2015). All these can be collected without anaesthesia only if the beaver is properly restrained within a cloth sack, by experienced personnel to minimize handling times and keep both the beaver and handlers safe. When collecting various samples from the back end of a beaver, its hind region should be raised upwards but without lifting it off the ground. The tail of course will need to be moved slightly out of the way, but it is best to handle the tail from its base, where it is strongest, and ensure never to apply any force to a bent tail; this would cause both discomfort and potential injury, as well as often provoking more of a reaction from the animal while it is being sampled (Campbell-Palmer and Rosell, 2013).

Numerous projects often take photographs of the full tail of each beaver they handle (this could also be collected from remote cameras if image quality is good enough). Distinctive markings may be present, often as a result of previous territorial inter-

action, and can aid in individual identification (Pollock et al., 2015).

10.4.1.7 Transportation

Animal transport regulations will vary between countries, method, and distance of transportation. The International Air Transport Association (IATA) Live Animal Regulations (LAR) are the global standard on moving animals by air and, along with all animals, provide crating and handling information for beavers. Moving beavers shorter distances by road does not require such stringent regulations or robust crates, but there are a few key considerations. Suitable beaver crates include those obviously resistant to chewing, so having flush internal surfaces, such as metal sheets, hard plastics, or smooth wood that can be lined with fine mesh, so there is no leverage for grip. Ventilation is key as beavers can easily become overheated and prone to heat stress; multiple ventilation holes or slants should be distributed throughout the crate top and sides, without being large enough to enable teeth to get a grip (Figure 10.13). Provisioning crates with a deep layer of straw and/or sawdust is important for comfort, to absorb moisture, and to prevent the animal slipping with vehicle motion. Keeping crates as dark as possible will also aid in reducing stress, along with avoiding sudden motion, lights, and noise as far as possible. To avoid injuries beavers should be transported in separate travel crates, unless moving youngsters with a parent or older

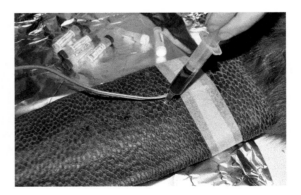

Figure 10.12 Blood sample collection in either anaesthetized or appropriately restrained animals is possible and best achieved via the ventral mid-tail blood vessels. (Photo supplied courtesy of Jon Cracknell.)

sibling, though two compatible animals have been moved together for shorter journeys in North American projects (K. Woodruff, pers. comm.).

10.4.2 Remote monitoring and tagging

Nocturnal and semi-aquatic lifestyles make monitoring more difficult. Of course, visual observations are possible, especially once emergence times and general behavioural patterns become more familiar. In general, beavers tend to become active outside of their daytime retreats between 18:00 and 20:00 hours, returning to their shelters around 05:00–06:30 hours (Hodgdon and Lancia, 1983; Stocker, 1985; see section 6.1). Therefore, lodge or burrow observations can be targeted. Monitoring food taken and fresh feeding signs are important methods. Remote cameras correctly positioned (at freshest feed stations, along worn forage trails, or on lodges) can offer an effective, non-invasive means to monitor activity and body condition and document rarer behaviours such as mating attempts. Individual recognition through natural markings alone is problematic due to the lack of variation in coloration and markings amongst beavers. Patterns of scaring and notching on the tail (often caused through previous territorial fighting) may enable individual identification, though it is difficult to employ from a distance.

10.4.2.1 PIT/microchip tagging

The use of radio-frequency identification (RFID) to detect passive integrated transponder (PIT) tags, more commonly known as microchipping, is widely employed across a wide range of companion animals, wildlife, and livestock (Boarman et al., 1998). Such tags are recommended as these are minimally invasive to apply, have low incidences of failure rates, and provide permanent identification throughout an individual's lifetime. Several beaver population monitoring studies, in both species, use PIT tagging and hand-held readers to identify animals at close range (e.g. Rosell and Hovde, 2001; Goodman et al., 2012; Havens et al. 2013). As best for many mammals, loose skin along the dorsal midline around the shoulders and lower neck region can be lifted up and a PIT tag can be inserted with a 12-gauge needle (Sharpe and Rosell, 2003). Beaver skin is very tough,

so appropriate pressure should be applied by experienced personnel, though excessive force should be avoided as this may cause injury.

One fundamental drawback of this tagging method is that the chip reader or scanner must come into fairly close contact with the animal in order to determine its unique ID code, which may often require a recapture event. Developments in RFID technology are enabling PIT tags to be detected from further away, and fixed-position tag readers are enabling less-invasive monitoring (Boarman et al., 1998).

Given that beaver lodges act as very suitable focal points for family groups, use of this technology has recently been trialed in Norway (Briggs et al., 2021; Farsund et al., in prep). In one study eight lodges were monitored (Briggs et al., 2021), with the reader antennas fixed at the opening to each entrance tunnel and attached using cable ties. Although it was determined that the number of beavers recorded via the RFID was significantly lower than known family size, this still marks the start of a promising non-invasive monitoring method. Further refinement, particularly with antenna placement and greater coverage within an active territory so that multiple readers are operating simultaneously were developed across ten beaver family groups (34 PIT-tagged individuals), focusing antenna placement on well used forage trails (Farsund et al., in prep; Figure 10.14). This study identified 47.1% of the tagged beavers, with a 72.6% success rate, and validated these results through use of camera traps.

10.4.2.2 Ear tagging

Ear tagging is a useful management tool (Sharpe and Rosell, 2003; Figure 10.15a and b), though retention rate can be an issue. It is a favoured external marking method in mammals as ears are often prominent and easy to view, it does not seem to affect behaviour, and as ears are mostly cartilage over vascular tissue they can heal quickly with presumably less pain and tissue damage and can provide in effect a double-marking system (Windels, 2013). As highly sociable animals and diligent groomers, external devices are often subject to increased destructive attention. As noted by Windels (2013), mutual grooming by another family member, although >90% of which tends to be directed away from the head, involves more nibbling (Patenaude

Figure 10.13 Metal transport crate, crucially beaver proof, with smooth insides and small mesh panel ends to provide adequate ventilation and prevent injuries. (Photo supplied courtesy of Derek Gow.)

and Bovet, 1984) and therefore is more likely to be a potential cause of tag loss than self-grooming. Visibility of tags can be enhanced by using bright, contrasting colours or adding reflective tape or fluorescent spray paint. It should be noted that the addition of glue, tape, or spray paint to tags may interfere with correct tag closure or hinder their fit into applicators.

Ensure the applicator and ear flaps are clean and antiseptic is used to prevent any infections. Position the beaver's head in one corner of the hessian/cloth sack. Feeling for each ear in turn, carefully use scissors to make a small hole in the sack big enough for the beaver's ear to be exposed. One person should restrain the beaver with a firm grip around the head by holding behind the jaws with their thumbs. Individual reactions to ear tagging across both species range from inertia to flinching and head jerking, but excessive struggling, vocalizations, or bleeding should not occur. The key to minimizing both beaver reactions and any potential risk of ear damage is secure and appropriate restraint (Windels, 2013). Note ear flaps are delicate in

Figure 10.14 Experimental set up in Norwegian trials of using radio-frequency identification (RFID) to detect passive integrated transponder (PIT) tags through antennae loops placed on used forage trails, with camera trap validation of known individuals across 10 families groups. (Photo supplied courtesy of USN/Peter Farsund.)

beavers and may be easily split if the tag is not applied appropriately. Tags should not be used on infected or cut ears or ears with multiple splits. Besides potential ear splitting, piercing too close to the inner ear has been reported with associated ear flap bending (Windels, 2013), though more recent use of non-looped ear tags is recommended to prevent tags being caught on vegetation for example. Each ear tag should be inserted with any piercing end pointing upwards and away from the beaver's head and any sharp edges rounded off. Plastic tags are often preferential as if they are groomed off by other family members an intact hole is often left in the ear flap, enabling new tags to be inserted easily.

Retention rates of ear tags are highly variable— with several long-term population studies finding individual differences (some animals just don't like to be tagged!) and even family differences, with some members being more diligent groomers and ear-tag removers than others, and of course there are loss rate differences according to tag design. Actual quantification of ear tag loss rates in beaver projects are few, but Windels (2013) reports on the use of 'self-piercing' metal in the North American beaver population and behavioural studies of 627 individual beavers. From this study, some interesting trends emerge. Most noteworthy was that tag

loss did not differ between the sexes or age class at time of trapping. Out of those recaptured, 6.3% (15 out of 237) had lost one ear tag, and thereby were still identifiable and not presumed to be 'new' animals into a population, which may be the case when both tags are lost in a population study. The probability of ear-tag loss was low in this study, though comparisons with other projects demonstrate this is not always the case. Loss rates of up to 18.4% ($n=87$) have been reported in Idaho (Leege, 1968), 14.3% ($n=64$) in Michigan (Shelton, 1966), and 5.9% ($n=27$) in New Mexico (Berghofer, 1961). These finding raise questions on tag design, application competency, and general use of this as a suitable long-term ID method for beavers.

10.4.2.3 External tagging

Various studies and population restoration programmes have required the implantation and/or attachment of various devices (e.g. dataloggers, global positioning system tags) for animal management and ecological and behavioural research purposes (e.g. Graf et al., 2015; Figure 10.16). Attaching an external tag is problematic owing to the behaviour, habitat, and fusiform body shape of the beaver, with neck collaring next to impossible as they readily slip off (Arjo et al., 2008). Harnesses may prove more reliable but could

Figure 10.15a and b Ear tagging can be a useful visual tool to aid individual identification from a distance and through a combination of colours, standard livestock tags have been used though great care should be taken in the application as beaver ears are small and easily ripped but tags that are too heavy for the ear, and/or have large surface area that may get caught. Non-looped tags should be used. (Photos supplied courtesy of Frank Rosell.)

Figure 10.16 External tags can include tags attached to the ear, tail, or glued to the fur and all have pros and cons, along with various functions and application. The beaver depicted in this photo is carried both a back tag, which is glued using expoy resin, and a tail tag, which has been mounted and inserted through a bolt and not attachment passing through the tail itself (Photos supplied courtesy of Frank Rosell.)

cause drowning if they get snagged on submerged vegetation, or straps could be gnawed through by the beaver.

Because they are highly sociable animals and diligent groomers, external devices are often subject to increased attention by the animal itself or by other family members. Therefore, it is difficult to make an external tag 'beaver proof', but tags can be made more robust if encased in metal or hardened plastic or covered with epoxy or araldite, especially around any weak or delicate points. Beavers' semi-aquatic lifestyle also requires tags to be waterproof, and any connections on the tag should be sealed to avoid corrosion. Tags may catch on branches, debris, rocks, etc., so it is important to make any tag as small and as streamlined as possible to try to avoid the tag being

damaged or ripped off. Biotelemetry/biologging devices may be lost inside burrows, in vegetation, and water, so the attachment of a small VHF (very high frequency) transmitter can help with their retrieval (Graf et al., 2016). Although beavers are large and strong animals, best practice suggests the weight of any tag and its attachment mechanism should be no more than 1% of an individual's total body weight. Heavier tags can be supported, but it should be noted that recommendations differ according to animal ethical guidelines for various scientific journals and between countries. The advantage of external tags can be twofold: they can be attached relatively quickly without surgery; and enable an antenna whip to be attached so that transmission is further compared to internal tags (Windel and Belant, 2016). For example, beaver dispersal patterns have often been studied through radio telemetry, which can be effective if tags stay on and beavers remain within receiving range, though this is often not the case (Sun et al., 2000).

Tail tagging has been used widely for beavers as a way of attaching external tags (Rothmeyer et al., 2002; Baker, 2006; Arjo et al., 2008). Tag placement has been at the base of the tail, though there are increased welfare concerns, as this is certainly more painful. More recent and common placement is several centimetres away from the spine and midway along the axis of the tail, to avoid nerve and bone damage (Arjo et al., 2008). Veterinary supervision is still highly recommended, and the attachment area should be cleaned thoroughly and injectable analgesic and hydrophilic antibiotic ointments applied. Belt pliers or drills are often used to make a hole in the tail, though it should be emphasized drills are not recommended and several studies have determined that a highly sharpened, stainless-steel, hollow leather punch generates a cleaner cut with less blood loss (Windels and Belant, 2016). Through the hole a short length of plastic tubing, a nylon sleeve, is inserted to prevent the hole enlarging over time and reduce rubbing injuries (Arjo et al., 2008). The transmitter can then be attached using a bolt washer (~25 mm) and nut secured with thread glue (Sharpe and Rosell, 2003; Campbell et al., 2005; Rosell and Thomsen, 2006; Windels and Belant, 2016).

Retention rates in tail tagging have been variable, largely according to tag placement, and can be low (Rothmeyer et al., 2002). More recently, Windels and Belant (2016) investigated beaver tail tagging, deploying 63 transmitters on 58 animals (5 lost tags being replaced—either on the opposite side of the tail via a newly cut hole or reusing an intact hole). These authors found that retention rates declined steadily over the first 90 days, with only 26% of tags still attached after ≥ 180 days and only a further 7% still attached after ≥ 360 days. The authors found no difference in tags deployed in either the autumn or spring or between the sexes, with mean retention times of 133 days (n=59, range 18–401 days). A clearer picture is given on discussion of examination of retrapped beaver tails (n=24); there is clear evidence that the majority of these animals (n=15, 63%) had split tails at the site of attachment, indicating that tags had been torn out, most likely either via becoming tangled or by the animal itself; a further 6 animals had widened attachment holes in which tags had pulled out; and a final 3 individuals (13%) were found to have intact holes, most likely indicating an unscrewing of the bolt. Other projects have reported the pulling out of tags and splitting of tails (in Massachusetts, DeStefano et al., 2006; Scottish Beaver Trial, Jones and Campbell-Palmer, 2014), along with varying mean retention times of 154 days in Colorado (Baker, 2006), 295 days in Illinois (Bloomquist and Nielsen, 2010), and 343 days in Arizona (Arjo et al., 2008). Given these typical retention rates, potential injuries, and possible secondary infections, their use should be carefully considered.

Gluing of tags to the outer guard hair of the lower back is a recognized, short-term attachment method in beavers (Graf et al., 2016). Any tag units should be attached to the fur of the lower back, approximately 15 cm from the base of the tail, using a two-component epoxy resin (e.g. 5-minute quick-cure epoxy or high-tech Araldite). Note glue should only be applied to the outer fur and not the skin, where it may cause chemical burns; also avoid excessive amounts of glue and pressing too hard. Coarse-meshed polyester or any light-weight, flexible material should be used as a base to increase the surface area of attachment and act as an additional barrier against glue reaching the skin. The antenna of any tag should face backwards towards the tail.

10.4.2.4 Intraperitoneal tagging

Radio transmitters have been implanted intraperitoneally in wild beavers through ventrolateral (Wheatley, 1997), paralumbar (Guynn et al., 1987), and midline-ventral (Ranheim et al., 2004) surgical incisions. Though such methods raise welfare issues in relation to pain and risk of infection, they have been deemed successful in both European and North American studies (Herr and Rosell, 2004; Smith et al., 2016). Typical transmitters are approximately 33 mm wide × 110 mm long, weighing around 100 g, with expected transmission life of 3 years (Breck et al., 2001). Long-term behaviour and movements are reportedly unaffected by these procedures or implants, apart from the first few days after implantation when the animals tend to spend more time within their lodges. Haemorrhage, post-operative infections, and damage caused by free-floating tags are all possible complications of implanting intraperitoneal transmitters (Arjo et al., 2008). Deaths have been recorded in beavers with abdominally implanted transmitters (Davis et al., 1984; Ranheim et al., 2004). Current veterinary standards advise the use of subcutaneous implants.

Given these potential risks, consideration for the requirement of such tags should be undertaken, especially if alternative monitoring techniques are more practical. Any trapping and tagging can be stressful events which can impact on welfare and behaviours. This has been investigated in wild Eurasian beavers in Norway, live trapped and fitted with GPS and acceleration data loggers (n=14) glued onto their backs (Graf et al., 2016). Individual differences were noted in some animals returning to their lodge for the rest of the night, but overall this study found no significant short-term effects on activity periods and distances moved in the first week of tagging. Similarly, tail-tagged and implant-tagged North American beavers showed no real significant differences in short- and long-term survival when compared to those with only ear tags (Smith et al., 2016). Ear-tagged individuals lost the least weight, body condition, and tail area over winter compared with those both ear-tagged and equipped with transmitters. Other studies, for both species, record reduced activity and increased lodge time in implant-tagged individuals (Sharpe and Rosell, 2003; Ranheim et al., 2004), suggesting these animals were reacting to these processes.

10.4.2.5 Remote cameras

Remote sensing technology is a popular tool providing vital information on the more secretive aspects of beaver behaviour, ecology, and life history traits. Early studies on beaver behaviour within the lodge used audio-taping or constructed viewing shelters to enable direct observation (Bloomquist and Nielsen, 2009; see Section 6.8 for activities within the lodge). These viewing devices involve an observer lying inside a specially constructed plywood box, with a window panel dug into the lodge where an observer can watch directly or use cameras. According to these Bloomquist and Nielsen, (2010). the beavers would leave the lodge as soon as digging to construct the device started, but would return and often build mud and sticks against the viewing panel, which were easily cleared (Potvin and Bovet, 1975; Patenaude-Pilote et al., 1980). Remote cameras and use of videography can prove invaluable, though placement is key. More invasive videography surveys were undertaken by Bloomquist and Nielsen (2009), comprising the insertion of fiber-optic cameras with lighting, encased in PVC tubes, into the roof of lodges and bank burrows. The behaviours of individually tail-tagged individuals could then be recorded within the shelters. From 1,080 hours of footage, 300 hours of beaver activity were captured, with 86% of this activity able to be classified into one of 12 behavioural categories. The marked beavers seen could be identified 97% of the time, though only 8 out of the 26 marked animals were seen. The beavers blocked the cameras through building activity around 10% of the viewing time. Overall it could be argued this system was expensive to install and time consuming to process the data; however, secretive behaviours were recorded, it was successful for counting kits (this could provide important information on natal fatalities), beavers did not appear to react to the researchers' presence, and no lodges were abandoned. Of note was the authors' observation that this method did not successfully capture complete family counts and there was a low rate of observations of marked individuals.

10.4.2.6 Aerial surveys and drones

Meaningful analyses have been undertaken from aerial and satellite surveillance of beaver-generated landscapes, especially when it comes to colonization extent, landscape alteration, and tracking changes over time (e.g. Townsend and Butler, 1996; Cunningham et al., 2006; Polvi and Wohl, 2012; Malison et al., 2014; Figure 10.17 a and b). Early records of beaver aerial surveys by airplane over West Virginia praise its usefulness for saving time and money and producing reliable territory counts (Swank and Glover, 1948). However, these authors do highlight potential sources of error, such as an inability to determine numbers of bank-dwelling beavers and difficulty in distinguishing territorial boundaries on stream systems with numerous dams situated close together. In more modern times helicopters are used to monitor beaver abundance and distribution (Beck et al., 2008), particularly the presence of food caches (typically October/early November as leaf fall ensures visibility whilst being prior to ice coverage) to determine active territories (Payne, 1981; Swenson et al., 1983). Monitoring to focus on food caches ensures accurate counts of families, as they typically only construct a single food cache over winter, thereby ignoring the multiple lodges that may exist within a territory (Hay, 1958; Novak, 1987). These techniques have been successfully employed to monitor population trends at a national forest scale, the Black Hills National Forest, Dakota and Wyoming (Beck et al., 2010). One helicopter flyover along randomly selected watersheds was typically sufficient to observe food caches, with detection probabilities of 0.89 relatively constant among helicopter surveys (Swenson et al., 1983; Osmundson and Buskirk 1993). Beck et al. (2010) recommend repeating such surveys every 6 years to determine abundance and presence. They also found no differences in food cache numbers between medium- and high-quality habitats (0.021 ± 0.008 and 0.030 ± 0.007 per km, respectively). In Kansas, where most beavers were living along river-constructed bank lodges and burrows, only 41 out of 146 territories located by ground surveys were recorded using aerial surveying (Robel et al., 1993). Overall, results from aerial surveys vary and depend greatly on topography,

Figure 10.17a and b Drones (a) can be useful survey techniques in circumstances such as determining presence of beavers, their distribution and extent of impounded water (b). (Photos supplied courtesy of Frank Rosell and Jorn Van Den Bogaert).

vegetation, beaver activity, and type of aircraft used (Hill, 1982; Novak, 1987).

Utilizing existing aerial photos of an area over time can provide invaluable documentation of beaver recolonization but also the artefacts of past activity. A charming case based around a map drawn by Lewis Morgan in 1868 depicting a new railroad providing access to Lake Superior before roads there, along with the surrounding watercourses, including multiple beaver dams, spurred a comparative analysis. Using established aerial photo interpretation methods, along with orthoimagery and Google Earth data over a 10-year period, a comparison was created with the 150-year-old map of this area of Michigan (Johnston and Naiman, 1990; Johnston, 2015). A key finding was that 46 out of the 64 beaver dams and ponds mapped in 1868 were still recognizable on aerial imagery from 2014. Eighteen original beaver sites were completely gone through building development and mining. Beaver canals on the north end of Grass Lake from Morgan's day were still visible, demonstrating the durability of beaver features such as dams, meadows, and ponds in a landscape context. Despite earlier validations on cost saving, in reality aerial-based survey can be costly and resolution on images may be poor, even being as high as 30 m (Townsend and Butler, 1996), though more recently <3 m; Malison et al., 2014).

As technology develops, drones (uninhabited aerial vehicles) and satellite imagery will most likely provide cheaper and less time-consuming survey methods, especially compared to ground-based surveying (Colomina and Molina, 2014). Beaver

dam building and recent activity on the Lower Noatak River and Wulik and Kivalina River watersheds (comprising 18,293 km²) were surveyed by analysing images from Landsat Thematic Mapper (TM), acquired from the United States Geological Survey (USGS) (Tape et al., 2018). This has enabled the identification of North American beaver expansion to be documented remotely in the Arctic tundra. Similarly, Henn et al. (2016) used satellite imagery to quantify non-native beaver impacts in Tierra del Fuego, facilitating eradication efforts. They point out the limitations of such a survey, including beaver detection in areas of open water, especially when trying to eradicate; an underestimation of beaver impacts; and a lack of verifiable population estimates.

More recently, digital cameras have been attached to lightweight multirotor drones to gather orthophoto and digital surface model data on reintroduced beaver-generated habitat in a headwater area, Devon, England (Puttock et al., 2015). Flight plans were designed so that at least 10 photos of every area of interest were taken and stitched together using reconstruction programs, the images then manually examined to identify beaver activity and digitized using geographic information system (GIS) programs. The authors ascertain that this methodology enables the collection of high spatial resolution imagery of ~0.01 m ground sampling distance, which surpasses that generated by both satellite- and aircraft-based surveys. Additionally, such survey methods can occur rapidly to more regularly monitor the dynamic changes beaver activities can generate over a short space of time, reducing costs and potential risk when surveying wetland environments. The generation of and rapidly developing field of 3D modelling makes this an exciting technique for hydrological modelling (Puttock et al., 2015).

10.4.2.7 Hair trapping

Live trapping for sample collection can be resource intensive, requiring specialized equipment and experienced handlers. Remote hair sampling of wild beavers has been trialled using barbed wire (Herr and Schley, 2009). For this trial, barbed wire was suspended 20–25 cm from the ground, across two appropriately situated woody stems, either side of active forage trails and feeding stations, approximately 1–2 m from the water's edge. The authors successfully collected beaver hair (0.26 samples per control day), with only one by-catch from a non-target species, identified as badger (*Meles meles*). Underfur was collected in all the 27 samples, but more importantly at least 1 guard hair with follicle was obtained for 70% of the samples, enabling DNA (for individual, sex, and species) analysis.

10.4.2.8 Questionnaires

Often the higher costs associated with some of the population counts and monitoring techniques discussed above have led to alternative survey methods such as citizen science. Return data forms are often deployed in beaver conflict and mitigation assessment (see Chapter 11), though they can be used in population assessment and distribution if designed appropriately and targeting those with experience of beavers such as hunters. Asking moose hunting team members to gather population information is a survey method that has been widely used in Nordic countries (Parker et al., 2002). The degree of accuracy of responses to these surveys varies greatly. The main reason for inaccurate answers is most likely a lack of motivation on the part of the respondents. In one survey moose hunters overestimated the number of occupied lodges by about 160%, while they underestimated the number by 25% in another survey. The underestimation would probably have been greater if it hadn't been for one person familiar with the area

who knew where the beaver population was highest. It is highly unlikely that the use of moose hunting teams to estimate the number of active beaver families (based on evidence of food caches outside lodges) will develop into a reliable method covering larger areas. This is because the terrain is often covered in an unsatisfactory manner, because the moose hunt occurs before the food caches are easily seen by the hunters, and they are, naturally enough, more interested in the hunt than in finding beaver families (Parker et al., 2002).

The government of Finland began nationwide beaver censuses in 1965; these have been followed up every fifth year. The census is based on surveys and copies of maps that are sent to hunting organizations and forest owner associations. Between 1995 and 2013 volunteers were also asked to count and record the number of active lodges during winter every 3 years (Brommer et al., 2017). Hunters in Finland make up around 6% of the whole population and already use an existing information system through the state wildlife agency, as they belong to hunting associations, so it was easy to ask them to mark territories on maps and location of active lodges. These surveys cover almost the entire country and are considered cost effective, though it is generally accepted that voluntary reporting is likely to result in underreporting, with a high degree of variability in data quality (Sutherland, 1996). Experience has shown that these surveys illustrate (1) the beaver's regional distribution, and (2) an estimate of population size which is probably somewhere between the maximum and minimum numbers. Issues with this technique are likely to increase in areas where the two beaver species overlap, where species identification will be increasingly required (Brommer et al., 2017).

10.4.3 Breeding and sex determination

Obviously, the emergence of small, buoyant balls of fluff floating around the entrance of a natal lodge or burrow will confirm breeding, though other methods may help confirm breeding status before this. Confirmation of pregnancy may be difficult, as typically adult beavers in varying body condition still exhibit rounded bellies and are often deemed 'fat', given their extensive digestive system which is evolved for consuming a high cellulose and fibrous diet. Therefore, confirming pregnancy through rotund body shape is problematic. In later stages of pregnancy (~2 months after mating) or lactation (~1 month after birth), females exhibit very prominent nipples (>0.5 cm) which are visible between their fore legs (four), with the upper two providing 50–75% of milk (Bradt, 1938; Osborn, 1955). Outside of these periods nipple size contracts and are not typically visible outside the fur. Observation of nipples during this time is indicative that breeding will or has occurred, ahead of any kits being visible. Some field researchers have assessed breeding status using sound recording equipment at natal lodges or burrows to listen for kit vocalizations (Nolet and Baveco, 1996; see Section 7.4.1).

Often when watching beavers with those unfamiliar with the species or listening to their descriptions of observations, a large beaver will typically be described as a male. In fact, adults of both sexes are incredibly similar in body sizes and weights; if anything, females can be 1.0–1.5 kg heavier. Sex determination is not straightforward in beavers due to lack of external genitalia and obvious morphological differences (apart from enlarged nipples. In both sexes the urinogenital and anal tracts open into a common cloaca. Both sexes also possess anal glands and castor sacs, which should not be confused with testes. The most practical method is to examine the coloration and viscosity of the anal gland secretion (AGS), which is yellowish and more liquid in male Eurasian beavers compared to females' AGS which tends to be grey/white and more thick/paste-like (Rosell and Sun, 1999; Figure 10.18a and b). In North American beavers the AGS is brown and viscous in males and whitish or light-yellow and runny in females (Schulte et al., 1995). Observational tests on inexperienced volunteers proved people can reliably distinguish colour and viscosity and assign sex and beaver species if they knew what they were looking for (Rosell and Sun, 1999). In some areas, beaver species may need to be confirmed (see Section 1.4 for differences between the species); this is most likely to be achieved through genetic analysis (e.g. McEwing et al., 2014), though examination of AGS if sex is established can provide a quicker and cheaper in-field test to allow species identification (Rosell and Sun, 1999). AGS can be collected whilst a beaver is safely restrained in a cloth sack with the tail region exposed. By palpating the anal gland and papillae, AGS can be examined through noting the colour and consistency of the secretions only coming from the papillae and not to be confused with any

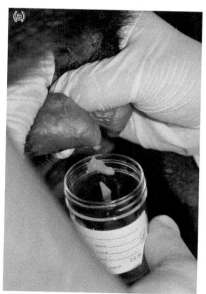

Figure 10.18a and b Sexing Eurasian beavers via colour and consistency of AGS: female (left) and male (right). (Photo supplied courtesy of Jon Cracknell.)

castoreum leaking from the cloaca area (Rosell and Sun, 1999). Additionally, trials with dogs have demonstrated species discrimination abilities through the presentation of castoreum as a potential future field technique (Rosell et al., 2019).

In dead and/or immobilized animals, the baculum (penis bone) or os penis will readily show up on X-ray (Wilsson, 1971) and can be felt for by palpating the abdomen or inguinal region, but this may be difficult in young individuals or next to impossible in unsedated adults (Bradt, 1938; Osborn, 1955). During beaver reintroduction to the Netherlands, this technique was shown to be prone to error, with 13% of the beavers involved being mis-sexed, later confirmed via X-ray and post-mortem examination (Nolet and Baveco, 1996). On beaver fur farms, Zaniewski (1965) gave beavers combelen, an anaesthetic which causes the penis to emerge through the cloaca. Visual examination of the cloaca openings (one in males and two in females) has been proposed as a sexing technique, but again this can be difficult and unreliable, especially in young animals, and has not been widely adopted (Young, 1936).

Cadavers can also be examined for either testicles (small, white) or uterus, again without confusing the paired castoreum and anal gland which lie in close proximity. Sexual dimorphism of the anal glands has been reported for Eurasian beavers in Russia (Shchennikov, 1986), though this has not provided a reliable sexing method in the field, noting species differences have been discussed (see Figure 3.21). Fetuses can generally be sexed through the examination of the gonads from 30+ days (Gunson, 1970). Sex can also be determined through more complicated laboratory processes, if time and money permit. These include examination of the sexual dimorphism of blood neutrophils (Larson and Knapp, 1971; Patenaude and Genest, 1977) or looking for Barr bodies (Barr and Bertram, 1949). Sex can also be determined genetically using polymerase chain reaction (PCR) amplification and then comparing differences in the sex-determining region on the Y chromosome (Kühn et al., 2002) or zinc-finger protein genes on the X and Y chromosomes (Zfx and Zfy, respectively) (Williams et al., 2004). In comparing the reliability of sexing via palpitation for an os penis and examination of anal glands, genetic examination for either Zfx (larger) or Zfy for confirmation of 96 North American beavers found that the field methods for sexing had an accuracy of 89.6% (Williams et al., 2004). Obviously, sexing of beavers can be prone to some degree of error and varies greatly with experience and field circumstances. Genetic tests can confirm sex, especially when establishing new pairs where sex determination is vital, though it may take several lab days to process.

10.4.4 Age determination

Age determination is important in life-history monitoring and can therefore inform population management and conservation strategies (Morris, 1972). It can be difficult to age beavers in the field, as the different age classes can look remarkably similar, especially if their fur is wet and they are seen from any distance, as they can look smaller than they really are. Generally, four age classes are defined: kits—offspring from that year; yearlings—offspring from the previous breeding season; subadults—between yearling and adult; and adults—≥3 years (Campbell, 2010; see Section 1.3). Female beavers have been determined to increase in body length and mass until around 3 years of age, when these dimensions generally flatten, allowing mass gain for pregnancies (Parker et al., 2017). Swimming position in the water can help assign age class: individuals having much of their back and even upper hind quarters above the surface of the water as they swim are denoted as kits or yearlings, whereas subadults and adults generally swim with only their heads exposed above the water surface (Figure 10.19). Older adults can be noticeably grey around the muzzle, noted in Eurasian but not so far in North American beavers, and although still large, their body weight can start to hang differently around their pelvis as they start to lose muscle mass, as typical of other mammals.

There are various ageing techniques, according to ageing a live or deceased individual, and there has been much discussion on the reliability of these methods within the academic literature. As a fur-bearing species across much of its range, pelt size has historically been investigated as a potential ageing technique in relation to body weight. A curvilinear regression of pelt size against weight has been developed (Buckley and Libby, 1955); other studies

Figure 10.19 Age determination based on approximate degree of body exposed above water while swimming. (Illustration provided courtesy of Rachael Campbell-Palmer, adapted from Long, 2000.)

have found considerable overlap and variation between age classes and not found this a reliable technique, especially in older animals (Larson and Van Nostrand, 1968). Studies on skeletal analysis and development in beavers have identified fairly constant trends, across both sexes, especially in the degree and sequence of epiphyseal fusion (Fandén, 2005). This method appears to reliably classify beavers into five age classes: juveniles (<1 year), subadults (1–2.5 years), young adults (3–5 years), middle-aged adults (6–9 years), and old adults (>12 years). One of the most reliable ageing methods employed for both species is the determination of cementum and dentine annuli in molars (van Nostrand and Stephenson, 1964; Larson and Van Nostrand, 1968; Klevezal and Kleinenberg, 1969; Boyce, 1974); this is based on tooth morphology and extent of basal closure (degree of eruption) and cementum annuli analysis (tooth sectioning) of pre-molars—developed independently (van Nostrand and Stephenson, 1964) and in Eurasian beavers (Kleinenberg and Klevezal, 1966). Of course, this is not an acceptable field technique for live animals and more recently it has been deemed practically impossible to age live beavers through tooth morph-ology using non-invasive methods (Stefen, 2009). The extent of the closure of pulp cavities and state of tooth eruption is a reliable ageing method for up

to 2 years (Ognev, 1963). Kits have incisors at birth, deciduous molars appear around 1 month, with the 1st, 2nd, and 3rd molars erupting by 6 months, and lastly the premolars appear by 10–12 months (Mayhew, 1978; Mayhew, 1979). Dental X-ray ana-lysis of the lower jaw is an ageing method that can be applied to living beavers, though it is question-able as a practical field technique (Hartman, 1992; Table 10.3).

Numerous studies have looked at head measure-ments, obviously on skulls of known ages, but with a mind for field application potential. Breadth of the zygomatic arches (skull width) has been a fairly reli-able method, given the thin layer of skin and fur over the bone here, and can be applied to live beavers, and of course skull size doesn't vary with seasonal weight fluctuations (Patric and Webb, 1960). Other authors are less keen on this method, and large error rates are still found (Baker and Hill, 2003). As with many of these techniques, regression modelling and estimat-ing the age of younger animals is relatively easy, but as beavers age, distinct classification categories become less defined. For example, various body and mandible measurements in Austrian beavers deter-mined a linear relationship with age and 'tooth-index' (crown width × body length / tooth length) (Mangen et al., 2018). The authors concluded that ageing beavers up to 5 years was best undertaken

Table 10.3 Criteria for determining age in beavers through lower jaw X-ray analysis (M_1 premolar and M_2 molar)

Beaver age	Description
1 year	Roots of M_1 and M_2 fully open. Primary premolar potentially visible as a cap over erupting permanent premolar.
2 years	M_2 thickened lip of deposited cementum on rim around open-ended root (posterior side). Sharp and pointed.
3 years	M_2 posterior lip thicker and irregular, with more rounded tip. Anterior lip club-shaped.
4 years	M_2 posterior lip even thicker. Anterior lip club-shaped and even more pronounced. Basal cavity can appear closed but looks open on X-ray.
5 years	Basal opening of M_2 fine and narrow.
6 years	Cementum and dentine deposition on M_2 appear as one solid tip.
≥7 years	Relations between solid cementum and dentine tip and non-cementum parts of M_1 and M_2 become main age criteria. Solid tip increases in size now with age, 10–15% to ≥50% in older individuals.

Source: After Hartman (1992).

through mandible measurements and inspection of basal tooth openings, whereas cementum annuli and tooth-index was more accurate in older animals.

For practical application of ageing beavers in the field various studies have analysed a range of weight versus body dimensions in both species to try to establish robust age estimates (Cook and Maunton, 1954; Buckley and Libby, 1955; Larson and Van Nostrand, 1968; Van Deelen, 1994). Body mass can help assign age class and is the most commonly used criterion. Eurasian beaver classifications of body masses of ≥17 kg and ≤19.5 kg indicate a minimum age of ≥2 years, adults of at least 3 years being ≥19.5 kg (Rosell et al., 2010; Mayer et al., 2017a; Mayer et al., 2017b). After the age of ~3 years beavers are virtually impossible to age beyond 'adult' as adult body length, skull dimensions, and weight are generally reached around this point and growth plates are closed (Rosell et al., 2006). The greatest variation in body weight typically occurs between kit and yearling age classes (Campbell et al., 2013). Kits are typically born ranging from 400 to 650 g but remain

in the lodge at least 4–6 weeks before emerging, so they should weigh several kilograms before first capture event (Parker and Rosell, 2001). Therefore, weight of beavers up to 1 year old is <10 kg, with yearlings (12–24 months) typically between 10 and 15 kg (Hartman, 1992; Rosell and Pedersen, 1999). More recently, Rosell et al. (2010) aged beavers (via counts of cementum annuli zones) and then correlated various body morphometries such as skull and tail dimensions, weight, and body and hind foot length to determine which, if any, could be used as age predictors. From these, only skull length and width proved reliable (probability >50%) but then only for beavers ≤2 years. Remaining measurements displayed curvilinear relationships, which are particularly ineffective for older animals. These authors concluded that if separated into age class (kit > yearling > subadult > adult) then skull length and width can be practical, otherwise body measurements are unsuitable for accurate ageing. Therefore, although age class assignment is generally possible in the field, accurate determination of age remains difficult in live beavers and requires further investigation, i.e. comparison studies with known age individuals.

10.5 Monitoring population dynamics

Many of the techniques described above for the captive setting, also have application in wider field to help answer the question of how to monitoring wild animals and determine their population dynamics.

The unique field signs of the beaver put them in a class of their own when it comes to opportunities for monitoring and measuring populations. Whilst it is often fairly straightforward to locate an active family in a given area, it may not always be so easy to detect early colonization and low population density. In northerly climates with winters reducing ground vegetation growth, autumn field signs of beaver activity are obvious and straightforward to monitor. Measuring active families at other times of the year and/or in regions with prolonged/year-round growth seasons increases the chance for incorrect estimations (Rosell and Parker, 1995).

The most commonly used technique when assessing beaver numbers in an area is to multiply the number of active families by the average number of beavers per family. Thus far there have been a number of different methods for attempting to estimate the number of active families and number of beavers per family; these have all given very different results (Rosell and Parker, 1995).

10.5.1 Identifying potential habitat and key signs of activity

Several authors have discussed key factors in identifying suitable beaver habitat and selection (Hartman, 1996; Fustec et al., 2001; Maringer and Slotta-Bachmayr, 2006; Pinto et al., 2009; Swinnen et al., 2017). As a starting point of any surveying, a detailed map will aid in pinpointing areas of freshwater, small wooded ponds, and small backwaters worth the effort of ground-truthing through a foot- or water-based survey. However, it should be noted, especially as densities increase, that beavers are perfectly adapted to reside in areas that may initially not appear suitable as part of a GIS map or modelling approach, and the degree of woodland coverage often quoted as a key requirement can be met by a scattering of trees not showing up as woodland and / or arable crops, so such areas should not be ruled out.

Some important tracks and signs can be used to evaluate the presence of beavers in an area (Novak, 1987; Rosell and Parker; 1995; Rosell and Pedersen, 1999; see also Chapters 4, 5 and 7):

- A newly constructed lodge and / or obvious resurfacing of a lodge with fresh materials (mud and branches). Note not all inhabited lodges will have new mud and branches in the autumn. It may be difficult to determine the age of the construction materials later in the autumn.
- Dams that have been rebuilt or that are newly constructed.
- Food caches outside the lodge.
- Newly felled trees and fresh teeth marks on stems.
- Newly debarked twigs—often at regular feeding sites.
- Fresh scent markings, often not always with the distinctive scent of castoreum.
- Active canals and ditches.
- Fresh forage trails.
- Faeces in still waters around the lodge or dam, potentially even frozen.

- Air bubbles trapped under the ice outside the lodge's entrance. Air bubbles may be generated from air trapped in the pelt or breathed out by a beaver swimming under the ice. Bubbles can also be generated from the fermentation of stored vegetation or sediments, but these will not be concentrated in the area just outside the lodge. In populated locations, there can be accumulations of bubbles outside several lodges, even though only one of the lodges is used as a permanent home. This is most likely a result of the beavers visiting older lodges during winter. Families can also occasionally split up and use several lodges; the reasons for this are unknown.
- Melted openings on the roof or rime around the ventilation opening of lodges or burrows. May not always be a very useful indicator because melted openings can form even when the animals are not in the lodge.
- Fresh footprints in areas without lodges. This can be explained in three ways:

  1. Overwintering in a burrow at this location; or
  2. Beavers have abandoned the area in the late autumn; or
  3. The area is being visited by animals from nearby locations.

It should be highlighted that families in underground burrows are often overlooked and under-counted. These burrowing families are typically found where the waterline is steep and the bank substrate is easy to excavate. Points 2 and 3 are heavily dependent upon the weather. If the water has frozen, migration/relocation is impossible until the ice breaks, but if the water remains unfrozen, the beaver can be mobile, over a wider area year-round. Exploratory movements outside the territory for shorter intervals can also occur (Mayer et al., 2017b; see also Section 6.9.2).

10.5.2 Determining number of active families

A number of different methods have been used to count the number of active beaver families in a given area. The first counts have been done on foot, on horseback, or by boat. In North America, lodges would often be marked on a map by wildlife managers, park rangers, or hunters. With such accurate censuses of lakes and waterways in an area it is easier to locate the families found there. Physical surveys could also serve as references for testing less-accurate population measurements. This method is time-consuming and expensive, and wildlife managers were quick to see the potential in the use of airplanes for population estimation.

Planes make it possible for wildlife managers to assess large, inaccessible areas for a relatively low price. Today the majority of aerial assessments occur just after the leaves have fallen and before the ice has set on the lakes. Visibility is better after the leaves have fallen, and the food caches are either under construction or are finished and are easily seen from the air. The primary criterion for recording that a lodge is active is the presence of a food cache, but remember that not all families have food caches (see Section 4.3). Other criteria such as muddy dams and lodges, newly debarked branches, high water levels or muddy water, freshly felled trees, and new trails can be used to draw the attention of observers looking for food caches. The effectiveness of aerial censuses can vary significantly. Use of a helicopter allows for more families to be observed, but this method is more expensive.

It is important to remember that a beaver family can have several lodges (as many as six), but the beavers will usually only build a single food cache for the entire family. It is therefore important to separate the total number of inhabited lodges from the number of repaired lodges. When assessing the number of families one should also estimate the size of the typical territory in the area so as to be aware of the possibility that there can be several lodges within the territorial borders, which will therefore belong to the same family. There is normally at least 400 m between lodges of active families, but they can be as close as 100 m.

10.5.3 Determining family group size

To estimate the population size within an area, the average family size is required in addition to the number of active families (see Section 10.5.2). This number is also important in connection with the establishment of hunting quotas. Beaver family group composition and size varies (see Section 6.3); therefore determining actual numbers within a family is more difficult than counting the number of families within an area. There are three main methods used for estimating family group size and the composition of individuals in a beaver family (Buech, 1985; Rosell et al., 2006). These are: (1) complete removal by dead-trapping (Hay, 1958); (2) mark and release live trapping (Busher et al., 1983; Briggs et al., 2020; see section 10.4.2); and (3) nocturnal observational counts using night-vision binoculars (Svendsen, 1980; McTaggart and Nelson, 2003). Unfortunately, the methods used to determine family group size, family composition, and litter size are often poorly described (Svendsen, 1980; Rosell et al., 2006). Undoubtedly, a range of many different techniques are used, which causes considerable variation in results (Novak, 1987; Rosell et al., 2006). Family group size and composition in beavers should therefore be viewed with caution if not obtained by methods tested for both precision and accuracy (see Section 6.3 and Chapter 7).

Attempting to determine family members using live traps or killing traps presents problems, as it can be difficult to capture all of the beavers; it is also time-consuming and work-intensive. If you capture them with killing traps it can be slightly easier to capture all of the members of a family than if live traps are used. Experience using live trapping shows that the effectiveness of capture is quickly reduced after the first animal is captured. It appears that the remaining animals learn to avoid the traps their compatriots have been taken in. On some occasions the captured animal can leave the family. Live trapping also varies according to the type of trap (the Hancock trap is usually better than the Bailey trap), bait (aspen or castoreum; castoreum will attract more males than females in the spring), which animal is captured (some are smarter than others), weather (beaver activity is reduced in poor weather), location (it is easier to trap beavers in grain fields), the trapper (some are more skilled than others and more enthusiastic about capturing animals), and the beaver species.

Direct counts of individuals outside a lodge can be one method to achieve an average number of animals per family, but this method is also highly unreliable (Rosell et al., 2006). If you are going to count animals near a lodge, you must find a place near the lodge and observe animals that appear in the evening. It is very difficult to get a good idea of how many animals are in a family. The number is dependent on the location, and in creeks and rivers it is nearly impossible to accurately estimate family size. It is easier to obtain a good estimate in smaller lakes and ponds, where one can observe the entire area from a single observation point. Once the animals emerge from the lodge they may not return before dark, but they will do so on occasion. It is very uncommon for several members of a family to leave the lodge at the same time. It is also easy to count the same animal several times. To be completely certain you should use the number of animals seen at the same time, but this number will almost certainly be lower than the real number. There are as a rule more animals in a family than one believes, especially in established families.

For example, in Telemark in Norway, family members in 19 families were repeatedly counted during dusk and dawn and during the months August and September (Rosell et al., 2006). No difference was found between dusk and dawn, but family group size was significantly greater during August than September. Significantly, all family members were rarely seen during a single dusk or dawn count, and successive counts often provided a better estimate for the family group size. The mean adjusted family group size increased during six successive counts over 7 weeks from 2.4 to 3.8. Overall, family composition was estimated as 54% adults (>27 months), 26% yearlings, and 19% kits. Note that kits are often underestimated, probably because many leave the lodge after total darkness. A review of 19 studies in various areas in North America found that on average the family composition was 38% (range 8–66%) for adults (≥2.5 years old), 26% (range 10–43%) for yearlings, and 35% (range 16–66%) for kits (Novak, 1987). A huge variation (range) was found between the different studies as many factors can affect the family composition (see factors mentioned in Section 6.7). There is a clear need for more studies in this area, performed using standard counting methods (to make the studies comparable) and in different phases of population development (increasing, stable, or decreasing) (Rosell and Parker, 1995; see Section 6.3).

The year's new kits can cause a number of problems under observation, because they tend to come out late in the evening and do not become active outside the lodge before late in the autumn when it is dark early. It becomes necessary to repeat the counts because it is very rare to observe the total number in the first evening. If you have a lot of experience you can separate the three age classes from each other in the autumn, these being the year's newest kits, the previous year's kits, and the animals that are 2 years old or older. It is very difficult to separate 2-year-olds from those even older. After several observations, the family size can be 'constructed' by adding together the different individuals from each age class that have been observed during the same period (Rosell et al., 2006). Results from Telemark, Norway, indicate that the last half of August is the best time to estimate family size, and that it is somewhat easier to count the number of animals that leave the lodge in the evening (Rosell et al., 2006). It can also be possible to count the number of animals later in the autumn when they are actively collecting food caches. It becomes dark earlier, so the use of a spotlight and binoculars is necessary. There is also a good deal of frost condensation over the waters that can make it difficult to see the individual animals. Family size can be easier to estimate if some animals are captured and tagged/marked before the observations begin; information on the number of tagged and untagged animals can be collected. It is then easier to gain information about how many animals there are within a given age class.

Some researchers believe the number of animals within a lodge can be estimated via the size of the food cache going into winter. There are a number of factors that govern the size of the food cache (see Section 4.3.3). The consensus is that this is not a method that can be used, though it can give a coarse indicator as to how many animals can be found in the lodge in some areas—it can indicate whether there is one animal, a few, or very many.

A completely accurate overview of the population in an area is difficult to attain for the following reasons:

- The number of beavers in an overwintering family can vary greatly from family to family.
- Single beavers or floaters (see Section 6.9.5) can overwinter (or be 'cruising' around a population)

in burrows in the same area where family lodges are in use, especially when they are near large lakes or rivers.

- Floaters can overwinter in burrows in areas where there are no other beavers.

Single individuals are therefore a confounding factor when estimating beaver populations within an area, mainly as they are difficult to discover and count. It is important to remember that the size of the family can change from one year to the next, for example because of hunting or changes in reproductive success, so changes in the total number of families will not necessarily reflect any changes in the population size.

By nature, humans, especially various government or landowning entities, are driven by numbers, and there is often a pressure to place figures on 'how many beavers are there'. Often it is far more practical, as there will undoubtedly be imprecision, to invest efforts in evaluating habitat quality, likely colonization and occupation, potential management issues (such as potential for damming or not), and local tolerance of beaver presence (see Chapter 11). The impact of beavers and general population status (expanding, contacting, at carrying capacity, etc.), coupled with proactive, robust, and measured management, should be the focus of long-term beaver presence and identifying numbers of active territories rather than getting hung-up on actual population numbers.

References

Aeschbacher, A. and Pilleri, G. 1983. Observation on the building behaviour of the Canadian beaver (*Castor canadensis*) in captivity. In: Pilleri, G. (ed.) *Investigations on Beaver*. Berne: Brain Anatomy Institute.

Aleksiuk, M. 1970. The function of the tail as a fat storage depot in the beaver (*Castor canadensis*). *Journal of Mammalogy*, 51, 145–148.

Arjo, W. M., Joos, R. E., Kochanny, C. O., Harper, J. L., Nolte, D. L., and Bergman, D. L. 2008. Assessment of transmitter models to monitor beaver *Castor canadensis* and *C. fiber* populations. *Wildlife Biology*, 14, 309–317.

Arnemo, J. M., Ahlqvist, P., Andersen, R., Berntsen, F., Ericsson, G., Odden, J., Brunberg, S., Segerström, P., and Swenson, J. E. 2006. Risk of capture-related mortality in large free-ranging mammals: experiences from Scandinavia. *Wildlife Biology*, 12, 109–113.

Backhouse, F. 2018. *Once They Were Hats: In Search of the Mighty Beaver*. Toronto: ECW Press.

Bailey, V. 1927. Beaver habits and experiments in beaver culture. *USDA Technical Bulletins*, 21, 1–39.

Baker, B. W. 2006. From the field: efficacy of tail-mounted transmitters for beaver. *Wildlife Society Bulletin*, 34, 218–222.

Baker, B. and Hill, E. 2003. Beaver (*Castor canadensis*). In: Feldhamer, G. A., Thompson, B. C., and Chapman, J. A. (eds.) *Wild Mammals of North America: Biology, Management and Conservation*. Baltimore: Johns Hopkins University Press.

Barr, M. L. and Bertram, E. G. 1949. A morphological distinction between neurones of the male and female, and the behaviour of the nucleolar satellite during accelerated nucleoprotein synthesis. *Nature*, 163, 676–677.

Becciolini, V., Lanini, F., and Ponzetta, M. P. 2019. Impact of capture and chemical immobilization on the spatial behaviour of red deer *Cervus elaphus* hinds. *Wildlife Biology*, 2019, 1–8.

Beck, J. L., Dauwalter, D. C., Staley, D. M., and Hirtzel, S. R. 2008. *USDA forest service region 2 monitoring protocol for American beaver* (Castor canadensis): *examples from the Bighorn and Black Hills National Forests*. Denver, CO: USDA Forest Service, Rocky Mountain Region.

Beck, J. L., Dauwalter, D. C., Gerow, K. G., and Hayward, G. D. 2010. Design to monitor trend in abundance and presence of American beaver (*Castor canadensis*) at the national forest scale. *Environmental Monitoring Assessment*, 164, 463–479.

Bejenaru, L., Stanc, S., Popovici, M., and Balasescu, A. 2015. Beavers (*Castor fiber*) in the past: Holocene archaeological evidence for beavers in Romania. *International Journal of Osteoarchaeology*, 25, 375–391.

Belovsky, G. E. 1984. Summer diet optimization by beaver. *American Midland Naturalist*, 111, 209–222.

Berghofer, C. B. 1961. Movement of beaver. *Annual Conference Western Association State Game and Fish Commissioners*, 1961. 181–184.

Bloomquist, C. K. and Nielsen, C. K. 2009. A remote videography system for monitoring beavers. *Journal of Wildlife Management*, 73, 605–608.

Bloomquist, C. K. and Nielsen, C. K. 2010. Demography of unexploited beavers in southern Illinois. *Journal of Wildlife Management*, 74, 228–235.

Boarman, W. I., Beigel, M. L., Goodlett, G. C., and Sazaki, M. 1998. A passive integrated transponder system for tracking animal movements. *Wildlife Society Bulletin*, 26, 886–891.

Boyce, M. S. 1974. Beaver population ecology in interior Alaska. MSc thesis, University of Alaska.

Bradt, G. W. 1938. A study of beaver colonies in Michigan. *Journal of Mammalogy*, 19, 139–162.

Breck, S. W. and Gaynor, J. S. 2003. Comparison of isoflu-rane and sevoflurane for anesthesia in beaver. *Journal of Wildlife Diseases*, 39, 387–392.

Breck, S. W., Wilson, K. R., and Andersen, D. C. 2001. The demographic response of bank-dwelling beavers to flow regulation: a comparison on the Green and Yampa rivers. *Canadian Journal of Zoology*, 79, 1957–1964.

Brenner, F. J. 1962. Foods consumed by beavers in Crawford county, Pennsylvania. *Journal of Wildlife Management*, 26, 104–107.

Brick, P. and Woodruff, K. 2019. The Methow beaver project: the challenges of an ecosystem services experiment. *Case Studies in the Environment*, 3, 1–14.

Briggs, A. J. A., Robstad, C. A., and Rosell, F. 2021. Using radio-frequency identification technology to monitor Eurasian beavers. Wildlife Society Bulletin, 45, 154–161.

Brommer, J. E., Alakoski, R., Selonen, V., and Kauhala, K. 2017. Population dynamics of two beaver species in Finland inferred from citizen-science census data. *Ecosphere*, 8, e01947.

Buckley, J. L. and Libby, W. 1955. Growth rates and age determination in Alaskan beaver. *Transactions of the North America Wildlife Conference*, 20, 495–507.

Buech, R. R. 1983. Modification of the Bailey live trap for beaver. *Wildlife Society Bulletin*, 11, 66–68.

Buech, R. R. 1985. Methodologies for observing beavers (*Castor canadensis*) during the activity period. In: Brooks, R. P. (ed.) *Nocturnal Mammals*. State College: Pennsylvania State University.

Busher, P. E. 1975. Movements and activities of beavers, *Castor canadensis*, on Sagehen Creek, California. MSc thesis, San Francisco State University.

Busher, P. E., Warner, R. J., and Jenkins, S. H. 1983. Population density, colony composition, and local movements in two Sierra Nevadan beaver populations. *Journal of Mammalogy*, 64, 314–318.

Campbell, R. 2010. Demography and life history of the Eurasian beaver *Castor fiber*. PhD thesis, University of Oxford.

Campbell-Palmer, R. and Rosell, F. 2013. *Captive Management Guidelines for Eurasian Beavers* (Castor fiber*). Edinburgh: Royal Zoological Society of Scotland.

Campbell-Palmer, R. and Rosell, F. 2015. Captive care and welfare considerations for beavers. *Zoo Biology*, 34, 101–109.

Campbell, R. D., Rosell, F., Nolet, B. A., and Dijkstra, V. A. A. 2005. Territory and group sizes in Eurasian beavers (*Castor fiber*): echoes of settlement and reproduction? *Behavioral Ecology and Sociobiology*, 58, 597–607.

Campbell, R. D., Newman, C., Macdonald, D. W., and Rosell, F. 2013. Proximate weather patterns and spring green-up phenology effect Eurasian beaver (*Castor fiber*) body mass and reproductive success: the implications of climate change and topography. *Global Change Biology*, 19, 1311–1324.

Campbell-Palmer, R., Gow, D., Campbell, R., Dickson, H., Girling, S., Gurnell, J., Halley, D., Jones, S., Lisle, S., Parker, H., Schwab, G., and Rosell, F. 2016. *The Eurasian Beaver Handbook: Ecology and Management of* Castor fiber. Exeter: Pelagic Publishing.

Campbell-Palmer, R., Rosell, F., Naylor, A., Cole, G., Mota, S., Brown, D., Fraser, M., Pizzi, R., Elliott, M., Wilson, K. and Gaywood, M., 2021. Eurasian beaver (*Castor fiber*) health surveillance in Britain: Assessing a disjunctive reintroduced population. *Veterinary Record*, p.e84.

Campbell-Palmer, R., Senn, H., Girling, S., Pizzi, R., Elliott, M., Gaywood, M., and Rosell, F. 2020. Beaver genetic surveillance in Britain. *Global Ecology and Conservation*, 24, e01275.

Cave, A. J. E. 1984. *Dentitional Anomalies in the Beaver and Some Other Mammals*. Berne: Brain Anatomy Institute.

Charlevoix, P. F. X. 1744. *Journal d'un Voyage dans l'Amérique Septentrionale*. Paris: Chez Rollin Fils.

Coles, B. J. 2019. *Avanke, Bever,* Castor: *The Story of Beavers in Wales*. Wetland Archaeology Research Project.

Collins, T. C. 1976. Population characteristics and habitat relationships of beaver, *Castor canadensis*. Thesis, University of Wyoming.

Colomina, I. and Molina, P. 2014. Unmanned aerial systems for photogrammetry and remote sensing: a review. *ISPRS Journal of Photogrammetry and Remote Sensing*, 92, 79–97.

Cook, A. H. and Maunton, E. R. 1954. A study of criteria for estimating the age of beavers. *New York Fish and Game Journal*, 1, 27–46.

Cunningham, J. M., Calhoun, A. J. K., and Glanz, W. E. 2006. Patterns of beaver colonisation and wetland change in Acadia National Park. *Northeastern Naturalist*, 13, 583–596.

Currier, A., Kitts, W. D., and Cowan, I. M. 1960. Cellulose digestion in the beaver (*Castor canadensis*). *Canadian Journal of Zoology*, 38, 1109–1116.

Davis, J. R., von Recum, A. F., Smith, D. D., and Guynn, D. C. 1984. Implantable telemetry in beaver. *Wildlife Society Bulletin*, 12, 322–324.

de Almeida, M. H. 1987. Nuisance furbearer damage control in urban and suburban areas. In: Novak, M., Baker, J. A., Obbard, M. E., and Malloch, B. (eds.) *Wild Furbearer Management and Conservation in North America*. Toronto: Queens Printer for Ontario.

Destefano, S., Koenen, K. K., Henner, C., and Strules, J. 2006. Transition to independence by subadult beavers (*Castor canadensis*) in an unexploited, exponentially growing population. *Journal of Zoology*, 269, 434–441.

Dezhkin, V. V. and Romashov, V. A. 1976. Characteristics of some interior features of beavers in connection with the peculiarities of their biology. *Proceedings of the Voronezh State Nature Reserve*, 21, 69–77.

Dezhkin, V. V., D'yakov, Y. V., and Safonov, V. G. 1986. *Bobr [The Beaver]*. Moscow: Agropromizdat.

Dollinger, P., Baumgartner, R., Isenbugel, E., Pagan, N., Tenhu, H., and Weber, F. 1999. Husbandry and pathology of rodents and lagomorphs in Swiss zoos. *Erkrankungen der Zoo- und Wildtiere*, 39, 241–254.

Fandén, A. 2005. Ageing the beaver (*Castor fiber* L.): a skeletal development and life history calendar based on epiphyseal fusion. *Archaeofauna*, 14, 199–213.

Farsund, P., Robstad, C.A., and Rosell, F. 2021. Monitoring Eurasian beavers with fixed PIT-tag readers on land and detection success rates. In prep.

Fish, F. E., Smelstoys, J., Baudinette, R. V., and Reynolds, P. S. 2002. Fur doesn't fly, it floats: buoyancy of pelage in semi-aquatic mammals. *Aquatic Mammals*, 28, 103–112.

Flemming, M. W. 1977. Induced sterility of the adult male beaver (*Castor canadensis*) and colony fecundity. MSc thesis, University of Massachusetts.

Fustec, J., Lodé, T., le Jacques, D., and Cormier, J. 2001. Colonization, riparian habitat selection and home range size in a reintroduced population of European beavers in the Loire. *Freshwater Biology*, 46, 1361–1371.

Girling, S. J., Campbell-Palmer, R., Pizzi, R., Fraser, M. A., Cracknell, J., Arnemo, J., and Rosell, F. 2015. Haematology and serum biochemistry parameters and variations in the eurasian beaver (*Castor fiber*). *PLoS ONE*, 10, e0128775.

Goodman, G., Girling, S., Pizzi, R., Meredith, A., Rosell, F., and Campbell-Palmer, R. 2012. Establishment of a health surveillance program for reintroduction of the Eurasian beaver (*Castor fiber*) into Scotland. *Journal of Wildlife Diseases*, 48, 971–978.

Gow, D. 2002. *The transport, quarantine and captive management of European beaver* Castor fiber. Inverness: Scottish Natural Heritage.

Graczyk, R. 1985. *Restitution of the European Beaver* (Castor fiber Linnaeus, 1758) in the Wielkopolska Region—Introduction, Population Size and Distribution. Wielkop: Kronik.

Graf, P. M., Wilson, R. P., Qasem, L., Hackländer, K., and Rosell, F. 2015. The use of acceleration to code for animal behaviours; a case study in free-ranging Eurasian beavers *Castor fiber*. *PLoS ONE*, 10, e0136751.

Graf, P. M., Hochreiter, J., Hackländer, K., Wilson, R. P., and Rosell, F. 2016. Short-term effects of tagging on activity and movement patterns of Eurasian beavers (*Castor fiber*). *European Journal of Wildlife Research*, 62, 725–736.

Grasse, J. E. and Putnam, E. F. 1950. Beaver management and ecology in Wyoming. *Wyoming Game Fish Committee Bulletin*, 6, 1–52.

Greene, S. A., Keegan, R. D., Gallagher, L. V., Alexander, J. E., and Harari, J. 1991. Cardiovascular effects of halothane anesthesia after diazepam and ketamine administration in beavers (*Castor canadensis*) during spontaneous or controlled ventilation. *American Journal of Veterinary Research*, 52, 665–668.

Gunson, J. R. 1970. Dynamics of the beaver of Saskatchewan's northern forest. MSc thesis, University of Alberta.

Guynn Jr, D. C., Davis, J. R., and von Recum, A. F. 1987. Pathological potential of intraperitoneal transmitter implants in beavers. *Journal of Wildlife Management*, 51, 605–606.

Halley, D. J. 2011. Sourcing Eurasian beaver *Castor fiber* stock for reintroductions in Great Britain and western Europe. *Mammal Review*, 41, 40–53.

Hammerson, G. A. 1994. Beaver (*Castor canadensis*): ecosystem alterations, management, and monitoring. *Natural Areas Journal*, 14, 44–57.

Harcourt, R. G., Turner, E., Hall, A., Waas, J. R., and Hindell, M. 2010. Effects of capture stress on free-ranging, reproductively active male Weddell seals. *Journal of Comparative Physiology A*, 196, 147–154.

Harper, J. A. 1969. The role of predation in vegetation diversity. *Brookhaven Symposium of Biology*, 22, 48–62.

Hartman, G. 1992. Age determination of live beaver by dental X-ray. *Wildlife Society Bulletin*, 20, 216–220.

Hartman, G. 1996. Habitat selection by European beaver (*Castor fiber*) colonizing a boreal landscape. *Journal of Zoology*, 240, 317–325.

Havens, R. P., Crawford, J. C., and Nelson, T. A. 2013. Survival, home range, and colony reproduction of beavers in east-central Illinois, an agricultural landscape. *American Midland Naturalist*, 169, 17–30.

Hay, K. G. 1958. Beaver census methods in the Rocky Mountain region. *Journal of Wildlife Management*, 22, 395–402.

Henderson, K., Spicklemier, G., and Ness, T. 2015. *Minnesota Zoological Garden, Minnesota Trial*. https://www.zoolex.org/gallery/show/1598/.

Henn, J. J., Anderson, C. B., and Pastur, G. M. 2016. Landscape-level impact and habitat factors associated with invasive beaver distribution in Tierra del Fuego. *Biological Invasions*, 18, 1679–1688.

Herr, J. and Rosell, F. 2004. Use of space and movement patterns in monogamous adult Eurasian beavers (*Castor fiber*). *Journal of Zoology*, 262, 257–264.

Herr, J. and Schley, L. 2009. Barbed wire hair traps as a tool for remotely collecting hair samples from beavers (*Castor sp.*). *Lutra*, 52, 123–127.

Hill, E. P. 1982. Beaver (*Castor canadensis*). In: Chapman, J. A. and Feldhamer, G. A. (eds.) *Wild Mammals of North America—Biology, Management and Economics*. Baltimore: Johns Hopkins University Press.

Hodgdon, H. E. 1978. Social dynamics and behavior within an unexploited beaver (*Castor canadensis*) population. PhD thesis, University of Massachusetts.

Hodgdon, H. E. and Lancia, R. A. 1983. Behaviour of the North American beaver, *Castor canadensis*. *Acta Zoologica Fennica*, 174, 99–103.

Hosey, G., Melfi, V., and Pankhurst, S. 2009. *Zoo Animals*. Oxford: Oxford University Press.

Jackson, M. D. 1990. Beaver dispersal in western Montana. MSc thesis, University of Montana.

Jenkins, S. H. 1981. Problems, progress, and prospects in studies of food selection by beavers. *Proceedings of the Worldwide Furbearers Conference*, Frostburg, MD, pp. 559–579.

Johnston, C. A. 2015. Fate of 150 year old beaver ponds in the Laurentian Great Lakes region. *Wetlands*, 35, 1013–1019.

Johnston, C. A. and Naiman, R. J. 1990. Browse selection by beaver: effects on riparian forest composition. *Canadian Journal of Forest Research*, 20, 1036–1043.

Jones, S. and Campbell-Palmer, R. 2014. *The Scottish Beaver Trial: the story of Britain's first licensed release into the wild. Final report*. Argyll: Royal Zoological Society of Scotland.

Kawata, K. 2008. Zoo animal feeding: a natural history viewpoint. *Der Zoologische Garten*, 78, 17–42.

Kim, J.-H., Lee, J. Y., Han, T.-S., Han, K.-B., Kang, S. S., Bae, C. S., and Choi, S. H. 2005. A case of maloccluded incisor teeth in a beaver (*Castor canadensis*). *Journal of Veterinary Science*, 6, 173–175.

Kleinenberg, S. and Klevezal, G. 1966. Age determination in mammals by the structure of tooth cement. *Zoologicheskii Zhurnal*, 45, 717–724.

Klevezal, G. A. and Kleinenberg, S. E. E. 1969. *Age Determination of Mammals from Annual Layers in Teeth and Bones*. Translated from Russian by J. Salkind. Jerusalem: Israel Program for Scientific Translations.

Koenen, K., Destefano, S., Henner, C., and Beroldi, T. 2005. Capturing beavers in box traps. *Wildlife Society Bulletin*, 33, 1153–1159.

Kühn, R., Schwab, G., Schröder, W., and Rottmann, O. 2002. Molecular sex diagnosis in Castoridae. *Zoo Biology*, 21, 305–308.

Larson, J. S. and Knapp, S. J. 1971. Sexual dimorphism in beaver neutrophils. *Journal of Mammalogy*, 52, 212–215.

Larson, J. S. and van Nostrand, F. 1968. An evaluation of beaver aging techniques. *Journal of Wildlife Management*, 32, 99–103.

Lavrov, L. S. 1969. Indigenous river beaver colonies in Eurasia: their condition, significance and ways for their preservation. *Proceedings of the Voronezh State Nature Reserve*, 16, 168–177.

Lavsund, S. 1977. Olika bäverjaktmetoder i Sverige. In: Lavsund, S. (ed.) *Proceedings of the Nordic Symposium on the Beaver 1975*, Institute of Forest Zoology, Research Notes No. 26.

Leege, T. A. 1968. Natural movements of beavers in southeastern Idaho. *Journal of Wildlife Management*, 32, 973–976.

MacArthur, R. A. and Dyck, A. P. 1990. Aquatic thermoregulation of captive and free-ranging beavers (*Castor canadensis*). *Canadian Journal of Zoology*, 68, 2409–2416.

Malison, R. L., Lorang, M. S., Whited, D. C., and Stanford, J. A. 2014. Beavers (*Castor canadensis*) influence habitat for juvenile salmon in a large Alaskan river floodplain. *Freshwater Biology*, 59, 1229–1246.

Mangen, T., Parz-Gollner, R., Böhm, J., Hölzler, G., and Graf, P. 2018. Tooth development and mandible measurements for age determination in the Eurasian beaver (*Castor fiber*). *Proceedings of the 8th International Beaver Symposium*, 18–20 September, Denmark.

Maringer, A. and Slotta-Bachmayr, L. 2006. A GIS-based habitat-suitability model as a tool for the management of beavers *Castor fiber*. *Acta Theriologica*, 51, 373–382.

Martin, T. H. 1892. *Castologia. History and Traditions of the Canadian Beaver*. Montreal: W.M. Drysdale and Company.

Marcuzzi, G. 1986. The relations man–beaver. In: Pilleri, G. (ed.) *Investigations on Beavers*. Bern: Brain Anatomy Institute Bern, Vammalan Kirjapaino.

Mason, C. E., Gluesing, E. A., and Arner, D. H. 1983. Evaluation of snares, leg-hold, and Conibear traps for beaver control. *Proceedings of the Annual Conference of the Southeast Fish and Wildlife Agencies*, 1983, pp. 201–209.

Mayer, M., Künzel, F., Zedrosser, A., and Rosell, F. 2017a. The 7-year itch: non-adaptive mate change in the Eurasian beaver. *Behavioral Ecology and Sociobiology*, 71, 1–9.

Mayer, M., Zedrosser, A., and Rosell, F. 2017b. When to leave: the timing of natal dispersal in a large, monogamous rodent, the Eurasian beaver. *Animal Behaviour*, 123, 375–382.

Mayhew, D. F. 1978. Age structure of a sample of subfossil beavers (*Castor fiber* L.). In: Butler, P. M. and Joysey, K. A. (eds.) *Development, Function and Evolution of Teeth*. London: Academic Press.

Mayhew, D. F. 1979. Evolution of a dental character in the beaver *Castor fiber* L. (Mammalia: Rodentia). *Zoological Journal of the Linnean Society*, 65, 177–184.

McEwing, R., Frosch, C., Rosell, F., and Campbell-Palmer, R. 2014. A DNA assay for rapid discrimination between beaver species as a tool for alien species management. *European Journal of Wildlife Research*, 60, 547–550.

McKinstry, M. C. and Anderson, S. H. 1998. Using snares to live-capture beaver, *Castor canadensis*. *Canadian Field Naturalist*, 112, 469–473.

McKinstry, M. C. and Anderson, S. H. 2002. Survival, fates, and success of transplanted beavers, *Castor canadensis*, in Wyoming. *Canadian Field-Naturalist*, 116, 60–68.

McNew, J., Lance B., and Woolf, A. 2005. Dispersal and survival of juvenile beavers (*Castor canadensis*) in southern Illinois. *American Midland Naturalist*, 154, 217–228.

McNew Jr, L. B., Nielsen, C. K., and Bloomquist, C. K. 2007. Use of snares to live-capture beavers. *Human–Wildlife Conflicts*, 1, 106–111.

McTaggart, S. T. and Nelson, T. A. 2003. Composition and demographics of beaver (*Castor canadensis*) colonies in central Illinois. *American Midland Naturalist*, 150, 139–150.

Milishnikov, A. N., Likhnova, O. P., Nikonova, O. A., Lavrov, V. L., and Orlov, V. N. 1994. Allozyme variability of the European beaver *Castor fiber* 1758 (Castoridae, Rodentia) from the Voronezh State Nature Reserve. *Russian Journal of Genetics*, 30, 529–534.

Mills, E. A. 1913. *In Beaver World*. Boston: Houghton Mifflin.

Morris, P. 1972. A review of mammalian age determination methods. *Mammal Review*, 2, 69–104.

Mortensen, R. M. and Rosell, F. 2020. Long-term capture and handling effects on body condition, reproduction and survival in a semi-aquatic mammal. *Scientific Reports*, 10, 17886.

Müller-Schwarze, D. and Haggart, D. P. 2005. From the field: a better beaver trap—new safety device for live traps. *Wildlife Society Bulletin*, 33, 359–361.

Müller-Schwarze, D. 2011. *The Beaver: Its Life and Impact*. New York: Cornell University Press.

Nikulin, V. N. 1954. On the age-related characteristics of the gastrointestinal tract of the beaver. *Proceedings of the Voronezh State Nature Reserve*, 5, 51–55.

Nolet, B. A. and Baveco, J. M. 1996. Development and viability of a translocated beaver *Castor fiber* population in the Netherlands. *Biological Conservation*, 75, 125–137.

Nolet, B. A. and Rosell, F. 1994. Territoriality and time budgets in beavers during sequential settlement. *Canadian Journal of Zoology*, 72, 1227–1237.

Nolet, B. A., Hoekstra, A., and Ottenheim, M. M. 1994. Selective foraging on woody species by the beaver *Castor fiber*, and its impact on a riparian willow forest. *Biological Conservation*, 70, 117–128.

Novak, M. 1987. Beaver. In: Novak, M., Baker, J. A., Obbard, M. E., and Mallock, B. (eds.) *Wild Furbearer Management and Conservation in North America*. Toronto: Queens Printer for Ontario.

O'Brien, D. F. 1938. A qualitative and quantitative food habit study of beavers in Maine. MSc thesis, University of Maine.

Ognev, S. I. 1963. *Mammals of the Union of Soviet Socialist Republics. Volume V—Rodents*. Tel Aviv: Israeli Programme of Scientific Translations.

Osborn, D. J. 1955. Techniques of sexing beaver, *Castor canadensis*. *Journal of Mammalogy*, 36, 141–143.

Osmundson, C. L. and Buskirk, S. W. 1993. Size of food caches as a predictor of beaver colony size. *Wildlife Society Bulletin (1973–2006)*, 21, 64–69.

Owesen, A. 1979. *I beverskog*. Oslo: Gylendal Norsk Forlag.

Parker, H. and Rosell, F. 2001. Parturition dates for Eurasian beaver *Castor fiber*: when should spring hunting cease? *Wildlife Biology*, 7, 237–241.

Parker, H., Haugen, A., Kristensen, Ø., Myrum, E., Kolsing, R., and Rosell, F. 2001. Landscape use and economic value of Eurasian beaver (*Castor fiber*) on a large forest in southern Norway. *Proceedings of the 1st European–American Beaver Congress*, Kazan, Russia, pp. 77–95.

Parker, H., Rosell, F., and Gustavsen, Ø. 2002. Errors associated with moose-hunter counts of occupied beaver *Castor fiber* lodges in Norway. *Fauna Norvegica*, 22, 23–31.

Parker, H., Zedrosser, A., and Rosell, F. 2017. Age-specific reproduction in relation to body size and condition in female Eurasian beavers. *Journal of Zoology*, 302, 236–243.

Parker, J. D., Caudill, C. C., and Hay, M. E. 2007. Beaver herbivory on aquatic plants. *Oecologia*, 151, 616–625.

Patenaude, F. and Bovet, J. 1984. Self-grooming and social grooming in the North American beaver, *Castor canadensis*. *Canadian Journal of Zoology*, 62, 1872–1878.

Patenaude, R. P. and Genest, F. B. 1977. The hematology and chromosomes of the Canadian beaver (*Castor canadensis*). *Journal of Zoo Animal Medicine*, 8, 6–9.

Patenaude-Pilote, F., Oertli, E. F., and Bovet, J. 1980. A device for observing wild beavers in their lodge. *Canadian Journal of Zoology*, 58, 1210–1212.

Patric, E. F. and Webb, W. L. 1960. An evaluation of three age determination criteria in live beavers. *Journal of Wildlife Management*, 24, 37–44.

Payne, N. F. 1981. Accuracy of aerial censusing for beaver colonies in Newfoundland. *Journal of Wildlife Management*, 45, 1014–1016.

Piechocki, R. 1962. Die Todesursachen der Elbe-Biber (*Castor fiber albicus* Matschie 1907) unter besonderer beriicksichtigung funktioneller Wirbelsäulen-Störungen. *Nova Acta Leopoldina*, 25, 1–75.

Pilleri, G., Gihr, M., and Krause, C. 1985. Vocalization and behaviour in the young beaver (*Castor canadensis*). In: Pilleri, G. (ed.) *Investigations on Beavers*. Berne: Brain Anatomy Institute.

Pinto, B., Santos, M. J., and Rosell, F. 2009. Habitat selection of the Eurasian beaver (*Castor fiber*) near its carrying capacity: an example from Norway. *Canadian Journal of Zoology*, 87, 317–325.

Poliquin, R. 2015. *Beaver*. London: Reaktion Books.

Pollock, M. M., Lewallen, G., Woodruff, K., Jordan, C. E., and Castro, J. M. 2015. *The Beaver Restoration Guidebook: Working with Beaver to Restore Streams, Wetlands, and Floodplains*. Portland, OR: United States Fish and Wildlife Service.

Polvi, L. E. and Wohl, E. 2012. The beaver meadow complex revisited—the role of beavers in post-glacial floodplain development. *Earth Surface Processes and Landforms*, 37, 332–346.

Potvin, C. L. and Bovet, J. 1975. Annual cycle of patterns of activity rhythms in beaver colonies (*Castor canadensis*). *Journal of Comparative Physiology*, 98, 243–256.

Puttock, A., Cunliffe, A., Anderson, K., and Brazier, R. E. 2015. Aerial photography collected with a multirotor drone reveals impact of Eurasian beaver reintroduction on ecosystem structure. *Journal of Unmanned Vehicle Systems*, 3, 123–130.

Ranheim, B., Rosell, F., Haga, H. A., and Arnemo, J. M. 2004. Field anaesthetic and surgical techniques for implantation of intraperitoneal radio transmitters in Eurasian beavers *Castor fiber*. *Wildlife Biology*, 10, 11–15.

Richard, P. B. 1975. The beaver in captivity. *International Zoo Yearbook*, 15, 48–52.

Richards, D. 1983. *Beaversprite: My Years Building an Animal Sanctuary*. Interlaken, NT: Heart of the Lakes Publishing.

Robel, R. J., Fox, L. B., and Kemp, K. E. 1993. Relationship between Habitat Suitability Index values and ground counts of beaver colonies in Kansas. *Wildlife Society Bulletin (1973–2006)*, 21, 415–421.

Rosell, F. and Bjørkøyli, T. 2002. A test of the dear enemy phenomenon in the Eurasian beaver. *Animal Behaviour*, 63, 1073–1078.

Rosell, F. and Hovde, B. 2001. Methods of aquatic and terrestrial netting to capture Eurasian beavers. *Wildlife Society Bulletin*, 29, 269–274.

Rosell, F. and Kvinlaug, J. K. 1998. Methods for live-trapping beaver (*Castor* spp.). *Fauna Norvegica, Series A*, 19, 1–28.

Rosell, F. and Nolet, B. A. 1997. Factors affecting scent-marking behavior in Eurasian beaver (*Castor fiber*). *Journal of Chemical Ecology*, 23, 673–689.

Rosell, F. and Parker, H. 1995. *Beaver Management: Present Practice and Norway's Future Needs*. Bø i Telemark: Telemark University College.

Rosell, F. and Parker, H. 1997. *Levendefangst av Beaver (Castor fiber)*. Bo in Telemark: Telemark University College.

Rosell, F. and Pedersen, K. V. 1999. *Bever [The Beaver]*. Oslo: Landbruksforlaget.

Rosell, F. and Sun, L. X. 1999. Use of anal gland secretion to distinguish the two beaver species *Castor canadensis* and *C. fiber*. *Wildlife Biology*, 5, 119–123.

Rosell, F. and Thomsen, L. R. 2006. Sexual dimorphism in territorial scent marking by adult Eurasian beavers (*Castor fiber*). *Journal of Chemical Ecology*, 32, 1301–1315.

Rosell, F., Parker, H., and Steifetten, Ø. 2006. Use of dawn and dusk sight observations to determine colony size and family composition in Eurasian beaver *Castor fiber*. *Acta Theriologica*, 51, 107–112.

Rosell, F., Zedrosser, A., and Parker, H. 2010. Correlates of body measurements and age in Eurasian beaver from Norway. *European Journal of Wildlife Research*, 56, 43–48.

Rosell, F., Cross, H. B., Johnsen, C. B., Sundell, J., and Zedrosser, A. 2019. Scent-sniffing dogs can discriminate between native Eurasian and invasive North American beaver. *Scientific Reports*, 9, 1–9.

Rothmeyer, S. W., McKinstry, M. C., and Anderson, S. H. 2002. Tail attachment of modified ear-tag radio transmitters on beavers. *Wildlife Society Bulletin*, 30, 425–429.

Ryden, H. 1990. *Lily Pond: Four Years with a Family of Beavers*. New York: HarperCollins.

Sainsbury, A. W. 2003. Rodentia (rodents). In: Fowler, M. E. and Miller, R. E. (eds.) *Zoo and Wild Animal Medicine*, 5th edn. St Louis: Saunders.

Saunders, R. 2016. *Other Insectivores and Rodents. BSAVA Manual of Wildlife Casualties*. Gloucester: British Small Animal Veterinary Association.

Schulte, B. A. 1998. Scent marking and responses to male castor fluid by beavers. *Journal of Mammalogy*, 79, 191–203.

Schulte, B. A., Müller-Schwarze, D., and Sun, L. 1995. Using anal gland secretion to determine sex in beaver. *Journal of Wildlife Management*, 59, 614–618.

Sharpe, F. and Rosell, F. 2003. Time budgets and sex differences in the Eurasian beaver. *Animal Behaviour*, 66, 1059–1067.

Shchennikov, G. N. 1986. On the sexual dimorphism of the anal glands of the European beaver. *Abstracts IV Congress of the All-Union Theriological Society*, 27–31 January 1985, Moscow, 2, pp. 119–120.

Shelton, P. C. 1966. Ecological studies of beavers, wolves, and moose in Isle Royale National Park, Michigan. PhD thesis, Purdue University.

Shepherdson, D. J., Mellen, J. D., and Hutchins, M. 1998. *Second Nature: Environmental Enrichment for Captive Animals*. Washington: Smithsonian Institution Press.

Simonsen, T. A. 1973. *Beverens næringsøkologi i Vest-Agder*. Trondheim: Direktoratet for jakt, viltstell og ferskvannsfiske.

Smith, D. W. and Jenkins, S. H. 1997. Seasonal change in body mass and size of tail of northern beavers. *Journal of Mammalogy*, 78, 869–876.

Smith, D. W. and Peterson, R. O. 1988. *The Effects of Regulated Lake Levels on Beaver in Voyageurs National Park, Minnesota, USA*. Omaha: US Department of the Interior, National Park Service, Midwest Regional Office.

Smith, J. B., Windels, S. K., Wolf, T., Klaver, R. W., and Belant, J. L. 2016. Do transmitters affect survival and body condition of American beavers *Castor canadensis*? *Wildlife Biology*, 22, 117–123.

Stefen, C. 2009. Intraspecific variability of beaver teeth (Castoridae: Rodentia). *Zoological Journal of the Linnean Society*, 155, 926–936.

Stegeman, L. C. 1954. The production of aspen and its utilization by beaver on the Huntington forest. *Journal of Wildlife Management*, 18, 348–358.

Stocker, G. 1985. The beaver (*Castor fiber* L.) in Switzerland. Biological and ecological problems of re-establishment. *Swiss Federal Intitute of Forestry Research Reports*, 242, 1–49.

Sun, L., Müller-Schwarze, D., and Schulte, B. A. 2000. Dispersal pattern and effective population size of the beaver. *Canadian Journal of Zoology*, 78, 393–398.

Sutherland, W. J. 1996. *Ecological Census Techniques: A Handbook*. Cambridge: Cambridge University Press.

Svendsen, G. E. 1980. Population parameters and colony composition of beaver (*Castor canadensis*) in southeast Ohio. *American Midland Naturalist*, 104, 47–56.

Svendsen, G. E. 1989. Pair formation, duration of pair-bonds, and mate replacement in a population of beavers (*Castor canadensis*). *Canadian Journal of Zoology*, 67, 336–340.

Swank, W. G. and Glover, F. A. 1948. Beaver censusing by airplane. *Journal of Wildlife Management*, 12, 214.

Swenson, J. E., Knapp, S. J., Martin, P. R., and Hinz, T. C. 1983. Reliability of aerial cache surveys to monitor beaver population trends on prairie rivers in Montana. *Journal of Wildlife Management*, 47, 697–703.

Swinnen, K. R. R., Strubbe, D., Matthysen, E., and Leirs, H. 2017. Reintroduced Eurasian beavers (*Castor fiber*): colonization and range expansion across human-dominated landscapes. *Biodiversity and Conservation*, 26, 1863–1876.

Taber, R. D. and McCowan, T. 1969. Capturing and marking wild animals. In: Giles, H. (ed.) *Wildlife Management Techniques*, 3rd edn. Washington, DC: The Wildlife Society.

Tape, K. D., Jones, B. M., Arp, C. D., Nitze, I., and Grosse, G. 2018. Tundra be dammed: beaver colonization of the Arctic. *Global Change Biology*, 24, 4478–4488.

Tournay, A. 2003. *Beaver Tails*. Boston: Boston Mills Press.

Townsend, P. A. and Butler, D. R. 1996. Patterns of landscape use by beaver on the lower Roanoke River floodplain, North Carolina. *Physical Geography*, 17, 253–269.

van Deelen, T. R. 1991. Dispersal patterns of juvenile beavers in western Montana. MSc thesis, University of Montana.

van Deelen, T. 1994. A field technique for aging live beavers, *Castor canadensis*. *Canadian Field-Naturalist*, 108, 361–363.

van der Ouderaa, A. 1985. Uitzetting van bevers (*Castor fiber* L.) in de Biesbosch. *Huid en Haar*, 4, 31–39.

van Nostrand, F. C. and Stephenson, A. B. 1964. Age determination for beavers by tooth development. *Journal of Wildlife Management*, 28, 430–434.

Vantassel, S. 2006. The Bailey beaver trap: modifications and sets to improve capture rate. *Proceedings of the 22nd Vertebrate Pest Conference*, University of California, Davis, pp. 171–173.

Veasey, J. 2006. Concepts in the care and welfare of captive elephants. *International Zoo Yearbook*, 40, 63–79.

Weaver, K. M. 1986. Dispersal patterns of subadult beavers in Mississippi as determined by implant radio-telemetry. MSc thesis, Mississippi State University.

Webster, J. P. 1994. *Animal Welfare: A Cool Eye Towards Eden*. Oxford: Blackwell Science.

Wheatley, M. 1997. A new surgical technique for implanting radio transmitters in beavers, *Castor canadensis*. *Canadian Field-Naturalist*, 111, 601–606.

Williams, C. L., Breck, S. W., and Baker, B. W. 2004. Genetic methods improve accuracy of gender determination in beavers. *Journal of Mammalogy*, 85, 1145–1148.

Wilson, R. P. and McMahon, C. R. 2006. Measuring devices on wild animals: what constitutes acceptable practice? *Frontiers in Ecology and the Environment*, 4, 147–154.

Wilsson, L. 1968. *My Beaver Colony*. New York: Doubleday.

Wilsson, L. 1971. Observations and experiments on the ethology of the European beaver (*Castor fiber* L.). *Viltrevy*, 8, 160–203.

Windels, S. K. 2013. Ear-tag loss rates in American beavers. *Wildlife Society Bulletin*, 38, 122–126.

Windels, S. K. and Belant, J. L. 2016. Performance of tail-mounted transmitters on American beavers *Castor canadensis* in a northern climate. *Wildlife Biology*, 22, 124–129.

Wingfield, J. C. 2005. The concept of allostasis: coping with a capricious environment. *Journal of Mammalogy*, 86, 248–254.

Young, F. W. 1936. The identification of the sex of beavers. *Michigan State College, Agricultural Bulletin*, 279, 8.

Zaniewski, L. 1965. Sex determination in the European beaver *Castor fiber* Linnaeus, 1758. *Acta Theriologica*, 10, 297–301.

Zbigniew, J., Bogdan, K., Andrzej, S., and Pawel, J. 1995. Hand-rearing of young beavers in captivity. *Proceedings of the International Union of Game Biologists, 22nd Congress 'The Game and the Man'*, 4–8 September, Sofia, Bulgaria, pp. 233–236.

Żurowski, W. and Krzywinski, A. 1974. Abnomalies in the wear of incisors in the European beaver. *Acta Theriologica*, 25, 367–370.

Żurowski, W., Kisza, J., Kruk, A., and Roskosz, A. 1974. Lactation and chemical composition of milk of the European beaver (*Castor fiber* L.). *Journal of Mammalogy*, 55, 847–850.

CHAPTER 11

Living with beavers: an 'adorable nuisance'?

11.1 Why should we live with the beaver?

Humans have been modifying, degrading, infilling, and draining wetlands for centuries. This has been documented in China for at least 2,000 years (An et al., 2007), in North America from the seventeenth century (Dahl, 1990), and undertaken at scale by the Romans throughout Europe (Davidson et al., 1991). Escalating in modern times, riparian landscapes across the northern hemisphere have seen widespread and prolonged decline due to intensive farming practices such as land drainage, grazing, monoculture arable planting, heavy pesticide use, and water abstraction; increasing urbanization and infrastructure; and hydrological engineering schemes (Davidson, 2014). The majority (> 90%) of all European rivers and significant tributaries are regulated by humans, including widespread disconnection from their floodplains (Harvolk et al., 2014). Average global long-term wetland loss has been estimated at 54–57%, with inland natural wetland exhibiting greater loss than coastal wetlands (Davidson, 2014). Wetlands have disappeared 3.7 times faster over the last hundred years than ever before, with an estimated 64–71% now gone from what was present in 1900 and with other regions such as parts of Asia with even higher loss (Davidson, 2014). In the western USA alone an estimated 90% of riparian ecosystems have been significantly degraded or lost completely due to human impacts (Ohmart and Anderson, 1986).

The conservation value of beaver activities in restoring lost wetlands could be significant in the long term. In addition to the biodiversity benefits, beavers are increasingly being proposed as key agents in nature-based solutions for river catchment management (Pollock et al., 2014; Bridgewater, 2018; Brazier et al., 2021). Climate change is causing the frequency and severity of flood events to increase in many part of the world (Kundzewicz et al., 2014). For example, severe flood events in Germany over the last decades, especially in 1999, 2005, and 2013, have generated flood mitigation strategies that integrate nature-based solution with technical processes (Neumayer et al., 2020). This is generating a global shift away from hard-engineering and towards developing a range of adaptive, resilient, long-term flood-management strategies (Morrison et al., 2018; Ciotti et al., 2021).

Already in many parts of North America, Europe, and Mongolia the translocation of beavers is an increasingly popular activity, given the positive ecological and socioeconomic benefits they can bring. This has also been accompanied by human–wildlife conflicts. If beaver presence causes issues it is generally not as straightforward to remove them, and so a wide range of management techniques have been devised and documented (Campbell-Palmer et al., 2016). People's perceptions and opinions on beavers can vary greatly, from conservation superhero to outright vermin. As we live alongside beavers once more, the future of our modern landscapes, perceptions of traditional woodland, and freshwater management will be challenged, away from uniform, tidy forests and canalized waterways. While it is typically much easier to focus on and calculate the costs of beavers, land managers and decision-makers involved with such conflicts should be encouraged to develop a more holistic

Beavers: Ecology, Behaviour, Conservation, and Management. Frank Rosell and Róisín Campbell-Palmer, Oxford University Press.
© Frank Rosell and Róisín Campbell-Palmer 2022. DOI: 10.1093/oso/9780198835042.003.0011

approach which considers the wider society bene-fits of beavers on a catchment, landscape, and regional scale (Pilliod et al., 2018).

The recoveries of both beaver species from near-global extinction have followed remarkably similar stories of decline, realization, active protection, and restoration, and can somewhat arguably be described as one of the greatest conservation success stories. Today, under the International Union for Conservation of Nature (IUCN) both species are classified as of least concern (LC), meaning popula-tions are stable and beavers are common and wide-spread (Batbold et al., 2016; Cassola, 2016; Halley et al., 2020). The Eurasian beaver alone has increased in abundance by > 14,000% since 1955 (Deinet et al., 2013), although Asian populations are small, iso-lated, and under threat, urgently requiring conser-vation interventions (Chu and Jiang, 2009).

As human populations become more urbanized there is also a growing wish to connect with nature, for instance through the growing popularity of 'rewilding' initiatives and species conservation pro-jects. Resource managers, even as early as the 1930s, hit on translocating beavers to degraded areas in order to improve floodplain connectivity, restore watersheds, reduce soil and water loss, and improve vegetation and biodiversity (Scheffer, 1938). Contrastingly, a minority of people typically must deal with the consequences the restoration of cer-tain species can bring, and this has generated deep-seated opposition. Additionally, some systems may not be able to support significant beaver damming and conflict with other land-use practices; therefore the beaver's capacity to restore floodplain connect-ivity may raise unrealistic expectations (Macfarlane et al., 2017).

So the joy of beaver return is not a shared con-sensus. The media may still often portray the prob-lems beavers can cause, even when the positive effects on ecosystems are emphasized (Kaphegyi et al., 2015). The impact of beavers on any environ-ment is strongly correlated with length of occupa-tion and population density (Wajdzik et al., 2013). Fundamental to beaver restoration and persistence in a landscape appears to be the need for both nat-uralization of the riparian zone and active man-agement, which often includes lethal control in

time. This is often a difficult concept in a world where beavers are viewed as anything from 'the greatest original wetland conservationist'(s) to blatant vermin leading to livelihood and property damage (Müller-Schwarze 2011).

11.2 Beaver restoration and human–wildlife conflict management

Land managers in areas where beavers have gener-ally been for some time usually understand and accept their vital role in restoring riparian zone func-tioning and encouragement of biodiversity hotspots. An increasing number of private landowners are allowing beaver restoration on their lands in both North America and Europe (Pollock et al., 2015). However, conflicts can arise and beaver activities may be incompatible with other land-uses. Beavers have the ability to colonize even marginal quality land (see Section 4.1) in human-dominated areas, especially as preferred habitat becomes occupied, which often sees reported conflicts increase (Nolet and Rosell, 1998; Müller-Schwarze and Schulte, 1999).

A study of 97 beaver translocations undertaken in western USA between 1950 and 2016, deter-mined that the main aim of such projects was to relocate 'nuisance beavers', with the next greatest project goal to increase water storage (Pilliod et al., 2018). Public opinion is a key consideration when implementing mitigation tools (Baker and Hill, 2003). People's responses to beaver activities, their perception of damage, and acceptability of various mitigation responses vary greatly, from complete opposition to any interference to requests for ongoing lethal control. Questionnaires sent to 5,563 residents in Massachusetts to assess public attitudes towards beaver management in their state (response rate 47.3%) found that the percep-tion of the species' status could influence the peo-ple's reaction towards beavers and acceptance of them (Jonker et al., 2009). Taking no action to vari-ous beaver impacts was not an option accepted by many people. Implementing non-lethal control was generally accepted, though flow device func-tion for example was not always understood. Lethal control acceptance increased if non-lethal mitigation was perceived to fail and when the perception that beaver damage was increasing. The authors stress that for beaver management actions and programmes to be successful there needs to be a full understanding of the types of beaver conflicts occurring, the type of mitigation offered, and the perception of the animal itself, along with perceived preference of population levels in the future.

Similar conclusions were made following interviews with local people in Hungary and Romania, who were relatively well informed about beaver biology, legal status, and behaviours, though they held more negative views on their ecological services capabilities and had concerns on management (Ulicsni et al., 2020). This study aptly concluded that beaver conflicts, as with other reintroduced species, are generally increasing, though this involves a complexity of perception, resource, and management issues that will take reciprocal learning among the local community, conservation/nature organizations, and government agencies, so as to develop adaptive management strategies. Local perceptions are important but highly variable—ranging from 'beaver slow water down and foul it up' to 'they have an important role in nature', from 'I like seeing them, they have cute faces' to 'there are too many', and 'they make the place look messy and chaotic' (Ulicsni et al., 2020). These views are generally repeated time and time again, thus demonstrating there is a knowledge gap that needs breaching, and therefore there is an opportunity to encourage greater cohabitation.

In Colorado, beavers remained historically after the Fur Trade Wars but were drastically reduced by 1850. From 1900 to 1960 trapping regulations resulted in successful population recovery (Rutherford, 1964). Commercial harvest and lethal control for conflict management were permitted again from 1956, with annual lethal control numbers varying from 5,000 to 20,000 right up until 1995 (Boyles and Owens, 2007). This changed following a citizen referendum in 1996 which voted for lethal trapping to be prohibited in Colorado (Boyles and Owens, 2007). Elsewhere, suburban New York residents more readily accepted trapping and lethal control if

beavers were causing local damage and economic impacts, as opposed to reasons of public health and safety (Loker et al., 1999; Figure 11.1). Experience of beaver impacts also influenced the likelihood of people to pursue non-lethal over lethal solutions. For example, highway maintenance supervisors attempted non-lethal (and therefore longer-term) solutions over lethal more readily than members of the public (Enck et al., 1997). In Wyoming, public land managers and 45% of private landowners recognized the importance of beavers in riparian management (McKinstry and Anderson, 1999). Experiences gathered in both America and Europe demonstrate that fast, proactive, evidence-based, and flexible responses are required for any beaver management programme (Campbell-Palmer et al., 2016).

11.2.1 Conflicts with humans

Somewhat humorously Ben Goldfarb describes the antics of beavers behaving badly in his book *Eager* (2018), including cutting off internet services to residents of Taos, New Mexico for 24 hours, thieving from supermarkets, and ruining golf courses. Of course, these are unusual and even one-off examples, but living alongside beavers in modern modified landscapes can be a challenge, and human–wildlife conflicts are one of the most significant challenges to conservation biology. Treves

Figure 11.1 Trapping for lethal control and translocation is widely implemented in established populations, but various welfare concerns should be addressed. (Photo supplied courtesy of Frank Rosell.)

et al. (2006) define such conflicts as when the presence of wildlife and their activities negatively impact on humans. These can range from the direct killing of people, to damage to property (Figure 11.2), to a social dislike of a species.

Fortunately, dangerous conflicts with beavers are rare, though they can be headline grabbing. In Belarus, a beaver attacked a 60-year-old fisherman. The fisherman spotted the beaver along the side of the road and tried to grab it to take a picture. The beaver bit the man, the bite cut a major artery in his leg, and he unfortunately bled to death (Scott, 2016). Several beaver attacks on humans have been associated with rabid individuals which generally act much more aggressively and have inflicted severe bite wounds requiring treatment on swimmers and kayakers (Morgan et al., 2015). Another more serious and well-publicized incident occurred in 1985 when a train derailed after heavy rain caused a series of beaver dam collapses, leading to a flash flood which washed the embankment away. Five people died and 26 were injured (Müller-Schwarze,

1996). Disease wise, beavers present no greater risk generally to humans compared to other wildlife (see Section 8.5), though they can transmit tularemia to humans on direct contact.

Rare beaver attacks on dogs have been recorded in various states and countries, all of which seem to occur when an overexcited or curious dog approaches a beaver. Typically, injuries may result as beavers defend themselves, especially a threatened adult during the breeding season. At least two attacks are known from almost opposite ends of Britain; in both cases dogs, off the lead, jumped into the water after a swimming beaver and both, although undoubtedly shocked, required some stitches. In 2010, a lethal beaver attack on a husky was recorded in Red Deer, Alberta (Backhouse, 2018). Otherwise, beaver attacks on pets and livestock are generally unknown.

The vast majority of beaver conflicts are related to their activities, predominantly damming and flooding areas, but also burrowing and destabilization of banks, and, to a lesser extent, feeding on crops and specimen trees (Schwab and Schmidbauer, 2003; Jonker et al., 2006). Watercourses with culverts or pinch points are more likely to lead to conflict if dammed by beavers (Jakes et al., 2007). In Alberta the probability of beaver damming of culverts was more generally related to additional factors such as their proximity to beaver ponds but also stream gradient, broadleaf availability, and stream order (Flynn, 2006). Therefore, costs of beaver management may be poorly documented and vary according to a wide range of factors including regional management strategy, surrounding land-use, beaver population densities, and stage of recolonization.

Many studies recording the costs of beaver damages often use extrapolated figures; for example estimated beaver impacts on water management structures at $100,000 and crop and forestry damage at $400,000 annually for Nebraska (Johnston and Timm, 1987). Beaver flooding and resultant damage to agriculture was estimated at $2.5 million in Mississippi in 1978 (Arner and Dubose, 1979). By 1985, timber damage (non-damming) by beavers alone was estimated at $25–118 per hectare, giving annual figures of $215 million (Bullock and Arner,

Figure 11.2 An unfortunate tree-felling incident! (Photo supplied courtesy of Elisabeth Finnekåsa.)

1985). Beaver-related damage on a national scale for the USA was estimated to exceed $75 million in the early 1980s (Miller, 1983), costing around $100 million in damage to public and private property in southeastern USA annually (West and Godwin, 2003). Culvert blockages, often leading to road flooding and damage, are some of the most commonly reported beaver-related sources of conflict in the USA, accounting for 27–40% of all complaints (Hodgdon and Hunt, 1953; Peterson and Payne, 1986). Between 48% and 94% of town highway departments have reported road damage due to beaver activity in central New York (Purdy and Decker, 1985; Enck et al., 1988). An estimated $2.5 million is thought to be spent annually on beaver management and damage repair, based on responses from the 52 counties/municipalities to a survey in 2013 (Hood, 2015). The town highway departments in New York are recorded as allocating between 5 and 25 staff days annually, with repair costs ranging from $543 to $4,900 per beaver-obstructed culvert (Purdy and Decker, 1985; Enck et al., 1988). It has been cited that beaver damage to infrastructure costs are high due to increasing populations (Fortin and Lizotte, 2007), but often such costs are more readily priced than placing values on ecosystem services, habitat creation, and biodiversity restoration on which millions are also spent.

Beavers do not like to move far from water, concentrating the vast majority of their activities close to the water's edge. Therefore, conflicts tend to be along this water–land fringe (Table 11.1). Habitat suitability declines with distance from freshwater for beavers, and most beaver foraging is constantly recorded within a 10 - 30m range (Macdonald et al., 1995; Jones et al., 2009; Swinnen et al., 2017). This is a strategy to both minimize energy expenditure and avoid predators (Jenkins, 1980; Fryxell and Doucet, 1991). European floodplains are typically very low relief (1–2 m) with very shallow channel depths (1–4 m) (Howard et al., 2008; Di Baldassarre et al., 2010). The Netherlands is densely populated with extensive water management networks, and even with low beaver populations, after their reintroduction in 1988 it had significant issues with them burrowing into dykes (Pillai and Heptinstall, 2013).

Beaver damming, depending on locations and numbers, can aid in flood alleviation and can decrease peak discharges (Harthun, 2000), though

Table 11.1 Potential sources of conflict with beavers.

Type of activity	Potential conflict	Potential solutions	
Foraging	Loss of crops Loss of ornamental vegetation	Temporary/deterrent fencing Permanent/deterrent fencing Planting unpalatable species Tree guards Anti-game/sand paint	Create more wetlands and naturalized riparian zones
Burrowing	Bank erosion Undermining of infrastructure	Riparian buffer zone Greenbank protection/reinforcement Livestock exclusion/grazing regimes Mesh facing Metal piling Hardcore infrastructure—stone facing	
Damming	Loss of crops Loss of trees Damage to infrastructure Downstream effects of dam failures	Removal Notching Flow devices Beaver dam analogues Culvert protection Building on higher ground/out of flood zone Use oversized culverts/larger bridge arches	

the failure of mature dams can cause infrequent but significant impacts. Dam failures/collapses can produce outburst floods and have been held responsible for deaths associated with a train derailment (as mentioned), and numerous other human injuries and damage to private property have been reported in the USA and Canada (Butler and Malanson, 2005). Sudden breaches in beaver dams can present health and safety issues and have caused major damage to homes, railways, and roads, and even an aircraft runway (Taylor and Singleton, 2014). One such dam in Alberta generated a flood wave 3.5 times greater than the typical maximum discharge, as recorded over 23 years of monitoring (Hillman, 1998). More often, the impoundment of water, especially in low-gradient landscape, can lead to flooding of crops and forestry. For example, in North Carolina > 35,000 ha of timber was flooded by beavers with dams < 0.5 m high (Woodward et al., 1985). However, it should be noted that this has also been used by some as a timber management tool, letting beavers dam and flood vegetation before removing beavers and draining ponds, to provide sites for planting commercial timber without having to pay land clearance costs (Houston et al., 1992). A recent study of 74 beaver dams in a cascade system across Kananaskis Country, Alberta, demonstrated that 68% remained intact or partially intact after a major flood event (Westbrook et al., 2020). These authors conclude that beaver dams are actually highly resistant to flood events, though the opposite is often reported in the literature.

11.3 Beaver introductions, coexistence of the two species, and eradication programs

The classic 'Who would win in a fight?' is the question along the fronts where both beaver species are meeting, namely in parts of Russian Karelia and Finland, primarily the regions of Pirkanmaa and Etelä-Pohjanmaa (Alakoski et al., 2019). The competitive interactions between the two species have long been debated (Parker and Rosell, 2012). The historic presumption was that the North American beaver, thought to be bigger, would outcompete the Eurasian beaver C. fiber (Danilov, 1976), as this originally appeared to occur in parts of Finland (Ermala et al., 1989). Interestingly, it was found that in nearly all areas of southern Karelia where North American beavers had been released and been resident for some time, they have now been replaced by Eurasian beavers (Danilov et al., 2011). Only one known area (Arkhangelsk region, northeast Karelia) has seen the North American beaver colonize former Eurasian, recently occupied range (Danilov et al., 2011). Population monitoring data in Karelia suggest the Eurasian outcompetes the North American beaver during the initial stages of competition (Danilov and Fyodorov, 2007), though the North American is believed to have slightly greater fecundity (Müller-Schwarze, 2011). Lodge and dam building activities were determined to be similar for both species under the same habitat conditions in south Karelia (Danilov and Fyodorov, 2015). Debates still occur as to which has the competitive edge; as hybridization is not an issue, will one species oust the other or is coexistence ever possible? Most recent evaluations have suggested competitive exclusion is certainly possible, with the North American traits of slightly higher fecundity and family size and earlier sexual maturation likely to give them the edge (Petrosyan et al., 2019).

Efforts to remove the North American beaver in Eurasia have been variable. For example, recent counts of active winter lodges estimate 1,172 Eurasian beaver lodges compared to an estimated 3,673 North American ones in Finland in 2018 alone (Alakoski et al., 2019; Alakoski et al., 2020; Figure 11.3). Therefore, measures to monitor and constrain North American populations in Eurasia are called for, with some modelling proposing that no action could see the Eurasian beaver eventually become significantly reduced or even locally extinct across much of its restored range, even as early as later this century (Petrosyan et al., 2019). In Scandinavia (Norway, Sweden, and Finland), hunting of both species is permitted; therefore some control occurs (Parker and Rosell, 2003; Ministry of Agriculture and Forestry Finland, 2016). North American populations in Germany have arisen from escapes from zoological collections (see section 10.2.1.5); the remaining captive animals were later replaced by Eurasian individuals (G. Schwab, pers. comm.).

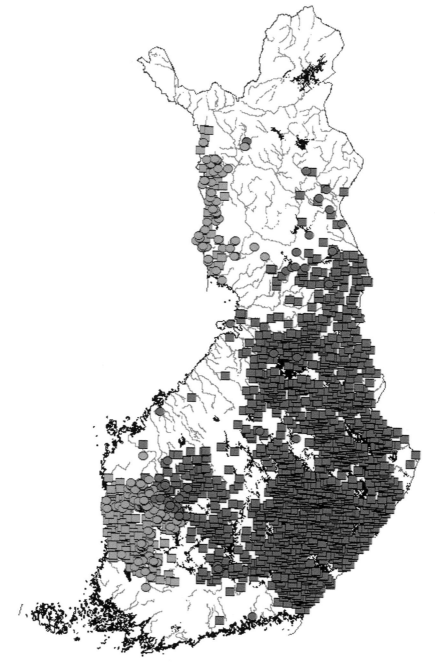

Figure 11.3 Present distribution of the Eurasian and North American beaver in Finland. (Reprinted from *Biological Conservation*, Alakoski, R., Kauhala, K., Tuominen, S., & Selonen, V, Environmental factors affecting the distributions of the native Eurasian beaver and the invasive North American beaver in Finland, 108680, Copyright 2020, with permission from Elsevier.)

Recent efforts have seen 66 North American individuals trapped and removed, sterilized, or transferred to zoos, largely due to public pressure against lethal control (Dewas et al., 2012; Frosch et al., 2014). Some modelling suggests this could constitute as low as 12% of the estimated population (Petrosyan et al., 2019), though this is contested by others as regular monitoring is still undertaken. In other countries such as Finland and Russia, North American beavers now exist in their thousands, so removal will be an astronomical task without greater appetite for coordinated efforts and resources investment. A similar immense task is being attempted in South America (Choi, 2008); though theoretically possible, estimated costs of eradication have been placed in the region of > £26/$30 million (Parkes et al., 2008; Box 11.1).

The introduction of North American beavers to South America has caused major alteration to habitats here and on ecosystems that never evolved with beavers. Some authors have argued that the extent and impact of these changes on native habitats are comparable to those of the last Ice Age ~10,000 years ago together with extensive livestock grazing on the nearby steppe ecosystems (Anderson et al., 2009). Control of beavers began in 1981 in Argentina, with the government permitting recreational hunting; commercial hunting was permitted

---

Box 11.1 Why are beavers so destructive in South America?

The presence of beavers has been described as threatening the biodiversity of a globally significant ecosystem, namely temperate forests and Holocene peat bogs, in this area of southern Patagonia (Malmierca et al., 2011). But how and why is this so significant?

Both beaver species are not native to the southern hemisphere; therefore any presence here constitutes an introduction of an exotic invader (e.g. American mink [*Neovison vison*] to the UK), rather than a reintroduction for restor-

Boxed Figure 11.1 The introduced North American beaver has had negative impacts on native Southern American riparian habitats which have not evolved alongside beaver activities; therefore the same biodiversity benefits generated in the northern hemisphere are not seen. (Photo supplied courtesy of Pablo Jusim.)

ation, and is of serious conservation concern. Native animal and plant species have evolved here along with specific ecosystems, which seem particularly at risk to introduced species, especially an ecological engineer like the beaver (Malmierca et al., 2011).

The Antarctic, southern beech (*Nothofagus*) forests have been significantly impacted by both felling and flooding; these native trees have no natural evolutionary adaptations against beavers and therefore die as a result of these impacts, and seedlings struggle to regenerate in areas where sediment accumulates as soil characteristics have altered (Anderson et al., 2006; Anderson et al., 2009). Beaver foraging and tree felling have been demonstrated to extend 30–60 m from their pool edges here (Anderson et al., 2009). This foraging removes tree biomass and significantly prevents the growth of saplings, whilst their dams alter the hydrology of these riparian systems by flooding areas, causing sediment and organic matter accumulation, changing water flow patterns, flow rates, and depths, and changing nutrient cycles—ultimately changing a lotic system into a lentic environment (Lizarralde et al., 1996; Jaksic, 1998; Lizarralde et al., 2004; Malmierca et al., 2011; Boxed Figure 11.1). Limited seedling recruitment has occurred in some previously beaver-flooded areas even 20+ years after abandonment (Martínez Pastur et al., 2006). Instead beaver meadows can become habitats encouraging the colonization and presence of non-native plant species, especially due to changes in soil moisture and light availability, which in turn also negatively impacted on transplanted beech seedling survival (Wallem et al., 2010; Henn et al.,

2016). Habitat suitability modelling has estimated that 14% of the Beech forests of Tierra del Fuego in Chile are vulnerable to beaver colonization, equivalent to 18,384 ha, which could result in up to 10% of the total stored carbon by these forests lost (Papier et al., 2019). Though the authors acknowledge beavers are unlikely to colonize all the modelled potential usable areas present, they argue that if they even occupy half of it, this is still a significant loss, exasperated by low forest regeneration potential.

Along with beavers, the muskrat (*Ondatra zibethicus*) and mink were also introduced by the fur industry to Tierra del Fuego, becoming abundant (Lizarralde and Escobar, 2000; Jaksic et al., 2002). In the northern hemisphere these three species interact, a cycle now replicated in these introduced environments. Beavers facilitate and encourage the presence and survival of muskrat through their damming activities; muskrat in turn are an important prey item for mink, thus facilitating their spread and predation on native species (Crego et al., 2016).

from 1997 and then a national management plan announced by 1999 with an elimination rate set at 21–23% per year (= 30,000–50,000 animals) (Malmierca et al., 2011). In 1993, the Chilean government passed legislation whereby beavers were officially deemed a nuisance species and could therefore be hunted all year round under D.S. 133 Chilean Hunting Law (Anderson et al., 2009). However, this approach recognizes that not only should beavers be eliminated but also, given the sensitivities of the native riparian ecosystem and its flow dynamics, all dams should be removed. In areas where beaver activities have caused sediment accumulation and thus impacted on the regeneration of native forest seedlings, transplanting of tree seedlings may be required (Anderson et al., 2009).

To incentivize local trappers a beaver fur market has been developed, along with promotion of consumption of beaver meat, and from 2001 the authorities pay a bounty for every beaver tail handed in (Malmierca et al., 2011). However, in both Argentina and Chile these initial efforts failed and elimination rates were not achieved. Today national control programmes are in place in both countries to trial as a trans-border approach, and hunting is actively encouraged, with even ecotourism hunting trips organized to save this unique habitat (Choi, 2008). Pilot eradication trials have highlighted the trapping effort required; practicalities including efficient trap types, reducing trap shy animals; and need to influence societal and political attitudes towards sustainable eradication strategies (Jusim et al., 2020). In Finland, although beaver eradication is considered feasible there are a number of realistic constraints including long-term costs, scale (including guaranteed and repeated access to all riparian areas), detection and monitoring efforts, political

support (including bi-national collaboration), and long-term capacity (including equipment, training, capacity building) (Malmierca et al., 2011).

11.4 Techniques for effective management

The Czech Republic beaver *Handbook* for coexisting describes beavers as 'persistent and tenacious animals' (Vorel et al., 2016). This can certainly be true, especially if they are determined to occupy an area they have decided will be their new home. With room to move on naturalized systems, beaver presence can be generally unobtrusive. Often humans and beavers seek to occupy and utilize the same resources; numerous stories describe ongoing battles of patience, endurance, and increasingly inventive measures to curtail any unwanted efforts. Effective and appropriate management actions are key in increasing the acceptance and tolerability of beavers. Any site may have several sources of conflict and/or require a combination of management solutions. Beaver mitigation programmes can be expensive, so such costs should be measured against the costs of no mitigation or setting land aside for nature conservation. This can be hard to quantify, but one estimation study of a Beaver Control Assistance Program in Mississippi run between 2005 and 2009 indicated every dollar spent saved between $39.67 and $88.52 for forestry and the state economy (Shwiff et al., 2011). Table 11.2 lists some management options.

Some of the first legal protection of remaining beaver populations occurred in Norway in 1845, followed by Finland and Sweden, though unfortunately by then the beaver was lost from these latter two countries (Nolet and Rosell, 1998). Under

Table 11.2 Management options.

Low	Level of human intervention			High
Information to general public and landowners to increase beaver acceptance	Protective measures, e.g. tree protection, fencing	Dam removal, flow devices	Live trapping and translocation	Local eradication of beavers
Low	Perceived seriousness of damage			High

Source: Modified from Sjöberg and Belova (2019).

European law, the Eurasian beaver is strictly protected across much of its range, even though it is now classed as *least vulnerable* and listed in the Habitat Directive Annex IV, prohibiting capture, killing, or possession without specific permissions. This protection can also extend to beaver structures including some dams, lodges, and burrows. Legal status of beaver varies in North America from state to state. A range of permitted management and levels of protection exists—from being deemed a pest that can be taken year-round to strict protection outside of furbearer seasons (Taylor et al., 2017). Under the Nature Conservancy, the North American beaver in Canada and the USA is classed nationwide as 'N5: demonstrably widespread, abundant, and secure', with wildlife state agencies having primary management authority ((Boyles and Owens, 2007; see Section 11.7). Within certain states the beaver has two further classifications: in Kansas, Nebraska, and South Dakota it is ranked statewide as 'S5: demonstrably widespread, abundant, and secure' and in Colorado and Wyoming as 'S4: apparently secure' (NatureServe, 2005). Beavers are further generally classed as furbearers in most states and therefore managed for sustainable hunting and fur harvest, and lethal control permitted with some form of regulation, to mitigate any damage conflicts (Boyles and Owens, 2007).

11.4.1 Scaring devices, unpalatable and scent deterrents

There are contradicting ideas of the repelling powers of various scents, including human odours, on beaver activities, perhaps more accurately reflecting the beaver's powers of threat level deduction and habituation. Indeed, many scaring devices (flashing lights, ultrasonic noise devices) and repellents are generally ineffective, especially over time

(Nolte et al., 2003). Hilfiker (1991) describes a more comical event (or sign of intelligence!). In an attempt to dissuade an industrious family of beavers from damming around a Boy Scout camp headquarters in Camp Massaweepie, New York, a heavily human-scented pair of old dungarees was hung at the destroyed dam location. This successfully held off the beaver building efforts for two nights until they pulled the offending item down and buried it, only to proceed with dam construction! Whilst indeed Rosell and Czech (2000) note a slight reaction on feeding of human-scented food items, it really cannot be recommended as a deterrent. More cryptic substances have been painted on trees in an attempt to deter beavers. Backhouse (2018) describes how the US Department of the Interior's Wildlife Research Laboratory developed 'Repellent 96—A-modified'; whatever the ingredients, had no impact whatsoever. Chemical repellents are not registered specifically for beavers and they appear unbothered by general herbivore repellent products (Taylor et al., 2017). Various commercial scent repellents have been trialed to create scent fences or protect trees from beavers in the Czech Republic, including 'Hagopur', 'Morsuvin', 'Antifer', and '*Pacholek*', none of which proved effective or was recommended (Vorel et al., 2016). Typically beavers may be initially cautious of 'new' scents but can learn fairly rapidly to habituate. Similarly, American experiments failed to find any effective and practical chemical repellent (Taylor et al., 2017).

Like other mammals, beavers seem to avoid plants treated with commercial herbivore deterrents containing egg or blood components (DuBow, 2000). Kimball and Perry (2008) demonstrated how captive beavers display a strong avoidance response to food treated with casein hydrolysate. The authors propose that treatment of desirable plants in a beaver-occupied territory could have a management application and even encourage foraging on invasive species. It has also been demonstrated that Eurasian beavers react to foodstuffs treated with predator scent (Rosell and Czech, 2000) and are less likely to over-mark beaver scent mounds that have been contaminated with a predator scent (Rosell and Sanda, 2006). Another study went a step further in proactively attempting to train beavers to avoid certain food items by giving an intraperitoneal injection of lithium chloride (LiCl) ahead of food presentation (Harper et al., 2005). This is a recognized method for the generation of food aversions in rodents as LiCl causes gastro-intestinal distress and if paired with an often-novel food item can produce a conditioned response (Nolte and Mason, 1998). Although they successfully produced an aversion to corn in these test beavers, the beavers compensated for this by increasing consumption of other offered food items, as successful generalist feeders. In conclusion, the application of this technique as a management strategy is questionable, as the authors point out the extensive resource investment and time needed to generate these aversions, which would need to be applied to all family members in a territory, in addition to any new colonizing individuals, all of which makes it impractical.

11.4.2 Tree protection and deterrent fencing

Beaver impact on trees varies with factors such as tree species and distance from a watercourse (Müller-Schwarze, 2011; Campbell-Palmer et al., 2016). Therefore, beaver tree impacts over the large scale are not often documented. One Swedish study on 3,500 ha of coniferous forest in Telemark, Norway, determined that beaver activity (foraging and flooding) had an impact rate of 2.6%. With the volume loss of flooded area estimated at 3.3 m^3/ha/year, overall total tree damage due to beaver activities is determined as insignificant on a national scale (Rosell and Parker, 1995).

Trees can be protected from beaver foraging impacts either individually or as stands of trees, mainly through fencing or application of protective, anti-game paints. The selection of which trees to protect can be due to a range of reasons—for example some personal to the landowner such as a favoured specimen for sentimental reasons; those that risk falling

and causing damage; crop trees; important specimens of conservation value; or trees that are part of an important collection. Investigations of impacts of North American beavers in urban areas have suggested that paths closely associated with beaver areas may in themselves provide a form of foraging deterrent to beavers through both creating an open, bare surface and their constant use by humans, and dogs especially, which may create an odour fence (Loeb et al., 2014). This study concluded that less tree and sapling damage was experienced in areas uphill of regularly utilized paths, on those that were macadam (stone-bound tar or bitumen) or wood chip covered, or paths consisting of raised wood boardwalks, whereas less-used paths, bare earth paths, and deer tracks did not seem to have this deterrent effect. Regardless, any protection method will have to take account of resources available, practicalities with access, and the importance of aesthetics.

Fencing can be very effective through directly preventing beavers from physically accessing the tree itself. Any protection should of course also cover any buttress roots and low-hanging limbs and not negatively impact on the tree's growth (Vorel et al., 2016). Stands of trees along the riparian zone, most likely within 0–40 m from a watercourse, can be protected using three-sided fencing as opposed to individually wrapping each tree (Kamczyc et al., 2016). 'Rabbit' or 'chicken mesh', light gauge, and highly flexible mesh wiring is often used or promoted for tree protection, probably as it is readily accessible and easy to work with (Figure 11.4). This should be used with caution and many examples of poor application exist. Though cheap and easy to work with, its high degree of flexibility means beavers can pull it loose or push underneath it, creating a gap for access easily unless either many layers are wrapped around the tree (which may restrict growth) or tension is put on the wiring using well-staked posts (Olson and Hubert, 1994). Fences should also be tall enough so they protect against beavers standing on their back legs, with provision for any regular snow which may provide a height advantage. The aesthetics of individual tree protection may be accompanied by a negative perception from some, especially in urban areas (Loeb et al., 2014). Therefore, in gardens or ornamental parks higher-specification or even custom-made fencing could be used or anti-game paint applied instead. Most importantly, the

Figure 11.4 Trees can be protected from beaver felling through wire mesh and anti-game paint successfully, though any mitigation is only as good as its application. (Photo supplied courtesy of Roisin Campbell-Palmer.)

mesh gap diameter should not allow a beaver to squeeze its head through for access.

Anti-game paint works on the principle that beavers can reach and access the tree trunk but because of the abrasive properties combined in the paint (which is typically quite glue-like in application), this produces a discomfort to their teeth and so has a dissuasive effect. Commercial and homemade versions are available and have been used successfully (Campbell-Palmer et al., 2016). 'Sand painting' can be both cheap and viewed as more attractive by some, but importantly the equipment is easily accessible, using latex paint and fine mason sand (20 oz of sand to 1 gallon of paint, or 140 g of sand per litre). Key factors in the successful application of this technique is to ensure all surfaces, including buttress roots, are covered up to an appropriate height, and rough bark will take much more paint and is more open for beavers to exploit any uncovered areas. Long-term efficiency of anti-game painting has been quoted at around 10 years by the manufacturers and with homemade latex paint and fine sand versions, but, in reality, trees should probably be checked annually to account

for growth, any peeling, or signs of trauma that may warrant fresh application (Vorel et al., 2016).

Larger stands of trees close to a water's edge or areas requiring prevention of beaver access can sometimes more readily be fenced off rather than spending time and effort on individual protection, if applicable. A significant issue with fencing along watercourses is that during high water, fences may become clogged with debris or washed out. Various designs exist, depending on the topography of a site, and a range of materials could be used, provided beavers cannot squeeze through. Large-scale fencing projects can be expensive and may not complement other land-use, so generally any fencing is targeted and/or involves other mitigation techniques (Nolte et al., 2005). Again, this principle works on preventing beavers access to resources, and their drive will vary with 'attractiveness' of species, alternative forage in the area, and distance from the watercourse. An area can be protected on three sides, i.e. the side running along the watercourse and the two adjoining sides. If carried out correctly this should turn the beaver away and back into the watercourse, provided they cannot easily walk around either edge. Taking fencing back into the stand, often at a wider angle, causes the beaver to have to move further and further away from the watercourse, something they are typically reluctant to do. Also, critical for this mitigation to work successfully is the principle that a beaver cannot simply dig under a fence when they meet it. To prevent this, extra fencing should either be buried, which can be more expensive and not always practical, or extended on the surface facing the direction of travel of the beaver and pegged into place.

Electric fencing has been used quite arguably with varying success for a reactive shorter-term protection. This works on the principle of dissuading beavers from accessing an area for a temporary period, such as crop ripening. Strong electrical outputs are not required, with suggested currents of 2,500 volts have been recommended—standard livestock fencing with low power (Vorel et al., 2016). As with any electrical fencing this should be viewed as a deterrent and should not hurt the animal; if this is not the case it should be removed and another mitigation method sought. It should be noted several cases of beaver biting electric fences after a shock, and subsequently being found dead on the lines have been witnessed. Additionally, camera trap footage has also demonstrated beavers accessing supposedly protected areas of high value food items, whilst obviously experiencing painful shocks. Therefore, alternative fencing is recommended for longer-term protection and less maintainence requirements. Two wire or tape strands set at 15–20 cm and then 30–40 cm are proposed to prevent animals from digging under or pushing through, with all vegetation removed under and around these wires to ensure circulation (Campbell-Palmer et al., 2016). In Bavaria, beaver managers have found that in general direct crop loss through beaver foraging is low and tolerated by farmers, but the flooding of crops through damming and/or erosion due to bank burrow collapse is a much more significant issue (Schwab and Schmidbauer, 2003).

11.4.3 Buffer zones and provision of alternative resources

It is consistently reported that the vast majority of beaver 'damage' occurs close to the water's edge, within but not limited to 20 m (Nolet and Rosell, 1998). Since their reintroduction, analysis of beaver conflict areas in Bavaria has concluded that these occur in areas where humans intensively utilize and/or have modified the narrow margin of freshwater shoreline (Schwab and Schmidbauer, 2003). The more naturalized this riparian edge is, the fewer the beaver impacts, for several reasons.

Naturalized riparian strips have a mixed range of plant species from which beavers can selectively forage and therefore have varying foraging pressure, allowing regrowth and encouraging biodiversity; the root systems stabilize banks more effectively than crops, grass, or low numbers of single mature trees, thereby making banks more resilient to burrowing; and reclaiming human land-uses such as agriculture back from the riparian edge decreases the direct and conflicting impact of beaver foraging, digging, or risk of flooding. Continuous and heavy grazing and high stocking densities will promote erosion, soil compaction, and degradation of riparian vegetation. Exclusion fencing provides the most immediate action to enable riparian vegetation recovery (Smith and Prichard, 1992). If this is not possible, alternative management strategies can be practiced, recognizing the varying

needs of both landowners and the riparian system in question (Chaney et al., 1990; Collins, 1993). Implementing shorter grazing periods, providing effective plant rest periods during the growing season, spreading the grazing load, regularly moving livestock, and avoiding grazing during vulnerable periods (e.g. waterlogged banks) can significantly improve riparian vegetation and minimize stream damage such as bank erosion and soil compaction (Olson and Hubert, 1994; Fitch et al., 2003). Also relocating feed stations, mineral licks, shelter provisions, shade trees, and chemical/fertilizer treatments away from riparian zones (Collins, 1993) are helpful. The preservation and restoration of these 'buffer zones' are therefore typically the most long-term and durable solution to reduce beaver conflicts (Campbell-Palmer et al., 2016). However, especially in highly urbanized and/or agricultural areas these buffers zones are not possible or practical to create and will be unrealistic to achieve, regardless of multiple potential ecological benefits (Swinnen et al., 2017). It is worth noting that huge amounts of money are spent throughout Europe and North America on river restoration and enhancement schemes; for example figures of $3 billion annually on a global scale have been quoted (Roni and Beechie, 2013). Annual spends of £6–10 million ($7.7–12.8 million) have been documented in England alone (Department of Environment Food and Rural Affairs, 2015). Therefore, more protective measures and management techniques will be required in certain areas.

Planting of unpalatable tree species has at times been suggested as a mitigation technique, and in theory this could provide sacrificial food and building sources to reduce foraging focus on vegetation of concern. Concerns with this as a sole protection strategy are apparent. Planting only such species, given the range of species beavers will consume, could lead to lower diversity in urban areas (Belovsky, 1984). The wide diet range of beavers is impressive and they readily adapt to available resources; therefore known unpalatable species over the long term is questionable and percentage of planted vegetation required and space available for this to be effective are unknown. Additionally, these species may not align with other conservation and planting plans for a specific area and may in fact reduce tree diversity in an area (Loeb et al., 2014).

Almost the reverse strategy could be applied, by planting favourable trees! Both species display strong tree species preferences, with willow (*Salix* spp.) being high on the consume list (Haarberg and Rosell, 2006). Some landowners may view such planting as merely providing more beaver food and thereby encouraging their ongoing presence. It should be noted that beavers will use an area regardless of presence of trees; therefore it would often be more beneficial for tree roots to be stabilized on the banks rather than leave them with little vegetation. Thus, selecting tree species that react to beaver foraging by becoming shrubby in structure not only provides ongoing food resources but also can protect commercial or ornamental species planted behind. Selecting these species next to roads, for example, for planting schemes and allowing beavers to forage reduces the chances of trees being felled onto roads. Leaving a beaver-felled tree in place is key, if possible, as its removal only serves to accelerate further felling.

11.4.4 Dam prevention, manipulation, and removal

One of the most commonly reported types of beaver damage is flooding caused by dam building and the blocking of culverts (Baker and Hill, 2003). Any mitigation method involving dam removal, alteration of the watercourse, and/or installation of any devices in it should be subject to local legalities and checked in advance of any application. Though beaver dams may be a natural occurrence, devices attached to them as part of any mitigation may be viewed as obstructions in the watercourse by others. Dams may be protected features in some countries, particularly those immediately supporting water levels around natal lodges or burrows, creating welfare issues when entrances are suddenly exposed during the breeding or winter seasons. Any dam removal as a management strategy will display immediate results and can quickly become a cycle of ongoing effort and expense, as beavers have both the determination and physical abilities to re-dam surprisingly quickly—in some cases reportedly reinstating a dam, with the associated extent of flooding, in as little as 24 hours (Taylor et al., 2017). Repeated dam removal will also

see an escalation of tree felling in the general area, as beavers tend not to reuse material from any dismantled dams. There are two main methods to control beaver flooding which involve building constructions; these can generally be split into exclusion or fence systems to prevent damming, and deception or pipe systems (Taylor and Singleton, 2014). Flow devices and culvert protection are discussed in Section 11.4.6, while general dam management is addressed here.

Preventative measures are often suggested, but these can be difficult to implement without making an area generally unattractive to biodiversity. Dredging of watercourses is often proposed, especially in more agriculture-dominated landscapes, as a method to deepen channels, thereby reducing the need for damming. However, this can be especially expensive, significantly impact on other biodiversity, and change hydrology (Schwab, 2014). In some areas, road ditches (acting as intermittent watercourses) have been filled with coarse stony materials, typically a thin layer of fine aggregate, with the remainder coarse aggregate of grain size 20–50 cm (Vorel et al., 2016). This enables water to drain as and when it is needed but prevents beavers from creating dams and impounding water on or along these embankments. Other less widely used options have included placing objects in the watercourse to physically prevent dams from being properly constructed. Floating objects or buoys (suspended from an overhanging beam or tied to a fence line) have been placed in watercourses at locations where beavers are likely to dam or block a culvert, the principle being that their constant movement in the water prevents beavers from building against them (Vorel et al., 2016).

A Canadian woodland forum magazine in Ontario 1997 determined that up to 48% of dams were removed using explosives. Whilst this may certainly look impressive there are obvious risks involved, including the sudden release of large amounts of water and downstream surges that could cause unexpected consequences without a full risk assessment in advance. Dam removal should really consider the time of year and function of the dam. Natal dams directly influencing water levels around breeding lodges or burrows will have welfare implications for pregnant females and any newborn kits should they suddenly be exposed through dam removal. Typically, most dam removal is undertaken using machinery, though removal by hand may be the only option in certain areas; though time consuming this often means less impact on riparian vegetation and compaction of banks.

Dam notching/breaching again provides instant results but does not necessarily prevent rebuilding, often only allowing partial and often temporary lowering of water levels behind dams, as these gaps are often rapidly fixed by the resident animals. As a mitigation technique, notching alone is not so regularly employed, as usually longer-term solutions are sought. Material should be removed from near the centre of the dam, as opposed to the littoral edges, as this is likely to increase bank erosion, which should be avoided (Vorel et al., 2016). A number of fishery authorities, including the Gitanyow Fisheries Authority in British Columbia, undertake beaver dam breaching programmes ahead of spawning migration to facilitate salmon passage (Kingston, 2004). For example, the Miramichi Salmon Association breached over 100 beaver dams to service a watershed of 13,552 km² ahead of the salmon run (Parker et al., 2013). Notching can be labour intensive and even appear a thankless task, especially over a catchment scale, but has been recommended as a practical tool for habitat managers, permitting dams to remain in place and therefore generate positive benefits to fish (see Section 9.3.3) by facilitating migratory fish movement as required and potentially alleviating concerns (Taylor et al., 2010). In some circumstances mitigation measures can be implemented to lower the height of the dam, deterring beavers from increasing height and impounding more water; this again is a compromise in accepted water levels.

In other sites and locations, dam management may actually involve encouraging or maintaining dams in specific locations. In areas where streams are severely incised, with narrow, deep channels and high velocity flow causing significant erosion, it is difficult for beavers to establish and function as a restorative tool without help (Demmer and Beschta, 2008; Pollock et al., 2015; Wheaton et al., 2019). Beaver dam analogues (BDAs), a collective term, are support structures, based on vertical posts

driven into the channel bed, spanning the watercourse, with finer material, such as willow whips, woven horizontally between the posts (though variations exist) (Pollock et al., 2015; Figure 11.5). These either mimic or reinforce existing beaver dams, by encouraging beavers to build against them, enabling beavers to establish dams in watercourses which may be difficult for them to and/or reinforce dam structures that may regularly wash out (Pollock et al., 2015). BDAs have been used to create pools, improve floodplain connectivity, encourage riparian vegetation, increase stream sinuosity, create multi-thread channels, increase wetland habitat, create sediment traps and sources, reduce bank erosion, recharge groundwater, encourage beaver colonization and range expansion, and create release sites for beavers (Pollock et al., 2015).

BDA construction materials come in a large range—stones, sand, clay, vegetation stalks, branches, shrubs, and conifer posts (Pollock et al., 2015). Past constructions have used rock dams with mortar, boulders, gabion mesh baskets filled with stones, and logs felled across the watercourse (Flosi et al., 2002; Cramer, 2012). Use of such hardcore infrastructure may not always be permitted and in some areas is actively discouraged. How these structures affect the hydrogeomorphology of a watercourse should be understood in advance of

Figure 11.5 A beaver dam analogue created with wooden posts and interwoven willow whips places in a spatey stream within an enclosed beaver restoration project to encourage the newly released pair to dam and create ponds. (Photo supplied courtesy of Alan Puttock.)

Box 11.2 Use of BDAs in Bridge Creek

Beavers and salmonids have a long evolutionary history, and both are documented as coexisting in high densities during expeditions of the Pacific Northwest in 1805 (Chapman, 1986). Bridge Creek watershed (710 km²), is located in semi-arid eastern Oregon and since this time has been heavily utilized for cattle ranching, resulting in significant overgrazing and a degraded riparian zone, with stream channel incision and disconnection from the floodplain (Demmer and Beschta, 2008; Weber et al., 2017). Middle Columbia River steelhead trout (*Oncorhynchus mykiss*) are listed as threatened. Therefore, the watershed has been identified as a restoration priority (Columbia-Blue Mountain Resource Conservation and Development Area [CBMRC], 2005).

A long-term restoration project along the lower 32 km was established to promote riparian stream ecology and increase habitat complexity and reconnection of the floodplain by encouraging beaver activities to restore geomorphic, hydro-

logical, and ecological functions (Bouwes et al., 2016; Boxed Figure 11.2a and b). A small beaver population was present, though commonly many dams were washed out seasonally during high discharge events in this incised system. Few beaver territories persisted longer than 2 years outside of the core beaver-occupied stretch, with most dams lasting less than a year, so little population expansion occurred (Demmer and Beschta, 2008). The observation of 161 beaver dams between 1988 and 1993 over a 25.4-km lower stretch of Bridge Creek saw 30% washed away and 32% breached in the centre of the dam, while 38% eroded the bank at one end of the dam (Demmer and Beschta, 2008). Therefore, a solution was sought to increase the longevity of beaver dams, which in turn would see the wider establishment and population growth of the current population, with the associated habitat restoration abilities to encourage steelhead productivity (Pollock et al., 2012).

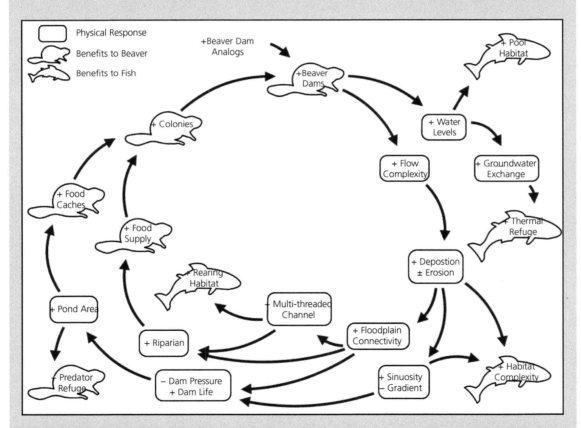

Boxed Figure 11.2a and b A constructed BDA installed to encourage beaver damming as a conservation strategy to increase habitat for threatened steelhead. Expected associated geomorphic and hydraulic changes, along with responses of the fish population are depicted. Reprinted from Scientific Reports, Bouwes, N., Weber, N., Jordan, C. et al., 2016, Ecosystem experiment reveals benefits of natural and simulated beaver dams to a threatened population of steelhead (*Oncorhynchus mykiss*). Sci Rep 6, 28581.

Boxed Figure 11.2a and b Continued.

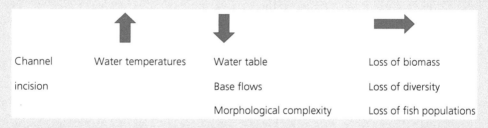

Boxed Figure 11.2c Summary of channel incision issues (Cluer and Thorne, 2014).

BDAs can be temporary features and biodegradable, and function like actual beaver dams, sequences of which can result in significant changes to a watercourse by facilitating channel aggradation and floodplain reconnection (Pollock et al., 2012). Lines of wooden posts (typically untreated lodgepole pine fence posts ~2 m long, 7–10 cm diameter, 0.3–0.5 m apart) are embedded across stream channels, often with wicker weaves intertwined, though other materials including sediment, clay, gravel, and vegetation stalks can also be used, allowing water to flow through the structure and providing a basis for beavers to build against. If not utilized by beavers, these can trap sediment and assist in functioning as a temporary dam. In other situations, the installation of only posts means they only become functional if beavers dam against them (Pollock et al., 2012). Additionally, active or intact abandoned beaver dams were further stabilized through the addition of posts along their lengths, with posts specifically installed in any gaps.

To monitor the impact of addition of BDAs and encourage beaver damming for steelhead, 35,867 juveniles were passive integrated transponder (PIT) tagged between 2007 and 2013 (Bouwes et al., 2016). This long-term monitoring determined both adult and juvenile steelhead were found throughout the restored stream section (Pollock et al., 2012). More than 1,000 tagged juveniles migrated downstream, with upstream migration also evident through more than 40 dams upstream (Bouwes et al., 2016). No impact on upstream adult spawners was found, with some individuals documented as having passed through more than 200 dams and BDAs (Bouwes et al., 2016). Therefore, fish passage is not presumed to be an issue with dams, and BDAs do not act as barriers to either juvenile or spawners. Additionally, steelhead density, survival, and production of juveniles all increased; this was directly attributed to the increased quantity and complexity of habitat, providing greater feeding, resting, and predator avoidance sites, and shelters against high-flow events (Bouwes et al., 2016). Boxed Figure 11.2c depicts some of the key channel incision issues.

their implementation (Pollock et al., 2015). Overall, if installed properly these can encourage beavers to dam in sites selected by humans, as opposed to giving beavers free-rein, and can be used to successfully manipulate damming behaviours. They can also greatly assist in enabling and encouraging beavers to establish in areas and on watercourses where they may otherwise struggle. Box 11.2 describes the use of BDAs in a long-term restoration project in eastern Oregon, USA.

11.4.5 Flow devices

In certain circumstances, flow devices (or pipe systems) can be installed that are designed to control water levels caused by beaver dams through the principle of deception (Lisle, 2003; Figure 11.6a and b). They work on the basis of allowing beavers to maintain a dam, but the level of impounded water is controlled by an installed pipe—beavers believe they are impounding water, whilst being 'deceived' by the level permitted. Flow devices or pond levelling devices can be highly successful and cost-effective means to reduce flooding, under appropriate circumstances, especially in light of

on-going dam or beaver removal and road maintenance (Simon, 2006; Boyles and Savitzky, 2009). They were first developed in North America as a non-lethal control response. The terminology varies and the design has developed and evolved over many years, across various states and across both continents (Wood and Woodward, 1992; Lisle, 2001; Lisle, 2003; Taylor and Singleton, 2014; Campbell-Palmer et al., 2016).

In fairness, these are visually and technically simple mechanisms, though there is a science to their design; however, there are many examples of failing, poor replicas that are readily made redundant or circumnavigated by industrial beavers. One of the most common potential reactions by beavers to any of these flow devices, and the ultimate failing of this mitigation technique, occurs when they take exception to dam interference. Beavers may create a subsidiary dam(s) below the flow device, potentially rendering it null and void. Whilst proving an incredibly reliable technique and a relatively low-cost and long-term solution, it should be noted that beavers may have multiple dams within a territory, which may cause conflicts and, depending on the hydrology of a system, may be set on creating a

Figure 11.6a and b Flow devices (a) and culvert protection (b) have been developed in North America as a successful means to regulate impounded water levels in beaver dams; this can be highly successfully at sites where beaver presence can be tolerated along with some increased water level rises. (Photos supplied courtesy of Roisin Campbell-Palmer.)

series of tiered ponds, so the practically of potentially installing multiple devices should be considered.

Another significant cause of failure is the installing of pipes with insufficient flow capacity, which can be easily rectified with greater-diameter pipes (Table 11.3). Dropping water levels too fast and below those acceptable by that beaver family can trigger subsidiary dam building, which, especially on a narrow, linear system, will readily put the flow device out of action. There is varying evidence of flow device failure rates, which experienced practitioners often question as being overrecorded, especially when installed by novice public relations (PR) actioners, but published data suggest that most flow device failures occur in the first 2–12 months after installation, highly related to the experience of the personnel involved and increasing with inappropriate site placement or poor design (Callahan, 2003; Czech and Lisle, 2003; Lisle, 2003).

There are some key considerations in the practical construction of flow devices. The pipe inflow must be protected through mesh caging, to prevent beavers blocking the device with sticks and other materials. Mesh size should roughly be 10 × 10 cm— any larger risks beavers getting trapped; any smaller increases material collection on the mesh, reducing functionality. The inflow of the pipe must be well away from the dam structure itself, at least a minimum of 3–5 m, to prevent the risk of the pipe

becoming part of the dam structure and/or becoming blocked with debris (Vorel et al., 2016). Pipe placement low in a dam, which causes the beaver pond to basically be drained, is not a function of this device. Although some specialists argue that this is possible in some locations largely to render the area unattractive to beavers, this technique works on the basis of compromise. Water levels should be set (by either increasing pipe flow capacity or pipe level placement within the dam), so that the conflicting issue caused by the impounded water levels is resolved while ensuring that beavers still have a pond to utilize. Otherwise this technique (and current dam) quickly becomes obsolete as beaver re-dam in a different position. The pipe outflow should not be flush with the downstream dam face as it could become blocked and incorporated in the dam itself, but neither does it need to be covered with a protective cage such as the inflow. A small degree of overhang seems to generally be sufficient for beavers to ignore any damming attempts, instead concentrating their efforts on the upstream face (see Section 4.4).

11.4.6 Culvert protection

Culverts and/or small bridges provide the ideal location to rapidly produce a pond where humans have helpfully caused a narrowing in the watercourse and therefore requiring less material and effort to plug. Culverts are often associated with water passing under roads or other infrastructure, including residential areas. Therefore, any obstruction can fairly rapidly cause back-flooding and associated conflict.

Various methods have been tried with varying success to protect culverts, including covering with grates or wire mesh, using electric breach guards, and installing T-culverts, pitchfork guards, and water level control devices (Roblee, 1987; Wood and Woodward, 1992; Hamélin et al., 1997). The first documented use of culvert guards was of wire-mesh fencing to dissuade beaver damming, but these guards often saw debris collection, requiring increased maintenance to keep clear (Laramie, 1963). Of course, such protection could simply consist of nothing more than a metal grill or welding rods across the culvert mouth. This will prevent

Table 11.3 Capacity of different pipes at free-flowing water rates.

Diameter of pipe (mm)	Gradient of pipe (%)	Capacity flow of pipe (m³/s)
200	1	0.03
200	2	0.04
200	3	0.05
300	1	0.08
300	2	0.12
300	3	0.15
400	1	0.18
400	2	0.26
400	3	0.31
500	1	0.33
500	2	0.46
500	3	0.57

Source: Reproduced from Vorel et al. (2016).

entry but can still provide a structure for beavers to actually dam against, or natural debris to collect, so requiring ongoing maintenance to keep clear (Taylor and Singleton, 2014). The 'culvert cleaner' is basically a heavy-duty grill to which a chain is attached and lifted by pulling upwards (a vehicle tow bar is often used) to dislodge any damming or collected debris regularly (Olson and Hubert, 1994). Evolving designs have included liftable grate sections to assist clearance (Taylor and Singleton, 2014). T-culverts were developed as a type of culvert extension, in which a horizontal section of pipe was fixed creating a T-shape so water dissipated out of the culvert and then from both sides of the pipe so that water drained out over a wider area (Roblee, 1987). This technique has been relatively successful under certain conditions including moderate flow, solid substrate and when both ends of the pipe rested in still water that was around 1.2–1.8 m deep (Roblee, 1987). 'Beavercones' have been installed on culverts in Canada and some American states; these are premade and retrospectively fitted metal cones pointing outwards from the culvert inlet. The shape may make damming more difficult, though not impossible, they prevent beavers accessing, and debris can be easily removed (BeaverCone, 2020).

Most long-term, successful protection commonly used today is often referred to as the 'beaver deceiver' and generally follows the same principles as flow devices as it involves exclusion fencing to prevent beavers entering and damming the culvert itself, which may or may not include a drainage pipe and protective cage surrounding the inflow (Lisle, 2001; Figure 11.6b). This fencing is typically attached to the culvert inlet or placed in the watercourse just upstream of the inlet and can vary in shape, ranging from rectangular, to semi-circular, to trapezoidal (Taylor and Singleton, 2014). The beaver deceiver is a type of culvert fence, first documented at the Penobscot Indian Nation, Maine (Lisle, 2001). By taking the fence line further out from the culvert itself and increasing and complicating the shape, the aim is to make damming more difficult. These devices can be effective in reducing culvert blockages, though they can require ongoing maintenance to prevent excessive debris accumulation. Sources of failure have generally been associated with damming of the fencing by beavers, lack of maintenance, or beavers damming downstream (Callahan, 2003; Callahan, 2005), though the author notes high success rates. Culvert fencing can be combined with a pipe to become a flow device if damming becomes a persistent issue (Taylor and Singleton, 2014).

Several management recommendations during the design and installation of culverts themselves have been investigated. The area of culvert inlet being the most significant factor as to whether beaver will plug them or not, oversized culverts in areas with repeated issues or when building new roads in beaver areas have been installed (Jensen et al., 2001). These authors found that to reduce the probability of beavers blocking a culvert, by 50%, the inlet opening should be oversized to a minimum of 2.1 m^2 for a 0% gradient stream and at least 0.8 m^2 for streams with gradients up to 3%. Stream gradient is a significant determinant in the appearance of dams at culverts, with those > 3% being less likely. Others determined that if the land immediately surrounding the culvert, for approximately 200 m either way, was devoid of vegetation, this could discourage beaver damming (Curtis and Jensen, 2004). Even with these factors, additional culvert protection measures such as fencing or flow devices may be required to alleviate all issues. Although expensive, bridges can provide a more sympathetic choice of infrastructure, as culverts can increase water velocity (due to lack of friction and channeling of water), especially if improperly sized, thereby increasing the potential of downstream erosion (Fitch et al., 2003).

In a similar vein, overflow devices in ponds or small reservoirs such as overspill pipes, monks, or sluice boards can all be blocked by beavers to increase the area of their pond. Wooden structures may be chewed and burrowing may occur which increases outflows above those set by humans. In all such cases, ideally beavers should be physically prevented from reaching such devices through fencing, metal grills, rods, and/or meshing of earthen banks. Fencing and grills alone may need regular debris removal, or they could be coupled with pipes to construct a flow device; often, if water levels are to be maintained until a time of letdown for example, fencing the beavers out is the recommended course of action. Along the same principle as the flow device, fencing or grills should be taken as far away as possible from the overflow device or spillway to attempt to dissociate this area with the beaver's perception of where water may be exiting from. Wooden structures, such as monks or sluice boards, could be reinforced to prevent gnawing if exclusion is not possible or practical. The base of such structures could be made from concrete and the wooden planks covered in either metal sheeting or weld mesh, or replaced with reinforced plastic or aluminium planks (Vorel et al., 2016).

11.4.7 Bank protection and restoration

Beavers will readily construct two shelter types, lodges or burrows, which are in fact a continuum of the same feature, involving varying degrees of burrowing into a bank and/or building upwards. Repeated climbing up a bank may also result in beavers wearing haul-out points and forage trails into a bank, which can be accompanied by some digging and canal formation (see Section 4.2). Additionally, dams in the watercourse can cause water to erode around one or both ends of the dam and cut into banking. Any heavy use and digging can be a real issue, especially in unstable banks prone to erosion due to sandy soils and/or lack of root structure, and burrows can collapse and banks give way (see Figure 11.7a and b). Infrastructure lying immediately

next to watercourses, particularly tracks, roads, and railway lines, may be undermined by burrowing, which should immediately be filled in to prevent collapse. There are various bank treatments by which beavers may be prevented from burrowing, but these can be costly solutions and require extensive coverage. Often an erroneous belief is that stripping the banks of vegetation will prevent beaver impacts; whilst these may be initially less attractive, or see less visual impacts of beaver presence, this does not prevent burrowing and beavers can still forage on any recolonizing vegetation. Almost more importantly, such action is often likely to only increase bank erosion rates, along with having additional impacts on biodiversity.

Techniques such as rock-facing/riprap or covering banks with metal mesh can provide a physical barrier against digging. Medium to coarse stone size is recommended, to a thickness of at least 40 cm, which can overlay buried mesh to give extra protection (Vorel et al., 2016). Mesh can of course be used on its own and may have the benefit that if buried it can permit some vegetation taking root, as opposed to bare stone. Such mesh must be resistant to corrosion, consisting of approximately 10 × 10 cm, with a minimum gauge of 2 mm (Vorel et al., 2016). Any covering must extend well below the waterline to prevent beavers extending burrows from below the protection point. An alternative

Figure 11.7a and b Digging impacts and collapsed burrows can lead to conflicts in agricultural areas and/or with flood bank if these then breach and flood often high value protected land laying behind (a). Regular use of bank access can also lead to eroded sections which may or may not be a significat issue (b). (Photo supplied courtesy of Roisin Campbell-Palmer.)

technique is to insert metal or reinforced plastic piles vertically into a bank, as opposed to covering the surface. This permits beavers (and other burrowing animals) to extend any digging only so far, without destabilizing the bank itself. This has been recently assessed on the Capolne River, Romania, in which vertical mesh netting was installed along both sides of vital dikes, to a depth of 1.7 m for a stretch of 500 m (25,500 €/km) (Pasca et al., 2019).

Greenbank protection solutions as part of river restoration techniques are gaining greater momentum as the realization dawns that hard infrastructure methods are proving ineffective over the long term, especially in light of climatic change. In the UK, various 'greener' river restoration techniques have been implemented across different catchments, though their resilience to beaver activities is yet to be tested. Bank protection methods including willow spiling, willow/brushwood mattress or coir matting revetment, and tree root wad protection offer more natural and living solutions (River Restoration Centre, 1997). Planting live willow or cottonwood stakes and willow spiling can prove cheap and effective measures, holding eroding banks together, especially effective on steep banks (Smith and Prichard, 1992). Planting schemes have been refined to encourage establishment plants, including recommendations for planting to occur November to March, when soil is moist and soft, avoiding hot dry periods, using freshly cut woody stakes between 50 and 100 mm thick, and protecting from herbivores. Willow can also be used to create living brash bundles, which can then be staked into place along the face of a bank. 'Juniper treatment' generally consists of using cut trees, even discarded Christmas trees, anchored into the banks to reduce erosion and bank sloughing and act as sediment traps (Smith and Prichard, 1992). A well-rooted vegetated bank is far more resilient to water erosion, as the increased friction slows water down. Even a 5-cm-deep root mass can resist erosion up to 20,000 times more effectively than bare soil shorelines (Fitch et al., 2003). Removing livestock or preventing them from intensively grazing bankside vegetation can quickly lead to the regeneration of a bankside vegetation protective layer, speeded by tree planting, and can be one of the fastest and cheapest management methods to reduce bank erosion (see Section 11.4.3). Placing shelters and feeding trough/salt licks outwith of riparian areas and creating off-stream watering sites can encourage livestock away and ease pressure on riparian areas (Fitch et al., 2003).

Infilling of canals or burrows, either whole or partially collapsed, is an immediate solution to repair that damage, as opposed to a preventative or protection strategy. Substances to infill with are dependent on the type of bank or structures being undermined. Rock or gabions have been used. Roads and tracks, for example, would require some form of gravel or hardcore, whereas gardens or agricultural areas can be backfilled with soil. In more naturalized banks, juniper treatment, brash bundles, or tree roots could be anchored into the depression. Infilling materials need to be stabilized and-or compacted; otherwise they can be washed out or depressions formed over time (Vorel et al., 2016). The impacts of beaver slides (see Figure 11.7b) in unwanted bank sections can be reduced by either preventing entry through fencing (which may be impractical) or mesh overlaid on the slide to reduce erosion. Meshing may prevent further digging and hold the bank section together to allow vegetation to grow through in time. However, it should be noted that these trails can often be related to foraging and can be temporary in nature so perhaps not worth of heavily resourced investment unless likely to lead to significant damage.

11.4.8 Fertility control

One advantage of sterilization is that it can enable beavers to remain as a family unit, preventing colonization by other beavers for as long as they can maintain that territory. Both sexes can be permanently surgically sterilized via minimally invasive surgery or laparoscopy ('keyhole surgery' Figure 11.8), via tubal ligation in females or vasectomy in males, across all age classes, with full recovery and maintenance of behaviours observed (Brooks et al., 1980; Pizzi, 2014; Campbell-Palmer et al., 2016). More traditional sterilization, through gonad removal (ovariectomy or ovariohysterectomy in females, or castration in males), has successfully reduced fecundity, though it can alter behaviours and cause family structure breakdown, including

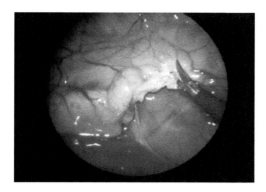

Figure 11.8 Beaver sterilization has been undertaken via keyhole surgery techniques. These laparoscopic camera views show the spermatic cord being cut and cauterized so that minimal surgical incision is required and quick recovery and healing times are ensured. (Photo supplied courtesy of Romain Pizzi.)

dispersal (Brooks et al., 1980; Taylor et al., 2017). The advantage of tubal ligation is the preservation of gonads and normal sex hormone production, and thus retention of social and hierarchical integrity (Brooks et al., 1980; Campbell-Palmer et al., 2016). Additionally, less pain, less risk of postoperative infections and complications, and faster recovery times have been reported (Pizzi, 2014).

Sterilization has been deployed more extensively as a management intervention for wild North American beavers in the Rhineland-Palatinate region of Germany. After huge conservation efforts for Eurasian beavers, campaigning, and public engagement in other parts of Germany, the first beavers to arrive in this region turned out to be North American rather than the native Eurasian (Schulte and Müller-Schwarze, 1999; Zahner et al., 2005). Public outcry against lethal control, along with confused messaging welcoming beavers back but at the same time these being are the wrong ones, resulted in large resource investment to trap and sterilize ~60 animals over 6 years. Wildlife managers considered this a balanced approach—preventing a non-native species from breeding, whilst enabling ecological benefits to establish and introduce concepts of coexistence with local people (S. Venske, pers. comm.). Over time as these individuals come to their end of life, they should readily be replaced by their Eurasian counterparts through natural dispersal.

Beaver sterilization has also been successfully employed in Canada at Montreal Airport, where

damming encouraged wetland habitats, thereby increasing bird strike concerns, and again at Farnham Garrison, where increased water levels interfered with military operations for the Canadian Armed Forces (Fortin, 2011). Beaver pairs remaining together for several years and no indication of lactation or new individuals were observed. Whilst sterilization of beavers as a theoretical means of population regulation proved successful at particular sites (Brook et al., 1980), widespread application is questionable and significant resource investment is required (Schulte and Müller-Schwarze, 1999).

11.4.9 Hunting, trapping, lethal control, and euthanasia

The lethal control of beavers should be teased apart in its role as a management tool and/or for recreational hunting. Any trapping, even if it involves live trapping, has the same management effect as lethal control, i.e. the beaver is removed from the system. Removing beavers can of course immediately reduce the source of damming and felling conflicts (burrowing impacts remain), but as a long-term management strategy it requires ongoing resource and time investment as empty territories act like vacuums, eventually filled by other beavers.

Beaver hunting and trapping are both considered to be population regulation tools, but also important cultural pastimes, especially in historical fur-trapping regions, with an estimated 176,573 trappers in the USA alone (Responsive Management, 2016). The majority of trapping in the USA (60% based on a survey in 2015) is associated with the trapping and removal of nuisance animals (primary species listed as raccoons [*Procyon lotor*], coyote [*Canis latrans*], muskrat, and beaver), apart from in Alaska in which trapping is a more significant source of income (Responsive Management, 2016). Spring hunting of beavers is more effective in reducing population density, as pregnant females are often shot first as they typically forage earlier in the evening (Parker et al., 2000; Parker and Rosell, 2001). When populations were shot at rates of 22–26% of the estimated spring population, a 46% decrease in occupied territories occurred within 3 years (Parker et al., 2002), though interestingly, this same population responded to this control through experiencing higher rates of dispersal into empty territories and increased fecundity once hunting was reduced, seeing a rebound of 93% (Parker and Rosell, 2014). Exploited beaver populations can of course be over-harvested and numerous regulations exist to prevent this, from hunting quotas to trapping seasons. As a mitigation tool, beavers causing conflict can often lead to the targeted removal of that whole family group as a reactive solution to damage. Shooting is legal in some but not all American states and European countries; local regulations should be checked and any permits or licences secured (Taylor et al., 2017).

Beaver pelt prices are generally low, but there is still a lucrative market. Some fur wholesalers have witnessed rises from 5,000 to 30,000 pelts a year, given increasing demand from Russia, Turkey, China, and South Korea, at prices of up to $35 per processed pelt, some parts of which may end up as dyed pieces on high-end fashion items costing thousands of dollars (Galbraith, 2006). More generally, beaver pelt prices at North American Fur Auctions for 2017 averaged between $8 and $13 (Wood, 2017). Trapping is commonly employed not only as a means of mitigation but also to utilize beaver fur and meat, often for a non-commercial market. Beaver castoreum and castor sacs are also harvested for the trapping market for use as lures (Dobbins, 2004; see section 10.4.1.3).

Trapping regulations vary, and local by-laws should be consulted both in advance and regularly. The Association of Fish and Wildlife Agency has published Best Management Practices for Trapping in the United States (2006). Trap line management advice cards give general ecological information, methods to estimate population size, harvest levels, and timing to maximize pelt quality. The Yukon trap line management series, for example, recommends trapping a whole family only from every second or third territory, rather than continuously trapping individuals over multiple families, to maximize trapping and maintain healthy populations; then to wait 3–4 years before re-trapping a removed territory; and only use quick-kill traps to ensure humane trapping practices (Yukon Renewable Resources, 1992).

Hunting in the Nordic countries is generally permitted for beaver management but in accordance with facilitating local and sustainable management. Commercial and recreational utilization of beavers as a game resource is recognized, alongside as a means to prevent damage and 'inconvenience to other societal interest' (Norwegian Nature Diversity Act). The Game Act in Sweden addresses both hunting rights and wildlife conservation through hunting seasons, fire-arms regulations, and legislation on wildlife disturbance including protection of active lodges. In some states and countries like Norway and Latvia, beaver meat is consumed generally only by local hunters, some restaurants, and specialty shops, though demand is low, with market values given as from $6.40 to $12.60/kg for choice cuts (Rosell and Parker, 1995; Parker and Rosell, 2001). Various lethal control methods are permitted in Estonia, including shooting, capture with dogs, nets, and even bow and arrow, according to the Estonian Hunting Rule and Hunting Act, within permitted dates. Beaver hunting in Russia (both species) was recorded as 16,968 animals between 2012 and 2013 and 14,429 in 2013 and 2014, which is still considered to be < 3% of the harvestable total (Sjöberg and Belova, 2019).

Live trapping for population monitoring and translocation is largely addressed in Chapter 10. Translocation, especially as a conflict resolution strategy, can work well until habitat availability and population density become the limiting factor, as beavers should not be moved into beaver-occupied territories without seriously questioning welfare implications. It should also be noted that, typically, dispersal distances from the release site (see Section 11.6.2) mean any translocation is effectively releasing beavers into a whole catchment area, so that they can easily choose to reside in areas not targeted for their occupations and this could include a non-supportive landowner.

Recreational and/or lethal trapping tends to involve either killing traps (typically body-grip type) or restraining traps (typically leg-hold traps). Note leg-hold traps are banned in over 80 countries, including the EU, on the basis of animal welfare grounds (Fox, 2004). The 330 Conibear trap is one of the most commonly deployed beaver kill traps, typically used in water, either in open water, placed in beaver canals, or in breaches made in dams (Murphey, 1996; Dobbins, 2004; Responsive Management, 2016).

Table 11.4 Beaver harvesting in Baltic region countries.

Country	Beaver hunting bag (number of animals)			Hunting season
	2015	2014	2013	
Sweden	12,928	8,448	8,210	01/10–10/05 (south) or 15/05 (north)
Norway	1,640	1,770	1,880	01/10–30/04 (hunting season can be extended for up to 15 days if ice and snow conditions persist)
Finland				
C. fiber	235	191	231	20/08–30/04 (both species)
C. canadensis	5,300	6,700	4,200	
Estonia	6,557	5,572	5,700	01/08–15/04 (with some exceptions)
Latvia	24,248	31,376	24,711	15/07–15/04
Lithuania	19,544	21,749	11,778	01/08–15/04
Poland	133	93	38	01/10–15/03 (partially protected under EU legal acts)
Republic of Karelia	238	165	150	01/10 to 28–29/02

Source: Reprinted from Sjöberg, G., Ulevicius, A., and Belova, O. 2020. Good practices for management of beavers in the Baltic Sea region, pp. 102–121. In: Sjöberg, G. & Belova, O. (eds.) Beaver as a Renewable Resource. A Beaver Dam Handbook for the Baltic Sea Region. Swedish Forest Agency.

Experienced trappers will also commonly set traps on underwater runs or dig out coves in a bank to attract beavers into, place traps along forage trails and slides, or create artificial scent mounds using castoreum as a lure (Murphey, 1996; Massaro and Massaro, 2020). They are generally effective for all beaver sizes, producing an extreme pressure when triggered (Taylor et al., 2017). Leg-hold traps, with padding often recommended for humane reasons, can be accompanied by a submersion set so that the animal is drowned on returning to the water (Massaro and Massaro, 2020). Obviously, the use of these trapping methods is considered controversial by some and risk accidentally trapping other species.

The welfare of trapped animals will vary according to trap type, expertise of trap management, level of mis-strikes, fail and escape rates, capture of non-target species, time to unconsciousness (kill traps), and degree of level of trauma experienced (Iossa et al., 2007). In a 2015 survey of trappers in the USA the most common secondarily caught species when trapping for beavers were muskrat followed by otter and raccoon predominantly, though the list also included low incidents of mink, opossum

(*Didelphis virginiana*), coyote, skunk (*Mephitis mephitis*), fox species, squirrel (*Sciurus carolinensis*), badger (*Taxidea taxus*), lynx (*Lynx canadensis*), and wolf (*Canis lupus*) (Responsive Management, 2016). Submersion or drowning traps are regularly used for beavers, especially in North America (Massaro and Massaro, 2020). The physiological adaptations beavers possess for diving (see Section 3.4.2) mean that death by drowning-induced hypoxia takes an average of 9 minutes and often is accompanied by very obvious signs of struggling and indicators of distress (Gilbert and Gofton, 1982). Drowning is not considered an acceptable means of euthanasia (Ludders et al., 1999; Beaver et al., 2001). Reported time to death or unconsciousness using 330 Conibear traps ranges from 1.5 to 9.25 minutes (Novak, 1987); others have reported death to be instant in the vast majority of times (Taylor et al., 2017).

Shooting beavers is often undertaken during the early morning or early evening hours to be most productive, unless thermal equipment is permitted for use at night. It has been estimated that the majority (80–90% of all beavers harvested) in Scandinavia are shot with centre-fire rifles (Parker and Rosell, 2003). Typical annual harvests in 2018 for Sweden range from 8,000 to 9,000 and in Finland are 6,200 (Natural Resources Institute Finland [LUKE], 2019; Table 11.4). Efforts are often focused on animals leaving or returning to a known lodge or burrow or following breaching of dams and waiting for animals to return to fix. Weapon type and caliber permitted will vary, though it is thought shotguns at close range are often used as an acceptable dispatch method (Taylor et al., 2017). The loss of a wounded animal, especially escaping into deeper water, is a real animal welfare issue. Head shots may offer an outright kill and are often practiced by North American hunters (Eastland, 2000), though body shots are encouraged for most Nordic hunters to reduce the chances of missing the head region (Parker et al., 2006). Body shots reportedly cause instantaneous unconsciousness in the vast majority of cases, thought to be due to pressure waves from impacting projectiles (Harvey et al., 1962). High immobilization rates and low loss rates determined in a Norwegian study concluded that centre-fire rifles produce a relatively humane death, with riffle ammunition delivering ≥ 980 joules of energy at 100 m (Parker et al., 2006).

Blunt force trauma so that the brain is destroyed is accepted as a humane lethal control for either free-ranging or confined beavers in some regions (American Veterinary Medical Association, 2013), though note this has practical limitations and is likely to be socially sensitive. Humane euthanasia is typically carried out using injectable barbiturates, such as sodium pentobarbital (80–160 mg/kg), following sedation. Beaver physiological adaptations for its semi-aquatic lifestyle, including breath holding, compacted chest cavity, and lack of neck region, mean inhalant agents, cervical dislocation, exsanguination, and thoracic compression are not practical or acceptable means of euthanasia (Taylor et al., 2017).

11.4.10 Exclusion zones

The incredible adaptability of beavers to live in habitats ranging from boreal forests to deserts is impressive (Naiman et al., 1988; Andersen and Shafroth, 2010). Sometimes suggestions for management plans can refer to 'exclusion zones'. This term can be somewhat misleading as there are no realistic means of completely excluding beavers

from larger areas once they are within its catchment. This would require every watercourse to be devoid of vegetation (not just trees) and completely covered in hard infrastructure, a state most people can probably agree is not desirable or acceptable (Schwab, 2014). It is more likely in zones where alternative methods to mitigate beaver presence are not possible that trap and removal or lethal control are more readily implemented. However, it should be noted such measures will not exclude beavers, just reduce their presence, as they will continually colonize, and therefore these will be ongoing management measures. Any removed animals must then either be translocated, if release areas exist, or be humanely euthanized. Both trap and removal and lethal control present their own animal welfare, legality, and resource implications (see Chapter 10).

11.4.11 Preventative management

Landowners and users in areas already or likely to be occupied by beavers can be engaged through local stakeholder groups—a useful tool to encourage people to feel included, disseminate information, and as a forum to provide mitigation advice. Areas with vulnerable banks, high-value trees, and crops lying within accessible range, and locations where dams are likely and would resultantly cause significant damage can all be more regularly monitored so that a rapid and appropriate response can be mounted.

One aspect of ongoing monitoring is of course population dynamics and distribution spread (see Chapter 10), which in turn can help set conservation and trapping targets. However, having the knowledge and resources to mitigate against beaver conflicts if, as, and when they occur is probably more relevant to individual land managers as opposed to knowing how many beavers are in an area. You are more likely to be focusing on that beaver family that has moved into yours! Therefore, opportunities for both preventing future issues and realizing when current activity may cause an issue are key components of reducing expenses down the line. New developments and planting schemes in and around freshwater areas should take beavers into account. Such measures, where possible, could include the following:

- Naturalized buffer strips of native riparian vegetation allowed to develop. Livestock grazing, gardens, and development pushed back from the water's edge.
- Exclusion fencing and/or tree protection installed ahead of planting schemes in riparian zones.
- Reinforcing artificial banks at time of construction to account for burrowing animals (stone facing, mesh fronting, sheet piling, gravel cores, greenbank protection, etc.).
- Outflows protected so beavers cannot enter with materials.
- Culverts made oversized and/or protected with metal grilles, welding rods, or fencing.
- Infrastructure taken away from water bodies, or embankments protected on water side to prevent burrowing.
- In areas prone to flooding, infrastructure built on raised foundations and/or associated protective embankments built to prevent waterlogging.
- Flood banks set back from the river's edge and resistant to burrowing.

11.4.12 Public education and outreach programmes

Given the potential for beaver–human conflicts, especially in highly urbanized areas, one of the key measures to encourage ongoing tolerance by humans and public support needs to incorporate engagement with local stakeholders, education programmes and campaigns, and good-quality information dissemination. Lack of beaver knowledge and experience of living alongside them can generate negative perceptions and lead to conflicts. In fact, the opposite has been proven, with projects providing high-quality beaver ecology and management information, in an objective and evidence-based manner, associated with increased tolerance of beavers and their activities (Parker and Rosell, 2012; Jones and Campbell-Palmer, 2014; Brazier et al., 2020). Well-informed, open engagement can go a long way in encouraging people to implement ways to live alongside beavers as opposed to viewing trap and removal or lethal control as the 'go to' option. For example, the Yakima Basin Beaver Reintroduction Project in Washington state provided an outreach programme for neighbouring school districts, landowners, and local agencies, engaging with 33,000 local residents through a range of presentations, festivals, newsletters, and media (Babik and Meyer, 2015). This is especially vital to encourage communities to embrace more naturalized and climate-resistant strategies.

11.4.13 Management of captive collections

Beavers are notorious escape artists, capable of burrowing, climbing, gnawing, or even swimming out of enclosures when opportunities present themselves. Zoo escapes are not uncommon, and during times when the Eurasian beaver was more difficult to source, numerous zoos throughout Europe kept the more sourceable North American species. Evidence of escapes has been documented and mapped (see Chapter 2.8). The risk of escapes of these non-native species into the wild is plausible still, even with precautions in place. This has resulted in calls from some authorities, conservationists, and European Beaver Management plans to question the desire and need to retain this species in European collections, and instead encourage their replacement with the native Eurasian (Vorel et al., 2003).

A slightly different variation of this has occurred in Britain. After its extinction, small numbers of beavers have been intermittently imported by private collectors and zoos, from around the eighteenth century onwards, with some misguided projects releasing the easier to source North American beaver. The early 'reintroductions' involving some associated captive time were haphazard to say the least and unlikely to be considered real attempts at restoring the species; in fact, North American beavers were often used. One of the first documented beaver releases was in Sotterley Park, Suffolk, England in the 1860s. These North American beavers soon made dams (considered 'eyesores' at the time), survived, and bred a few years and then seem to have died out and disappeared (Conroy et al., 1998). Two pairs of beavers were released on the Isle of Bute, Scotland, around 1874, with conflicting sourcing claims from both France and North America (Conroy et al., 1998). The Marquis of Bute employed a 'keeper of beavers' who published an account of his observations, including descriptions of dam and lodge building and seasonal variations in feeding behaviours; however, he mistakenly thought beavers had two litters per year of only a single kit at a time (Black, 1880). These beavers are thought to have died out by the late 1890s (Coles, 2006). One of these dams was still visible in 1918, though no beavers themselves were thought to be still present (Gibson, 1980).

None of these can be considered as serious attempts to re-establish the species or resulted in free-living populations. Eurasian beaver reintroduction feasibility and desirability really began in Scotland during the early 1990s, with a key driver being the EU Habitats Directive (Gaywood, 2017). Various people began to import Eurasian beavers for captive collections before any official restoration processes resulted. The official reintroduction process has taken decades and currently has still not resulted in full acceptance or restoration of the species. This accusation of official processes taking too long, being overly expensive, and believed by some to be loaded in favour of politically astute opponents has gained momentum over the years (Arts et al., 2014). Anecdotal evidence suggests that mounting frustration with this situation is resulting in unofficial and illegal animal releases as a counter action. Additionally, a range of fencing specifications and husbandry practices have existed for private collections and escapees, without doubt contributing to the unofficial wild populations manifested today. Where the official response has been to remove beavers, public debate has commonly forced a change of approach and resulted in them remaining. Following the granting of European Protected Species status in Scotland, the statutory authorities also permitted lethal control licences, especially on prime agricultural land experiencing beaver conflicts, with 87 individuals dispatched in the first year.

11.5 Role as a charismatic flagship species for conservation projects

Reintroductions are viable and increasingly being implemented as a tool in reversing biodiversity loss and encouraging ecological restoration, though such projects can be emotive, generating strong support or objections depending on the species (or perception of) involved (Douglas and Veríssimo, 2013). Overall, wildlife restoration projects tend to capture the public's imagination, are popular, and are often easier to fundraise for compared with other ecological restoration measures such as invasive species removal. The success of many conservation and reintroduction projects is highly dependent on the species involved, with iconic flagship species inspiring the public and generating funds in clever marketing ploys. Arts et al. (2014) even go as far as to say 'the reintroductions of charismatic animals symbolize Western conservation ideologies', recognizing that the reintroduction of charismatic species tends also to be the most sociopolitically contentious, whilst releases of other animals such as fish, insects, or plants are often overlooked.

Fortunately, beavers are charismatic mammals (Figure 11.9). Their restoration has received much media attention, influenced by public enthusiasm, academic investigations, and political discussion. Under the EU Habitats Directive, member states are legally obligated to investigate the feasibility of native species reintroductions, and even promote their implementation (Rees, 2001). A quick search online demonstrates beavers' popularity—from

Figure 11.9 Eurasian beaver stamp released by The Royal Mail as part of a series recognizing reintroduced wildlife to Britain. Copy Right Stamp design ©Royal Mail Group Limited.

passionate rehabilitation folk to those dedicated to reducing their conflicts via non-lethal solutions. Families can readily become tolerant to their human observers, lending themselves readily to wildlife watching and photography. People in numerous states and countries have fallen for the charms of beavers and their curious lifestyle and actively defended their attempted removal or lethal control. For example, when beavers appeared in the Devon countryside unofficially, attempts by the local authorities to remove them from the River Otter were actively stopped by local residents refusing to give permission for traps to be set on their land. For any such project there are multiple ecological, social, and political factors to be balanced, with any management being both practical and generally accepted (Arts et al., 2014).

11.5.1 Wildlife tourism

The return of charismatic species has seen wildlife tourism development (Johansson et al., 2016). Beavers with the wildlife watching and habitat immersion opportunities they can provide are generally popular with people across their range. In more recent years, political interest in beaver restoration has risen, in conjunction with growing social awareness on the impacts of climate change and biodiversity loss. Typically, public opinion surveys reveal strong support for beavers, e.g. 86% of 1,944 responses regarding beaver reintroduction to Scotland were supportive—especially those engaged with conservation and academic sectors, with fishing/angling interests displaying least support (Scott Porter Research & Marketing Ltd, 1998).

Beavers can contribute to wildlife tourism and seemingly be quite lucrative for well-organized facilities with regular successful viewing. Beavers can readily lend themselves to tourism opportunities, being territorial, often with identifiable homes which they regularly utilize at routine times, enabling observations (exiting early evening and returning early morning). This makes it feasible to organize regular walks, place information displays, and facilitate viewing through hides or platforms. For example, such infrastructure was invested by the Scottish Beaver Trial during the first official reintroduction of beavers to Britain (https://www.

scottishbeavers.org.uk), after which local community-focused wildlife business took on visitor education and wildlife watching opportunities (https://www.argyllbeavercentre.co.uk). Typical beaver emergence timings also encourage overnight stays, providing further spend within the local community (Campbell et al., 2007).

In Belgium, the conservation non-government organization (NGO) bringing back the beaver developed a marketing strategy 'Beavers, Beer and Castles' to promote their restoration and generate funds for local landowners, guest houses, pubs, and guides. Websites such as Nordic Discovery promote 'Beaver Safari' packages in which people can immerse themselves in nature, canoeing with an experienced guide at dusk to look for emerging beavers, and stay in local accommodation. One wildlife tourism company in Sweden focusing on beaver tours discusses how they organize 'beaver evenings' during which they provide informative talks on the species and encourage customers to find evidence of beaver activities so that it's not just about seeing the animals (Margaryan and Wall-Reinius, 2017). A single beaver family alone, taking up residency in a highly accessible location (with riverside path, car parking, and refreshment opportunities nearby) in Devon, England, saw increased footfall with beaver presence, especially during the summer months when kits were easily observable (Auster et al., 2020a). Local businesses reported a largely positive response, including increased custom and selling opportunities for beaver-related products, with the average 'willing-to-pay' for a beaver watching experience value suggested by local residents to be £7.74 ($10.34/€8.67) (Brazier et al., 2020). At the first official beaver reintroduction in Scotland, guided walks given by project staff were used as a proxy for 'wildlife experiences' with an estimated value over 5 years of £355,000–£520,000 (Moran and Lewis, 2014).

11.6 Beaver restoration

The now infamous story of how the Idaho Fish and Game Department restored beaver populations would most certainly not get past any animal welfare laws today, but surprisingly and undoubtedly resulted in a long-term population. In the 1940s

Idaho still remained a fairly inaccessible wilderness with few suitable roads, accessible only by foot and pack horses. Initially, attempts to use mules and horses to carry beavers to the planned release areas were thwarted as the carrying beasts took objection to carrying unsettled beavers, which in turn did not fare well during long, jolting journeys in hot crates. A plan was hatched to fly and then drop (from heights of 150–250 m) beavers in purposely built crates that would break open safely on impact and free the beavers. Geronimo, an adult male, was the test pilot making several successful aerial trips and eventually released; 76 beavers where dropped in 1948, with only one death recorded escaping from the travel crate mid-drop (Heter, 1950). Many of these early projects document large numbers of animals being released as part of what must have been very impressive strives to restore this species (see Chapter 2). Many beaver populations successfully established in time, while others failed drastically, and early records lack details of survival rates and project success.

Reintroductions and/or conservation transloca-tions are viable means of reversing biodiversity loss and encouraging ecological restoration, increas-ingly seen as important conservation tools (Donlan et al., 2006; Figure 11.10). Depending on the species in question and length of time of absence, reintro-ductions are generally popular with the public and can be used to promote conservation projects. Today the beaver can certainly be valued as an iconic, flagship species for the promotion of wet-land conservation and receives high accolade for its ecological restoration services (Figure 11.11). Analysis of reintroduction efforts throughout Europe deter-mined that the Eurasian beaver was involved in > 40% of all reintroduction attempts (Thévenin et al., 2020), though this was not always the case, and the first reintroductions were often undertaken to support the fur industry.

Beavers have been promoted as tools with restorative powers. Numerous land managers and riparian restoration projects report beaver activities as producing positive resource-saving impacts, including reed-cutting, natural tree-regeneration thinning, creation of open water, and willow scrub removal. All these activities are common manage-ment practices in wetland landscapes to maintain

Figure 11.10 Beavers are popular and charismatic animals, often generating a range of merchandise, wildlife watching, and cultural references for both species throughout their respective ranges. Reintroduction and release programmes generate much interest (Photo supplied courtesy of Chris Parkes.)

favorable habitat types for biodiversity. Beavers have also been widely endorsed as positive agents for improving degraded freshwater systems, to restore hydrologic and geomorphology regimes, and to enhance habitats for other plants and ani-mals (Bailey et al., 2019), such as fish (Pollock et al., 2007; Eaton et al., 2013; see Chapter 9). Beaver habi-tat engineering as a habitat restoration tool in degraded agricultural environments has been docu-mented, through their creation of 'a mosaic of aquatic, semi-aquatic and terrestrial habitat patches' (Law et al., 2017). Reintroductions in semi-arid regions of North America have successfully improved habitat condition and restored riparian zones (Brayton, 1984; Albert and Trimble, 2000; McKinstry et al., 2001).

Generally, three approaches to beaver restoration typically occur: (1) passively enabling beavers to colonize and increase in an area through lethal

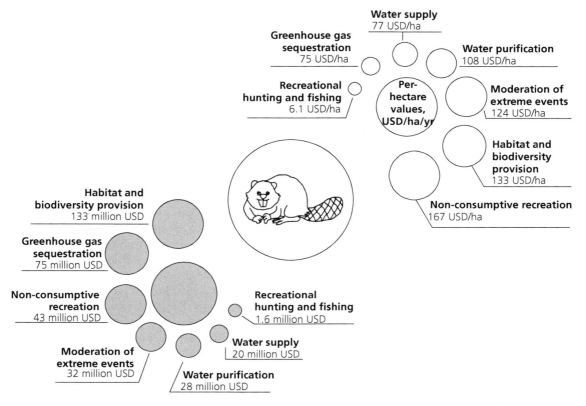

Water supply
77 USD/ha

Greenhouse gas
sequestration
75 USD/ha

Water purification
108 USD/ha

Recreational
hunting and fishing
6.1 USD/ha

Per-
hectare
values,
USD/ha/yr

Moderation of
extreme events
124 USD/ha

Habitat and
biodiversity
provision
133 USD/ha

Non-consumptive recreation
167 USD/ha

Habitat and
biodiversity provision
133 million USD

Greenhouse gas
sequestration
75 million USD

Non-consumptive
recreation
43 million USD

Recreational
hunting and fishing
1.6 million USD

Moderation of
extreme events
32 million USD

Water supply
20 million USD

Water purification
28 million USD

Figure 11.11 Ecological services generated by beavers and their activities in per hectare values (white circles) and aggregated over the Northern hemisphere beaver range (grey circles) per year. (Reprinted from Thompson, S., Vehkaoja, M., Pellikka, J., and Nummi, P. *Mammal Review* © 2020 Wiley Periodicals, Inc.)

control restrictions and/or bringing agriculture and development back from freshwater, enabling riparian vegetation to recover; (2) more actively employing habitat manipulation techniques to improve habitat and encourage beavers to colonize, including tree plantation schemes and/or increasing wetland areas; and (3) actively translocating beavers into suitable areas and increasing connectivity of populations (Pollock et al., 2015).

Buckley et al. (2011) investigated the economic value of restoring beavers to carrying capacity within a Utah watershed, concluding that the annual value of their presence could amount to tens to hundreds of millions of dollars. Beaver restoration projects are generally not just about beaver conservation but tend to incorporate related goals, including wetland and incised stream restoration; water management, quality improvement, and storage; conflict management; habitat creation especially for species of conservation value; and restoration of the riparian zone (Pollock et al., 2015). However, as a number of authors warn, beavers should not be viewed as the 'silver bullet' to wetland restoration and biodiversity recovery, as there will be varying impacts on certain species, sensitive habitats may not thrive with their activities, and their benefits may not become evident for some time (Law et al., 2017). Any sensible beaver restoration projects should come with access to a range of pragmatic mitigation tools and a wider national, long-term management strategy. Numerous regional management documents have been produced and are available online.

Both beaver species are still in the process of recovery after widespread extirpation, though where many are reintroduced or appear again, they tend to successfully recolonize. Beaver recovery in northern Minnesota was documented and examined through aerial photographs of the boreal forests, ~300 km², over 46 years (Johnston and Naiman, 1990a; Johnston and Naiman, 1990b). They showed rapid expansion over the first 20 years, which slowed as habitat became less available—an increase of 71 to 835 territories over this period. It has been estimated that the recovery of beavers globally from near extinction in the early 1900s has been accompanied by the creation of 9,500–42,000 km² of shallow aquatic habitat with an associated increase in > 200,000 km of riparian shoreline (Whitfield et al., 2015).

Beaver-related restoration projects may have a wide range of goals, making comparisons between projects difficult, but recommendations for better guidelines have been proposed, accompanied by research focusing on best practices, including social along with ecological components (Pilliod et al., 2018). This wider societal component is most strongly demonstrated in relation to unofficial beaver releases, which have occurred in several European countries including Spain, Belgium, and Britain (Dewas et al., 2012). Identifying project goals and strategies, collaborations, and resource availability both in advance and over the long term, along with continual landowner and stakeholder engagement are key criteria for success (Pollock et al., 2015), even without the ecological requirements of the animals themselves. Often the effectiveness of beaver translocation in achieving the restoration goals of a project may be hard to assess, particularly due to a wide range of impacting factors and the time frame of the project.

Future research areas for consideration to improve beaver restoration include long-term genetic management, especially of small reintroduced populations with restricted genetic flow. Britain for example as an island presents little opportunity for new genetic exchange unless animals are proactively translocated with consideration of genetic diversity. Beavers exhibit low major histocompatibility complex (MHC) polymorphism after recent bottlenecks caused by the fur trade, and the majority of restoration programmes have involved small numbers of founder animals with little genetic consideration (Ellegren et al., 1993; Babik et al., 2005).

The susceptibility of populations with limited MHC diversity to disease and potential extinction is controversial, especially as many populations, including other mammalian species, have recovered well without apparent deleterious effects (Ellegren et al., 1993; Edwards and Potts, 1996; Mikko et al., 1999). Ultimately, reintroductions, translocations, and release of animals from conflict zones or rehabilitation facilities should seek to result in genetically diverse populations, which minimize inbreeding and therefore help ensure long-term survival and robustness to long-term environmental changes (Senn et al., 2014; Iso-Touru et al., 2020). The perception exists that, even today, the majority of beaver moves often occur without full consideration of genetics, disease and parasites (Campbell-Palmer et al., 2020; 2021), and long-term monitoring and management resources, and that clear guidelines for best practice are lacking (Pilliod et al., 2018). Restoration projects should strive to demonstrate best practice and build on the knowledge base to inform future work. For instance, although so much science has been published over the decades, a recent evaluation determined only around 22% of projects documented post-release monitoring (Pilliod et al., 2018).

11.6.1 Release site suitability assessment and release techniques

Of course, not all reintroduction projects and every translocation are a success, with some projects having quite mixed results and providing important protocol refinement details for others. Like any animal translocation, factors encouraging site fidelity, resource availability, animal welfare, and ability to choose are more likely to be successful than humans determining where would be beneficial for the animal to remain. Numerous studies have investigated and highlighted different variables as being key habitat characteristics for beaver colonization, with a range of conclusions, useful in identifying release sites and potential future management issues (Touihri et al., 2018). Habitat quality and food availability are key determinants in species distribution which can be readily modelled (see Chapter 4); however, human disturbance can influence both whether a species will occupy an area and/or be tolerated, and so this should be considered when planning reintroduction and species management (Alakoski et al., 2020).

Recent studies modelling the distribution of both beaver species in Finland determined that aquatic habitat and temperature were the most significant factors. Human population and distance from agricultural areas influenced Eurasian beaver distribution, contributing 12.5% and 9.2% of the full modelling respectively, but appeared to have little effect on North American beavers (both contributing < 4%) (Alakoski et al., 2020), with the greatest Eurasian beaver habitat suitability next to agriculture but in areas with low numbers of humans. Water-related variables (such as stream gradient and width) seem to be more significant to long-term colonization of sites in comparison to food-associated ones (such as percentage of shrub cover for example) (Howard and Larson, 1985), with other studies finding shrub and canopy cover being more significant (Cox and Nelson, 2009).

Beaver site selection and the habitats they create are variable, often changing precise damming location seasonally and/or with population density and often colonizing sites not requiring dams first if given a choice (Campbell et al., 2005). Two initial beaver relocation projects in Oregon evaluated beaver translocations: one as a tool to increase habitat quality for salmon (Petro et al., 2015), the other to examine beaver–habitat relationships within a watershed (Oregon Department of Fish and Wildlife, 2012). GIS modelling was used to identify sites where beaver damming would increase habitat quality for salmon, and habitat surveys were used to identify key features for beaver occupation. Both projects experienced high levels of mortality and dispersal from the release site. As the Alsea Basin project modelled where beaver damming would most likely improve salmon habitat, those beavers that didn't dam either dispersed fairly rapidly, so some degree of restocking was required, and/or were predated by cougars (*Puma concolor*), due to lack of deeper water (Petro et al., 2015). The seasonal washout of dams in some watersheds and therefore the ability of beavers to maintain dams over the long term was also most likely an unpredicted factor. Beaver relocation to the Umpqua watershed was viewed

with mixed success. The Oregon Department of Fish and Wildlife (ODFW) selected release sites based on evaluation of typical beaver habitat features, releasing only animals in good body condition. There was a varied outcome, with mortality rates > 50% (predation, roadkill, drowning, and water-fall associated deaths), whilst others remained at their release site, and others still dispersed up to 8 miles (~13 km) before settling (Oregon Forest Resources Institute, 2016).

Some translocations fail or suffer significant setbacks as beavers may die in the process, for example 15 beavers died in trapping and 13 beaver during transport out of 277 animals, or disperse from the release site, for example 51% beavers emigrated >10km within 180 days of release (McKinstry and Anderson, 2002). Post-release survival rates vary greatly; for example, 57% of predator mortality occurred in the first week of release in an Oregon project (Petro et al., 2015). In Switzerland, > 50 of the 141 animals released died, though populations slowly developed but were vulnerable and fragmented. Following active riparian habitat restoration, beaver populations expanded both in numbers and range (Angst, 2010; Dewas et al., 2012). First-year mortality rates in an Illinois study of beavers released into a predominantly agricultural region saw 74% of animals dying through disease (mainly tularemia) or trapping/lethal control, but both these factors were significantly reduced in the following years, when no predation was recorded (Havens et al., 2013).

Beavers have large territories and can disperse great distances through catchments before settling, though some catchments may be difficult to colonize without assistance. In France, 273 beavers were translocated over 26 release projects, mainly into river systems where natural colonization was unlikely due to significant natural and artificial obstacles, including hydroelectric dams (Rouland and Migot, 1990). Initial failures were experienced, but fairly high release site retention resulted in successful population expansion (Dewas et al., 2012). Challenges to beaver restoration in Scotland include project costs, animal mortality during quarantine, small population vulnerability, and the complication of unofficial beaver releases elsewhere in the country (Campbell-Palmer et al., 2016).

The identification of suitable release sites does not guarantee animals will remain there and be successful. Numerous studies exist, and developing methods of predicting habitat assessment and suitability in itself suggests no single consensus (Pollock et al., 2015). GIS-based studies and population modelling have often been applied to identify suitable release sites, but these should be treated with caution. Although they can give a very broad indication of areas, they often lack the resolution given by field-work surveys; also they may give the impression that beavers may not occupy certain stretches, while, for example, in real life we see beavers readily colonizing areas lacking significant woodland or highly urbanized areas. Early methodology to identify potential beaver release sites in Scotland initially used GIS and woodland landcover maps based on the interpretation of aerial photography at a scale of 1:25,000 to generally define areas as either suitable for beaver breeding and beaver dispersal or assumed not to be used by beavers (Macdonald et al., 1995; South et al., 2000). These methods enabled potential release areas to be identified and subject to field-based assessment using a standardized survey protocol to assess habitat quality factors, such as degree and type of vegetation cover of river bank, bank substrate, slope and height of bank, and water width and depth. Gurnell (1998) concluded that the combination of water flow and regime, water channel characteristics, and food availability were key habitat criteria for beaver colonization and long-term residency. Others have determined that accessibility and the creation of a series of connecting subpopulations is key for wider and long-term restoration (Żurowski and Kasperczyk, 1988; Nolet and Baveco, 1996).

In existing beaver populations, particularly those reaching carrying capacity, habitat selection operates slightly differently, as often the best and most suitable areas are already occupied; therefore beavers may 'lower their standards' and occupy areas that need more work to make it suitable for them. In such circumstances, beavers will occupy areas we humans may feel are unsuitable for them and less conforming to our ideas of suitable release sites. These can include narrower and shallower watercourses, deciduous forests, low-gradient bank slopes (Pinto et al., 2009), and even highly urban areas. Some studies have determined that a minimum distance of 0.48 km typically exists between active territories on Alaskan rivers (Boyce, 1981), though often in other areas with high food abundance and good quality habitat, beaver territories may border each other, with this border shifting slightly between years as families gain or lose neighbouring habitat (Campbell et al., 2005; see Chapter 7). Intraspecific aggression in both species increases with population density in well-connected landscapes, less so in more isolated water bodies (Mayer et al., 2020). Therefore, this may have release implications (and consequently dispersal patterns, mate acquisition, and future colonization), suggesting gaps should be left between active territories within connected habitats and release sites to enable animals to settle without experiencing immediate territorial conflict, with associated welfare challenges on release.

How habitat is managed can have important consequences for restoring beaver populations. Of course, planting willow and aspen along riparian margins will greatly increase the carrying capacity, with higher beaver fecundity reached in these areas (Heidecke, 1991). However, interactions with large herbivores also accessing such browse need careful monitoring and management, which may include changes in farming practices, as discussed in section 11.1, and/or increased deer culling. For example, cattle grazing at Forest Service allotments in New Mexico had limiting factors on beaver restoration as willow scrub was largely removed, with beavers rarely occupying these areas (Small et al., 2016). In such areas, changes in livestock practices and planting schemes, along with BDAs to encourage reconnection of streams with floodplains, may all be required to promote beaver recovery. Forestry practices may impact on beaver populations as they favour stands of young to middle-aged broadleaved species that are typically the result of clear-felling practices which are not then later thinned excessively (Solbraa, 1996). Similarly, any mature stands that have some degree of broadleaf within ~50 m of freshwater shorelines appear to support greater beaver population densities (Härkonen, 1999). Overall, Parker et al. (1999) concluded, in their study of Norwegian forest management, that beavers were a positive force for biodiversity, with restorative influences on forestry.

The gradient and composition of banks influences burrow and lodge construction but may also limit the suitability of a site for territory selection (Dieter and McCabe, 1989). Beavers tend not to be found in gradients > 15%, with optimum gradients usually placed around < 3% (Hodgdon and Hunt, 1966). Even though beavers can access and reside in steep-gradient streams with good vegetation, their dams tend to get washed out seasonally and this limits population expansion and longevity in such areas (see Section 4.4.5). Oppositely, low-order gradient streams are often preferred and readily colonized, though these areas tend also to overlap with areas of prime agricultural lands, for example, if we think of nutrient-rich floodplains.

The creation of artificial burrows has been promoted in a small number of beaver manuals—potentially leading to the confusion by some that this is a release requirement. The vast majority of releases, across numerous countries, do not create any shelter provisions on release. This is because release sites in which suitable banks enabling shelter construction by the beavers themselves should

be selected. Beavers readily and quickly build their own shelter, not often using artificial ones. In circumstances where hard infrastructure prevents burrowing, there may be a need for artificial burrow provision, but then this probably raises wider questions on release site suitability. The most common use of artificial burrows/lodges is as an aid to increase release site fidelity, to hold the animals for a day or two to acclimatize them to the area and let them relax (Schwab, 2014). For these to be effective for this use and worth the effort to construct, they need to be robust, as beavers will readily push, chew, and dig their way out surprisingly quickly. The Methow Beaver and Yakima Beaver Projects in Washington state routinely construct artificial lodges and food supplies at release sites, typically cut aspen (Babik and Meyer, 2015; Woodruff, 2015). See Table 11.5 for release site criteria.

The translocation of established pairs/family units and/or the release of beavers in pairs goes some way to increase release site fidelity and

Table 11.5 Release site criteria.

Habitat feature	Criteria range	Release considerations	References
Cause of local extinction	Removed or mitigated	Avoid releasing directly into conflict areas or where persecution/trapping likely	
Bank slope	> 36° appears rarely used		Dieter and McCabe (1989); Pinto et al. (2009)
Bank substrate	Finer preferable, silt-soil selected	Majority of substrate present should provide digging and building opportunities Avoid areas with majority sand or stone banks Avoid majority rocky cliff banks	Howard and Larson (1985); McComb et al. (1990); Pinto et al. (2009); Dittbrenner et al. (2018)
Channel confinement		May not support beaver colonization as modification difficult	Pollock et al. (2015)
Community support	Increased success of restoration for ongoing beaver presence, expansion, and engagement with local authorities	Gauge level of support in advance, invest time and resources in information dissemination Maintain ongoing dialogue and support	Jones and Campbell-Palmer (2014); Pollock et al. (2015); Auster et al. (2020a,b)
Deciduous tree cover	Essential—range of hardwood species	Avoid dominant conifer tree cover. Preferred woody species, such as willows and cottonwoods, over others	Macdonald et al. (1995); Hartman (1996); Campbell et al. (2005); Pinto et al. (2009); Macdonald et al. (2000)

continued

Table 11.5 Continued.

Habitat feature	Criteria range	Release considerations	References
Woody or herbaceous vegetation	Available year round, < 6 m from shoreline and continuous minimum of 400 m up- or downstream		Macdonald et al. (1995); Macdonald et al. (2000)
Landowner tolerance	Can range from n/a, indifference, tolerant, to activity wanted	Areas with open hostility may result in persecution. Any release should work with landowners, with mitigation support to increase tolerance	Jones and Campbell-Palmer (2014); Pollock et al. (2015); Brazier et al. (2020)
Stream gradient	1–10% (possible), with 0–6% optimal and < 3% preferred	High stream gradient (> 10%) hinders colonization and increases dam washouts. Low gradient can cause increased management conflicts, especially with agriculture	Allen (1982); Beier and Barrett (1987); Cotton (1990), Suzuki and McComb (1998); Jakes et al. (2007); Macfarlane et al. (2017)
Water supply	Permanent and preferably stable throughout the year	Do not release in areas subject to drought or excessive water levels fluctuations	Slough and Sadlier (1977); Allen (1982)
Mean water depth	> 1 m or readily damable	2–4 m preferable, encouraging aquatic plant growth and burrow/lodge building into deeper water	Howard and Larson (1985); Suzuki and McComb (1998); Pollock et al. (2004); Macfarlane et al. (2015)
Water speed	Slower to medium flow preferable	Flow should enable typical swimming expenditure and dam establishment. Presence of numerous sheltered bays with higher flows acceptable. Rapids/excessive water speeds and seasonal washouts should be avoided	Slough and Sadlier (1977); Pinto et al. (2009); Pollock et al. (2014)
Watercourse width	> 2 m wide	Avoid < 2 m or > 300 m	Macdonald et al. (2000)
Habitat connectivity	Part of a network to permit population expansion	Not isolated catchments	Macdonald et al. (1995); Macdonald et al. (2000)
Habitat size	No general consensus	Connectivity and habitat quality more important considerations	Macdonald et al. (1995)
Disturbance	Low to medium	Beavers are readily tolerant and can live in close proximity to humans and various land-use, though disturbance factors on release should be as low as possible. Avoid transport links, bankside infrastructure, riparian zones cleared of vegetation, and low-gradient prime agricultural land	Slough and Sadlier (1977); Macdonald et al. (2000)
Future management	Plans and resource availability post release		Jones and Campbell-Palmer (2014); Pollock et al. (2015); Auster et al. (2020b)

improve restoration success overall, but most importantly each group has its own release site away from others to prevent fighting (Schwab, 2014; Campbell-Palmer et al., 2016). A small number of early translocations recognized increased pair and site fidelity success when beavers where released within artificial lodges surrounded by a wooden

fence, the beavers freeing themselves once they had chewed their way out (Couch, 1942; Żurowski and Jaczewski, 1974). These studies also supplied additional food, mainly willow, poplar, and bread placed along the shoreline of the release lake. Other studies cite numerous instances of pair separation soon after release, so holding a pair together on site for a few days may reduce this (Myrberget, 1967; Helminen and Lahti, 1970). In trying to encourage the survival of a local beaver family threatened by artificial water regimes washing away winter food caches, Lars Wilsson movingly describes in his book *My Beaver Colony* (Wilsson, 1968) tying bundles of brash and branches together and sinking this outside beaver lodges to sustain them on a watercourse.

Water regimes are also vital for long-term beaver recovery, with natural regimes being the most constructive. Where water levels are artificially managed, it is vital that floodplains are inundated regularly and/or for sustained periods so that 'beaver hills' are created so that islands of dry land are available (Hinze, 1953). Artificial water regimes, especially in northern regions where beavers are more dependent on food caches, sudden water draw-downs, or rapid water rises may cause caches to be washed out; also extended periods of drying out or flooding will affect resources and shelters (Smith and Peterson, 1991).

11.6.2 Release site fidelity and dispersal

Dispersal distances following release and fidelity to a release site vary greatly. Numerous projects have implemented habitat suitability models to identify favourable release sites, only to experience high rates of dispersal (McKinstry and Anderson, 2002; Methow Beaver Project, 2014; Babik and Meyer, 2015; Pollock et al., 2015). Perhaps on the extreme end, one translocated beaver was recorded as dispersing 238 km (148 miles) (Hibbard, 1958). A more typical average dispersal distance of 17 translocated individuals was 14.6 km (9 miles), according to Hibbard (1958), while Denney (1952) reported an average of 16.7 km (10.4 miles) for 26 beavers; Hodgdon and Larson (1985) 11.3 km (7 miles) in 12 Maine beavers; and Courcelles and Nault (1983) 18 km (11 miles) in 18 individuals translocated to northern Quebec. More recently, McKinstry and Anderson (2002) determined that 51% of the 114 beavers translocated in a project in Wyoming dispersed > 10 km from their initial release site. Differences between the sexes were discovered by Berghofer (1961) in New Mexico, with females moving less than males: 9.7 km (6 miles) and 12.9 km (8 miles) on average, respectively.

To try to encourage beavers to remain at a release site, a few techniques have been used, including installing BDAs and digging release pools so beavers have immediate access to deeper water, reducing predation risk, and facilitating building burrows into the associated banks for shelter. Artificial lodges have been constructed by some projects, either to release the beavers directly into and potentially hold for a few hours to allow them to settle or just to have a shelter if needed (Jones and Campbell-Palmer, 2014; Scottish Beaver Trial). Habitat quality can be improved by planting beaver-favoured tree species, such as willow, aspen, and cottonwood species which are all readily sourced, propagated, and planted cheaply and which establish quickly (Hall et al., 2015). Any planting should obviously be associated with the riparian strip, taking into account areas likely to be flooded by beavers and where new resultant water levels are likely to lie (Pollock et al., 2015). New planting should be protected from beavers and herbivores while it establishes, but, in general, grazing by large herbivores especially can generate competition for vegetation (see Section 9.3.7). Competition for food resources has been documented between beavers and large herbivores, such as white-tailed deer (*Odocoileus virginianus*), significantly impacting on hardwood regeneration in beaver meadows, and moose (*Alces alces*) intensively grazing on regenerating shoots of beaver-coppiced trees (Busher and Lyons, 1999; Hartman, 2003). Heavy livestock grazing (sheep and cows) can vastly reduce woody vegetation available to beavers and also prevent its regrowth (Parker et al., 2000). If significant, this may lead to site abandonment by beavers and increasing dispersal. Further research on this interaction and the level of additional herbivory that riparian woodland can tolerate before it becomes unsuitable for beaver colonization is required (Baker et al., 2005). The fastest way to improve habitat for beavers is to allow riparian strip to recover by preventing livestock from grazing to the water's edge. In some parts, wildlife competitors can include deer species, elk, and moose; these can be fenced out or subjected to increased population control, or in some projects top predators have been reintroduced (see Box 11.3).

Box 11.3 Wolf reintroduction to Yellowstone National Park

The release of 31 gray wolves in 1995/1996 was a symbolic conservation moment for large carnivore restoration in one of the first national parks of North America (Bangs, 1996). Grey wolf–prey relationships with large herbivores such as various deer species, elk (*Cervus canadensis*), moose (*Alces americanus*), bighorn sheep (*Ovis canadensis*), and wild goat species are well documented, used time and again to justify the release of large predators, often the subject of myths and folklore and widespread persecution. Over the last 100 years, wolf persecution saw escalated ungulate population rises and a shift in the plant community to a shrub-grass dominated system with significant elk populations (Wolf et al., 2007).

Although some beaver trapping undoubtedly occurred in the region during the height of the fur-trading decades, it seems Yellowstone retained good beaver populations, with 'numerous beaver dams' reported in 1863 by Walter DeLacy and 'streams full of beaver' in 1874 stated by the Earl of Dunraven (Goldfarb, 2018). By the 1950s, however, beavers had all but disappeared, including from streams where they had been well documented, where dams had been measured and photographed (Warren, 1926). This drastic loss of beaver was an unpredicted consequence of wide-scale predator control policies implemented to protect livestock (Goldfarb, 2018). Without wolf predation the large herbivore populations,

particularly elk, went unchecked, leading to overgrazing of vegetation, incised waterways, and bank erosion. This competition for food resources saw the beavers lose out and disappear, especially through selective grazing of beaver-coppiced shoots by the elk.

Beavers had fairly wide- but low-density and patchy distribution across Yellowstone, though following wolf reintroduction curious changes occurred (Smith and Tyers, 2012). Beavers were actively reintroduced upstream of Yellowstone, between 1986 and 1999, and therefore were posed to recolonize when conditions permitted (Smith and Tyers, 2012). However, compared to previous historic levels, beaver populations remained low to absent, especially in northern Yellowstone first- to fourth-order streams, which generally constitute elk winter range, and their overgrazing has left the land devoid of shrubs and favoured tree species such as willow and aspen (Chadde, 1991), until the wolf returned. Though the wolf has undoubtedly assisted beaver recovery in some parts, enabling beavers to once again restore the degraded waterways within the park, not all the valleys and waterways are being recolonized, as degradation is suspected to be too advanced. The complexity of these relationships has often been oversimplified, grabbing headlines given the Park's status, often neglecting the years of beaver restoration work preceeding the wolf's return (Goldfarb, 2018).

11.7 Long-term management strategies and future planning

The mass extermination of beavers began across Eurasia during the mediaeval period, spreading throughout North America as beaver populations crashed behind this wave of exploitation. This, in combination with the rise in human-induced changes such as forest clearance, intensified arable cultivation, channel abandonment, and flow confinement, increased sediment erosion and dead wood removal; this was later followed by significant infrastructural changes during the industrial revolution, all of which led to the drastically altered riverine systems of today and were largely responsible for the general state of degradation seen (Brown et al., 2018). Beavers assisted in the mitigation of climate change through the creation of wetlands and maintenance of open water, resisting

both drought and flooding (Hood and Bayley, 2008). Such is the believed importance of this role that ecologists have proposed that the removal of beavers from riparian systems is equivalent to known wetland degradations such as infilling of ponds and ground water abstraction (Hood and Bayley, 2008). Similarly, the beaver's key role in freshwater carbon cycling is considered so significant that their elimination has been suggested as potentially having both local and global-scale impacts (Gatti et al., 2018).

The Eurasian beaver is Europe's most widely reintroduced species—a conservation success; however, its strict protection reveals tensions between species management and socio-economic issues (Pillai and Heptinstall, 2013; Auster et al., 2020b). In most European countries, the Eurasian beaver is protected under a series of legislation, including Annex III (protected species) to the Treaty on Conservation of European Wild Flora, Fauna and

Natural Habitats (Bern Agreement); the European Council Directive (92/43/EEC) on the conservation of wild habitats, wild fauna, and wild flora (Annex II and Annex IV); and the Ramsar Convention on wetlands, which provides indirect protection to beavers in certain wetlands. Management plans are developed within each member state but generally comprise elements of legislative implementation, administration of management options, and information dissemination to landowners and the public. A number of EU countries have varying derogations to this protective legislation in order to manage beavers, noting that a derogation is only permissible if there is 'no satisfactory alternative' (Habitats Directive, Article 16). Between 2001 and 2008, 845 derogations relating to beavers were reported by the European Commission, including Sweden, Poland, Latvia, Germany, and Lithuania (Pillai et al., 2012; Pillai and Heptinstall, 2013).

Hunting is the main form of management in Fennoscandia, with heavy harvesting permitted in areas most likely to see conflicts with humans (Hartman, 1999; Parker and Rosell, 2001). Norway, which sits outside of the EU, has traditionally operated a quota-based management of beavers since 1855 (Parker and Rosell, 2012). This was recently updated so that it is mainly up to the individual municipalities ('*kommunes*') if and how they permit hunting, within a legal hunting season running from 1 Oct to 30 April. Very occasionally some kommunes will decide hunting zones: such as Stjørdal near Trondheim does not permit hunting on the main river, where little conflict occurs, and landowner discretion elsewhere (D. Halley, pers. comm.). Such strategies give management powers to the landowners, can generate direct economic incentives, and are incorporated into land management practices, and therefore are largely self-financing (Halley and Rosell, 2002).

In other European countries, hunting is not permitted. In Bavaria, Germany, for example, a number of derogations are given to permit a range of mitigation methods—from scaring techniques to lethal control by wildlife managers—following the development and implementation of a management programme 30 years after first releases (Schwab and Schmidbauer, 2003). Support for landowners to access mitigation options and public engagement

and education are viewed as key for successful long-term beaver presence. Annually, 1,900 beavers are now lethally controlled in Bavaria (G. Schwab, pers. comm.), following extensive translocation strategies to other parts of Europe (see Section 2.8.1), after they were reintroduced from 1966 to the 1970s (following complete expatriation in 1867). Lethal control measures, as determined by the authorities, were brought in in 2002, though these measures are typically viewed as a last resort due to lethal control generally being an ineffective long-term beaver management strategy (Schwab and Schmidbauer, 2003). The management plan is well developed, with dedicated beaver managers and a team of trained volunteers to implement the plan. The role of beaver management is more focused on preventive advice to conflict resolution, information dissemination (from individual land managers to local school groups), development planning, assistance with the installation of beaver mitigation measures, community engagement, and beaver monitoring. Lethal control is made as a case-specific decision in consultation among the beaver consultant, landowner, and beaver manager. Some compensation is paid if direct crop damage, for example, is documented and the application process is made and accepted.

Latvia was one of the first countries to bring back the beaver, with first releases occurring in the 1930s; they have taken a different approach as it is a bit of a unique case. Beavers established very well here, with population estimates of up to 150, 000 (Halley et al., 2020). Therefore, when Latvia joined the EU in 2003, it already had a large self-sustaining population of beavers and therefore asked for a permanent exception from the Habitats Directive for this species (Pillai and Heptinstall, 2013). To manage conflicts, the beaver is classed as a game species under the Hunting Act 2003, in which they may be hunted under licence, during specific time periods. However, there is no tangible national strategy, and beavers provide poor value to hunters as there is no strong commercial market for their products. To try to encourage further hunting the government has extended the hunting season, one of the longest for mammals in Latvia, and has set unlimited quotas (Pillai and Heptinstall, 2013).

The Czech Republic has a very different strategy: a 10- to 15-year Eurasian beaver management plan primarily focused on the differentiation of protection based around 'zones' (Vorel et al., 2003). This aims to balance the existence, development, and protection of beaver populations, whilst also ensuring this population remains within socially accepted ranges and levels (Vorel et al., 2016). Three zones are recognized by this management plan, each consisting of varying levels of beaver protection. Zone A represents the smallest areas within the country, linking important conservation sites, including the Natura 2000 network, where beavers receive full protection. These areas ensure the long-term presence and existence of beavers within the country; moving outside these zones means other interests such as infrastructure, agriculture, and fisheries are prioritized. Zone B covers 85.5% of the Czech Republic; here a series of management measures are actively employed, including dam management and lethal control if non-lethal methods fail. Therefore, beavers can exist here within limits and provide dispersal corridors between A zones. Beaver presence in Zone C is deemed as high risk, including damage to extensive, historic fish ponds and infrastructure; therefore their presence is not tolerated. Zone C represents 13.3% of the country. Beavers are considered as protected and cannot therefore be hunted, without exemptions granted by the state hunting management authority. As a protected animal, financial compensation is available for specific cases of damage, mainly in relation to forestry and agriculture.

Britain represented one of the last EU member states to see the reintroduction of the beaver; this has been a long, uncoordinated affair over many decades. There is a strong drive for full beaver reintroduction and it is increasingly evident that more beavers are now present through unofficial processes then official ones. None of the devolved nations has, as yet, actively permitted their full restoration and further wild releases. The Eurasian beaver is now fully protected as a European Protected Species (EPS) under the Habitats Regulations in Scotland within a range stated by the Scottish government, with those appearing outside of this area subject to trap and removal. Through this legislation, lethal control licences are also permitted and exercised. The Scottish government announcement in 2019 of accepting both the official

trial reintroduced population and the substantially larger unofficially released population was actually an historic occasion as it represented the first formally approved reintroduction of any mammal species anywhere in the UK (Gaywood, 2017).

In North America and Canada beaver management varies greatly between states, from protected to game animal, and national long-term management plans appear lacking. Beaver trapping is permitted in 42 Californian counties, yet is restricted in Arizona and New Mexico as small populations sizes place beavers here as vulnerable and at risk of local extinction (Gibson and Olden, 2014). In the USA, jurisdictional authority in beaver management is varied and state specific, with state, federal, and even tribal governments interacting; thus, whichever agency has the regulatory authority changes, though typically it tends to lie with state legislation and state fish and wildlife departments (Pollock et al., 2015). Much management, especially in urban areas, is reactive and generally related to the mitigation of damming of watercourses (Taylor et al., 2008). However, in North America, beavers are typically managed as furbearers, which includes setting sustainable harvest levels that are deemed scientifically determined, as many populations are considered restored, while recognizing that incentives of economic gain should not lead to overharvesting (White et al., 2015). To achieve this, beavers are classed along with other wildlife to be a public resource administrated by the government (Sax, 1970); thus they are managed primarily by the state and provincial agencies who regulate responsible harvesting, managed by wildlife law enforcement officers (White et al., 2015).

As beavers recover and human urbanization expands, recent studies are looking to future incorporation of beavers in designed green spaces, as a countermeasure to their traditional view as a nuisance species. Today, more than ever, we are looking at our urban spaces to become more holistic, providing health and well-being for their human inhabitants along with increasingly recognizing the potential for these areas to function as habitat corridors, for example, or provide ecosystem services such as water storage and quality improvement. Beavers have been proposed to be incorporated by landscape architects to enhance green spaces and utilize their activities to 'better mimic natural systems', as opposed to only being agents of conflict, as beavers will readily colonize urban green spaces (Bailey, 1919). In fact, planning wetland green spaces, especially those on low-lying flat areas, which may become accessible by beavers will probably do little more than exasperate beaver conflict issues. If general public opinion is against lethal control, a repeated cycle of damage, trap, and removal, then recolonization typically ensues.

Designing future sites with the forethought of beaver colonization, and also allowing the flexibility to enable their activities, whilst mitigating any

extremes, will only serve to meet complementary objectives such as increasing ecological benefits and recreational opportunities (Bailey et al., 2019). The potential ecological services provided by beavers have been estimated in a hope to facilitate decision-making and management decisions, in the hope to enable landowners to be compensated (mitigation resource investment and/or direct monetary value) for beaver impacts whilst also recognizing wider societal benefits accrued through their presence (Figure 11.12). An increasing number of case studies now exist: for example, Magnuson Park, Seattle, a 142-ha public park set in a decommissioned

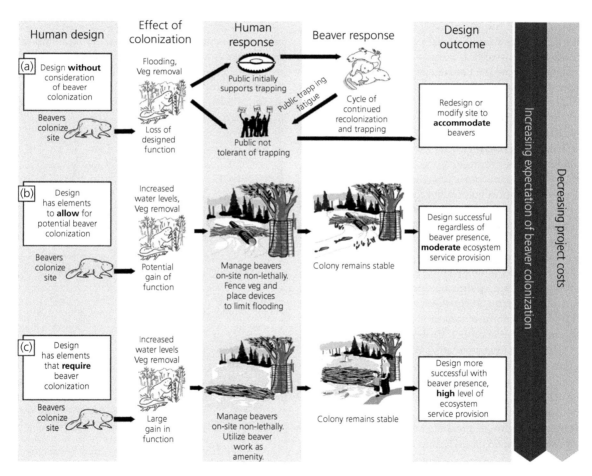

Figure 11.12 Depicted pathway scenarios for integrating beavers into an urban landscape. These authors predict that urban areas designed to anticipate beaver colonization are less costly over the long-term and receive greater ecological benefits and demonstrated ecological services. Reprinted from Bailey, D. R., Dittbrenner, B. J., and Yocom, K. P. 2018. Reintegrating the North American beaver (*Castor canadensis*) in the urban landscape. *Wiley Interdisciplinary Reviews.* (© 2018 Wiley Periodicals, Inc.)

military airfield, redesigned by a team including ecologists, hydrologists, and architects to promote wildlife habitat, especially birds and amphibians, store and filter surface water run-off, and provide recreational opportunities. Given the extent and variation of wetland habitats developed, including open ponds and wet meadow areas, it was anticipated that beavers would readily colonize. To plan for this, multiple features to reduce negative impacts were implemented, including setting pathways back from areas likely to be flooded, minimizing hydrological pinch-points, and installing beam-style weirs to function like beaver dams. Beavers have now colonized this site successfully, building dams, increasing surface water storage by 30% and diversifying the bankside vegetation, resulting in both public and park designers seeing them as a benefit. Beaver management is designed around keeping the beavers present, implementing tree protection, and flow devices where required (Bailey et al., 2019). This provides an inspiring vision of what could be—how relatively straightforward but well thought-out planning can ease conflict and positively utilize the impressive restoration abilities of beaver to restore even seemingly degraded urban spaces, increasing their ecological value and improving human quality of life.

Beaver presence and restoration are not new concepts, and even if still absent from some areas they are certainly recolonizing their former ranges. The Eurasian beaver is now recolonizing intensively modified and densely populated, low-relief regions such as Britain, Belgium, the Netherlands, and northwest Germany (Halley et al., 2020). However, their long-term presence, across both species' range, is dependent on conflict resolution and coexistence with humans (Swinnen et al., 2017). Wetland restoration and even rewilding projects are growing in both public appeal and transcending into wider applications in some land-use practices. Beavers have often been promoted as a vital species in such processes, with a growing trend towards their wider restoration (see Chapter 9), and with this, the need for management programmes comes hand in hand. Obviously, with beavers in more naturalized systems, in which rivers have space to move, riparian vegetation develops unhindered by grazing for example, and catchment-based understanding is

incorporated into management practices, less beaver mitigation will be required. In reality, many modern rivers remain densely populated, heavily modified and managed for often competing purposes. Beavers have been reintroduced in countries with some of the densest human populations, such as the Netherlands, and are dispersing into city centres, such as Vienna, Munich (there are ~20–25 active territories now within Munich) (Halley and Rosell, 2002), and New York City after an absence of 200 years (O'Connor, 2007). Beavers now occupy 86% of the 60 km of suitable habitat waterways within Seattle's city limits, whilst they also increasingly attempt to modify marginal waterways here through damming and tree felling (Bailey et al., 2019). Therefore, local and national beaver management programmes are required to balance the restorative powers of beavers with the conflicts they can generate. As recently surmised, the predictable 'disturbance mosaic' effects beavers generate on abiotic regimes and their biotic consequences make them an entirely suitable driver of riparian habitat restoration (Kivinen et al., 2020).

References

Alakoski, R., Kauhala, K., and Selonen, V. 2019. Differences in habitat use between the native Eurasian beaver and the invasive North American beaver in Finland. *Biological Invasions*, 21, 1601–1613.

Alakoski, R., Kauhala, K., Tuominen, S., and Selonen, V. 2020. Environmental factors affecting the distributions of the native Eurasian beaver and the invasive North American beaver in Finland. *Biological Conservation*, 248, 108680.

Albert, S. and Trimble, T. 2000. Beavers are partners in riparian restoration on the Zuni Indian Reservation. *Ecological Restoration*, 18, 87–92.

Allen, A. W. 1982. *Habitat Suitability Index Models: Beaver*. Washington, DC: US Fish and Wildlife Service.

An, S., Li, H., Guan, B., Zhou, C., Wang, Z., Deng, Z., Zhi, Y., Liu, Y., Xu, C., and Fang, S. 2007. China's natural wetlands: past problems, current status, and future challenges. *AMBIO: A Journal of the Human Environment*, 36, 335–342.

Andersen, D. C. and Shafroth, P. B. 2010. Beaver dams, hydrological thresholds, and controlled floods as a management tool in a desert riverine ecosystem, Bill Williams River, Arizona. *Ecohydrology*, 3, 325–338.

Anderson, C. B., Griffith, C. R., Rosemond, A. D., Rozzi, R., and Dollenz, O. 2006. The effects of invasive North

American beavers on riparian plant communities in Cape Horn, Chile: do exotic beavers engineer differently in sub-Antarctic ecosystems? *Biological Conservation*, 128, 467–474.

Anderson, C. B., Pastur, G. M., Lencinas, M. V., Wallem, P. K., Moorman, M. C., and Rosemond, A. D. 2009. Do introduced North American beavers *Castor canadensis* engineer differently in southern South America? An overview with implications for restoration. *Mammal Review*, 39, 33–52.

Angst, C. 2010. *Vivre avec le* Castor. *Recensement National de 2008. Perspectives pour la Cohabitation avec le* Castor *en Suisse*. Berne: Office Federal de l'Environnement et Centre Suisse de Cartographie de la Faune.

Arner, D. and Dubose, J. 1979. The impact of the beaver on the environment and economics in the southeastern United States. *Proceedings of the International Wildlife Congress*, Dublin, Ireland.

Arts, K., Fischer, A., and van der Wal, R. 2014. Political decision making, governance shifts and Scottish animal reintroductions: are democratic principles at stake? *Journal of Environmental Planning and Management*, 57, 612–628.

Auster, R. E., Barr, S. W., and Brazier, R. E. 2020. Wildlife tourism in reintroduction projects: Exploring social and economic benefits of beaver in local settings. *Journal for Nature Conservation*, 58, 125920.

Auster, R. E., Barr, S. W., and Brazier, R. E. 2020b. Improving engagement in managing reintroduction conflicts: learning from beaver reintroduction. *Journal of Environmental Planning and Management*, 1-22.

American Veterinary Medical Association 2013. *Guidelines for the Euthanasia of Animals, 2013 edition*. Schaumburg, IL: American Veterinary Medical Association.

Babik, W. and Meyer, W. 2015. *Yakima Basin Beaver Reintroduction Project. 2011–2015 Progress report*. Ellensburh: Washington Department of Wildlife.

Babik, W., Durka, W., and Radwan, J. 2005. Sequence diversity of the MHC DRB gene in the Eurasian beaver (*Castor fiber*). *Molecular Ecology*, 14, 4249–4257.

Backhouse, F. 2018. *Once They Were Hats: In Search of the Mighty Beaver*. Toronto: ECW Press.

Bailey, D. R., Dittbrenner, B. J., and Yocom, K. P. 2019. Reintegrating the North American beaver (*Castor canadensis*) in the urban landscape. *Wiley Interdisciplinary Reviews: Water*, 6, e1323.

Bailey, V. 1919. A new subspecies of beaver from North Dakota. *Journal of Mammalogy*, 1, 31–32.

Baker, B. W. and Hill, E. P. 2003. *Beaver (*Castor canadensis*)*, 2nd edn. Baltimore: Johns Hopkins University Press.

Baker, B. W., Ducharme, H. C., Mitchell, D. C., Stanley, T. R., and Peinetti, H. R. 2005. Interaction of beaver and elk

herbivory reduces standing crop of willow. *Ecological Applications*, 15, 110–118.

Bangs, E. E. 1996. Reintroducing the gray wolf to central Idaho and Yellowstone National Park. *Wildlife Society Bulletin*, 24, 402–413.

Batbold, J., Batsaikhan, N., Shar, S., Amori, G., Hutterer, R., Kryštufek, B., Yigit, N., Mitsain, G., and Muñoz, L. 2016. *Castor fiber*. www.iucnredlist.org.

Beaver, B. V., Reed, W., Leary, S., McKiernan, B., Bain, F., Schultz, R., Bennett, B. T., Pascoe, P., Shull, E., Cork, L. C., Francis-Floyd, R., Amass, K. D., Johnson, R., Schmidt, R. H., Underwood, W., Thornton, G. W., and Kohn, B. 2001. Report of the AVMA panel on euthanasia. *Journal of the American Veterinary Medical Association*, 218, 669–696.

BeaverCone. 2020. The beaver expert for flood control. www.beavercone.com.

Beier, P. and Barrett, R. H. 1987. Beaver habitat use and impact in Truckee river basin, California. *Journal of Wildlife Management*, 51, 794–799.

Belovsky, G. E. 1984. Summer diet optimization by beaver. *American Midland Naturalist*, 111, 209–222.

Berghofer, C. B. 1961. Movement of beaver. *Proceedings of the Annual Conference of Western Association State Game and Fish Commissioners*, pp. 181–184.

Black, J. S. 1880. A short account of how the Marquis of Bute's beavers have succeeded in the Isle of Bute, Scotland. *Journal of Forestry & Estates Management*, 3, 695–698.

Bouwes, N., Weber, N., Jordan, C. E., Saunders, W. C., Tattam, I. A., Volk, C., Wheaton, J. M., and Pollock, M. M. 2016. Ecosystem experiment reveals benefits of natural and simulated beaver dams to a threatened population of steelhead (*Oncorhynchus mykiss*). *Scientific Reports*, 6, 28581.

Boyce, M. S. 1981. Beaver life-history responses to exploitation. *Journal of Applied Ecology*, 18, 749–753.

Boyles, S. and Owens, S. 2007. *North American Beaver (*Castor canadensis*): A Technical Conservation Assessment*. Lakewood, CO: USDA Forest Service, Rocky Mountain Region.

Boyles, S. L. and Savitzky, B. A. 2009. An analysis of the efficacy and comparative costs of using flow devices to resolve conflicts with North American beavers along roadways in the coastal plain of Virginia. *Proceedings of the 23rd Vertebrate Pest Conference*, 23, 47–52.

Brayton, D. S. 1984. The beaver and the stream. *Journal of Soil and Water Conservation*, 39, 108–109.

Brazier, R. E., Elliott, M., Andison, E., Auster, R. E., Bridgewater, S., Burgess, P., Chant, J., Graham, H., Knott, E., Puttock, A. K., Sansum, P., and Vowles, A. 2020. *River Otter Beaver Trial: Science and Evidence Report*. Exeter: University of Exeter.

Brazier, R. E., Puttock, A., Graham, H. A., Auster, R. E., Davies, K. H., and Brown, C. M. 2021. Beaver: Nature's ecosystem engineers. *Wiley Interdisciplinary Reviews: Water*, 8(1), e1494.

Bridgewater, P. 2018. Whose nature? What solutions? Linking ecohydrology to nature-based solutions. *Ecohydrology & Hydrobiology*, 18, 311–316.

Brooks, R. P., Fleming, M. W., and Kennelly, J. J. 1980. Beaver colony response to fertility control: evaluating a concept. *Journal of Wildlife Management*, 44, 568–575.

Brown, A. G., Lespez, L., Sear, D. A., Macaire, J.-J., Houben, P., Klimek, K., Brazier, R. E., Van Oost, K., and Pears, B. 2018. Natural vs anthropogenic streams in Europe: history, ecology and implications for restoration, river-rewilding and riverine ecosystem services. *Earth-Science Reviews*, 180, 185–205.

Buckley, M., Souhlas, T., Niemi, E., and Reich, S. 2011. *The economic value of beaver ecosystem services: Escalante River basin*. Utah: Report for the Grand Canyon Trust.

Bullock, J. F. and Arner, D. H. 1985. Beaver damage to non-impounded timber in Mississippi. *Southern Journal of Applied Forestry*, 9, 137–140.

Busher, P. E. and Lyons, P. J. 1999. Long-term population dynamics of the North American beaver *Castor canadensis* on Quabbin Reservation, Massachusetts, and Sagehen Creek, California. In: Busher, P. E., and Dzieciolowski, R. (eds.) *Beaver Protection, Management, and Utilization in Europe and North America*. New York: Kluwer Academic/Plenum Publishers.

Butler, D. R. and Malanson, G. P. 2005. The geomorphic influences of beaver dams and failures of beaver dams. *Geomorphology*, 71, 48–60.

Callahan, M. 2003. Beaver management study. *Association of Massachusetts Wetland Scientists (AMWS) Newsletter*, 44, 12–15.

Callahan, M. 2005. Best management practices for beaver problems. *Association of Massachusetts Wetland Scientists (AMWS) Newsletter*, 53, 12–14.

Campbell, R. D., Rosell, F., Nolet, B. A., and Dijkstra, V. A. A. 2005. Territory and group sizes in Eurasian beavers (*Castor fiber*): echoes of settlement and reproduction? *Behavioral Ecology and Sociobiology*, 58, 597–607.

Campbell, R. D., Dutton, A., and Hughes, J. 2007. *Economic impacts of the beaver*. Report for the Wild Britain Initiative. Tubney, Oxon: WildCRU.

Campbell-Palmer, R., Gow, D., Campbell, R., Dickinson, H., Girling, S., Gurnell, J., Halley, D., Jones, S., Lisle, S., Parker, H., Schwab, G., and Rosell, F. 2016b. *The Eurasian Beaver Handbook: Ecology and Management of* Castor fiber. Exeter: Pelagic Publishing.

Campbell-Palmer, R., Rosell, F., Naylor, A., Cole, G., Mota, S., Brown, D., Fraser, M., Pizzi, R., Elliott, M., Wilson, K. and Gaywood, M., 2021. Eurasian beaver (*Castor fiber*) health surveillance in Britain: Assessing a disjunctive reintroduced population. *Veterinary Record*, p.e84.

Campbell-Palmer, R., Senn, H., Girling, S., Pizzi, R., Elliott, M., Gaywood, M., and Rosell, F. 2020. Beaver genetic surveillance in Britain. *Global Ecology and Conservation*, 24, e01275.

Cassola, F. 2016. *Castor canadensis*. The IUCN Red List of Threatened Species 2016: e.T4003A22187946. https://www.iucnredlist.org/species/4003/22187946.

Chadde, S. W. 1991. Tall-willow communities on Yellowstone's northern range: a test of the 'natural-regulation' paradigm. In: Keiter, R. B. and Boyce, M. S. (eds.) *The Greater Yellowstone Ecosystem*. Binghamton, NY: Yale University Press.

Chaney, E., Elmore, W., and Platts, W. S. 1990. *Livestock Grazing on Western Riparian Areas*. Washington, DC: US Environmental Protection Agency.

Chapman, D. W. 1986. Salmon and steelhead abundance in the Columbia River in the nineteenth century. *Transactions of the American Fisheries Society*, 115, 662–670.

Choi, C. 2008. Tierra del Fuego: the beavers must die. *Nature*, 453, 968–969.

Chu, H. and Jiang, Z. 2009. Distribution and conservation of the Sino-Mongolian beaver *Castor fiber* birulai in China. *Oryx*, 43, 197–202.

Citti, D.C., McKee J., Pope, K.L., Kondolf, G.M. and Pollock, M.M. 2021. Design criteria for process-base restoration of fluvial systems. Bioscience, doi.org/10.1093/biosci/biab065

Cluer, B. and Thorne, C. 2014. A stream evolution model integrating habitat and ecosystem benefits. *River Research and Applications*, 30, 135–154.

Coles, B. 2006. *Beavers in Britain's Past*. Oxford: Oxbow Books.

Collins, T. 1993. *The role of beaver in riparian habitat management. Habitat Extension Bulletin*. Cheyenne: Wyoming Game and Fish Department.

Columbia-Blue Mountain Resource Conservation and Development Area (CBMRC) 2005. *John Day subbasin plan*. Prepared for Northwest Power and Conservation Council, Portland, OR.

Conroy, J. W. H., Kitchener, A. C., and Gibson, J. A. 1998. The history of the beaver in Scotland and its future reintroduction. In: Lambert, R. A. (ed.) *Species History in Scotland*. Edinburgh: Scottish Cultural Press.

Cotton, F. E. 1990. Potential beaver colony density in parts of Quebec. MSc thesis, Blacksburg, Virginia Polytechnic Institute and State University.

Couch, L. K. 1942. Trapping and transplanting live beavers. *Conservation Bulletin*, 30, 1–20.

Courcelles, R. and Nault, R. 1983. Beaver programs in the James Bay area, Quebec, Canada. *Acta Zoologica Fennica*, 174, 129–131.

Cox, D. R. and Nelson, T. A. 2009. Beaver habitat models for use in Illinois streams. *Illinois State Academy of Science*, 102, 55–64.

Cramer, M. L. 2012. *Stream Habitat Restoration Guidelines*. Olympia, WA: Co-published by the Washington Departments of Fish and Wildlife. Natural Resources, Transportation and Ecology, Washington State

Recreation and Conservation Office, Puget Sound Partnership, and the US Fish and Wildlife Service.

Crego, R. D., Jiménez, J. E., and Rozzi, R. 2016. A synergistic trio of invasive mammals? Facilitative interactions among beavers, muskrats, and mink at the southern end of the Americas. *Biological Invasions*, 18, 1923–1938.

Curtis, P. D. and Jensen, P. G. 2004. Habitat features affecting beaver occupancy along roadsides in New York state. *Journal of Wildlife Management*, 68, 278–287.

Czech, A. and Lisle, S. 2003. Understanding and solving the beaver (*Castor fiber* L.)–human conflict: an opportunity to improve the environment and economy of Poland. *Denisia*, 9, 91–98.

Dahl, T. E. 1990. *Wetland Losses in the United States 1780s to 1980s*. Washington, DC: US Department of the Interior, Fish and Wildlife Service.

Danilov, P. I. 1976. *The History of Beaver Spread in Karelia. Ecology of Birds and Mammals in the North-West of USSR*. Petrozavodsk: Institute of Biology Karelian Research Centre of RAS.

Danilov, P. I. and Fyodorov, F. V. 2007. *Rechnye Bobry Evropeyskogo Severa Rossii*. Moskva: Nauka.

Danilov, P. I. and Fyodorov, F. V. 2015. Comparative characterization of the building activity of Canadian and European beavers in northern European Russia. *Russian Journal of Ecology*, 46, 272–278.

Danilov, P., Kanshiev, V., and Fyodorov, F. 2011. Characteristics of North American and Eurasian beaver ecology in Karelia. In: Sjøberg, G. and Ball, J. P. (eds.) *Restoring the European Beaver: 50 Years of Experience*. Sofia-Moscow: Pensoft.

Davidson, N. C. 2014. How much wetland has the world lost? Long-term and recent trends in global wetland area. *Marine and Freshwater Research*, 65, 934–941.

Davidson, N. C., Laffoley, D. A., Doody, J. P., Way, L. S., Gordon, J., Key, R., Drake, C. M., Pienkowski, M. W., Mitchell, R. M., and Duff, K. L. 1991. *Nature Conservation and Estuaries in Great Britain*. Peterborough: Nature Conservancy Council.

Deinet, S., Ieronymidou, C., McRae, L., Burfield, I. J., Foppen, R. P., Collen, B., and Böhm, M. 2013. *Wildlife Comeback in Europe: The Recovery of Selected Mammal and Bird Species. Final Report to Rewilding Europe by ZSL*. London: ZSL: BirdLife International and the European Bird Census Council.

Demmer, R. and Beschta, R. L. 2008. Recent history (1988–2004) of beaver dams along Bridge Creek in central Oregon. *Northwest Science*, 82, 309–318.

Denney, R. N. 1952. *A Summary of North American Beaver Management, 1946–1948*. Denver: Colorado Game and Fish Department.

Department of Environment Food and Rural Affairs 2015. *Catchment Restoration Fund: Environmental Agency Final Annual Report 2014–2015*. London: DEFRA.

Dewas, M., Herr, J., Schley, L., Angst, C., Manet, B., Landry, P., and Catusse, M. 2012. Recovery and status of native and introduced beavers *Castor fiber* and *Castor canadensis* in France and neighbouring countries. *Mammal Review*, 42, 144–165.

Di Baldassarre, G., Schumann, G., Bates, P. D., Freer, J. E., and Beven, K. J. 2010. Flood-plain mapping: a critical discussion of deterministic and probabilistic approaches. *Journal of Hydrological Science*, 55, 364–376.

Dieter, C. D. and McCabe, T. R. 1989. Factors influencing beaver lodge-site selection on a prairie river. *American Midland Naturalist*, 122, 408–411.

Dittbrenner, B. J., Pollock, M. M., Schilling, J. W., Olden, J. D., Lawler, J. J., and Torgersen, C. E. 2018. Modeling intrinsic potential for beaver (*Castor canadensis*) habitat to inform restoration and climate change adaptation, *PLoS ONE*, 13, e0192538.

Dobbins, C. L. 2004. *Beaver and Otter Trapping: Open Water Techniques*. North Carolina: Bullet Printing.

Donlan, J. C., Berger, J., Bock, C. E., Bock, J. H., Burney, D. A., Estes, J. A., Foreman, D., Martin, P. S., Roemer, G. W., and Smith, F. A. 2006. Pleistocene rewilding: an optimistic agenda for twenty-first century conservation. *The American Naturalist*, 168, 660–681.

Douglas, L. R. and Veríssimo, D. 2013. Flagships or battleships: deconstructing the relationship between social conflict and conservation flagship species. *Environment and Society*, 4, 98–116.

DuBow, T. J. 2000. Reducing beaver damage to habitat restoration sites using palatable tree species and repellents. MSc thesis, Utah State University.

Eastland, W. 2000. Spring beaver hunting. *Fur Taker Magazine*, 36(2), 22–27.

Eaton, B., Fisher, J. T., Muhly, T., and Chai, S. L. 2013. Potential impacts of beaver on oil sands reclamation success—an analysis of available literature. ORSIN Report. Edmonton: Oil Sands Research and Information Network, University of Alberta.

Edwards, S. T. and Potts, W. K. 1996. Polymorphism of genes in the major histocompatibility complex (MHC): implications for conservation genetics of vertebrates. In: Smith, T. B. and Wayne, R. K. (eds.) *Molecular Genetic Approaches in Conservation*. New York: Oxford University Press.

Ellegren, H., Hartman, G., Johansson, M., and Andersson, L. 1993. Major histocompatibility complex monomorphism and low levels of DNA fingerprinting variability in a reintroduced and rapidly expanding population of beavers. *Proceedings of the National Academy of Sciences*, 90, 8150–8153.

Enck, J. W., Purdy, K. G., and Decker, D. J. 1988. *Public Acceptance of Beavers and Beaver Damage in Wildlife Management Unit 14 in DEC region 4*. Ithaca, NY: Human Dimensions Research Unit, Cornell University.

Enck, J. W., Connelly, N. A., and Brown, T. L. 1997. Acceptance of beaver and actions to address nuisance beaver problems in New York. *Human Dimensions of Wildlife*, 2, 60–61.

Ermala, A., Helminen, M., and Lahti, S. 1989. Some aspects of the occurrence, abundance and future of the Finnish beaver population. *Suomen Riista*, 35, 108–118.

Fitch, L., Adams, B., and O'Shaughnessy, K. 2003. *Caring for the Green Zone: Riparian Areas Grazing Management.* Lethbridge, AB: Cows and Fish Program.

Flosi, G., Downie, S., Bird, M., Coey, R., and Collins, B. 2002. *California Salmonid Stream Habitat Restoration Manual.* Sacramento: California Department of Fish and Wildlife.

Flynn, N. J. 2006. Spatial associations of beaver ponds and culverts in boreal headwater streams. MSc thesis, University of Alberta.

Fortin, C. and Lizotte, M. 2007. *Castors*, routes et chemins de fer: une problématique méconnue. *Vivo*, 27, 8–10.

Fortin, M. A. 2011. Beaver sterilization project, *Castor canadensis*, in an urban setting. *Proceedings of the 2011 Bird Strike North America Conference*, Niagara Falls.

Fox, C.H. and Papouchis, C.M. 2004. Cull of the Wild. A contemporary analysis of wildlife trapping practices in the United States. Bang Publishing, Brainerd.

Frosch, C., Kraus, R. H. S., Angst, C., Allgöwer, R., Michaux, J., Teubner, J., and Nowak, C. 2014. The genetic legacy of multiple beaver reintroductions in central Europe. *PLoS ONE*, 9, e97619.

Fryxell, J. M. and Doucet, C. M. 1991. Provisioning time and central-place foraging in beavers. *Canadian Journal of Zoology*, 69, 1308–1313.

Galbraith, K. 2006. Back in style: the fur trade. *New York Times*, 24 December.

Gatti, R. C., Callaghan, T. V., Rozhkova-Timina, I., Dudko, A., Lim, A., Vorobyev, S. N., Kirpotin, S. N., and Pokrovsky, O. S. 2018. The role of Eurasian beaver (*Castor fiber*) in the storage, emission and deposition of carbon in lakes and rivers of the River Ob flood plain, western Siberia. *Science of the Total Environment*, 644, 1371–1379.

Gaywood, M. J. 2017. Reintroducing the Eurasian beaver *Castor fiber* to Scotland. *Mammal Review*, 48, 48–61.

Gibson, J. A. 1980. Supplementary notes on Bute mammals. *Transactions of the Buteshire Natural History Society*, 21, 93–95.

Gibson, P. P. and Olden, J. D. 2014. Ecology, management, and conservation implications of North American beaver (*Castor canadensis*) in dryland streams. *Aquatic Conservation: Marine and Freshwater Ecosystems*, 24, 391–409.

Gilbert, F. F. and Gofton, N. 1982. Heart rate values for beaver, mink and muskrat. *Comparative Biochemistry and Physiology Part A: Physiology*, 73, 249–251.

Goldfarb, B. 2018. Why two countries want to kill 100,000 beavers. https://www.washingtonpost.com/news/ animalia/wp/2018/08/09/why-two-countries-want-to-kill-100000-beavers/ [Accessed 10/09/2018].

Gurnell, A. M. 1998. The hydrogeomorphological effects of beaver dam-building activity. *Progress in Physical Geography*, 22, 167–189.

Haarberg, O. and Rosell, F. 2006. Selective foraging on woody plant species by the Eurasian beaver (*Castor fiber*) in Telemark, Norway. *Journal of Zoology*, 270, 201–208.

Hall, J. E., Pollock, M., Hoh, S., Volk, C., Goldsmith, J., and Jordan, C. E. 2015. Evaluation of deep-planting and herbivore protection methods to restore riparian vegetation in a semi-arid watershed without irrigation. *Ecosphere*, 6, 263.

Halley, D. and Rosell, F. 2002. The beaver's reconquest of Eurasia: status, population development and management of a conservation success. *Mammal Review*, 32, 153–178.

Halley, D., Saveljev, A., and Rosell, F. 2020. Population and distribution of Eurasian beavers (*Castor fiber*). *Mammal Review*, 51(1), 1–24.

Hamelin, D., Dougherty, D., Fuerst, G., Jenks, D., Raffaldi, T., Gilligan, V., Golja, G., and Tuller, B. 1997. *Beaver Damage Control Techniques Manual.* Albany: New York State Department of Environmental Conservation.

Härkonen, S. 1999. Forest damage caused by the Canadian beaver (*Castor canadensis*) in South Savo, Finland. *Silva Fennica*, 33, 247–259.

Harper, J., Nolte, D. L., DeLiberto, T. J., and Bergman, D. 2005. Conditioning beaver to avoid desirable plants. *Wildlife Damage Management Conference Proceedings*, 122, 354–362.

Harthun, M. 2000. Influence of the damming-up by beavers (*Castor fiber albicus*) on physical and chemical parameters of Highland brooks (Hesse, Germany). *Limnologica*, 30, 21–35.

Hartman, G. 1996. Habitat selection by European beaver (*Castor fiber*) colonizing a boreal landscape. *Journal of Zoology*, 240, 317–325.

Hartman, G. 1999. Beaver management and utilization in Scandanavia. In: Busher, P. E. and Dzięciołowski, R. M. (eds.) *Beaver Protection, Management, and Utilisation in Europe and North America.* New York: Kluwer Academic/ Plenum Press.

Hartman, G. 2003. Irruptive population development of European beaver (*Castor fiber* L.) in southwest Sweden. *Lutra*, 46, 103–108.

Harvey, E.N., MacMillen, J.H., Butler, E.G. and Puckett, W.O. 1962. Mechanism of wounding. In: Wound Ballistics, pp. 143-235 eds Coates, B. and Beyer, J.C. Office of the Surgeon General, Department of the Army, Washington, D.C. USA.

Harvolk, S., Symmank, L., Sundermeier, A., Otte, A., and Donath, T. W. 2014. Can artificial waterways provide a refuge for floodplain biodiversity? A case study from north western Germany. *Ecological Engineering*, 73, 31–44.

Havens, R. P., Crawford, J. C., and Nelson, T. A. 2013. Survival, home range, and colony reproduction of beavers in east-central Illinois, an agricultural landscape. *American Midland Naturalist*, 169, 17–30.

Heidecke, D. 1991. Zum Status des Elbebibers sowie ethoökologische Aspekte. *Seevögel*, 12, 33–38.

Helminen, M. and Lahti, S. 1970. A beaver (*Castor fiber*) with a life span of at least 30 years in Hyvinkää, southern Finland. *Luonnon Tutkija*, 74, 11–14.

Henn, J. J., Anderson, C. B., and Pastur, G. M. 2016. Landscape-level impact and habitat factors associated with invasive beaver distribution in Tierra del Fuego. *Biological Invasions*, 18, 1679–1688.

Heter, E. W. 1950. Transplanting beavers by airplane and parachute. *Journal of Wildlife Management*, 14, 143–147.

Hibbard, E. A. 1958. Movements of beaver transplanted in North Dakota. *Journal of Wildlife Management*, 22, 209–211.

Hilfiker, E. L. 1991. *Beavers, Water, Wildlife and History*. Interlaken, NY: Windswept Press.

Hillman, G. R. 1998. Flood wave attenuation by a wetland following a beaver dam failure on a second order boreal stream. *Wetlands*, 18, 21–34.

Hinze, G. 1953. *Unser Biber*. Leipzig: Die Neue Brehm-Bücherei, Akademische Verlagsgesellschaft Geest and portig K.G.

Hodgdon, H. E. and Larson, J. S. 1985. Vocal communication outside the lodge by Canadian beavers (*Castor canadensis*). In: Pilleri, G. (ed.) *Investigations on Beavers*. Berne: Brain Anatomy Institute.

Hodgdon, K. W. and Hunt, J. H. 1953. *Beaver Management in Maine*. Augusta: Maine Department of Inland Fisheries & Game.

Hodgdon, K. W. and Hunt, J. H. 1966. Beaver management in Maine. *Game Division Bulletin*, 3.

Hood, G. A. 2015. *Mitigating Human–Beaver Conflicts Through Adaptive Management*. Final Report. Edmonton: University of Alberta.

Hood, G. A. and Bayley, S. E. 2008. Beaver (*Castor canadensis*) mitigate the effects of climate on the area of open water in boreal wetlands in western Canada. *Biological Conservation*, 141, 556–567.

Houston, A. E., Buckner, E. R., and Rennie, J. C. 1992. Reforestation of drained beaver impoundments. *Southern Journal of Applied Forestry*, 16, 151–155.

Howard, A. J., Brown, A. G., Carey, C. J., Challis, K., Cooper, L. P., Kincey, M., and Toms, P. 2008. Archaeological resource modelling in temperate river valleys: a case study from the Trent valley, UK. *Antiquity*, 82, 1040–1054.

Howard, R. J. and Larson, J. S. 1985. A stream habitat classification system for beaver. *Journal of Wildlife Management*, 49, 19–25.

Iossa, G., Soulsbury, C., and Harris, S. 2007. Mammal trapping: a review of animal welfare standards of killing and restraining traps. Animal Welfare, 16, 335–352.

Iso-Touru, T., Huitu, O., Tapio, M., Kučinskienė, J., Ulevičius, A., Bukelskis, E., Tirronen, K., Fyodorov, F., Panchenko, D., Saarma, U. and Valdmann, H., 2020. Low genetic polymorphism in the re-introduced Eurasian beaver (*Castor fiber*) population in Finland: implications for conservation. *Mammal Research*, 65, 331-338.

Jakes, A. F., Snodgrass, J. W., and Burger, J. 2007. *Castor canadensis* (beaver) impoundment associated with geomorphology of southeastern streams. *Southeastern Naturalist*, 6, 271–282.

Jaksic, F. M. 1998. Vertebrate invaders and their ecological impacts in Chile. *Biodiversity & Conservation*, 7, 1427–1445.

Jaksic, F. M., Iriarte, J. A., Jiménez, J. E., and Martínez, D. R. 2002. Invaders without frontiers: cross-border invasions of exotic mammals. *Biological Invasions*, 4, 157–173.

Jenkins, S. H. 1980. A size–distance relation in food selection by beavers. *Ecology*, 61, 740–746.

Jensen, P. G., Curtis, P. D., Lehnert, M. E., and Hamelin, D. L. 2001. Habitat and structural factors influencing beaver interference with highway culverts. *Wildlife Society Bulletin*, 29, 654–664.

Johansson, M., Dressel, S., Kvastegård, E., Ericsson, G., Fischer, A., Kaltenborn, B. P., Vaske, J. J., and Sandström, C. 2016. Describing human–wildlife interaction from a European perspective. *Human Dimensions of Wildlife*, 21, 158–168.

Johnston, C. A. and Naiman, R. J. 1990a. Aquatic patch creation in relation to beaver population trends. *Ecology*, 71, 1617–1621.

Johnston, C. A. and Naiman, R. J. 1990b. The use of a geographic information system to analyze long-term landscape alteration by beaver. *Landscape Ecology*, 4, 5–19.

Johnston, R. J. and Timm, R. M. 1987. Wildlife damage to agriculture in Nebraska: a preliminary cost assessment. *Proceedings of the 3rd Eastern Wildlife Damage Control Conference*, 18–21 October, Gulf Shores, Alabama.

Jones, K., Gilvear, D., Willby, N., and Gaywood, M. 2009. Willow (*Salix* spp.) and aspen (*Populus tremula*) regrowth after felling by the Eurasian beaver (*Castor fiber*): implications for riparian woodland conservation in Scotland. *Aquatic Conservation: Marine and Freshwater Ecosystems*, 19, 75.

Jones, S. and Campbell-Palmer, R. 2014. *The Scottish Beaver Trial: The Story of Britain's First Licensed Release into the Wild*. Final report. Knapdale, Argyll: Scottish Beaver Trial.

Jonker, S. A., Muth, R. M., Organ, J. F., Zwick, R. R., and Siemer, W. F. 2006. Experiences with beaver damage and attitudes of Massachusetts residents toward beaver. *Wildlife Society Bulletin*, 34, 1009–1021.

Jonker, S. A., Organ, J. F., Muth, R. M., Zwick, R. R., and Siemer, W. F. 2009. Stakeholder norms toward beaver management in Massachusetts. *Journal of Wildlife Management*, 73, 1158–1165.

Jusim, P., Goijman, A. P., Escobar, J., Carranza, M. L., and Schiavini, A. 2020. First test for eradication of beavers (*Castor canadensis*) in Tierra del Fuego, Argentina. *Biological Invasions*, 22, 3609-3619.

Kamczyc, J., Bielachowicz, M., and Pers-Kamczyc, E. 2016. Damages caused by European beaver (*Castor fiber* L., 1758) in broadleaved stands. *Forestry Letters*, 109, 7–10.

Kaphegyi, T. A. M., Christoffers, Y., Schwab, S., Zahner, V., and Konold, W. 2015. Media portrayal of beaver (*Castor fiber*) related conflicts as an indicator of changes in EU-policies relevant to freshwater conservation. *Land Use Policy*, 47, 468–472.

Kimball, B. A. and Perry, K. R. 2008. Manipulating beaver (*Castor canadensis*) feeding responses to invasive tamarisk (*Tamarix* spp.). *Journal of Chemical Ecology*, 34, 1050–1056.

Kingston, D. 2004. *The 2003 Upper Kitwanga Beaver Dam Breaching Program*. Kitwanga, BC: Gitanyow Fisheries Authority.

Kivinen, S., Nummi, P., and Kumpula, T. 2020. Beaver-induced spatiotemporal patch dynamics affect landscape-level environmental heterogeneity. *Environmental Research Letters*, 15, 094065.

Kundzewicz, Z. W., Kanae, S., Seneviratne, S. I., Handmer, J., Nicholls, N., Peduzzi, P., Machler, R., Bouwer, L., Arnell, N., Mach, K., Muir-Wood, R., Brakenridge, G. R., Kron, W., Beniti, G., Honda, Y., Takahashi, K., and Sherstyukov, B. 2014. Flood risk and climate change: global and regional perspectives, *Hydrological Sciences Journal*, 59, 1–28.

Laramie Jr, H. A. 1963. A device for control of problem beavers. *The Journal of Wildlife Management*, 471–476.

Law, A., Gaywood, M. J., Jones, K. C., Ramsay, P., and Willby, N. J. 2017. Using ecosystem engineers as tools in habitat restoration and rewilding: beaver and wetlands. *Science of the Total Environment*, 605, 1021–1030.

Lisle, S. 2001. Beaver management at the Penobscot nation, USA: using flow devices to protect property and create wetlands. *Proceedings of the European Beaver Symposium*, 2, 147–156.

Lisle, S. 2003. The use and potential of flow devices in beaver management. *Lutra*, 46, 211–216.

Lizarralde, M. and Escobar, J. M. 2000. Mamíferos exóticos en la Tierra del Fuego. *Ciencia Hoy*, 10, 52–63.

Lizarralde, M., Deferrari, G., Alvarez, S. E., and Escobar, J. M. 1996. Effects of beaver (*Castor canadensis*) on the nutrient dynamics of the southern beech forest of Tierra del Fuego (Argentina). *Ecología Austral*, 6, 101–105.

Lizarralde, M., Escobar, J., and Deferrari, G. 2004. Invader species in Argentina: a review about the beaver (*Castor canadensis*) population situation on Tierra del Fuego ecosystem. *Interciencia*, 29, 352–356.

Loeb, R. E., King, S., and Helton, J. 2014. Human pathways are barriers to beavers damaging trees and saplings in urban forests. *Urban Forest and Urban Greening*, 13, 290–294.

Loker, C. A., Decker, D. J., and Schwager, S. J. 1999. Social acceptability of wildlife management actions in suburban areas: 3 cases from New York. *Wildlife Society Bulletin*, 27, 152–159.

Ludders, J. W., Schmidt, R. H., Dein, F. J., and Klein, P. N. 1999. Drowning is not euthanasia. *Wildlife Society Bulletin (1973–2006)*, 27, 666–670.

Macdonald, D. W., Tattersall, F. H., Brown, E. D., and Balharry, D. 1995. Reintroducing the European beaver to Britain: nostalgic meddling or restoring biodiversity? *Mammal Review*, 25, 161–200.

Macdonald, D. W., Tattersall, F. H., Rushton, S., South, A. B., Rao, S., Maitland, P., and Strachan, R. 2000. Reintroducing the beaver (*Castor fiber*) to Scotland: a protocol for identifying and assessing suitable release sites. *Animal Conservation*, 3, 125–133.

Macfarlane, W. W., Wheaton, J. M., Bouwes, N., Jensen, M. L., Gilbert, J. T., Hough-Snee, N., and Shivik, J. A. 2017. Modeling the capacity of riverscapes to support beaver dams. *Geomorphology*, 277, 72–99.

Malmierca, L., Menvielle, M. F., Ramadori, D., Saavedra, B., Saunders, A., Soto Volkart, N., and Schiavini, A. 2011. Eradication of beaver (*Castor canadensis*), an ecosystem engineer and threat to southern Patagonia. In: Veiitch, C. R., Clout, M. N., and Towns, D. R. (eds.) *Island Invasives: Eradication and Management*. Gland: IUCN.

Margaryan, L. and Wall-Reinius, S. 2017. Commercializing the unpredictable: perspectives from wildlife watching tourism entrepreneurs in Sweden. *Human Dimensions of Wildlife*, 22, 406–421.

Martínez Pastur, G., Lencinas, M. V., Escobar, J., Quiroga, P., Malmierca, L., and Lizarralde, M. 2006. Understorey succession in *Nothofagus* forests in Tierra del Fuego (Argentina) affected by *Castor canadensis*. *Applied Vegetation Science*, 9, 143–154.

Massaro, P. P. and Massaro, P. J. 2020. *Trapping 101: A Complete Guide to Taking Furbearing Animals*. New York: Skyhorse Publishing.

Mayer, M., Estalella, C. A., Windels, S. K., and Rosell, F. N. 2020. Landscape structure and population density affect intraspecific aggression in beavers. *Ecology and Evolution*, 10(24), 13883–13894.

McComb, W. C., Sedell, J. R., and Buchholz, T. D. 1990. Dam-site selection by beavers in an eastern Oregon basin. *Great Basin Naturalist*, 50, 273–281.

McKinstry, M. C. and Anderson, S. H. 1999. Attitudes of private- and public-land managers in Wyoming, USA, toward beaver. *Environmental Management*, 23, 95–101.

McKinstry, M. C. and Anderson, S. H. 2002. Survival, fates, and success of transplanted beavers, *Castor canadensis*, in Wyoming. *Canadian Field-Naturalist*, 116, 60–68.

McKinstry, M. C., Caffrey, P. and Anderson, S. H. 2001. The importance of beaver to wetland habitats and waterfowl in Wyoming. *Journal of the American Resources Association*, 37, 1571–1577.

Methow Beaver Project 2014. *Methow Beaver Project Successes and Challenges. Report to National Forest Foundation*. Winthrop, WA: Methow Beaver Project.

Mikko, S., Roed, K., Schmutz, S., and Anderson, L. 1999. Monomorphism and polymorphism at MHC DRB loci in domestic and wild ruminants. *Immunological Reviews*, 167, 169178.

Miller, J. E. 1983. Control of beaver damage. *Proceedings Eastern Wildlife Damage Control*, 1, 177–183.

Ministry of Agriculture and Forestry Finland. 2016. Hunting and game management. http://mmm.fi/en/wildlife-and-game/hunting-and-game-management [Accessed 25 November 2016].

Moran, D. and Lewis, A. R. 2014. *The Scottish Beaver Trial: socio-economic monitoring, final report*. Scottish Natural Heritage Commissioned Report No. 799. Inverness: Scottish Natural Heritage.

Morgan, S. M. D., Pouliott, C. E., Rudd, R. J., and Davis, A. D. 2015. Antigen detection, rabies virus isolation, and Q-PCR in the quantification of viral load in a natural infection of the North American beaver (*Castor canadensis*). *Journal of Wildlife Diseases*, 51, 287–289.

Morrison, A., Westbrook, C. J., and Noble, B. F. 2018. A review of flood risk management governance and resilience literature. *Flood Risk Management*, 11, 291–304.

Müller-Schwarze, D. 1996. *Proactive management: avoiding conflict by knowing beaver behaviour*. Living with Wildlife Report. North Grafton, MA: Tufts Centre for Animals and Public Policy.

Müller-Schwarze, D. 2011. *The Beaver: Its Life and Impact*, 2nd edn. New York: Cornell University Press.

Müller-Schwarze, D. and Schulte, B. A. 1999. Behavioral and ecological characteristics of a 'climax' population of beaver (*Castor canadensis*). In: Busher, P. E. and Dzięciołowski, R. M. (eds.) *Beaver Protection, Management, and Utilization in Europe and North America*. Boston: Springer.

Murphey, W. 1996. *Conibear Beaver Trapping in Open Water: Mastering Beaver Trapping Techniques*. La Pine, OR: Lost Creek Books.

Myrberget, S. 1967. The beaver in Norway. *Acta Theriologica*, 2, 17–26.

Naiman, R. J., Johnston, C. A., and Kelley, J. C. 1988. Alteration of North American streams by beaver. *BioScience*, 38, 753–762.

Natural Resources Institute Finland (LUKE). 2019. *Beavers*. https://www.luke.fi/en/natural-resources/game-and-hunting/beavers/.

NatureServe. 2005. Explorer. *An Online Encyclopedia of Life*. Arlington, VA: NatureServe. http://www.natureserve.org/explorer.

Neumayer, M., Teschemacher, S., Schloemer, S., Zahner, V., and Rieger, W. 2020. Hydraulic modeling of beaver dams and evaluation of their ompacts on flood events. *Water*, 12, 1–23.

Nolet, B. A. and Baveco, J. M. 1996. Development and viability of a translocated beaver *Castor fiber* population in the Netherlands. *Biological Conservation*, 75, 125–137.

Nolet, B. A. and Rosell, F. 1998. Comeback of the beaver *Castor fiber*: an overview of old and new conservation problems. *Biological Conservation*, 83, 165–173.

Nolte, D. L. and Mason, J. R. 1998. Bioassays for mammals and birds. In: Millar, J. and Haynes, K. (eds.) *Methods in Chemical Ecology Volume 1*. London: Hall Publishers.

Nolte, D. L., Lutman, M. W., Bergman, D. L., Arjo, W. M., and Perry, K. R. 2003. Feasibility of non-lethal approaches to protect riparian plants from foraging beaver in North America. In: Singleton, G. R., Hinds, L. A., Krebs, C. J., and Spratt, D. M. (eds.) *Rats, Mice and People: Rodent Biology and Management*. Canberra: Australian Centre for International Agricultural Research.

Nolte, D. L., Arner, D. H., Paulson, J. C., Jones, J., and Trent, A. 2005. *How to keep beavers from plugging culverts*. US Department of Agriculture Forest Service. Technical Report 0571-2830-MTDC. Missoula, MT: Missoula Technology and Development Center.

Novak, M. 1987. Traps and trap research. In: Novak, M., Baker, J. A., Obbard, M. E., and Mallock, B. (eds.) *Wild Furbearer Management and Conservation in North America*. Toronto: Queens Printer for Ontario. pp. 941-969.

O'Connor, A. 2007. After 200 years, a beaver is back in New York City. *The New York Times*, 23 February.

Ohmart, R. D. and Anderson, B. W. 1986. Riparian habitats. In: Cooperrider, A.Y., Boyd, R. J., and Stuart, H. R. (eds.) *Inventory and Monitoring of Wildlife Habitat*. Denver: Department of the Interior, Bureau of Land Management.

Olson, R. A. and Hubert, W. A. 1994. *Beaver: Water Resources and Riparian Habitat Manager*. Laramie: University of Wyoming.

Oregon Department of Fish and Wildlife 2012. *Guidelines for Relocation of Beaver in Oregon*. Salem: ODFW.

Oregon Forest Resources Institute 2016. *Wildlife in Managed Forests: The American Beaver*. Portland: OFRI.

Papier, C. M., Poulos, H. M., and Kusch, A. 2019. Invasive species and carbon flux: the case of invasive beavers (*Castor canadensis*) in riparian *Nothofagus* forests of Tierra del Fuego, Chile. *Climatic Change*, 153, 219–234.

Parker, H. and Rosell, F. 2001. Parturition dates for Eurasian beaver *Castor fiber*: when should spring hunting cease? *Wildlife Biology*, 7, 237–241.

Parker, H. and Rosell, F. 2003. Beaver management in Norway: a model for continental Europe? *Lutra*, 46, 223–234.

Parker, H. and Rosell, F. 2012. *Beaver Management in Norway—A Review of Recent Literature and Current Problems 2012*. Porsgrunn: Telemark University College.

Parker, H. and Rosell, F. 2014. Rapid rebound in colony number of an over-hunted population of Eurasian beaver *Castor fiber*. *Wildlife Biology*, 20, 267–270.

Parker, H., Haugen, A., Kristensen, Ø., Myrum, E., Kolsing, R., and Rosell, F. 1999. Landscape use and economic value of Eurasian beaver (*Castor fiber*) on large forest in southeast Norway. *Proceedings of the 1st Euro-American Beaver Congress*, 24–27 August.

Parker, H., Rosell, F., Hermansen, T. A., Sørløkk, G., and Stærk, M. 2000. Can beaver *Castor fiber* be selectively harvested by sex and age during spring hunting? In: Czech, A. and Schwab, G. (eds.) *The European Beaver in a New Millennium*. 2nd European Beaver Symposium, Białowieża, Poland. Kraków: Carpathian Heritage Society, pp. 164–169.

Parker, H., Rosell, F., Hermansen, T., Sørløkk, G., and Stærk, M. 2002. Sex and age composition of spring-hunted Eurasian beaver in Norway. *Journal of Wildlife Management*, 66, 1164–1170.

Parker, H., Rosell, F., and Danielsen, J. 2006. Efficacy of cartridge type and projectile design in the harvest of beaver. *Wildlife Society Bulletin*, 34, 127–130.

Parker, H., Steifetten, Ø., Uren, G., and Rosell, F. 2013. Use of linear and areal habitat models to establish and distribute beaver *Castor fiber* harvest quotas in Norway. *Fauna Norvegica*, 33, 29–34.

Parkes, J. P., Paulson, J., Donlan, J., and Campbell, K. 2008. *Control of North American beavers in Tierra del Fuego: feasibility of eradication and alternative management options*. Landcare Research Contract Report: LC0708/084 prepared for Comité Binacional para la Estrategia de Erradicación de Castores de Patagonia Austral.

Pasca, C., Popa, M., Gridan, A., Ionescu, G., Visan, D., and Ionescu, O. 2019. Solutions for protection of hydrotechnical dikes in the areas populated with beavers. *Revista de Silvicltura si Cinegetica*, 24(45), 91–96.

Peterson, R. P. and Payne, N. F. 1986. Productivity, size, age, and sex structure of nuisance beaver colonies in Wisconsin. *Journal of Wildlife Management*, 50, 265–268.

Petro, V. M., Taylor, J. D., and Sanchez, D. M. 2015. Evaluating landowner-based beaver relocation as a tool to restore salmon habitat. *Global Ecology and Conservation*, 3, 477–486.

Petrosyan, V. G., Golubkov, V. V., Zavyalov, N. A., Khlyap, L. A., Dergunova, N. N., and Osipov, F. A. 2019. Modelling of competitive interactions between native Eurasian (*Castor fiber*) and alien North American (*Castor canadensis*) beavers based on long-term monitoring data (1934–2015). *Ecological Modelling*, 409, 1–15.

Pillai, A. and Heptinstall, D. 2013. Twenty years of the Habitats Directive: a case study on species reintroduction, protection and management. *Environmental Law Review*, 15, 27–46.

Pillai, A., Heptinstall, D., Hammond, M., Redpath, S., and Saluja, P. G. 2012. *Derogations for reintroduced protected species*. Inverness: Scottish Natural Heritage.

Pilliod, D. S., Rohde, A. T., Charnley, S., Davee, R. R., Dunham, J. B., Gosnell, H., Grant, G. E., Hausner, M. B., Huntington, J. L., and Nash, C. 2018. Survey of beaver-related restoration practices in rangeland streams of the western USA. *Environmental Management*, 61, 58–68.

Pinto, B., Santos, M. J., and Rosell, F. 2009. Habitat selection of the Eurasian beaver (*Castor fiber*) near its carrying capacity: an example from Norway. *Canadian Journal of Zoology*, 87, 317–325.

Pizzi, R. 2014. Minimally invasive surgery techniques. In: Miller, R. E. and Fowler, M. E. (eds.) *Zoo and Wild Animal Medicine*. St Louis, MO: Elsevier Saunders.

Pollock, M. M., Pess, G. R., Beechie, T. J., and Montgomery, D. R. 2004. The importance of beaver ponds to coho salmon production in the Stillaguamish river basin, Washington, USA. *North American Journal of Fisheries Management*, 24, 749–760.

Pollock, M. M., Beechie, T. J., and Jordan, C. E. 2007. Geomorphic changes upstream of beaver dams in Bridge Creek, an incised stream channel in the interior Columbia River basin, eastern Oregon. *Earth Surface Processes and Landforms*, 32, 1174–1185.

Pollock, M. M., Wheaton, J. M., Bouwes, N., Volk, C., Weber, N., and Jordan, C. E. 2012. *Working with beaver to restore salmon habitat in the Bridge Creek intensively monitored watershed: design rationale and hypotheses. NOAA Technical Memorandum NMFS–NWFSC*. Washington, DC: NOAA.

Pollock, M. M., Beechie, T. J., Wheaton, J. M., Jordan, C. E., Bouwes, N., Weber, N., and Volk, C. 2014. Using beaver dams to restore incised stream ecosystems. *BioScience*, 64, 279–290.

Pollock, M. M., Lewallen, G., Woodruff, K., Jordan, C. E., and Castro, J. M. 2015. *The Beaver Restoration Guidebook: Working with Beaver to Restore Streams, Wetlands, and Floodplains*. Portland, OR: United States Fish and Wildlife Service.

Purdy, K. G. and Decker, D. J. 1985. *Central New York Beaver Damage Tolerance Study*. Ithaca: Human Dimensions Research Unit, Cornell University.

Rees, P. A. 2001. Is there a legal obligation to reintroduce animal species into their former habitats? *Oyrx*, 35, 216–223.

Responsive Management 2016. *Trap Use, Furbearers Trapped, and Trapper Characteristics in the United States in 2015*. Harrisonburg, VA: Responsive Management National Office.

River Restoration Centre 1997. *Manual of River Restoration Techniques.* Cranfield: River Restoration Centre.

Roblee, K. J. 1987. The use of the T-culvert guard to protect road culverts from plugging damage by beavers. *Proceedings of the 3rd Eastern Wildlife Damage Control Conference,* 18–21 October, Gulf Shores, Alabama, pp. 23–33.

Roni, P. and Beechie, T. 2013. *Stream Watershed Restoration: A Guide to Restoring Riverine Processes and Habitats (Advancing River Restoration and Management).* Chichester: Wiley-Blackwell.

Rosell, F. and Czech, A. 2000. Responses of foraging Eurasian beavers *Castor fiber* to predator odours. *Wildlife Biology,* 6, 13–21.

Rosell, F. and Parker, H. 1995. *Beaver Management: Present Practice and Norway's Future Needs.* Porsgrunn: Telemark University College.

Rosell, F. and Sanda, J. 2006. Potential risks of olfactory signaling: the effect of predators on scent marking by beavers. *Behavioral Ecology,* 17, 897–904.

Rouland, P. and Migot, P. 1990. La réintroduction du castor (*Castor fiber*) en France. Essai de synthése et réflexions. *Revue d'Ecologique la Terre et la Vie,* 5, 145–158.

Rutherford, W. H. 1964. *The Beaver in Colorado: Its Biology, Ecology, Management and Economics.* Colorado Game, Fish and Parks Technical Publication. Denver: Colorado Parks & Wildlife.

Sax, J. L. 1970. The public trust doctrine in natural resource law: effective judicial intervention. *Michigan Law Review,* 68, 471–566.

Scheffer, P. M. 1938. The beaver as an upstream engineer. *Soil Conservation,* 3, 178–181.

Schulte, B. A. and Müller-Schwarze, D. 1999. Understanding North America beaver behaviour as an aid to management. In: Busher, P. E. and Dzieciolowski, R. M. (eds.) *Protection, Management, and Utilization in Europe and North America.* New York: Kluwer Academic/Plenum Press.

Schwab, G. 2014. *Handbuch für den Biberberater.* Nuremberg: Bund Naturschutz in Bayern.

Schwab, v. G. and Schmidbauer, M. 2003. Beaver (*Castor fiber* L., Castoridae) management in Bavaria. *Denisia,* 9, 99–106.

Scott, N. 2016. Beaver attacks Latvian man, who couldn't be helped because police thought his report was a prank call. *USA Today,* 21 April. https://ftw.usatoday.com/2016/04/beaver-attack-latvia.

Scott Porter Research & Marketing Ltd. 1998. Re-introduction of the European beaver to Scotland: results of a public consultation. Scottish Natural Heritage Research, Survey and Monitoring Report No 121.

Senn, H., Ogden, R., Frosch, C., Syrůčková, A., Campbell-Palmer, R., Munclinger, P., Durka, W., Kraus, R. H. S., Saveljev, A. P., Nowak, C., Stubbe, A., Stubbe, M., Michaux, J., Lavrov, V., Samiya, R., Ulevicius, A., and Rosell, F. 2014. Nuclear and mitochondrial genetic structure in the Eurasian beaver (*Castor fiber*)—implications for future reintroductions. *Evolutionary Applications,* 7, 645–662.

Shwiff, S. A., Kirkpatrick, K. N., and Godwin, K. C. 2011. *Economic Evaluation of Beaver Management to Protect Timber Resources in Mississippi.* USDA National Wildlife Research Center Staff Publications, 1367. Washington, DC: USDA National Wildlife Research Center.

Simon, L. J. 2006. Solving beaver flooding problems through the use of water flow control devices. *Proceedings of the 22nd Vertebrate Pest Conference,* University of California, Davis, pp. 174–180.

Sjöberg, G. and Belova, O. 2019. *Beaver as a Renewable Resource. A Beaver Dam Handbook for the Baltic Sea Region.* Sweden: Water Management in Baltic Forests.

Sjöberg, G., Ulevicius, A., and Belova, O. 2020. Good practices for management of beavers in the Baltic Sea Region. In: Sjöberg, G. and Belova, O. (eds.) *Beaver as a renewable resource. A beaver dam handbook for the Baltic Sea Region.* Swedish Forest Agency. https://www.skogsstyrelsen.se/en/.

Slough, B. G. and Sadleir, R. M. F. S. 1977. A land capability classification system for beaver (*Castor canadensis* Kuhl). *Canadian Journal of Zoology,* 55, 1324–1335.

Small, B. A., Frey, J. K., and Gard, C. C. 2016. Livestock grazing limits beaver restoration in northern New Mexico. *Restoration Ecology,* 24, 646–655.

Smith, B. and Prichard, D. 1992. *Riparian Area Management: Management Techniques in Riparian Areas. Technical Reference.* Denver: US Department of the Interior.

Smith, D. W. and Peterson, R. O. 1991. Behavior of beaver in lakes with varying water levels in northern Minnesota. *Environmental Management,* 15, 395–401.

Smith, D. W. and Tyers, D. B. 2012. The history and current status and distribution of beavers in Yellowstone National Park. *Northwest Science,* 86, 276–288.

Solbraa, K. 1996. *Veien til et bærekraftig skogbruk.* Oslo: Universitetsforlaget.

South, A., Rushton, S., and Macdonald, D. 2000. Simulating the proposed reintroduction of the European beaver (*Castor fiber*) to Scotland. *Biological Conservation,* 93, 103–116.

Suzuki, N. and McComb, W. C. 1998. Habitat classification models for beaver (*Castor canadensis*) in the streams of the central Oregon Coast Range. *Northwest Science,* 72, 102–110.

Swinnen, K. R. R., Strubbe, D., Matthysen, E., and Leirs, H. 2017. Reintroduced Eurasian beavers (*Castor fiber*): colonization and range expansion across human-dominated landscapes. *Biodiversity and Conservation,* 26, 1863–1876.

Taylor, B. R., MacInnis, C., and Floyd, T. A. 2010. Influence of rainfall and beaver dams on upstream movement of spawning Atlantic salmon in a restored brook in Nova Scotia, Canada. *River Research and Applications*, 26, 183–193.

Taylor, J., Bergman, D., and Nolte, D. 2008. If you build it they will come—management planning for a suburban beaver population in Arizona. In: Timms, R. M. and Madon, M. B. (eds.) *Proceedings of the 23rd Vertebrate Pest Conference*, University of California, Davis.

Taylor, J. D. and Singleton, R. D. 2014. The evolution of flow devices used to reduce flooding by beavers: a review. *Wildlife Society Bulletin*, 38, 127–133.

Taylor, J. D., Yarrow, G. K., and Miller, J. E. 2017. *Beavers*. Wildlife Damage Management Technical Series, 11. Fort Collins, CO: USDA, APHIS, National Wildlife Research Center.

Thévenin, C., Morin, A., Kerbiriou, C., Sarrazin, F., and Robert, A. 2020. Heterogeneity in the allocation of reintroduction efforts among terrestrial mammals in Europe. *Biological Conservation*, 241, 108346.

Thompson, S., Vehkaoja, M., Pellikka, J., & Nummi, P. 2021. Ecosystem services provided by beavers Castor spp. *Mammal Review*, 51, 25–39.

Touihri, M., Labbé, J., Imbeau, L., and Darveau, M. 2018. North American beaver (*Castor canadensis* Kuhl) key habitat characteristics: review of the relative effects of geomorphology, food availability and anthropogenic infrastructure. *Ecoscience*, 25, 9–23.

Treves, A., Wallace, R. B., Maughton-Treves, L., and Morales, A. 2006. Co-managing human–wildlife conflicts: a review. *Human Dimensions in Wildlife*, 11, 383–396.

Ulicsni, V., Babai, D., Juhász, E., Molnár, Z., and Biró, M. 2020. Local knowledge about a newly reintroduced, rapidly spreading species (Eurasian beaver) and perception of its impact on ecosystem services. *PLoS ONE*, 15, e0233506.

Vorel, A., Šíma, J., Uhlíková, J., Peltánová, A., Mináriková, T., and Švanyga, J. 2003. *Management Plan of the Eurasian Beaver in the Czech Republic*. Prague: Ministry of the Environment of the Czech Republic.

Vorel, A., Dostál, T., Uhlíková, J., Korbelová, J., and Koudelka, P. 2016. *Handbook for Coexisting with Beavers*. Prague: Czech University of Life Sciences.

Wajdzik, M., Kubacki, T., and Tomek, A. 2013. Szkody wyrzadzane przez bobra europejskiego *Castor fiber* w gospodarce rolnej, lesnej I rybackiej w Malopolsce. Rogów: Studia I Materialy CEPL w Rogowie.

Wallem, P. K., Anderson, C. B., Martínez Pastur, G., and Lencinas, M. V. 2010. Using assembly rules to measure the resilience of riparian plant communities to beaver invasion in subantarctic forests. *Biological Invasions*, 12, 325–335.

Warren, E. R. 1926. A study of beaver in the Yancey region of Yellowstone National Park. *Roosevelt Wild Life Annals*, 1–2, 17.

Weber, N., Bouwes, N., Pollock, M. M., Volk, C., Wheaton, J. M., Wathen, G., Wirtz, J., and Jordan, C. E. 2017. Alteration of stream temperature by natural and artificial beaver dams. *PLoS ONE*, 12, e0176313.

West, B. C. and Godwin, K. 2003. *Managing Beaver Problems in Mississippi*. Mississippi State: Mississippi State University and US Department of Agriculture. http://www.aphis.usda.gov/wildlife_damage/state_office/mississippi_info.shtml.

Westbrook, C. J., Ronnquist, A., and Bedard-Haughn, A. 2020. Hydrological functioning of a beaver dam sequence and regional dam persistence during an extreme rainstorm. *Hydrological Processes*, 34(18), 3726–3737.

Wheaton, J. M., Bennett, S. N., Bouwes, N., Maestas, J. D., and Shahverdian, S. M. 2019. *Low-Tech Process-Based Restoration of Riverscapes: Design Manual*. Logan: Utah State University Restoration Consortium.

White, H. B., Decker, T., O'Brien, M. J., Organ, J. F., and Roberts, N. M. 2015. Trapping and furbearer management in North American wildlife conservation. *International Journal of Environmental Studies*, 72, 756–769.

Whitfield, C. J., Baulch, H. M., Chun, K. P., and Westbrook, C. J. 2015. Beaver-mediated methane emission: the effects of population growth in Eurasia and the Americas. *Ambio*, 44, 7–15.

Wilsson, L. 1968. *My Beaver Colony*. New York: Doubleday.

Wolf, E. C., Cooper, D. J., and Hobbs, N. T. 2007. Hydrologic regime and herbivory stabilize an alternative state in Yellowstone National Park. *Ecological Applications*, 17, 1572–1587.

Wood, G. W. and Woodward, L. A. 1992. The Clemson beaver pond leveller. *Proceeding of the Annual Conference of the Southeast Association of Fish and Wildlife Agencies*, 46, 179–187.

Wood, J. 2017. Fur prices: NAFA July auction results. *Trapping Today*, 12 July. http://trappingtoday.com/2017-fur-prices-nafa-july-auction-results.

Woodruff, K. 2015. *Methow Beaver Project: Accomplishments 2014*. Twisp, WA: Methow Salmon Recovery Foundation.

Woodward, D. K., Hazel, R. B., and Gaffney, B. P. 1985. Economic and environmental impacts of beavers in North Carolina. *Proceedings of the 2nd Eastern Wildlife Damage Control Conference*, 22–25 September, North Carolina State University.

Yukon Renewable Resources 1992. Managing your trapline. *Yukon Trapline Management Series*. Whitehorse: Government of Yukon.

Zahner, V., Schmidbauer, M., and Schwab, G. 2005. *Der Biber: Die Rückkehr Der Burgherren*. Amberg: Buch und Kunstverlag Oberpfalz.

Żurowski, W. and Jaczewski, Z. 1974. Observations on the releasing of the European beaver. *Acta Theriologica*, 25, 370–374.

Żurowski, W. and Kasperczyk, B. 1988. Effects of reintroduction of European beaver in the lowlands of the Vistula basin. *Acta Theriologica*, 33, 325–338.

Index

nutritional 38, 63, 120, 148, 150, 152–153, 156–159, 162, 168, 171, 347, 352

O

Oak (*Quercus* spp.) 103, 121, 144, 147, 149, 159, 169, 349
oak, Black (*Quercus velutina*) 147
oak, Red (*Quercus rubra*) 121, 147
oak, Shingle (*Quercus imbricaria*) 147
Oats (*Avena sativa*) 98, 144, 150, 161, 164
obstruction 91, 130, 387, 396, 401, 415
occipital 6
occlusal 25, 352
ocular 101
oesophagus 78
offspring 85, 116, 343–344, 347, 349, 370
Ohio 53, 58, 147, 157, 160, 162–163, 170–171, 303, 382
oil 36, 94, 427
olfaction 91, 97–98, 101, 152–153, 170, 432
Oligocene 8–9, 23, 25
Ontario 12, 18, 24, 51–53, 60, 62, 84, 100, 119, 121, 129, 131, 133, 137, 147, 151, 164, 168, 377, 380, 396, 431
Opossum (*Didelphis virginiana*) 405, 407
oral 67, 91, 350
Oregon 14, 23, 28, 52, 58, 63, 126, 136–138, 398, 400, 415, 427, 430–433
Organic matter accumulation 318
organisation 27, 42, 62, 78, 97, 385, 391, 411
Oribatidae 318
Oslo 25, 63–64, 100–101, 136, 138, 170, 380–381, 433
Osprey (*Pandion haliaetus*) 324
Ottawa 58
otter, Eurasian (*Lutra lutra*) 18–20, 22, 25–26, 34, 38–39, 41, 47, 58, 93, 97, 99, 111, 131, 137, 140, 266–267, 270, 326, 344, 353, 355, 407, 411, 425, 427
otter, North American (*Lutra canadensis*) 19, 43, 89, 266, 290, 328, 351
outbreeding 15, 17
ovariectomy 404
ovariohysterectomy 404
ovary 83, 86, 100
overbank flows 304
overexploitation 41–42
overgrazing 398, 420
overkill 24, 62
overwintering 373, 375
Owls (Strigiformes) 324

oxbow 57–58, 167, 426
oxygen 79, 93, 96, 99–100, 359
oxygen storage capacity 93

P

Pacific 398
Pair 4, 16, 41, 55, 71, 74, 81, 85, 116, 120, 130, 140, 246, 342–344, 370, 382, 392, 397, 405, 409, 417–418
Palaearctic 11, 23, 25, 99
Palaeocastor 8–10, 24
Palaeocastor barbouri 10
Palaeocastor fossor 10
Palaeocastor magnus 10
Palaeocastorine 8–9
Palaeocene 9
palaeoclimate, palaeoenvironment 21–22, 24, 26, 65
palaeodietary 12
palaeoecology 22, 24–26
palaeogeography 22, 24, 26
palaeolithic 28–29, 59, 65
Palaeomys 9
palaeontology 23–25
palatability 146, 166, 170, 427
Paleocene 21
Paleoindians 21
Paleolithic 58, 61
palpitation 370
papillae 81, 86, 88, 360, 369
paralumbar 365
parasite 12, 70, 286, 359, 414
pars cervicalis 80
pars thoracalis 80
parturition 94, 380, 431
Patagonia 58, 62–63, 104, 134, 137, 148, 170, 390, 430, 432
pathogens 102, 352, 378
PCR 370, 430
Pea (*Pisum sativum*) 144, 161
pedunculate 169
pelage 3, 5, 98, 106, 378
pelagic 72, 77, 112, 134, 145, 167, 377, 426
pelvis 75, 86, 351, 370
peninsula 58–59, 65, 135
penis 83, 370
Pennsylvania 52–54, 57, 167, 377
perception 343, 348, 383, 385, 392–393, 403, 409–410
perennial 108, 125
perirenal 95
peritoneal 365
persecution 18, 417–418, 420
Perth 163
pest 382, 392, 425, 433
pesticide 383

pet 27, 99, 342, 353–354, 386
petroglyphs 29
phenolic 165–167, 169, 377
phenotype 101
pheromone 91, 97
Philadelphia 57–59, 66, 100
philopatry 16
philosopher 24–25, 27
phloem 140, 158
Phocidae 34
Phoenicians 38
phosphorus 82, 157–158, 160, 164–165, 310, 347–348, 350
photogrammetry 377
photograph 2, 5, 12, 15, 19, 28, 31, 37–39, 68, 70, 73–75, 79, 83–85, 104, 106, 109, 112, 114–115, 118, 120, 123, 128, 131, 133, 146, 152, 154–158, 160–162, 164, 343, 346–347, 350, 352, 356–360, 362–364, 367, 369, 381, 384–386, 390, 394, 397, 400, 403, 405, 411–412, 414, 416, 420
phyletic 13
phylogenetics 7, 18, 20–21, 23, 25, 97
phylogeography 22, 24, 54
physeal 98, 378
physiology 12, 67, 76, 84–85, 90–91, 95, 97–101, 129, 157, 165–166, 353, 378, 381, 407–408, 428
pigment 89, 95, 164
Pike (*Esox lucius*) 267
Pilewort (*Ficaria verna*) 144
Pin cherry (*Prunus pensylvanica*) 147
Pine (*Pinus* spp.) 38, 104, 121, 143–144, 149, 159, 166, 353, 399, 431
Pine marten (*Martes martes*) 38–39, 43, 267, 326, 328, 353
pine, Jeffrey (*Pinus jeffreyi*) 143, 166
pine, Scots (*Pinus sylvestris*) 149
pinniped 67, 99
Pinus 121, 143–144, 149, 159
piscivorous 89
pituitary 85, 98
placenta 85, 100, 193, 199
planetary 23
plant defences, toxins 140, 165–167, 255
plantation 413
plasma 85, 98
plastic 116, 358, 360, 363–364, 403–404
play (fun, amusement) 245
Pleistocene 8–9, 11–13, 18, 21–23, 25–26, 28, 32, 62–64, 427
Pliny 27
Pliocene 8–9, 11, 26
Po river 59, 90, 104, 122, 139